Communications in Computer and Information Science 1037

Commenced Publication in 2007
Founding and Former Series Editors:
Phoebe Chen, Alfredo Cuzzocrea, Xiaoyong Du, Orhun Kara, Ting Liu, Krishna M. Sivalingam, Dominik Ślęzak, Takashi Washio, and Xiaokang Yang

More information about this series at http://www.springer.com/series/7899

K. C. Santosh · Ravindra S. Hegadi (Eds.)

Recent Trends in Image Processing and Pattern Recognition

Second International Conference, RTIP2R 2018 Solapur, India, December 21–22, 2018 Revised Selected Papers, Part III

Editors
K. C. Santosh
Department of Computer Science
University of South Dakota
Vermillion, SD, USA

Ravindra S. Hegadi
Solapur University
Solapur, India

ISSN 1865-0929 ISSN 1865-0937 (electronic)
Communications in Computer and Information Science
ISBN 978-981-13-9186-6 ISBN 978-981-13-9187-3 (eBook)
https://doi.org/10.1007/978-981-13-9187-3

This Springer imprint is published by the registered company Springer Nature Singapore Pte Ltd.
The registered company address is: 152 Beach Road, #21-01/04 Gateway East, Singapore 189721, Singapore

Preface

It is our great pleasure to introduce the collection of research papers in the *Communication in Computer and Information Science* (CCIS) Springer series from the second Biennial International Conference on Recent Trends in Image Processing and Pattern Recognition (RTIP2R). The RTIP2R conference event took place at the Solapur University, Maharastra, India, during December 21–22, 2018, in collaboration with the Department of Computer Science, University of South Dakota (USA) and Universidade de Evora (Portugal). Further, the conference had a very successful workshop titled Pattern Analysis and Machine Intelligence (PAMI): Document Engineering to Healthcare, with more than 70 participants.

As announced in the Call For Paper, RTIP2R attracted current and/or recent research on image processing, pattern recognition, and computer vision with several different applications, such as document understanding, biometrics, medical imaging, and image analysis in agriculture. Altogether, we received 371 submissions and accepted 173 papers based on our thorough review reports. We followed a double-blind submission policy and therefore the review process was extremely solid. On average, there were at least three reviews per paper except the few that had desk rejections, and therefore we had 859 review reports. We also made the authors aware of plagiarism, and rejected a few of them even after making review reports.

During the event, we hosted more than 200 participants from more than 29 different countries, such as USA, Vietnam, Australia, Russia and Sri Lanka (not just limited to India). In brief, the event was found to be a great platform bringing together research scientists, academics, and industry practitioners throughout the world. We categorized the papers into five different tracks: (a) computer vision and applications; (b) machine learning and applications; (c) document analysis; (d) healthcare and medical imaging; (e) biometrics and applications; (f) data mining, information retrieval and applications; (g) image processing; and (h) image analysis in agriculture.

We also selected the best papers based on the review reports, review scores, and presentations at the conference, and provided authors an opportunity to publish their extended works in the following journals: (a) *Multimedia Tools and Applications* (Springer); (b) *International Journal of Ambient Computing and Intelligence* (IGI Global); and (c) *Journal of Healthcare Informatics Research* (Springer).

The conference event was full of new ideas, including keynote speeches that were from (a) Sameer Antani, National Institutes of Health; (b) Mohan Gundeti, University of Chicago Medicine; and (c) Ernest Cachia, University of Malta.

April 2019

K. C. Santosh
Ravindra S. Hegadi

Organization

Patron

M. M. Fadnavis

Organizing Chairs

V. B. Ghute
V. B. Patil
B. C. Shewale

Honorary Chairs

P. Nagabhushan	IIIT, Allahabad, India
P. S. Hiremath	KLE University, Hubballi, India
B. V. Dhandra	Symbiosis University, India

General Chairs

Jean-Marc Ogier	University of la Rochelle, France
Laurent Wendling	University of Paris Descartes, France
Sameer Antani	US National Library of Medicine, USA
D. S. Guru	University of Mysore, India

Conference Chairs

Ravindra Hegadi	Solapur University, India
Teresa Goncalves	Universidade de Evora, Portugal
K. C. Santosh	University of South Dakota, USA

Area Chairs

Szilard Vajda	Central Washington University, USA
Mickael Coustaty	University of La Rochelle, France
Nibaran Das	Jadavpur University, India
Nilanjan Dey	Techno India College of Technology, India
Jude Hemanth	Karunya University, India

Publicity Chairs

Hubert Cecotti	California State University, USA
Odemir Martinez Bruno	University of Sao Paulo, Brazil
Alba Garcia Seco de Herrera	University of Essex, UK
Sheng-Lung Peng	National Dong Hwa University, Taiwan
Do T. Ha	VNU University of Science, Vietnam
B. Uyyanonvara	Thammasat University, Thailand
Sk Md. Obaidullah	University of Evora, Portugal
V. Bevilacqua	Polytechnic of Bari, Italy
R. S. Mente	Solapur University, India
Pratim P. Roy	Indian Institute of Technology (IIT), India
Manjunath T. N.	BMSIT, Bangalore, India
Nadra Ben Romdhane	University of Sfax, Tunisia
M. A. Jabbar	Vardhaman College of Engineering, India

Finance Chairs

A. R. Shinde	Solapur University, Solapur, India
S. D. Raut	Solapur University, Solapur, India

Advisory Committee

Daniel P. Lopresti	Lehigh University, USA
Rangachar Kasturi	University of South Florida, USA
Sargur N. Srihari	CEDAR, USA
K. R. Rao	University of Texas at Arlington, USA
Ishwar K. Sethi	Oakland University, USA
G. K. Ravikumar	CVS Health/Wipro, Texas, USA
Jose Flores	University of South Dakota, USA
Rajkumar Buyya	University of Melbourne, Australia
Arcot Sowmya	UNSW, Sydney, Australia
Antanas Verikas	Halmstad University, Sweden
Diego Liberati	Politecnico di Milano, Italy
B. B. Chaudhuri	Indian Statistical Institute, Kolkata, India
Atul Negi	University of Hyderabad, India
Arun Agarwal	University of Hyderabad, India
Hemanth Kumar	University of Mysore, India
K. V. Kale	Dr. BAMU, Aurangabad, India
B. V. Pawar	NMU, Jalgaon, India
R. R. Deshmukh	Dr. BAMU, Aurangabad, India
Karunakar A. K.	MIT, Manipal, India
Suryakanth Gangashetty	IIIT Hyderabad, India
Kaushik Roy	West Bengal University, India
Mallikajrun Hangarge	KASCC, Bidar, India
T. Devi	Bharathiar University, Coimbatore, India

G. R. Sinha	IIIT, Bangalore, India
U. P. Kulkarni	SDMCE, Dharwad, India
Rajendra Hegadi	IIIT, Dharwad, India
S. Basavarajappa	IIIT, Dharwad, India
B. P. Ronge	SVERI'S College of Engineering, India

Technical Program Committee (Country-Wise)

Randy C. Hoover	South Dakota School of Mines and Technology, USA
Sivarama Krishnan Rajaraman	US National Library of Medicine, NIH, USA
Yao-Yi Chiang	University of Southern California - LA, USA
Ullas Bagci	University of Central Florida, USA
Yuhlong Lio	University of South Dakota, USA
Eugene Borovikov	Intelligent Automation Inc., USA
Szilard Vajda	Central Washington University, USA
Hubert Cecotti	California State University, USA
Sema Candemir	US National Library of Medicine, NIH, USA
Md Mahmudur Rahman	Morgan State University, USA
Gabriel Picioroaga	University of South Dakota, USA
Peter Dolan	University of Minnesota Morris, USA
Michael Clement	York University, Canada
Alba Garca Seco de Herrera	University of Essex, UK
Nico Hochgeschwender	University of Luxembourg, Luxembourg
Benoit Naegel	University of Strasbourg, France
Vincent Bombardier	CRAN, University of Lorraine, France
Isabelle Debled-Rennesson	LORIA, University of Lorraine, France
Camille Krutz	University Institutes of Technology (IUT de Paris), France
Jean Cousty	University Paris-Est, France
Jonathan Weber	University of Haute-Alsace, France
Sabine Barrat	University of Tours, France
Muhammad Muzzamil Luqman	University of La Rochelle, France
Mickael Coustaty	University of La Rochelle, France
Jean-Pierre Salmon	University of Bordeaux Montaigne, France
Victor Codocedo	University de Lyon, CNRS, INSA-Lyyon, France
Diego Liberati	Politecnico di Milano, Italy
Vitoantonio Bevilacqua	Polytechnic of Bari, Italy
Salim Jouili	Euro Nova, Belgium
Paulo Quaresma	University of Evora, Portugal
Luis Rato	University of Evora, Portugal
Joao Barroso	University of Tras-os-Montes e Alto Douro, Portugal
Vitor M Filipe	University of Tras-os-Montes e Alto Douro, Portugal
Mohamed-Rafik Bouguelia	Halmstad University, Sweden
Marcal Rusinol	Universitat Autonoma de Barcelona, Spain

Name	Affiliation
Margit Antal	Sapientia University, Romania
Laszlo Szilagyi	Sapientia University, Romania
Srikanta Pal	Griffith University, Australia
Alireza Alaei	Griffith University, Australia
M. Cerda Villablanca	University of Chile, Chile
B. Uyyanonvara	SIIT, Thammasat University, Thailand
V. Sornlertlamvanich	Thammasat University, Thailand
S. Marukatat	Thammasat University, Thailand
I. Methasate	NECTEC, Thailand
C. Pisarn	Rangsit University, Thailand
Makoto Hasegawa	Tokyo Denki University, Japan
P. Shivakumara	University of Malaya, Malaysia
Sophea Prum	National R&D Center in ICT, Malaysia
Lalit Garg	University of Malta, Malta
Nadra Ben Romdhane	University of Sfax, Tunisia
Nafaa Nacereddine	Centre de Recherche en Techno. Industrielles (CRTI), Algeria
Aicha Baya Goumeidane	Centre de Recherche en Techno. Industrielles (CRTI), Algeria
Ameni Boumaiza	Qatar foundation, Qatar
Nguyen Thi Oanh	Hanoi University of Science Technology, Vietnam
Do Thanh Ha	VNU University of Science, Vietnam
Tien-Dat Nguyen	FPT Corp., Vietnam
T. Kartheeswaran	University of Jaffna, Sri Lanka
Shaikh A. Fattah	Bangladesh University of Engineering and Technology, Bangladesh
Pratim P. Roy	Indian Institute of Techno (IIT), India
Surekha Borra	KS Institute of Technology, (KSIT), India
Ajit Danti	JNN College of Engineering, Shimoga, India
Lalita Rangarajan	University of Mysore, Mysore, India
Manjaiah D. H.	Mangalore University, Mangalore, India
V. S. Malemath	KLE Engineering College, Belagavi, India
B. H. Shekar	Mangalore University, Mangalore, India
G. Tippeswamy	BMSIT, Bangalore, India
Aziz Makandar	Akkamahadevi Women's University Karnataka, Vijayapura, India
Mallikarjun Holi	BDT College of Engineering, Davangere, India
S. S. Patil	Agriculture University, Bangalore, India
H. S. Nagendraswamy	University of Mysore, Mysore, India
Shivanand Gornale	Ranichannamma University, Belagavi, India
S. Shivashankar	Karnatak University, Dharwad, India
Ramesh K.	Akkamahadevi Women's University Karnataka, Vijayapura, India
H. L. Shashirekha	Mangalore University, Mangalore, India
Dayanand Savakar	Ranichannamma University, Belagavi, India
S. B. Kulkarni	SDM College of Engineering, Dharwad, India

M. T. Somashekhar	Bangalore University, Bangalore, India
Manjunath Hiremath	Christ University, Bangalore, India
Sridevi Soma	PDA College of Engineering, Gulbarga, India
V. M. Thakare	SGB Amravati University, Amaravati, India
G. V. Chaudhari	SRTM University, Nanded, India
R. K. Kamat	Shivaji University, Kolhapur, India
Ambuja Salgaonkar	University of Mumbai, India
Praveen Yannavar	Dr. BAM University, India
R. R. Manza	Dr. BAM University, Aurangabad, India
A. S. Abhyankar	SP Pune University, India
V. T. Humbe	SRTMU Sub-Centre, Latur, India
P. B. Khanale	SRTMU, Nanded, India
M. B. Kokre	GGSIET, Nanded, India
Gururaj Mukrambi	Symbiosis International University, Pune, India
S. R. Kolhe	North Maharashtra University, Jalgaon, India
M. Sundaresan	Bharathiar University, Coimbatore, India
C. P. Sumathi	SDNBV College for Women, Chennai, India
J. Satheeshkumar	Bharathiar University, Coimbatore, India
Britto Ramesh Kumar	St. Joseph's College, Tiruchirappalli, India
Neeta Nain	Malaviya National Institute of Technology (MNIT), Jaipur, India
A. A. Desai	Veer Narmad South Gujarat University, Gujarat, India
Chandra Mouli P. V. S. S. R.	VIT University, Vellore, India
Nagartna Hegde	Vasavi Eng. College, Hyderabad, India
B. Gawali	Dr. BAM University, Aurangabad, India
K. T. Deepak	IIIT, Dharwad, India
P. M. Pawar	SVERI'S College of Eng., India
S. R. Gengaje	Walchand Inst. of Technology, Solapur, India
B. Ramadoss	National Inst. of Technology, Tamil Nadu, India

Local Organizers

P. Prabhakar
S. S. Suryavanshi
V. B. Ghute
R. B. Bhosale
B. J. Lokhande
G. S. Kamble
J. D. Mashale
C. G. Gardi
P. M. Kamble
D. D. Sawat
A. B. Jagtap
D. D. Ruikar
P. P. Gaikwad

Additional Reviewers

Abdullah Mohammed Kaleem
Abhinav Muley
Addepalli Krishna
Adithya Pediredla
Aditya Patil
Ajay Nagne
Ajeet A. Chikkamannur
Ajit Danti
Ajju Gadicha
Akbaruddin Shaikh
Alba García Seco De Herrera
Alessia Saggese
Alexandr Ezhov
Almas Siddiqui
Ambika Annavarapu
Amol Vibhute
Amruta Jagtap
Anagha Markandey
Anderson Santos
Andrés Rosso-Mateus
Aniket Muley
Anita Dixit
Anita Khandizod
Anitha H.
Anitha J.
Anitha N.
Ankita Dhar
Anupriya Kamble
Archana Nandibewoor
Arjun Mane
Arunkumar K. L.
Ashish Mourya
Atish Patel
Aznul Qalid Md Sabri
Balachandran K.
Balaji Sontakke
Balamurugan Karnan
Basavaprasad B.
Basavaraj Dhandra
Bb Patil
Benoit Naegel
Bharath Bhushan
Bharathi Pilar
Bharatratna Gaikwad
Bhausaheb Pawar
Bindu V. R.
Brian Keith
C. Namrata Mahender
C. P. Sumathi
Camille Kurtz
Chandrashekhara K. T.
Chetan Pattebahadur
Daneshwari Mulimani
Daniel Caballero
Darshan Ruikar
Dattatray Sawat
Dericks Shukla
Diego Bertolini
Diego Liberati
Dnyaneshwari Patil
E. Naganathan
Ebenezer Jangam
Evgeny Kostyuchenko
G. P. Hegde
G. R. Sinha
G. S. Mamatha
Ganesh Janvale
Ganesh Magar
Ganga Holi
Gireesh Babu
Girish Chowdhary
Gururaj Mukarambi
H. L. Shashirekha
Hajar As-Suhbani
Hanumant Gite
Haripriya V.
Harshavardhana Doddamani
Hayath Tm
Hemavathy R.
Himadri Mukherjee
Hubert Cecotti
Ignazio Gallo
Jayendra Kumar
João Cardia
Jonathan Weber
Joseph Abraham Sundar K.

Jude Hemanth
Jyoti Patil
K. K. Chaturvedi
K. C. Santosh
Kalman Palagyi
Kalpana Thakare
Kapil Mehrotra
Kartheeswaran Thangathurai
Kasturi Dewi Varathan
Kaushik Roy
Kavita S. Oza
Kiran Phalke
Kwankamon Dittakan
Laszlo Szilagyi
Latchoumi Thamarai
Lingdong Kong
Lorenzo Putzu
Lp Deshmukh
Lucas Alexandre Ramos
Luis Rato
M. T. Somashekhar
Madhu B.
Mahesh Solankar
Mahmudur Rahman
Mainak Sen
Maizatul Akmar Ismail
Mallikarjun Hangarge
Mallikarjun Holi
Manasi Baheti
Manisha Saini
Manjunath Hiremath
Manjunath T. N.
Manohar Madgi
Manoj Patil
Mansi Subhedar
Manza Ramesh
Marçal Rusiñol
Margit Antal
Masud Rana Rashel
Md Obaiduallh Sk
Md. Ferdouse Ahmed Foysal
Md. Rafiqul Islam
Michael Clement
Midhula Vijayan
Miguel Alberto Becerra Botero
Mikhail Tarkov
Minakshi Vharkate
Minal Moharir
Mohammad Idrees Bhat Bhat
Mohammad Shakirul Islam
Mohan Vasudevan
Mohd. Saifuzzaman
Monali Khachane
Muhammad Muzzamil Luqman
Mukti Jadhav
Nadra Ben Romdhane
Nafis Neehal
Nagaraj Cholli
Nagaratna Hegde
Nagsen Bansod
Nalini Iyer
Nico Hochgeschwender
Nita Patil
Nitin Darkunde
Nitta Gnaneswara Rao
P. P. Patavardhan
Pankaj Agrawal
Parag Bhalchndra
Parag Kaveri
Parag Tamhankar
Parashuram Bannigidad
Parashuram Kamble
Parminder Kaur
Paulo Quaresma
Peter Dolan
Pooja Janse
Poonam Ghuli
Poornima Patil
Prabhakar C. J.
Pradeep Udupa
Prajakta Dhamdhere
Prakash Hiremath
Prakash Khanale
Prakash Unki
Praneet Saurabh
Prasanna Vajaya
Prasanth Vaidya
Pratima Manhas
Praveen K.
Pravin Metkewar
Pravin Yannawar
Prema T. Akkasaligar

Priti Singh
Pushpa Patil
Pushpa S. K.
Qazi Fasihuddin
Rafaela Alcântara
Rajendra Hegadi
Rajesh Dhumal
Rajivkumar Mente
Rajkumar Soundrapandiyan
Rajkumar Yesuraj
Rakesh K.
Ramya D.
Rashmi Somshekhar
Ratnadeep Deshmukh
Ratnakar Ghorpade
Ravi Hosur
Ravi M.
Ravindra Babu Tallamraju
Ravindra Hegadi
Rim Somai
Ritu Prasad
Rodrigo Nava
Rohini Bhusnurmath
Rosana Matuk Herrera
Rupali Surase
S. Basavarajappa
S. Ramegowda
S. B. Kulkarni
Sachin Naik
Sahana Das
Sameer Antani
Sanasam Inunganbi
Sangeeta Kakarwal
Sanjay Jain
Santosh S. Chowhan
Sarika Sharma
Satish Kolhe
Sema Candemir
Shajee Mohan
Shankru Guggari
Shanmugapriya Padmanabhan
Shanthi D. L.
Sharath Kumar
Shaveta Thakral
Sheikh Abujar
Shilpa Bhalerao
Shiva Murthy Govindaswamy
Shivani Saluja
Shivashankar S.
Shridevi Soma
Shrikant Mapari
Siddanagouda Patil
Siddharth Dabhade
Sivarama Krishnan Rajaraman
Slimane Larabi
Smriti Bhandari
Srikanta Pal
Sudha Arvind
Suhas Sapate
Sunanda Biradar
Suneeta Budihal
Sunil Nimbhore
Swapnil Waghmare
Szilard Vajda
Tejaswi Potluri
Thanh Ha Do
Ujwala Suryawanshi
Ulavappa B. Angadi
Umakant Kulkarni
Urmila Pol
Usha B. A.
Vaibhav Kamble
Veerappa Pagi
Víctor Codocedo
Vidyagouri Hemadri
Vijay Bhaskar Semwal
Vijaya Arumugam
Vikas Humbe
Vilas Naik
Vilas Thakare
Vinay T. R.
Vincent Bombardier
Virendra Malemath
Vishal Waghmare
Vishweshwarayya Hallur
Yao-Yi Chiang
Yaru Niu
Yoanna Martínez-Díaz
Yogesh Gajmal
Yogesh Rajput
Yogish H. K.
Yuhlong Lio
Zati Hakim Azizul Hasan

Contents – Part III

Image Analysis in Agriculture

Data Mining, Information Retrieval and Applications

Document Image Analysis

Text Extraction Using Sparse Representation over Learning Dictionaries

Thanh-Ha Do[1(✉)], Thi Minh Huyen Nguyen[1], and K. C. Santosh[2]

[1] Department of Informatics, VNU University of Science, Hanoi, Vietnam
hadt_tct@vnu.edu.vn, huyenntm@hus.edu.vn

[2] Department of Computer Science, University of South Dakota, Vermillion, USA
santosh.kc@ieee.org

Abstract. This paper presents a new approach for text detection using sparse representation over learned dictionaries. More specifically, the K-SVD algorithm is used for constructing two dictionaries, one for the background and one for the text. Then, text detection is done by comparing the error constructions of each patch of image over two dictionaries. Results on ICDAR dataset present that proposed method is competitive related to state-of-the-art methods.

Keywords: Text extraction · Sparse representation · Learning dictionary

1 Introduction

Texts in scene images contain lots of semantic information that can be very useful in retrieval systems. However, the differences in text fonts, colors, and sizes, as well as geometric distortions, partial occlusions, different lighting conditions and image resolutions, make detecting the text regions is more challenging [35].

Many efforts have been made in the image processing community to address problems of text detection [6,19,20,23], and these can be divided into four directions being edge-based, texture-based, connected component based and hybrid based.

The first direction is the approaches based on the edge images obtained by analyzing the contrast difference between the text and the background. Because the effect of the edge-based methods [16,17,41] depends on the effect of edge detection approaches and therefore they are not work well when images contain the shadow or highlight. The second direction is based the analysis textures properties over images [3]. In texture-based approaches [3,31], features a certain region is calculated and then the classifier techniques are used to identify the text regions. These methods can handle well with noise images because of using the

This research is funded by the Vietnam National University, Hanoi (VNU) under project number QG.18.04.

K. C. Santosh—IEEE senior member

K. C. Santosh and R. S. Hegadi (Eds.): RTIP2R 2018, CCIS 1037, pp. 3–13, 2019.
https://doi.org/10.1007/978-981-13-9187-3_1

textural properties of regions. However, it has the issue of computing time. The third direction is connected component based methods [4,18,24], which group the small connected components to a larger ones. These segment the candidate text regions by clustering the color or utilizing the edge information. In general, the connected component based methods are effect in computing time. However the approaches fails when the text regions over images are not dominant. The last direction [5,32,34,43] is developed that utilize a combination of the above, therefore, they can deal with the large number of possible variations in text.

Recently, the use of a sparse representation has take a lot of attentions of researchers, such as face recognition [39], signal classification [1] and texture classification [21]. *Pan et al.* [30] proposed the use of a sparse representation over one learned dictionary for text detection. However, because of using only one dictionary, therefore the performance is not good in the case existing the confusion between the backgrounds with text-like regions. In [44], the authors overcome the deficiency of *Pan et al.* by using the discriminate dictionary. Generally, this method provides a high precision and recall rate. However, its performance depends on the size of the dictionary. This disadvantage can be overcome by using the multi-learned dictionaries [10].

In this paper, we also focuses on the direction of using sparse representation and learning dictionaries for text regions extraction. Specifically, two learning dictionaries are constructed by K-SVD algorithm. Then, the text/background separation is done by comparing reconstruction errors of each patch of edge image over learned dictionaries. Finally, text detection is efficiently performed by applying the adaptive run-length projection profile analysis [22] and the grouping text components into text strings [38].

The remainder of this paper is organized as follows. We briefly review the meaning of sparsity in Sect. 2. Learning algorithm and how to construct two dictionaries for text regions extraction is presented in Sect. 3. Section 4 shows the main steps of proposed method. Afterwards, the experimental results is shown in Sect. 5 and finally, we give our conclusions in Sect. 6.

2 Sparse Representation

A signal $h \in \mathbb{R}^L$ is considered as sparse vector if most of its entries are equal to zero. A $k-$ sparse signal is a signal that has exactly k nonzero entries. A signal can be presented as a linear combination of k columns (or atoms) of a given over-complete matrix (or dictionary) A, such as

$$h = Ax = \sum_{i=1}^{k} a_i x_i \tag{1}$$

Mathematically, if $A = \{a_1, a_2, \ldots, a_M\} \in \mathbb{R}^{L\times M}, a_i \in \mathbb{R}^L, i = 1, ..., M; M \gg L$, is a full−rank matrix, then the under-determined linear system of equations (1) will have infinitely many solutions that satisfy it simultaneously.

From the application point of view, we needs a solution x that represents signal h well comparing with others. In fact, this task is equivalent to finding the best subset columns from the dictionary. As the results, to gain this solution, a function $f(x)$ is added to evaluate the "good" of x, with smaller values being preferred:

$$(P_f) : \min_x f(x) \text{ s.t } Ax = h \tag{2}$$

If $f(x)$ is the l_0 pseudo-norm $\|x\|_0$, then the problem (P_f) becomes finding the sparsity vector x of the signal h satisfying:

$$(P_0) : \min_x \|x\|_0 \text{ s.t } Ax = h \tag{3}$$

In general, solving Eq. (3) is often difficult (NP-hard problem) and, therefore, it is necessary to find algorithms computing sub-optimally, but '*good enough*' solution. One of the choices is the greedy algorithms, such as Matching Pursuit (MP) [27], Orthogonal-MP (OMP) [33], Weak-MP [37].

Another strategy is using other norm instead of l_0-norm, such as l_1-norm proposed by Donoho *et al.* [12]:

$$(P_1) : \min_x \|W^{-1}x\|_1 \text{ s.t } Ax = h \tag{4}$$

The matrix W is a diagonal positive-definite matrix in which $w(i,i) = 1/\|a_i\|_2$. Let $\tilde{x} = W^{-1}x$, then Eq. (4) is re-written as:

$$(P_1) : \min_{\tilde{x}} \|\tilde{x}\|_1 \text{ s.t } h = AW\tilde{x} = \tilde{A}\tilde{x} \tag{5}$$

where $\tilde{A}$ is the normalized matrix of matrix A. Eq. (5) is defined as the basis pursuit (BP) problem, and its solution x can be found by using some numerical algorithms, named Basis Pursuit by Linear Programming [7], IRLS (*Iterative Reweighed Least Squares*) (for $p = 1$) [9]. Following [26], the basis pursuit is more computing time than matching pursuit, but its solution is more sparser.

In some problems, the exact constraint $h = Ax$ is changed by using the quadratic penalty function $Q(x) = \|Ax - h\|_2 \le \epsilon$ $(\epsilon \ge 0)$. Therefore, the (P_0) is changed to following problem:

$$(P_0^\epsilon) : \min_x \|x\|_0 \text{ subject to } \|Ax - h\|_2 \le \epsilon \tag{6}$$

In (P_0^ϵ), the l_2-norm can be replaced by others, such as l_1, l_2, or l_∞.

$$(P_1^\epsilon) : \min_x \|x\|_1 \text{ subject to } \|Ax - h\|_2 \le \epsilon \tag{7}$$

Naturally, Eq. (7) needs a proper choice of the dictionary A. Our working hypothesis is that if the dictionary A is constructed from training database then its columns are adapted better to characteristics of documents. We will see details how to construct the learned dictionary A the next Section.

3 Learning Dictionaries

This section introduces the core idea of the learned dictionary and the algorithm K-SVD used to build these dictionaries from training set being the patches of images.

3.1 The Core Idea of Learning Dictionary and K-SVD Algorithm

The optimal solution of (P_1^ϵ) depends on the matrix A. In fact, using sparse representation with a specific choice of A has taken a lot of research attentions in the past decade. Some translation-invariant redundant dictionaries can be named as the curvelets, the contourlets, the wedgelets, the bandelets, and the steerable wavelets. These dictionaries perform reasonably well, however they are limited to images that they are designed to handle and they are difficult to be utilized for a new type of images. Therefore, an ambitious goal still continue. This ambitious goal leads to build a dictionary adapted to the image by training it directly from the images, therefore it is better adapted to document characteristics.

To take this advantage of learning dictionary, we decide to use one of learning algorithms to construct the learned dictionary A. In a learning dictionary algorithms, a training database $\{h_j\}_{j=1}^S$ is needed to construct the dictionary A that ensures all the $h_j, j = 1, ..., S$ to have these optimally sparse approximations in A.

$$\min_{A, x_j} \sum_{j=1}^{S} \|x_j\|_1 \text{ s.t } \|h_j - Ax_j\|_2 \leq \epsilon, j = 1, .., S \tag{8}$$

The dictionary A is obtained by using learning dictionary algorithm. There are four well-known algorithms including K-SVD [2], MOD [14], ODL [29] and RLS−DLA [36]. In these four learning algorithms, the dictionary A is obtained by iterative process via two main steps: *sparse representation step* and *update dictionary step*. In the sparse representation step, the dictionary A is fixed and finding all sparse representations $X = \{x_j\}_{j=1}^S \in \mathbb{R}^{M \times S}$ of $H = \{h_j\}_{j=1}^S \in \mathbb{R}^{L \times S}$ over A. In the update dictionary step, using the updating rule to optimize solutions obtained by sparsity step.

In [13], some numerical experiments have been performed with the purpose of evaluating the complexity, finding out advantages as well as disadvantages of algorithms used in the sparse representation step and the update dictionary step. Basing on the performance and the computing time, we set the choice for this paper is using OMP algorithm for the sparse coding stage and the updating rule in K-SVD algorithm for update dictionary's columns. Particularly, in the update dictionary step, each columns of A is updated such that the residual error (9) is minimized, where X and $\{a_1, ...a_{j_0-1}, a_{j_0+1}, ..., a_M\}$ are fixed,

$$\begin{aligned} \|H - AX\|_F^2 &= \|H - \sum_{j=1}^{M} a_j x_j^T\|_F^2 \\ &= \|(H - \sum_{j \neq j_0} a_j x_j^T) - a_{j_0} x_{j_0}^T\|_F^2 \\ &= \|E_{j_0} - a_{j_0} x_{j_0}^T\|_F^2 \end{aligned} \tag{9}$$

In the description (9), $x_{j_0}^T \in \mathbb{R}^S$ is the j_0-*th* row in X. Because X and all the columns of A are fixed excepted a_{j_0}, thus $E_{j_0} = H - \sum_{j \neq j_0} a_j x_j^T$ is fixed. It means that the minimum error $\|H - AX\|_F^2$ depends only on the optimal a_{j_0} and $x_{j_0}^T$. This is the problem of approximating a matrix E_{j_0} with another one which has a rank 1 based on minimizing the *Frobenius* norm. The optimal solutions $\tilde{a}_{j_0}, \tilde{x}_{j_0}^T$ can be given by the SVD (*Singular Value Decomposition*) of E_{j_0}. More detail about the K-SVD algorithm can be found in [2,11].

3.2 Learning Dictionaries for Text Detection

In proposed paper, we use K-SVD algorithm to build two over-complete dictionaries, one for the text and one for the background. Text dictionary is trained using dataset including images of 26 english letters and 10 numbers with difference fonts, sizes and and orientations. As for the background dictionary, the images of GREC'11 dataset are used to construct the training set. Afterwards, a small sliding window scans images into patches and patches having less than 16 pixels will be not included in training set. The size of the sliding window is very important because it effect to both the precision rate and computing time. Therefore, we use a good tradeoff size being 16×16 pixels. Finally, a dataset includes 37.667 text patches and 3.423 background patches are generated to train two dictionaries using K-SVD algorithm mentioned in Sect. 3.1.

4 Text Detection by Two Learning Dictionaries

A flowchart of the proposed text detection algorithm is shown in Fig. 1. There are three key steps, each of which will be described in details below.

4.1 Edge Detection

In fact, Canny operator is used in many text detection methods. However, this paper we use wavelets transform to detect edge image because this transform is multiscale and *Mallat et al.* [28] proved that it is equivalent to the multiscale Canny edge detection. Figure 2 presents the results obtained by wavelet transform and Canny edge operator. Obviously, the wavelet is better than Canny edge detection method.

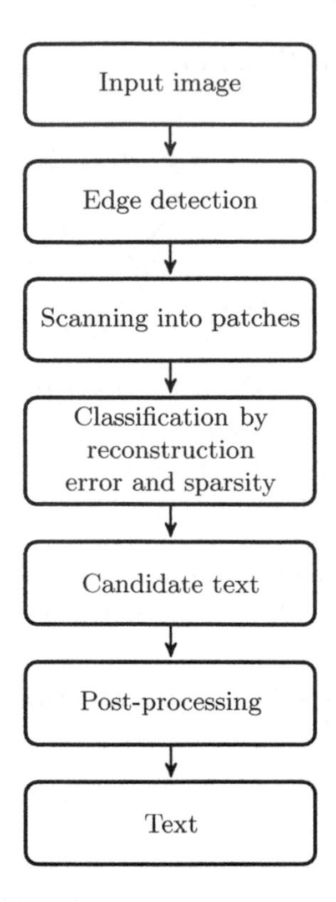

Fig. 1. Flowchart for text region extraction.

4.2 Text Classification

In this step, all candidate text edges is found by performing classification based on the sparsity. The main idea to compare the reconstruction errors of candidate patches over two dictionaries. More detail, we use these learned dictionaries combined with OMP algorithm to find the sparse representations of all patches $\{p_i\}_{i=1}^{K}$ of the image. Let $\{\bar{q}_{a,i}\}_{i=i}^{K}$ being optimal sparse representations of $\{p_i\}_{i=1}^{K}$ over learned dictionary a, where a stands for text and background dictionary.

$$\bar{q}_{a,i} = \arg\min p_i ||p_i - aq_{a,i}||_2 \text{ subject to } ||p_i||_0 \leq s \tag{10}$$

Next, the patch p_i is classified into text edge or background edge by comparing its reconstruction errors over text and background dictionaries using Eq. (11):

$$\epsilon_{a,i} = ||p_i - a\bar{q}_{a,i}||_2 \tag{11}$$

If the construction error of p_i in the text dictionary is smaller than its reconstruction error in the background dictionary, it indicates that the representation of p_i in the text dictionary is sparse and not sparse (or at least not enough sparse) in the background dictionary, therefore p_i is consider as a text patch. Finally, all text edges is reconstructed into edge image for the post processing step.

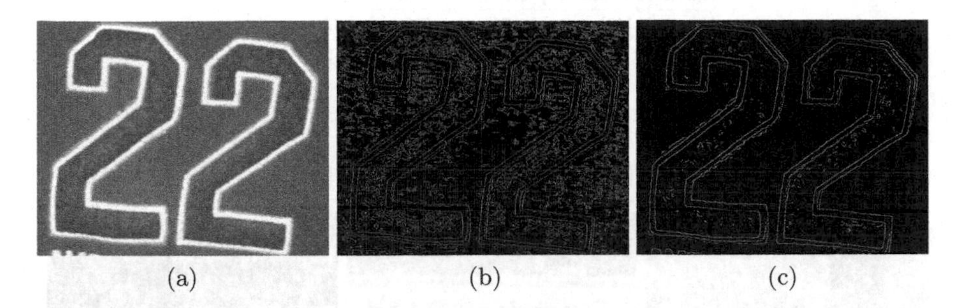

(a) (b) (c)

Fig. 2. Edge detection results for the word "22": (a) Original image; (b) result obtained by Canny operator; (c) result obtained wavelet transform.

4.3 Post Processing

In the post processing step, we further filter out some text candidate regions by verifying the behavior of the sparse representation of candidate text as suggestion in [10]. Afterward, the adaptive runlength projection profile analysis [22] is applied to decide the the true texts. Finally, text candidates are grouped using some heuristic criteria, such as as neighborhood, inter-distance, orientation, and overlapping [38].

5 Experiment Results

In this section, experiments are performed on images from ICDAR'03 text locating test set[1]. These images include the texts with variety of color and fonts with complex background and various orientations. Moreover, these images are different to the images used in the training process to create the learned dictionaries.

Normally, the results of methods using sparse representation depend on two main parameters: size of the patch, size of the dictionary (number of columns M). Because finding the best values of M, s takes exhausted search, therefore we fix the value of $M = 512$, and do the experiments with value of s from 8 to 22. The best tradeoff between the complexity and the performance with $s = 16$.

Measure used to evaluated methods is the same measure proposed by *C. Yao et al.* [40], in which the definitions of precision and recall rate as following:

$$\text{Precision} = \frac{\text{Number of true positive regions}}{\text{Number of estimated rectangles}} \tag{12}$$

[1] http://www.iapr-tc11.org/mediawiki/index.php/ICDAR_2003_Robust_Reading_Competitions.

Table 1. Performance of different approaches for text region extraction.

Method	Precision	Recall	F-measure
TD-Mixture [40]	0.69	0.66	0.67
Epshtein et al. [15]	0.73	0.60	0.66
Yi et al. [42]	0.71	0.62	0.62
Becker et al. [25]	0.62	0.67	0.62
Chen et al. [8]	0.60	0.60	0.58
Proposed method	0.76	0.88	0.82

Fig. 3. Some text detection results using the proposed method.

$$\text{Recall} = \frac{\text{Number of true positive regions}}{\text{Number of ground truth rectangles}} \tag{13}$$

In which, the estimated rectangle is defined as the minimum rectangles around the detected text regions. The true positive region is the estimated region satisfies the conditions of the angle between this region and the ground true rectangle is less than $\pi/8$ and their overlap ration exceeds 0.5.

The F-measure being a combination of the precision and recall rate is defined as following:

$$f = \frac{2 \times \text{Precision} \times \text{Recall}}{(\text{Precision} + \text{Recall})} \tag{14}$$

Table 1 presents the quantitative comparisons of different methods mentioned in [44]. Obviously, the precision and recall rate of proposed method are higher than that of *Epshtein et al*'s method. It indicates that using sparse representation over two learned dictionaries achieves much better performance than the state-of-the-art methods when dealing with scene images having different text font, color, and size (see Fig. 3).

6 Conclusion

In this paper, we use a sparse representation over two learned dictionaries for text detection over scene images. Firstly, edges of an input image is extracted by the wavelet transform. Then comparing their reconstruction errors over learned dictionaries to classify the text and non-text regions. Experimental results present that the proposed text detection method outperforms the state-of-arts. In the future work, we will verify the performance of our approach in both precision and computing time when detecting the text regions over videos.

Acknowledgements. This research is funded by the Vietnam National University, Hanoi (VNU) under project number QG.18.04.

References

1. Aerschot, W., Jansen, M., Bultheel, A.: Normal mesh based geometrical image compression. Image Vis. Comput. **27**(4), 459–468 (2009)
2. Aharon, M., Elad, M., Bruckstein, A.: K-SVD: An algorithm for designing overcomplete dictionaries for sparse representation. Sig. Process. **54**(11), 4311–4322 (2006)
3. Angadi, S., Kodabagi, M.: A texture based methodology for text region extraction from low resolution natural scene images. In: Advance Computing Conference, pp. 121–128 (2010)
4. Belaid, A., Santosh, K., D'Andecy, V.P.: Handwritten and printed text separation in real document. In: The Thirteenth International Conference on Machine Vision Applications (2013)

5. Bui, T., Pan, W., Suen, C.: Text detection from natural scene images using topographic maps and sparse representations. In: The IEEE International Conference on Image Processing (2009)
6. Chen, D., Jean-Marc, O., Herve, B.: Text detection and recognition in images and video frames. Pattern Recogn. **37**(3), 595–608 (2004)
7. Chen, S.S., Donoho, D.L., Saunders, M.A.: Atomic decomposition by basis pursuit. SIAM J. Sci. Comput. **20**(1), 33–61 (1998)
8. Chen, X., Yuille, A.: Detecting and reading text in natural scenes. In: Proceeding of CVPR (2004)
9. Daubechies, I., Devore, R., Fornasier, M., Gunturk, C.: Iteratively reweighted least squares minimization for sparse recovery. Commun. Pure Appl. Math. **63**(1), 1–38 (2009)
10. Do, T.H., Tabbone, S., Terrades, O.R.: Text/graphic separation using a sparse representation with multi-learned dictionaries. In: The International Conference on Pattern Recognition, pp. 689–692 (2012)
11. Do, T.H., Tabbone, S., Terrades, O.R.: Document noise removal using sparse representations over learned dictionary. In: ACM Symposium on Document Engineering, pp. 161–168 (2013)
12. Donoho, D., Elad, M.: Optimally sparse representation in general (nonorthogonal) dictionaries via ell1 minimization. PNAS **100**(5), 2197–2202 (2003)
13. Elad, M.: Sparse and Redundant Representation: From Theory to Applications in Signal and Images Processing. Springer, New York (2010). https://doi.org/10.1007/978-1-4419-7011-4
14. Engan, K., Skretting, K., Husoy, J.H.: Family of iterative LS-based dictionary learning algorithm, ILS-DLA, for sparse signal representation. Digit. Signal Process. **17**(1), 32–49 (2007)
15. Epshtein, B., Ofek, E., Wexler, Y.: Detecting text in natural scenes with stroke width transform. In: Proceedings of the CVPR (2010)
16. Ezaki, N., Bulacu, M., Schomaker, L.: Text detection from natural scene images: towards a system for visually impaired persons. In: Proceedings of the 17th International Conference on Pattern Recognition, vol. 2, pp. 683–686 (2004)
17. Jain, A., Yu, B.: Automatic text location in images and video frames. Pattern Recogn. **31**(12), 2055–2076 (1998)
18. Jiang, R., Qi, F., Xu, L., Wu, G.: Using connected components' features to detect and segment text. J. Image Graph. **11**, 1653–1656 (2006)
19. Kim, K., Jung, K., Kim, J.: Texture-based approach for text detection in images using support vector machines and continuously adaptive mean shift algorithm. IEEE Trans. Pattern Anal. Mach. Intell. **25**(12), 1631–1639 (2003)
20. Kumar, S., Gupta, R., Khanna, N., Chaudhury, S., Joshi, S.: Text extraction and document image segmentation using matched wavelets and MFR model. IEEE Trans. Image Process. **16**(8), 2117–2128 (2007)
21. Lee, T.W., Lewicki, M.: Unsupervised image classification, segmentation and enhancement using ICA mixture models. IEEE Trans. Image Process. **11**(3), 270–279 (2002)
22. Lienhart, R., Wernicke, A.: Localizing and segmenting text in images and videos. IEEE Trans. Circuits Syst. Video Technol. **12**, 256–268 (2002)
23. Lim, J., Park, J., Medioni, G.: Text segmentation in color images using tensor voting. Image Vis. Comput. **25**(5), 671–685 (2007)
24. Liu, Z., Sarkar, S.: Robust outdoor text detection using text intensity and shape features. In: The 19th International Conference on Pattern Recognition, pp. 1–4 (2008)

25. Lucas, S.M.: ICDAR 2005 text locating competition results. In: Proceedings of the ICDAR (2005)
26. Mallat, S.: Geometrical grouplets. Appl. Comput. Harmonic Anal. **26**(2), 161–180 (2009)
27. Mallat, S.G., Zhang, Z.: Matching pursuits with time-frequency dictionaries. Sig. Process. **41**(12), 3397–3415 (1993)
28. Mallat, S., Zhong, S.: Characterization of signals from multiscale edges. IEEE Trans. Pattern Anal. Mach. Intell. **11**(7), 710–732 (1992)
29. Marial, J., Bach, F., Ponce, J., Sapiro, G.: Online dictionary learning for sparse coding. In: Proceedings of the 26th Annual International Conference on Machine Learning, pp. 689–696 (2009)
30. Pan, W., Bui, T., Suen, C.: Text detection from scene images using sparse representation. In: Proceedings of the 19th International Conference on Pattern Recognition (ICPR 2008), pp. 1–5 (2008)
31. Pan, Y., Liu, C., Hou, X.: Fast scene text localization by learning-based filtering and verification. In: The 17th IEEE International Conference on Image Processing, pp. 2269–2272 (2010)
32. Park, J., Chung, H., Seong, Y.: Scene text detection suitable for parallelizing on multi-core. In: IEEE International Conference on Image Processing, pp. 2425–2428 (2009)
33. Pati, Y., Rezaiifar, R., Krishnaprasad, P.: Orthogonal matching pursuit: recursive function approximation with applications to wavelet decomposition. In: Proceedings of the 27th Annual Asilomar Conference on Signals, Systems, and Computers, pp. 40–44 (1993)
34. Santosh, K.C.: g-DICE: graph mining-based document information content exploitation. IJDAR **18**(4), 337–355 (2015)
35. Santosh, K.C.: Document Image Analysis. Current Trends and Challenges in Graphics Recognition. Springer, Singapore (2018). https://doi.org/10.1007/978-981-13-2339-3
36. Skretting, K., Engan, K.: Recursive least squares dictionary learning algorithm. Sig. Process. **58**(4), 2121–2130 (2010)
37. Temlyakov, V.N.: Weak greedy algorithms. Adv. Comput. Math. **12**(2–3), 213–227 (2000)
38. Hoang, T.V., Tabbone, S.: Text extraction from graphical document images using sparse representation. In: Proceedings of the 9th International Workshop on Document Analysis Systems (2010)
39. Wright, J., Ganesh, A., Yang, A., Ma, Y.: Robust face recognition via sparse representation. IEEE Trans. Pattern Anal. Mach. Intell. **31**(2), 210–227 (2009)
40. Yao, C., Bai, X., Liu, W., Ma, Y., Tu, Z.: Detecting text of arbitrary orientations in natural images. In: Proceedings of CVPR (2012)
41. Ye, Q., Jiao, J., Huang, J., Yu, H.: Text detection and restoration in natural scene images. J. Vis. Commun. Image Represent. **18**(6), 504–513 (2007)
42. Yi, C., Tian, Y.: Text string detection from natural scenes by structure-based partition and grouping. In: Image Processing (2011)
43. Yi, C., Tian, Y.: Text string detection from natural scenes by structure-based partition and grouping. IEEE Trans. Image Process. **20**(9), 2594–2605 (2011)
44. Zhao, M., Li, S., Kwok, J.: Text dectection in images using sparse representation with discriminative dictioanries. Image Vis. Comput. **28**, 1590–1599 (2010)

Word Level Plagiarism Detection of Marathi Text Using N-Gram Approach

Ramesh R. Naik(✉), Maheshkumar B. Landge(✉), and C. Namrata Mahender(✉)

Department of CS & IT, Dr. B. A. M. University, Aurangabad, (MS), India
ramesh.naik31@yahoo.com,
Maheshkumar.landge@gmail.com, nam.mah@gmail.com

Abstract. Plagiarism is increasing day by day. Plagiarism detection is one of the most complex, but a must requirement. This paper deals with word level plagiarism detection for Marathi text by using N-gram language model and a Marathi corpus. This is most simple in form still provides good depth for understanding and emphasing copy-paste and paraphrased plagiarism detection. It forms basis for sentence as well as paragraph level processing

Keywords: Plagiarism detection · N-gram · Marathi language

1 Introduction

"Plagiarism is defined as the integration of someone other's work without providing proper cite" [1] Plagiarism in a text papers by observing similarity between it and other written document is called external plagiarism detection, to determine the writing fashion of author is called the intrinsic plagiarism. n-gram approach to plagiarism detection is commonly used by many of the existing plagiarism detection tools. n-gram can be defined as a sequence of N words. In [3, 4] an ENCOPLOT organisation is described which uses 16-g and encoplot diagram to identify plagiarized written document. Different approach is used in [5] where suspicious document is schism into sentences and these sentences are rip into n-grams. Compared n-grams to identify duplication and reused papers on the web [6].

1.1 Types of Plagiarism

Copy and paste plagiarism It's most commonly done plagiarism and simplest one. Here an author copies contain from viable source and paste in its own work as if it is his own work sometimes with or without appropriately citing the source or not providing any credit to the original author [7].

K. C. Santosh and R. S. Hegadi (Eds.): RTIP2R 2018, CCIS 1037, pp. 14–23, 2019.
https://doi.org/10.1007/978-981-13-9187-3_2

Paraphrasing plagiarism: Paraphrasing word itself expresses that produce some given thing in another form. That is write the same information in different style using word order, synonyms of words, changing tense of sentences etc., and marks such writing as one's own [8].

Metaphor plagiarism: Here authors presents some one's idea or work in their own way, that is in more asthetics fashion that makes it to appear as it is his own work, but in reality it is not [9].

Idea plagiarism: It deficit that authors idea or a solutions are copied from sources, as they are available in public domain, and asserted as if those are relevant general knowledge for particular domain. Thus are free to be copied and used [10].

Mosaic plagiarism: Topic search is done and content is collected from different sources the collected information in aggregated, rephrased and recorded in collectively new form. This is finally presented as one's own work i.e. without citing the sources [7, 8].

Self-plagiarism: Author reuses his work to generate a new aspect of his previous work, which seems to be new [11].

2 Text Corpus

For this work poem is considered as the basic unit of writing as poem are generally defined as "a composition inverse, especially one that is characterized by highly developed artistic form and by the of use heightened language and rhythm to express an intensely imaginative interpretative of the subject or a piece of writing in which the expression of feelings and ideas is given instantly by particular attention to diction rhythm and imaginary". The main purpose of using poem is that when people are asked to explore and summarize the poem in their own words, it means the users have to express the summary in a way that the concept of poem to be written, where in finding how the paraphrase is done and which are those words were much level of paraphrasing was done or in other way to see how many terms or words were used literally as it is.

Poem 1	Poem 2	Poem 3
आनंदाने नाचूया !	पावसा, पावसा ये ! ये!	गवताचं पात वारयावर
चला गड्यानो खेळूया !	झरझर झरझर आला वारा,	डोलत, डोलताना म्हणत
आनंदाने नाचूया !	टपटप टपटप पडल्या	खेळायला चला.
बाग चिमुकली	गारा.	झरयातलं पाणी खळाखळा
फुले उमलली	धड्म धड्म ढगांचा बाजा,	हसत,हसताना म्हणत
सुंदर फुलली	कोकीळ म्हणतो	पळायला चला.
फुलासारखे होऊया	पाऊसराजा.	निळनिळ पाखरू आंब्यावर
फुलाभोवती नाचूया ॥	पावशा म्हणतो पाणी	गात,गाताना म्हणत
फुलाभोवती	पाणी,	नाचायला चला.
भुंगे फिरती	धरती म्हणते द्या	झिम्मड पावसात गारांची
गाणी गाती	न्हाऊनी.	बरसात,बरसात म्हणते
तसेच आपण गाऊया	नदी-नाल्यांना आला पूर,	वेचायला चला.
ताल धरोनी नाचूया ॥	केरकचरा गेला दूर.	छोटासा मोती लपाछपी
आभाळाचे तहेतहेचे	माती झाली चिंब चिंब,	खेळतो,धावताना म्हणतो
रंग मजेचे	आता उगवतील हिरवे	शिवायला चला.
चला चला रे पाहूया	कोंब.	मनीच पिल्लू पायाशी
आनंदाने नाचूया ॥	गडबड करते वीजराणी,	लोळत,लोळताना म्हणत
-- के. नारखेडे	आम्ही गातो पाऊसगाणी.	जेवायला चला.
	पावसा, पावसा ये ! ये !	--कुसुमाग्रज
	पीकपाणी खूप दे!	
	--उषाकिरण आत्राम	

Fig. 1. Three Marathi poems for database

For the development of corpus, three poems of Marathi usually most popular and small are selected. The poem names are “Anandane nachuya”, “Pavsa pavsa ye ye”, and “Gavtache pate” and having length of “17”, “16”, “24” lines. The summary for these poems were collected from 300 users each.

The text corpus contains 900 summary files from 300 authors for three Marathi poems. This 990 KB of information is available in the database in the format of text file for the further processing.

We have considered three Marathi poems namely “Anandane nachuya”, “Pavsa pavsa ye ye”, and “Gavtache pate” (Figs. 1 and 2).

कवी कुसुमाग्रज यांनी या कवितेत निसर्गाचे वर्णन करून निसर्गाशी संवाद साधण्याचा प्रयत्न केला आहे.
या कवितेत गवताचे पण, पक्षी, वर, मोती, मांजरीच पिल्लू सर्वजण त्यांच्यासोबत सहभागी होऊन
त्यांच्यासारखे कार्य करण्यास सांगतात.

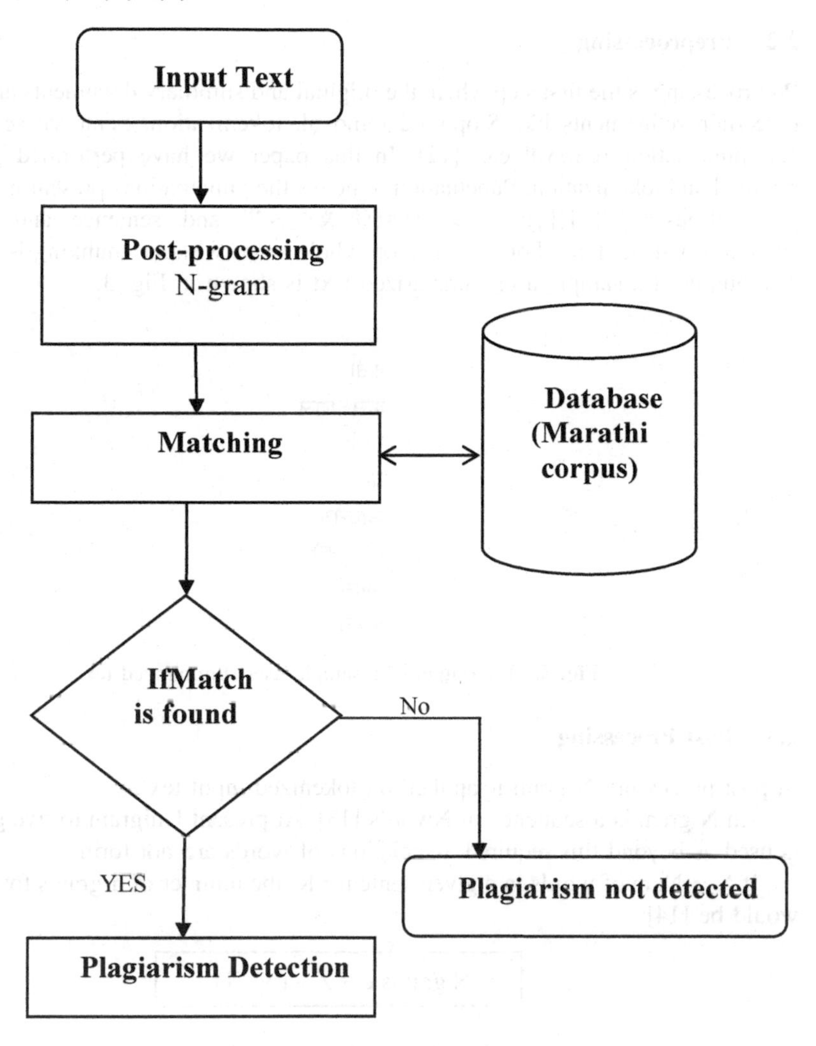

Fig. 2. Block diagram of plagiarism detection

2.1 Input

First the system reads two files. Main file of poem and summary of written by users file. The file format is .txt.

2.2 Preprocessing

Preprocessing is the first step where the original and summary documents are subjected to certain refinements like Stopword removal, tokenization, Sentence segmentation, and punctuation removal etc. [12]. In this paper we have performed punctuation removal and tokenization. Punctuation removes the punctuations present in file for e.g. punctuations = “‘‘!()-[]{};:’”\, <>./?@#$%^&*_~’’’ and sentence and word tokenization will be done. For tokenization whole word without chunking is considered. The output of a sample user summarized text is shown in Fig. 3.

कवी
कुसुमाग्रज
यांनी
या
कवितेत
निसर्गाचे
वर्णन
करून

Fig. 3. The output of a sample user summarized text

2.3 Post Processing

In post processing N-gram is applied on tokenized input text.

An N-gram is a sequence of Nwords [13]. At present Unigram to five gram model is used as beyond this required associations of words are not form.

If X = Num of words in a given sentence K, the number of n-grams for sentence K would be [14]:

$$\text{N grams}_K = X - (N - 1)$$

Unigram- It is a one word sequence of words like e.g.

पात

Bigram- It is a two-word sequence of words like example

डोलत डोलताना

Trigram- It is a three-word sequence of words like e.g.

वारयावर डोलत डोलताना

Fourgram- It is a four-word sequence of words like e.g.

वारयावर डोलत डोलताना म्हणत

Fivegram- It is a five-word sequence of words like e.g. (Table 1).

वारयावर डोलत डोलताना म्हणत खेळायला

Following table shows n-gram representation of sentence

Table 1. N-gram representation of sentence.

Sentence: गवताचं पात वारयावर डोलत, डोलताना म्हणत खेळायला चला.

Unigram	Bigram	Trigram	Fourgram	Fivegram
गवताचं	गवताचं पात	गवताचं पात वारयावर	गवताचं पात वारयावर डोलत	गवताचं पात वारयावर डोलत डोलताना
पात	पात वारयावर	पात वारयावर डोलत	पात वारयावर डोलत डोलताना	पात वारयावर डोलत डोलताना म्हणत
वारयावर	वारयावर डोलत	वारयावर डोलत डोलताना	वारयावर डोलत डोलताना म्हणत	वारयावर डोलत डोलताना म्हणत खेळायला
डोलत	डोलत डोलताना	डोलत डोलताना म्हणत	डोलत डोलताना म्हणत खेळायला	डोलत डोलताना म्हणत खेळायला चला
डोलताना	डोलताना म्हणत	डोलताना म्हणत खेळायला	डोलताना म्हणत खेळायला चला	
म्हणत	म्हणत खेळायला	म्हणत खेळायला चला		
खेळायला	खेळायला चला			
चला				

3 Analysis

In above Tables 2, 3, and 4 shows that we have compared main file of poems with summary of poem written by different users. Firstly we have calculated unigram, bigram, trigram, Fourgram, Fivegram of main file of poem and summary of poem written by different users. Then we compared main file of unigram with users unigram file and likewise we matched bigram, trigram, Fourgram, Fivegram files and calculated the matching percentage for each file.

Table 2. Plagiarism percentage of main file of poem with summary of poem written by different users. (Gavtach Pat)

	1	2	3	4	5	6	7	8	9	10	11	12	13	14	15
Unigram_1	18.66	2.70	7.59	0.0	0.0	12.65	22.80	8.57	0.0	3.84	10.95	6.01	15.94	0.0	0.0
Bigram_2	16.17	2.98	8.33	0.0	0.0	13.88	22.42	9.01	0.0	3.09	13.91	5.55	16.03	0.0	0.0
Trigram_3	14.20	3.33	7.17	0.0	0.0	11.28	18.77	9.19	0.0	2.27	10.23	5.63	14.53	0.0	0.0
Fourgram_4	12.03	3.77	6.89	0.0	0.0	10.34	16.02	9.48	0.0	2.15	6.74	6.73	12.5	0.0	0.0
Fivegram_5	11.06	3.47	6.27	0.0	0.0	11.02	10.56	4.39	0.0	2.93	6.71	0.0	8.82	0.0	0.0

Table 3. Plagiarism percentage of main file of poem with summary of poem written by different users. (Anadane Nachuya)

	1	2	3	4	5	6	7	8	9	10
Unigram_1	5.33	4.70	12.19	11.49	12.19	7.31	2.12	5.40	4.54	9.09
Bigram_2	3.44	4.76	8.13	9.37	8.13	4.87	1.48	5.29	3.10	6.20
Trigram_3	2.58	3.63	6.17	7.18	6.17	3.70	1.14	4.38	2.38	4.76
Fourgram_4	2.08	2.97	5.02	5.88	5.02	3.01	0.94	3.77	1.95	3.90
Fivegram_5	2.80	2.67	4.52	5.30	4.52	2.71	0.85	3.48	1.76	3.52

Table 4. Plagiarism percentage of main file of poem with summary of poem written by different users. (Pavsa pavsa)

	1	2	3	4	5	6	7	8	9	10
Unigram_1	4.54	2.19	13.00	21.53	4.30	5.26	6.06	15.38	6.83	4.93
Bigram_2	3.70	2.32	15.78	24.34	3.44	5.82	6.72	16.82	7.54	3.89
Trigram_3	1.68	1.58	16.36	22.49	2.35	5.94	6.26	16.50	7.69	2.66
Fourgram_4	0.0	0.0	13.77	20.67	0.63	6.36	10.47	17.46	8.0	0.0
Fivegram_5	0.0	0.0	10.43	19.39	0.0	7.79	8.87	20.47	8.31	0.0

Table 5. Average percentage of copy & paste plagiarism for three poems

N-gram files	Average of copy & paste plagiarism (Gavatach Pat)	Average of copy & paste plagiarism (Anandane nachuya)	Average of copy & paste plagiarism (pavsa pavsa)
Unigram_1	7.31	11.01	8.402
Bigram_2	7.42	9.70	9.037
Trigram_3	6.44	8.35	8.351
Fourgram_4	5.77	7.58	7.736
Fivegram_5	4.34	7.09	7.526

Above Table 5 shows average percentage of copy & Paste plagiarism for three Marathi poems namely ‘Gavtach pat’, ‘Anandane nachuya’, and ‘Pavsa pavsa’. The average plagiarism percentage for poem Gavtach pat is highest using bigram representation i.e. 7.42, likewise average plagiarism percentage for poem Anandane nachuya is highest using unigram i.e.11.01 and the poem pavsa pavsa is having highest percentage using bigram i.e. 9.037.

In above Tables 6, 7 and 8 show that we have compared main file of poems with summary of poem written by different users. Firstly we have calculated unigram, bigram, trigram, Fourgram, Fivegram of main file of poem and summary of poem written by different users. Then we compared main file of unigram with users uni-gram file and likewise we matched bigram, trigram, Fourgram, Fivegram files and calculated the matching percentage for each file. Also we matched user summary file.

Table 6. Paraphrasing plagiarism percentage of main file of poem with summary of poem written by different users. (Gavtach Pat)

	1	2	3	4	5	6	7	8	9	10	11	12	13	14	15
Unigram_1	16.27	4.70	6.66	0.0	0.0	13.33	20.8	7.94	0.0	3.47	10.19	5.55	14.76	4.87	3.33
Bigram_2	13.75	3.79	7.14	0.0	0.0	14.28	21.84	8.95	0.0	3.66	12.59	5.79	15.38	3.94	2.63
Trigram_3	9.90	2.73	5.12	0.0	0.0	11.96	20.25	10.66	0.0	3.30	10.23	6.81	15.24	2.85	1.95
Fourgram_4	8.08	2.23	4.16	0.0	0.0	10.41	16.90	10.30	0.0	3.59	6.81	6.79	12.20	2.34	1.75
Fivegram_5	7.09	1.96	3.63	0.0	0.0	10.33	10.02	3.44	0.0	4.02	6.99	0.0	8.64	2.06	1.69

Table 7. Paraphrasing plagiarism percentage of main file of poem with summary of poem written by different users. (Pavsa pavsa)

	1	2	3	4	5	6	7	8	9	10
Unigram_1	2.77	1.34	10.16	15.73	4.0	4.81	5.49	11.76	4.73	2.85
Bigram_2	3.25	1.56	11.53	17.83	3.100	4.13	4.96	13.42	5.40	3.36
Trigram_3	3.63	2.06	14.67	20.16	4.095	5.27	5.65	14.73	6.85	4.56
Fourgram_4	0	0	13.77	20.67	0.63	6.36	10.47	17.46	8.0	0
Fivegram_5	0	0	10.43	19.39	0	7.79	8.87	20.47	8.31	0

Table 8. Paraphrasing plagiarism percentage of main file of poem with summary of poem written by different users. (Anandane nachuya)

	1	2	3	4	5	6	7	8	9	10
Unigram_1	4.49	4.16	4.25	4.08	4.25	4.25	1.88	6.53	3.96	8.08
Bigram_2	3.52	3.22	3.33	3.17	3.33	3.33	1.96	6.20	3.09	6.31
Trigram_3	2.89	2.59	1.80	2.53	1.80	8.45	2.32	15.53	1.64	5.06
Fourgram_4	2.23	2.0	1.38	11.84	1.38	8.33	9.67	12.24	1.29	3.89
Fivegram_5	8.97	5.86	12.84	10.74	12.84	9.28	11.20	8.64	2.94	5.69

With wordnet for identifying paraphrasing plagiarism in which user uses synonyms for words.

Above Table 9 shows average percentage of paraphrasing plagiarism for three Marathi poems namely 'Gavtach pat', 'Anandane nachuya', and 'Pavsa pavsa'. The average plagiarism percentage for poem 'Gavtach pat' is highest using unigram representation i.e. 11.01, likewise average plagiarism percentage for poem 'Anandane nachuya' is highest using Fivegram i.e. 8.9 and the poem 'pavsa pavsa' is having highest percentage using trigram i.e. 8.16.

Table 9. Average percentage of Para-phrasing plagiarism for three poems

N-gram files	Average percentage of Paraphrasing plagiarism (Gavtach Pat)	Average percentage of Para-phrasing plagiarism (Anandane nachuya)	Average percentage of Paraphrasing plagiarism (Pavsa pavsa)
Unigram_1	11.01	4.593	6.364
Bigram_2	9.70	3.746	6.854
Trigram_3	8.35	4.461	8.1675
Fourgram_4	7.58	5.425	7.736
Fivegram_5	7.09	8.9	7.526

As discussed through definition of poem, in poems words are makeup, decorated in such way its asthetics creates a so called beautiful environment enhancing the overall view of the concept, the same when is asked to be explained in a summarized form, accepted is that the representation will be simple and clear to express the terms and concepts. Thus majorly we expect that copy to paste types of form will be not be much identical is both but our study using N-gram model express that bi-gram representation shows minimum of 7.42 the maximum of 9.037% of plagiarized data for copy-paste. And for paraphrased detection it has given results from 8.16 for trigram and the 8.9 for Fivegram. It shows that these results further can be utilized on sentence and paragraph based plagiarism detection.

4 Conclusion

The act of plagiarizing is when you use someone else's words and ideas without giving due credits to the original writer and regard the work as your own. We created Marathi text corpus using three poems and collected summaries from different users. We have applied punctuation removal and tokenization on input text. We have calculated n-gram for main file of poem and summary written by users. We compared n-gram of main file and summary file. Average percentage for copy & paste plagiarism and paraphrasing plagiarism is higher using unigram i.e.11.01. Unigram gives better results for paraphrasing and copy & paste plagiarism detection

Acknowledgement. Authors would like to acknowledge and thanks to CSRI DST Major Project sanctioned No.SR/CSRI/71/2015(G), Computational and Psycholinguistic Research Lab Facility supporting to this work and Department of Computer Science and Information Technology, Dr. Babasaheb Ambedkar Marathwada University, Aurangabad, Maharashtra, India.

References

1. University of Melbourne (2005). What is plagiarism? https://services.unimelb.edu.au/__data/assets/pdf_file/0004/821668/5297-Avoiding-PlagiarismWEB.pdf. Accessed 27 June 2018
2. Paul clough, plagiarism in natural and programming languages an overview of current tools and technologies, Technical report, University of Sheffeld, Sheffeld, UK, June 2000
3. Grozea, C., et al.: ENCOPLOT: pairwise sequence matching in linear time applied to plagiarism detection. In 3rd PAN Workshop. Uncovering Plagiarism, Authorship, and Social Software Misuse, p. 10 (2009)
4. Grozea, C., Popescu, M.: Who's the thief? automatic detection of the direction of plagiarism. In: Gelbukh, A. (ed.) CICLing 2010. LNCS, vol. 6008, pp. 700–710. Springer, Heidelberg (2010). https://doi.org/10.1007/978-3-642-12116-6_59
5. Barrón-Cedeño, A., Rosso, P.: On automatic plagiarism detection based on *n*-grams comparison. In: Boughanem, M., Berrut, C., Mothe, J., Soule-Dupuy, C. (eds.) ECIR 2009. LNCS, vol. 5478, pp. 696–700. Springer, Heidelberg (2009). https://doi.org/10.1007/978-3-642-00958-7_69
6. Chiu, S., Uysal, I., Croft, B.W.: Evaluating text reuse discovery on the web. In: Proceedings of the Third Symposium on Information Interaction in Context, pp. 299–304 (2010)
7. Weber Wulff, D.: Copy, Shake, and Paste- A blog about plagiarism from a German professor, written in English. http://copy-shake-paste.blogspot.com. Accessed 28 June 2018
8. Lancaster, T.: Effective and efficient plagiarism detection. Ph.D. thesis, school of computing, information systems and mathematics south bank university (2003)
9. Barnbaum, C.: Plagiarism: A Student's Guide to Recognizing It and Avoiding It. Valdos Ta state university. http://www.valdosta.edu/cbarnbau/personal/teaching_MISC/plagiarism.htm. Accessed 28 June 2018
10. Maurer, H., et al.: Plagiarism-a survey. J. Univ. Comput. Sci. **12**, 1050–1084 (2006)
11. Bretag, T., Mahmud, S.: Self-plagiarism or appropriate textual re-use. J. Acad. Ethics **7**, 193–205 (2009)
12. Vani, K., Gupta, D.: Using k-means cluster based techniques in external plagiarism detection. In: 2014 International Conference on Contemporary Computing and Informatics (IC3I), pp. 1268–1273. IEEE 2014
13. Jurafsky, D., Martin, J.H.: Text book on "Speech and Language Processing", Copyright c 2016. All rights reserved (2017)
14. What-are-n-grams.html. http://text-analytics101.rxnlp.com/2014/11/. Accessed 18 Aug 2018

UHTelPCC: A Dataset for Telugu Printed Character Recognition

Rakesh Kummari(✉) and Chakravarthy Bhagvati(✉)

School of Computer and Information Sciences,
University of Hyderabad, Hyderabad 500046, India
rakeshkummarics@gmail.com, chakcs@uohyd.ernet.in

Abstract. This paper describes how UHTelPCC, a dataset for Telugu printed character recognition, is created and its characteristics. The dataset is created from characters extracted from images of printed Telugu texts from the period 1950–1990. Thus, it is hoped that the dataset provides the basis for developing practical Telugu OCR systems. UHTelPCC is to provide a standard benchmark for comparing different algorithms for Telugu OCR and helps in research and development of Telugu OCR systems. UHTelPCC contains 70K samples of 325 classes, and these samples are divided into 50K, 10K, 10K training, validation, and test sets respectively. It is hoped that UHTelPCC serves like MNIST, a dataset for handwritten digit recognition, for Telugu printed character recognition. The baseline performances on the test set using KNN, MLP, and CNN are 98.85%, 99.52%, and 99.68% respectively. UHTelPCC is available at http://scis.uohyd.ac.in/~chakcs/UHTelPCC.html.

Keywords: Optical Character Recognition · OCR · Printed Telugu OCR · UHTelPCC · Telugu dataset · OCR dataset · Telugu character dataset

1 Introduction

Optical Character Recognition (OCR) is a well-known problem in the pattern recognition community. The steps involved in the OCR system and the research progress in OCR systems can be found in [2,3,5,7,11,19,28,30,33]. The primary approach in classifier development for OCR systems is supervised learning with training done on a labeled dataset of several thousand characters. Dataset plays a vital role in the design of supervised learning systems.

The following are available datasets for OCR systems of different languages. *NIST-MPDB* (NIST machine printed database) is in grayscale and binary images. It contains approx. 3 million English characters from 360 document pages. It is also known as *Special Database 8*. *Letter Recognition Dataset* from the UCI machine learning repository contains 20K samples of 26 classes (uppercase English alphabet). These 20K samples are obtained from 20 different fonts and random distortions. *InftyCDB-1* contains 688,580 characters from 476 pages

K. C. Santosh and R. S. Hegadi (Eds.): RTIP2R 2018, CCIS 1037, pp. 24–36, 2019.
https://doi.org/10.1007/978-981-13-9187-3_3

of 30 mathematics articles in English. *InftyCBD-2* contains 662,142 characters from English articles, 37,439 characters from French articles and 77,812 characters from German articles. *InftyCDB-3* is divided into two parts *InftyCDB-3A* and *InftyCDB-3B*. *InftyCDB-3A* contains 188,752 characters extracted from 300 sources which includes Japanese documents. *InftyCDB-3B* contains 70,637 characters from 20 articles of *InftyCDB-1*. All three *InftyCDB* datasets are available at www.inftyproject.org/en/database.html. *ETL Character Database* contains 1.2 million samples that include handwritten and machine printed numerals, symbols, Latin alphabet, and Japanese characters. This dataset is created by Electrotechnical laboratory. It is available at etlcdb.db.aist.go.jp. *APTI Database* [31] (Arabic Printed Text Image Database) is a synthetically generated dataset using 10 different fonts in 10 different sizes with 4 styles. It is available at http://diuf.unifr.ch/diva/APTI/download.html. It contains 250 million characters and 45 million words. *UW-I* contains 1147 pages of English technical journals. *UW-II* contains 623 pages of English technical journals, 63 pages of English memorandum and 477 pages of Japanese technical journals. *UW-III* contains line and word bounding boxes for the English pages in UW-I [9].

The research on OCR for Indic scripts is started from 1970 [8]. Pal and Chaudhuri [23], and Srinivas *et al.*[32] are presented a review of Indic OCR, Govindaraju and Setlur [8] edited a book on OCR for Indic scripts. Some of the early works on handwritten OCR for Indic scripts are present in [4,10,12–17,24,29]. The datasets for Indic OCR systems are developed as part of Development of Robust Document Analysis and Recognition System for Indian Languages (OCR) project. Broadly Indic scripts are divided into two categories: scripts with *shirorekha*(Devanagari, Bangla, Gurmukhi) and scripts without *shirorekha* (Telugu, Tamil, Kannada, Malayalam). Zone-based OCR system for Indic scripts with *shirorekha* and connected component based OCR system for Indic scripts without *shirorekha* are used. Telugu OCR is using connected component approach.

Telugu is one of the official languages in India and it is primary (native or mother tongue) language in Telangana and Andhra Pradesh states. Approximately 80 million people speak Telugu around the world [26]. The details of Telugu script can be found in [22]. The research progress on Telugu OCR can be found in [22,27,32]. A standard dataset is needed to compare one approach against another. Though research on Telugu OCR started in the 1970s, there is no publicly available standard dataset for Telugu OCR till 2015. Achanta *et al.* created a synthetic Telugu glyph dataset consisting of 73K glyph samples from 460 classes and made it publicly available [1]. Prakash *et al.* also published a synthetic dataset with 17387 classes [26]. Synthetic datasets may not be a good representative of real data and the real dataset is highly valued in the OCR community. As there is no real dataset for Telugu, we are attempting to create such a dataset called UHTelPCC. This paper describes UHTelPCC, a real dataset for Telugu printed OCR.

1.1 Terminology

Table 1 shows basic terms and their definitions or meanings.

Table 1. Basic terms and their definitions or meanings

Term	Definition or meaning
Achchu	A vowel
Akshara	A C*V syllable where C, V are consonant, vowel sounds respectively [20]
Connected Component(CC)	A basic recognizable unit by OCR [22] and it is also referred as glyph
Gunitham	Consonant and its vowel modified variants
Hallu	A consonant
Maatra	Vowel sound that combine with consonants [20]
Pure consonant	Consonant ending with *halanth*(consonant without vowel sound)
Voththu	A half consonant [22]

The organization of remainder of this paper is as follows: Sect. 2 describes connected components. Section 3 describes how the UHTelPCC is created and the details of the UHTelPCC. Section 4 describes experiments using UHTelPCC and baseline accuracies on the test set of UHTelPCC using KNN, MLP, and CNN. In the last section, the summary of the paper and conclusion are provided.

2 Connected Components

Telugu alphabet consists of 16 vowels, 36 consonants. In addition to vowels and consonants, 16 *maatras*, 36 *gunithams* and each *gunitham* consists of 16 *aksharas*. Theoretically, Telugu script has over 21K *aksharas* and these includes vowels, consonants, *maatras*, *gunithams*, *voththus* and other *aksharas* that can be formed by assembling consonants, *maatras*, and *voththus* in suitable manner. Practically, all possible *aksharas* are not used in the language and it is estimated that 5K–10K *aksharas* are frequently used in the language [22]. All possible *aksharas* can be formed by assembling around 400 distinct connected components. The exact number of distinct connected components is font-dependent and varies between 350 and 400. The UHTelPCC contains 325 distinct frequently used connected components.

Connected components may belongs to one of the following categories: (1) a vowel, (2) a consonant, (3) a *maatra*, (4) a *voththu*, (5) part of a vowel or a consonant or a *maatra*, (6) combination of a consonant and a *maatra*, (7)*halanth*, (8) a pure consonant, (9) a punctuation mark and (10) a number.

Connected components are grouped into 38 clusters based on the structure and location of the component relative to the line of text. These clusters include vowels, pure consonants, *gunithams*, *maatras*, *voththus*, Arabic numbers, *halanth* and punctuations. Table 2 shows clusters and their size and are sorted in descending order of cluster size. Romanization is used to represent *gunitham* cluster names [21]. Figure 1 shows vowels, Fig. 2 shows *maatras*, Fig. 3 shows pure consonants, Fig. 4 shows *voththus*, Fig. 5 shows Arabic numbers. Figure 6 shows *k gunitham* and its 11 distinct connected components. All other *gunithams* have 11 distinct connected components except *gh,p,ph,s,Sh*, and *h* have only 7 distinct connected components because few *maatra* are disconnected. Figure 7 shows *p gunitham* and its has only 7 distinct connected components. Characters *D* and *Dh* have two distinct representations per each of them. Figure 8 shows different connected components of *D*.

Table 2. Clusters and their size

Clusters	Number of glyphs
voththus	27
D	20
vowels	13
pure consonants	12
k,g,d,n,bh,m,y,r,l	11
Arabic numbers,c,j,T,t,b,sh	10
L,v	9
kh	8
maatras,dh,p	7
N	6
h	5
jh,ph,rZ,s	4
Sh	3
gh,Dh	2
halanth,punctuations,nY	1

Fig. 1. Vowels

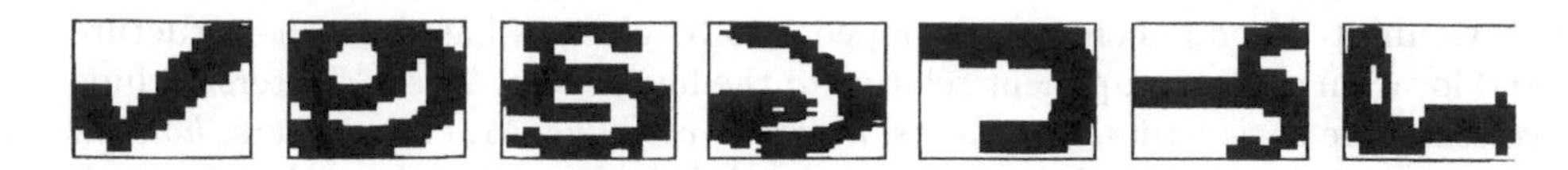

Fig. 2. *Maatras*

Fig. 3. Pure consonants

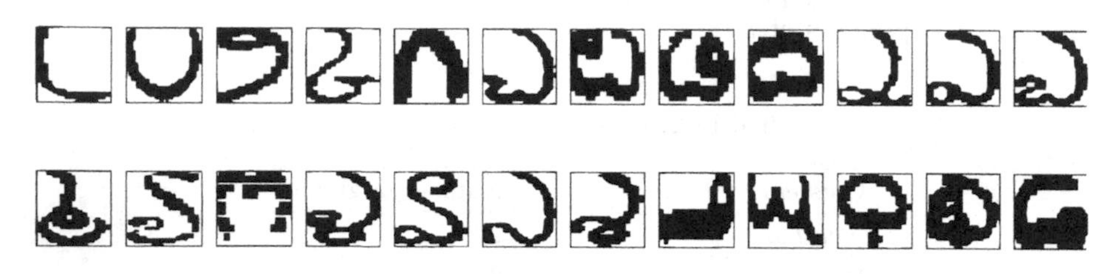

Fig. 4. *Voththus*

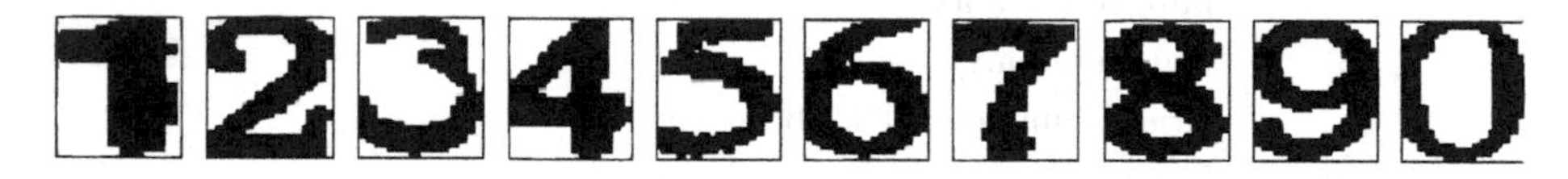

Fig. 5. Arabic numbers

Fig. 6. *k gunitham*

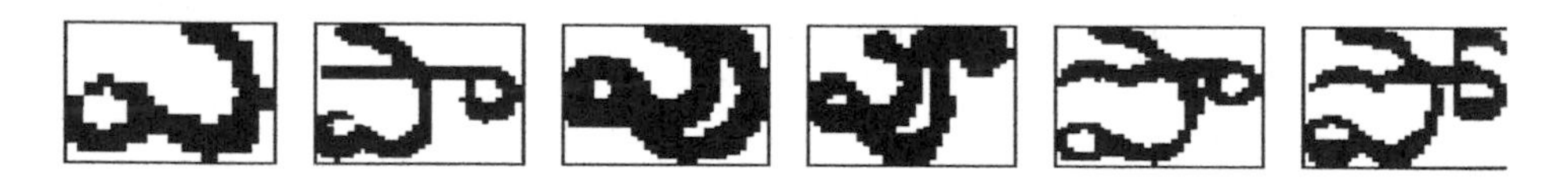

Fig. 7. *p gunitham*

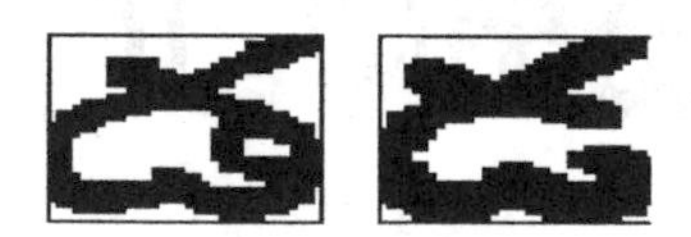

Fig. 8. Two different representations for *D*

3 UHTelPCC Creation and Its Characteristics

3.1 UHTelPCC Creation

UHTelPCC is created from Telugu printed fiction books scanned at 300 dpi. These books are printed in different fonts during the years 1950–1990. The steps in UHTelPCC creation are pre-processing, Connected Component(CC) extraction, normalization, and labeling. The block diagram of UHTelPCC creation is depicted in Fig. 9. First, scanned Telugu document page image is binarized using global thresholding method, then noise removal and skew correction are applied on binarized document page image obtained after pre-processing step. Then connected components are extracted using standard connected component extraction algorithm [6] and then the extracted connected components are normalized to 32×32. Finally, normalized connected components are labeled manually. Figure 10 is a sample Telugu document page image and Fig. 11 is a portion of a sample Telugu document page image with CC bounding boxes.

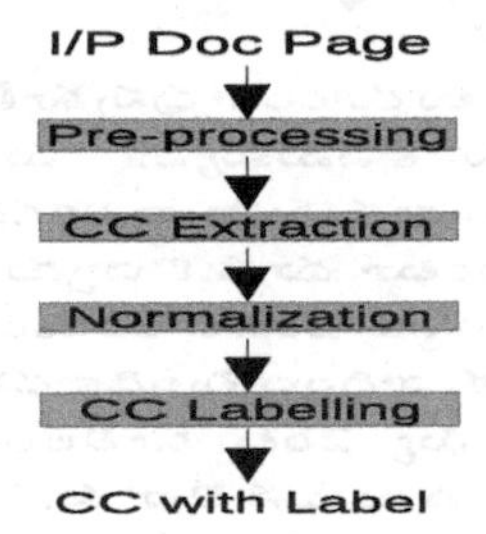

Fig. 9. Block diagram of UHTelPCC creation

3.2 Sample Distribution

UHTelPCC contains 70K samples (connected components) of 325 classes (distinct connected components). Each sample is a connected component. The number of samples in each class ranges from one to 5958. Figure 12 is a bar chart with the number of samples range on X-axis and number of classes having samples in particular range on Y-axis. In Fig. 12, green, blue, cyan, magenta and red colors indicate the number of classes in the range 11–20, 21–30, 31–40, 41–50 and 51–60 respectively. In UHTelPCC, 33 classes have either one or two samples, 22 classes have samples in the range 3–5, 30 classes are having samples in the range 6–10, 15 classes are having samples in the range 11–15, 14 classes are having samples in the range 16–20, 41 classes are having samples in the range 21–50, 48 classes are having samples in the range 51–100, 29 classes are having samples in the range 101–200, 53 classes are having samples in the range 201–500, 28 classes have samples in the range 501–1000 and 12 classes are having samples in the range 1001–6000.

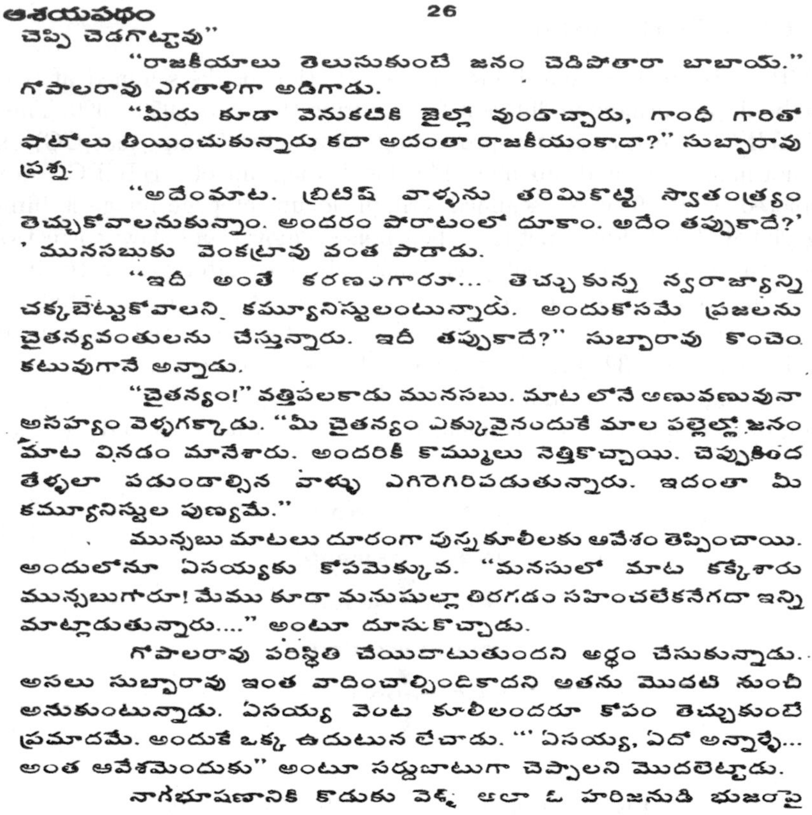

ఆశయపథం 26

చెప్పి చెడగొట్టావు"

"రాజకీయాలు తెలుసుకుంటే జనం చెడిపోతారా బాబాయ్." గోపాలరావు ఎగతాళిగా అడిగాడు.

"మీరు కూడా వెనుకటికి జైల్లో వుండొచ్చారు, గాంధీ గారితో ఫొటోలు తీయించుకున్నారు కదా అదంతా రాజకీయంకాదా?" సుబ్బారావు ప్రశ్న.

"అదేంమాట. బ్రిటిష్ వాళ్ళను తరిమికొట్టి స్వాతంత్ర్యం తెచ్చుకోవాలనుకున్నాం. అందరం పోరాటంలో దూకాం. అదేం తప్పుకాదే?' ' మునసబుకు వెంకట్రావు వంత పాడాడు.

"ఇదీ అంతే కరణంగారూ... తెచ్చుకున్న స్వరాజ్యాన్ని చక్కబెట్టుకోవాలని కమ్యూనిస్టులంటున్నారు. అందుకోసమే ప్రజలను చైతన్యవంతులను చేస్తున్నారు. ఇదీ తప్పుకాదే?" సుబ్బారావు కొంచెం కటువుగానే అన్నాడు.

"చైతన్యం!" వత్తిపలకాడు మునసబు. మాట లోనే అణువణువునా అసహ్యం వెళ్ళగక్కాడు. "మీ చైతన్యం ఎక్కువైనందుకే మాల పల్లెల్లో జనం మాట వినడం మానేశారు. అందరికీ కొమ్ములు నెత్తికొచ్చాయి. చెప్పుకింద తేళ్ళులా పడుండాల్సిన వాళ్ళు ఎగిరెగిరిపడుతున్నారు. ఇదంతా మీ కమ్యూనిస్టుల పుణ్యమే."

మున్సబు మాటలు దూరంగా వున్న కూలీలకు ఆవేశం తెప్పించాయి. అందులోనూ ఏసయ్యకు కోపమెక్కువ. "మనసులో మాట కక్కేశారు మున్సబుగారూ! మేము కూడా మనుషుల్లా తిరగడం సహించలేకనేగదా ఇన్ని మాట్లాడుతున్నారు...." అంటూ దూసుకొచ్చాడు.

గోపాలరావు పరిస్థితి చేయిదాటుతుందని అర్థం చేసుకున్నాడు. అసలు సుబ్బారావు ఇంత వాదించాల్సిందికాదని అతను మొదటి నుంచీ అనుకుంటున్నాడు. ఏసయ్య వెంట కూలీలందరూ కోపం తెచ్చుకుంటే ప్రమాదమే. అందుకే ఒక్క ఉదుటున లేచాడు. "' ఏసయ్య, ఏదో అన్నారే... అంత ఆవేశమెందుకు" అంటూ సర్దుబాటుగా చెప్పాలని మొదలెట్టాడు.

నాగభూషణానికి కొడుకు వెళ్ళి ఆలా ఓ హరిజనుడి భుజంపై

Fig. 10. Sample book page

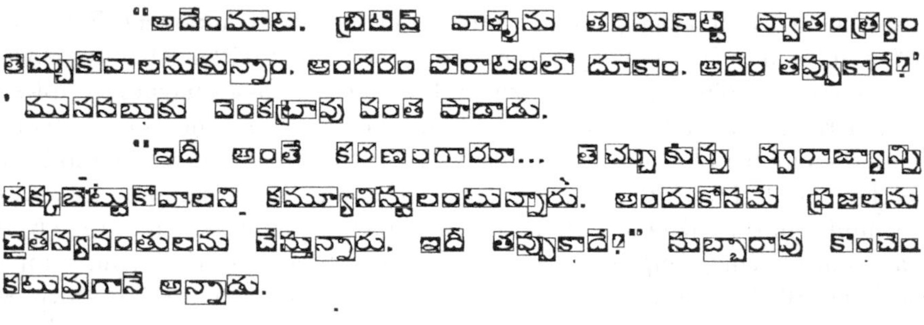

"అదేంమాట. బ్రిటిష్ వాళ్ళను తరిమికొట్టి స్వాతంత్ర్యం తెచ్చుకోవాలనుకున్నాం. అందరం పోరాటంలో దూకాం. అదేం తప్పుకాదే?' ' మునసబుకు వెంకట్రావు వంత పాడాడు.

"ఇదీ అంతే కరణంగారూ... తెచ్చుకున్న స్వరాజ్యాన్ని చక్కబెట్టుకోవాలని కమ్యూనిస్టులంటున్నారు. అందుకోసమే ప్రజలను చైతన్యవంతులను చేస్తున్నారు. ఇదీ తప్పుకాదే?" సుబ్బారావు కొంచెం కటువుగానే అన్నాడు.

Fig. 11. A portion sample book page with CC bounding boxes

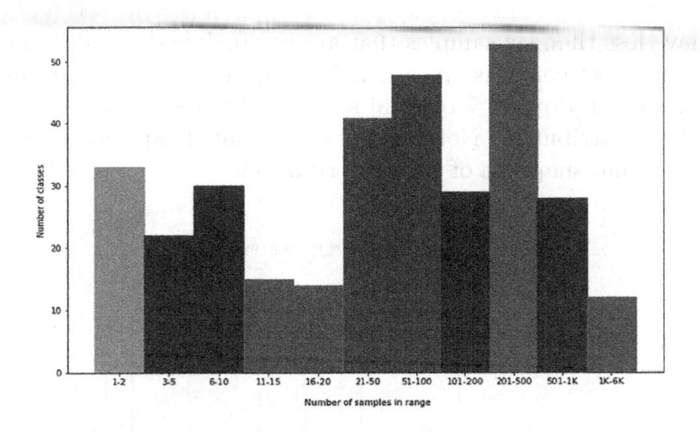

Fig. 12. Sample distribution (Color figure online)

Figure 13 shows sample coverage by the number of classes where X-axis is the number of classes and Y-axis is the percentage of samples covered by the number of classes. From Fig. 13, only 106 classes cover 1% of samples, 10% of samples covered by only 215 classes. 300 classes cover 50% of samples and rest 25 classes covers 50% of the samples.

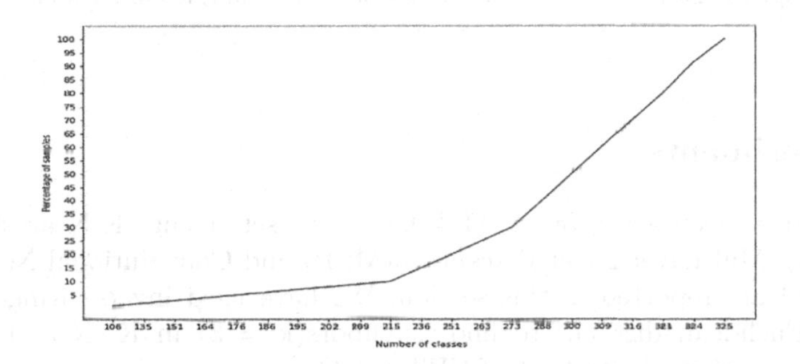

Fig. 13. Sample coverage

3.3 Data Splits

As mentioned in Sect. 2, UHTelPCC contains 70K binary samples of size 32×32. Each sample can have two values either zero (foreground pixel) or 255 (background pixel). The 70K samples are divided into three sets (training, validation and test sets) like MNIST. The training set contains 50K samples of 325 classes, validation and test set contains 10K samples each from 263 classes. The training set is used for training classifier, the validation set is used for tuning hyper-parameters, and the test set is used for evaluating the performance of the model.

62 classes have less than six samples that are not included in validation and test sets. The training set contains approx. 72% of total samples, validation set, and test sets contains approx. 14% of total samples. Figure 14 shows training, validation and test distribution. Note that the class labels are not continuous and it is to give a visual snapshot of class distribution.

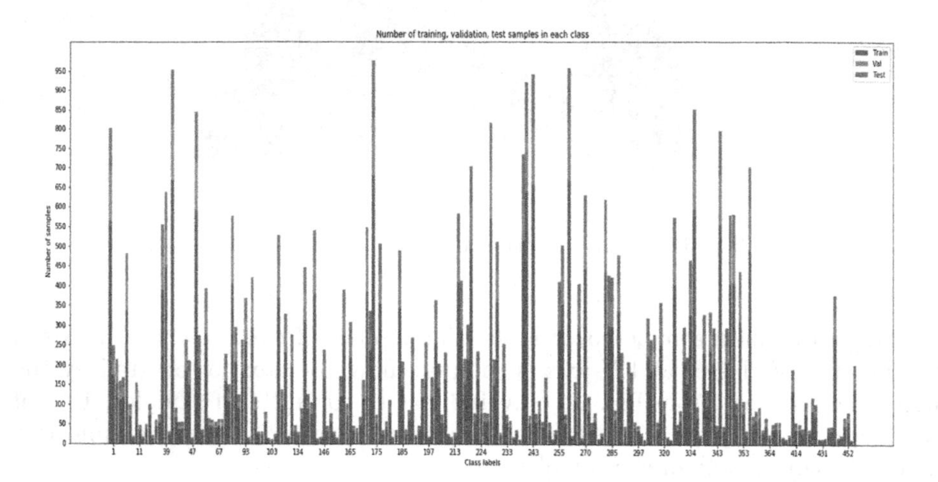

Fig. 14. Number of training, validation, test samples in each class

4 Experiments

The baseline accuracies on UHTelPCC test set using K-Nearest Neighbors(KNN), Multilayer Layer Perceptron(MLP) and Convolutional Neural Network(CNN) are reported in this section. We have used inverse fringes [25] as features, Euclidean distance to find neighbors(k = 3) in KNN and achieved 98.85% accuracy on the test set of UHTelPCC.

Table 3. KNN, MLP and CNN accuracy and loss on test set of UHTelPCC

Model	Accuracy	Loss
KNN	98.85%	-
MLP	99.55%	0.029
CNN	99.75%	0.009

Figure 15 shows MLP architecture that contains 1024 neurons in input layer, 512 neurons in each of the hidden layers and 508 neurons in output layer. *Sigmoid* activations are used in hidden layers and *softmax* activations are used in output layer. The *adam* optimizer and categorical cross-entropy loss function are used. The left plot in Fig. 16 shows training, validation accuracies for 10 epochs and the right plot in Fig. 16 shows training, validation loss for 10 epochs. Table 3 shows the model loss and accuracy.

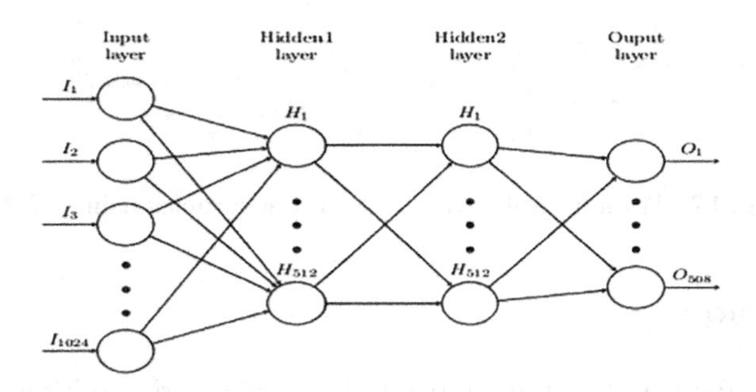

Fig. 15. MLP architecture

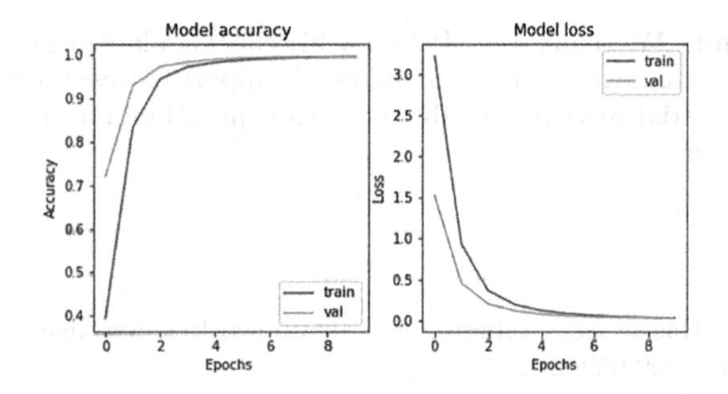

Fig. 16. Training, validation loss and accuracies using MLP

The CNN architecture is similar to LeNet5 [18]. The input size is 32 × 32, the number of nodes in each fully connected layer is 1000, 750, 508 are used in this architecture. *ReLU* activations are used in convolutional, hidden layers and *softmax* activations are used in output layer. The *adam* optimizer and categorical cross-entropy loss function are used. The plot on the left in Fig. 17 shows training and validation accuracies for ten epochs and the plot on the right in Fig. 17 shows training and validation loss for ten epochs. Table 3 shows the model loss and accuracy. The dataset may be used for developing Telugu OCR systems or for comparing the performance of different OCR algorithms provided a reference is given to this paper.

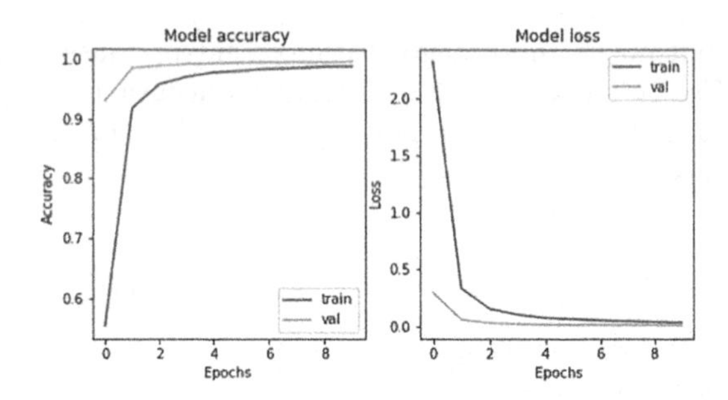

Fig. 17. Training, validation loss and accuracies using CNN

5 Conclusion

UHTelPCC, a dataset for Telugu printed character recognition is proposed. UHTelPCC contains 70K binary connected components of size 32×32 from 325 classes. These 70K samples are divided into the training set(50K), the validation set(10K) and the test set(10K). The baseline accuracies on test data are 98.85% using KNN, 99.55% using MLP and 99.75% using CNN.

Acknowledgment. We thank Amit Patel for his efforts in labeling connected components. The first author acknowledges the financial support received from the Council of Scientific and Industrial Research (CSIR), Government of India in the form of a Junior Research Fellowship.

References

1. Achanta, R., Hastie, T.: Telugu OCR framework using deep learning. arXiv preprint arXiv:1509.05962 (2015)
2. Balm, G.: An introduction to optical character reader considerations. Pattern Recogn. **2**(3), 151–166 (1970)
3. Casey, R.G., Lecolinet, E.: A survey of methods and strategies in character segmentation. IEEE Trans. Pattern Anal. Mach. Intell. **18**(7), 690–706 (1996)
4. Dongre, V.J., Mankar, V.H.: Development of comprehensive Devnagari numeral and character database for offline handwritten character recognition. Appl. Comput. Intell. Soft Comput. **2012**, 29 (2012)
5. Fujisawa, H., Nakano, Y., Kurino, K.: Segmentation methods for character recognition: from segmentation to document structure analysis. Proc. IEEE **80**(7), 1079–1092 (1992)
6. Gonzalez, R.C., Woods, R.E., et al.: Digital Image Processing (2002)
7. Govindan, V., Shivaprasad, A.: Character recognition - a review. Pattern Recogn. **23**(7), 671–683 (1990)
8. Govindaraju, V., Setlur, S.: Guide to OCR for Indic Scripts. Springer, London (2009)
9. Guyon, I., Haralick, R.M., Hull, J.J., Phillips, I.T.: Data sets for OCR and document image understanding research. In: Handbook of Character Recognition and Document Image Analysis, pp. 779–799. World Scientific (1997)

10. Hegadi, R.S., Kamble, P.M.: Recognition of Marathi handwritten numerals using multi-layer feed-forward neural network. In: 2014 World Congress on Computing and Communication Technologies (WCCCT), pp. 21–24. IEEE (2014)
11. Impedovo, S., Ottaviano, L., Occhinegro, S.: Optical character recognition: a survey. Int. J. Pattern Recogn. Artif. Intell. **5**(01n02), 1–24 (1991)
12. Jayadevan, R., Kolhe, S.R., Patil, P.M., Pal, U.: Offline recognition of Devanagari script: a survey. IEEE Trans. Syst. Man Cybern. Part C (Appl. Rev.) **41**(6), 782–796 (2011)
13. John, J., Pramod, K., Balakrishnan, K.: Offline handwritten Malayalam character recognition based on chain code histogram. In: 2011 International Conference on Emerging Trends in Electrical and Computer Technology (ICETECT), pp. 736–741. IEEE (2011)
14. Kamble, P.M., Hegadi, R.S.: Handwritten Marathi character recognition using r-hog feature. Procedia Comput. Sci. **45**, 266–274 (2015)
15. Kamble, P.M., Hegadi, R.S.: Comparative study of handwritten Marathi characters recognition based on KNN and SVM classifier. In: Santosh, K.C., Hangarge, M., Bevilacqua, V., Negi, A. (eds.) RTIP2R 2016. CCIS, vol. 709, pp. 93–101. Springer, Singapore (2017). https://doi.org/10.1007/978-981-10-4859-3_9
16. Kannan, R.J., Prabhakar, R.: An Improved Handwritten Tamil Character Recognition System Using Octal Graph (2008)
17. Kannan, R.J., Prabhakar, R., Suresh, R.: Off-line cursive handwritten Tamil character recognition. In: International Conference on Security Technology, 2008. SECTECH 2008, pp. 159–164. IEEE (2008)
18. LeCun, Y., Bottou, L., Bengio, Y., Haffner, P.: Gradient-based learning applied to document recognition. Proc. IEEE **86**(11), 2278–2324 (1998)
19. Mantas, J.: An overview of character recognition methodologies. Pattern Recogn. **19**(6), 425–430 (1986)
20. Murthy, K.N.: Natural Language Processing: An Information Access Perspective. Ess Ess Publications for Sarada Ranganathan Endowment For Library Science (2006)
21. Murthy, K.N., Srinivasu, B.: Roman transliteration of Indic scripts. In: 10th International Conference on Computer Applications, University of Computer Studies, Yangon, Myanmar, 28–29 February 2012 (2012)
22. Negi, A., Bhagvati, C., Krishna, B.: An OCR system for Telugu. In: Sixth International Conference on Document Analysis and Recognition, 2001. Proceedings, pp. 1110–1114. IEEE (2001)
23. Pal, U., Chaudhuri, B.: Indian script character recognition: a survey. Pattern Recogn. **37**(9), 1887–1899 (2004)
24. Pal, U., Jayadevan, R., Sharma, N.: Handwriting recognition in Indian regional scripts: a survey of offline techniques. ACM Trans. Asian Lang. Inf. Process. (TALIP) **11**(1), 1 (2012)
25. Patel, A., Sukumar, B., Bhagvati, C.: SVM with inverse fringe as feature for improving accuracy of Telugu OCR systems. In: Sa, P.K., Sahoo, M.N., Murugappan, M., Wu, Y., Majhi, B. (eds.) Progress in Intelligent Computing Techniques: Theory, Practice, and Applications. AISC, vol. 518, pp. 253–263. Springer, Singapore (2018). https://doi.org/10.1007/978-981-10-3373-5_25
26. Prakash, K.C., Srikar, Y., Trishal, G., Mandal, S., Channappayya, S.S.: Optical character recognition (ocr) for telugu: Database, algorithm and application. arXiv preprint arXiv:1711.07245 (2017)
27. Rajasekaran, S., Deekshatulu, B.: Recognition of printed Telugu characters. Computer Graph. Image Process. **6**(4), 335–360 (1977)

28. Santosh, K.C.: Character recognition based on DTW-radon. In: 2011 International Conference on Document Analysis and Recognition (ICDAR), pp. 264–268. IEEE (2011)
29. Santosh, K.C., Wendling, L.: Character recognition based on non-linear multi-projection profiles measure. Front. Comput. Sci. **9**(5), 678–690 (2015)
30. Singh, S.: Optical character recognition techniques: a survey. J. Emerg. Trends Comput. Inf. Sci. **4**(6), 545–550 (2013)
31. Slimane, F., Ingold, R., Kanoun, S., Alimi, A.M., Hennebert, J.: A new Arabic printed text image database and evaluation protocols. In: 10th International Conference on Document Analysis and Recognition, 2009. ICDAR 2009, pp. 946–950. IEEE (2009)
32. Srinivas, B.A., Agarwal, A., Rao, C.R.: An overview of OCR research in Indian scripts. IJCSES **2**(2), 141–153 (2008)
33. Trier, O.D., Jain, A.K., Taxt, T., et al.: Feature extraction methods for character recognition-a survey. Pattern Recogn. **29**(4), 641–662 (1996)

On-Line Devanagari Handwritten Character Recognition Using Moments Features

Shalaka Prasad Deore[1,3](✉) and Albert Pravin[2]

[1] Sathyabama Institute of Science and Technology, Chennai, India
shalakadeore@gmail.com

[2] Department of Computer Science and Engineering, Sathyabama Institute of Science and Technology, Chennai, India

[3] Department of Computer Engineering, M.E.S. College of Engineering, S. P. Pune University, Pune, Maharashtra, India

Abstract. Now a days recognizing the handwritten character is receiving high significance because of numerous applications like Educational Software, On-line Signature Verification, Bank Cheque Processing, postal code recognition, Electronic library etc. Very less work is accounted in the research of Devanagari handwritten character recognition (HWDCR), so that there is a large scope of research in this area. In this paper we proposed a HWDCR system that recognizes Devanagari handwritten characters, the most popular script in India. Using pen tablet handwritten character is inputted and its on-line features are extracted like sequence of (x, y) coordinates, stroke and pressure information which are passed to classifier for classification. We have used MLP-BP Neural Network Classifier for classification. The average recognition accuracy is achieved by the proposed HWDCR system is 90% using on-line data.

Keywords: Devanagari handwritten character recognition · On-line features · MLP-BP Neural Network · Classifier

1 Introduction

Character recognition is winding up increasingly vital and huge in the advanced world. It encourages people to carry out their jobs effortlessly.Handwritten character recognition (HCR) is most challenging and demanding research area as far as image processing is concerned. Goal of this research is to facilitate automation in order to minimize human efforts. Handwritten character recognition (HCR) mainly divided into two classes: On-line HCR and Off-line HCR. In off-line HCR method the image of the written character is detected from a paper by optical scanning called scanner. In on-line HCR method input is detected by movement of pen tip called Digitizer. Handwritten character recognition is very complex

K. C. Santosh and R. S. Hegadi (Eds.): RTIP2R 2018, CCIS 1037, pp. 37–48, 2019.
https://doi.org/10.1007/978-981-13-9187-3_4

due number of reasons [20]. Firstly, shape similarity between characters. Secondly, handwriting of a person is different in various circumstances. Thirdly, noise present while collection of data. With this we also have to consider large variations in stroke primitives due to different handwriting styles. In English language there are only 26 characters but Indian scripts consist of vowels and consonants total as well as Compound characters so recognition of Indian languages is challenging as compared to English language. The proposed work done for Devanagari scripts mainly adopted for different languages such as Marathi, Hindi, Sanskrit, Nepali etc. Devanagari scripts consist of 12 vowels and 36 basic consonants and 10 numeral characters. So there is challenge to create a software system that will recognize handwritten Devanagari characters of different font size, font style and shape.

In Fig. 1 shows different samples of Devanagari Handwritten characters. We can observe that for same characters also there exist variations in each sample due different handwriting styles and their number of strokes. We can also observed shape similarity for some characters.

Fig. 1. Handwriting samples by five writers

The paper covers literature review in Sect. 2 and proposed work in Sect. 3. Section 4 is about Data Collection followed by the Preprocessing in Sect. 5. Section 6 describes methods for feature extraction followed by classifier used for the recognition in Sect. 7. Section 8 elaborates the results followed by conclusion in Sect. 9.

2 Literature Review

Digitizers are mostly electromagnetic-electrostatic tablets, which send the coordinates of the pen tip to the computer at regular intervals. The on-line handwriting recognition provides more discriminate properties compared to off-line recognition resulting in improvement in accuracy. There are many advantages of on-line HCR over off-line HCR like real time processing, it is very adaptive in real time and the main advantage of on-line HCR is very little per-processing is required compared to off-line HCR. The operations, such as smoothing, de- skewing, finding of number of strokes and loops are efficient using the pen tablet compared to scan images [12]. When the user will write character using pen tablet, that image can also store as .bmp format in specified folder and the same image we can used as off-line images. By doing this we required very little preprocessing due to less noise is present. Not many attempts have been found in recognizing the Devanagari Handwritten Characters. Shelke and Apte, proposed Fuzzy based classification schema for recognition of Handwritten Devanagari character [1]. Wang, Y. et al. [2] presented topic Language adaption model focuses on specific topic of Chinese text. In the first stage author used bi-gram model to recognize topic of text image at the character level and then text topic is coordinated adaptively. The implemented model reduces error rate by 11.94%. In paper [3], author explores 2-D moments feature extraction techniques on Devanagari Handwritten characters. In this paper experiments are conducted using separate as well as combination of different moment methods and got improved results using all combination (central+ Geometric+ Complex+ Zernike) of moment methods. Omer, M. et al. [4] presented new handwritten Arabic character dataset. Using stroke information collected from real time samples implemented online Arabic handwriting recognition system for isolated characters. Jayashree et al. [5] suggested method of pattern matching in which unique features are extracted and its pattern is created for classification. Venkatesha et al. [6] proposed orthogonal moments based approach for verifying the on-line handwritten signature with reduced computational resources. In paper [7], handwritten Gurmukhi characters recognition system is implemented in two stages. In the first stage stroke recognition is done and in next stage based on recognized stroke character is assessed with good accuracy. More and Rege, they have used Geometric Moment and Zernike Moment to recognize off-line Devanagari Numerals [8].

In paper [9,11], authors explained moment theory and how efficiently it can be used as feature extraction technique. In paper [10], author used stroke based information to recognize on-line Devanagari characters. Agui and Takahashi used moments for feature extraction to recognize Katakana off-line characters with 3-layered feed forward NN [12]. Santosh, K.C. et al. [13–15] presented an established as well as validated approach of using combination of stroke and spatial relationship between them in stroke clustering format. Isolated offline character recognition is done using radon features in paper [16]. In paper [17], author presented the Dynamic time wrapping method to manage dissimilar shapes and different sizes while addressing the multi-class similarities and distortions. Santosh et al. [18] elaborated the handwritten Nepali characters by using structural approach of identifying the Stroke Number and Order Free handwriting. In paper [19] author presented an Ensembling Model to recognize Devanagari handwritten characters. Authors used very famous SVM, K-NN and NN classifiers for Ensembling and achieved good accuracy.

3 Proposed Work

We implemented a system which is used to recognize 12 vowels of Handwritten Devanagari character. In India the Devanagari script used by masses especially in Gujarat, Maharashtra, North India, Madhya Pradesh and West Bengal. Devanagari script is read and written from left to right direction. Here input image is captured using a pen tablet and while writing we have recorded its x and y coordinates and pressure information. Then captured on-line information of image is stored in text file. This information is passed to feature extraction phase for calculating features of different characters, which is implemented using various non-orthogonal Moments.Then neural network is applied to classify characters. Figure 2 depicts architecture of proposed Character Recognition system.

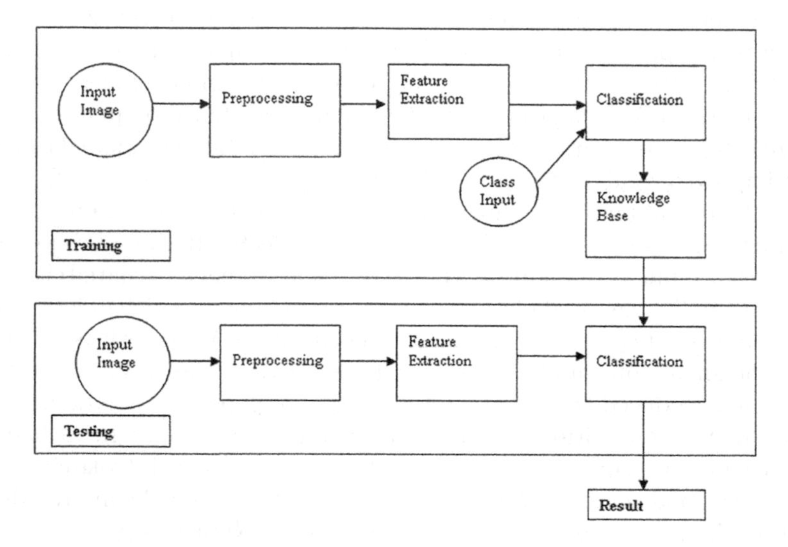

Fig. 2. Architecture of proposed Devanagari handwritten character recognition System

4 Dataset Creation

On-line handwritten data is inputted with a digitizing pen tablet and special stylus. These pen movements gives us (x, y) pixel coordinates of a character image. These coordinates, along with the pressure information are stored in text file. Out of which only x and y coordinates information is used by our on-line HCR system. Figure 3 shows GUI of our developed application and Fig. 4 shows contents of text file. No restriction was forced on the content or style of writing, the only constraint is the requirement of isolated characters. We have collected 100 samples of each 12 characters from different persons. We have created our own on-line data set of 1200 characters.

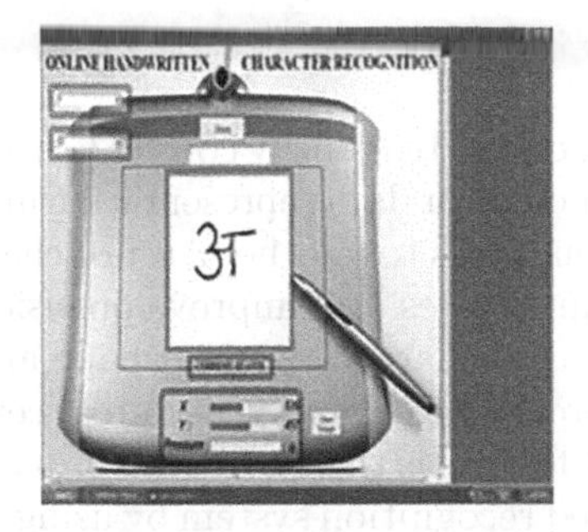

Fig. 3. GUI of HWDCR

X Coordinate	Y Coordinate	Pressure information
415	495	89
403	495	36
403	494	74
411	496	231
412	497	232
.	.	.
418	501	230
419	501	229

Fig. 4. On-line data of handwritten character

5 Preprocessing

Preprocessing is one of the essential phase of handwriting recognition. It removes the noise introduced due to software and device constraints thus improving overall recognition rate. When user writes a character its on-line information (i.e. x and y coordinate information) is captured in timely basis form. We are storing N no. of samples in text file. We are taking on-line data for feature calculation in which very less noise is present. So that not much preprocessing is required for on-line data set. We have done only sample normalization and thinning operation on image. Figure 5(a) shows original input character and (b) shows same processed character.

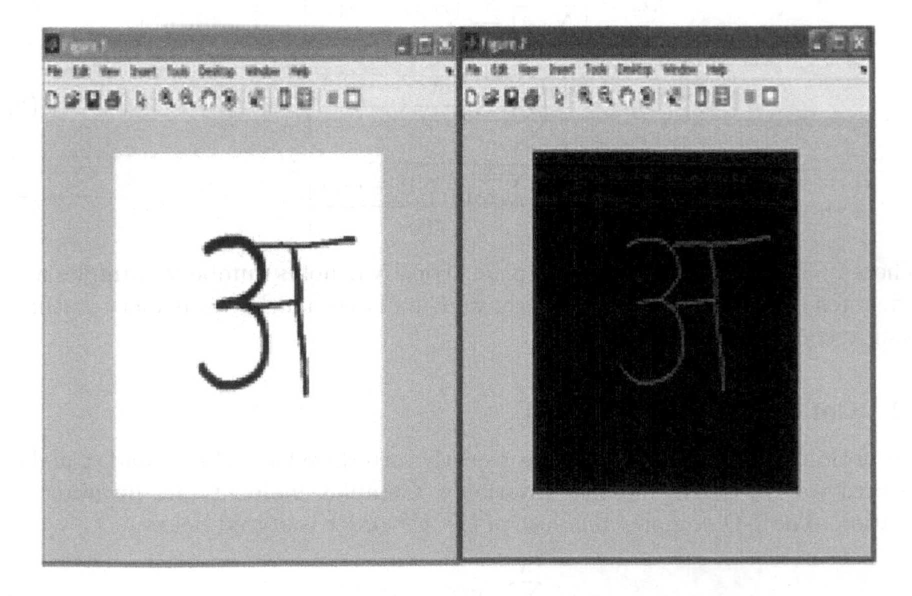

Fig. 5. (a) Original input character (b) Same processed character

6 Feature Extraction

In the process of handwriting recognition, it is important to identify correct feature set. Feature extraction is very important phase for efficient data representation and for further processing. To differentiate between one class to another a set of characteristics are mined. This would minimize the intricacies and improve precision of algorithm. Moments are shape descriptors used to characterize the shape and size of the image. Moments provide important properties of image like area, centroid, orientation etc. Main advantage of moment feature is their ability to provide invariant measures of shape. We have implemented recognition system by using 1-D Moments as feature extraction method using on-line data set.

6.1 Geometric Moment (1-D)

Geometric moments (1-D) are basically projections of the image function onto the monomials, i.e., x^p, the pth order geometric moment, m_p, is stated below:

$$m_p = \sum_{i=1}^{N} X[i]^p \tag{1}$$

Using on-line x coordinate information we calculated 4 moment features as:

$$m_1 = \frac{1}{N}\sum_{i=1}^{N} X[i] \tag{2}$$

$$m_2 = \left[\frac{1}{N-1}\sum_{i=1}^{N}(X[i]-m_1)^2\right]^{0.5} \tag{3}$$

$$m_3 = \left[\frac{1}{N}\frac{\sum_{i=1}^{N}(X[i]-m_1)}{m_2}\right]^3 \tag{4}$$

$$m_4 = \left[\frac{1}{N}\frac{\sum_{i=1}^{N}(X[i]-m_1)}{m_2}\right]^4 \tag{5}$$

Where i = 1..N, X(i) denotes the x pixel array, N denotes number of samples and p denotes order of moment. Similarly we calculated another 4 moment features using y coordinate information.

6.2 Complex Moments (1-D)

The notion of complex moments was recently introduced as a simple and straightforward way to derive moment invariants. Complex moments are invariant to rotation. The 1-D complex moment of the pth order is stated below:

$$c_p = \sum_{i=1}^{N}(X[i] + iY[i])^p \tag{6}$$

Where i = 1..N, X(i) denotes the x pixel array, Y(i) is y pixel array, N denotes number of samples and p denotes order of moment. We calculated 4 features of complex moment.

6.3 Central and Hu's Moments (1-D)

Geometric moments cannot handle the shift translation, change in rotation and difference in size. To make it invariant to translation we calculated Central moments and to handle the rotation and size, Hu's moments are computed. Below listed Hu's moments can tackle the translation, rotation and size of the image. The 1-D central moment of the pq^{th} order is expressed as:

$$\mu_{pq} = \sum_{i=1}^{N} (X - \bar{X})^p (Y - \bar{Y})^q \tag{7}$$

Where i = 1..N, X(i) denotes the x pixel array, Y(i) is y pixel array, N denotes number of samples and p denotes order of moment. Where,

$$\bar{X} = \frac{m_{10}}{m_{00}} \quad \text{and} \quad \bar{Y} = \frac{m_{01}}{m_{00}} \tag{8}$$

To handle deviation in size of character image the central moments is normalized using below expression:

$$\eta_{pq} = \frac{\mu_{pq}}{{\mu_{00}}^{\gamma}} \quad \text{and} \quad \gamma = \frac{p+q}{2} + 1 \tag{9}$$

For p, q = 0,1,.. and for (p+q) = 2,3,..

As per above transformation done in central moment, Hu's moments can be derived as follows [12]:

$$\phi_1 = \eta_{20} + \eta_{02} \tag{10}$$

$$\phi_2 = (\eta_{20} - \eta_{02})^2 + 4\eta_{11}^2 \tag{11}$$

$$\phi_3 = (\eta_{30} - 3\eta_{12})^2 + (3\eta_{21} + \eta_{03})^2 \tag{12}$$

$$\phi_4 = (\eta_{30} + \eta_{12})^2 + (\eta_{21} + \eta_{03})^2 \tag{13}$$

$$\begin{aligned} \phi_5 &= (\eta_{30} - 3\eta_{12})(\eta_{30} + \eta_{12})\left[(\eta_{30} + 3\eta_{12})^2 - 3(\eta_{21} + \eta_{03})^2\right] \\ &+ (3\eta_{21} - \eta_{03})(\eta_{21} + \eta_{03})\left[3(\eta_{30} + \eta_{12})^2 - (\eta_{21} + \eta_{03})^2\right] \end{aligned} \tag{14}$$

$$\phi_6 = (\eta_{20} - \eta_{02})[(\eta_{30} + \eta_{12})^2 - (\eta_{21} + \eta_{03})^2] + 4\eta_{11}(\eta_{30} + \eta_{12})(\eta_{21} + \eta_{03}) \tag{15}$$

$$\begin{aligned} \phi_7 &= (3\eta_{12} - \eta_{30})(\eta_{30} + \eta_{12})[(\eta_{30} + \eta_{12})^2 - 3(\eta_{21} + \eta_{03})^2] \\ &+ (3\eta_{21} - \eta_{03})(\eta_{21} + \eta_{03})[3(\eta_{30} + \eta_{12})^2 - (\eta_{21} + \eta_{03})^2] \end{aligned} \tag{16}$$

7 Character Classification and Recognition

Character Recognition is done based on a popular method called Multilayer Neural Network as depicted in Fig. 6. Feed forward is an acyclic network in which the flow is unidirectional from the input to output nodes via hidden nodes. We have considered leveraging the Back propagation algorithm for training the classifier. In network, calculated mean squared error information is propagated back to hidden layer and weights are adjusted to get desired output. Neural network is consist of one hidden layer trained using sigmoid activation function. We have 12 character samples of 100 different peoples. Out of which 80 character samples are trained and 20 are used for testing. Results are different for different features. Characters are normalized before passing to Neural Network. Using following equation they are normalized:

$$\sum_{i=1}^{N}\left[\frac{(feature(i) - mean(i))}{\sigma(i)}\right] \tag{17}$$

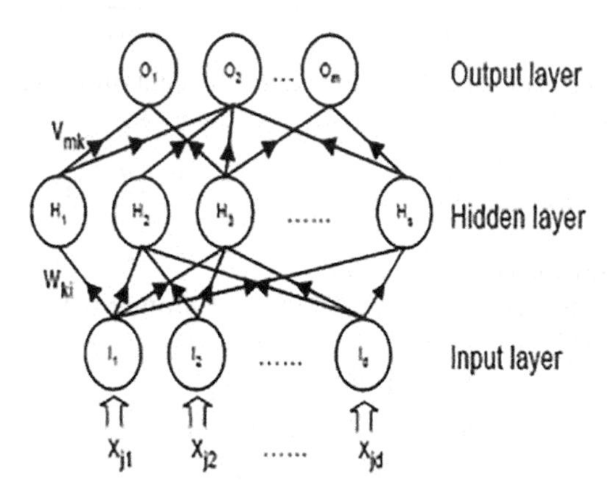

Fig. 6. Architecture of MLP neural network

8 Experiments and Results

8.1 Individual Moment-Based Results

Data used for the present work were collected from different individuals. Class 1 consists of samples of character अ of 100 different peoples; class 2 consists of samples of character आ and so on. Like this we have total 12 classes for 12 Devanagari characters so we have total 1200 samples of Devanagari basic characters (only vowels) for the experiment of the proposed work. Table 1 show

Table 1. Devanagari character recognition results using online features

Character classes	Geometric moment(%)	Central moment(%)	Hu's moment(%)	Complex moment(%)
1	45	70	60	40
2	70	70	25	05
3	80	65	55	10
4	85	90	60	35
5	70	55	45	05
6	55	65	50	05
7	60	70	50	35
8	70	85	35	05
9	65	75	65	15
10	50	80	40	30
11	35	65	15	10
12	45	60	40	15
Average recognition rate	**61**	**71**	**45**	**18**

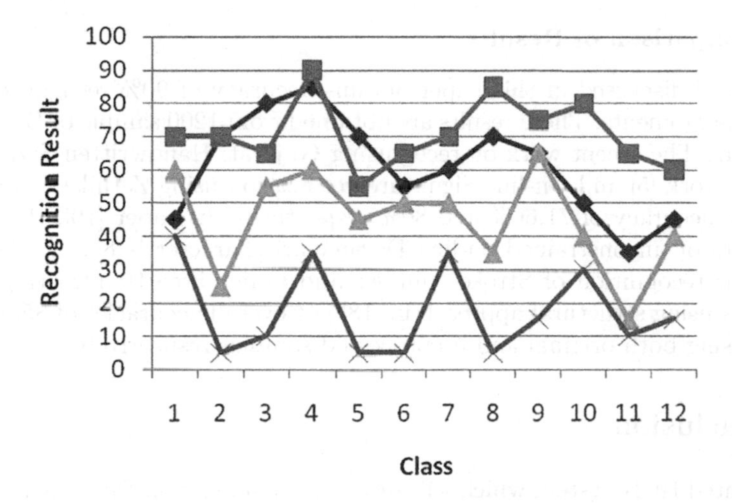

Fig. 7. On-line character recognition result

recognition rate of each class from 1 to 12 determined using different moment features which used on-line information for its calculation.

From Table 1 we can observe that Central moment have good recognition results among all other moments because of CM are invariant to translations. We also observed that features calculated using Complex moment (COM) have very much variation so that its recognition result is very poor.

Figure 7 shows recognition rate of different on-line moments shown in Table 1 for 12 class characters. From Fig. 7 we can observe that graph of all features is non-linear so that we can combined two different features. These two combined features are passed to neural network to see its recognition result. Table 2 depicts how after combining features we are getting improved recognition result. Overall average recognition rate is increased as compared to using them individual. When we combined all four features it increases recognition rate of each individual class characters and average rate is also 90%.

Table 2. Average character recognition result of 1-D combined moments

Online moments (combined)	Average recognition rate
CM & GM	84%
CM & COM	73%
GM & COM	70%
Combined all 4 1-D moments (GM,CM,HuM,COM)	90%

8.2 Comparison of Results

The method discussed in this paper obtains accuracy of 90% recognition rates for On-line Moments. These results are obtained from 1200 sample of Devanagari Characters. The recent work on recognizing Gujarati Handwritten Script using neural network [5] and On-line Signature verification using Zernike moments [6] shows the accuracy of 71.66% and 80% respectively. In paper [10], the classification rate of unconstrained on-line Devanagari character is 86.5%. Presented system for recognition of Stroke Number and Order Free Handwriting Nepali characters using structural approach in [18] got overall accuracies of 85.87% and 88.59% using both original and preprocessed samples respectively.

9 Conclusion

Implemented HCR system which will recognize handwritten Devanagari characters using on-line data and moments as feature. We collected 100 samples of each 12 characters from different persons so our data set consists of total 1200 characters. Out of 100 samples we used 80 samples for training Neural Network and 20 samples for testing. We observed that: (i) Central Moment (CM) gives the highest recognition rate 71% and Complex Moment (COM) gives lowest recognition rate 18% for on-line features compared to other moments when each moment is applied separately (ii) Combination of any two moments improves the recognition rate. (iii) Combination of all 1-D moments gives recognition rate to 90% (iv) 1-D Moments can also use as feature for on-line system which is not influenced by stroke sequence. For future work we plan to recognize the character set of "Devanagari Consonants" and "Consonant-Vowel combination (Compound Character)" using moments. And also wants check results with different types of features.

References

1. Shelke, S., Apte, S.: A fuzzy based classification scheme for unconstrained handwritten Devanagari character recognition. In: International Conference on Communication, Information and Computing Technology (ICCICT), pp. 1–6. IEEE Press, Mumbai (2015). https://doi.org/10.1109/ICCICT.2015.7045738
2. Wang, Y., Ding, X.: Topic language model adaption for recognition of homologous on-line handwritten Chinese text image. IEEE Signal Process. Lett. **21**, 550–553 (2014). https://doi.org/10.1109/LSP.2014.2308572
3. Deore, S., Ragha, L.: Moment based online and online handwritten character recognition. CiiT Int. J. Biometrics Bioinf. **3**, 111–115 (2011). BB032011004
4. Omer, M., Ma, S.: Online Arabic handwriting character recognition using matching algorithm. In: 2nd International Conference on Computer and Automation Engineering (ICCAE), pp. 259–262. IEEE Press, Singapore (2010) . https://doi.org/10.1109/ICCAE.2010.5451492
5. Prasad, J., Kulkarni, U., Prasad, R.: Offline handwritten character recognition of Gujarati script using pattern matching. In: 3rd International Conference on Anti-counterfeiting, Security, and Identification in Communication, pp. 611–615, IEEE Press, Hong Kong (2009). https://doi.org/10.1109/ICASID.2009.5276999
6. Radhika, K., Venkatesha, M., Shekar, G.: On-line signature authentication using Zernike moments. In: 3rd International conference on Biometrics: Theory, applications and systems, pp. 109–112. IEEE Press, Washington (2009). https://doi.org/10.1109/BTAS.2009.5339022
7. Sharma, A., Kumar, R.: On-line handwritten Gurmukhi character recognition using elastic matching. In: Congress on Image and Signal Processing (CISP), pp. 391–396. IEEE Press, Sanya (2008). https://doi.org/10.1109/CISP.2008.297
8. More, V., Rege, P.: Devnagari handwritten numeral identification based on Zernike moments. In: IEEE Region 10 Conference (TENCON), pp. 1–6. IEEE Press, Hyderabad (2008). https://doi.org/10.1109/TENCON.2008.4766863
9. Shu, H., Luo, L.: Moment-based approaches in imaging part 1 basic features. IEEE Eng. Med. Biol. Mag. **26**, 70–74 (2007)
10. Connell, S.D., Sinha, R., Jain, A.: Recognition of unconstrained on-line Devanagari characters. In: 15th International Conference on Pattern Recognition, pp. 368–371. IEEE Press, Barcelona (2000). https://doi.org/10.1109/ICPR.2000.906089
11. Liao, S., Pawlak, M.: On image analysis by Moments. IEEE Trans. Pattern Anal. Mach. Intell. **18**, 254–266 (1996). https://doi.org/10.1109/34.485554
12. Agui, T., Takahashi, H., Nagahashi, H.: Recognition of handwritten katakana in a frame using moment invariants based on neural network. In: IEEE International Joint Conference on Neural Networks, pp. 659–664. IEEE Press, Singapore (1991). https://doi.org/10.1109/IJCNN.1991.170475
13. Santosh, K., Nattee, C., Lamiroy, B.: Relative positioning of stroke based clustering: a new approach to on-line handwritten Devangari character recognition. Int. J. Image Graph. (IJIG) **12**, 1–24 (2012). https://doi.org/10.1142/S0219467812500167
14. Santosh, K., Iwata, E.: Stroke-Based Cursive Character Recognition. IntechOpen (2012). https://doi.org/10.5772/51471

15. Santosh, K., Nattee, C., Lamiroy, B.: Spatial similarity based stroke number and order free clustering. In: 12th International Conference on Frontiers in Handwriting Recognition (ICFHR), pp. 652–657. IEEE Press, Kolkata (2010). https://doi.org/10.1109/ICFHR.2010.107
16. Santosh, K.: Character recognition based on DTW-radon. In: 11th International Conference on Document Analysis and Recognition (ICDAR), pp. 264–268. IEEE Press, Beijing (2011). https://doi.org/10.1109/ICDAR.2011.61
17. Santosh, K., Wendling, L.: Character recognition based on non-linear multi-projection profiles measure. Front. Comput. Sci. **9**, 678–690 (2015). https://doi.org/10.1007/s11704-015-3400-2
18. Santosh, K.C., Nattee, C.: Stroke number and order free handwriting recognition for Nepali. In: Yang, Q., Webb, G. (eds.) PRICAI 2006. LNCS (LNAI), vol. 4099, pp. 990–994. Springer, Heidelberg (2006). https://doi.org/10.1007/978-3-540-36668-3_120
19. Deore, S.P., Pravin, A.: Ensembling: model of histogram of oriented gradient based handwritten Devanagari character recognition system. Traitement du Signal **34**, 7–20 (2017). https://doi.org/10.3166/ts.34.7-20
20. Jagtap, A.B., Hegadi, R.S.: Offline handwritten signature recognition based on upper and lower envelope using eigen values. In: World Congress on Computing and Communication Technologies (WCCCT), pp. 223–226. IEEE (2017)

Artistic Multi-character Script Identification Using Iterative Isotropic Dilation Algorithm

Mridul Ghosh[1(✉)], Sk Md Obaidullah[2], K. C. Santosh[3], Nibaran Das[4], and Kaushik Roy[5]

[1] Department of Computer Science, Shyampur Siddheswari Mahavidyalaya, Howrah, India
mridulxyz@gmail.com

[2] Department of Computer Science and Engineering, Aliah University, Kolkata, India
sk.obaidullah@gmail.com

[3] Department of Computer Science, University of South Dakota, Vermillion, SD, USA
santosh.kc@ieee.org

[4] Department of Computer Science and Engineering, Jadavpur University, Kolkata, India
nibaranju@gmail.com

[5] Department of Computer Science, West Bengal State University, Barasat, India
kaushik.mrg@gmail.com

Abstract. In this work, a new problem of script identification named artistic multi-character script identification has been addressed. Two types of datasets of artistic documents/images prepared with Bangla, Devanagari and Roman script have been used: one is real life artistic multi-character script image and another is synthetic artistic multi-character script image. After binarization using Otsu's algorithm, some character images found to be broken into components. To overcome this, a novel iterative isotropic dilation algorithm is proposed here to convert the components into a single component object. Then two types of features, namely shape based and texture based features have been considered. Discrete Gabor wavelet has been exploited with 2 scales and 4 orientations for texture feature extraction and PCA is used to reduce the dimensionality of the texture feature space. The performance of the proposed algorithm has been tested with different machine learning classifiers and promising accuracy has been observed.

Keywords: Script identification · Multi-character script · Otsu's binarization method · Random forest · Multilayer perceptron

---

K. C. Santosh—IEEE Senior Member

K. C. Santosh and R. S. Hegadi (Eds.): RTIP2R 2018, CCIS 1037, pp. 49–62, 2019.
https://doi.org/10.1007/978-981-13-9187-3_5

1 Introduction

Since last decade Script identification [1–8] has come up with a very interesting research topic which is the precursor to OCR [9], is one of the promising domain of research in the field of image processing and pattern recognition. Many methods have been proposed for automatic identification of script identification in different phases like block level, word level, line level, etc. and many combinations like bi-script [10], tri-script [11] and multi-script [12,13] scenario. Different characters in different scripts have attributes which causes difficulty in character recognition of multi-script word and it is not possible to devise a universal approach and technique to get better performance in OCR. The countries like India, where multilingual with a multi-script written people live, developing a generalized OCR for all Indus Script is not possible due to different graphic symbols in the script. Moreover, script identification, i.e. a particular language belongs to which script is itself a complicated work. So, designing a preprocessor of script identification system first, then script specific OCR can be applied. However, in the literature, we find a generic shape descriptor-based character/text recognition for different types of scripts [38,39], which are based on [40] and [41].

Script identification has progressed, but is not the same when we talk about artistic documents. It is one of the interesting problems which yet to get proper attention from the researchers, though the presence of not only mixing of scripts along with wide variation in style, size, color texture etc. but mixing of scripts individual word makes it a more interesting problem. For example, see Fig. 2.

Section 2 describes the review of literature in script identification. Section 3 describes data collection and preprocessing. In Sect. 4 feature extractions has been discussed and in Sect. 5 experimental results have been shown. Contribution in this work includes by collecting the multi-script images from different places and broken component within images have been connected by our algorithm iterative isotropic dilation.

2 Literature Review

In [14] Dhanya et al. classify the scripts at the word level, in a bilingual document consisting of Roman and Tamil scripts with various fonts having 4 different patterns for each words. This problem has been treated in two ways. In the first way, distinct spatial zones have been considered for each word. The spatial extend in different areas and with the solidity of a word has been considered and in second way by texture analysis using Gabor filters used to identify the script.

In [15] Dhanya et al. proposed a hierarchical feature extraction technique for acknowledgment of printed Tamil and Roman text. Features have been computed by extraction of some spatial structural features from the geometric moments, DCT based features and Wavelet transform based features. A linear transformation is used on the subsets for feature space representation by PCA and maximization of Fisher's ratio and maximization of divergence measure. This method obtains good accuracy as a result of the transformations.

Pati et al. [16], have used a Gabor function based multichannel directional filtering approach for both text area separation and script identification at the word level (Fig. 1).

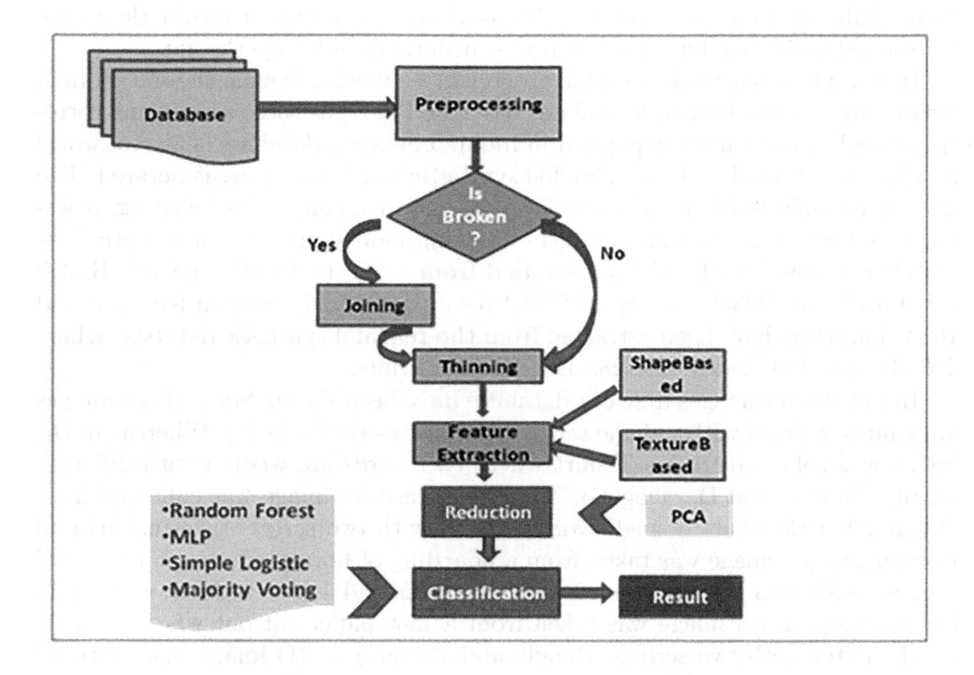

Fig. 1. Flowchart of our proposed work.

Mohanty et al. [17] proposed a work to discriminate Roman and Oriya scripts. They used dynamic thresholding technique for binarization and histogram analysis through horizontal and vertical projection used for segmentation. By chain code method the geometrical features are extracted and by the classification is done using.

Pal et al. discussed in [18] the problem of segmenting and recognizing the complex documents of Bangla and Devanagari with text line having curved and with different orientations. They used convex hull and reservoir principle for segmenting characters with the help of candidate envelope points and after dividing the character into different zones, circular and convex hull features are extracted and obtained a very good recognition rate after feeding it to SVM classifier.

3 Data Collection and Preprocessing

Data collection is very important in any pattern recognition problem as from developing the algorithm to validation of the system all depends on the quality and variety of the data. So, for the current problem we need a database which is not only very good in size, but it contains all possible variations from not only font, size and color variations but their combinations considering texture and scripts considered for the present work. But, the hardcore reality is they are very difficult to collect. To overcome the same in the proposed work we aren't only planning to collect data from all possible sources to cover the above mentioned variations but also generate synthetic data bridge the gap.

In the proposed work, we plan to consider Bangla, Devnagari and Roman scripts under the tri-script formula as described in [1] as these scripts ans corresponding language are very popular in India. Till now a database of 30 real world images are collected and a total of 300 synthetic word images are generated. The images are collected from hoarding, festoon, placard, banner, wall writing, newspaper advertisement, t-shirts graffiti etc. using mobile phones, camera etc. The mobile cameras used for the same varied from 8 MP to 16 MP and DSLR still camera (Nikon D3300) having 24.2 MP resolution and 18–55 mm lens. Around 1033 characters have been extracted from the real and synthetic database where 306 Bangla, 439 Roman and 288 Devanagari scripts.

In Fig. 2 some images from our database have been shown. Some of the images are simply written with a single script (Roman) as shown in (a). Whereas in (b) image was collected from a T-shirt where two words are written with different scripts (Roman and Devanagari), in another case (c) image was collected from of a movie title where a single word written with two scripts (Devanagari and Roman), in (d) image was taken from a hoarding of library where a upper word written with two scripts (Bangla and Roman) and lower word is written in Roman only, in (e) image was taken from a newspaper cut out where a single word written with two scripts (Bangla and Roman), in (f) image was captured from hoarding of a Biriyani shop where a single word written with two scripts (Bangla an Roman) and other line is written with single script (Bangla) and also with different size, in (g) image was collected from a movie title of a movie where upper part is written with a single script (Bangla) and lower part is written with two different script (Bangla an Roman) and also there is variations in color between the scripts, in (h) the image was captured from logo of company where the image is not only written with different scripts (Devanagari, Roman) but also it is written artistic way.

From these images it has been seen that the images have, not only in variations in color, size, fonts, textures, but also the backgrounds have wide variation like bag, T-shirt, wall etc. In this scenario, detecting region of interest automatically is a very complicated task. So, to make this step simple, we detect the ROI manually.

After collecting the images we have used Otsu [19] based binarization method to convert into a binary image after converting to gray image. Prior to that we have extracted the ROI and segmented the images into characters. In our database, the words are formed by bi-script characters. Considering 3 different scripts Bangla, Devanagari and Roman, the words are formed by Bangla with Devanagari, Devanagri with Roman and Roman with Bangla or in other words,

(a) (b) (c) (d) (e) (f) (g) (h)

Fig. 2. Shows some sample images from our database.

there are $\binom{3}{2}$ combinations of script in word formation. After manual segmentation, we separated each character into its corresponding script class.

During binarization, we found that some of the images are very well binarize or converted to two tone images, but in some cases it is found that the characters are broken into parts which needs to be connected to get proper results. Here we have used novel iterative isotropic dilation algorithm to join the broken parts. Isotropic dilation is the technique of thresholding the distance transform. This algorithm uses connected component labelling [20,21] and flood fill algorithm [22] to find out number connected component object and Euclidean distance transform [23] algorithm to make the components join with changing the threshold value (Fig. 3).

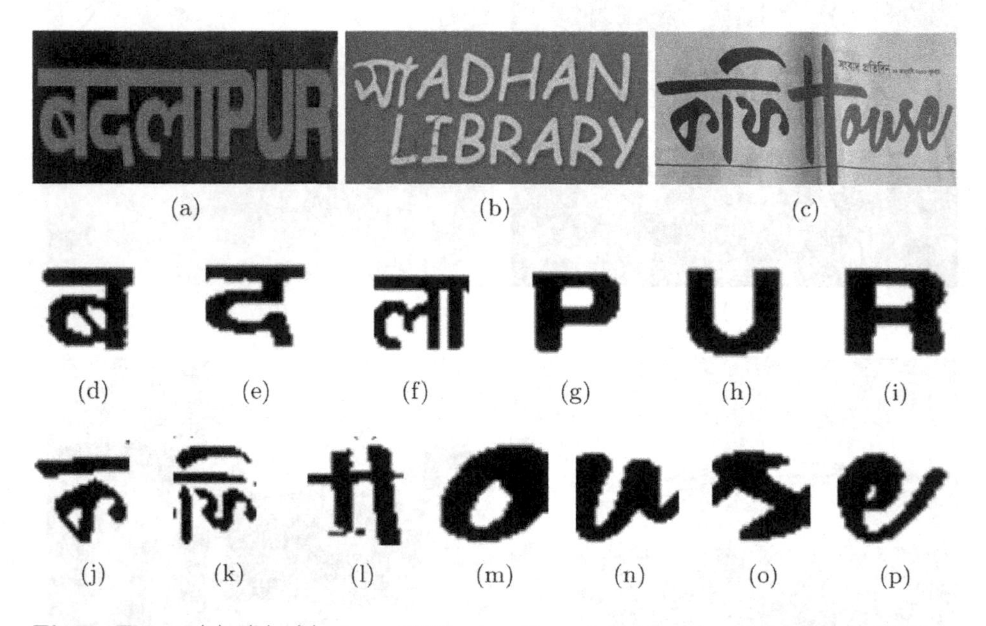

Fig. 3. Figure (a), (b), (c) shows ROI extraction of real images and (d), (e), (f), (g), (h), (i), (j), (k), (l), (m), (n) are the binarized and segmented character images.

From Figs. 4(c) and (d), we can see the broken parts of the characters are joined by only a single pixel. In this algorithm, by changing the threshold value t by 0.1, the Euclidean distance transformed value also changed and at some value of the threshold the number of components will be 1 and the iterative isotropic dilation algorithm will stop executing and we get joined images as shown in Figs. 4(c) and (d).

Next, we have normalized the characters into a predefined size of [32, 32]. This was done to avoid size variation of the images as explained in Sect. 3. Next, the thinning of the character images has been done using Zhang-Suen thinning algorithm [24]. This thinning algorithm consists of 2 phases. First phase

Algorithm 1. Iterative Isotropic Dilation

Input: Binary_ Image, Number_of_connected_component
i. N= Number_of_connected_component
ii. Initialize t=0
iii. If (N¿1) then go to step (iv)
iv. Repeat Step (v) to (vi) until N= =1
v. Binary_Image = Euclidean distance transform ¡=t
vi. t=t+0.1
vii. update N
Output: Joined Binary_Image.

Fig. 4. Figure (a), (b) are the broken binarized character images and figure (c), (d) are the images after the iterative isotropic dilation algorithm has been applied

to remove south-west pixels and second phase to remove north-west pixels and these phases are iterated if any pixels remain to remove.

4 Feature Extraction

Feature extraction is an important pattern recognition task. Features should be robust enough to differentiate the scripts and at the same time it should be easily computable. In this paper, we have used feature based on the shape and texture to identify the scripts.

4.1 Shape Based Features

The shape based properties [25] used in shape perceptive along with their details are reviewed. Shape feature is extremely useful in many image databases and pattern matching applications.

The simple geometric features can categorize shapes with large differences. The features computed based on shape are area, major axis length, minor axis length etc. which are explained below. These features are selected with the assumption that they will be simple to compute at the same time we will be able to distinguish the inputs. To compute the features we have separated the images into 4 parts from the Center of Mass [26]. The Center of Mass is selected as a segmentation point, so the every quadrant contains sufficient numbers of object pixels, so that we can extract proper feature from the images.

a. Area: This scalar feature measures number of pixels in the region of the whole image.

b. Major Axis Length: Returns a scalar value as the major axis of the ellipse that has the same normalized second central moments as the region, formed around the object.
c. Minor Axis Length: Returns a scalar value as the minor axis of the ellipse that has the same normalized second central moments as the region, formed around the object.
d. Eccentricity: The eccentricity which has the same second-moments as the region is the ratio of the distance between the foci of the ellipse and its major axis length.
e. Orientation: This scalar value represents the angle between the major axis and x axis of the ellipse that has the same second-moments as the region.
f. Convex Area: this scalar quantity denotes the number of pixels in convex image.
g. Euler Number: Euler number is a scalar value calculated as the number of objects in the region minus the number of holes in those objects.
h. Equiv Diameter: Returns a scalar can be represented as the diameter of a circle with the same area as the region.
i. Solidity: Returns a scalar representing the proportion of the pixels in the convex hull that are also in the region.
j. Extent: A scalar value which can be calculated as the Area/ the area of the bounding box.
k. Perimeter: A scalar represents the distance between each neighboring pair of pixels around the border of the region.
l. Circularity ratio: This is the ratio of the area of a shape to the area of the bounding circle (Fig. 5).

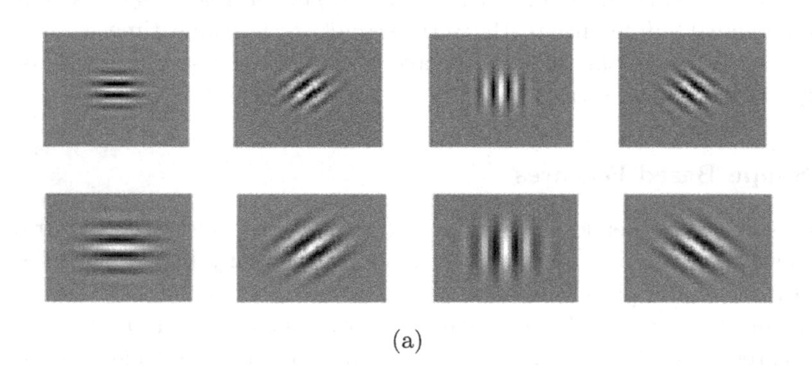

(a)

Fig. 5. The real part of Gabor wavelet having 2 scales and 4 orientation of 0°, 45°, 90°, 145°.

4.2 Texture Feature

Texture features describe the contents of an image or a region of an image and give important low level features and mean oriented energy. Discrete Gabor wavelet is a oriented filter and returns mean oriented energy as a texture feature [27,28].

The discrete Gabor wavelet transform on an image I(x, y) having dimension m X n can be represented as

$$G(x,y) = \sum_{s}\sum_{t} I(x-s, y-t)\psi^{*}_{\mathrm{pq}}(s,t) \tag{1}$$

Where s, t are the filter mask sizes. ψ^* is the complex conjugate of ψ which is a self similar function generated from dilation and rotation of the mother wavelet ψ, can be represented as

$$\psi(x,y) = (\frac{1}{2\pi\sigma_x\sigma_y})\exp(\frac{-1}{2}(\frac{x^2}{\sigma_x^2} + \frac{y^2}{\sigma_y^2}))\exp(2\pi jWx) \tag{2}$$

The self-similar Gabor wavelets are obtained through the generating function

$$\psi_{pq}(x,y) = a^{-p}\psi(x', y') \tag{3}$$

Where p denotes scale and q denotes orientations and

$$x' = a^{-p}(x\cos\theta + y\sin\theta) \tag{4}$$

and

$$y' = (-x\sin\theta + y\cos\theta) \tag{5}$$

for a¿1 and

$$\theta = \frac{\pi q}{Q} \tag{6}$$

After applying Gabor filters on the image with different orientation at different scale, the energy content is calculated using

$$E(p,q) = \sum_{x}\sum_{y} |G_{pq}(x,y)| \tag{7}$$

2-D Gabor filter with modulation frequency W can be represented as

$$G(x,y) = g_\sigma \exp(2\pi jW(x\cos\theta + \sin\theta)) \tag{8}$$

where

$$g_\sigma(x,y) = (\frac{1}{2\pi\sigma_x\sigma_y})\exp(\frac{-1}{2}(\frac{x^2}{\sigma_x^2} + \frac{y^2}{\sigma_y^2})) \tag{9}$$

The parameters θ and σ denotes orientation and scale of Gabor filters. For different values of these parameters features can be obtained.

A total of 512 Features has been extracted using Gabor wavelet. But to use this huge dimension of feature is very challenging an many of the features are redundant. So to minimize the feature dimension and preserve the diversity of feature intake we have reused the feature dimension.

Feature dimensionality reduction is one of the challenging problems in pattern recognition and there are a number of algorithms like PCA [29], kernel PCA [30], LDA [31], GDA [32] etc. for the same.

Here we have used PCA as it uses eigenvectors of the covariance matrix of the feature set which are easily computable. Here we have retained our Gabor feature set of 40. From shape based feature, 12 features are obtained and as we have divided the characters into 4 parts from the center of mass point as discussed in Sect. 4.1, the features are extracted from the 4 parts of a character and from the character itself, makes $5 \times 12 = 60$ shape based features. So, in total 40 (Gabor wavelet) + 60 (shape based) = 100 features are extracted.

5 Experiments

5.1 Evaluation Protocol

In this work, for evaluating the performance of our method, 10 fold cross validation (CV) has been used. In this 10-fold method, we divide the dataset into 10 equal sized parts. From 10 parts, training and testing sets are chosen. One part is chosen as the testing and remaining 9 parts chosen as the training and this process is repeated for all parts and result is obtained by averaging all results. The performance parameter metrics have been used are Correct Recognition Rate (CRR), kappa static, Root Mean Square error (RMSE) can be defined as follows.

$$CRR = \frac{(\#correctly\ identified\ character\ images)}{(\#total\ character\ images)} * 100\% \tag{10}$$

The Kappa statistic compares an observed accuracy with an expected accuracy. Expected accuracy can be defined as the accuracy that any random classifier would be anticipated to realize based on the confusion matrix.

$$Kappa = \frac{(observed\ accuracy - expected\ accuracy)}{(1 - expected\ accuracy)} \tag{11}$$

Another metric Root mean square error (RMSE) can be defined as

$$RMSE = \sqrt{\frac{\sum_{m=1}^{M} \sum_{n=1}^{N} I(x-s, y-t)}{MN}} \tag{12}$$

Where I (x, y) and T (x, y) stand for reference and target image of size M X N respectively.

5.2 Result and Analysis

During experimentation, performance of different well known classifiers namely random forest [33], majority voting [34], simple logistic [35] and multilayer perceptron (MLP) [36,37] were considered.

Table 1. Classifiers with different parameters.

Classifier	CRR	Kappa statistic	Root mean square error
Random Forest	93.18	0.8959	0.2731
Majority Voting	92.20	0.8815	0.2089
Simple Logistic	90.90	0.862	0.214
MLP	90.90	0.8618	0.2462

From the table, we found that the best result is obtained using random Forest of 93.18% recognition rate followed by Majority Voting, MLP, Simple Logistic and it is observed that MLP and Simple Logistic and gives the same recognition rate. The other parameter values for Random Forest i.e. Kappa Statistic and Root Mean Square Error are 0.8959 and 0.2731 respectively.

From Table 1 it can be observed that using Random Forest, Majority Voting, MLP and Simple Logistic the mis-classification rates are obtained 6.81,7.79, 9.09, 9.09 respectively and RMS errors we get 0.2731, 0.2089,0.214,0.2462 respectively. In Table 2 we tabulate the confusion matrix of the highest performer classifier on our dataset.

Table 2. Confusion Matrix for the highest accuracy rate of Bi-script identification using Random Forest.

	Bangla	Devanagari	Roman
Bangla	291	3	12
Devanagari	32	252	4
Roman	18	4	417

From Table 2 it has been seen that 291 characters out of 306 has been correctly classified as Bangla characters, 3 characters as Devanagari and 12 as Roman. In case of devanagari script, 252 characters out of 288 have been correctly identified as Devangari but 32 have been identified as Bangla script and 3 as Roman script. For Roman characters, 417 out of 439 have been correctly classified as Roman but 22 characters have been misclassified and out of 22, 18 characters have been seen in Bangla class and 4 in Devanagari class.

6 Conclusion

A new problem of script identification, from artistic multi-script characters has been addressed in this work. Such problem has inherent complexities in terms of character size, color, texture, pattern, orientation and sometimes heterogeneous background making the problem really challenging. Using some structural and texture based features we found an average accuracy of 93.18% applying random forest classifier. The result is promising observing the inherent complexity of the said problem.

Availability of standard dataset is an issue for this type of work. At present we have considered both synthetic and natural images for the experiment. In future, we would like to extend our dataset by incorporating more natural artistic images. Proposing some novel script dependent features to improve the recognition accuracy is also in our plans.

References

1. Ghosh, D., Dube, T., Shivaprasad, A.: Script recognition-a review. IEEE Trans. Pattern Anal. Mach. Intell. **32**(12), 2142–2161 (2010)
2. Obaidullah, S.M., Santosh, K.C., Das, N., Halder, C., Roy, K.: Handwritten Indic script identification in multi-script document images: a survey. Int. J. Pattern Recognit. Artif. Intell. **32**, 1856012 (2018)
3. Obaidullah, S.M., Bose, A., Mukherjee, H., Santosh, K.C., Das, N., Roy, K.: Extreme learning machine for handwritten Indic script identification in multi-script documents. J. Electron. Imaging **27**(5), 051214 (2018)
4. Obaidullah, S.M., Halder, C., Santosh, K.C., Das, N., Roy, K.: Automatic line-level script identification from handwritten document images-a region-wise classification framework for Indian subcontinent. Malaysian J. Comput. Sci. **31**(1), 63–84 (2018)
5. Obaidullah, S.M., Santosh, K.C., Halder, C., Das, N., Roy, K.: Automatic Indic script identification from handwritten documents: page, block, line and word-level approach. Int. J. Mach. Learn. Cybern. **10**, 1–20 (2017)
6. Obaidullah, S.K., Santosh, K.C., Halder, C., Das, N., Roy, K.: Word-level multi-script Indic document image dataset and baseline results on script identification. Int. J. Comput. Vis. Image Process. (IJCVIP) **7**(2), 81–94 (2017)
7. Obaidullah, S.M., Halder, C., Santosh, K.C., Das, N., Roy, K.: PHDIndic_11: pagelevel handwritten document image dataset of 11 official Indic scripts for script identification. Multimedia Tools Appl. **77**(2), 1643–1678 (2018)
8. Obaidullah, S.M., Halder, C., Das, N., Roy, K.: Bangla and Oriya script lines identification from handwritten document images in tri-script scenario. Int. J. Serv. Sci. Manag. Eng. Technol. (IJSSMET) **7**(1), 43–60 (2016)
9. Mori, S., Suen, C.Y., Yamamoto, K.: Historical review of OCR research and development. Proc. IEEE **80**(7), 1029–1058 (1992)
10. Rajput, G.G., Anita, H.B.: Handwritten script identification from a bi-script document at line level using Gabor filters. In: Proceedings of SCAKD, pp. 94–101 (2011)
11. Aithal, P.K., Rajesh, G., Acharya, D.U., Krishnamoorthi, M., Subbareddy, N.V.: Script identification for a tri-lingual document. In: Das, V.V., Stephen, J., Chaba, Y. (eds.) CNC 2011. CCIS, vol. 142, pp. 434–439. Springer, Heidelberg (2011). https://doi.org/10.1007/978-3-642-19542-6_82
12. Pal, U., Sinha, S., Chaudhuri, B.B.: Multi-script line identification from Indian documents. In: Seventh International Conference on Document Analysis and Recognition, 2003. Proceedings, pp. 880. IEEE (2003)
13. Pati, P.B., Ramakrishnan, A.G.: Word level multi-script identification. Pattern Recogn. Lett. **29**(9), 1218–1229 (2008)

14. Dhanya, D., Ramakrishnan, A.G.: Script identification in printed bilingual documents. In: Lopresti, D., Hu, J., Kashi, R. (eds.) DAS 2002. LNCS, vol. 2423, pp. 13–24. Springer, Heidelberg (2002). https://doi.org/10.1007/3-540-45869-7_2
15. Dhanya, D., Ramakrishnan, A.G.: Optimal feature extraction for bilingual OCR. In: Lopresti, D., Hu, J., Kashi, R. (eds.) DAS 2002. LNCS, vol. 2423, pp. 25–36. Springer, Heidelberg (2002). https://doi.org/10.1007/3-540-45869-7_3
16. Pati, P.B., Raju, S.S., Pati, N., Ramakrishnan, A.G.: Gabor filters for document analysis in Indian bilingual documents. In: Proceedings of International Conference on Intelligent Sensing and Information Processing, 2004, pp. 123–126. IEEE (2004)
17. Mohanty, S., Dasbebartta, H.N., Behera, T.K.: An efficient bi-lingual optical character recognition (English-Oriya) system for printed documents. In: Seventh International Conference on Advances in Pattern Recognition, 2009. ICAPR 2009, pp. 398–401. IEEE (2009)
18. Pal, U., Roy, P.P., Tripathy, N., Lladós, J.: Multi-oriented Bangla and Devanagari text recognition. Pattern Recogn. **43**(12), 4124–4136 (2010)
19. Otsu, N.: A threshold selection method from gray-level histograms. IEEE Trans. Syst. Man Cybern. **9**(1), 62–66 (1979)
20. Chang, F., Chen, C.J., Lu, C.J.: A linear-time component-labeling algorithm using contour tracing technique. Computer Vis. Image Underst. **93**(2), 206–220 (2004)
21. Samet, H., Tamminen, M.: Efficient component labeling of images of arbitrary dimension represented by linear bintrees. IEEE Trans. Pattern Anal. Mach. Intell. **10**(4), 579 (1988)
22. Silvela, J., Portillo, J.: Breadth-first search and its application to image processing problems. IEEE Trans. Image Process. **10**(8), 1194–1199 (2001)
23. Breu, H., Gil, J., Kirkpatrick, D., Werman, M.: Linear time Euclidean distance transform algorithms. IEEE Trans. Pattern Anal. Mach. Intell. **17**(5), 529–533 (1995)
24. Zhang, T.Y., Suen, C.Y.: A fast parallel algorithm for thinning digital patterns. Commun. ACM **27**(3), 236–239 (1984)
25. Yang, M., Kpalma, K., Ronsin, J.: A survey of shape feature extraction techniques. In: Yin, P.-Y. (ed.) Pattern Recognition, pp. 43–90. In-Tech (2008)
26. Harmsen, J.J., Pearlman, W.A.: Steganalysis of additive-noise modelable information hiding. In: Security and Watermarking of Multimedia Contents V, vol. 5020, pp. 131–143. International Society for Optics and Photonics (2003)
27. Ma, W.Y., Manjunath, B.S.: Texture features and learning similarity. In: 1996 IEEE Computer Society Conference on Computer Vision and Pattern Recognition, 1996. Proceedings CVPR 1996, pp. 425–430. IEEE (1996)
28. Zhang, D., Wong, A., Indrawan, M., Lu, G.: Content-based image retrieval using Gabor texture features. IEEE Trans. PAMI 13–15 (2000)
29. Jolliffe, I.T.: Principal Component Analysis. Springer Series in Statistics, 2nd edn, p. XXIX, 487. Springer, New York (2002). https://doi.org/10.1007/b98835
30. Schölkopf, B., Smola, A., Müller, K.R.: Nonlinear component analysis as a kernel eigenvalue problem. Neural Comput. **10**(5), 1299–1319 (1998)
31. Martinez, A.M., Kak, A.C.: PCA versus LDA (PDF). IEEE Trans. Pattern Anal. Mach. Intell. **23**(2), 228–233 (2001)
32. Baudat, G., Anouar, F.: Generalized discriminant analysis using a kernel approach. Neural Comput. **12**(10), 2385–2404 (2000)
33. Rodriguez-Galiano, V.F., Ghimire, B., Rogan, J., Chica-Olmo, M., Rigol-Sanchez, J.P.: An assessment of the effectiveness of a random forest classifier for land-cover classification. ISPRS J. Photogrammetry Remote Sens. **67**, 93–104 (2012)

34. Franke, J., Mandler, E.: A comparison of two approaches for combining the votes of cooperating classifiers. In: 11th IAPR International Conference on Pattern Recognition. Vol. II. Conference B: Pattern Recognition Methodology and Systems, pp. 611–614. IEEE (1992)
35. Gardezi, S.J.S., Faye, I., Eltoukhy, M.M.: Analysis of mammogram images based on texture features of curvelet sub-bands. In: Fifth International Conference on Graphic and Image Processing (ICGIP 2013), vol. 9069, p. 906924. International Society for Optics and Photonics (2014)
36. Alkan, A., Koklukaya, E., Subasi, A.: Automatic seizure detection in EEG using logistic regression and artificial neural network. J. Neurosci. Methods **148**(2), 167–176 (2005)
37. Orhan, U., Hekim, M., Ozer, M.: EEG signals classification using the K-means clustering and a multilayer perceptron neural network model. Expert Syst. Appl. **38**(10), 13475–13481 (2011)
38. Santosh, K.C., Wendling, L.: character recognition based on non-linear multi-projection profiles measure. Frontiers Comput. Sci. **9**(5), 678–690 (2015)
39. Santosh, K.C.: Character recognition based on DTW-radon. In: ICDAR, pp. 264–268 (2011)
40. K.C., S., Lamiroy, B., Wendling, L.: DTW for matching radon features: a pattern recognition and retrieval method. In: Blanc-Talon, J., Kleihorst, R., Philips, W., Popescu, D., Scheunders, P. (eds.) ACIVS 2011. LNCS, vol. 6915, pp. 249–260. Springer, Heidelberg (2011). https://doi.org/10.1007/978-3-642-23687-7_23
41. Santosh, K.C., Lamiroy, B., Wendling, L.: DTW-radon-based shape descriptor for pattern recognition. IJPRAI **27**(3) (2013)

Recognition of Meitei Mayek Using Statistical Texture and Histogram Features

Sanasam Inunganbi[1(✉)] and Prakash Choudhary[2]

[1] National Institute of Technology, Manipur, Imphal, India
inung.sam@gmail.com
[2] National Institute of Technology, Hamirpur, India
choudharyprakash87@gmail.com

Abstract. Recognition of handwritten characters is one of the challenging and interesting problems in recent trends considering the empirical aspect. The research work on Meitei Mayek (Manipuri Script) recognition is still in the infant stage due to its intricate patterns and being a regional language. Moreover, the language has been reinstated recently, and there is no standard database available for research work. Therefore, we attempt to develop a recognition system on our developed dataset of Meitei Mayek using statistical texture and histogram features. A total of 4,900 samples of 35 letters of Meitei Mayek have been collected in 140 pages from 90 different people of varying age group. Feature extraction technique like Uniform Local Binary Pattern (ULBP) and Projection Histogram (PH) have been considered to evaluate and validate the developed dataset. Finally, recognition is performed using the K-nearest neighbor (KNN) classifier. The combination of ULBP and PH has given the recognition rate of 97.85%.

Keywords: Handwritten character recognition · Meitei Mayek · Texture · Projection Histogram

1 Introduction

Handwritten character recognition is a notable area in the field of pattern recognition drawing lots of attention in academic as well as commercial interest. It is a challenging task due to its divergence in writing styles and occurrence of similar looking characters. A recognition system generally follows the following procedure: image acquisition, preprocessing, feature extraction and recognition. Handwritten character recognition finds its application in various areas such as digitization of documents, automation in mailing system and bank data processing. The result of OCR systems can act as an intermediate step in translation or speech synthesis. Meitei Mayek is already encoded into Unicode. Therefore, the OCR output of text as Unicode would help to maintain a common standard for further processing in translation to other languages.

K. C. Santosh and R. S. Hegadi (Eds.): RTIP2R 2018, CCIS 1037, pp. 63–71, 2019.
https://doi.org/10.1007/978-981-13-9187-3_6

Manipuri is the official language of Manipur, a state in north-eastern India. It belongs to Tibeto-Burman branch of the Sino-Tibetan language family and is the primary language for communication in Manipur which has also spread in some part of Bangladesh and Myanmar. The Meitei Mayek (Manipuri Script) has been reinstated recently, and that is why only a few researchers have been accomplished. The current script is the construction of the ancient Meitei Mayek script. The current script consists of 27 alphabets (Eeyek Eepee), 8 letters with short-ending (Lonsum Eeyek derived from Eeyek Eepee), 8 vowel signs (Cheitap Eeyek), 10 digits (Cheising Eeyek), 3 punctuation marks (Khudam Eeyek). Interestingly the alphabets of Meitei Mayek Scripts are named after parts of the human body. The 27 alphabets and 8 short ending letters of the Meitei Mayek Scripts are given in Fig. 1.

Fig. 1. Meitei Mayek alphabets

Various Character Recognition System has been developed for different languages in the world. To mention a few, Rectangle Histogram Oriented Gradient

(R-HOG) had been used as the feature for recognition of 8000 samples of Marathi character [1]. The authors had used Support Vector Machine (SVM) and Feed-Forward Artificial Neural Network (FFANN) for classification achieving an accuracy of 95.64% and 97.15% respectively. In [2], Kannada numerals classification had been performed using the multi-layer neural network on 50 images. All numerals are normalized to 7 $\times$ 5 pixels whose one-dimensional representation were fed to the neural network for classification. Radon transformed based features with dynamic time wrapping (DTW) had been used for pattern recognition and retrieval method [3–6]. The DTW algorithm had been used to match corresponding pairs of the histogram at every projection angle for Radon transformed. The feature is more focus on shape descriptor for pattern recognition due to its statistical properties. However, very few research paper existed in literature for recognition based on Meitei Mayek Script. In 2010, identification of the handwritten character of Meitei Mayek script using Artificial Neural Network was proposed by [7]. The authors have used 594 samples images, out of which 495 have been trained, and the remaining 135 had been used for testing. Probabilistic and fuzzy feature and their combination had been used as the feature vector in their work. In 2013, an OCR system of Meitei Mayek Script on printed documents was proposed using Support Vector Machine (SVM) in [8]. They had scanned 100 pages at 300 dpi from textbook, magazine, and newspaper and stored in "tiff" or "png" format. Chain code and directional distribution had been used as local features, aspect ratio and longest vertical run had been used as the global feature for classification. In 2014, another OCR system for Meitei Mayek Script using Neural Network (NN) with back propagation had been proposed in [9]. The authors had used 1000 samples of images, 500 images were used for training while the remaining 500 were used for testing. They had used the binary pattern as the feature for their model. Recognition of Meitei Mayek numerals or digits had also been carried out using SVM in [11,12] and NN in [10]. Various work performed on Meitei Mayek script has been summarized in Table 1.

Table 1. Various work performed on Meitei Mayek script.

Database	Method	Accuracy
594 sample images [7]	ANN with probabilistic and fuzzy feature	90.03%
100 pages printed documents [8]	SVM with RBF kernel chain code and directional feature with aspect ratio and longest vertical run	96%
1000 samples [9]	NN with back propagation	80%
1000 samples [10]	NN using pixel density of binary pattern as feature	85%
800 samples [12]	SVM using Gabor filter	89.58%

Since there is no standard database for Meitei Mayek (Manipuri script) available for research work we have collected our handwritten database from various people. To validate the collected database, we have performed experiments using statistical texture and projection histogram feature. Extraction of a distinctive feature is a vital stage in the recognition process. Extracted features should enhance interclass difference and decrease intraclass difference. Texture represents an essential property of an object that does not change with the change of illumination conditions. Texture gives the spatial arrangement or structure of color or pixel intensities over a local area or of the whole image. It defines how a pixel intensities correlate with their neighboring pixels. Combining this feature with the histogram on pixels arrangement of character image will promote distinctive feature. The extracted features are then fed to KNN classifier for recognition.

The rest of the paper is organized as follows: Sect. 2 presents the image acquisition stating the dimension of the dataset followed by Sect. 3 explaining the preprocessing technique follow in our character recognition system. Section 4 describes the feature extraction technique to be used for recognition in Sect. 5. This section also presents the experimental results and lastly Sect. 6 concludes the paper stating the key finding and future scope.

2 Image Acquisition

The first and foremost important step in developing an OCR system is to design and acquire databases for evaluation and validation of the proposed algorithms or techniques. So, for development and evaluation of an efficient character recognition system of Meitei Mayek, we have collected 4 sets of the 35 alphabets (27 Eeyek Epee with 8 Lom Eeyek) in 140 pages. Each alphabet has 35 samples collected per page so that one set comprises of 1225 (35×35) isolated characters. So, in total there are 4,900 samples of the alphabets written by 90 different people of various age group. The image acquisition is associated with demographic information of the writer such as name, address, occupation, qualification signature, etc. So that this information can be made available for other application like signature verification as well. The collected image has been scanned using a scanner at 300 dpi and stored in "png" file format for further processing.

3 Preprocessing

Preprocessing is vital for further processing and crucial for maximizing recognition accuracy. It is the procedure performed to enhance image properties and suppress undesirable noise. The use of a preprocessing technique improved document image and prepared it for the next stage in the character recognition system. So, it is essential to have an active preprocessing step to achieve higher recognition rate. Besides, it makes recognition system more robust through accurate enhancement, noise removal, image thresholding, character segmentation, character normalization, additional space removal around the character. Segmentation is the main preprocessing perform in our methodology so that only identifiable elements have been retained and unrelated information can be avoided.

The characters have been collected in tabular form where each character occupies one slot, and every character needs to be isolated for further processing. Firstly, the character table (the section where the writer replicates characters in their writing style) has been separated from the rest of the part by cropping as illustrated in Fig. 2. Since the table has been appropriately aligned, every character has been isolated by diving the table into an appropriate number of rows and columns, and further processing has been performed to eliminate extra space other than the alphabets.

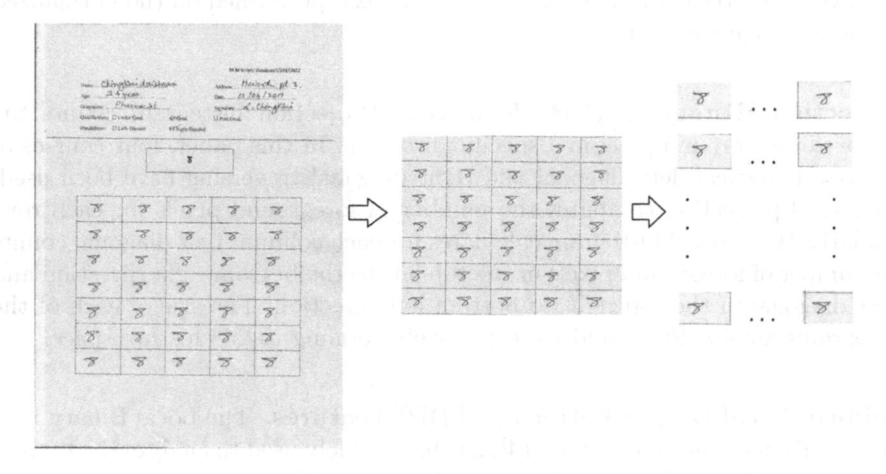

Fig. 2. Preprocessing performed in Meitei Mayek dataset

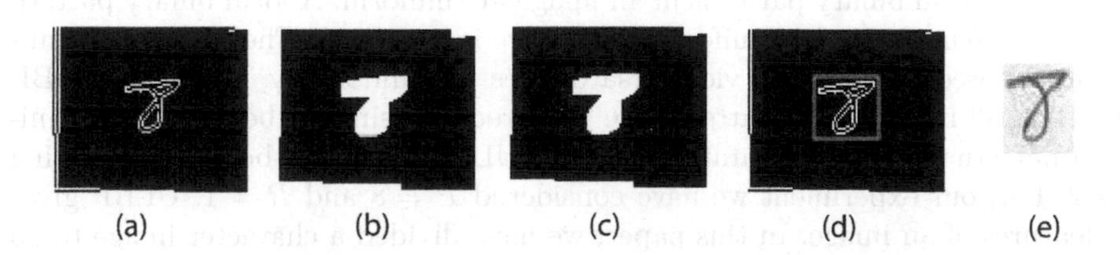

Fig. 3. Process for obtaining minimum bounding rectangle of a character in Meitei Mayek dataset (a) Edge image (b) Dilate image (c) Filled image (d) Minimally bounded box (e) Cropped image

From the isolated image, the edge of the alphabet is determined using the Sobel operator. To thicken the border, we have operated the morphological dilation using structural element disk of size three. Image filling operation is performed on the binary image to fill any hole exists in the picture. Then, the connected component analysis is implemented on the filled image to obtain properties like area and bounding box. The bounding box corresponding to the most significant object (here the alphabet) is taken into account, and the image is cropped accordingly. The whole procedure, adopted from [13], is depicted diagrammatically in Fig. 3.

4 Feature Extraction

For evaluation and validation, statistical texture and projection histogram features have been used for experimentation on the self-collected dataset. All feature extraction techniques in this paper have been performed on the normalized character image size of 40×40.

Projection Histogram (PH) Features. Projection histogram counts the number of foreground pixel in a specific direction. In this paper, four courses of horizontal, vertical, left diagonal and right diagonal traversing have been used. Horizontal projection computes the number of foreground pixels for each row. Similarly, the vertical histogram calculates for each column. Left diagonal counts the number of foreground pixels in the top left to the bottom right direction and right diagonal in the top right to bottom left direction. The size of each of the histograms are 40, 40, 79 and 79 respectively totaling to 238 feature space.

Uniform Local Binary Pattern (ULBP) Features. The Local Binary Pattern (LBP), introduced by Ojala [14], has been widely used to analyze the texture of an image. It may be defined as an ordered set of binary numbers formed by thresholding between central and its neighbor pixels later converted to decimal. Mathematically, LBP can illustrated using Eq. 1 where g_c is the center pixel of 3×3 block and g_p are the neighboring pixels. Further, Ojala et al. noticed that 90% of the local binary patterns in an image are uniform. A local binary pattern is said to be uniform if its uniformity measure is at most 2. The number of transitions between 0 and 1 or vice versa defines the uniformity. In Uniform LBP (ULBP), all non-uniform patterns are gathered in a single label while each uniform pattern is given a separate label. The ULBP value can be evaluated using Eq. 2. For, our experiment we have considered $P = 8$ and $R = 1$. ULBP gives 58 features of an image. In this paper, we have divided a character image to 25 zones by horizontal and vertical division giving a total of 1450 ULBP features.

$$LBP_{P,R} = \sum_{P=0}^{P-1} s(g_p - g_c)2^P,\ s(x) = \begin{cases} 1, & if\ x \geq 0\ or \\ 0, & Otherwise \end{cases} \tag{1}$$

$$\begin{aligned} U(LBP_{P,R}) = {} & |s(g_{p-1} - g_c) - s(g_0 - g_c)| \\ & + \sum_{p=1}^{p-1} |s(g_p - g_c) - s(g_{p-1} - g_c)| \end{aligned} \tag{2}$$

5 Recognition and Experimental Results

The K-Nearest Neighbor (KNN) classifier is used for experimentation which is one the most basic yet efficient algorithm for classification and regression [15] for recognition of Meitei Mayek alphabets. It is a supervised learning method, and it relates unknown pattern to the known according to some distance or some other similarity function. This classifier is a non-parametric algorithm which does not lay any underlying assumption for distribution of feature. The decision is taken based on the statistic gathered during training, to what group its nearest neighbor belong to and give a label to an unknown image. We have used the distance function to determine the closest neighbor as 'Cityblock' distance metric with the value of $k = 1$. The value of k should be odd to get an unbiased majority decision. Increasing the value of k does not give significant improvement in the recognition accuracy, that is why $k = 1$ is fixed for our experiment.

Considering the above two types of features, we have performed three experiments. Fourfold cross-validation has been implemented so that every sample contribute to building the recognition system and their average is considered as the accuracy. In Table 2, the recognition accuracy using KNN classifier with the two features and the combination of both have been shown. It can be observed from the table that the combined feature vector yields a higher recognition rate than each of the feature considered alone. A recognition rate of 97.85% is achieved by the combination of the ULBP and histogram feature. Combining both the features increases the discriminative properties, and hence the number of test images are correctly classified.

Table 2. Recognition Rate (RR) using texture and histogram feature

Experiment no	Feature(s) used	RR
1	Projection Histogram (PH)	94.18%
2	ULBP	88.49%
3	PH + ULBP	97.85%

Table 3. Comparison of handwritten character recognition rate (RR) with the existing methods in literature.

Papers	Recognition methods	Recognition rate
[7]	Probabilistic and fuzzy feature with ANN	90.3%
[9]	Binary pattern as vector and NN with back propagation	80%
Proposed method	PH + ULBP and KNN classifier	**97.85%**

Comparison of some of the existing work in literature for recognition of handwritten Meitei Mayek has been performed with the experimental results of our proposed work, and the results have been summarized in Table 3. It can be observed from the table that our proposed model for recognition has achieved a higher recognition rate as compared to the one existing in the literature.

6 Conclusions and Future Work

In this paper, we have proposed a character recognition technique on a self-collected database of size 4,900 sample of Meitei Mayek alphabets. Till date, this is one of the largest available datasets for Meitei Mayek Script. The database has been collected in four sets and does have the variation of above 80%. Some individual has written with a pen while some, with a pencil. Moreover, each has their unique characteristic of writing. Hence, it can be concluded that the proposed handwritten character recognition system is robust to the writing style of the individual.

This paper only deals with the recognition of isolated Meitei Mayek alphabets using projection histogram and uniform LBP. The spatial arrangement of the pixels has been explored in this paper. Further, it can be extended to transform domain as well as other spatial structure and even deep neural network. In future, identification of dataset having complete sentences of Meitei Mayek can be performed and focus can be made on the full-length dataset of both printed and handwritten, encoding the corpus, benchmarking of the dataset on different standards.

References

1. Kamble, P.M., Hegadi, R.S.: Handwritten Marathi character recognition using R-HOG Feature. Procedia Comput. Sci. **45**, 266–274 (2015)
2. Hegadi, R.S.: Classification of Kannada numerals using multi-layer neural network. In: Aswatha Kumar, M., Selvarani, R., Suresh Kumar, T.V., et al. (eds.) Proceedings of International Conference on Advances in Computing. Advances in Intelligent Systems and Computing, pp. 963–968. Springer, New Delhi (2013). https://doi.org/10.1007/978-81-322-0740-5_116
3. Santosh, K.C.: Character recognition based on DTW-radon. In: 2011 International Conference on Document Analysis and Recognition (ICDAR). IEEE (2011)
4. Santosh, K.C., Lamiroy, B., Wendling, L.: DTW for matching radon features: a pattern recognition and retrieval method. In: Blanc-Talon, J., Kleihorst, R., Philips, W., Popescu, D., Scheunders, P. (eds.) ACIVS 2011. LNCS, vol. 6915, pp. 249–260. Springer, Heidelberg (2011). https://doi.org/10.1007/978-3-642-23687-7_23
5. Santosh, K.C., Lamiroy, B., Wendling, L.: DTW-radon-based shape descriptor for pattern recognition. Int. J. Pattern Recognit. Artif. Intell. **27**(03), 1350008 (2013)
6. Santosh, K.C., Wendling, L.: Character recognition based on non-linear multi-projection profiles measure. Front. Comput. Sci. **9**(5), 678–690 (2015)
7. Thokchom, T., et al.: Recognition of handwritten character of Manipuri script. JCP **5**(10), 1570–1574 (2010)

8. Ghosh, S., et al.: An OCR system for the Meetei Mayek script. In: 2013 Fourth National Conference on Computer Vision, Pattern Recognition, Image Processing and Graphics (NCVPRIPG). IEEE (2013)
9. Laishram, R., et al.: A neural network based handwritten Meitei Mayek alphabet optical character recognition system. In: 2014 IEEE International Conference on Computational Intelligence and Computing Research (ICCIC). IEEE (2014)
10. Laishram, R., et al.: Simulation and modeling of handwritten Meitei Mayek digits using neural network approach. In: Proceedings of the International Conference on Advances in Electronics, Electrical and Computer Science Engineering-EEC (2012)
11. Kumar, C.J., Kalita, S.K.: Recognition of handwritten numerals of Manipuri script. Int. J. Comput. Appl. **84**(17) (2013)
12. Maring, K.A., Dhir, R.: Recognition of Cheising Iyek/Eeyek-Manipuri digits using support vector machines. IJCSIT **1**(2) (2014)
13. Inunganbi, S., Choudhary, P.: Recognition of handwritten Meitei Mayek script based on texture feature. Int. J. Nat. Lang. Comput. (IJNLC) **7**(5), 99–108 (2018)
14. Ojala, T., Pietikäinen, M., Harwood, D.: A comparative study of texture measures with classification based on featured distributions. Pattern Recogn. **29**(1), 51–59 (1996)
15. Cover, T., Hart, P.: Nearest neighbor pattern classification. IEEE Trans. Inf. Theor. **13**(1), 21–27 (1967)

Symbolic Approach for Word-Level Script Classification in Video Frames

C. Sunil[1(✉)], K. S. Raghunandan[2], H. K. Chethan[1], and G. Hemantha Kumar[2]

[1] Department of Computer Science and Engineering, Maharaja Research Foundation, Maharaja Institue of Technology, Mysore, Karnataka, India
sunilchaluvaiah87@gmail.com, hkchethan@gmail.com

[2] Department of Studies in Computer Science, University of Mysore, Manasagangotri 570006, Mysore, India
raghu0770@gmail.com, ghk.2007@yahoo.com

Abstract. In recent years, addiction towards internet and digital world has made difficult for people to understand multilingual scripts in various circumstances. In this work, we proposed a model for classification of South Indian multilingual word script extracted from video frames namely, Kannada, Tamil, Telugu, Malayalam and English. Firstly, we extracted Local Binary Pattern (LBP), Histogram of Oriented Gradients (HoG) and Gradient Local Auto-Correlation (GLAC) features for each multilingual word script. The multilingual word script consists of five classes and each class of images are clustered by implementing k-means clustering technique. Further, we proposed symbolic representation to capture intra-class variations between each clusters and symbolic classifier is employed for classification. For experimentation, we have extracted 600 word images from each script and total of 3000 word images from video frames. Further, we have made comparative study to show the robustness of symbolic representation and classifier with SVM and ANN classifiers.

Keywords: Multilingual scripts identification · Interval data · K-means clustering · Symbolic representation · Symbolic classification

1 Introduction

From recent survey it is evident that in today's world everyone are dependent on multiple languages around them due to diversification of religion and culture, so we come across multilingual scripts from videos and images. This has been demanding a major challenging task for all researchers in the domain of image processing and pattern recognition to analyze the multilingual scripts from videos and images. Nowadays OCR engines are designed to identify single scripts than multiple scripts and recent survey shows that there is no such OCR engines which recognizes more than one languages. The task of automatic recognition of the characters available in an images and video frames is called as Optical Character Recognition (OCR). The ultimate goal of an machine translation is

K. C. Santosh and R. S. Hegadi (Eds.): RTIP2R 2018, CCIS 1037, pp. 72–84, 2019.
https://doi.org/10.1007/978-981-13-9187-3_7

to convert all the multilingual scripts into machine readable form which leads to design an effective OCR system [1].

The research on script identification and classification in videos is not explored much when differentiated with printed and handwritten documents as it focus on uncomplicated backgrounds and high resolution which is vital for OCR. The challenges associated with the identification of scripts in video frames are low resolution, blurriness, noise, orientations, fonts and complex backgrounds. Samples of multilingual word script images from the video frames are shown in Fig. 1. In Fig. 1(a)–(e) we can see that word images extracted from video frames contains multiple scripts such as Kannada, Telugu, Tamil, Malayalam and English with various size, fonts, orientations and some with blur and low resolution. Figure 1 clearly shows the essentiality and challenges present in the identification of multilingual scripts and this motivates us to take up this problem. In this work, we propose the most efficient classification method based on symbolic representation and classifier for extracted multilingual word script from video frames [2].

Fig. 1. Shows samples of (a) Kannada, (b) Telugu, (c) Tamil, (d) Malayalam, (e) English word multi-script images extracted from video frames

Subsequently, the organization of proposed work is as follows, Sect. 2 gives a brief explanation on related work, Sect. 3 we discuss proposed model of multilingual word script classification based on symbolic representation and classifier, Sect. 4 illustrate experimentation and results and Sect. 5 concludes the multilingual script classification and contributes for future to build an effective Multilingual-OCR for South Indian Scripts.

2 Related Works

Recently some methods which are adopted for identification and classification of scripts. Pan et al. [3] proposed an average angle feature for classification using KNN classifiers for Chinese and Tamil, Tamil and English, English and Chinese for combined script. Rani et al. [4] considered four classes multi Support Vector Machine classifier is adopted classification of scripts namely, Gurumukhi, English character and numeral. Pal et al. [5] reconstructed Quadratic Classifier is used for classification of Indian numeral scripts such as Tamil, Bangla, Devanagari, Oriya, Telugu and Kannada. Sarkar et al. [6] Multi Layer Perceptron classifier is used for script classification of words from handwritten document of Bangla and Devanagari varied with Roman scripts.

Santosh et al. [7] for identification of Indic script in handwritten documents they have considered same documents with different levels of script identification such as page, line, word and block. Two features such as script dependent and script independent for computation and classification such as Multilayer perceptron and Random forest method are adopted. Among all line based gives better results compared to block, word and page. Obaidullah et al. [8] for identification of eleven official Indic script based on page-level in handwritten documents they have created their own handwritten document image datasets considering 11 official Indic scripts such as Roman, Urdu, Oriya, Arabic, Bangla, Devanagari, Gurumukhi, Telugu, Tamil, Kannada, Gujarati and Malayalam and for evaluation they have consider some of state-of-the-art features and classifiers such as Multilayer perceptron and Simple logistic and combination of both for recognition of scripts based on Bi-script, Tri-script and multi-script. Obaidullah et al. [9] proposed a novel technique which segregate the Indic scripts with "matra" that will act as precursor that approaches an effective way for the identification of handwritten scripts in multi-script documents. To segregate they have consider Bangla and Devanagari as positive samples and Roman and Urdu as negative samples. Canny edge detector, Fractal geometry analysis and morphological linear transform are the methods used for extraction of features and classifiers such as Bayesnet, Random Forest and Multilayer perceptron and also combination of features and classifiers are adopted for computation purpose and drawback is some scripts are misclassified. Obaidullah et al. [10] proposed an word-level document image datasets for identification of multi scripts which considers 13 various Indic languages from 11 official scripts which includes 3000 words per languages. In this model they have adopted various classifier such as fuzzy unordered rule induction algorithm, Bayesnet, liner, multilayer perceptron and simple logistic. Random forest, wavelet energy and spatial energy features are used and also some of their combinations and MLP performs better among all. Obaidullah et al. [11] In this work they have undergone an survey based on various techniques used for identification of Indic scripts in handwritten documents and which also includes various pre-processing, feature extraction and classifiers. They have also conducted survey on different datasets available for Indic Scripts. Obaidullah et al. [12] proposed an novel method for identification of Indic scripts in multi script documents based on extreme learning machine

(ELM) which furnish better performance of neural network with lesser user intervention and provides faster training. They also study some various activations functions along with ELM such as sine, hard limiter, sigmoidal, radial and triangular basis and for computation both script dependent and independent features are used and sigmoidal function gives better results.

Gomez et al. [13] proposed novel Ensembles Conjoined based networks classification of scene text word scripts such as Arabic, Russian, Cambodian, Kannada, Korean and Mongolian. This method make use of mutual way of learning discriminative stroke-part representation and relative importance of patch based classification through ensembles conjoined convolutional networks for an efficient classification of scripts. Bharath et al. [14] proposed a novel approach for classification of text documents which consists of two text representation model that is Sentence-Vector space model and Unigram representation model. For classification of text documents two classifiers are used one is symbolic classifier and neural network classification for effective capturing of semantic information of text data. Score level fusion technique is used to calculate the accuracy of different text representation models. Shi et al. [15] proposed a novel method for identifying the scripts in both documents which consists of word images of eleven various scripts and as well as in videos which consists of ten Indian scripts from television videos. Discriminative convolution neural network(DisCNN) is an novel deep learning approach which is the combination of deep features with discriminative mid level representations. This DisCNN optimized model helps for differentiating among various scripts. Jamil et al. [16] proposed a novel method for multilingual text identification and extraction in video images. The method uses combination of both supervised and unsupervised techniques for detection of text regions and classification of text into English, Arabic, Urdu and Chinese by determining Local Binary Pattern histogram and which classifies using the artificial neural network classifier and fed to appropriate recognition engines. Singh et al. [17] proposed a novel method of word level script identification for six different handwritten Indic scripts such as Roman, Oriya, Telugu, Devanagari, Bangla etc. This model has produced a new technique of combining elliptical and polygonal approximation based feature method and these independent features have been implemented on various classifiers for the purpose of classification and have produce better results. Angadi et al. [18] proposed a novel fuzzy technique for identifying word wise script in display boards of low resolution images. In this model five Indian scripts such as Tamil, Hindi, English, Kannada and Malayalam have been considered and to discriminate between these classes wavelet features and horizontal run statistics has been adopted. Crisp vector representation and crisp sets of knowledge bases is constructed for all four scripts. To evaluate the performance overall accuracy as well as separate accuracy of all scripts have been calculated.

3 Proposed Model

In this work, we propose the concept of symbolic representation and classifier for classification of multilingual word scripts such as Tamil, Kannada, Malayalam,

English and Telugu in video frames. Various steps involved during the process of proposed multilingual word scripts classification are shown in the Fig. 2 and further explanations is given in sub sections below.

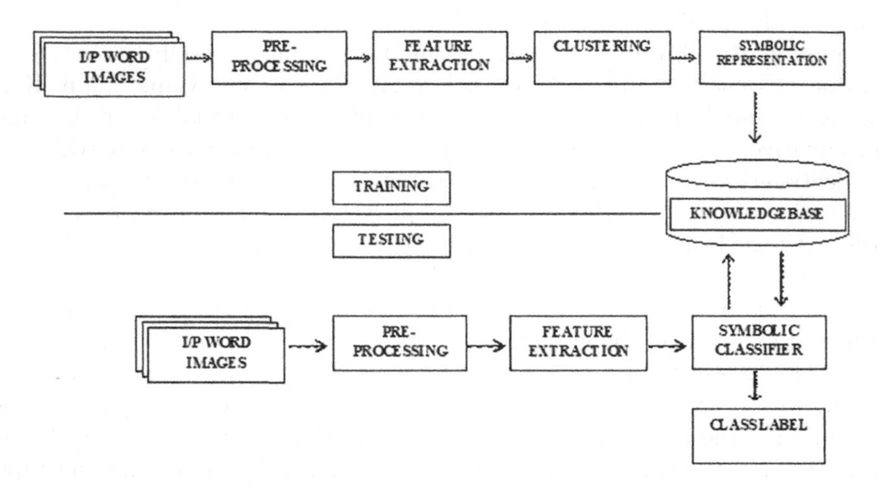

Fig. 2. Shows block diagram of proposed model

While Pre-processing we extract frames from video using key frame extraction technique and to maintain uniformity in the dimensions we resize all the frames of dimension P × Q to p × q and prior dimension relates to the original frame and last dimension relates to the resized video frame. Further, we transform color images to gray images and this transformation helps to extract better features from gray scale images.

3.1 Feature Extraction

For extracting features we adopt feature extraction technique namely, one is Local Binary Pattern feature based on texture and other two are gradient based features likewise Histogram of oriented gradients and Gradient Local Auto-Correlation. In following sub sections we have explained in brief about the feature extraction technique which is used in the proposed model.

Local Binary Pattern. The basic idea of Local Binary Pattern is that it illustrates the texture of images by employing two kind of measures such as gray scale contrast and local spatial patterns. The basic definition of $LBP_{P,R}$ is given below [19],

$$LBP_{P,R}(x_c, y_c) = \sum_{p=0}^{P-1} S(g_p - g_c)2^P \tag{1}$$

where,

$$S(x) = \begin{cases} 1, & \text{if } x >= 1 \\ 0, & \text{otherwise} \end{cases}$$

From the above equation S(x) denotes thresholding function,(x_c, y_c) indicates that it is the center pixel of 8 pixel neighborhood, g_c and g_p are the gray levels of center pixel, around the radius R, P is the adjacent pixels in circular neighborhood.

Histogram of Oriented Gradients. The normal idea of HoG descriptor, it is gradient based depiction that is invariant to local geometric and photometric changes i.e. shape and appearance of the different script of word images which can be represented through intensity gradient distribution or the edge directions [20]. Surrounded by cells an histogram of gradient directions of the pixels are studied and images are divided into small spatial regions called cells that determines HoG descriptors.

In the present model we make use of block size 3×3 in script images of size 16×16 and orientation of histogram comprised of 9 bins given $81(3 \times 3 \times 9)$ features for HoG descriptor. In our experiment HoG feature is well suited as it utilizes the localized cells and will be effective in illustrating appearance and shape of the object, which is an word in present circumstances.

Gradient Local Auto-correlation. GLAC is essentially the continuation of HoG or SIFT from first order statistics i.e. histograms to second order statistics i.e. auto-correlations. For feature extraction GLAC adopt spatial and orientation of auto correlations of local gradients. These features capture information about the gradients, but also it is invariant to structure of rotations that appears in text, and it is expressed with regard to magnitude and orientation [21].

GLAC can be illustrated as, the image region is considered as I and $r = (x, y)^t$ let it be a position vector in I. The image gradient $(\frac{\partial I}{\partial x}, \frac{\partial I}{\partial y})^t$ at each pixel is constituted with regard to magnitude $n = \sqrt{\frac{\partial I^2}{\partial x} + \frac{\partial I^2}{\partial y}}$ and the orientation angle $\theta = arctan(\frac{\partial I}{\partial x}, \frac{\partial I}{\partial y})$.

3.2 Clustering and Symbolic Representation

At this stage, word script images which is identical are grouped together to their respective classes and to cluster these images we adopted partitional clustering approach as it make easy for clustering by employing feature matrix rather than using proximity matrix as in hierarchical clustering. K-means clustering technique is adopted to cluster particular script to their respective classes by varying k.

We have used interval data representation for clustered script images, it captures the variations in intra-class [22]. We consider a sample $Y_i = (z^1, z^2, z^3, \ldots, z^b)$ $(Y_i \in s^b)$ those lies to i^{th} class which include b features. For total number of P samples let there be q number of classes. Then, we apply clustering for the samples which lies to i^{th} class from which number of clusters achieved from each class is k and total number of samples j^{th} cluster which lies to class i is $p_j^i j = 1, 2, \ldots, k$ and $i = 1, 2, .., q$. Mean-standard deviation interval representation is used to capture the intra-class differences from each cluster and the mean and standard deviation is evaluated for the clustered samples is specified in Eqs. (2) and (3).

$$\mu_{j_i}^l = \frac{1}{p_j^i} \sum_{h=1}^{p_j^i} z_h^l \tag{2}$$

$$\sigma_{j_i}^l = \sqrt{\frac{1}{(p_j^i - 1)} \sum_{h=1}^{p_j^i} (z_h^l - \mu_{j_i}^l)} \tag{3}$$

Where, $\mu_{j_i}^l$ and $\sigma_{j_i}^l l$ the mean and standard deviation value of lth feature belongs to j^{th}cluster corresponding to class i respectively. The mean and standard deviation is evaluated for all features, those lies to j^{th} cluster corresponding to i^{th} class.

Then we have combined mean and standard deviation to obtain interval cluster representative that belongs to each class. Then, the difference of mean and standard deviation and sum of mean and standard deviation signifies the lower bound and upper bound of an interval. Eventually, from each class we get k number of such cluster interval representatives. Cluster representative is by,

$CS_j^i = [(\mu_{(j_i)}^1 - \sigma_{(j_i)}^1), (\mu_{(j_i)}^1 + \sigma_{(j_i)}^1)], [(\mu_{(j_i)}^2 - \sigma_{(j_i)}^2), (\mu_{(j_i)}^2 + \sigma_{(j_i)}^2)], ..,$ $[(\mu_{(j_i)}^b - \sigma_{(j_i)}^b), (\mu_{(j_i)}^b + \sigma_{(j_i)}^b)]$

$CS_j^i = [f_1^-, f_1^+], [f_2^-, f_2^+], .., [f_b^-, f_b^+]$

where,

$$f_l^- = (\mu_{(j_i)}^l - \sigma_{(j_i)}^l) \, and \, f_l^+ = (\mu_{(j_i)}^l + \sigma_{(j_i)}^l)$$

3.3 Script Classification

In this section, we used symbolic classifier in our proposed classification method to check the efficacy for classification of word scripts from video frames. Therefore, we represent the word script images in the form of interval data which is explained above. We consider a test sample $U_t = u^1, u^2, \ldots, u^b$, contains b number of features and the test sample U_t must be categorized into any one member of the class from the five classes. For test samples and all the reference samples similarity is calculated at feature level. For single valued feature and reference interval feature we calculate the similarity which is as shown.

When the single value lies between the upper bound and lower bound, then the value of similarity is 1 or else 0. Likewise, the similarity between U_t and rest of the samples are computed. If U_t is consider to be member of any one of the five classes, at that time the value of acceptance count $AC_q^{j_i}$ will be too high concerning to the cluster representative that lies to a specific class. Therefore, acceptance count $AC_q^{j_i}$ for a test sample which correlates to j^{th} cluster of ith class is given by:

$$AC_q^{j_i} = \sum_{l=1}^{h} Sim(U_{(t)}, CS_j^i) \quad (4)$$

where,

$$Sim(U_t, CS_j^i) = \begin{cases} 1, & \text{if } u^l \geq f_l^- \, and \; u^l \leq f_l^+ \\ 0, & \text{otherwise} \end{cases} \quad (5)$$

i = 1,2,...,q; j = 1,2,...,k; and l = 1,2,...,b;

Finally, we get the interval-valued feature matrix and it used as a reference matrix for classification.

4 Experimentation and Results

In this segment we discuss the experimentation and results that acquired from all the feature extraction technique, symbolic representation and symbolic classifiers. In our proposed method, to analyze the performance of the features and classifiers, we have build our own word scripts dataset that is extracted from the multilingual video frames, as there is no readily available standard datasets. Our dataset consists of 600 word script images from each scripts and total of 3000 word script images extracted from video frames.

Table 1. Confusion matrix for symbolic classifier

Classes	Kannada	Tamil	Telugu	Malayalam	English
Kannada	88.7	1.2	5.7	3.2	1.2
Tamil	1.7	85.2	2.1	7.8	3.2
Telugu	9.3	2.6	80.7	3.8	3.6
Malayalam	2.4	10.8	1.7	83.8	1.3
English	1.8	2	3.2	2.5	90.5

Table 2. Confusion matrix for SVM classifier

Classes	Kannada	Tamil	Telugu	Malayalam	English
Kannada	82.3	2.9	9.3	3.8	1.7
Tamil	5.3	78.6	3.4	10.2	2.5
Telugu	12.7	3.8	75.3	6.3	1.9
Malayalam	11.2	4.2	5.7	76.8	2.1
English	4.2	3.9	4.5	3	84.4

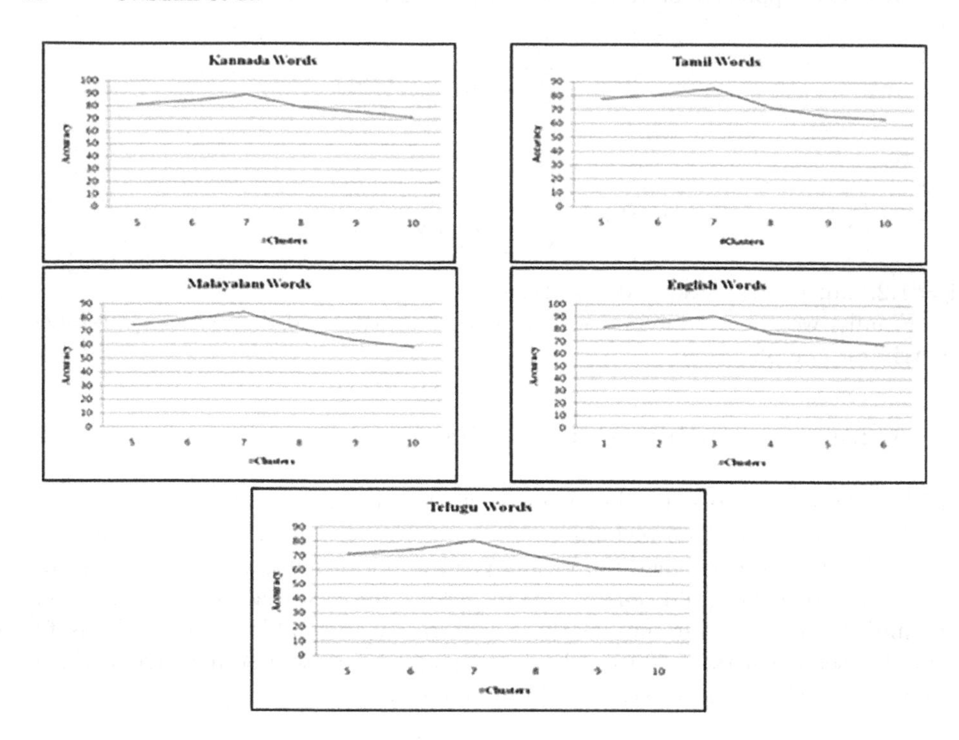

Fig. 3. Graph shows the average results of ten-fold cross validation for different k values of cluster with respect to Kannada, Tamil, Telugu, Malayalam and English Words

Table 3. Confusion Matrix for ANN Classifier

Classes	Kannada	Tamil	Telugu	Malayalam	English
Kannada	84.8	3.3	6.1	3.5	2.3
Tamil	4.6	82.3	3.9	7.3	1.9
Telugu	12.4	5.2	78.4	2.6	1.4
Malayalam	5.3	8.1	3.7	80.5	2.4
English	3.7	2.3	3.5	2.9	87.6

For classification we varied cluster value from 5 to 10 clusters and for each cluster we used 10 fold cross validation. To preserve intra-class variation we adopt symbolic representation. The classification results for the proposed model is done by obtaining confusion matrix from the average of ten-fold cross validation during the classification, where Table 1 shows the superior results of confusion matrix with cluster value 7 for symbolic classifier. When we varied the cluster value from 5 to 10, for cluster value 5 and 6 we achieved poor results for classification because from those cluster we are not able to capture intra-class variation between the classes such as, Kannada and Telugu, Tamil

Fig. 4. Successfully classified multilingual word images (a) Kannada script word images (b) Tamil script word images (c) Malayalam script word images (d) Telugu script word images (e) English script word images

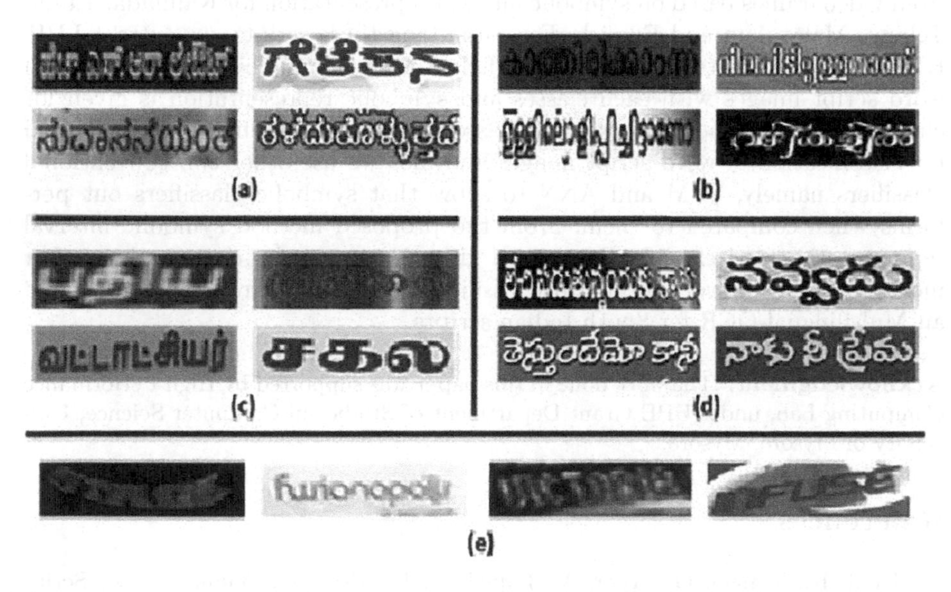

Fig. 5. Misclassified multilingual word images (a) Kannada as Telugu script word images (b) Tamil as Malayalam script word images (c) Malayalam as Tamil script word images (d) Telugu as Kannada script word images (e) English as (left-right) Kannada, Tamil, Telugu, Malayalam script word images

and Malayalam but we got good results for English. When we have fixed the cluster value to 7 we have achieved the better results compared to other cluster values as shown in the Fig. 3. For comparison in ANN we adopted feed forward Multi-Layered Perceptron (MLPs) which is also trained with flexible back propagation algorithm. For illustration we have altered the number of inputs and hidden units and reconstructed the architectures. We have considered six scripts as output units. Throughout the ANN examination the number of hidden units has been set to 10–60 units and iterations extended to1000 till 6000. Every ANNs has been trained with 0.01 learning rate and 0.9 momentum rate. For classification using SVM we have adopted RBF kernel and to calculate the accuracy 5-fold cross validation have been used with various kernel parameters γ and cost parameters c for each binary classifier $\gamma = [(2^{-15}, 2^{-14}, 2^{-13}, \ldots, 2^{15}]$ and $c = [(2^{-15}, 2^{-14}, 2^{-13}, \ldots, 2^{15}]$ and which gives better accuracy among the parameters is adopted. Where, Table 2 shows the confusion matrix for SVM classifier and Table 3 shows confusion matrix for ANN classifier and from the classification results we can see symbolic classification outperforms compared to other conventional classifiers such as SVM and ANN. From the Figs. 4 and 5 shows the successful and unsuccessful results of proposed method.

5 Conclusion and Future Work

In this paper, we proposed the classification of multilingual word script extracted from video frames based on symbolic interval representation for Kannada, Tamil, Telugu, Malayalam and English. For classifying these scripts, we extract LBP, HoG and GLAC features. Then we used clustering technique to cluster identical word script images with each classes and symbolic representation is given for clusters belong to the corresponding classes. Symbolic classifier is then adopted for classification of word script images and also we use other two conventional classifiers namely, SVM and ANN to show that symbolic classifiers out performs when compared to them. From the proposed method symbolic interval representation and classification model gives better results compared to other methods. In future work, using this classification model we are planning to build an Multilingual-OCR for South Indian scripts.

Acknowledgment. The work done in this paper was supported by High Performance Computing Lab, under UPE Grant Department of Studies in Computer Science, University of Mysore, Mysore.

References

1. Ubul, K., Tursun, G., Aysa, A., Impedovo, D., Pirlo, G., Yibulayin, T.: Script identification of multi-script documents. IEEE Access **5**, 6546–6559 (2017)
2. Pati, P.B., Ramakrishnan, A.G.: OCR in Indian scripts: a survey. J. IETE Tech. Rev. **22**, 217–227 (2015)
3. Pan, T.Q., Shivakumara, P., Ding, Z., Lu, S., Tan, C.L.: Video script identification based on text lines. In: International Conference on Document Analysis and Recognition (ICDAR 2011). IEEE (2011)

4. Rani, R., Dhir, R., Lehal, G.S.: Script identification of pre-segmented multi-font characters and digits. In: International Conference on Document Analysis and Recognition (ICDAR 2013). IEEE (2013)
5. Pal, U., Sharma, N., Wakabayashi, T., kimura, F.: Handwritten numerical recognition of six popular Indian scripts. In: International Conference on document Analysis and Recognition (ICDAR 2007). IEEE (2007)
6. Sarkar, R., Das, N., Basu, S., Kundu, M., Nasipuri, M., Basu, D.K.: Word level script identification from Bangla and Devanagari handwritten texts mixed with roman scripts. J. Comput. **2**, 103–108 (2010)
7. Obaidullah, S.M., Santosh, K.C., Halder, C., Das, N., Roy, K.: Automatic Indic script identification from handwritten documents: page, block, line and word-level approach. Int. J. Mach. Learn. Cybern. **10**, 87–106 (2017)
8. Obaidullah, S.M., Santosh, K.C., Halder, C., Das, N., Roy, K.: PHDIndic_11: page level handwritten document image dataset of 11 official Indic scripts for script identification. Multimedia Tools Appl. **77**, 1643–1678 (2017)
9. Obaidullah, S.M., Goswami, C., Santosh, K.C., Halder, C., Das, N., Roy, K.: Separating Indic Scripts with matra for effective handwritten script identification in multi-scripts documents. Int. J. Pattern Recognit. Artif. Intell. **31**(5), 1753003 (2017)
10. Obaidullah, S.M., Santosh, K.C., Halder, C., Das, N., Roy, K.: Word-level multi script Indic document image dataset and baseline results on script identification. Int. J. Comput. Vis. Image Process. **7**(2), 81–94 (2017)
11. Obaidullah, S.M., Santosh, K.C., Halder, C., Das, N., Roy, K.: Handwritten Indic script identification in multi-script images: a survey. Int. J. Pattern Recognit. Artif. Intell. **32**(10), 1856012 (2018)
12. Obaidullah, S.M., Bose, A., Mukherjee, H., Santosh, K.C., Das, N., Roy, K.: Extreme learning machine for handwritten Indic script identification in multi script documents. J. Electron. Imaging **27**(5), 051214 (2018)
13. Gomez, L., Nicolau, A., Karatzas, D.: Improving patch-based scene text script identification with ensembles of conjoined networks. Pattern Recogn. **67**, 85–96 (2017)
14. Bharath Bhushan, S.N., Danti, A.: Classification of text documents based on score level fusion approach. Pattern Recogn. Lett. **94**, 118–126 (2017)
15. Shi, B., Bai, X., Yao, C.: Script identification in the wild via discriminative convolution neural network. Pattern Recogn. **52**, 448–458 (2016)
16. Jamil, A., Batool, A., Malik, Z., Mizar, A., Siddiqi, I.: Multilingual artificial text extraction and script identification from video images. Int. J. Adv. Comput. Sci. Appl. **7**(4) (2016)
17. Singh, P.K., Sarkar, R., Nasipuri, M., Doermann, D.: Word-level script identification for handwritten Indic scripts. In: International Conference on Document Analysis and Recognition (ICDAR 2015) (2015)
18. Angadi, S.A., Kodabagi, M.M.: A fuzzy approach for word level script identification of text in low resolution display board images using wavelet features. In: International Conference on Advances in Computing, Communications and Informatics (ICACCI 2013). IEEE (2013)
19. Ojala, T., Pietikainen, M.: Multiresolution gray-scale invariant texture classification with local binary patterns. IEEE Trans. Pattern Anal. Mach. Intell. **24**(7), 971–981 (2002)
20. Dalal, N., Triggs, B.: Histograms of oriented gradients for human detection. In: IEEE Computer Society Conference on Computer Vision and Pattern Recognition (CVPR 2005) (2005)

21. Kobayashi, T., Otsu, N.: Image feature extraction using gradient local autocorrelations. In: Forsyth, D., Torr, P., Zisserman, A. (eds.) ECCV 2008, Part I. LNCS, vol. 5302, pp. 346–358. Springer, Heidelberg (2008). https://doi.org/10.1007/978-3-540-88682-2_27
22. Guru, D.S., Vinay Kumar, N.: Symbolic representation and classification of logos. In: Proceedings of International Conference on Computer Vision and Image Processing (CVIP 2016) (2016)

Development of Inter-primitive Grammar for Construction of Kannada Language Vowels and Consonants Based on Their Hierarchical Structures

Basavaraj S. Anami and Deepa S. Garag(✉)

K.L.E. Institute of Technology, Hubballi 580030, Karnataka, India
anami_basu@hotmail.com, deepagarag@rediffmail.com

Abstract. Kannada, 2500 years old, the administrative and official language of Karnataka state, India. The language script comprises of 16 vowels and 34 consonants, which are formed with primitives, connection points and relative positions of other primitives. The writing skill of the language needs teaching, the teachers and parents put efforts on children to make them learn the good writing skills. Here lies the scope for automation. Robots assist children in constructing the characters of the language and improve their handwriting skills. Till date, formalism has been applied to languages to check their syntax and semantics to frame words, sentences and paragraphs. But not for the character construction, which needs a formal approach. This paper presents the development of Inter-Primitive Grammar for Construction of Kannada Language Vowels and Consonants based on their Hierarchical Structures. The unambiguous Context Free Grammar (CFG), consisting of a combination of a set of primitives written in specific sequence for construction of Kannada vowels and consonants is devised. Since two primitives are connected at a time, productions are written in Chomsky Normal Form (CNF). To corroborate the grammar, a given string of primitives as input to the tools, the corresponding transliteration code for the given character is generated. Lex and Yacc tools are used, to verify the completeness and soundness of the grammar.

Keywords: Kannada · Vowels · Consonants · Primitive · Grammar · Context free grammar · Chomsky normal form · Position label · Connecting point

1 Introduction

India is a country of diverse languages, culture and tradition. Kannada is the administrative and official language of Karnataka state, India, which is estimated to be 2500 years old. Significant numbers of Kannada speaking people have migrated from India to USA, UAE, Singapore, Australia and UK, where the parents teach the children to read and write Kannada. On an average, there

K. C. Santosh and R. S. Hegadi (Eds.): RTIP2R 2018, CCIS 1037, pp. 85–95, 2019.
https://doi.org/10.1007/978-981-13-9187-3_8

are about 60 million Kannadigas i.e. the Kannada speaking people in the world, making it the 27th most spoken language in the world. Since kannada script is evolved from stone carving, most of the characters are round and symmetric in nature.

The script of Kannada language is syllabic consisting of more than 250 basic, modified and compound character shapes with $(544 * 34) + 15 = 18511$ distinct characters. The language uses 49 phonemic letters, which are segregated into two groups-vowels called as Swaras and consonants called as Vyanjanas in Kannada. The Kannada vowels and consonants along with their transliteration code in Nudi are given in Table 1. Nudi is Kannada software used for Desktop publishing.

Table 1. Transliteration code of Kannada vowels and consonants.

Vowels (Swaras)	Code	Consonants (Vyanjanas)	Code	Consonants (Vyanjanas)	Code
ಅ	a	ಕ	k	ದ	d
ಆ	A	ಖ	K	ಧ	D
ಇ	i	ಗ	g	ನ	n
ಈ	I	ಘ	G	ಪ	p
ಉ	u	ಙ	Z	ಫ	P
ಊ	U	ಚ	c	ಬ	b
ಋ	R	ಛ	C	ಭ	B
ಎ	e	ಜ	j	ಮ	m
ಏ	E	ಝ	J	ಯ	y
ಐ	Y	ಞ	z	ರ	r
ಒ	o	ಟ	q	ಲ	l
ಓ	O	ಠ	Q	ವ	v
ಔ	V	ಡ	w	ಶ	S
ಅಂ	aM	ಢ	W	ಷ	x
ಅಃ	aH	ಣ	N	ಸ	s
		ತ	t	ಹ	h
		ಥ	T	ಳ	L

Compound characters are formed by combining two or more basic characters. The way children learn writing the alphabet in a language, specifically in kindergarten schools, involved practice of writing the characters and thereby memorize them. Every character writing involves combining the primitives, in a definite sequence, which in turn depends upon the type of the writer, whether left-hand-writer or right-hand-writer. Hence, the character construction and writing skills form an important activity in Education. Researchers from the University of Leeds, Bradford Institute of Health Research and University of Indiana (US) have developed a robotic arm. It helps children to practice hand and wrist movements commonly made to improve their handwriting and other manual hand coordination tasks as shown in Fig. 1. This is extended with formalism in the background, one can teach the children, through the Robot by holding their hands, the definite sequence of primitives to write a character.

Every character has a definite way of writing it and is combination of character-parts called primitives. This combination of primitives is a systematic approach in generating or constructing the characters. It is important to note that while reading, we read the whole character and while writing, we write primitives in an order by joining them at appropriate connection positions. Since writing of characters are also governed by the rules and hence an attempt is made to apply formalism for Kannada characters, which finds applications in teaching and learning other languages, television and film industry etc. In order to know the state of the art in this area we have carried out a literature survey. The gist of papers cited during the survey is given in Sect. 2.

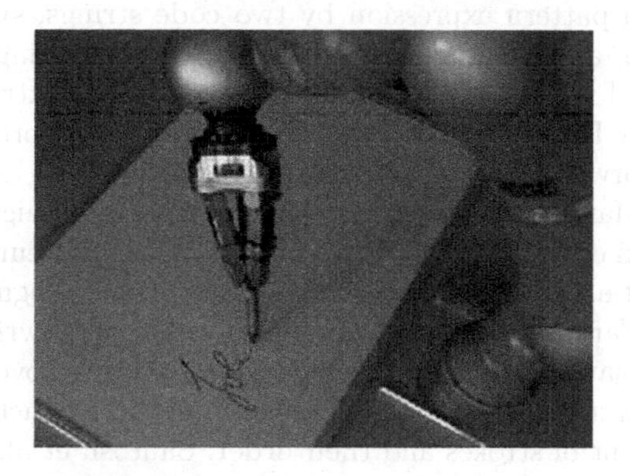

Fig. 1. A robotic arm that helps children to improve handwriting.

2 Literature Review

Prasanna and Kumar [1] have presented a combination of a syntax analysis module with an offline recognition system for handwritten Kannada sentences. As a measure of the syntactical quality, the probability is computed by a parser using a stochastic context free grammar for general Kannada sentences. Ota et al. [2] have describing the structure of Kanji using Stochastic Context Free Grammar (SCFG) to recognize on-line handwritten Kanji character. The work is extended to recognize Kanji strings. They have described Kanji with a few stroke shape and relative position labels. The method enables efficient training and thus robust recognition. Wang et al. [3] have presented a unified approach for modeling and synthesizing Chinese characters. Character is decomposed into basic components, which forms the stroke database and radical database. In the synthesis process, wavelet-based approach is used to select proper strokes and radicals. Genetic algorithm is employed to search for the optimal results. Ota et al. [4] have focused on hierarchical structure to recognize online handwritten Kanji characters. To improve the recognition accuracy, they have used stochastic context free grammar (SCFG) in combination with Hidden Markov Models (HMM) Kim et al. [7] have presented an extended handwritten Korean character (Hangul) recognition algorithm by stroke extraction which is based on the structure of the character and writing habits. To solve the problem, 26 primitive strokes are defined for handwritten Hangul and build stroke separation rules.

Wang et al. [8] have presented an algorithm for the automated generation of the Chinese character structure data based on extracting stokes. The Gothic styles are used as inputs of Chinese characters. Ohmori [9] have presented a new approach for online handwritten Kanji recognition, using hierarchically represented knowledge. It has achieved high recognition rate in reduced processing time. Nagahashi et al. [10] have presented a pattern description and generation method of structural Chinese and Korean character patterns. In this method, composite patterns called as patterns blocks are constructed by several simpler subpatterns. To encode a pattern expression by two code strings, syntactic grammar is defined. Indira et al. [11] have presented a review of existing work on printed Kannada script and their results. Kamble et al. [16] have extraction of features using the Rectangle Histogram Oriented Gradient and Support Vector Machines (SVM) and feed-forward Artificial Neural Network (FFANN) classifiers to classify handwritten Marathi characters. Results demonstrate high performance of these features when classified using feed-forward Artificial Neural network classification. Santosh et al. [21] have constructed a prototype recognizer that uses the "Dynamic Time Warping" (DTW) algorithm to align handwritten strokes with stored stroke templates and determine their similarity. A novel real time handwriting recognition system on Nepalese alphanumeric characters is introduced which is independent of strokes and their order. Santosh et al. [18] have done a comprehensive survey on on-line handwriting recognition system along with the real application by taking Nepali natural handwriting (a real example of one of the cursive handwritings). Kamble et al. [19] have used connected pixel based features and k-nearest neighbor to recognize handwritten Marathi characters. Santosh et al. (2010) have presented an innovative approach to integrate spatial relations in stroke clustering for handwritten Devanagari character recognition. Santosh et al. (2010) have proposed spatial similarity based stroke clustering, using the dynamic time warping (DTW) algorithm to recognize Devanagari natural handwritten characters.

No work is cited in the literature on formalism for construction of Kannada characters namely, vowels and consonants. Kannada has its own unique features to construct the characters like Kanji, Korean, and Chinese etc. Most of the time Kannada character construction is considered complex than other languages as it consists of more curved shapes unlike straight lines [11,12]. This motivated the authors for the present proposed work.

In this paper, we propose formalism for construction of Kannada vowels and consonants characters using the primitives, connection points and relative position of other primitives. All characters are combination of certain primitives written in exact and specific sequence. The method takes a string of primitives as input and generates the corresponding transliteration code for the corresponding character. The rest of this paper is organized as follows: We propose a method to construct Kannada vowels and consonants based on Inter-primitive grammar in Sect. 3. Experimentation results are given in Sect. 4 and finally we conclude the paper in Sect. 5.

3 Construction of Kannada Vowels and Consonants Based on Interprimitive Grammar

3.1 Kannada Character Primitives

Kannada characters are curved in shape. Some regions are highly denser than others. Some shapes are longer than others and some are wider with some kind of symmetry observed in most cases. Initially, the characters are composed of some character-parts, i.e. combination of one or more primitives, and each of these smaller character-parts which can finally be divided into primitives. The simplest primitives are called terminals.

Table 2. Primitives constructing vowels and consonants.

Primitives used to Construct Vowels and Consonants							
t1	ಅ	t11	[illegible]	t21	ಸ	t31	▬
t2	ಆ	t12	ಣ	t22	[illegible]	t32	⌣
t3	ಇ	t13	ಬ	t23	[illegible]	t33	[illegible]
t4	ಉ	t14	ಲ	t24	[illegible]	t34	⌒
t5	ಊ	t15	○	t25	[illegible]	t35	(
t6	ಋ	t16	∩	t26	[illegible]	t36	⌣
t7	ಎ	t17	ಡ	t27	\	t37	ʔ
t8	ಏ	t18	[illegible]	t28	•	t38	ı
t9	ಐ	t19	ω	t29	[illegible]	t39	ಃ
t10	ಒ	t20	ತ	t30	ɔ		

With the help of language experts and domain knowledge, 39 primitives are identified to construct the Kannada vowels and consonants, as given in Table 2. For example, some possible primitives needed to construct certain characters and their corresponding transliteration codes are given in Table 3. To form the given characters, primitives are joined at appropriate positions.

3.2 Hierarchical Structure of Kannada Characters

In Kannada character construction, the primitives are connected in hierarchical manner. The hierarchical structure varies from character to character based on the positional labels, connecting points and number of primitives or character-parts connected with each other. For example, the hierarchical structure of Kannada consonant character "J" is shown in Fig. 2. Wherein a character-part is composed of two primitives, in turn a primitive connected with this small character-part generates a large character-part and finally three character-parts connected using five primitives, constructs the character "J".

Table 3. Sample primitives.

Character	Primitives Used	Transliteration code / Output generated
ಅ	ಅ	a
ಗ	∩ + ⌣	g
ಈ	○ + ⌣ + [illegible]	I
ಢ	ω + ⌣ + ı + •	T
ಝ	○ + ⌣ + ɔ + ı + ɔ	J

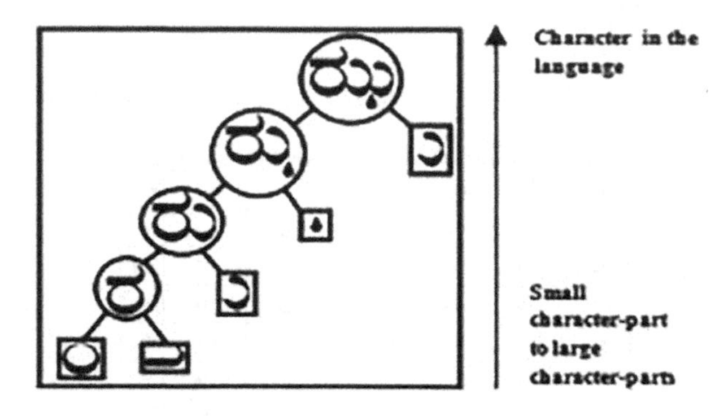

Fig. 2. An example of hierarchical structure of Kannada consonant "J".

3.3 Relative Position and Connecting Points of the Primitives

The primitive positions and connections between the primitives are essential to character construction. Practically, it is necessary to check whether the second primitive is located in the "top" or "bottom" or "right" position with regard to the first primitive or a character-part and at which connecting point the primitive needs to be connected. Hence, it is necessary to perform decision on the relative position and connecting points. On analysis, we have identified 13 relative position labels and assume that all positions are represented using these labels. Figure 3 shows identified relative position labels, where dots represented the connection points, the solid square represents first primitive or a character-part B and dotted square represent the second primitive to be connected. The dots in Fig. 4 show all possible connecting points at which the primitives are connected to each other.

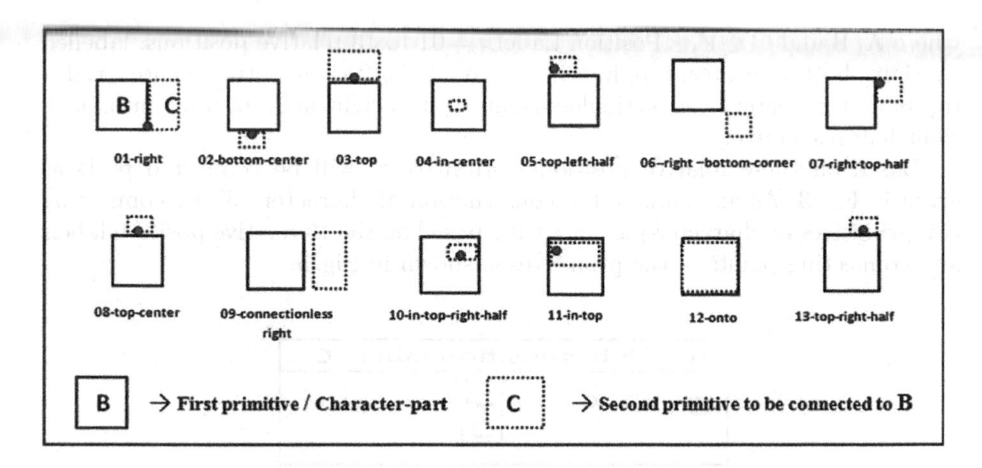

Fig. 3. Relative position labels and connection points

Connecting points of all primitives			
ಅ	[illegible]	[illegible]	[illegible]
ಆ	[illegible]	[illegible]	[illegible]
[illegible]	[illegible]	[illegible]	[illegible]
ಉ	ಲ	[illegible]	[illegible]
ಊ	[illegible]	[illegible]	[illegible]
[illegible]	[illegible]	[illegible]	[illegible]
ಎ	[illegible]	[illegible]	[illegible]
ಏ	[illegible]	[illegible]	[illegible]
ಐ	[illegible]	[illegible]	
[illegible]	[illegible]	[illegible]	

Fig. 4. Connecting points of all primitives

3.4 Inter-primitive Grammar

Inter-primitive grammar, G = (V, T, P, S), expressed as CFG, is used to construct the vowels and consonant characters. S is a start symbol consisting of all the transliteration codes of Kannada vowels and consonants, V is a set of non-terminal symbols consisting of all the transliteration codes of Kannada vowels and consonants, intermediate nodes representing character-parts and all primitives. T is a set of terminal symbols consisting of all primitives and 13 relative position labels. P is a set of production rules written in Chomsky Normal Form, which allows following two forms of rules.

(i) Rules for terminal symbols: $A \rightarrow a$, where A ∈ V and a ∈ T.
(ii) Rules for non-terminal symbols: $A \rightarrow B\langle Position\ Label\rangle C$,

where A, B and $C \in V$, <Position Label> = 01 to 13 relative positions, labelled as, right, bottom-center, top, in-center, top-left-half, right-bottom-center, right-top-half, top-center, connectionless-right, in-top-right-half, in-top, onto, top-right-half respectively.

Based on these relative positions, primitive C will be connected to B as given in Fig. 3. As an example the construction of character "J" by connecting two primitives or character-part at a time based on the 13 relative position labels and connecting points of the primitives is shown in Fig. 5.

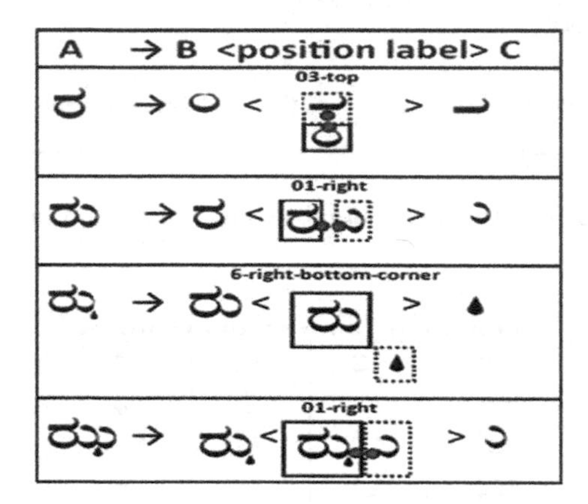

Fig. 5. Connecting primitives based on relative position labels and connecting points of the primitives.

4 Experimental Results

All the primitives are defined as terminal symbols or tokens in the lex program, as Lex and Yacc tools are used to verify the developed grammar [13–15]. Since two primitives are connected at a time, the parse tree for derivations is represented as a binary tree. Figure 6 gives the equivalent binary tree representation for the hierarchical structure of Ķannada consonant character "J" shown in Fig. 2.

The BNF representation of the production rules used to generate J is given in Table 4. Given a string of primitives as input, the corresponding transliteration code for the constructed character is generated. Certain example input/output generated by using lex and yacc tools are

Case1: Given a string of primitives as input, namely, t15 "03" t32 "01" t30 "06" t38 "01" t30, the corresponding transliteration code J for the character is generated.

Case2: Given input string: t15 "03" t32, the corresponding transliteration code r for the character is generated.

Case3: Given input string: t15 "03" t32 "01" t30, it displays invalid character as this character is neither vowel nor consonant.

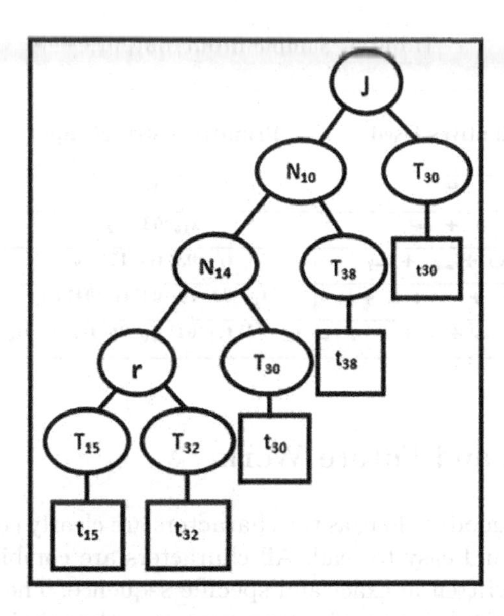

Fig. 6. An example of hierarchical structure of Kannada consonant character

Case 4: Given input string: t15 "03" t32 "01" t30 "06" t38, it displays invalid character as this sequence of primitives generates a character-part but not a complete character.

Table 4. Production rules for character J.

Production rules
<**J**> ::= <$\mathbf{N_{10}}$> 01-right <$\mathbf{T_{30}}$>
<$\mathbf{N_{10}}$> ::= <$\mathbf{N_{14}}$> 06-right-bottom-corner <$\mathbf{T_{38}}$>
<$\mathbf{N_{14}}$> ::= <**r**> 01-right <$\mathbf{T_{30}}$>
<**r**> ::= <$\mathbf{T_{15}}$> 03-top <$\mathbf{T_{32}}$>
<T_{30}> ::= t_{30}
<T_{38}> ::= t_{38}
<T_{15}> ::=t_1

Table 5 gives the input string and the transliterated code of the constructed character as the code for five Kannada and consonant characters.

The production rules written in the inter-primitive grammar recognize all and only 49 Kannada vowels and consonants. Hence the grammar is robust and complete. The experimental result reveals that no two characters are constructed from the same sequence of primitives. Hence, the grammar is sound.

Table 5. Sample input/output.

Character	Primitives Used	Primitives string/ Input	Transliteration code / Output generated
ಅ	ಅ	t_1	a
ಗ	∩ + ◡	t_{16} '03' t_{32}	g
ಈ	○ + ◡ + ⌐	t_{15} '03' t_{32} '12' t_{29}	I
ಥ	ω + ◡ + ı + •	t_{19} '03' t_{32} '02' t_{38} '04' t_{23}	T
ಝ	○ + ◡ + ↄ + ı + ↄ	t_{15} '03' t_{32} '01' t_{30} '06' t_{38} '01' t_{30}	J

5 Conclusion and Future Work

The handwriting is good, as long as the characters are clearly constructed, evenly spaced on the line and easy to read. All characters are combination of a set of certain primitives written in exact and specific sequence. The proposed character construction method helps novice learners to use the primitive sequence with the knowledge of connection points and relative position of other primitives to construct vowels and consonants of the Kannada characters. The method takes a string of primitives as input and generates the corresponding transliteration code for the character. Experimental results show that the inter-primitive grammar developed, is sound and complete. The future work can be extended for constructing Kagunitha and Vathakshara Kannada characters at which the authors are working. The extension to the work finds applications in teaching and learning other languages, handwriting improvement, advertisements in television and film industries calligraphy, robot generated document recognition & analysis, digitizing old manuscripts of the Kannada literature. The same with the help of archeological society of India can be extended to read halegannada, which is old Kannada.

References

1. Prasanna, K., Kumar, P.R.: Handwriting recognition of Kannada characters and context free grammar based syntax analysis. Int. J. Sci. Res. **1**(1), 24–29 (2012)
2. Ota, I., Yamamoto, R., Nishimoto, T., Sagayama, S.: On-line handwritten Kanji string recognition based on grammar description of character structures. In: 2008 19th International Conference on Pattern Recognition ICPR (2008)
3. Wang, Y., Wang, H., Pan, C., Fang, L.: Style preserving Chinese character synthesis based on hierarchical representation of character. In: IEEE International Conference on Acoustics, Speech and Signal Processing (2008)
4. Ota, I., Yamamoto, R., Sako, S., Sagayama, S.: Online handwritten Kanji recognition based on inter-stroke grammar. In: 9th International Proceedings on Document Analysis and Recognition, pp. 1188–1192. IEEE Computer Society, Washington, DC (2007)

5. Kim, H.J., Kim, S.K.: On-line recognition of cursive Korean characters using art-based stroke classification (recognition of cursive Korean characters). Int. J. Pattern Recogn. Artif. Intell. **10**(7), 791–812 (1996)
6. Kim, S.K., Kim, J.W., Kim, H.J.: On-line recognition of cursive Korean characters using neural networks. Neurocomputing **10**(3), 291–305 (1996)
7. Kim, P.K., Lee, J.K., Kim, H.J.: Handwritten Korean character recognition by stroke extraction and representation. In: Proceedings of TENCON 1993 in Computer, Communication, Control and Power Engineering (1993)
8. Wang, J.H., Ozawa, S.: Automated generation of Chinese character structure data based on extracting the strokes. In: Proceedings of the 2nd International Conference on Document Analysis and Recognition (1993)
9. Ohmori, K.: On-line handwriting Kanji character recognition using hypothesis generation in the space of hierarchical knowledge. In: 3rd International Workshop on Frontiers in Handwriting Recognition, pp. 242–251 (1993)
10. Nagahashi, H., Nakatsuyama, M.: A pattern description and generation method of structural characters. IEEE Trans. Pattern Anal. Mach. Intell. **8**(1), 112–118 (1986)
11. Indira, K., Selvi, S.S.: Kannada character recognition system: a review. Inter JRI Sci. Technol. **1**(2) (2009)
12. Pal, U., Chaudhuri, B.B.: Indian script character recognition: a survey. Pattern Recogn. **37**(9), 1887–1899 (2004)
13. Johnson, S.C.: Yacc: yet another compiler-compiler. Computing Science Technical report no. 32. Bell Laboratories, Murray Hill (1975)
14. Lesk, M.E., Schmidt, E.: Lex - a lexical analyzer generator. Computing Science Technical report no. 39. Bell Laboratories, Murray Hill (1975)
15. Levine, J.R., Mason, T., Brown, D.: Lex & Yacc. O'Reilly & Associates Inc., Sebastopol (1992)
16. Kamble, P.M., Hegadi, R.S.: Handwritten Marathi character recognition using R-HOG feature. In: International Conference on Advanced Computing Technologies and Applications (2015)
17. Santosh, K.C., Nattee, C.: Stroke number and order free handwriting recognition for Nepali. In: Yang, Q., Webb, G. (eds.) PRICAI 2006. LNCS (LNAI), vol. 4099, pp. 990–994. Springer, Heidelberg (2006). https://doi.org/10.1007/978-3-540-36668-3_120
18. Santosh, K.C., Nattee, C.: A comprehensive survey on on-line handwriting recognition technology and its real application to the Nepalese natural handwriting. Kathmandu Univ. J. Sci. Eng. Technol. **5**(1), 31–55 (2009)
19. Kamble, P.M., Hegadi, R.S.: Geometrical features extraction and KNN based classification of handwritten Marathi characters. In: World Congress on Computing and Communication Technologies (2017)
20. Santosh, K.C., Nattee, C.: Spatial similarity based stroke number and order free clustering. In: International Conference on Frontiers in Handwriting Recognition, Kolkata, India (2016)
21. Santosh, K.C., Nattee, C.: Relative positioning of stroke based clustering: a new approach to on-line handwritten Devanagari character recognition. Int. J. Image Graph. **12**(02), 1250016 (2012)

Automatic Recognition of Legal Amount Words of Bank Cheques in Devanagari Script: An Approach Based on Information Fusion at Feature and Decision Level

Mohammad Idrees Bhat(✉) and B. Sharada

Department of Studies in Computer Science, University of Mysore, Mysore 570006, Karnataka, India
idrees11@yahoo.com, sharadab21@gmail.com

Abstract. Legal amount word recognition is an essential and challenging task in the domain of automatic Indian bank cheque processing. Further intricacies get accumulated by inherent complexities in Devanagari script besides cursiveness present in handwriting. Due to segmentation ambiguity and variability of constituent parts present in handwritten word analytical approach is inadequate, in contrast, to the holistic paradigm, where the word is taken indivisible entity. Despite the proliferation of various feature representations, it still remains a challenge to get effective representation/description for holistic Devanagari words. In this paper, we made an attempt to exploit robust, most discriminative and computationally inexpensive Histogram of Oriented Gradients (HOG) and Local Binary Pattern (LBP) for effective characterization of Devanagari legal amount words taking into account different writing styles and cursiveness. Two models are proposed based on fusion strategies for word recognition. In the first model, LBP and HOG features are fused at feature level and in second, fused at decision level. In both models, recognition is performed by the nearest neighbour (NN) and support vector machine (SVM) classifiers. For corroboration of the results, extensive experiments have been carried out on ICDAR 2011 Devanagari Legal amount word dataset. Experimental results reveal that fusion based approaches are more robust than conventional approaches.

Keywords: Analytical and holistic word recognition · Writing styles · Cursiveness · Feature representation · Legal amount · Feature and decision level fusion

1 Introduction

In developing countries like India, paper cheques are still used for non-cash transactions despite the proliferation of credit cards, debit cards, charge cards,

K. C. Santosh and R. S. Hegadi (Eds.): RTIP2R 2018, CCIS 1037, pp. 96–107, 2019.
https://doi.org/10.1007/978-981-13-9187-3_9

pre-paid cards and store cards (plastic money) and many other electronic forms. In this traditional system, a bank employee has to manually enter the details, written by the customer, in a computer system, authenticate the signature, date etc. As millions of cheques are produced and processed on daily basis, an automatic recognition for such a system can save a considerable amount of time and human effort. Although significant amount of work can be found in the field of script identification [1–3], handwritten character, numeral recognition [4–7], the automatic recognition of handwritten fields of bank cheque is still a challenge [8]. The handwritten content of bank cheque includes legal amount, payee details, signature, courtesy amount and date as shown in Fig. 1. The amount of bank cheques are written in two ways (a) legal amount, written in textual manner (b) courtesy amount, written in a numeric manner. The practice of writing amount in textual format helps in to counter check the courtesy amount written by customer moreover, it also aids to detect fraud, as the courtesy amount is easy to tamper with. Therefore, legal amount words are an extremely important handwritten field in bank cheques and their automatic recognition can further accelerate the entire field of automatic bank cheque processing. Within this area, a number of studies have addressed the automatic recognition of Latin, Arabic and Chinese [8], legal amount words but work in Indian context i.e., recognition of legal amount words written in Indian scripts such as Devanagari is far way behind. Perhaps the main reason for lagging behind in its automatic recognition might be due to the complex shapes, the presence of inherent cursiveness [9], intra-class variations, and different writing styles (Fig. 2) thus making automatic recognition of legal amount words written in Devanagari script more challenging.

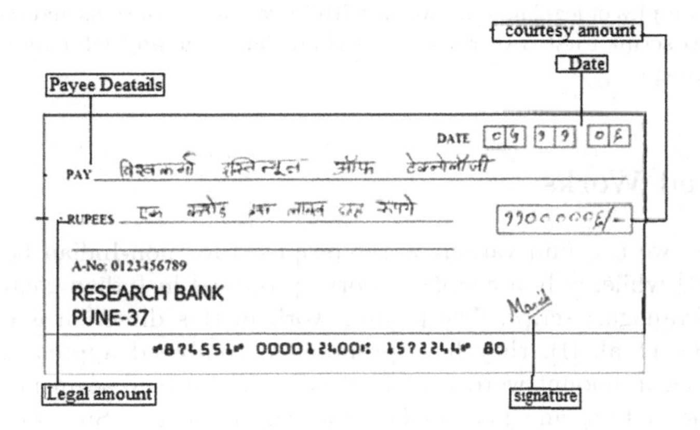

Fig. 1. Different handwritten fields in Bank Cheque. (Image taken from the dataset used in this paper)

State-of-the-art techniques [10], for recognition of handwritten words, are broadly classified into two approaches viz. Analytical and Holistic approaches. In Analytical approach, a word is segmented into simpler constituent units (such as characters, allograph, different zones, graphemes etc.) later these constituent units are successively used to recognize the entire word. On the other hand in Holistic approach, segmentation step is eliminated by treating the word as a single entity rather than decomposing it into constituent units. The holistic approach starts from feature extraction and later unknown word is recognized by comparing its representation with the representation of words stored in a knowledge base. It is tolerant towards deformations present in a handwritten word such as poor writing [10]. However, extraction of suitable features plays a significant role, in the domain of holistic approach. Furthermore, extracting only one type of features gives only one particular cue of information and thus may not be capable fully in characterizing the underlying pattern. As stated earlier, due to the presence of inherent complexities in Devanagari script, it fully qualifies for viewing it, from multiple cues of information (multiple features). With this motivation, in this paper, we address the recognition of legal amount words written in Devanagari script in a holistic manner by extracting different complementary features.

Fig. 2. Few samples of legal amount words with intra-class variations and style, written in Devanagari script First, two rows are in Hindi language and later rows are in the Marathi language

2 Related Works

In literature, we can find various works proposed for non-Indian bank cheque processing [1] while, only a couple of works proposed in Indian context particularly in Devanagari script. The pioneer work in this direction is carried out by Jayadevan et al. [1], they have proposed a combined approach, in which Devanagari legal amount word is divided into four vertical segments such that each segment contains an equal number of foreground pixels. Subsequently, word image is further divided into 8-segments horizontally by following the same procedure. Later from each segment ($4 \times 8 = 32$) combined features are extracted based on gradient structural cavity (GSC) and vertical projection profile features (VPP). GSC features are considered with binary vector matching (BVM) and VPP features are considered with dynamic time wrapping (DTW) and later recognition is performed by following a ranking scheme. Most of the other works which are proposed for Devanagari word recognition, in general, differ mainly in domain considered, feature extraction techniques and recognition classifiers employed. In [12], GSC and directional distance distribution features (DDD) are fused at the feature and decision level (classifier level) for the recognition of handwritten Devanagari city names later multiple support vector machines (SVM) are employed for recognition purpose. A novel method based on zone segmentation is proposed in [13]. First, three zones (upper, middle and lower) are segmented from the handwritten Devanagari word, than from each individual zone pyramid histogram of oriented gradient (PHOG) features are extracted. Upper and lower zones are recognized by SVM and for the middle zone, hidden markov model (HMM) is used. To corroborate the final result, individual results are combined. More recently, numerous architectures based on convolution neural networks (CNN) were presented for large-scale recognition of Indic words [9] For the extensive survey, readers are referred to [14].

In the existing works, a different number of features, their combinations, various recognition strategies have been proposed. The basic motivation behind using a holistic approach in word recognition is based on the psychological behaviour of human reading [4] i.e., humans try to read handwritten word based on its appearance and by employing various information cues. Furthermore, the inherent nature of Devanagari words offers stable structural pattern because of the presence of Shirorekha or headline, constituent character shapes [14]. In view of this, we have extracted two types of features from handwritten Devanagari words. The first method is based on the local binary pattern (LBP) [15], a texture descriptor which transforms word image into an image of integer labels describing low-level appearance. From various fields in pattern recognition, LBP has proven to be robust and discriminative in characterizing the respective images. Furthermore, its invariance towards monotonic gray level changes and computationally inexpensive nature makes it idle in demanding applications such as word recognition. Since LBP has limitations in capturing edge and directional information, to compensate, second feature extraction method is used, which is based on Histogram of oriented gradient (HOG) descriptors [16]. HOG extracts appearance and shape within the word image by describing them as the distribution of intensity and edge directions.

Overall, LBP and HOG capture different information making them complementary to each other. Finally, both these features are combined at feature and decision level in order to investigate improvement in recognition performance. The main aim of the paper is not to outperform any existing system but to observe the performance of LBP and HOG features. The rest of the paper is structured as follows: - In Sect. 3 brief description is provided for features considered, Sect. 4 elaborates end to end pipeline of the proposed methodology. The experimentation starts with dataset description followed by experimental setup and concluded with an experimental result. Finally, the conclusion is drawn in Sect. 5.

3 Description of the Features

Brief and concise definitions are given for LBP and HOG descriptors in this section. However, for comprehensive reading, readers are referred to [15,16].

3.1 Local Binary Pattern (LBP)

Basic LBP operator is a robust means of texture characterization. For each input image, LBP code assigns the labels to pixels of an image after thresholding every pixel in 3 × 3 neighbourhoods with their respective value of the centre pixel. The final result is considered as a binary number (Fig. 3). Thereafter, labels of pixels are represented by a histogram based on the LBP computation from each pixel of an input image. Subsequently, LBP was considered with different sizes of the neighbourhoods [11]. Exploiting circular neighbourhoods with bilinear interpolation of the pixel values provide usage of LBP with different possible radii R and neighbourhood P. Let P is 3 × 3 neighbourhood of each pixel and R is the unit radius from each centre pixel than LBP is computed from each pixel as follows:- Displayed equations are centered and set on a separate line.

$$LBP = \sum_{i=1}^{P} D * 2^{n-1} \tag{1}$$

$$where \begin{cases} 1, & if(U_i - U_c) \geq 0 \\ 0, & \text{Otherwise} \end{cases}$$

Where U_i and U_c describe the intensity values of the center pixel and neighborhood pixel respectively.

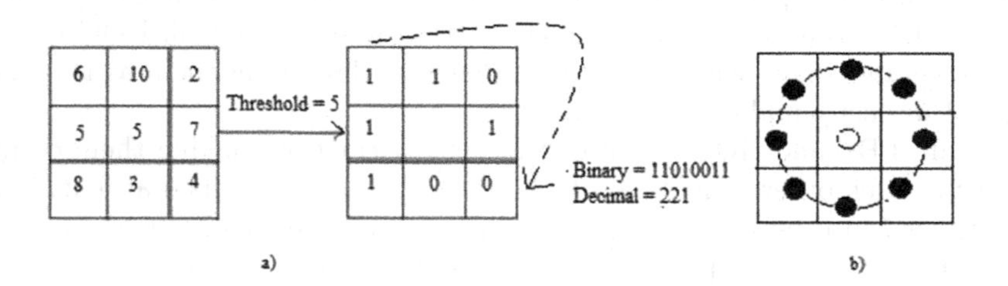

Fig. 3. (a) Computation of the basic LBP operator with radius R = 1 and neighbourhood P, 3 × 3. (b) Circular neighbourhood of 8 with radius R = 1.

Later different extensions of the basic LBP were proposed, one such extension is uniform LBP [17]. An LBP is a uniform if its computation consists of at most two circular bit-wise transitions. For example following LBP patterns are uniform 00000000, 11111111, 00111000 and 11011111 following are non-uniform 10101100, 01010000, and 01001110. The main motivation behind using uniform

LBP is twofold; first, it reduces total combinations of the patterns to $P(P-1)+3$ as compared to 2^P combinations in basic LBP (where P is the neighbourhood, when $2^8 = 256$) thus, reducing the dimensionality of the feature vector. Second, it detects most important and fundamental features such as spot, flat areas, corners, line endings and edges as shown in Fig. 4). These two characteristics make uniform LBP more precise, simpler and effective. With these motivations, Uniform LBP has been used for different tasks in document image processing such as word spotting [18], and many more.

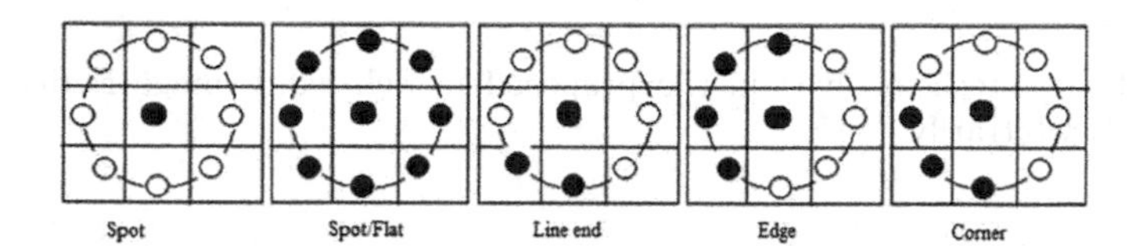

Fig. 4. Illustration of different low-level primitives detected by uniform LBP.

3.2 Histogram of Oriented Gradients (HOG)

HOG is an object appearance and shape descriptor, enumerates the occurrences of intensity gradients or directions of edges. It decomposes an underlying image into different cells and for each cell computes a histogram for oriented gradients. Later these histograms are combined for an overall descriptor for the image. HOG extracts discriminative information as the magnitude of gradients is more towards edges and corners. Let $\theta(x)$ and $h(x)$ represent orientation and magnitude of image gradient at pixel location (x, y) than:-

$$h(x,y) = \sqrt{(h^2x + h^2y)} \tag{2}$$

$$\theta = \arctan \frac{h_y}{h_x} \tag{3}$$

Finally, each pixel is discretized into different evenly spaced bins according to their orientation gradient. For HOG descriptor computation we have used vlfeat library [19], in contrast to original HOG implementation, it reduces the dimensionality of the feature vector (from 36 dimensions to 31) and enriches it with components that are both directed and in-directed to gradients. Further, it also computes texture energy features thus more discriminative than Dalal and Trigs [16]. A snapshot of HOG descriptor extracted from Handwritten Legal amount word '*Aik* (one) is shown in Fig. 6'. In literature, HOG has been used for several practical recognition applications such as word recognition [20]. Moreover, we also find applications like face recognition [21], word spotting [22], and many more where HOG descriptor has extensively combined with LBP feature descriptor, since, they extract complimentary information.

4 Proposed Model

Different steps present (end-to-end pipeline) in the proposed holistic recognition of Devanagari legal amount words are shown in Fig. 3. End to end pipeline is illustrated in following subsections:-

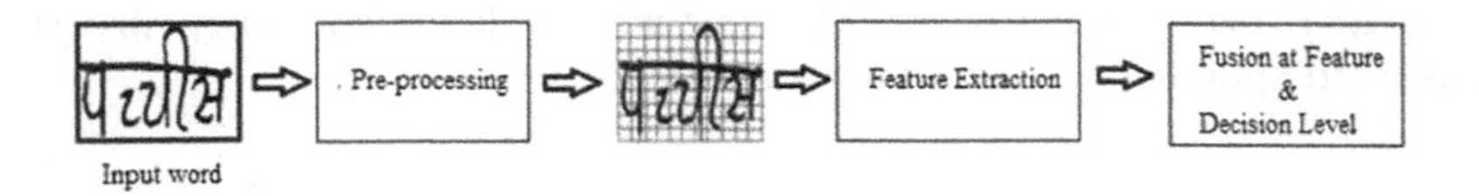

Fig. 5. Different steps involved in the proposed model, word image is divided into different cells arbitrarily.

4.1 Pre-procesing

Due to variations present in scanned images of handwritten legal amount words caused by noise, they are first filtered by Difference-of-Gaussian (DoG)-filtering. For convenience, simplicity and uniformity in feature extraction, we have resized the word images into 100×100 dimensions.

4.2 Feature Extraction

For each Input word image, a regular grid is overlaid, partitioning the image into multiple non-overlapping cells as shown in Fig. 5). The dimension of the grid is obtained by dividing the total number of columns and rows of a word image by a number chosen empirically (empirical value of b is 5 in this case).

$$c = \frac{Number\,of\,columns\,in\,word\,image}{b}, r = \frac{Number\,of\,rows\,in\,word\,image}{b}$$

Next, a word image is partitioned into $c \times r$ cells having equal size (20×20 nonoverlapping cells each having a dimension 5×5 in this case). Thereafter, each cell is encoded with uniform LBP and Histogram of Gradient (HOG) descriptors. The uniform LBP descriptors of each cell are combined and later normalized by Euclidean norm 2 (L_2 norm). A similar procedure is adopted for HOG descriptors. Finally, both the vectors are concatenated in order to obtain final single vector for feature fusion. The final feature vector $F \in T^v, where\ v = (31 + 58)20.20 = 35600$.

5 Experimentation

5.1 Dataset

For Experimentation, we have used Devanagari Legal amount word dataset [11]. It consists of 26,720 handwritten Devanagari legal amount words written in

Fig. 6. Distribution of direction of edges (legal amount word *Aik* in an image with cell size 5×5

Hindi and Marathi languages, respectively. Dataset is collected from specially designed bank forms, which contains different boxes where each writer has to write possible lexicon legal amount words by following a specific order. The writers were from all walks of life, different age groups; however, most writers were in the age group of 18 to 22. In a dataset words are with different writing styles, sizes and stroke widths and later skew corrected with radon transform, complete description of the dataset can be found in [11]. Figure 2 shows some samples from the dataset.

5.2 Experimental Setup

Feature extraction follows, representing features into a feature matrix in which rows, columns correspond word samples and features respectively. In this paper, two models are proposed one is based on feature level fusion and second, decision level fusion (Fig. 7). To design model first, we have used, as stated earlier, $J = 2$ for feature extraction viz. uniform-LBP and HOG and for classifiers, we have employed $J = 2$ classifiers viz. nearest neighbour with Euclidean distance and support vector machines (SVM) with a linear kernel. In order to choose optimized parameter K in the classifier, we have varied value of K from 3 to 9, the value of K which shows maximum accuracy in terms of *F-measure* is used. We have divided the feature matrix into training and testing ratios of 75:25 and 80:20 respectively (by following standard procedures in literature). To corroborate the final results, first, features are tested individually for recognition potential and later fused at feature level with KNN classifier and SVM separately. In the model second, individual features are fused at decision level by using simple majority vote scheme due to its simplicity, less computation costs, robustness and scalability [23]. First, Individual features are fused with KNN-classifier and later with SVM classifier (Table 1). It should be noted that for a combination of classifiers at decision level, each should be accurate and diverse [23]. The accuracy of the KNN classifiers and SVM's are validated by various applications in pattern recognition and diversity mean each classifier should have different decision boundaries or should make different errors. In this paper, diversity is achieved by means of extracting different types of features for the word images.

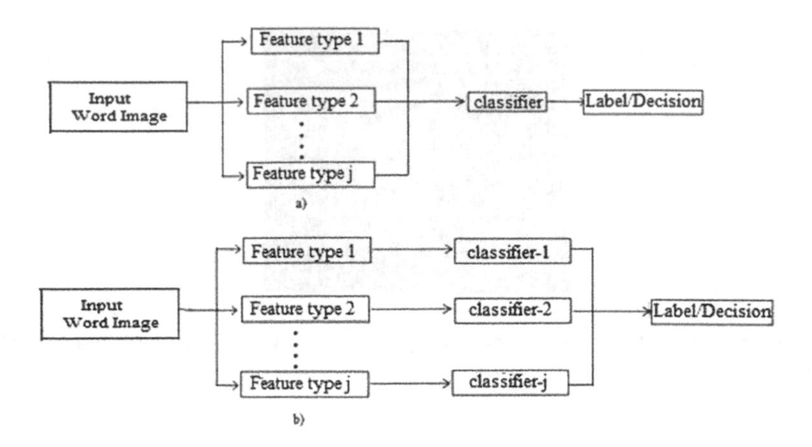

Fig. 7. Architectures of the models proposed for Holistic recognition of Legal amount words (a) Feature level Fusion (b) Decision level Fusion

5.3 Experimental Results

For corroboration of the efficacy of the proposed models, extensive experiments were carried out on the dataset and repeated for 20 random trails of training and testing in the ratios of 75:25 and 80:20 respectively. In each trial, recognition performance of the proposed models is recorded by computing *F-measure* and the average *F-measure* is computed from all 20 trails as shown in Table 1.

From Table 1 it can be observed that HOG performs better than LBP when compared individually. Individually these feature types (HOG and LBP) generate 65–78% average recognition rate. Since magnitude information is ignored in LBP computation, perhaps this could be the reason for its lowest recognition performance (65–68)%. In contrast, the HOG preserves magnitude and direction information hence it generates over 70% average recognition rates. As both features types contain complementary information, combining them at feature and decision level greater recognition results can be achieved. With information fusion at feature level, we achieved maximum average recognition rate 82% as shown in Table 1. Similarly, Information fusion at decision level maximum average recognition rate 84% is achieved. It can be also observed from the Table 1, as training samples are increased feature and decision fusion performs significantly better.

In order to choose optimal K in KNN classifier, value of K is varied from 3 to 9 (beyond 9 we didn't observe any significant recognition performance). Table 2 gives recognition accuracy in-terms of *F-measure* for various values of K. The value for K = 3 shows significant development, hence in this paper, we conducted experiments with K = 3 in KNN classifier. For SVM simple linear kernel is employed as our feature vector is large as compared to training samples.

Table 1. Devanagari script Legal amount word recognition accuracy (in terms of F measure) under two training and testing samples with individual, feature and decision level fusion

Type		Ratio's of training and testing	
Feature	Classifier	75:25	80:20
LBP	NN-classifier	65.0126 ± 1.1528	68.0307 ± 0.0427
HOG		75.8798 ± 1.1859	77.7689 ± 0.0751
LBP+ HOG (NN Classifier)	Feature level fusion	80.8596 ± 0.2589	82.8789 ± 0.1862
	Decision level fusion	81.9632 ± 0.8596	83.9898 ± 1.9623
LBP	SVM classifier	65.2528 ± 2.2649	68.1418 ± 1.1538
HOG		77.7778 ± 1.1758	78.8898 ± 2.2649
LBP+ HOG(SVM)	Feature level fusion	81.4528 ± 2.2589	82.9898 ± 3.3759
	Decision level fusion	82.1589 ± 2.5891	82.2589 ± 4.4868
NN+SVM	Decision level fusion	83.4788 ± 6.7896	84.3699 ± 5.5979

Table 2. Empirical evaluation of K in KNN Classifier

Value of K in KNN classifier	Feature type	75:25	80:20
3	LBP	65.0126 ± 1.1528	68.0307 ± 0.0427
	HOG	75.8798 ± 1.1859	77.7689 ± 0.0751
6	LBP	61.1127 ± 2.2639	66.1218 ± 0.1816
	HOG	73.7898 ± 1.1859	75.1884 ± 1.1661
9	LBP	60.8898 ± 1.1759	65.12547 ± 2.0821
	HOG	70.8421 ± 2.5871	73.2995 ± 3.32589

6 Conclusion and Future Work

In this paper, we have proposed fusion based approaches for holistic recognition of Devanagari legal amount words. We have divided the word image into non-overlapping cells of equal dimension. From each cell, a histogram of oriented gradient (HOG) and uniform local binary pattern (LBP) features are extracted. In the first model, two complementary features (LBP and HOG) are fused after L2 normalization. In the second model, individual features are fused at decision level by simple majority vote. After observing the results, we conclude that HOG features perform better than LBP features when compared individually for characterizing the legal amount words. We also observe decision level fusion perform better than feature level fusion which is also observed from numerous fields of pattern recognition. There are various issues that can be investigated in future studies. Compressing non-pen-stroke cells and giving more importance (weighted) to cells which are having pen-strokes overall recognition accuracy of LBP and HOG features can be improved. It would be interesting to observe the results with reduced dimension of features at feature level fusion by utilizing any Dimensional reductionality technique like principal component analysis (PCA). Finally, investigating different combinations of features proposed in the literature, with feature and decision level fusion seems to be an interesting direction.

Acknowledgement. We thank Prof. Jayadeyan of Department of Information Technology of Pune Institute of Computer Technology (IT-PICT) Pune, India for providing the legal amount word dataset.

References

1. Obaidullah, S.M., Halder, C., Santosh, K.C., et al.: Page-level handwritten document image dataset of 11 official Indic scripts for script identification. Multimed. Tools Appl. **77**(2), 1643 (2018). https://doi.org/10.1007/s11042-017-4373-y
2. Obaidullah, S., Goswamir, C., Santosh, K.C., et al.: Separating indic scripts with matra for effective handwritten script identification in multi-script documents. Int. J. Pattern Recogn. Artif. Intell. **31**(5), 1753003 (2017). https://doi.org/10.1142/S0218001417530032
3. Obaidullah, S.M., Santosh, K.C., Das, N., Halder, C., et al.: Handwritten indic script identification in multi script document images. Int. J. Pattern Recogn. Artifi. Intell. **32**(10), 1856012 (2018). https://doi.org/10.1142/S0218001418560128
4. Santosh, K.C., Wendling, L.: Character recognition based on non-linear multi-projection profiles measure. Front. Comput. Sci. **9**, 678 (2015). https://doi.org/10.1007/s11704-015-3400-2
5. Santosh, K.C.: Character recognition based on DTW-radon. In: International Conference on Document Analysis and Recognition, 18–21 September 2011. https://doi.org/10.1109/ICDAR.2011.61
6. Bhat, M.I., Sharada, B.: Recognition of handwritten Devanagari numerals by graph representation and lipschitz embedding. In: Santosh, K.C., Hangarge, M., Bevilacqua, V., Negi, A. (eds.) RTIP2R 2016. CCIS, vol. 709, pp. 102–110. Springer, Singapore (2017). https://doi.org/10.1007/978-981-10-4859-3_10
7. Bhat, M.I., Sharada, B.: Spectral graph-based features for recognition of handwritten characters: a case study on handwritten Devanagari numerals. J. Intell. Syst. (2018). https://doi.org/10.1515/jisys-2017-0448. Accessed 21 July 2018
8. Jayadevan, R., et al.: Automatic processing of handwritten bank cheque images: a survey. Int. J. Doc. Anal. Recogn. **15**(4), 99–110 (2012). https://doi.org/10.1007/s10032-011-0170-8
9. Dutta, K., Krishnan, P., Mathew, M., Jawahar, C.V.: Towards accurate handwritten word recognition for hindi and bangla. In: Rameshan, R., Arora, C., Dutta Roy, S. (eds.) NCVPRIPG 2017. CCIS, vol. 841, pp. 470–480. Springer, Singapore (2018). https://doi.org/10.1007/978-981-13-0020-2_41
10. Cheriet, M., El Yacoubi, M., Fujisawa, H., Lopresti, D., Lorette, G.: Handwriting recognition research: twenty years of achievement... and beyond. Pattern Recogn. **42**(12), 3131–3135 (2009)

11. Jayadevan, R., et al.: Database development and recognition of handwritten Devanagari legal amount words. In: 2011 International Conference on Document Analysis and Recognition (ICDAR), pp. 304–308 (2011). https://doi.org/10.1109/ICDAR.2011.69
12. Shaw, B., et al.: Offline handwritten Devanagari word recognition: a holistic approach based on directional chain code feature and HMM. In: Proceedings of the 11th International Conference on Information Technology, ICIT 2008, pp. 203–208 (2008). https://doi.org/10.1109/ICIT.2008.33
13. Roy, P.P., et al.: HMM-based Indic handwritten word recognition using zone segmentation. Pattern Recogn. **60**, 1057–1075 (2016). https://doi.org/10.1016/j.patcog.2016.04.012
14. Jayadevan, R., Kolhe, S.R., Patil, P.M., Pal, U.: Offline recognition of Devanagari script: a survey. IEEE Trans. Syst. Man. Cybern. Part C Appl. Rev. **41**, 782–796 (2011). https://doi.org/10.1109/TSMCC.2010.2095841
15. Ojala, T., et al.: A comparative study of texture measures with classification based on featured distributions. Pattern Recogn. **29**(1), 51–59 (1996). https://doi.org/10.1016/0031-3203(95)000674
16. Dalal, N., Triggs, B.: Histograms of oriented gradients for human detection. In: Proceedings of 2005 IEEE Computer Society Conference on Computer Vision and Pattern Recognition, CVPR 2005, vol. 1, pp. 886–893 (2005). https://doi.org/10.1109/CVPR.2005.177
17. Ojala, T., et al.: Multiresolution gray-scale and rotation invariant texture classification with local binary patterns. IEEE Trans. Pattern Anal. Mach. Intell. **24**(7), 971–987 (2002). https://doi.org/10.1109/TPAMI.2002.1017623
18. Dey, S., Nicolaou, A., Llados, J., Pal, U.: Local binary pattern for word spotting in handwritten historical document. In: Robles-Kelly, A., Loog, M., Biggio, B., Escolano, F., Wilson, R. (eds.) S+SSPR 2016. LNCS, vol. 10029, pp. 574–583. Springer, Cham (2016). https://doi.org/10.1007/978-3-319-49055-7_51
19. Vedaldi, A., Fulkerson, B.: VLFeat: an open and portable library of computer vision algorithms. In: Proceedings of the International Conference on Multimedia, pp. 1469–1472. ACM (2010). https://doi.org/10.1145/1873951.1874249
20. Ahmed, R., et al.: A survey on handwritten documents word spotting. Int. J. Multimed. Inf. Retrieval **6**(1), 31–47 (2017). https://doi.org/10.1007/s13735-016-0110-y
21. Xie, Z., et al.: Fusion of LBP and HOG using multiple kernel learning for infrared face recognition. In: Proceedings of the 16th IEEE/ACIS International Conference on Computer and Information Science, ICIS 2017, pp. 81–84 (2017). https://doi.org/10.1109/ICIS.2017.7959973
22. Kovalchuk, A., et al.: A simple and fast word spotting method. In: Proceedings of the International Conference Frontiers Handwriting Recognition, ICFHR 2014, pp. 3–8 (2014). https://doi.org/10.1109/ICFHR.2014.9
23. Kuncheva, L.I.: Combining Pattern Classifiers: Methods and Algorithms: Combining Pattern Classifiers: Methods and Algorithms (2005). ISBN 978-1-118-31523-1

OnkoGan: Bangla Handwritten Digit Generation with Deep Convolutional Generative Adversarial Networks

Sadeka Haque(✉), Shammi Akter Shahinoor, AKM Shahariar Azad Rabby(✉), Sheikh Abujar, and Syed Akhter Hossain

Department of Computer Science and Engineering, Daffodil International University, Dhanmondi, Dhaka 1205, Bangladesh
{sadeka15-5210,akter15-4970,azad15-5424,sheikh.cse}@diu.edu.bd, aktarhossain@daffodilvarsity.edu.bd

Abstract. From a very early age human achieve a precious skill that is a handwriting. After this invention, the ardor of it changed day by day. And every human has a different style of handwriting. So, facsimile anyone's handwriting is a difficult task and it needs the strong ability of brain and practice. This paper is about this mimicry where an artificial system will do this by using Generative Adversarial Networks (GANs) [1]. GANs used in unsupervised machine learning that is implemented by two neural networks. GANs has a generator which generates fake images and a discriminator which make a difference between a real image and a fake image. We trained our proposed DCGAN [2] (Deep convolutional generative adversarial networks) to achieve our goal by using the three most popular Bangla handwritten datasets CMATERdb [3], BanglaLekha-Isolated [4], ISI [5] and our own dataset Ekush [6]. The proposed DCGAN successfully generate Bangla digits which makes it a robust model to generate Bangla handwritten digits from random noise. All code and datasets are freely available on https://github.com/SadekaHaque/BanglaGan.

Keywords: DCGAN · GAN · Handwriting recognition · Deep learning · Bangla handwriting

1 Introduction

Generative Adversarial Networks (GANs) is an effective class of Artificial Intelligence algorithm which implemented by a system of two neural networks to compete with each other in a zero-sum framework. From these two network one generate candidates and another evaluate them. Basically it able to create fake images of a character that is used many field as hand written recognition, traffic number plate recognition, automatic postal code identification, automatic ID card reading, automatic reading of bank cheques and digitalization of documents etc.

The original version of this chapter was revised: The names of the two Authors have been corrected as "AKM Shahariar Azad Rabby" and "Syed Akhter Hossain". The correction to this chapter is available at https://doi.org/10.1007/978-981-13-9187-3_67

K. C. Santosh and R. S. Hegadi (Eds.): RTIP2R 2018, CCIS 1037, pp. 108–117, 2019.
https://doi.org/10.1007/978-981-13-9187-3_10

We use this algorithm on Bangla dataset CMATERdb, BanglaLekha-Isolated, ISI and our own dataset Ekush. If it has seen then there is not that much work on Generative Adversarial Networks (GANs) with Bangla. We preferred the Bangla handwriting digit for the proposed method.

Bangla is a language that acquired by many anticipations and sacrifices. When the time has come to exuberant the language by technology, Bangla is always lagging behind. There is few works on Bangla handwritten character but the number of works is not huge as expected. Giving an eye to other languages there is much work with artificial intelligence on those. Such as Latin, Chinese, Japanese have achieved a great success in machine learning and deep learning application. Handwriting digit or character recognition is always a big challenge because of the shape, type, size of characters have changed by people to people. Also sometimes some handwritten character looks like so clumsy that it is so difficult to recognize for a system. The number of work has been done so far face that same difficulty inasmuch as the result did not come as expected. To get an expected result there is one more difficulty with data. If there is fewer data to train then the result will not that much veritable. So, the amount of data should be huge but collecting data is troublesome. GANs can be the best way to solve this trouble.

Bangla is the native language of South Asian Bengal area which is divided into two different countries Bangladesh and West Bengal the state of India. Apart from this Asam, Tripura, Andaman Island also uses Bangla. This language occupies the sixth place among the world's most common languages. Around 30 million people speak in Bangla. Moreover, Bangla is the state language of Bangladesh.

Bangla is a compound language. Bangla takes new form at every step from its journey. After British came in Indian subcontinent, they left some impact of their own language in our language. Therefore, Bangla has some touch of English, Arabic, Hindi, Dutch, French etc. But Bangla has its own script which is come from Sanskrit script. Naturally, it's completely different from English, France or other popular scripts. There are 50 characters, 10 numerical digits, more than 200 compound characters. Here we use Bangla numerical data or digit. Figure 1 gives an example of Bangla digits.

2 Literature Review

Handwritten recognition is a popular work in machine learning area and dataset is the raw material to establish the model. There is many work on it in other languages. Such as "Character Recognition based on DTW–Radon" [17] they tries to recognize characters using the "DTW algorithm" [18] at every projecting angle. Another work is "Radon transform" [16] used to produce the non-linear multi- projection profile. Other work is "Character recognition based on non-linear multi-projection profiles measure" [15], they did their experiment on several languages like Roman, Japanese, Katakana, Bangla etc. There proposed method used dynamic programming to match the nonlinear multi-projection

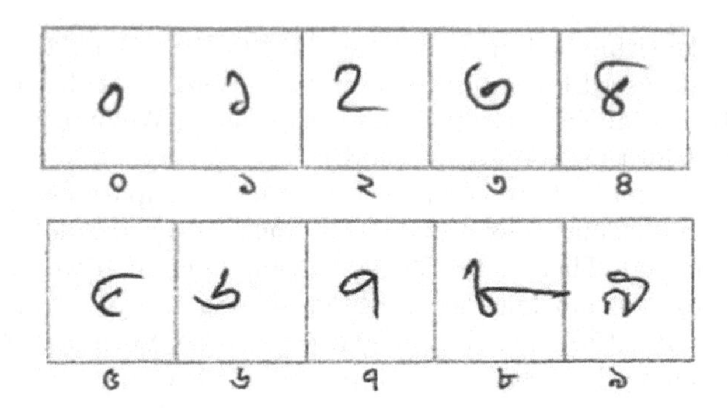

Fig. 1. Example of Bangla digit

profile which is used to recognize handwritten characters. "Deep Learning For Word-level Handwritten Indic Script Identification" [19] used multilevel 2D discrete Haar wavelet transform. They are worked on 11 different handwritten script where per script contain 1000 words.

As we know Bangla is sixth most common language in the world but the work in Bangla is less compared to other language. Recently we have known about quite much work on Bangla handwritten character using deep learning. We found some work on several algorithms such as CNN, SVM, MLP, and Regression. There are several works using these algorithms on Bangla handwritten. "Handwritten Numeral Databases of Indian Scripts and Multistage Recognition of Mixed Numerals" [5] is such kind of work where the dataset contains mixed handwritten numerals of three Indian scripts Devanagari, Bangla and English. The present databases include respectively 22,556 and 23,392 handwritten isolated numeral samples of Devanagari and Bangla. "Bangla Handwritten Digit Recognition Using Convolutional Neural Network" [6] prefers CNN model for classifying Bangla Handwriting Digits and gets validation accuracy of 99.74% on ISI handwritten character database and 99.42% on CAMTERdb dataset.

Before starting work, we focused on some other work on Bangla character and Generative Adversarial Networks related work. The work on Bangla with GANs is not that much available so we study the work of GANs on another language such as English or on images and also follow the work on Bangla handwritten character.

A few years ago, a related work has been reported as "Unsupervised Representation Learning with Deep Convolutional Generative Adversarial Networks" [2] which is mainly based on Deep Convolutional Generative Adversarial Networks (DCGAN) that refers the gap between supervised learning and unsupervised learning which help to bridge by using DCGAN. "Handwriting Profiling Using Generative Adversarial Networks" [7] prefers a system that tries to learn the handwriting of an entity using a modified architecture of DCGAN. "Bengali Handwritten Character Recognition Using Modified Syntactic Method" [8] introduces a new algorithm to develop the stroke of Bangla handwritten character, "Handwritten Bangla Digit Recognition Using Deep Learning" [9] acquired

98.78% recognition rate using CNN. "Generative Adversarial Nets" [1] basically proposed the authentic method for the area of artificial intelligence.

3 Proposed Methodology

The proposed method was used a DCGAN to create Bangla handwritten images which have many phases as below.

3.1 Dataset

The proposed method of DCGAN used 4 different datasets. CMATERdb, ISI Handwritten Dataset, BanglaLekha-Isolated Dataset and Ekush Dataset. All of these datasets contain both character and digits, we only choose digits for the proposed model.

ISI Handwriting Database has 23392 numerical images which is used for training the model of deep convolutional network of GAN. On this dataset, digits are almost correctly labeled and edge look smooth. The images have a white background where digits are written in black.

The CMATERdb dataset has total 6000 digits images in BMP format. Most of the images are noise free with correctly labeled and edge are bulkier.

BanglaLekha-Isolated datasets contain 19748 noisy images where some images are incorrectly labeled and edge are looking smooth. The images are grayscale and inverted where background as black and digit wrote as white.

Ekush Dataset has 30687 digits images inverted into the background as black and digits in white. All images are 28×28 pixel while preserving the aspect ratio of the images as well as edges looks smooth.

Finally, we merge this all dataset and make a final dataset which contains total 58,827 images. Figure 2 showing an example from all datasets.

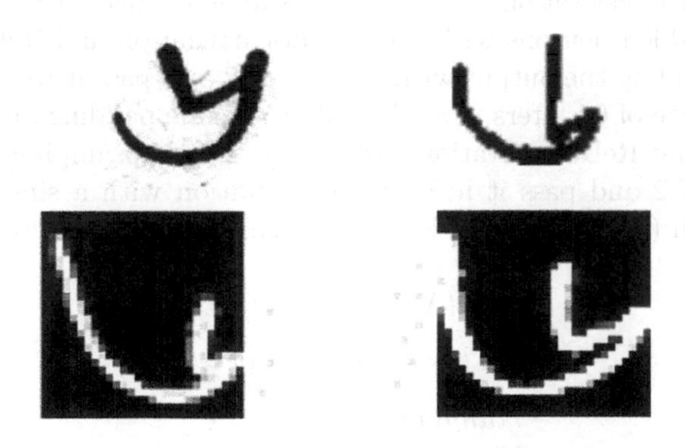

Fig. 2. Sample from different dataset.

3.2 Preparation of Dataset

Dataset preparation is the most important task in machine learning. To train our DCGAN with those datasets we first process all of our datasets as bellow point:

1. Converted all images into grayscale.
2. Resized all data into 28 × 28 pixels
3. Inverted ISI and CMATERdb dataset where background as black and digits as white to reduce the computational expenses.
4. Extract all pixel value and store them in a CSV file with the label.
5. Normalized all images (1). Where maximum value 127.5 and minimum value 0.

$$Z_i = \frac{X_i - \text{minimum}(X)}{\text{maximum}(X) - \text{minimum}(X)} \tag{1}$$

3.3 Proposed DCGAN

DCGAN is one the most successful architecture for GANs. It mostly grouping with traditional Convolutional Neural Network without max polling or fully connected layer. For the down-sampling and upsampling, it uses transposed Convolutional and convolutional stride. DCGAN has 2 sets of neural network names as Generator and Discriminator. The generator creates the fake images from random noise and discriminator maps the likelihood of the real and fake images. DCGAN use Batch normalization [10] to generate the image in the generator and the input of the discriminator. It uses ReLU (2) [11] in the generator and except the output layer which uses tanh and LeakyReLU (3) [12] in discriminator.

Our proposed model's generator is a multilayer convolutional neural network. The generator starts with a fully connected dense layer with 1024 hidden node with an input size of 100. Then the output is batch normalization layer. Again, pass with a ReLU activation. Later the output is connected to another dense layer of 6272 hidden neurons with a batch normalization and ReLU activation. Then we upsampling the output with a tuple of 2 and pass it to a convolutional layer with the size of 64 filters, 2 × 2 kernel, and same padding and added batch normalization and ReLU activation. And then, again upsampling of the output with the size of 2 and pass it in a new convolution with a single filter, 2 × 2 kernel with tanh (4) activation. Figure 3 showing the architect of the generator of DCGAN.

$$ReLU(X) = MAX(0, X) \tag{2}$$

$$f(x) = 1\,(x < 0)\,(\alpha x) + 1\,(x \geq 0)\,(x) \tag{3}$$

$$tanh(x) = 2\sigma(2x) - 1 \tag{4}$$

Discriminator classifies individual element as being fake or real produced by generator compare to training set on batch and loss function are computed based on that output. Mainly The discriminator takes an image as input, passes through convolution stacks and output a probability telling whether or not the

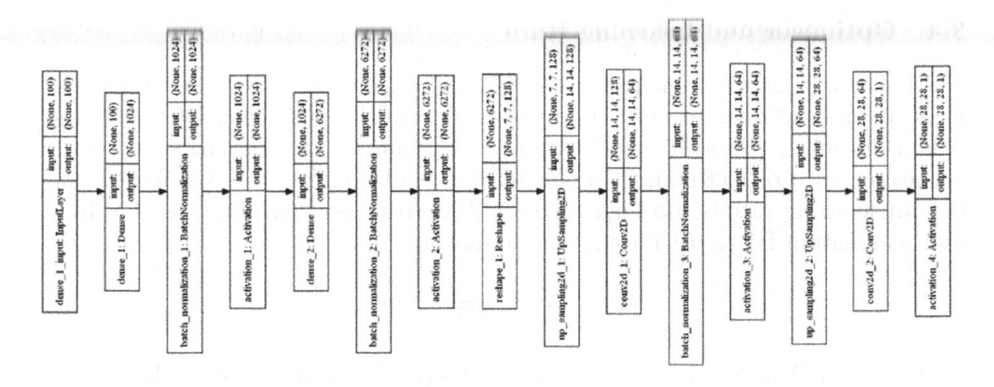

Fig. 3. Generator architecture.

image is real. Proposed DCGAN's discriminator start with a convolutional Neural network with a size of 64 filters, 5×5 kernel, and same padding. The layer uses LeakyReLU activation. The second layer uses 128 filters with a 5×5 kernel with LeakyReLU activation. The output later connects with a dense layer with 256 hidden units with 50% dropout. The final output layer has one node with sigmoid (5) [13] activation which gives the probability of the images being real or fake. Figure 4 showing the architect of the discriminator of DCGAN.

$$\sigma(x) = 1/(1+e-x) \tag{5}$$

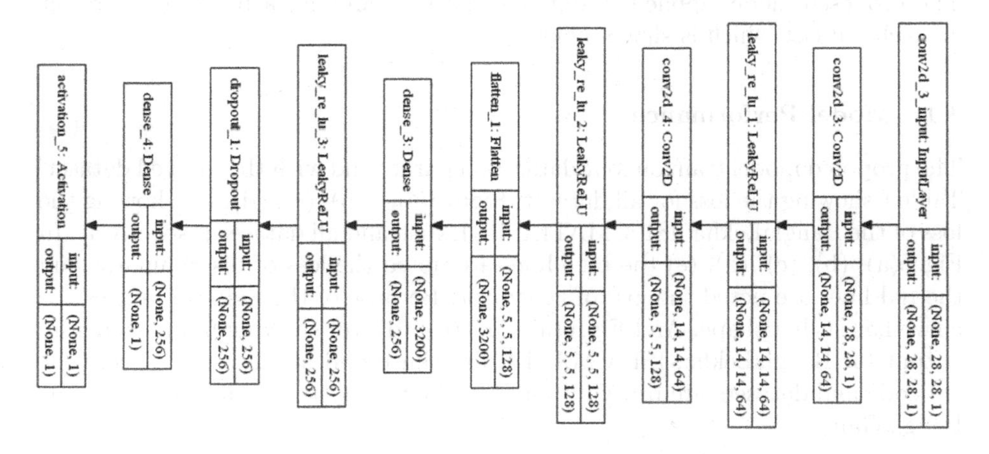

Fig. 4. Discriminator architecture.

3.4 Optimizer and Learning Rate

The Optimizer algorithm helps to make better performance and change the result in deep convolutional generative adversarial networks. Proposed DCGAN used Adam optimizer (6) with tuned parameters. Adam paper [14] suggested using a learning rate of 0.001 and beta as 0.9 for traditional CNN. We found that default learning rate is too high for DCGAN, so we set learning rate 1×10^{-5} and momentum B1 as 0.1 for the discriminator.

$$\theta_{t+1} = \theta_t - \frac{\eta}{\sqrt[1]{\hat{v}_t + \varepsilon}} \hat{m}_t \tag{6}$$

Proposed DCGAN uses Binary cross-entropy (7) to calculate the loss of the proposed model.

$$Binary\ Cross\ Entropy = -(ylog(p) + (1-y)log(1-p)) \tag{7}$$

3.5 Train the DCGAN

To train the model first we train it with all four datasets separately. And later merge the all dataset and train again. We use 50 epochs for each case with a batch size of 32. After 50 epochs model produced good output images which are almost similar to real digits.

4 Evaluate the Model

The proposed model applied to different datasets and get a pretty good result on each dataset which is shown below.

4.1 Model Performance

The proposed model train on four datasets separate and with the merged dataset. Table 1 showing the loss for all dataset. Figure 5(a), (b), (c), (d), (e) showing the loss of the BanglaLekha, ISI, CMATERdb, Ekush and all dataset respectively. In Fig. 6(a), (b), (c), (d), (e) the blue lines are noted the loss of discriminator and the red lines are noted the loss of generator. Because of the noise in images the curve has such a shape. And Fig. 6(a), (b), (c), (d), (e) showing 9 final generate output for BanglaLekha, ISI, CMATERdb, Ekush and all dataset respectively. All code and datasets are freely available on https://github.com/SadekaHaque/BanglaGan.

Table 1. DCGAN of different datasets

Dataset name	Loss
BanglaLekha Isolated	0.624247
ISI Handwritten Dataset	0.65499914
CMATERdb Dataset	0.66252947
Ekush	0.56648755
All merged	0.64720273

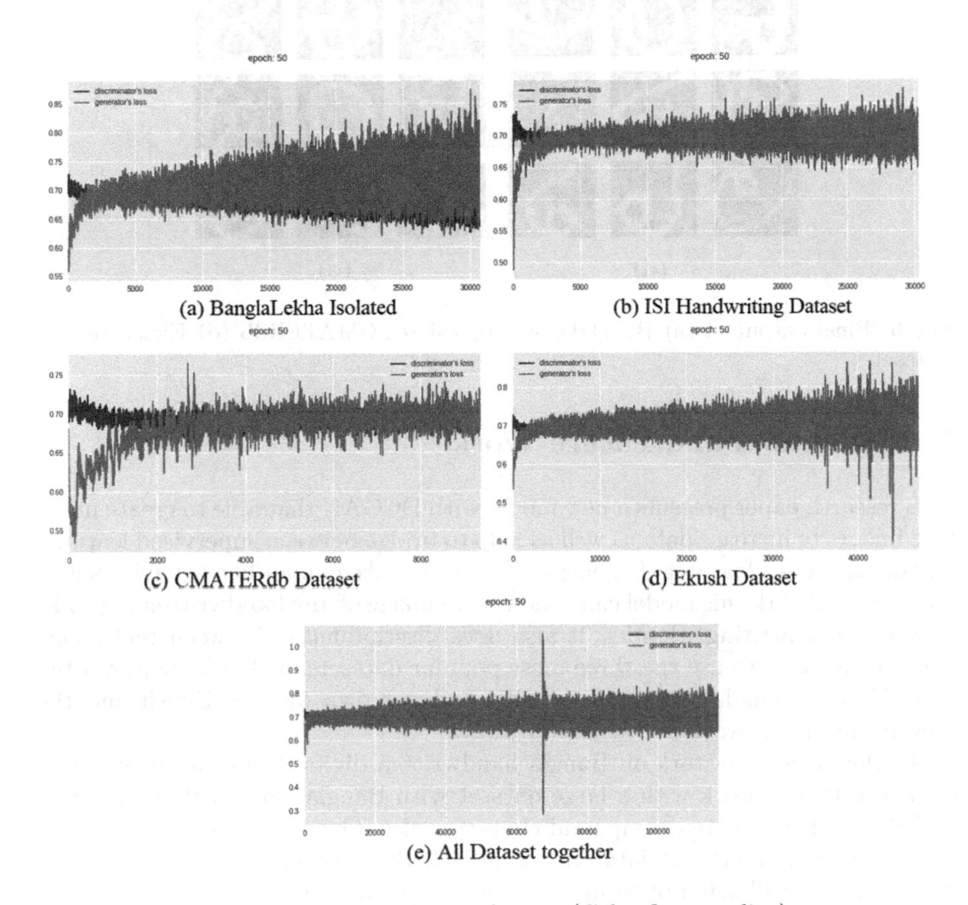

Fig. 5. Loss on a different dataset (Color figure online)

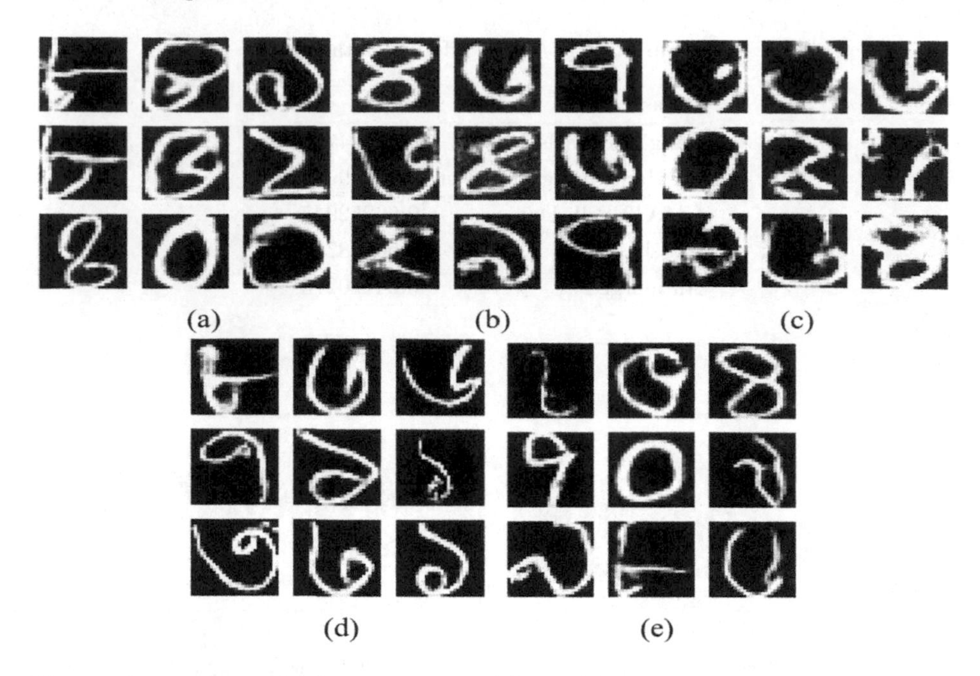

Fig. 6. Final output of (a) BanglaLekha (b) ISI (c) CMATERdb (d) Ekush (e) All Dataset.

5 Conclusion and Future Work

This research paper presents a new model with DCGAN that able to create many fake images to increase data as well as help to bridge between supervised learning and unsupervised learning. In general, GANs play like human-like behavior which means in our work this model can generate an image of any handwritten digit like a human handwriting. Besides, it also does discrimination between real image and fake image. We use the three most popular datasets of Bangla handwritten CMATERdb, BanglaLekha-Isolated, ISI and our own dataset Ekush and the final result of this work is reached far away.

In this paper, we work on Bangla handwritten digits. In future, it will be a great exertion to work with a large dataset with Bangla handwritten character and Bangla handwritten compound character also. Over and above this we will use any real-life image as data on this model. By resolving the bug or error of this model we will attempt to make a big Neural Network.

References

1. Goodfellow, I., et al.: Generative adversarial nets. In: Ghahramani, Z., Welling, M., Cortes, C., Lawrence, N.D., Weinberger, K.Q. (eds.) Advances in Neural Information Processing Systems 27, pp. 2672–2680. Curran Associates Inc., Red Hook (2014)

2. Radford, A., Metz, L., Chintala, S.: Unsupervised representation learning with deep convolutional generative adversarial networks. CoRR, abs/1511.06434 (2015)
3. Sarkar, R., Das, N., Basu, S., Kundu, M., Nasipuri, M., Basu, D.K.: CMATERdb1: a database of unconstrained handwritten Bangla and Bangla-English mixed script document image. Int. J. Doc. Anal. Recogn. (IJDAR) **15**(1), 71–83 (2012)
4. Biswas, M., et al.: BanglaLekha-Isolated: a multi-purpose comprehensive dataset of Handwritten Bangla Isolated characters. Data in Brief. **12**, 103–107 (2017). https://doi.org/10.1016/j.dib.2017.03.035
5. Bhattacharya, U., Chaudhuri, B.: Handwritten numeral databases of indian scripts and multistage recognition of mixed numerals. IEEE Trans. Pattern Anal. Mach. Intell. **31**, 444–457 (2009). https://doi.org/10.1109/TPAMI.2008.88
6. Rabby, AKM Shahariar Azad., Abujar, S., Haque, S., Hossain, S.A.: Bangla Handwritten Digit Recognition Using Convolutional Neural Network. In: Abraham, A., Dutta, P., Mandal, J.K., Bhattacharya, A., Dutta, S. (eds.) Emerging Technologies in Data Mining and Information Security. AISC, vol. 755, pp. 111–122. Springer, Singapore (2019). https://doi.org/10.1007/978-981-13-1951-8_11
7. Ghosh, A., Bhattacharya, B., Chowdhury, S.B.R.: Hand-writing profiling using generative adversarial networks. CoRR, abs/1611.08789 (2016)
8. Islam, M.B., Azadi, M.M.B., Rahman, Md.A., Hashem, M.M.A.: Bengali handwritten character recognition using modified syntactic method. NCCPB-2005 Independent University, Bangladesh
9. Alom, Md.Z., Sidike, P., Taha, T.M., Asari, V.: Handwritten Bangla digit recognition using deep learning (2017)
10. Ioffe, S., Szegedy, C.: Batch normalization: accelerating deep network training by reducing internal covariate shift. In: Bach, F., Blei, D. (eds.) Proceedings of the 32nd International Conference on Machine Learning. Proceedings of Machine Learning Research, Lille, France, 07–09 July 2015, vol. 37, pp. 448–456. PMLR (2015)
11. Ramachandran, P., Zoph, B., Le, Q.V.: Searching for activation functions. CoRR, abs/1710.05941 (2017)
12. Xu, B., Wang, N., Chen, T., Li, M.: Empirical evaluation of rectified activations in convolutional network. CoRR, abs/1505.00853 (2015)
13. Han, J., Moraga, C.: The influence of the sigmoid function parameters on the speed of backpropagation learning. In: Mira, J., Sandoval, F. (eds.) IWANN 1995. LNCS, vol. 930, pp. 195–201. Springer, Heidelberg (1995). https://doi.org/10.1007/3-540-59497-3_175
14. Kingma, D.P., Ba, J.: Adam: a method for stochastic optimization. CoRR, abs/1412.6980 (2014)
15. Santosh, K.C., Wendling, L.: Character recognition based on non-linear multi-projection profiles measure. Front. Comput. Sci. **9**(5), 678–690 (2015)
16. Deans, S.R.: Applications of the Radon Transform. Wiley Interscience Publications, New York (1983)
17. Santosh, K.C.: Character recognition based on DTW-Radon. In: 11th International Conference on Document Analysis and Recognition – ICDAR 2011, Beijing, China, September 2011, pp. 264–268. IEEE Computer Society (2011). https://doi.org/10.1109/ICDAR.2011.61. inria-00617298
18. Kruskall, J.B., Liberman, M.: The symmetric time warping algorithm: from continuous to discrete. In: Time Warps, String Edits and Macromolecules: The Theory and Practice of String Comparison, pp. 125–161. Addison-Wesley (1983)
19. Ukil, S., et al.: Deep learning for word-level handwritten Indic script identification. arXiv:1801.01627v1 [cs.CV], 5 January 2018

Breaking News Recognition Using OCR

Ahmed Ridwan[1(✉)], Ajit Danti[2], S. P. Raghavendra[2], Hesham Abdo Ahmed Aqlan[2], and N. B. Arunkumar[1]

[1] Kuvempu University, Shimoga, Karnataka, India
arash982@gmail.com, arunkumarnb93@gmail.com
[2] Department of Computer Science and Engineering, Christ (Deemed to be University), Bangalore, Karnataka, India
ajit.danti@christuniversity.in, raghusp.bdvt@gmail.com, heshamaklan1980@gmail.com

Abstract. Identifying and recognition of breaking news in most of the TV channels in different backgrounds with varying positions from a static image plays a significant role in journalism and multimedia image processing. Now a days it's very challenging to isolate only breaking news from headlines due to overlapping of many categories of news, keeping all this in mind, a novel methodology is proposed in this paper for detecting specific text as a breaking news from a given multimedia image. Basic digital image processing techniques are used to detect text from the images. The methods like MSER (Maximally Stable Extremal Regions) and SWT (Stroke Width Transform) are used for text detection. The proposed work focuses on extraction of text in breaking news images also discusses the different methods to overcome existing challenges in text detection along with different types of breaking news datasets collected from various news channels are used to identify text from images and comparative study of different text detection methods. The comparative study proves that MSER and SWT is a better technique to detect text in images. Finally using OCR (Optical Character Recognition) technique to extract the breaking news text from the detected regions will help in easy indexing and analysis for journalism and common people. Extensive experiments are carried out to demonstrate the effectiveness of the proposed approach.

Keywords: Image segmentation · Feature extraction · Text detection · OCR

1 Introduction

Recently, there is a noticeable rapid growing in multimedia libraries, which increase the necessity of browsing, retrieval and indexing the multimedia information. Many methods are introduced from time to time in literature reviews to retrieve the image data. All these techniques are based on colour, shape, texture, and relation between the parts of the objects. It is possible to provide

K. C. Santosh and R. S. Hegadi (Eds.): RTIP2R 2018, CCIS 1037, pp. 118–130, 2019.
https://doi.org/10.1007/978-981-13-9187-3_11

a vital tool from the Textual information embedded for indexing and retrieval in multimedia. The process of text extraction has many old problems because of the variation in sizes of letters font, colour, resolution and the backgrounds in which the text appears. So the text detection and determining its location are the most challenging stage of text's extraction process whereas extraction the results of the text are largely depends on these stages. As yearly technology develops, new applications are also designed and produced much. Along with this exploration of applications, the designers and experts face serious challenges and complexities. One of these complex problems in scene images is a text detection. The real applications all over the world require text detection in scene images, such as applications for blind people who try to know the breaking news and help journals to archive these breaking news and index them as headlines which requires text detection in scene images. Text detection is still an open issue in scene images due to unpredictable nature of text. According to articles none of these approaches suggest perfect solution, such as some new methods on scene images which may captured to recognize the text and detect it. In fact the rate of detection and recognition. In fact detection and recognition rate typically are not 100% but varies between 0% to almost 70%. It can be say, what is added in this article is using the MSER (Maximally Stable Extremal Regions) and SWT (Stroke Width Transform) methods to detect and capture the text in images and it is possible to split or crop this image to many parts as much as it required to concentrate the process only on the part of the image which contains the text to be detected whereas the process was working on the whole image and that will reduce the time complexity and the region of processing. The geometrical features of text information help to simplify the process of determining the shape of text in the image. It is not only determining the shape of the text but also it gives more accurate results and reduces the time complexity. Nowadays the text detection from the scene image becomes an important role in our daily life which helps the experts in vision based application to determine the text in a given input image. In the past, they were using only the camera to detect text from the images directly. OCR (Optical Character Recognition) was the technology to detect a text from scanned images as well as printed images. But it has many optical difficulties like the complexity of the text background and interference factors ... etc., which effects of the quality of the detected text. The technological development and the appearance of smartphones and mobile cells have contributed widely to the expansion of the media data which have shown another challenge in browsing information due to enlarging the huge amount of media databases and the different types of its management methods. The image processor considered promising technologies especially in the relation of determination and extraction of texts in scene images.

2 Related Work

(Obaidullah et al. 2015) The proposed of this approach is an automatic HSI technique for document images of six popular Indic scripts namely Bangla, Devanagari, Malayalam, Oriya, Roman and Urdu, based on (BRT, BDCT, BFFT and

BDT) are used for the same and initially 34-dimensional feature vector, textural and statistical techniques. Finally using a GAS (Greedy Attribute Selection) method 20 attributes are selected for learning process. In Ahuja et al. (2018) provide user with multiple functionalities in a single application such as text extraction from an image, translating the text present in an image, generating the text file from translated text and speech out the translated text for handicapped/illiterate people. Santosh et al. (2011) isolated used as method on offline character recognition using radon features is proposed. The key characteristic of the method is to use DTW algorithm to match corresponding pairs of radon histograms at every projecting angle. It avoids compressing feature matrix into a single vector which may miss information. Comparison has been made with the state–of–the–art of shape descriptors over several different character as well as numeral datasets from different scripts. This research paper provides a detailed information on the evolution of text detection in images which are taken from. This study aims to analyze, compare and discuss the different approaches of existing challenges in text detection to overcome them in the future. The current study viewing different types of datasets through which the users can determine the texts from natural images which are taken by cameras and smart cells and it is considered a comparative study of different types of text detection methods. In (Nadarajan et al. 2018) the comparative study proves that CNN technique is a better technique in the text detection in natural images. The paper main idea is to use dynamic time warping (DTW) algorithm to match corresponding pairs of the Radon features for all possible projections. By using DTW, we can avoid compressing feature matrix into a single vector which may miss information. It can handle character images in different shapes and sizes that are usually happened in natural handwriting in addition to difficulties such as multi-class similarities, deformations, and possible defects. Besides, a comprehensive study is made by taking a major set of state-of the-art shape descriptors over several character and numeral datasets from different scripts. The method shows a generic behavior by providing optimal recognition rates but, with high computational cost (Santosh and Wendling 2015). This paper presents a general overview of the approaches in which the participants used and the measurement to evaluate the methodology performance. It also presents the performance of all used methods for text localization and how can word recognition tasks be done by comparing performance results using standard approaches of the area precision, edit distance and recall the images (Shahab et al. 2011). This study proposed a large neural network to classify the high-resolution of 1.2 million images, in the ImageNet LSVRC-2010 contest, into the 1000 different classes (Krizhevsky et al. 2017). This proposed suggests that the novel image operator which tries to find a value of stroke width for each pixel in the image. After that, it is used to demonstrate the task of text detection in the natural image. After a lot of testing which showed that the suggested scheme outperforms the latest published algorithms. Its simplicity allows the algorithm to be able to detect texts in different fonts and various languages (Epshtein et al. 2010). Proposed reviewed and different analyzed methods are designed to find characters and strings from the natural images. The reviewers have reviewed different types of techniques like extraction of string regions and characters which are taken from the scenery images based on the characters' contours and thickness. There are enhancement technique and efficient binarization followed by a favorable connected component analysis. The proposed methods, which discussed, is outperforming the results of the public Robust Reading Dataset, which consists of text only in horizontal orientation. This proposed method which is used to detect traffic panels in roads-level images and to recognize the information which included in them that lead them to produce an application to intelligent transportation systems (ITS). The results of Experimentations on the real images from Google Street View prove that the efficiency of the proposed method and suggest many ways to use street-level images by using different applications on ITS (Gonzalez et al. 2014). The proposed methods are simultaneously processing multiple text line hypotheses, eliminating the need for time-consuming acquisition and labeling of real-world training data finally Maximally Stable Extremal Regions (MSERs) used as novel. The performance of the method is evaluated on two standard datasets (Neumann and Matas 2010).

Proposed of This paper is to present a new NN (neural network) based method for OCR (Optical Character Recognition) as well as HCR (Handwritten Character Recognition). Experimental results show that our proposed method achieves increased accuracy in optical character recognition as well as handwritten character recognition. All the algorithms describes more or less on their own. Handwritten character recognition is a very popular and computationally expensive task; described advanced approaches for handwritten character recognition. To compare the most important once out of the variety of advanced existing techniques, it leads to the behavior of the algorithms reaches to the expected similarities (Venkata Rao et al. 2016). Text detection and localization results are evaluated on two publicly available datasets namely ICDAR 2013 and IPC-Artificial text. Moreover, results are compared with state-of-the-art techniques and the Comparison demonstrates the superiority of the presented research (Ye and Doermann et al. 2015). Proposed methods are compares, and contrasts technical challenges, methods, and the performance of text detection and recognition research in color imagery. It summarizes the fundamental problems and enumerates factors that should be considered when addressing these problems in the field (Zhang et al. 2015). The proposed method recently, a variety of real-world applications have triggered huge demand for techniques that can extract textual information from natural scenes. Therefore, scene text detection and recognition have become active research topics in computer vision. In this work, they investigated the problem of scene text detection from an alternative perspective and propose a novel algorithm for it. The experiments on the latest ICDAR benchmarks demonstrate that the proposed algorithm achieves state-of-the-art performance. Moreover, compared to conventional approaches, the proposed algorithm shows stronger adaptability to texts in challenging scenarios (Zhao et al. 2010). In this work text detection is important in the retrieval of texts from digital pictures, video databases and webpages. The proposed classification-based algorithm for text detection using a sparse representation with discriminative dictionaries. Proposed method can effectively detect texts of various sizes, fonts and colors from images and videos.

3 Proposed Methodology

Text extraction in images passes through five stages, the important stage is the "text detection" and text localization which are closely related to each other from a given image or video frame. The early steps of obtaining the information from the text which will focus on the attention of most research scholars. Text variation that are related to size, style, alignment and direction, as well as low image contrast and complex backgrounds make the problem of automatic text detection extremely challenging. According to the survey of many of literature reviews, text localization methods is classified into two categories such as compressed domain. Each method in each category takes into account the different features depending on the different types of the images as well as its quality and the tackling thus above problems since several decades and hence has obtained impressive performance. Overview of the proposed methodology is given in the Fig. 1.

Let g(x, y) binary image of the breaking news image the proposed methodology will be exhibited as mentioned below:

- Select the bottom region of g(x, y) by applying grid, since any news channel contains the information related to the breaking news only be at bottom of the image g(x, y) so rest of the regions are ignored, divide news image into two equal halves upper half and lower one and ignore the upper region and retain below region by dividing the number of rows by 2.

$$t(x, y) = g\left(\frac{\sum_{i=1^x}^{m}}{2}, \sum_{i=1}^{n} y\right) \tag{1}$$

Using MSER in t(x, y) for detecting possible text regions, require converting a colour of the image into a binary image, MSER algorithm used to find text regions from everywhere of the regions within the image and trick the results in this step. To detect text regions, first, you have to MSER. Then, remove most of the non-text regions which can be detected in neighboring regions as what has been done in the third step.

- The Standard MSER algorithm is employed in order to eliminate non textual regions by using fundamental geometrical attributes also the algorithm sometimes misdetects some of stable zones. The proposed methodology is employed to recognize and eliminate several non-text zones located in the image. The process of filtering occurs by considering only text zones and eliminating rest. Subsequently this method uses a machine learning technique to train a classifier in order to distinguish text from non-text zones, the adequate blending of geometrical attributes along with a suitable machine learning methodology generates superior results.

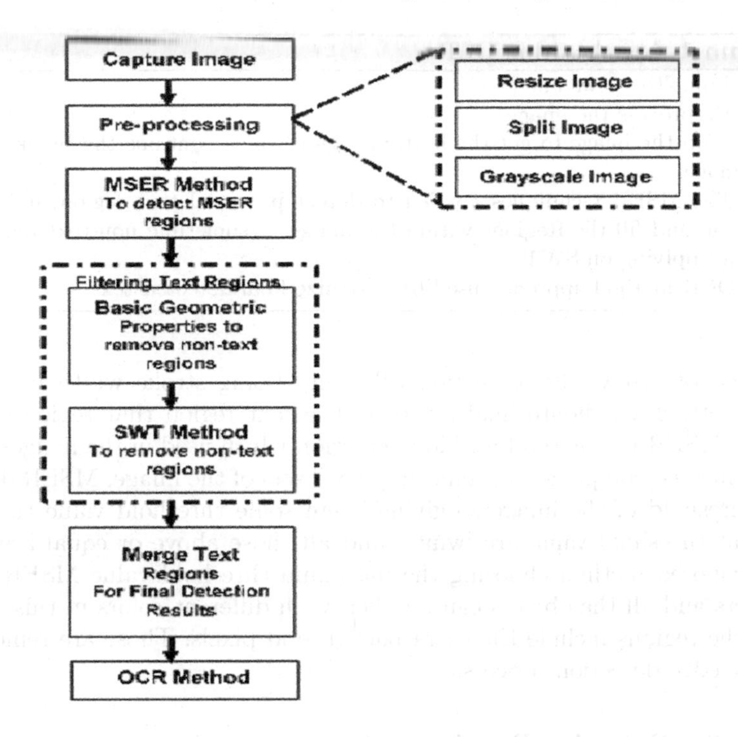

Fig. 1. Overview of the proposed system

- Remove Non-Text Regions Based on Stroke Width Variation, is another common metric which is used to make a decision between text and non-text is stroke width. To measure the width of the curves and lines, which makes up a character, it is suitable to use the stroke width variation properties. Actually, the text regions in the image have small stroke width variations, whereas non-text regions in the same image have larger ones. To help users to understand how they can use the stroke width to remove non-text regions from the image, they will be able to estimate the stroke width of the parts of the detected MSER regions.
- The individual characters in the news image can be extracted in this phase by merging text regions in order to form a word else lines of words as similar as the job done during optical character recognition stage, this technique facilitate the users to identify and recognize the definite words in the input image which has significant information, not only individual characters, such as recognizing the string 'HELP' vs. the set of individual characters {'L', 'H', 'P', 'E'} in which they lose the meaning without the correct ordering.

3.1 Extracting Maximally Stable Extremal Regions (MSER)

The MSER technique to detect the text regions. The text regions consist of some unnecessary information, to remove that unwanted information, the technique

Algorithm 1. My algorithm OCR

Step1: Input the news image.
Step2: Preprocess the image
step3: Crop the image to get the bottom region which contains the breaking news of the image.
step4: The MSER techniques applied to detect potential text region, it finds the text region and fill the Regions within the image, sometime non-text regions are appear it applying on SWT.
step5: OCR method approach used to recognize final detected text.

which used of canny edge detection Filtering. Using stroke width method to filter the letters' candidate finally getting the text region that region gives to the OCR. MSER is a method for blob detection which used in the images; it is a stable connected component of some gray level sets of the image. MSER depends on the threshold of the image so giving them some threshold value the pixels below that threshold value are 'white' and all those above or equal are black. In the proposed method choosing the minimum threshold value MSER detect the objects and all the objects can be filled with different colors in this process some of the regions include the extra background pixels. Those are removed in the canny edge detection process.

3.2 Region Detection Requirements

The requirements of region detection should be considered regarding to region detection invariance transformation such as illumination changes, rotation, translation, full affine transform and scale (i.e. a region should correspond to the same pre-image for many viewpoints. changing of Viewpoint can be done by affine transform if assuming locally orthographic camera and planar objects, ignoring perspective effects). Region detection must be stable and repeatable and having discriminate capability between all regions as show in Fig. 2.

Fig. 2. Steps of region detection processor

The following are the steps for MSER Extraction implements:

- Performing a simple luminance by sweeping threshold of intensity starting from black to white.

- (Extremal Regions) which is extract connected components.
- Detected the threshold while an Extremal region is "Maximally Stable", i.e. local lower of comparative growth of its square. Because of separated nature of image, the region above/below may be consonant with the current region, in the way that region is still considered as maximal.
- Optionally approximate a region an ellipse.
- Keeping regions descriptors as features.

Sometime the Extremal region may be rejected even if it's maximally stable:

- There is a parameter MaxArea (very big).
- There is a parameter MinArea (too small).
- There is a parameter MaxVariation (unstable)
- Its matching to its parent MSER.

Image I is a mapping I:$R \subset Z^2 \rightarrow S$. Extremal regions are will defined on images if:

1. S is totally ordered, i.e. reflexive, ant symmetric and transitive binary relation

   $\leq$ exists. In this paper only $S = \{0, 1, \ldots.., 255\}$ is considered, but Extremal

   regions can be defined on e.g. real-valued images ($S = R$).
2. An adjacency (neighborhood) relation

$$A \subset RXR$$

is defined. In this paper 4-neighborhoods are used, i.e.

$$p, q \in R \text{ are adjacent } (pAq) iff \sum_{f=1}^{d} |p1 - qi| \leq 1$$

Region Q is a contiguous subset of, i.e. for each p, q $\in$ Q there is a sequence

$$\text{p, } a_1, a_2, \ldots., a_{nq} \text{ and } pAa1, aiAai + 1, a_n Aq$$

(Outer) Region Boundary

$$\partial \mathbf{Q} = \{\mathbf{q} \in \mathbf{R} \backslash \mathbf{Q} : \exists \mathbf{p} \in \mathbf{Q} : \mathbf{qAp}\} \text{ , i.e. the boundary } \partial\mathbf{Q} \textbf{ of } \mathbf{Q}$$

is the set of pixels being adjacent to at least one pixel of Q but not belonging to Q. Extremal region

$$\mathbf{Q} \subset \mathbf{R} \text{ is a region such that for all } \mathbf{p} \in \mathbf{Q}, \mathbf{q} \in \partial\mathbf{Q} : I(\mathbf{p}) > I(\mathbf{q})$$

(maximum intensity region) or $I(\mathbf{p}) < I(\mathbf{q})$ (minimum intensity region).

Maximally Stable Extremal Region (MSER). Let

$$\mathbf{Q1}, ..., \mathbf{Qi} - \mathbf{1}, \mathbf{Qi}, ...$$

be a sequence of nested extremal regions, i.e. $\mathbf{i} \subset \mathbf{Qi} + \mathbf{1}$ Extremal region Qi. is maximally stable

$$\mathit{iff}\, \mathbf{q}(\mathbf{i}) = |\mathbf{Qi} + \triangle \backslash \mathbf{Qi} - \triangle / |\mathbf{Qi}|$$

has a local minimum at $i*$ ($|.|$denotes cardinality). $\triangle \in S$ is a parameter of the method.

4 Experimental Result

In the experimental results 500 images from the reputed international media channels are considered for segmentation and recognition, the success rate and

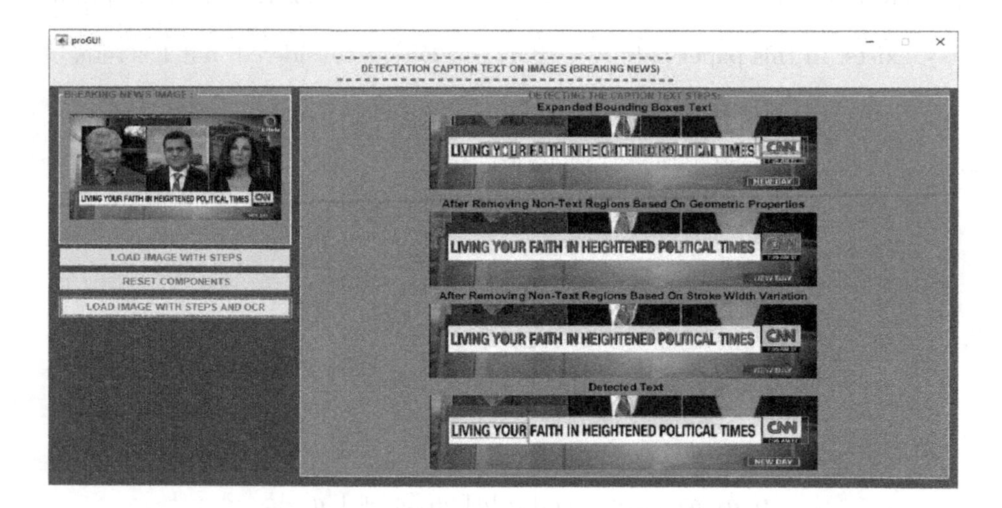

Fig. 3. Steps for breaking news detection

Fig. 4. Result for segmented text

analysis conducted will be illustrated as follows: The result of the recognized text in image with one Simple misdetection are shown in the Fig. 3.

To avoid the noise caused by the use of the OCR (Optical text recognition technology), OCR will recognize the text within each bounding box of MSER (Maximally stable Extremal regions technique), which show good results with high accuracy, less error and reducing time complexity, focusing on regions only without searching all regions in image (Figs. 4 and 5).

Fig. 5. Message box for OCR results

Fig. 6. Simple misdetection

Experimental have been done on six news channel as Aljazeera, BBC, CNBC, CNN, RT and Sky News. Analysis and evaluation result shown in Table 1 and Fig. 6. Some factors influenced the accuracy of the results derived from the channels. For example, Aljazeera has reached results accuracy up to 71% due to the existence of some obstacles such as font's types. It can also be noted that news channels such as the Russian TV reached the results of extracting texts using OCR technology more than 80%, because the quality and size of the fonts are more common, thus increasing results accuracy of these channels. The CNBC news channel is using popular font type which the proposed algorithm makes it easy extract high quality scores results. Compared to other methods in previous studies on lectures survey, the method which used in this experimental has more accuracy and less errors. The following table shows the percentage of results extracted from the data of each news channel separately (Figs. 7, 8 and Table 2).

Table 1. Average of percentage for each channel

No	T.V.channel	Total of images (Data sets)	Recognition rate (Success rate)
1	Aljazeera	500	71%
2	BBC	500	75%
3	CNBC	500	94%
4	CNN	500	81%
5	RT	500	84%
6	Sky News	500	78%

Table 2. Competitive analysis

Author	Paper title	Accuracy	Remarks
Julinda Gllavata	A text detection, localization and segmentation system for OCR in images	83.9%	Using color and edge detection the text will be selected based on horizontal projection
Iqbal, Khalid et al.	Text localization based on Bayesian network	72.44%	K2 Algorithm is used with Bayesian network using geometric features for text localization
Ikica, Andrej et al.	Text detection using edge profile in natural images	71%	Edge-based, connected-component based and texture-based methods
Yin, Xu-Cheng et al.	Text recognition in real images	76%	self-training distance metric learning algorithm, clustering threshold simultaneously
Our method	Novel technique for recognition of breaking news from television using OCR technique	94%	Fixed size for all images, splitting images, MSER

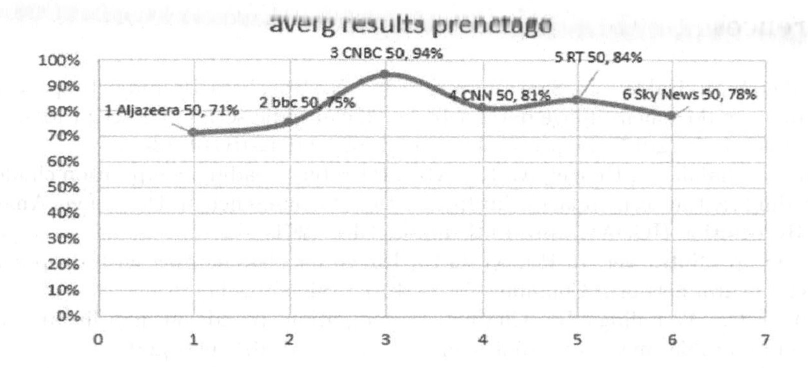

Fig. 7. Success rate comparison for each channel

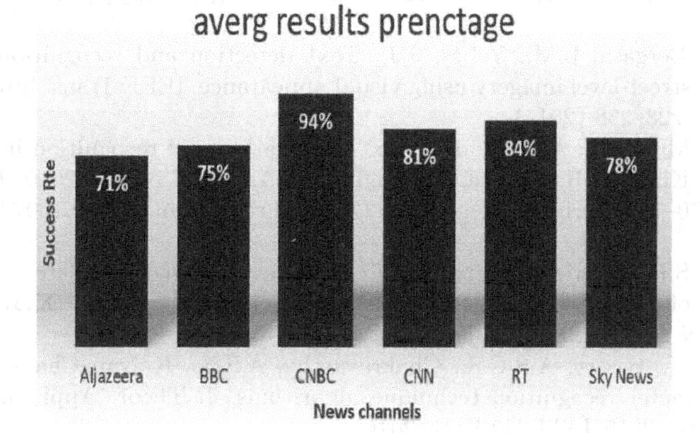

Fig. 8. Success rate comparison for each channel.

5 Conclusion

Algorithm proposed used in this work to detect the only breaking news caption from popular English news channels. After preprocessing resizing the image which gives the fixed size for all images that's make splitting images more elective to focus only on the breaking news line in the image – which comes on the bottom down of screen of the channels by default - decrease the time complexity by making the search area less. The performance of the proposed algorithm showing more efficiency after segments the caption text from the set of images with MSER and SWT techniques. The algorithm gives good results with overall accuracy of 80.5% with highest success rate of 94% and less error comparing with previous work obtained for CNBC news channel.

References

Obaidullah, S.M., Halder, C., Santosh, K.C., Das, N., Roy, K.: PHDIndic_11: page-level handwritten document image dataset of 11 official Indic scripts for script identification. 1643–1678 (2017). https://doi.org/10.1007/s11042-017-4373-y

Shahab, A., Shafait, F., Dengel, A.: ICDAR 2011 robust reading competition challenge 2: reading text in scene images. In: International Conference on Document Analysis and Recognition (ICDAR), pp. 1491–1496. IEEE (2011)

Krizhevsky, A., Sutskever, I., Hinton, G.E.: ImageNet classification with deep convolutional neural network. Commun. ACM **60**(6), 84–90 (2017)

Santosh, K.C., Wendling, L.: Character recognition based on non-linear multi-projection profiles measure. Front. Comput. Sci. **9**(5), 678–690 (2015)

Santosh, K.C.: Character recognition based on DTW-Radon. In: International Conference on Document Analysis and Recognition (ICDAR), pp. 264–268. IEEE (2011)

Epshtein, B., Ofek, E., Wexler, Y.: Detecting text in natural scenes with stroke width transform. Institute of Electrical and Electronics Engineers, pp. 2963–2970. IEEE, 13 June 2010

Gonzalez, A., Bergasa, L.M., Yebes, J.J.: Text detection and recognition on traffic panels from street-level imagery using visual appearance. IEEE Trans. Intell. Transp. Syst. **15**(1), 228–238 (2014)

Neumann, L., Matas, J.: A method for text localization and recognition in real-world images. In: Kimmel, R., Klette, R., Sugimoto, A. (eds.) ACCV 2010. LNCS, vol. 6494, pp. 770–783. Springer, Heidelberg (2011). https://doi.org/10.1007/978-3-642-19318-7_60

Raghavendra, S.P., Danti, A., Suresha, M.: Correlation based template matching for recognition of Bank Cheque number. Int. J. Comput. Eng. Appl. **XII**(III), 61–76 (2018). ISSN 2321-3469. www.ijcea.com

Venkata Rao, N., Sastry, A.S.C.S., Chakravarthy, A.S.N., Kalyan Chakravarthy, P.: Optical character recognition technique algorithms. J. Theor. Appl. Inf. Technol. **83**(2) © 2005–2015 JATIT & LLS (2016)

Ye, Q., Doermann, D.: Text detection and recognition in imagery: a survey. IEEE Trans. Pattern Anal. Mach. Intell. **37**(7), 1480–1500 (2015). https://doi.org/10.1109/tpami.2014.2366765

Zhang, Z., Shen, W., Yao, C., Bai, X.: Symmetry-based text line detection in natural scene. IEEE. INSPEC Accession Number 15524342, 7–12 (2015). http://rrc.cvc.uab.es

Zhao, M., Li, S., Kwok, J.: Text detection in images using sparse representation with discriminative dictionaries. Image Vis. Comput. **28**(12), 1590–1599 (2010)

Ahuja, D., Amesar, J., Gurav, A., Sachdev, S., Zope, V.: Text extraction and translation from image using ML in Android. Int. J. Innovative Res. Sci. Eng. Technol. (2018). ISSN 2319-8753, ISSN (Print) 2347-6710

Siamese Network for Learning Genuine and Forged Offline Signature Verification

Amruta B. Jagtap[1(✉)], Dattatray D. Sawat[1], Ravindra S. Hegadi[1], and Rajendra S. Hegadi[2]

[1] Department of Computer Science, Solapur University, Solapur 413255, Maharastra, India
amrutaj88@gmail.com, sawat.datta@gmail.com, rshegadi@gmail.com
[2] Indian Institute of Information Technology, Dharwad, Karnataka, India
rajendra.hegadi@gmail.com

Abstract. This work aims to use a Siamese network to verify between genuine and forged signatures by making signature embeddings more robust. Currently, the Siamese network is most widely used in many applications such as Dimensionality reduction, Learning image descriptor, Face recognition, Image ranking, etc. This network is termed as twin network since it consists of two similar neural networks which take two input images and shares same weights. The critical task in signature verification is to discriminate between genuine and skilled forger since forged signature differs by some precise kind of deformation. Embedding vector is generated by Siamese network and to make embedding vector more robust we propose to add statistical measures to it, which are calculated on the embedding vector itself. The contrastive loss function is then applied on the resultant embedding vector.

Keywords: Siamese network · Deep neural network · Signature verification · Statistical measures

1 Introduction

Machine Learning methods can learn data from the feature vector, where learning technique can be supervised or unsupervised [11]. The deep neural network is one of the super-fast machine learning algorithms that has multiple layers through which the data is transformed. To prevent falsification of offline signature verification, Convolutional deep neural network is broadly researched. The network has the capability to learn features from input images, to obtain effective features for classification purpose. To achieve high accuracy of the system, large size dataset is required during training of the convolutional neural network. In contrast to this, the Siamese neural network can train on one input image and one target image.

Recently, Siamese neural network is used to recognize between two similar or dissimilar images. This network has two identical networks which share the same

K. C. Santosh and R. S. Hegadi (Eds.): RTIP2R 2018, CCIS 1037, pp. 131–139, 2019.
https://doi.org/10.1007/978-981-13-9187-3_12

weights and parameter. To train network two distinct inputs are given, having energy function at the top. Later these input images are processed by series of convolutional layers, pooling layers, along with fully connected layer consisting the feature vector for each image, than these feature vector is transformed in to dense vector is called as embedding vector. To recognize the similarity and dissimilarity between given input images, the distance is calculated. If the distances are small, then images signify of the same class, and if it is large, images are from a different class.

The signature verification systems have genuine and forged signatures of the same subject. Such pair of signatures have a similar characteristic, so it became difficult to discriminate them. If the verification system does not have a mechanism to deal with this problem, then the system may result in high False Acceptance Ratio (FAR). To resolve this problem, several researchers designed various Siamese networks that reduced FAR significantly. Apart from FAR, False rejection Ratio (FRR) is one of the significant evaluation protocol that describes systems robustness. We aim to design a robust system for signature verification by designing a Siamese network by making embeddings more robust. We propose to calculate geometric mean and standard deviation of the embedding vector and add it to same vector. Methodology for the proposed technique described in the Sect. 3.

2 Related Works

In recent years, the task of offline handwritten signature has been broadly researched. By means of various techniques many researchers have performed different experiments, but still, the area of offline signature verification remains an open problem for researchers. The offline signature system is addressed with two approaches: Writer Dependent and Writer Independent [7]. As compared to writer independent, writer dependent is not a preferable approach, since it requires to retrain the system with every new signer. Whereas, in the case of writer-independent, the system is modeled to discriminate between genuine and forged signatures. For the training of writer-independent system, the dataset is divided into train and test sets, moreover to test the system same procedure of training is repeated.

Several hand crafted features have been proposed. Vahid Kiani et al. proposed a method based on Local Randon Transform and used randon transform locally as a feature extractor [12]. Various other features like global [20], statistical and geometrical such as area, centroid, skewness, kurtosis [3,15] were popular feature extraction methods for offline signature verification. Furthermore, many novel techniques proposed to verify signatures of individual, like pixel matching [1], structural features extracted from shape of signature using modified direction and its enhanced version [16], hybrid features [5] includes local and global characteristics and these features are combined to form visual codebook having correlated features, Applied eigen value technique for upper and lower envelop of signature [9,10]. Santosh et al. [18] developed a technique based on spatial similarity based stroke clustering for Devanagari natural handwritten character recognition. To determine their similarity dynamic time warping (DTW) algorithm is used. Ukil et al. [21] proposed approach is two-fold: Firstly for three different scales of input image, they used two and three layered CNNs. Secondly, for two different scales of transformed images same CNNs were used and later they merged these obtained features for identification of script.

Apart from these methods, nowadays most popular technique used is representation of feature learning using deep neural network for signature verification. Instead of using handcrafted features, Luiz G. Hafemann et al. [8] proposed to use deep convolutional neural network to learn features from input images and additionally this model was trained on another set of users. Further they developed a model for fixed sized representation of signatures, as input size of neural network is same, but the size of signature vary from one user to another. By addressing this problem they customized the network architecture via Spatial Pyramid Pooling. Besides this, twin like network called Siamese network are extremely popular for other verification tasks like face recognition, object recognition and online signature verification [2]. Sounak Dey et al. proposed a network called SigNet using convolutional Siamese network for writer independent offline signature verification. They addressed the network by observing similar and dissimilar pair of observations and for similar pair, they minimized the Euclidean distance while maximized distance for dissimilar pair. Experiments were conducted on different datasets such as GPDS 300, CEDAR, Bengali, Hindi, GPDS Synthetic and achieved good results as compared to state-of-the-art performance [4]. Siamese neural network can train on single input image for each new class known as one-shot network for recognition [13].

3 Methodology

In following sections we discuss the convolutional neural network which is used to obtain embedding vector. We focus to make embedding more robust using different statistical measures to discriminate between genuine and forged signature. We also discuss contrastive loss function, experiment design and robust embeddings in sections below.

3.1 CNN and Siamese Network

In Neural network, high dimensional image vector takes a huge amount of parameters to characterize the network. To address this problem, Deep Convolutional neural network are proposed. Sawat et al. [19] used CNN for deep features extraction. It is a multilayer architecture consisting of input, convolutional, max-pooling, rectified linear unit (ReLU) and fully connected layers. All layers are connected in chronological order, i.e output of one layer is given as input next layer. The first layer i.e Convolution is a building block of network, which performs the dot product between input and weights, and obtained matrix are integrated across the channel. The input is organized as $H \times W \times D$

where H is a height, W is a width and D is a depth of input image. Rectified linear unit (ReLU) applied activation function to increase non-linearity of the network. Convolutional layer generates activation maps, pooling layer carry out down sampling on these maps to reduces the window size. Lastly, fully connected layer vectorize the previous layer.

Siamese neural network is two branch network having identical CNNs which shares same parameters and shared weights. The framework is used for dimensionality reduction [6] having contrastive loss function at bottom of network. Equation 1, which computes similarity and dissimilarity between pair by using distance metric called Euclidean distance. The contrastive loss function is defined as

$$L(c, s_1, s_2) = (1 - c)\,\frac{1}{2}\,(D_W)^2 + (c)\frac{1}{2}\,\{max\,(0, m - D_W)\}^2 \tag{1}$$

3.2 Architecture

The CNN architecture used in the proposed work is inspired by [14] for recognition task and used in [4]. Input given to network is signature image of size 155 × 220. This input is passed to first convolutional layer which has 96 different filters with 11 × 11 convolution and 1 stride. To normalize local input regions, local Response Normalization is applied across the channel that performs a kind of “Lateral inhibition”. After normalization, the image input to the next layer known as pooling layer with size of 96 × 3 × 3 having parameter with stride equal to 2. The output of first convolutional layer, local normalization and pooling layer is fed as a input to second convolutional layer having 256 different filters with size of 5 × 5. The third and fourth convolutional layers are connected in parallel without disturbance of pooling and normalization layer. These layers have 384 and 256 different filters with size of 3 × 3 having stride equal to 1. In addition to this, dropout is used with combination of pooling layer and first fully connected layer to avoid unwanted units. The last fully connected layer represents embedding of input data which has 128 neurons.

The input given for training is in pair form $(s1, s2)$ which can be labeled in binary form as

$$c = \begin{cases} 1 \;...if(s1, s2)\;(Genuine, Genuine) \\ 0 \;...if(s1, s2)\;\;(Genuine, Forged) \end{cases} \tag{2}$$

3.3 Robust Embedding

Embedding vector is generated by Siamese network as mention in Sect. 3.2. To make this embedding vector more robust we propose to add statistical measures to it, which are calculated on the embedding vector itself. Following paragraph discusses the statistical measures to be used.

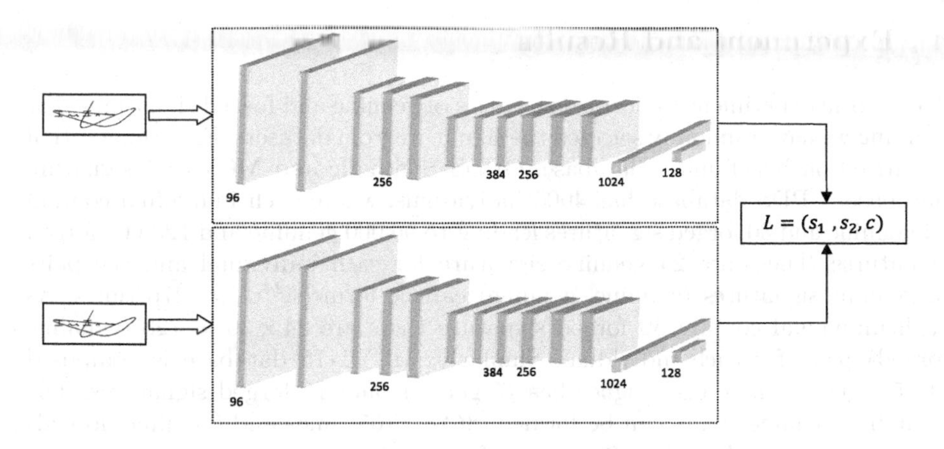

Fig. 1. Architecture of siamese network

Geometric mean: It is often used when each object has various properties with different numeric ranges. It is n^{th} root of the product of n number of pixels in signature, represented by $(s_1, s_2, s_3, ...s_n)$ and defined as:

$$(\prod_{i=1}^{n} s_i)^{\frac{1}{n}} = \sqrt[n]{s_1, s_2, ...s_n} \tag{3}$$

In proposed work, $s_1, s_2, s_3, ...s_n$ will be the elements of original embedding vector.

Standard deviation: It is one of the important measure of dispersion of data values. It is measured as a square root of variation. If $s_1, s_2, s_3, ...s_n$ are number of pixels in signature then standard deviation is denoted by σ_s and is given by

$$\sigma_s = \sqrt{\frac{1}{n}\sum_{i=1}^{n}(s_i - \overline{s})^2} \tag{4}$$

where, $\overline{s}$ is mean of $s_1, s_2, s_3, ...s_n$. High standard deviation represents higher dispersion or variation in data points.

The calculated geometric mean and standard deviation is concatenated with original embedding to generate robust embedding. After concatenation the embedding vector will be 1×130 in size. The contrastive loss function defined in Eq. 1 is then applied on the resultant embedding vector.

4 Experiment and Results

To perform experiment we formed the pairs of genuine and forged signatures. For training we have randomly selected the signature from datasets. Experimentation is carried on bench-mark database: GPDS Synthetic and MCYT-75 signature database. GPDS database has 4000 individuals, where each individual contain 24 genuine and 30 forged signatures leading to 96,000 genuine and 120,000 forged signatures. There are 24 genuine signature for each individual and the pairs of genuine signatures (genuine, genuine) can be formed ${}^{24}C_2 = 276$ times. As each individual contain 30 forged signature there are $24 \times 30 = 720$ (genuine, forged) pairs for each individual. Similarly, MCYT-75 database is composed of 75 signers, where each signer has 15 genuine and 15 forged signatures. The (genuine, genuine) pairs can be formed ${}^{15}C_2 = 105$ times and (genuine, forged) pairs are formed $15 \times 15 = 225$ times. Out of the signature pairs prepared by using above criteria, 80% are used for training, 10% for validation and 10% for testing. The result obtained using robust embedding shows improvement in accuracy for GPDS and MYCT database.

The proposed experiment is implemented using CUDA-9.0, CuDNN-7.0, Python-3.6, Tensorflow-1.9 as a backend and Keras-2.1.6 as front end tool for Siamese neural network. The model is trained for 10 epochs, using batch size set to 32. The initial learning rate is set to 0.0001 with momentum rate = 0.9 and $\epsilon = 1e-8$. The training was performed using NVIDIA GeForce 940 m GPU. Following table shows the comparison between proposed method and other state-of-art methods.

Databases	Methods	Signers	Accuracy
GPDS synthetic signature corpus	Dutta et al. [5]	4000	73.67
	Sounak et al. [4]	4000	77.76
	Proposed	4000	**84.58**
MCYT-75 signature corpus	Prakash et al. [17]	75	81.74
	Proposed	75	**85.38**

As shown in Fig. 2, the initial validation loss is around 0.560, after 4 epoch validation loss dropped to $\approx$0.2200. Figure shows validation loss for 10 epoch using batch size of 32.

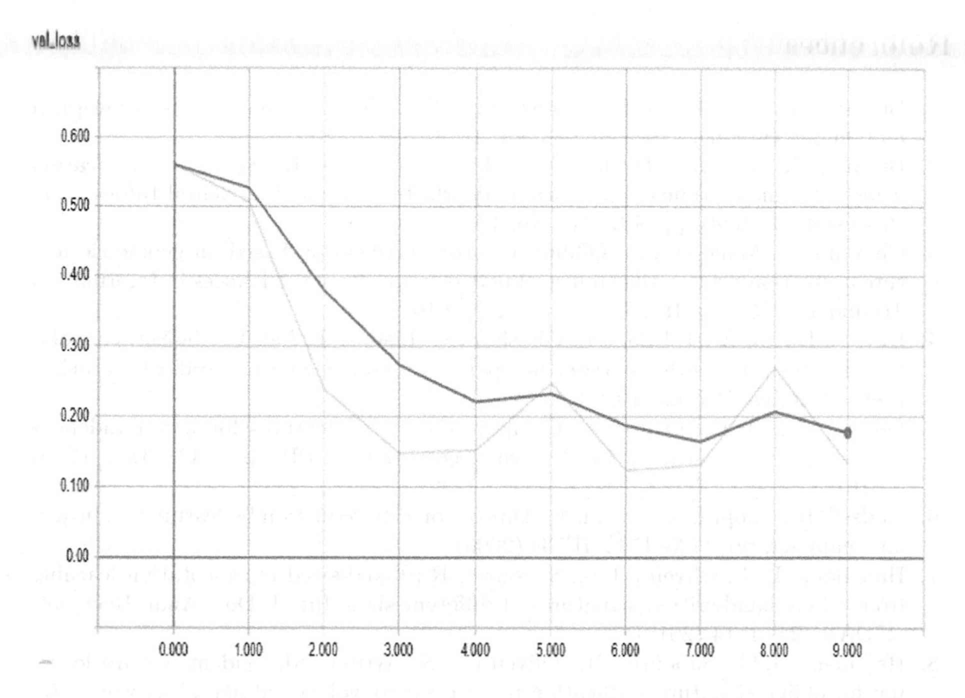

Fig. 2. Validation loss

5 Conclusion and Future Work

The proposed work focus on learning feature from deep neural network instead of hand-drafted features. To verify genuine and forged signature, embedding vector is generated by Siamese network and to make embedding vector more robust we proposed to add statistical measures to it, which are calculated on the embedding vector itself. The statistical measures used in this work are geometric mean and standard deviation are concatenated with original embedding to generate robust embedding vector and contrastive loss function is then applied on the resultant embedding vector. Experiment is carried on GPDS synthetic and MCYT-75 signature corpus. Results of proposed work on bench-mark database out performs the state-of-art methods. In future, we will develop more robust model which lead to recognize between genuine and skilled forgery, also increases verification rate, reduces false acceptance rate (FAR) and false rejection rate (FRR) of robust model.

Acknowledgments. Authors thank the Ministry of Electronics and Information Technology (MeitY), New Delhi for granting Visvesvaraya Ph.D. fellowship through file no. PhD-MLA\4(34)\2015-16 Dated: 05/11/2015.

References

1. Bhattacharya, I., Ghosh, P., Biswas, S.: Offline signature verification using pixel matching technique. Procedia Technol. **10**, 970–977 (2013)
2. Bromley, J., Guyon, I., LeCun, Y., Säckinger, E., Shah, R.: Signature verification using a "siamese" time delay neural network. In: Advances in Neural Information Processing Systems, pp. 737–744 (1994)
3. Chandra, S., Maheskar, S.: Offline signature verification based on geometric feature extraction using artificial neural network. In: Recent Advances in Information Technology (RAIT). IEEE, pp. 410–414 (2016)
4. Dey, S., Dutta, A., Toledo, J.I., Ghosh, S.K., Lladós, J., Pal, U.: SigNet: convolutional siamese network for writer independent offline signature verification. arXiv preprint arXiv:1707.02131 (2017)
5. Dutta, A., Pal, U., Lladós, J.: Compact correlated features for writer independent signature verification. In: Pattern Recognition (ICPR), pp. 3422–3427. IEEE (2016)
6. Hadsell, R., Chopra, S., LeCun, Y.: Dimensionality reduction by learning an invariant mapping, pp. 1735–1742. IEEE (2006)
7. Hafemann, L.G., Oliveira, L.S., Sabourin, R.: Fixed-sized representation learning from offline handwritten signatures of different sizes. Int. J. Doc. Anal. Recognit. (IJDAR) **21**, 1–14 (2018)
8. Hafemann, L.G., Sabourin, R., Oliveira, L.S.: Writer-independent feature learning for offline signature verification using deep convolutional neural networks. In: International Joint Conference on Neural Networks (IJCNN), pp. 2576–2583. IEEE (2016)
9. Jagtap, A.B., Hegadi, R.S.: Eigen value based features for offline handwritten signature verification using neural network approach. In: Santosh, K.C., Hangarge, M., Bevilacqua, V., Negi, A. (eds.) RTIP2R 2016. CCIS, vol. 709, pp. 39–48. Springer, Singapore (2017). https://doi.org/10.1007/978-981-10-4859-3_4
10. Jagtap, A.B., Hegadi, R.S.: Offline handwritten signature recognition based on upper and lower envelope using eigen values. In: World Congress on Computing and Communication Technologies (WCCCT). IEEE, pp. 223–226 (2017)
11. Jarad, M., Al-Najdawi, N., Tedmori, S.: Offline handwritten signature verification system using a supervised neural network approach. In: Computer Science and Information Technology (CSIT), pp. 189–195. IEEE (2014)
12. Kiani, V., Pourreza Shahri, R., Pourreza, H.R.: Offline signature verification using local radon transform and support vector machines. Int. J. Image Process. **3**, 184–194 (2009)
13. Koch, G., Zemel, R., Salakhutdinov, R.: Siamese neural networks for one-shot image recognition. In: ICML Deep Learning Workshop, vol. 2 (2015)
14. Krizhevsky, A., Sutskever, I., Hinton, G.E.: ImageNet classification with deep convolutional neural networks. In: Advances in Neural Information Processing Systems, pp. 1097–1105 (2012)
15. Majhi, B., Reddy, Y.S., Babu, D.P.: Novel features for off-line signature verification. Int. J. Comput. Commun. Control **1**, 17–24 (2006)
16. Nguyen, V., Blumenstein, M., Muthukkumarasamy, V., Leedham, G.: Off-line signature verification using enhanced modified direction features in conjunction with neural classifiers and support vector machines. In: Document Analysis and Recognition, ICDAR, vol. 2, pp. 734–738. IEEE (2007)

17. Prakash, H., Guru, D.: Offline signature verification: an approach based on score level fusion. Int. J. Comput. Appl. **1**, 0975–8887 (2010)
18. Santosh, K., Nattee, C., Lamiroy, B.: Relative positioning of stroke-based clustering: a new approach to online handwritten devanagari character recognition. Int. J. Image Graph. **12**(02), 1250016 (2012)
19. Sawat, D.D., Hegadi, R.S.: Unconstrained face detection: a deep learning and machine learning combined approach. CSI Trans. ICT **5**(2), 195–199 (2017)
20. Serdouk, Y., Nemmour, H., Chibani, Y.: Topological and textural features for off-line signature verification based on artificial immune algorithm. In: Soft Computing and Pattern Recognition (SoCPaR), pp. 118–122. IEEE (2014)
21. Ukil, S., Ghosh, S., Obaidullah, S.M., Santosh, K., Roy, K., Das, N.: Deep learning for word-level handwritten indic script identification. arXiv preprint arXiv:1801.01627 (2018)

Multi-scale Local Binary Patterns- A Novel Feature Extraction Technique for Offline Signature Verification

Bharathi Pilar[1(✉)], B. H. Shekar[2(✉)], and D. S. Sunil Kumar[2]

[1] Department of Computer Science, University College, Mangalore, Karnataka, India
bharathi.pilar@gmail.com

[2] Department of Computer Science, Mangalore University, Konaje, Karnataka, India
bhshekar@gmail.com, dssunil6@gmail.com

Abstract. This paper presents a powerful feature representation method called Multi-scale Local Binary Patterns for offline signature verification. The multi-scale representation oriented local binary patterns can be obtained by changing the radius R value of Local Binary Patterns(LBP) operator and combining the LBP features at different scales. In this proposed approach the LBP operator is applied at 3 different scales by varying the radius R value and at each scale equal number of pixels are considered for the processing. Finally, by cascading a group of LBP operators at 3 different scales over a signature image with fixed number of pixels at each scale and combining their results, a multi-scale representation LBP can be obtained. This essentially represents nonlocal information. Features fusion is performed by the linear combination of the histogram corresponding to 3 different radii results in a multi resolution (scale) feature vector. Support Vector Machine (SVM) is a well known classifier employed to classify the signature samples. Experimental results on standard datasets like CEDAR and a regional language datasets shows the proposed technique's performance. A comparative analysis with few well known methods is also presented to demonstrate the performance of proposed technique.

Keywords: Multi-scale Local Binary Patterns · Signature verification · Support Vector Machine · Local binary patterns

1 Introduction

Signatures are one of the biometric trait, globally accepted to authenticate individuals. Automatic Signature verification is categorized into two branches based on the way signatures are acquired, which are offline and online signature verification. Offline signature verification method takes the scan of signature which is written on a paper this is static by nature. Offline signature verification works on the static information mainly on texture such as vertical and horizontal projections, line width, orientation etc. Where as online signature verification method takes the signature using digital gadget like digitizing tablets etc

K. C. Santosh and R. S. Hegadi (Eds.): RTIP2R 2018, CCIS 1037, pp. 140–148, 2019.
https://doi.org/10.1007/978-981-13-9187-3_13

which is dynamic by nature. Online signature verification uses function features and parameter features like x-y coordinates, pen pressure, pen up, pen down, azimuth, pen inclination etc. Considering the nature of forging method forgeries are categorized into three categories namely: *simple, random and skilled forgery.* The simple forgery is one where the forger is aware of the signature name but not its pattern so that there will be variation in his/her own pattern of writing. Random forgery is another type where the forger is not aware of both the signature name and its pattern. Lastly the skilled forgery is one where the forger knows the genuine signature name and its pattern, practices for a while and then imitates. This almost resembles the original pattern and most challenging issue along with high intra class variations in offline signature verification system.

In this work, we made an attempt to exploit the extension of Local Binary Patterns, called Multi-scale Local Binary Patterns. The local binary patterns is a powerful texture descriptor, extending this descriptor to higher scales will helps in capturing nonlocal information. In the literature we found extensive work done on local binary patterns in texture discrimination. Multi-scale local binary patterns also performs well when there is a need for combined local and nonlocal features. Hence, we made an attempt to explore Multi Scale Local Binary Pattern for offline signature verification. The order of this paper with section wise is as follows: Sect. 2 presents a detailed literature of some of the state- of- the-art papers. The elaborated description of the proposed approach is in Sect. 3. Section 4 illustrates the experimental results followed by the discussion. Finally, Sect. 5 represents the conclusion part.

2 Review of the Literature

Biometric is extensively used in person authentication. Signature is one of the behavioural traits and can be used to represent a person. The different kinds of signature forgery raises the need for signature verification system which verifies the genuineness of the signature. There are several algorithms used to design offline signature verification system. We have listed few important methods from the literature. Hafemann et al. [3] presents learning features formulations for offline signature verification. Learning features are used to train writer independent classifier using convolution neural networks. Hadjadji et al.[2] proposes an open system based on one-class classifier using curvelet transform along with the principal component analysis. Grid based template matching method by Zois et al. [17] uses the geometric pattern of a signature, which is encoded by grid templates, apparently partitioned as subsets. Yesmine et al. [12] presents Artificial Immune Recognition System for offline signature verification. The proposed method uses two descriptors one is gradient local binary patterns to estimate gradient features from the neighboring local binary patterns and another is longest run feature to describe the signature topology. Score level fusion of classifiers [15] proposed by Mustafa et al., This approach extracts set of features namely, Scale Invariant Feature Transformation (SIFT), Histogram of Oriented Gradients (HoG) and Local Binary Patterns (LBP). Radhika et al. [10] proposed a combined approach of both offline and online signature verification.

Author extracts features such as pen tip tracking from online signatures. Gradient features and projection profile features are extracted from offline signatures. Experiments are conducted separately. The well known classifier Support Vector Machine is employed for classification. Results obtained are combined and verified. Yilmaz et al. [16] proposed a method where signature samples are partitioned in to different zones based on both polar and cartesian coordinate systems. From different zones of both coordinate system histograms are obtained. Classification has been done by employing Support Vector Machine (SVM).

3 Proposed Approach

In this work, we propose a Multi-scale Local Binary Patterns (MSLBP) to capture the local as well as global features from the signature image. The LBP features are extracted from signature image at different radii and are stored in the form of histograms. The fusion of histogram features at various scales are performed to form a single feature vector.

The LBP [9] is a gray-scale texture descriptor which describes the local spatial structure of the texture of an image. Based on central pixel value in an image, a code sequence is generated by keeping it as a threshold with its neighborhood pixel values.

$$LBP_{P,R} = \sum_{p=0}^{p-1} s(g_p - g_c)2^p \tag{1}$$

$$s(x) = \begin{Bmatrix} 1, & if\ x \geq 0 \\ 0, & if\ x < 0 \end{Bmatrix}$$

here g_c represents the central pixel gray value, g_p neighborhood value, P represents the number of neighborhood pixels involved and R represents the radius of the neighborhood pixels. The LBP values are computed for each pixel $P_{i,j}$ by considering N neighboring pixels. The N neighboring pixels of P are considered in clockwise direction resulting a binary stream S. In case of $N = 8$ with $R = 1$, the binary stream is defined as follows

$$S = \{P_{i,j+1}, P_{i+1,j+1}, P_{i+1,j}, P_{i+1,j-1}, P_{i,j-1}, P_{i-1,j-1}, P_{i-1,j}, P_{i-1,j+1}\} \tag{2}$$

Compute the decimal equivalent d_i of the binary sequence S. The resulting value d_i is variant to rotation. To make it invariant to rotation shift S one bit towards right side by applying right shift operator. This will results in another decimal equivalent d_j of S. Repeat the process for all the remaining bits in binary sequence S to obtain N decimal equivalent values.

$$D = d_1, d_2, \ldots, d_N \tag{3}$$

The minimum decimal value in the set D is taken as the value for pixel $P_{i,j}$. The value of radius R can be varied. If $R = 1$ then the neighboring 8 pixels at distance 1 are taken as the binary stream representing the pixel under consideration. The radius can be extended to 2, 3, etc. The neighboring pixels for a given pixel at R = 1,2 and 3 are shown in the Fig. 1.

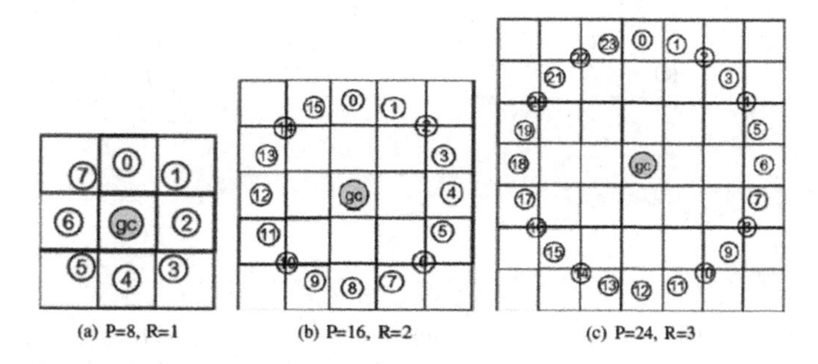

Fig. 1. (a) LBP with R = 1 and P = 8 (b) LBP with R = 2 and P = 16 (c) LBP with R = 3 and P = 24 (Image Curtsey : Rassem et al. [11]).

In the proposed approach we have used 3 scales, (R = 1, P = 8), (R = 2, P = 8), (R = 3, P = 8). The input image is applied with LBP(R = 1,P = 8) operator. The LBP transformed image is converted into an histogram H_1 and stored in a knowledge base. The process is repeated with LBP(R = 2, P = 8) and LBP(R = 3 and P = 8) giving rise to two LBP transformed images which are converted to histogram H_2 and H_3 correspondingly and are stored in knowledge base. The Histograms are combined to form a single histogram representing the resultant MSLBP feature vector for input signature image.

The Fig. 2 shows the input signature image and the transformed images after applying the LBP operators LBP(R = 1, P = 8), LBP(R = 2, P = 8), LBP(R = 3, P = 8), LBP(R = 4, P = 8) and LBP(R = 5, P = 8) respectively. The classification is done using SVM as follows.

3.1 Classification

In this work we have used the well known classifier Support Vector Machine(SVM) to classify signatures samples. The Support Vector Machine is intend to develop a model which learns from training samples based on the extracted features to predict target values of test samples [4,7]. The dataset is divided into training set as well as test set during the classification process. Every sample in the training set having its own target value called class label with set of features known as observed variables. Support vector machines are a bunch of supervised classification and regression algorithms. They are bi-linear by nature and used to classify two class objects. The Multi-SVM can be used to classify more than two class objects, in this case Multi-SVM uses one versus all strategy. The aim of SVM is to separate objects of different class by maximizing

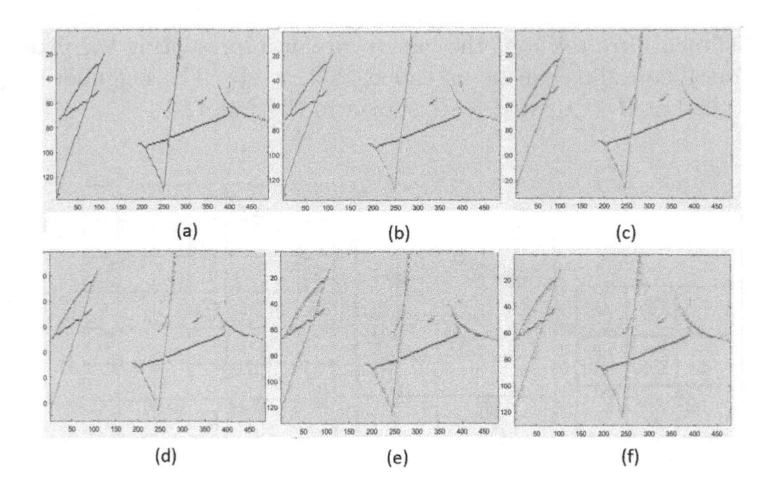

Fig. 2. (a) Original image; (b) LBP with r = 1 and n = 8; (c) LBP with r = 2, n =8; (d) LBP with r = 3, n = 8; (e) LBP with r = 4, n = 8; (f) LBP with r = 5, n = 8.

the margin of a hyper plane. The vectors which define the hyper plane are the support vectors.

$$m = 2/(||w||) \tag{4}$$

Where m is the margin and w is the width of the hyper plane.

$$w^T X + b = 1 \tag{5}$$

Equation 5 represents the upper boundary of the hyper plane.

$$w^T X + b = 0 \tag{6}$$

Equation 6 represents the center of the hyper plane.

$$w^T X + b = -1 \tag{7}$$

Equation 7 represents the lower boundary of the hyper plane.

For classification of N classes, N SVM classifiers are required. Therefore, in the proposed work we employed each SVM classifier for each writer. Here the SVM classifier uses one versus all strategy for classification of signatures.

The next section presents the experimental results in detail.

4 Experimentation Results and Discussion

The experiments are conducted on publicly available well-known dataset namely CEDAR (Center for Document Analysis and Recognition) database. Further we continued the experiments also on Local Regional Kannada dataset namely MUKOS (Mangalore University Kannada Offline Signature) corpus. The experiments are carried out using MATLAB tool with version 2016b on DEL laptop with 8 GB ram, 1tb Hard disk and i7 processor on windows 10 operating system (Table 1).

Table 1. Datasets details

Datasets	Number of signature contributors	Number of genuine signatures	Number of skilled forgeries	Total number of signatures
CEDAR	55	24	24	2640
MUKOS	30	30	15	1350

The features database contains MSLBP histograms, obtained from both datasets. The MSLBP features are extracted from genuine as well as skilled forgery signatures. The signature samples are categorized into training set and test set from both the dataset. Experiments are carried out in 4 epochs. In epoch-1, we choosen first 10 signatures of both genuine as well as skilled forgeries to train the model and remaining samples are used to test. In set-2, we considered the first 15 genuine samples along with first 15 skilled forgeries to train and remaining samples are used to test. For set-3, 10 genuine along with 10 skilled forgeries are chosen randomly for training and for testing remaining samples are used. Lastly for set-4, again randomly chosen 15 genuine signatures and 15 skilled signatures are considered to train and remaining samples used to test the proposed model. To avoid the impact of randomness, experiments are repeated for set-3 as well as set-4 and finally we considerd average results and are tabulated.

Experiments on CEDAR Dataset

CEDAR is a well-known publicly available database, stands for Center for Document Analysis and Recognition. The database consists 2640 signatures, which are collected from 55 contributors. Each contributor contributed 24 genuine signature samples. 20 randomly chosen contributors are allowed to practice the patterns of genuine signatures and from each of them 24 forge samples are collected. Finally the database contains 48 signature samples, among that 24 genuine and 24 forge. Experiments were started with set-1 along with set-3 where 10 genuine and 10 skilled forge signatures are considered for training and for testing around 14 genuine signatures with 14 skilled forgery signatures are considered. Experiments were continued with Set-2 and set-4 with numerical figures such as first 14 genuine signature samples with first 14 skilled forge signature samples were considered for training and tested against remaining both 9 genuine signature samples with 9 skilled forgery signatures. To overcome from the effect of randomness 5 times experiments are repeated for set-2 and set-4 finally average result is considered. The results are tabulated in Table 2, where FAR and FRR are the metrics.

Table 2. Obtained Experimental Results on CEDAR Dataset:

Experiment Set-up	Accuracy	FRR	FAR
Set-1	94.74	4.0	5.6
Set-2	92.72	9.29	5.25
Set-3	90.58	11.94	6.88
Set-4	91.05	10.60	9.2

From the literature, we found the experimental results of few well known approaches on CEDAR dataset. We made an compararision analysis presented in Table 3 shows the improvements in accuracy by the proposed approach.

Table 3. Experimentation results on CEDAR dataset

Proposed by	Classifier	Accuracy	FAR	FRR
Kalera et al. [5]	PDF	78.50	19.50	22.45
Chen and Shrihari [1]	DTW	83.60	16.30	16.60
Kumar et al. [8]	SVM	88.41	11.59	11.59
Pattern spectrum [14]	EMD	91.06	10.63	9.4
Surroundedness [8]	MLP	91.67	8.33	8.33
Inter point envelop [6]	SVM	92.73	6.36	8.18
Proposed Approach	**SVM**	**94.74**	**4.0**	**5.6**

Experiments on MUKOS Dataset

MUKOS (Mangalore University Kannada Offline Signature) database is a regional language Kannada dataset. The database consists 1350 signatures collected from 30 Signer. There are 30 genuine signature samples and 15 skilled forgeries from each Signer. Contributors had written the signature with black ink pen on a white paper representing 14 boxes. After acquiring genuine signatures, the contributors are allowed to practice the genuine signature pattern for a time interval and contributed skilled forgeries. All experiments are conducted in terms of set-1, set-2, set-3 and set-4. In set-1 and set-2 there are 10 genuine and 10 skilled forgeries are chosen for training and 15 genuine signature and 5 skilled forgeries are chosen for testing. Experiments on set-2 as well as on set-4 we considered 15 genuine signatures and 15 skilled forgeries for training and 15 genuine signatures and 15 skilled forgeries are tested against. Repeat the set-2 and also set-4 experiments 5 times to overcome from the effect of randomness and considering the average result. The metrics FAR and FRR used to measure the accuracy of the our approach. Results obtained for MUKOS dataset is tabulated below (Tables 4 and 5).

Table 4. Experimentation on MUKOS database.

Experimental Set-up	Accuracy	FRR	FAR
Set-1	97.73	0.53	4
Set-2	97.25	0.58	4.9
Set-3	97	1.2	4.8
Set-4	98.6	0.8	2.0

Table 5. Experiments outcome on MUKOS database- A comparative analysis

Method	Classifier	Accuracy	FAR	FRR
Shape based eigen signature [13]	Euclidean distance	93.00	11.07	6.40
Pattern spectrum [14]	EMD	97.39	5.6	8.2
Proposed Approach	**SVM**	98.6	0.8	2.0

5 Conclusion

In this work we tried to explore Multi-scale Local Binary Patterns for offline signature verification. The Local Binary Patterns is a well known powerful texture descriptor which captures local features. We made an attempt to explore Multi-scale Local Binary Pattern for the purpose of offline signature verification. The MSLBP can effectively represent both local information (micro structure) as well as global information of an image. We conducted extensive experiments on CEDAR database, which is a publicly available well known database. We also conducted experiments on MUKOS data corpus, which is a regional Kannada language offline signature database. Further we made a comparative analysis wherein we found that, the proposed approach performs better than some of the well known approaches on the CEDAR dataset. It is observed experimentally that, the implementation of MSLBP is simple yet gives high accuracy in signature verification task.

Acknowledgment. We acknowledge Bharathi R.K for providing a regional language kannada dataset namely MUKOS(Mangalore University Kannada Off-line Signature) dataset.

References

1. Chen, S., Srihari, S.: Use of exterior contours and shape features in off-line signature verification. In: Proceedings 2005 Eighth International Conference on Document Analysis and Recognition, pp. 1280–1284. IEEE (2005)
2. Hadjadji, B., Chibani, Y., Nemmour, H.: An efficient open system for offline handwritten signature identification based on curvelet transform and one-class principal component analysis. Neurocomputing **265**, 66–77 (2017)

3. Hafemann, L.G., Sabourin, R., Oliveira, L.S.: Learning features for offline handwritten signature verification using deep convolutional neural networks. Pattern Recogn. **70**, 163–176 (2017)
4. Hsu, C.W., Chang, C.C., Lin, C.J.: A Practical Guide to Support Vector Classification (2003)
5. Kalera, M.K., Srihari, S., Xu, A.: Offline signature verification and identification using distance statistics. Int. J. Pattern Recognit. Artif. Intell. **18**(07), 1339–1360 (2004)
6. Kumar, M.M., Puhan, N.: Inter-point envelope based distance moments for offline signature verification. In: 2014 International Conference on Signal Processing and Communications (SPCOM), pp. 1–6. IEEE (2014)
7. Kumar, R., Kundu, L., Chanda, B., Sharma, J.D.: A writer-independent off-line signature verification system based on signature morphology. In: Proceedings of the First International Conference on Intelligent Interactive Technologies and Multimedia, IITM 2010, pp. 261–265. ACM, New York (2010)
8. Kumar, R., Sharma, J., Chanda, B.: Writer-independent off-line signature verification using surroundedness feature. Pattern Recogn. Lett. **33**(3), 301–308 (2012)
9. Ojala, T., Pietikainen, M., Maenpaa, T.: Multiresolution gray-scale and rotation invariant texture classification with local binary patterns. IEEE Trans. Pattern Anal. Mach. Intell. **24**(7), 971–987 (2002)
10. Radhika, K., Gopika, S.: Online and offline signature verification: a combined approach. Procedia Comput. Sci. **46**, 1593–1600 (2015)
11. Rassem, T.H., Khoo, B.E., Makbol, N.M., Alsewari, A.A.: Multi-scale colour completed local binary patterns for scene and event sport image categorisation. IAENG Int. J. Comput. Sci. **44**(2), 197–211 (2017)
12. Serdouk, Y., Nemmour, H., Chibani, Y.: New off-line handwritten signature verification method based on artificial immune recognition system. Expert Syst. Appl. **51**, 186–194 (2016)
13. Shekar, B.H., Bharathi, R.K.: Eigen-signature: a robust and an efficient offline signature verification algorithm. In: 2011 International Conference on Recent Trends in Information Technology (ICRTIT), pp. 134–138, June 2011
14. Shekar, B.H., Bharathi, R.K., Pilar, B.: Local morphological pattern spectrum based approach for off-line signature verification. In: Maji, P., Ghosh, A., Murty, M.N., Ghosh, K., Pal, S.K. (eds.) PReMI 2013. LNCS, vol. 8251, pp. 335–342. Springer, Heidelberg (2013). https://doi.org/10.1007/978-3-642-45062-4_45
15. Yılmaz, M.B., Yanıkoğlu, B.: Score level fusion of classifiers in off-line signature verification. Inf. Fusion **32**, 109–119 (2016)
16. Yilmaz, M.B., Yanikoglu, B., Tirkaz, C., Kholmatov, A.: Offline signature verification using classifier combination of HOG and LBP features. In: 2011 international joint conference on Biometrics (IJCB), pp. 1–7. IEEE (2011)
17. Zois, E.N., Alewijnse, L., Economou, G.: Offline signature verification and quality characterization using poset-oriented grid features. Pattern Recogn. **54**, 162–177 (2016)

Ekush: A Multipurpose and Multitype Comprehensive Database for Online Off-Line Bangla Handwritten Characters

AKM Shahariar Azad Rabby(✉), Sadeka Haque, Md. Sanzidul Islam, Sheikh Abujar, and Syed Akhter Hossain

Department of Computer Science and Engineering, Daffodil International University, Dhanmondi, Dhaka 1205, Bangladesh
{azad15-5424,sadeka15-5210,sanzidul15-5223,sheikh.cse}@diu.edu.bd, aktarhossain@daffodilvarsity.edu.bd

Abstract. Ekush the largest dataset of handwritten Bangla characters for research on handwritten Bangla character recognition. In recent years Machine learning and deep learning application-based researchers have achieved interest and one of the most significant application is handwritten recognition. Because it has the tremendous application such in Bangla OCR. Also, Bangla writing script is one of the most popular in the world. For that reason, we are introducing a multipurpose comprehensive dataset for Bangla Handwritten Characters. The proposed dataset contains Bangla modifiers, vowels, consonants, compound letters and numerical digits that consists of 367,018 isolated handwritten characters written by 3086 unique writers which were collected within Bangladesh. This dataset can be used for other problems i.e.: gender, age, district base handwritten related research, because the samples were collected include verity of the district, age group and the equal number of male and female. It is intended to fabricate acknowledgment technique for hadn written Bangla characters. This dataset is unreservedly accessible for any sort of scholarly research work. The Ekush dataset is trained and validated with EkushNet and indicated attractive acknowledgment precision 97.73% for Ekush dataset, which is up until this point, the best exactness for Bangla character acknowledgment. The Ekush dataset and relevant code can be found at this link: https://github.com/ShahariarRabby/ekush.

Keywords: Bangla handwritten · Data science · Machine learning · Deep learning · Computer vision · Pattern recognition

1 Introduction

There are large numbers of research have been introduced for the handwritten recognition of Latin, Chinese, and Japanese text and characters.On the other

---

The original version of this chapter was revised: The names of the two Authors have been corrected as "AKM Shahariar Azad Rabby" and "Syed Akhter Hossain". The correction to this chapter is available at https://doi.org/10.1007/978-981-13-9187-3_67

K. C. Santosh and R. S. Hegadi (Eds.): RTIP2R 2018, CCIS 1037, pp. 149–158, 2019.
https://doi.org/10.1007/978-981-13-9187-3_14

hand, relatively few research has been done on Bangla Handwritten recognition due to its compliance of Bangla characters and limitation of Bangla datasets. Bangla language has 50 basic characters, 10 modifiers, 10 numerals and more than 300 compound characters. But till now Bangla has no complete dataset that contains all of these characters. So that, recognition of Bangla is at the beginning period contrasted with the techniques for recognition of Latin, Chinese, and Japanese text.

Handwritten character recognition is an imperative issues because of its numerous implementation as Optical Character Recognition (OCR), office robotization, bank check mechanization, postal computerization and also human-PC connections. Bangla Handwritten framework recognition is an extraordinary test since still a long ways behind the human acknowledgment capacity. In this manner, numerous datasets in handwritten recognition area have been accumulated and utilized in different dialects and applications are contrasted with our examination result. There are datasets in CEDAR [1] English words and characters, English sentence dataset IAM [2], Indian [3] for handwritten recognition applications. However, a few studies are attested on handwritten characters of Bangla scripts. Though Bangla is the seventh most spoken language in the world by population and major language in the Indian subcontinent as well as first language in Bangladesh. Though, several studies have been dealing for Bangla handwriting recognition, a robust model for Bangla numerals and Characters classification is still due. One of the main reasons for that the lack of a single comprehensive dataset which covers different types of the Bangla character. There are existing data sets which cover either just the Bangla numerals or just the Bangla characters without modifiers. While it is possible to combine them to form a unified data set, the inconvenience faced by the researchers stems from the lack of consistency in the data presentation of the different datasets. Ekush is the first of a chain of dataset being introduced which aims to Bangla handwriting related research.

In order to support research on recognition for Bangla handwriting, Machine learning, and deep learning applications, we have collected samples (isolated characters) that contain 122-character samples of 4 categories (10 modifiers, 11 vowels and 39 consonants, 52 frequently use compound letter and 10 numeral digit) written by 3086 unique persons include verity of the district, age group and the equal number of male and female. Figure 1 showing the sample of different character images of the Ekush dataset.

2 Literature Review

There are three open access datasets available for Bangla characters, these are the BanglaLekha-Isolated [4], the CMATERdb [5], and the ISI [3]. But every dataset has some drawback. BanglaLekha-Isolated dataset contains 166,105 squared images (while preserving the aspect ratio of the characters), where each sample have 84 different Bangla characters. Those characters have 3 categories which is 10 numerals, 50 basic characters, and 24 compound characters. Two others datasets CMATERdb and ISI where CMATERdb also has 3 categories for basic characters, numerals and compound characters and ISI dataset has two different categories for basic characters and numerals. A comparison between Ekush and with that three popular sources of Bangla handwriting related datasets (BanglaLekha-Isolated dataset, CMATERdb, and the ISI Handwriting dataset) are given in Table 1.

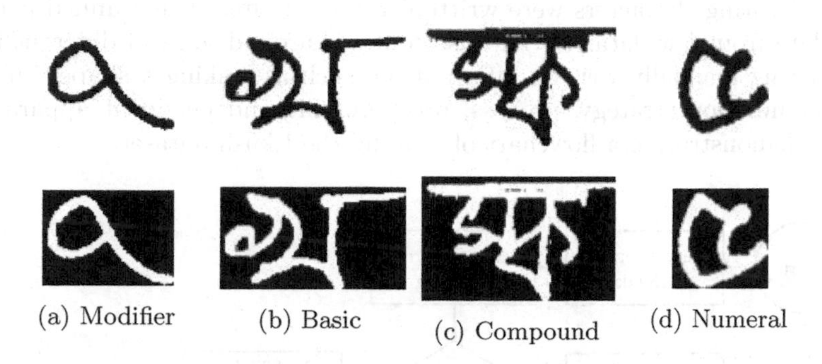

(a) Modifier (b) Basic (c) Compound (d) Numeral

Fig. 1. Sample images of Ekush dataset

Table 1. Number of images in different datasets

Dataset name	Modifiers	Basic characters	Compound characters	Numeral	Total
ISI	None	30,966	None	23,299	34,256
CMATERdb	None	15,103	42,248	6,000	63,351
BanglaLekha-Isolated	None	98,950	47,407	19,748	166,105
Ekush	30,667	154,824	150,840	30,687	367,018

The Ekush dataset consists of 367,018 images that contain 122 classes and it became the largest dataset for Bangla characters yet. For other language like DEVANAGARI there are some work like "Relative positioning of stroke based clustering: a new approach to on-line handwritten devnagari character recognition" [6] where author used to WCACOM table to collect the data. Same authour has other work like "Character recognition based on non-linear multi-projection profiles measure" [7], They did their experiment on several languages like Roman, Jap-anese, Katakana, Bangla etc. There proposed method used dynamic programming to match the nonlinear multiprojection profile which is used to recognize hand-written characters. "Radon transform" [8] used to produce the nonlinear multi-projection profile. On other paper "Character Recognition based on DTW–Radon" [9] author tries to recognize characters using the "DTW algorithm" [10] at every projecting angle.

3 The Ekush Dataset

Ekush is a dataset of Bangla handwritten characters which can be used as multi-purpose way. Ekush dataset of isolated Bangla handwritten characters structured and organized data was collected from 3086 peoples covering university, school, college students, where approximately 50% 1510 male and 50% 1576 female. The handwriting characters were written in a form after then scanned it to get image data from raw data. We likewise centered around some of different issues for gathering manually written information, such as making a shape, information accumulation strategy, process, programming and pertinent apparatuses. Figure 2 demonstrating a flowchart of making the Ekush dataset.

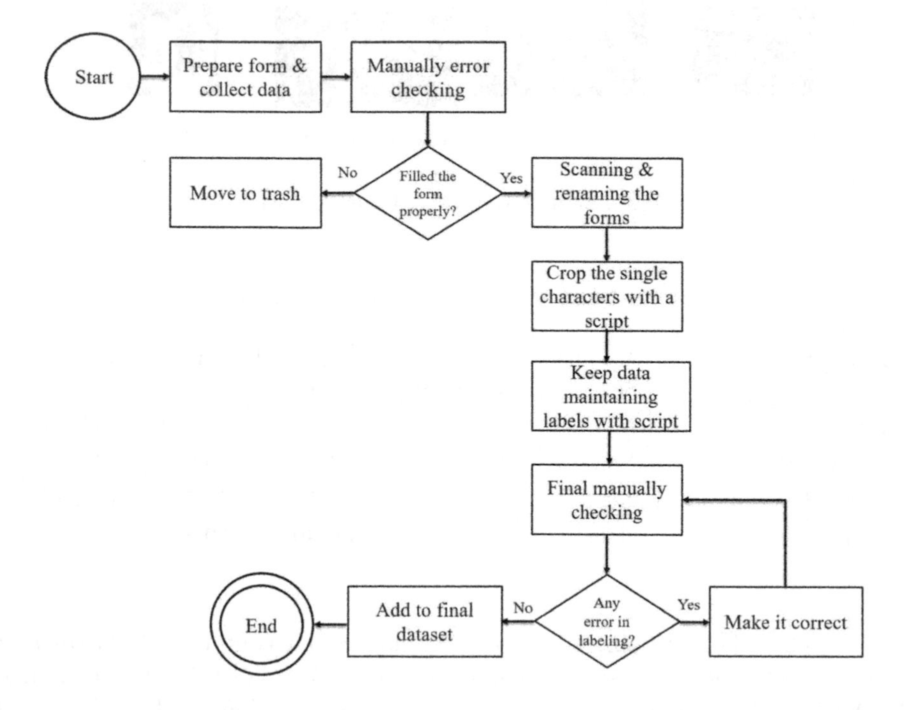

Fig. 2. A flow diagram of "Ekush" dataset

3.1 From Preparation and Collection

Initially creating the form was followed some steps as cell created with equal size, which letter helps to separate the character automatically and made border which help to crop the form equal size. There are lots of compound characters in Bangla language, so we selected most frequently used compound characters.

Than peoples were voluntarily filled up the form which letter scanned for further process.

3.2 From Processing

To process the form we follow the A Universal Way to Collect and Process Handwritten Data for Any Language [11]. During scanning the form, a big black boundary was created to detect the biggest couture. To find that we use canny edge detection algorithm. as same shape with same angle.

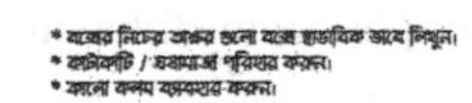

Fig. 3. Sample form for Ekush

After skew correction, first cropped the images by row where 12 characters contains each row. After that cropped the row by column. That cropping was separated all the characters and stored all the images in folder which was labeled by the character name. This process was separate all the 120 characters.

The OTSU [12] algorithm find the best threshold value instead of the specified thresh value. In our method used Gaussian blur for smoothing the images.

We removed all extra bit of information to provide efficient output and stored them. Then Adding white padding to preserve the aspect ratio of the images.

This stored image then inverted into character into white and background as black. All the processed can be done atomically by our designed GUI which able to process 100 scanned images per minute. A filled template used for collection of data is shown In Fig. 3.

3.3 Constructing Ekush

Ekush dataset of Bangla handwritten characters were collected from 3086 peoples where 50% male and 50% female. Each individual developed 120 characters including 50 Bangla basic characters, 50 frequent Bangla compound characters, 10 modifiers, 10 digits. But after processing the dataset we found more than 55 type compound characters because of Bangla compound characters are confusing and some of are similar. So when the writer wrote they made mistake to understand which character actually exist in the form. Then we manually checked and found 2 compound character most of the people were mistake to understand the proper character. So after that, we picked those 2 characters and added in our dataset and finally got 52 types of compound character. These writers were selected from various age, gender, and educational background groups. Figure 4 showing a bar chart of data samples from different ages base on gender.

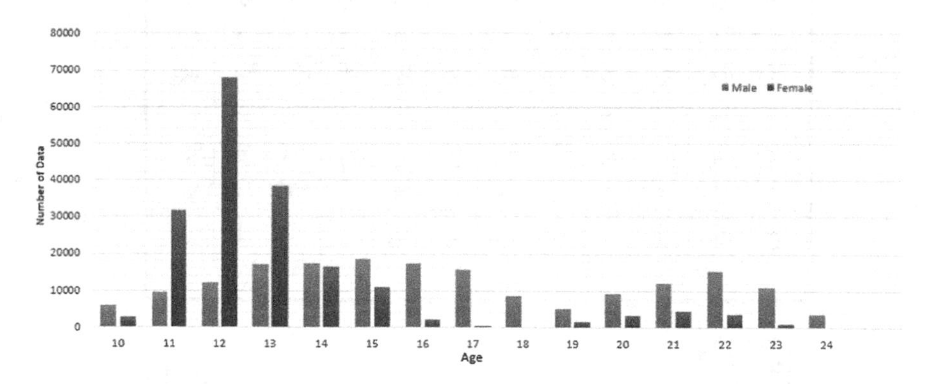

Fig. 4. A bar chart of data samples from different ages base on gender

The collected dataset is preprocessed in three different format:

1. Foreground and background are inverted so that images have a black background with the letter drawn in white.

2. Foreground and background filled with white so that images have a white background with the letter drawn in black
3. Removed noise and smoothing attempted using the Trash holding and Gaussian blur filter. The proceed dataset is further filtered and after necessary smoothing different format are created including CSV.

The Ekush dataset will available on the various format, depending on the user desired applications as well as will be available without extra information on character images, and with padding added to preserve aspect ratio and also in CSV format. Table 2 showing the details of Ekush dataset base on gender.

Table 2. Number of character in Ekush by gender

Gender	Modifiers	Basic characters	Compound characters	Numeral	Total	Total in Ekush
Female	15580	78615	76912	15622	186729	367018
Male	15087	76209	73928	15065	180289	

3.4 Visual Representation of the Ekush Dataset

The class breakdown of Ekush dataset shows in Fig. 5, where see that Fig. 5(a), (b) showing Bangla modifiers and digits class as well as seeing that all class are almost equally distributed. The number of the Ekush dataset Modifier and Digits has 30769 and 30687 respectively. From Fig. 5(c), (d) most of the classes are equally distributed, but there are few classes exist where number of images of that characters are not averagely equal to compare the other classes. While writing the compound characters people made mistake to write the proper character. Then we relabeled some compound character also deleted other as well as collected some incorrect compound characters and added them to the dataset in two new class which we are seeing at the last to Fig. 5(d) Compound character bar chart.

4 Data Labeling

The extracted image is saved as the JPEG format and each image has unique ID or name that represent writer gender, hometown, age, education level and serial number. In order to storing the writer information which helps to identify that person using an ID. So that this dataset not only use for handwritten recognition but also help to predict a person gender, age and his or her location as well as it can help investigators focusing more on a certain category of suspects and forensic purposes. This id or name fixed according to the following criteria, Its first one digit indicates writer gender. If it is 0 which means writer was male and

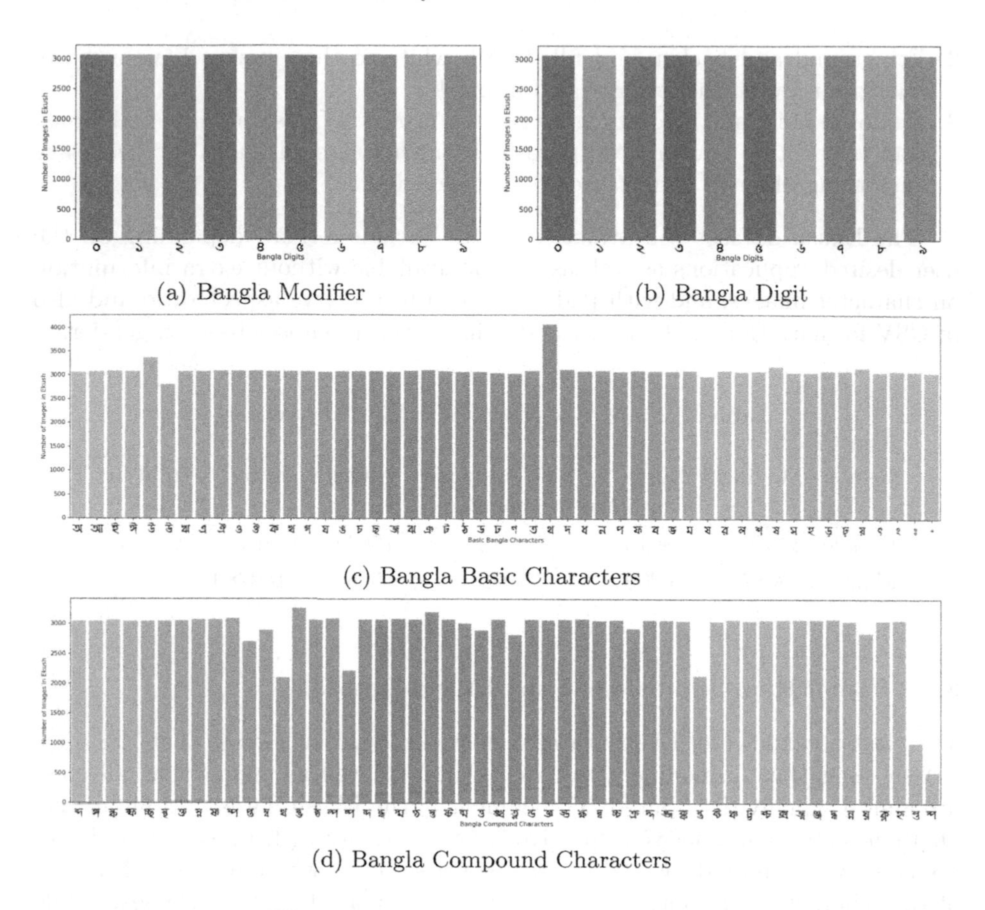

(a) Bangla Modifier

(b) Bangla Digit

(c) Bangla Basic Characters

(d) Bangla Compound Characters

Fig. 5. Visual represent of Ekush dataset

1 means writers were female. After that have writer home district names First 3 or 4 letter, the next one represents age and then their education or occupation level (0 means primary level, 1 high school level, 2 means college level, 3 means university and 4 means other occupation) and the last one is the serial number. And that information are separated by an underscore (_). Here an example.

0_Dha_20_3_00052

Here the first digit one so it was written by a male writer and he is from Dhaka district the next one is 20 which means his age 20 and he is a university student and the last one is the serial number of male data.

5 Possible Uses of Ekush Dataset

This dataset can be used for handwritten character recognition, Bangla OCR, Machine learning and Deep learning base research fields. The prediction of age, gender, the location from handwriting is a very interesting research field. These information's can also be used for forensic purposes, where it will help investigators focusing more on a certain category of suspects.

6 EkushNet

To test the performance of Ekush dataset we built a multilayer CNN model EkushNet [13] in order to classifying Bangla Handwritten Characters. Figure 6 showing the CNN architect.

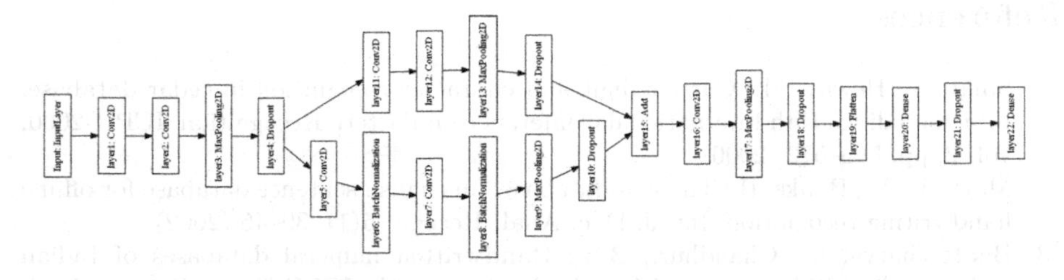

Fig. 6. The architect of EkushNet

We trained the EkushNet and got 96.90% accuracy on training and 97.73% accuracy on validation of Ekush datasets. After train the model we cross validate that with CMATERdb dataset and got satisfactory accuracy of 95.01%. Figure 7 showing the training and validation loss and accuracy on Ekush Dataset.

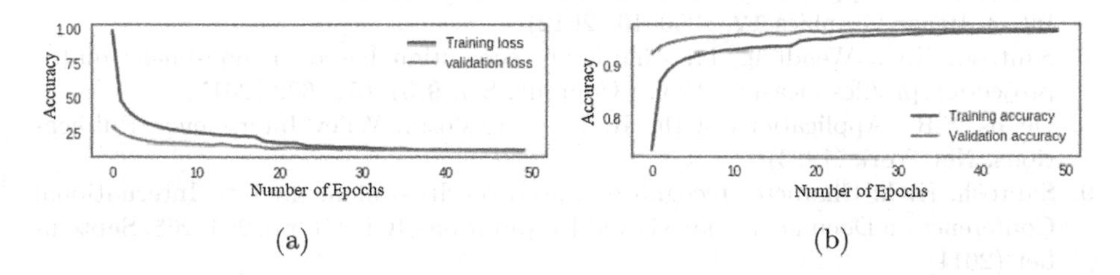

Fig. 7. (a) Training and validation loss. (b) Training and validation accuracy of Ekush

7 Conclusion and Future Work

This research formulate a diverse repository containing primarily for the Computer Vision and NLP research and called the dataset Ekush. In future, we will extend our dataset including all kinds of compound characters. Also, we will make a website where the user can download the form and upload the scan copy which will automatically process those data and added to the dataset after verifying. That website also gives the user to search and download character data by age, gender and districts.

Acknowledgement. I would like to express my deepest appreciation to all those who had provided us the possibility to complete this research under the Daffodil International University. A special gratitude we give to our university and Daffodil International University NLP and Machine Learning Research LAB for their instructions and support. Furthermore, I would also like to acknowledge that, this research partially supported by Notre Dame College, Mirpur Bangla School, Dhanmondi Govt. Girls' High School, Shaheed Bir Uttam Lt. Anwar Girls' College and Adamjee Cantonment Public School who gave permission to collect data from their institution. Any errors are our own and should not tarnish the reputations of these esteemed persons.

References

1. Singh, S., Hewitt, M.: Cursive digit and character recognition in cedar database. In: Proceedings 15th International Conference on Pattern Recognition, ICPR-2000, vol. 2, pp. 569–572 (2000)
2. Marti, U.-V., Bunke, H.: The iam-database: an english sentence database for offline handwriting recognition. Int. J. Doc. Anal. Recogn. **5**(1), 39–46 (2002)
3. Bhattacharya, U., Chaudhuri, B.B.: Handwritten numeral databases of Indian scripts and multistage recognition of mixed numerals. IEEE Trans. Pattern Anal. Mach. Intell. **31**(3), 444–457 (2009)
4. Biswas, M., et al.: Banglalekha-isolated: a multi-purpose comprehensive dataset of handwritten bangla isolated characters. Data in Brief, 12, 103–107 (2017)
5. Sarkar, R., Das, N., Basu, S., Kundu, M., Nasipuri, M., Basu, D.K.: Cmaterdb1: a database of unconstrained handwritten Bangla and Bangla-English mixed script document image. Int. J. Doc. Anal. Recogn. (IJDAR) **15**(1), 71–83 (2012)
6. Santosh, K.C., Nattee, C., Lamiroy, B.: Relative positioning of stroke-based clustering: a new approach to online handwritten devanagari character recognition. Int. J. Image Graphics **12**, 1250016 (2012)
7. Santosh, K.C., Wendling, L.: Character recognition based on non-linear multi-projection profiles measure. Front. Comput. Sci. **9**(5), 678–690 (2015)
8. Deans, S.R.: Applications of the Radon Transform. Wiley Interscience Publications, New York (1983)
9. Santosh, K.C.: Character recognition based on dtw-radon. In: 11th International Conference on Document Analysis and Recognition - ICDAR, pp. 264–268, September (2011)
10. Liberman, M., Kruskall, J.B.: The symmetric time warping algorithm: From continuous to discrete. In: Time Warps, String Edits and Macromolecules: The Theory and Practice of String Comparison, pp. 125–161. Addison-Wesley, Boston (1983)
11. Shahariar Azad Rabby, A.K.M., Haque, S., Shahinoor, S.A., Abujar, S., Hossain, S.A.: A universal way to collect and process handwritten data for any language. Procedia Comput. Sci. **143**, 502–509 (2018). 8th International Conference on Advances in Computing & Communications (ICACC-2018)
12. Otsu, N.: A threshold selection method from gray-level histograms. IEEE Trans. Syst. Man Cybern. **9**(1), 62–66 (1979)
13. Shahariar Azad Rabby, A.K.M., Haque, S., Abujar, S., Hossain, S.A.: Ekushnet: using convolutional neural network for bangla handwritten recognition. Procedia Comput. Sci. **143**, 603–610 (2018). 8th International Conference on Advances in Computing & Communications (ICACC-2018)

ShonkhaNet: A Dynamic Routing for Bangla Handwritten Digit Recognition Using Capsule Network

Sadeka Haque(✉), AKM Shahariar Azad Rabby(✉), Md. Sanzidul Islam, and Syed Akhter Hossain

Department of Computer Science and Engineering, Daffodil International University, Dhanmondi, Dhaka 1205, Bangladesh
{sadeka15-5210,azad15-5424,sanzidul15-5223}@diu.edu.bd, aktarhossain@daffodilvarsity.edu.bd

Abstract. In the present world, one of the most interesting topics is Handwritten Recognition due to its academic and commercial interest in different research fields. But deal with it a little bit tough because of different size and style. There are many works have been accomplished base in handwritten recognition including Bangla. Here proposed a model which is classified Bangla handwritten numeral using capsule net (a new type of neural network represents activity vector as parameters). The Model is trained and valid with ISI handwritten database [1], BanglaLekha Isolated [2], CMATERdb 3.1.1 [3] and all database together that was achieved 99.28% validation accuracy on ISI handwritten character database, 97.62% validation accuracy on BanglaLekha Isolated, 98.33% validation accuracy on CMATERdb 3.1.1 dataset and 98.90% validation accuracy combination mixed dataset. This model gives satisfactory recognition accuracy compared to other existing models.

Keywords: Bangla numeral · Bangla handwritten recognition · Pattern recognition · Capsule · CapsNet

1 Introduction

In recent decades, to make all the documents digital and portable is one of the greatest challenges. Because older form and information that was written by hand. So, the problem lies to convert them to follow tradition system on manual typing to copy the existing archive. Also, it takes lot of time to analyze the documents. Handwriting recognition can handle this problem and making all the document digital and reliable in short time. However, there are one of the most important challenges is to recognize handwritten characters because every person has a unique style to write and the scripts have different size. Bangla script contains 50 basic characters, 10 numerals and more than 300 compound characters having a very complex arrangement. About 300 million people overall

The original version of this chapter was revised: The names of the two Authors have been corrected as "AKM Shahariar Azad Rabby" and "Syed Akhter Hossain". The correction to this chapter is available at https://doi.org/10.1007/978-981-13-9187-3_67

K. C. Santosh and R. S. Hegadi (Eds.): RTIP2R 2018, CCIS 1037, pp. 159–170, 2019.
https://doi.org/10.1007/978-981-13-9187-3_15

in the world. It is the first language of Bangladesh and second popular language in India. But the research held on Bangla handwriting is very few compared with other languages like English, Arabic, Hindi, Chinese etc. Recognition of handwritten character mostly found other languages compare with Bangla for example Roman script [4] related to English, Chinese [5], Japanese [6], Korean [7]. Nowadays researchers want to make Bang-la OCR which has many application Bangla HLDR, such as, Bangla traffic number plate recognition, automatic postal code identification, extracting data from hard copy forms, automatic ID card reading, automatic reading of bank cheque and digitization of documents etc. Our goal to make a model that can recognize Bangla Handwritten digits using capsule net. Though several years CNN has been very successful for recognizing the handwritten characters, the performance of Capsule gains better when the dataset is highly overlapping [8]. Capsule net works very well for handwritten character recognition for its distinguished features. Capsule net is able to classify image and visual pattern from image pixel with less preprocessing, this ability adds a new dimension in image processing. The output of the capsule sent to its applicable parents in the previous layer to perform a dynamic routing. A capsule is built with many small groups of neurons which came from each layer. An active capsule is a correspondent to a parse tree of each node [9]. Figure 1 showing Bangla digits.

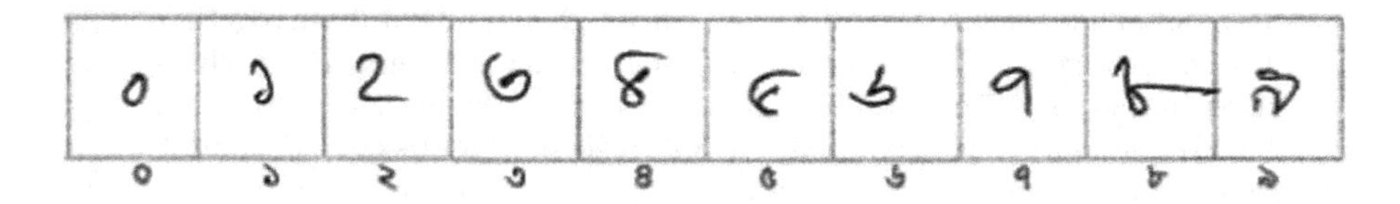

Fig. 1. Example of Bangla digits

2 Literature Review

Capsules perform pretty good to recognize English digits, but till now there is no work for Bangla language. In past studies, by using different supervised machine learning algorithm, some works are done based on Bangla handwriting digit recognition *"Handwritten numeral databases of Indian scripts and multistage recognition"* [1] of mixed numerals the main feature for their model includes matra, the upper part of the character, vertical line and double vertical line and for the MLF classifier, the feature is constructed from the stroke feature of the characters. Another work is *"Character recognition based on non-linear multi-projection profiles measure"* [21], They did their experiment on several languages like Roman, Japanese, Katakana, Bangla etc. There proposed method used dynamic programming to match the nonlinear multi-projection profile which is used to recognize handwritten characters. *"Radon transform"* [22] used to produce the nonlinear multi-projection profile. On other paper *"Character Recognition based on DTW–Radon"* [23] author tries to recognize characters using the *"DTW algorithm"* [24] at every projecting angle. *"Automatic Recognition of Unconstrained Off-Line Bangla Handwritten Numerals"* [10] is also one of them, their proposed method is to extract features from a concept called water reservoir. They used it for postal automation. Another work is *"Handwritten Bangla Digit Recognition Using Deep Learning"*. [11] They use a CNN model that achieves 98.78% accuracy on CMATERdb 3.1.1 datasets. Another work of author Shahariar Azad Rabby et al. [12] to recognize Bangla digit which accuracy trained and tested with ISI handwritten character database, BanglaLekha Isolated and CAMTERdb 3.1.1 and achieve 99.74%, 98.93%, 99.42% validation accuracy respectively and 99.43% validation accuracy on all three dataset combined data.

3 Propose Methodology

The proposed system uses Capsules which has many phases to accomplish the model such as Dataset preparation, model training etc.

3.1 Dataset

In our proposed model we used three datasets which are ISI handwritten character dataset, CMATERdb 3.1.1 and BanglaLekha-Isolated datasets. In ISI database has total 23392 numerical images, The CMATERdb 3.1.1 has the total 6000-digit images, BanglaLekha-Isolated datasets contain 19748 images where each dataset split as train, test, and validation 80%, 10%, and 10% respectively. After splitting each of the datasets as train, test, and validation, then combined as all training, testing and validation datasets. The training dataset contains 39279 (80% of total data) images, the testing dataset contains 4910 (10% of total data) images and 4911 validation datasets (10% of total data). BanglaLekha and ISI images have different pixel size with a smooth edge. In CMATERdb the images are 32 × 32 pixels, noise-free, almost correct labeling with no overwriting. Some of the BanglaLekha Isolated images have incorrect labeling and some images are overwritten with others. BanglaLekha Isolated and ISI images are grayscale and PNG format and the images of BanglaLekha Isolated are inverted. Figure 2 showing an example of different datasets and Table 1 shows the dataset size.

Fig. 2. Example of different dataset

Table 1. Number of digits in different datasets

Dataset	Numeric digits
CMATERdb 3.1.1	6000
ISI	23299
BanglaLekha Isolated	19748
Mixed dataset	49047

3.2 Dataset Preparation

Preparing a dataset is the first step in building any model. For Bangla, we build a larger dataset of 10 digits. For ISI and CMATERdb dataset are converted them to 1 channel grayscale to reduce computational expense. But we inverted to customized the datasets where black for background and white for foreground numeral digit. In BanglaLekha-Isolated we corrected some labeling and removed a few incorrect images. This dataset already contains inverted images. Then all images were resized to 28×28 pixel.

Then converted these datasets image into a 784 + (1 label) D matrix and storing them into a CSV file to make the computation efficient. Also, a MinMax (1) normalization was done on the dataset to reduce the effect of illumination differences. Moreover, the CNN converge faster on $[0 \ldots 1]$ than on $[0 \ldots 255]$.

$$Z_i = \frac{X_i - \text{minimum}(X)}{\text{maximum } (X) \ - \text{minimum}(X)} \tag{1}$$

Then convert the 10 labels of Bangla digit into one hot encoding.

[1, 0, 0, 0, 0, 0, 0, 0, 0, 0] = ০
[0, 1, 0, 0, 0, 0, 0, 0, 0, 0] = ১
[0, 0, 1, 0, 0, 0, 0, 0, 0, 0] = ২
[0, 0, 0, 1, 0, 0, 0, 0, 0, 0] = ৩
[0, 0, 0, 0, 1, 0, 0, 0, 0, 0] = ৪
[0, 0, 0, 0, 0, 1, 0, 0, 0, 0] = ৫
[0, 0, 0, 0, 0, 0, 1, 0, 0, 0] = ৬
[0, 0, 0, 0, 0, 0, 0, 1, 0, 0] = ৭
[0, 0, 0, 0, 0, 0, 0, 0, 1, 0] = ৮
[0, 0, 0, 0, 0, 0, 0, 0, 0, 1] = ৯

Then those 784 D matrix converted to 28×28 pixel images. Now the data is ready for making the model.

3.3 Capsule Network Between Dynamic Routing

Capsule network is one kind of neural networks tries to perform inverse graphics. A capsule is a function that tries to predict the presence and the instantiation parameters of a particular object of a given location. Capsule network also uses translated replied of learned feature detectors as CNN, though the scaler-output feature of CNN replacing by vector output of Capsule. The capsule net is better than convolutional neural networks because the Internal data representation of a CNN does not consider important spatial hierarchies between simple and complex objects. However capsule net tries to fix this problem and find a global solution.

In our network, we use dynamic routing which is a process that helps to change the configuration of a network which result makes automatic updates into the routing table. The similar approach applied the authors of *"Dynamic Routing Between Capsules"* [8] in that strong connection between capsules at different layers are encouraged. Procedure 1 showing the Routing algorithm from the *"Dynamic Routing Between Capsules"* paper.

Procedure 1: Routing algorithm

1: procedure ROUTING ($\hat{u}_{j|i}, r, l$)
2: for all capsule i in layer l and capsule j in layer $(l+1)$: $b_{ij} \leftarrow 0$.
3: for r iterations do
4: for all capsule i in layer l: $c_i \leftarrow \text{softmax}(b_i)$ SoftMax computes Eq. 5
5: for all capsule j in layer $(l+1)$: $s_j \leftarrow \Sigma_i c_{ij} \hat{u}_{j|i}$
6: for all capsule j in layer $(l+1)$: $v_j \leftarrow \text{squash}(s_j)$ squash computes Eq. 3
7: for all capsule i in layer l and capsule j in layer $(l+1)$: $b_{ij} \leftarrow b_{ij} + \hat{u}_{i|j} . v_j$
return v_j

The first line represents all procedure which takes capsule, and l represents lower level capsule and u and r represent output and routing iteration respectively. The last line return produce output of high level capsule which is represent by v_j. This algorithm uses forward pass calculation of the network. In the second line a new coefficient b_{ij} which is a temporary value that will be iteratively updated, and completing the procedure the value of b_{ij} stored in c_{ij}. Initially $b_{ij} = 0$. In line 3 use a loop that is make a boundary from steps 4–7 will be repeated r times. Line 4 calculates the value of lower level capsule vector c_i which is all routing weights. Here use an activation function SoftMax due to it will help that each weight c_{ij} is a non-negative number and their sum equals to one. After calculated all weights of c_{ij} for all lower level capsules, move on to line 5, where higher level capsules start. Here calculates a linear combination of input vectors, c_{ij} which is determined in the previous step. Then scaling down input vectors and adding them together, which produces output vector s_j and higher-level capsules calculation done. Line 6 last step are passed through the squash nonlinearity from vector. And 7 calculate Given a capsule i on layer l, capsule j on layer $l+1$, the routing weight b_{ij} updated using the formula $b_{ij} \leftarrow b_{ij} + \hat{u}_{i|j} . v_j$.

3.4 Proposed Model

This proposed model uses a multilayer neural network, mainly which are Convolution layer, Primary caps, digit caps and output capsule length layer. Here a total number of parameters of 6,815,744.

Convolution layer (Conv1) has 256 filters, 9×9 kernels with stride of 1×1, "valid" padding with the activation function ReLU (2). This layer reshapes 28×28 pixel intensities which are used as inputs to the primary capsules. Each kernel in a convolutional layer has bias term that is 1 so Conv1 has the total number of trainable parameters $((9 \times 9) + 1) \times 256 = 20992$.

$$\text{ReLU}(X) = \text{MAX}(0, X) \tag{2}$$

The primary capsule is the lower level capsule which has multidimensional entities. It has three-layer, Primary Capsule Conv which take Conv1 convert pixel as input and has 32 × 8 filter (32 primary capsule and 8D convolution unit) with a 9 × 9 kernel and a stride of 2 × 2. A total number of parameters are 5308672. Then reshape layer and finally primary Capsule output. This layer calculates Primary Capsule Conv output with squash (3) activation function that was used as digit capsule layer input. The squash function used to calculate the length of activity vector from the capsule.

$$v_j = \frac{\|s_j\|^2}{1 + \|s_j\|^2} \frac{s_j}{\|s_j\|} \tag{3}$$

Where v_j is the output of j capsule and s_j is total input of v_j. To calculate s_j we have to use the Eq. 4.

$$s_j = \sum_i c_{ij} \hat{U}_{j|i}, \quad \hat{U}_{j|i} = w_{ij} u_j \tag{4}$$

where the c_{ij} are coupling coefficients help to calculate iterative dynamic routing process and *"routing SoftMax"* [8] (5) whose primary logits b_{ij} is log prior probabilities that coefficients among capsule i, all the other capsules in above layer sum to 1 (Fig. 3).

$$c_{ij} = \frac{\exp\left(b_{ij}\right)}{\Sigma_k \, exp\left(b_{ik}\right)} \tag{5}$$

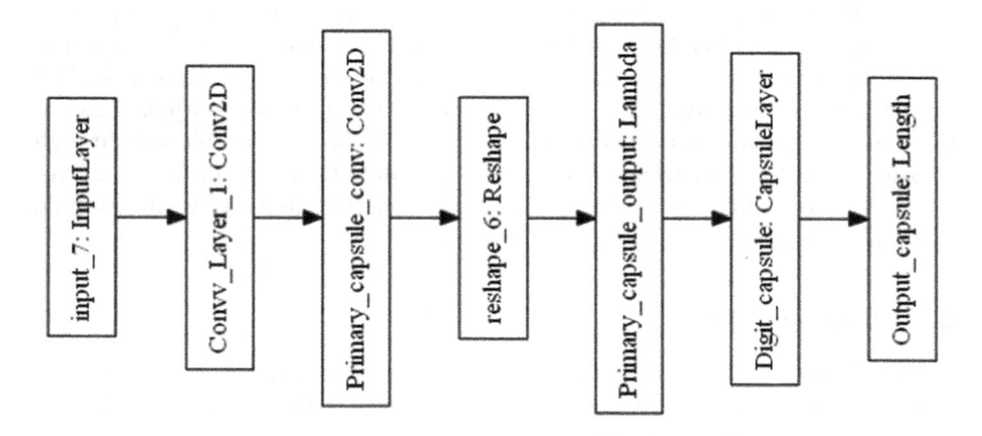

Fig. 3. The architecture of the proposed model.

Digit caps is the higher-level layer has which has digits capsules each of the 10 numerals. Each capsule takes as input a 6 × 6 × 8 × 32 =1152 as 8-dimensional vectors, as input. So, each capsule has 1152 matrices, and also 1152 c coefficients and 1152 b coefficients used in the dynamic routing. The total number of parameters are 1152 × 8 × 16 + 1152 + 1152, and get 149760 trainable parameters for each numeral capsule, then the last stage has to multiply by 10 to get the final number of parameters for this layer.

3.5 Optimizer and Learning Rate

The optimization algorithm plays an important rule to change the result and make it better in Deep Learning and computer vision work. Proposed method used ADAM (6) Optimizer [13] with learning rate 0.001.

$$\theta_{t+1} = \theta_t - \frac{\eta}{\sqrt[1]{\hat{v}_t} + \varepsilon}\hat{m}_t \tag{6}$$

To perform classification and prediction task using a neural network then calculate the error rate more important. And recent study [14] shows that cross entropy function performs better than classification error and mean square error. Proposed method used categorical cross entropy (7) as loss function.

$$\mathrm{L_i} = -\sum_{\mathrm{j}} \mathrm{t}_{\mathrm{i,\ \ j}} \log\left(\mathrm{p_{i,\ \ j}}\right) \tag{7}$$

4 Evaluate the Model

The proposed model of Capsules applied different dataset and achieved sufficient result on train, test and validation datasets.

4.1 Train, Test, Validation Datasets

Each dataset of ISI, BanglaLekha Isolated and CAMTERdb 3.1.1 used as train, test, and validation accordingly 80%, 10% and 10% of the total dataset. After that three dataset which is 39279 images on the training set, and 10% (4911) images on the validation set, remaining 10% (4910) images in test set.

4.2 Model Performance

After 30 epoch proposed model gets 98%, 96.81%, 95.71%, and 96.40% validation accuracy respectively for CMATERdb, ISI, BanglaLekha- Isolated dataset and mixed dataset. Also, all of this dataset cross-validate with each other and perform accurately. Figure 4 showing the accuracy and loss of training and validation set of ISI, BanglaLekha-Isolated, CMATERdb and mixed dataset respectively. Table 2 showing the details accuracy on different datasets.

Later the trained model is tested with the training dataset and got 99.55%, 99.07%, 99.74% and 99.57% accuracy on CMATERdb, ISI, BanglaLekha-Isolated and Mixed dataset test split respectively. Figure 5 showing the confusion Matrix for all of these datasets trained on mixed dataset model.

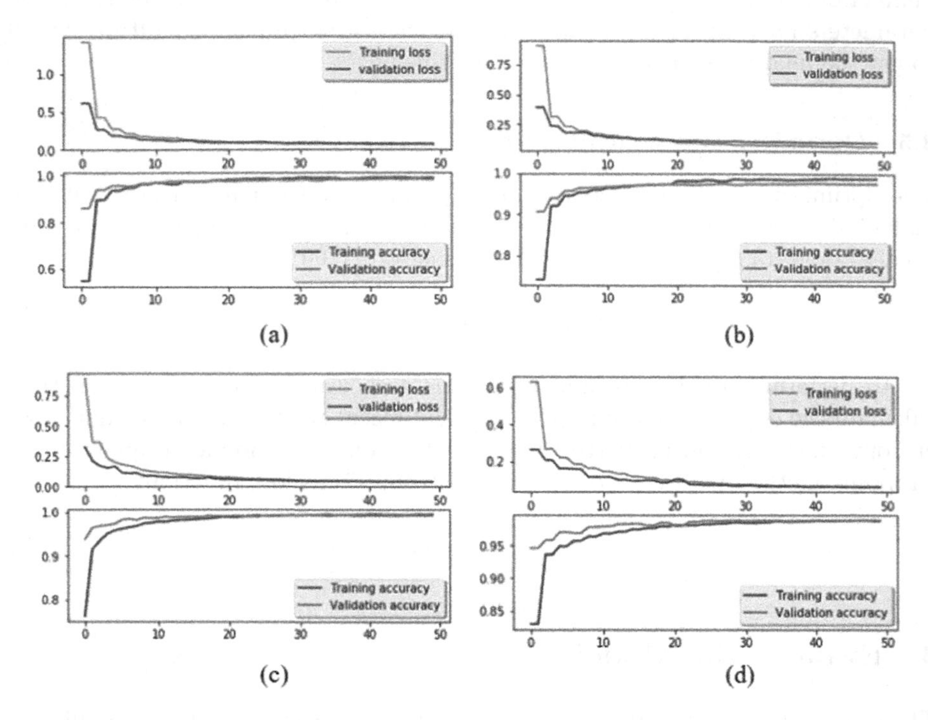

Fig. 4. Accuracy and loss on train and validation set for (a) ISI (b) BanglaLekha-Isolated (c) CMATERdb (d) Mixed dataset.

Table 2. Result comparison in different dataset.

Dataset name	Tr. loss	Val. loss	Tr. acc	Val acc
BanglaLekha	0.0694	0.0998	98.67%	97.62%
CMATERdb	0.0690	0.0791	98.96%	98.33%
ISI	0.0423	0.0421	99.54%	99.28%
Mixed dataset	0.0644	0.0637	98.78%	98.90%

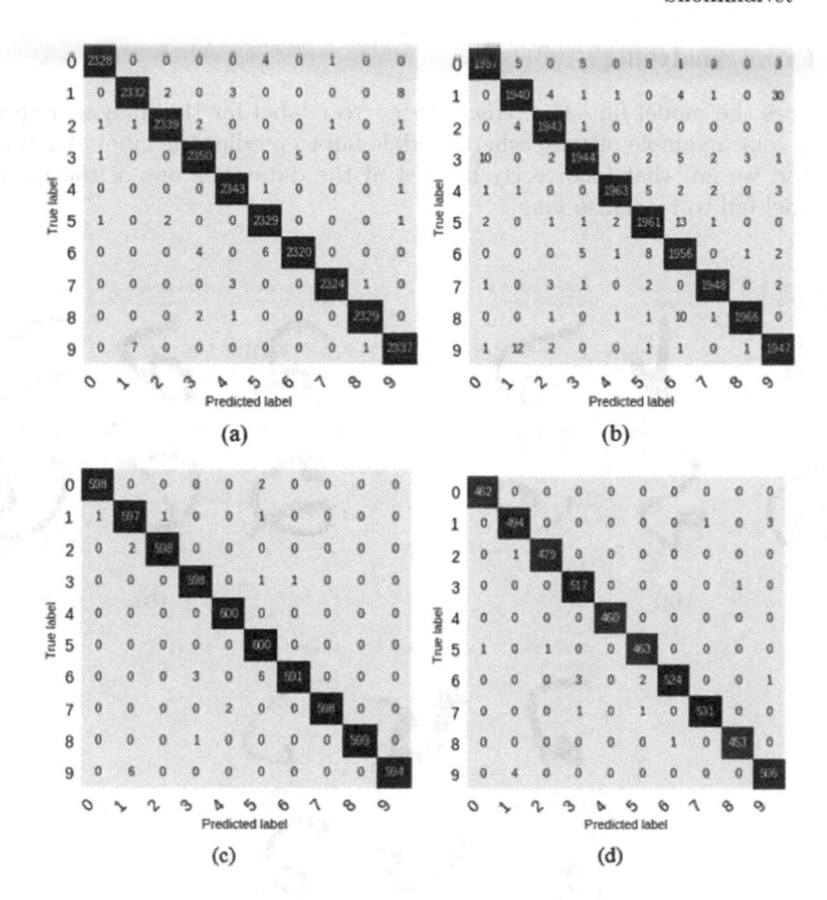

Fig. 5. Confusion matrix for, (a) CMATERdb (b) ISI (c) BanglaLekha (d) Mixed test set.

4.3 Result Comparison

Table 3 is showing Sonkhanet result comparison with some previous works.

Table 3. Result comparison with previous works.

Work	Accuracy	Work	Accuracy
Haider et al. [15]	94%	Sarkhel et al. [19]	98.23%
Wen and He [16]	96.91%	Hassan et al. [25]	96.7%
Nasir and Uddin [17]	96.80%	Basu et al. [20]	95.10%
Akhnad et al. [18]	97.93%	**Proposed model**	**98.90%**

4.4 Error Analyzing

Sometimes the model fails to predict the correct label for the images. Figure 6 showing some example of error where models fail to predict the label. Analyzing this error, we got that incorrectly labeled of the dataset is one of the reasons this model fail to recognize label.

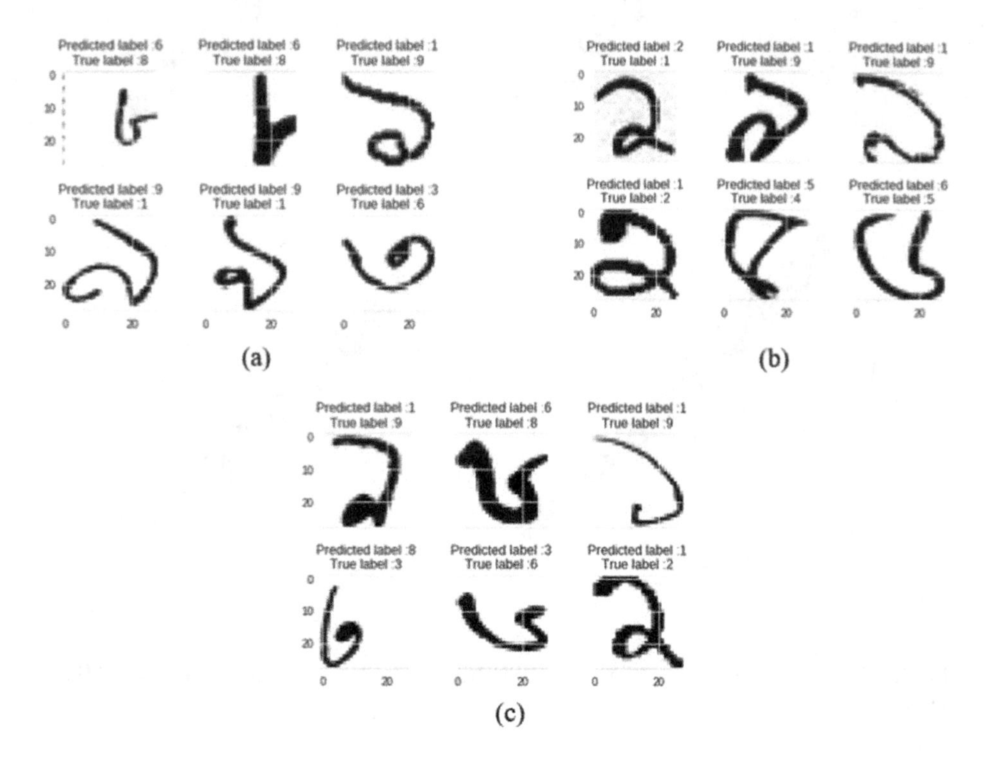

Fig. 6. Error from (a) Mixed train set (b) Mixed validation set (c) Mixed test set.

5 Conclusion

This proposed model introduced Capsules for the recognition of Bangla handwritten numerals. Capsule net works very well for handwritten character recognition for its distinguished features. The proposed scheme is more effective because of its performance over CNNs when the data is overlapping. The best recognition rate found by the system was 98.90%. The proposed recognition method has tested and validated on a large Bangla handwritten numeral dataset. In the future, it is possible to gain more accuracy by providing a better collection of the dataset.

References

1. Bhattacharya, U., Chaudhuri, B.: Handwritten numeral databases of Indian scripts and multistage recognition of mixed numerals. IEEE Trans. Pattern Anal. Mach. Intell. **31**, 444–457 (2009). https://doi.org/10.1109/TPAMI.2008.88
2. Biswas, M., et al.: BanglaLekhaIsolated: a multi-purpose comprehensive dataset of Handwritten Bangla Isolated characters. Data in Brief **12**, 103–107 (2017). https://doi.org/10.1016/j.dib.2017.03.035
3. Sarkar, R., Das, N., Basu, S., Kundu, M., Nasipuri, M., Basu, D.K.: CMATERdb1: a database of unconstrained handwritten Bangla and Bangla-English mixed script document image. Int. J. Doc. Anal. Recogn. (IJDAR) **15**(1), 71–83 (2012)
4. Cheriet, M., Yacoubi, M.E., Fujisawa, H., Lopresti, D., Lorette, G.: Handwritten recognition research: Twenty years of achievement... and beyond. Pattern Recogn. **42**, 3131–3135 (2009)
5. Dong, J., Krzyżak, A., Suen, C.Y.: An improved handwritten Chinese character recognition system using support vector machine. Pattern Recogn. Lett. **26**(12), 1849–1856 (2005)
6. Zhu, B., Zhou, X.-D., Liu, C.-L., Nakagawa, M.: A robust model for on-line handwritten Japanese text recognition. IJDAR **13**(2), 121–131 (2010)
7. Kim, H.J., Kim, P.K.: Recognition of off-line handwritten Korean characters. Pattern Recogn. **29**, 245–254 (1996)
8. Geoffrey, E.H., et al.: Dynamic Routing Between Capsules. 1710.09829v2 [cs.CV], 7 November 2017
9. Hinton, G.E., Krizhevsky, A., Wang, S.D.: Transforming auto-encoders. In: Honkela, T., Duch, W., Girolami, M., Kaski, S. (eds.) ICANN 2011. LNCS, vol. 6791, pp. 44–51. Springer, Heidelberg (2011). https://doi.org/10.1007/978-3-642-21735-7_6
10. Pal, U., Chaudhuri, B.B.: Automatic recognition of unconstrained off-line Bangla handwritten numerals. In: Tan, T., Shi, Y., Gao, W. (eds.) ICMI 2000. LNCS, vol. 1948, pp. 371–378. Springer, Heidelberg (2000). https://doi.org/10.1007/3-540-40063-X_49
11. Alom, M.Z., Sidike, P., Taha, T.M., Asari, V.: Handwritten Bangla Digit Recognition Using Deep Learning (2017)
12. Rabby, A.K.M.S.A., Abujar, S., Haque, S., Hossain, S.A.: Bangla handwritten digit recognition using convolutional neural network. In: Abraham, A., Dutta, P., Mandal, J.K., Bhattacharya, A., Dutta, S. (eds.) Emerging Technologies in Data Mining and Information Security. AISC, vol. 755, pp. 111–122. Springer, Singapore (2019). https://doi.org/10.1007/978-981-13-1951-8_11
13. Kingma, D.P., Ba, J.: Adam: A Method for Stochastic Optimization. arXiv:1412.6980 [cs.LG], December 2014
14. Janocha, K., Czarnecki, M.W.: On loss functions for deep neural networks in classification. arxiv, abs/1702.05659 (2017)
15. Khan, H.A., Al Helal, A., Ahmed, K.I.: Handwritten Bangla digit recognition using sparse representation classifier. In: 2014 International Conference on Informatics, Electronics & Vision (ICIEV), pp. 1–6. IEEE (2014)
16. Wen, Y., He, L.: A classifier for Bangla handwritten numeral recognition. Expert Syst. Appl. **39**(1), 948–953 (2012)
17. Nasir, M.K., Uddin, M.S.: Handwritten Bangla numerals recognition for automated postal system. IOSR J. Comput. Eng. **8**(6), 43–48 (2013)

18. Islam, S., Shill, P.C., Rahman, M.M., Akhand, M.A.H., Rahman, M.M.H.: Bangla handwritten character recognition using convolutional neural network. Int. J. Image Graphics Signal Process. (IJIGSP) **73**, 42–49 (2015)
19. Sarkhel, R., Das, N., Saha, A.K., Nasipuri, M.: A multi-objective approach towards cost-effective isolated handwritten Bangla character and digit recognition. Pattern Recogn. **58**, 172–189 (2016)
20. Basu, S., Sarkar, R., Das, N., Kundu, M., Nasipuri, M., Basu, D.K.: Handwritten *Bangla* digit recognition using classifier combination through DS Technique. In: Pal, S.K., Bandyopadhyay, S., Biswas, S. (eds.) PReMI 2005. LNCS, vol. 3776, pp. 236–241. Springer, Heidelberg (2005). https://doi.org/10.1007/11590316_32
21. Santosh, K.C., Wendling, L.: Character recognition based on non-linear multi-projection profiles measure. Front. Comput. Sci. **9**, 678–690 (2015)
22. Deans, S.R.: Applications of the Radon Transform. Wiley Interscience Publications, New York (1983)
23. Santosh, K.C.: Character recognition based on DTW-radon. In: 2011 International Conference on Document Analysis and Recognition. IEEE (2011)
24. Kruskall, J.B., Liberman, M.: The symmetric time warping algorithm: From continuous to discrete. In: Time Warps, String Edits and Macromolecules: The Theory and Practice of String Comparison, pp. 125–161. Addison-Wesley, Boston (1983)
25. Hassan, T., Khan, H.A.: Handwritten bangla numeral recognition using local binary pattern. In: 2015 International Conference on Electrical Engineering and Information Communication Technology (ICEEICT), pp. 1–4. IEEE (2015)

Reader System for Transliterate Handwritten Bilingual Documents

Ranjana S. Zinjore[1] and R. J. Ramteke[2(✉)]

[1] KCES's Institute of Management & Research, Jalgaon, India
rszinjore14@gmail.com
[2] School of Computer Sciences, North Maharashtra University, Jalgaon, India
rakeshramteke@yahoo.co.in

Abstract. India is a Multistate- Multilingual country. Most of the people in India used their state official language and English is treated as a binding language used for form filling or some official work. So there is a need to create a system which will convert the handwritten bilingual document into digitized form. This paper aims at development of reader system for handwritten bilingual (Marathi-English) documents by recognizing words. This facilitates many applications such as Natural language processing, School, Society, Banking, post office and Library automation. The proposed system is divided into two phases. The first phase focuses on recognition of handwritten bilingual words using two different feature extraction methods including combination of structural and statistical method and Histogram of Oriented Gradient Method. K-Nearest Neighbor classifier is used for recognition. This classifier gives 82.85% recognition accuracy using Histogram of Oriented Gradient method. The dataset containing 4390 words collected from more than 100 writers. The second phase focuses on digitization and transliteration of recognized words and conversion of transliterated text into speech, which is useful in the society for visually impaired people.

Keywords: Reader system · Transliteration · Handwritten bilingual document · Histogram of oriented gradient

1 Introduction

Handwriting is a media for communication. Most of the historical documents, books and official documents are in handwritten form and it is necessary to preserve for future generation [14]. The preservation is important because information technology and the era of digitization increases which reduces processing time. Handwritten recognition is the most challenging area of the image processing. Recognition process is divided into two categories as: Online Recognition and Offline Recognition [16]. This paper is based on recognition of Offline Handwritten bilingual documents. The document written with two languages is called bilingual document. In this case people use their state official language and

K. C. Santosh and R. S. Hegadi (Eds.): RTIP2R 2018, CCIS 1037, pp. 171–182, 2019.
https://doi.org/10.1007/978-981-13-9187-3_16

English is treated as binding language [12]. Marathi is the state official language of Maharashtra state. People use Marathi language for official work but some words-numbers are written in English. To produce such hand-written bilingual document system is a tedious job. This paper focuses on word level recognition of handwritten bilingual document which is useful in many application areas including Natural Language processing for processing of handwritten messages on social media, School form processing, Postal Automation, Bank Cheque processing, Library Automation and in society for reading advertisement and books for visually impaired people by attaching a voice synthesizer [13].

2 Literature Review

During literature review it is observed that most of the researcher put their contribution on single script or on printed documents. Whereas very less research work have been reported on handwritten bilingual/multilingual documents. Shaw et al. [18] suggested stroke based feature for recognition of handwritten Devanagari words and Hidden Markov Model is used for recognition. Singh et al. [19] proposed curvelet transform for recognition of handwritten Devanagari city name. Patil et al. [15] discussed different methodology for recognition of handwritten Devanagari words. Sandyal et al. [7] reported 92% recognition accuracy of hand-written Kannada words. Manoj Kumar et al. [8] used directional feature extraction and zoning for recognition of handwritten Malayalam words. Jino et al. [1] considered histogram of oriented gradient descriptor, number of black Pixels in the upper half and lower half as a feature for recognition of Malayalam district name. For recognition of Kannada Text authors suggest Zernike Moments as a feature extraction at character level and Support Vector Machine is working as a classifier [20]. Handwritten Devanagari words are recognition using Hidden Markov Model by considering combination of Gradient and structural feature [17]. Belaid et al. [2] suggested double smearing technique to extract the word for separation of handwritten and printed text from real documents. Obaidullah et al. [9] presented different script identification techniques for offline and on-line documents at page, block, line, word and Character level. Obaidullah et al. [10] discusses script dependent and script independent as feature extraction methods for identifying script from handwritten documents. In this paper Sect. 3 focuses on architecture of the proposed system. Section 4 describes the preprocessing techniques, Line and word segmentation methodology, feature extraction and word recognition approach. Finally recognized words are digitized and transliterated to produce speech. Section 5 reports comparative results of the proposed system.

3 Proposed System Architecture

Figure 1 depicts the architecture of the reader system for transliterated handwritten bilingual document. In this architecture the input sample is preprocessed

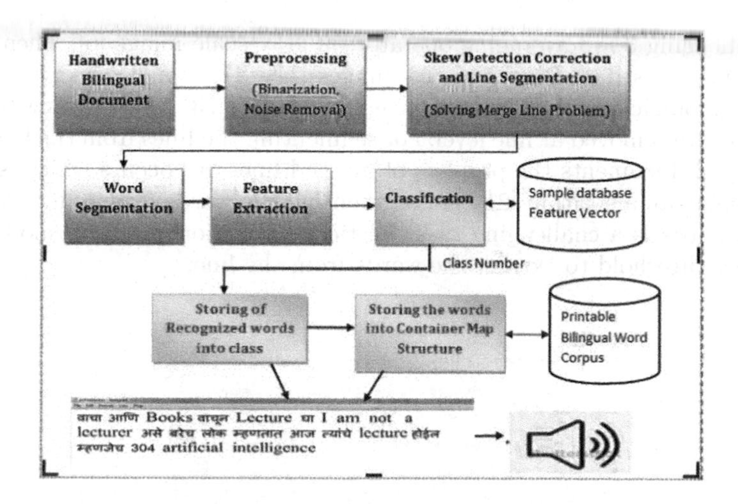

Fig. 1. Proposed system architecture

in the form of noise removal and skew detection and correction. During line segmentation problem of merged lines are solved and from individual segmented lines words are extracted. Two different feature extraction techniques are used for word recognition and K-Nearest Neighbor (K-NN) is used as a classifier. The container map structure is used for mapping the respective class number with printable bilingual word corpus. The result is digitized copy of handwritten bilingual document. Finally the transliterated copy of digitized document is created and passed to voice synthesizer for producing speech.

4 Methodology

This section describes the methodology used for development of the proposed system. The proposed system is developed in two different phases which is discussed in the next subsection.

4.1 Recognition of Words from Handwritten Bilingual Document

During recognition of handwritten bilingual word different phases of Optical Character Recognition is used. The phases are discussed as below.

Preprocessing and Segmentation. As no standard handwritten bilingual dataset is available for experiments, we have collected handwritten sample datasets from various sources like-books, news-papers and student notes. One sample database is shown in Fig. 2. The collected datasheets of handwritten bilingual documents were scanned using scanner with 300 dpi then all sheets are stored in image file. Preprocessing techniques are used to bridge the gap between documents using 3 × 3 averaging operator on gray scale image [6]. Then Otsu's thresholding is used to Binarized the images [11]. The small dots are removed using morphological opening operation as shown in Fig. 3. Also the skew from the documents are removed at line level. For segmenting the lines from the handwritten bilingual documents the problem of merged lines are occurs which is solved by developing an algorithm [21]. For segmenting the words from a line consist of bilingual words is a challenging task. Vertical projection profile is modified by calculating threshold to extract the words from the line.

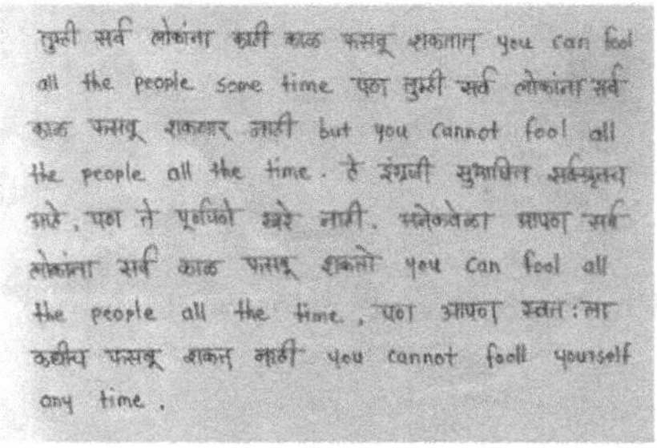

Fig. 2. Sample of handwritten bilingual document

Feature Extraction. After extracting the words from a line, features are extracted for recognizing the words. Two different feature extraction methods are used which is discussed in the next subsection.

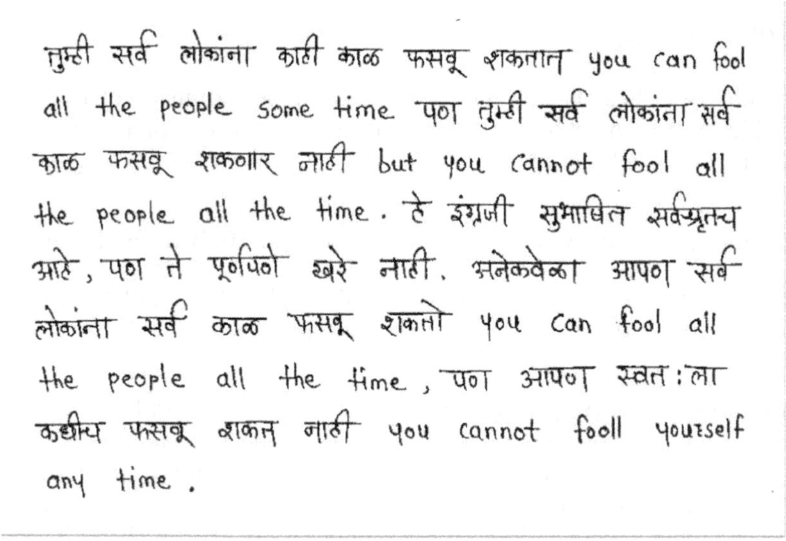

Fig. 3. Preprocessed document

Method 1: Structural and Statistical Feature
Structural features involved number of horizontal and vertical lines(Table 1), minimum length of horizontal line, minimum length of vertical line, maximum length of horizontal lines, maximum length of vertical line, total length of all horizontal and vertical lines, number of cavities profile in the word and some statistical features such as Aspect Ratio, Eccentricity and Extent [4]. We have developed an algorithm Cavity_Profile_Word, based on water reservoir principle to extract total number of cavities from the word (Table 2). Finally total twelve features are extracted using Method1.

Table 1. Results of algorithm used for Method1

Input Image	No. of Horizontal Lines	No. of Vertical Lines
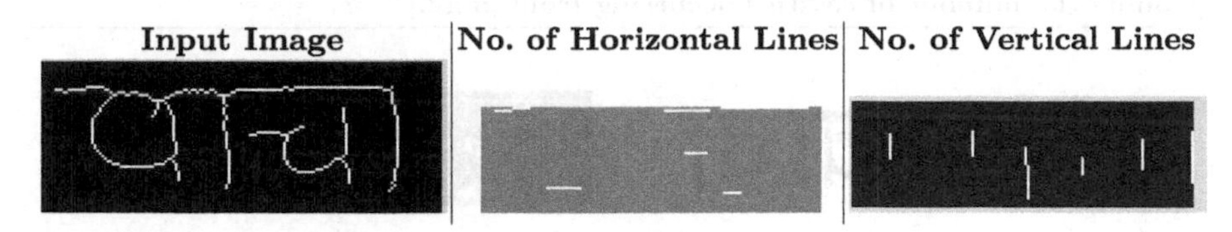		

Method 2: Histogram of Oriented Gradient
Histogram of Oriented Gradient (HOG) is proposed by Dalal and Triggs [3]. HOG counts occurrences of gradient orientation in localized portions of an image. For extracting the features using HOG in Matlab we have performed following steps: Step 1) Resized preprocessed segmented word into [40×80] and stored it into img1. Step 2) [fc1, visual] = extractHOGFeatures(img1); Where fc1 is a feature vector of size 1×1296 and visual represent HOG feature visualization shown in Fig. 4.

Algorithm 1. Extract_lines_bilingualwords(bw,fvector)

Preprocessed thin image: bw and Feature Vector: fvector

1. **[Extract Horizontal and Vertical lines from the words.]**

1.1 Find connected component of the binary image.

1.2 Use regionprop to set a shape measure as bounding box for each connected component object.

1.3 Find the sum of the height of each object.

$Sum\_ht = \Sigma_{i=1}^{stats} ht\_object$

where ht_object height of each connected component

1.4 Find horizontal line using morphological structuring elements.

1.5 Find vertical line using morphological structuring elements.

2. **[Find minimum, maximum and total number of horizontal and vertical lines.]**

3. **Compute the cavity profile of a word.**

no_cavity = Cavity_Profile_Word(bw);

4. **[Compute the aspect ratio, eccentricity and extent of all the connected components of an input image and obtain their average values.]**

5. fvector = [No_horz_line, min_horz_line, max_horz_line, total_horz_cnt, No_vert_line, min_vert_line, max_vert_line, total_vert_cnt, no_cavity, AAR, Avg_Eccen, Avg_Extent]

Algorithm 2.Cavity_Profile_Word(bw,no_cavity)

Preprocessed thin image: bw and Number of cavities no_cavity

1. **[Crop the image (bw) from top to its 50% height into img1 shown in Table2.]**
2. **[Generate a blank image img2 of size of img1.]**
3. **[Find Cavity Profile of a word.]**

Let [x1 y1] = size(img1);

3.1 Scan the img1 from top to bottom such that replace substitute pixel value with 1 in img2 in following cases:

Case 1) If the top pixel shows open area replaces all zeros with 1 till the edge pixel occurs at the bottom.

Case 2) If the top pixel is closed then scan the pixels in the same column till the next edge pixel occurs. Replace all zeros with 1 between these two edge pixels.

4. **[Counts the number of cavities occurring from img2.]**

Fig. 4. Left-sample image Right: HOG feature visualization

Table 2. Results of algorithm used for Method1

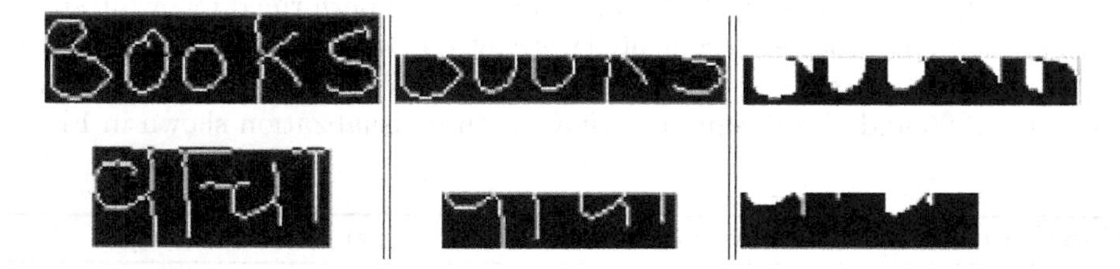

4.2 Word Recognition

K-Nearest Neighbor classifier is used for recognition of handwritten words. This classifier gives class number of the test samples by mapping the features with train samples. This class numbers are mapped with printable bilingual word corpus using container map data structure to generate the digitized form of handwritten bilingual document shown in Fig. 5. The corresponding algorithm is shown in below:

Algorithm 3 . Mixed_Word_Recognition(Mix_words, print_bilingual_doc)

1. Stored the class number of recognized word using K-NN classifier into a text file as classf1.
2. Open classf1 in read mode.
3. Use container map.
newMap1 = containers.Map;;
4. Open a standard printed bilingual dataset file in read mode and extract the words and stored it in newMap1
5. Generate printable bilingual document shown in Fig.5.

5 Text to Speech System

After receiving the printable bilingual document, the document is converted into transliterated document for producing speech from the text as shown in Fig. 6. The architecture shows the flow of the conversion of text to speech. Table 3 shows the transliterated samples of Marathi words. C#. NET is used as an interface to develop text to speech system as shown in Fig. 7.

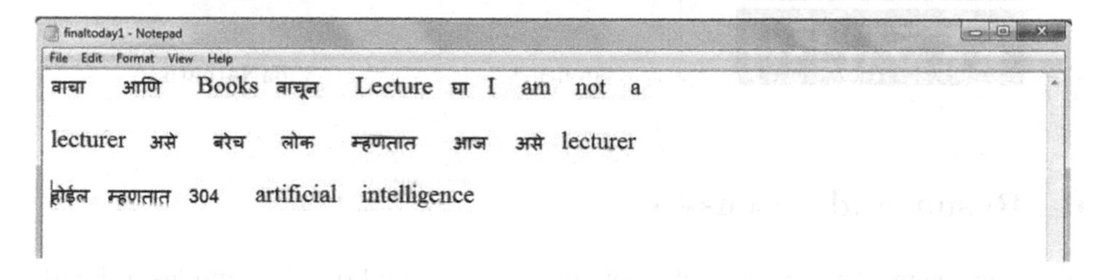

Fig. 5. Printable bilingual document

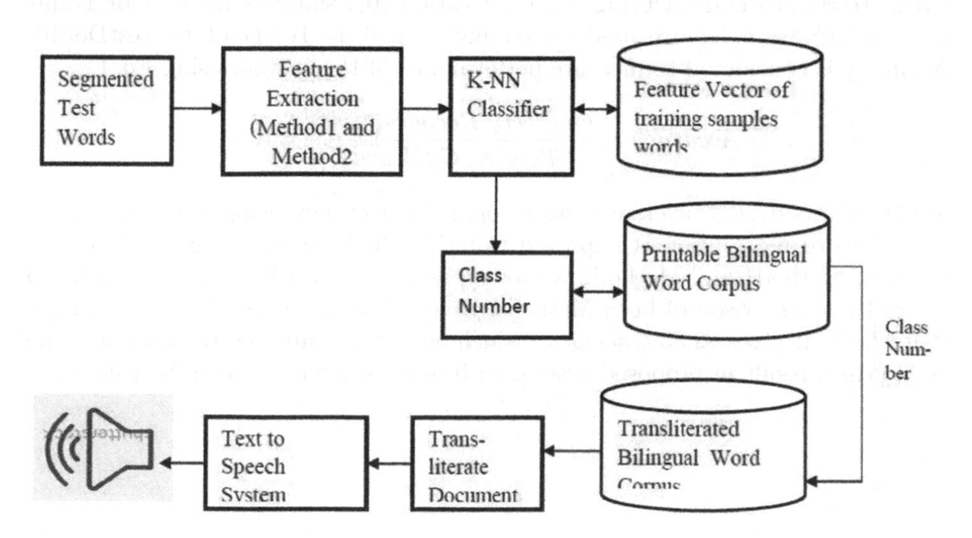

Fig. 6. Text to speech for transliterated text

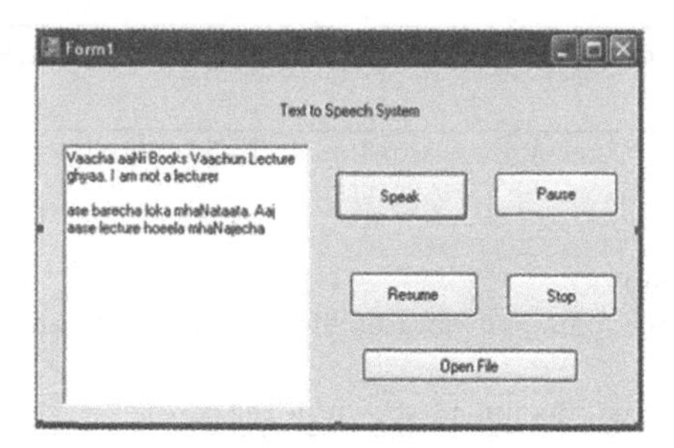

Fig. 7. Text to speech system

Table 3. Word image to Devnagari printed & Romanized Marathi text

Handwritten Marathi words	Printable Words	Transliterate Marathi Word
	वाचा	Vaacha
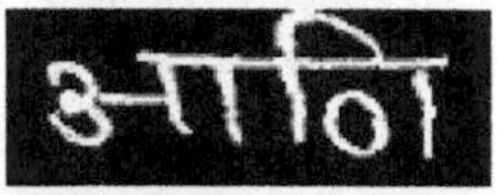	आणि	aaNi
	म्हणतात	mhaNataata

6 Result and Discussion

An experiment of the proposed algorithm is conducted on 100 samples of handwritten bilingual documents collected from 10 database samples named as HandBiDb1 to HandBiDb10. During experimentation 70% samples are used for Training and 30% samples are used for testing named as TestDoc1 to TestDoc10. Accuracy is considered to measure performance of the system using Eq. 1.

$$Accuracy = \frac{CorrectlyRecogSamples}{TotalNoOfWords} * 100 \tag{1}$$

Table 4 shows 57.72% average word recognition accuracy using Method1. Accuracy of word recognition is improved using Method2 as shown in Table 5. The result of Method1 and Method2 is shown graphically in Figs. 8 and 9. Figure 10 shows the comparison of both Methods. It could be understood from the graph that Method2 gives 82.85% accuracy which better in compared to Method1. The comparison result of proposed system with other system is shown in Table 6.

Table 4. Accuracy of mixed script word recognition (Method I)

Sr.No.	Database name	#Bilingual words	#Correctly recognized bilingual words	Accuracy of recognized (%)
(1)	HandBiDb1	77	47	61.03896
(2)	HandBiDb2	40	24	60
(3)	HandBiDb3	41	21	51.21951
(4)	HandBiDb4	75	54	72
(5)	HandBiDb5	32	19	59.375
(6)	HandBiDb6	18	9	50
(7)	HandBiDb7	42	15	52.38095
(8)	HandBiDb8	29	21	72.41379
(9)	HandBiDb9	40	20	50
(10)	HandBiDb10	41	12	48.78049
Average accuracy				**57.72%**

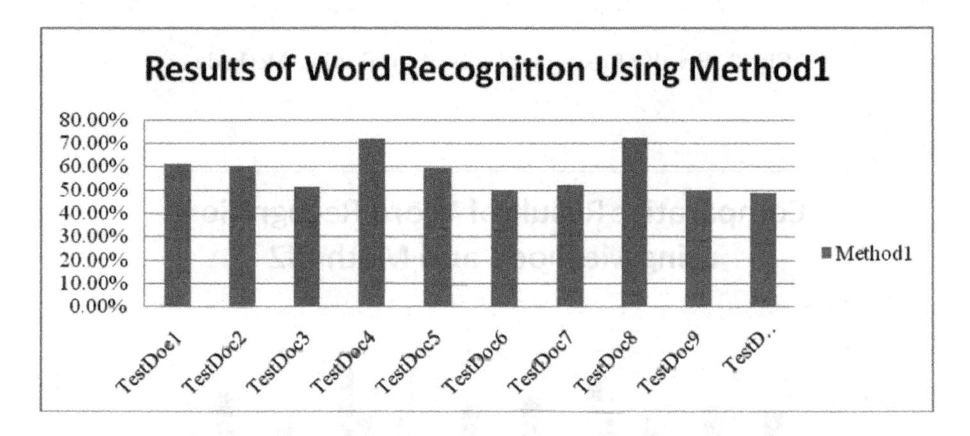

Fig. 8. Result of recognized mixed word using Method1

Table 5. Accuracy of mixed script words recognition (Method II)

Sr.No.	Database name	#Bilingual words	#Correctly recognized bilingual words	Accuracy of recognized words
(1)	HandBiDb1	77	58	75.32%
(2)	HandBiDb2	40	31	77.50%
(3)	HandBiDb3	41	30	73.17%
(4)	HandBiDb4	75	68	90.66%
(5)	HandBiDb5	32	26	81.25%
(6)	HandBiDb6	18	14	77.77%
(7)	HandBiDb7	42	42	100%
(8)	HandBiDb8	29	29	100%
(9)	HandBiDb9	40	28	70.00%
(10)	HandBiDb10	41	34	82.92%
Average accuracy				**82.85%**

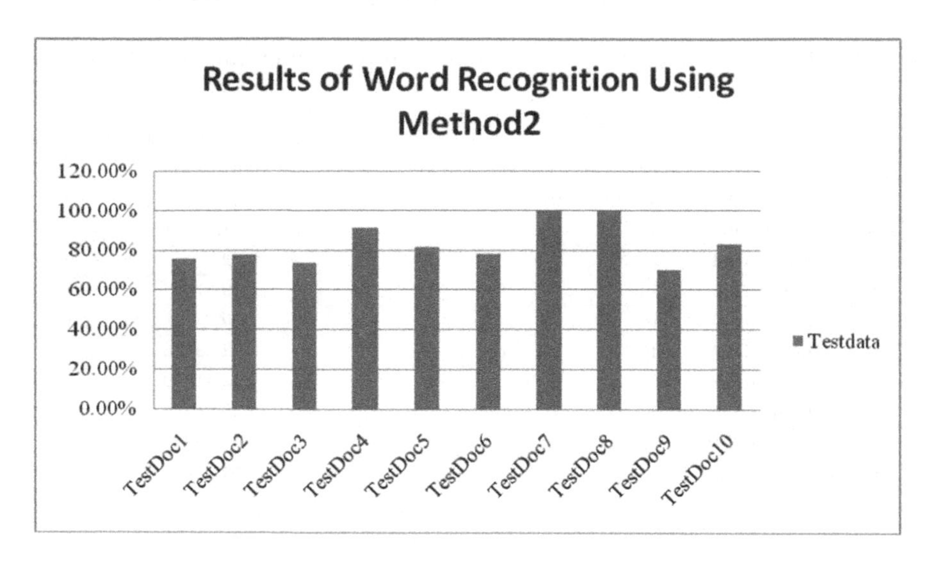

Fig. 9. Result of recognized mixed word using Method2

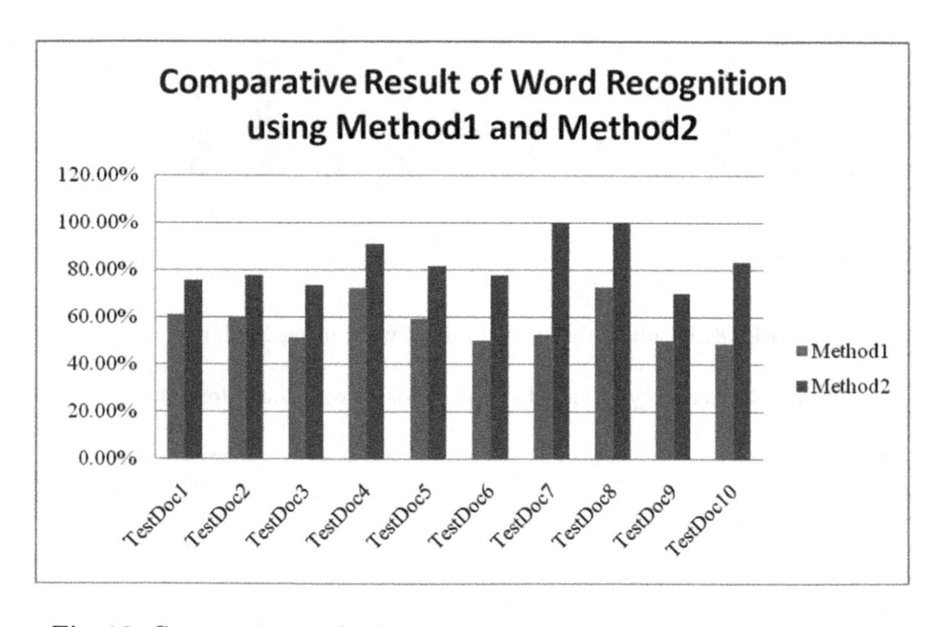

Fig. 10. Comparative result of word recognition using Method1 and Method2

Table 6. Comparison of proposed system with other method

Author	Document type	Feature extraction method	Script type	Recognition accuracy
Shaw et.al. [18]	Handwritten Devanagari word recognition	Stroke based feature	Single script	84.31%
Singh et.al. [19]	Handwritten Devanagari city name	Curvelet transform	Single script	93.21%
Gatos et.al. [5]	Handwritten Roman cursive word	Hybrid features	Single script	80.76%
Proposed system	Handwritten bilingual words	HOG	Bilingual script	82.85%

7 Conclusion

In this paper a system is proposed for conversion of handwritten bilingual document into speech. Handwritten bilingual words are recognized using two feature extraction methods including structural and statistical method and histogram of oriented gradient method. K- Nearest Neighbor is used as a classifier. It is observed that Method2 gives satisfactory results as compared to Method1 for word recognition to create digitized document. Finally the transliterated documents are created which is passed to voice synthesizer. This system could be useful for visually impaired people.

References

1. Balakrishnan, K., et al.: Offline handwritten recognition of Malayalam district name-a holistic approach. arXiv preprint arXiv:1705.00794 (2017)
2. Belaïd, A., Santosh, K.C., d'Andecy, V.P.: Handwritten and printed text separation in real document. arXiv preprint arXiv:1303.4614 (2013)
3. Dalal, N., Triggs, B.: Histograms of oriented gradients for human detection. In: 2005 IEEE Computer Society Conference on Computer Vision and Pattern Recognition, CVPR 2005, vol. 1, pp. 886–893. IEEE (2005)
4. Dhandra, B.V., Hangarge, M.: Morphological reconstruction for word level script identification. Int. J. Comput. Sci. Secur. (IJCSS) **1**(1), 41–51 (2007)
5. Gatos, B., Pratikakis, I., Kesidis, A.L., Perantonis, S.J.: Efficient off-line cursive handwriting word recognition. In: Tenth International Workshop on Frontiers in Handwriting Recognition, Suvisoft (2006)
6. Kamble, P.M., Hegadi, R.S.: Handwritten Marathi basic character recognition using statistical method (2014)
7. Sandyal, K.S., Patel, M.S.: Offline handwritten Kannada word recognition. pp. 19–22 (2014)
8. Manoj Kumar, P., Chandran, S.: Handwritten Malayalam word recognition system using neural networks. Int. J. Eng. Res. Technol. **4**, 90–99 (2015)

9. Obaidullah, S.M., Santosh, K.C., Das, N., Halder, C., Roy, K.: Handwritten Indic script identification in multi-script document images: a survey. Int. J. Pattern Recogn. Artif. Intell. **32**(10), 1856012 (2018)
10. Obaidullah, S.M., Santosh, K.C., Halder, C., Das, N., Roy, K.: Automatic Indic script identification from handwritten documents: page, block, line and word-level approach. Int. J. Mach. Learn. Cybern. **10**, 1–20 (2017)
11. Otsu, N.: A threshold selection method from gray-level histograms. IEEE Trans. Syst. Man Cybern. **9**(1), 62–66 (1979)
12. Pal, U., Chaudhuri, B.B.: Automatic separation of words in multi-lingual multi-script Indian documents. In: ICDAR, p. 576. IEEE (1997)
13. Patel, M.S., Kumar, R., Linga Reddy, S.C.: Offline Kannada handwritten word recognition using locality preserving projection (LPP) for feature extraction. IJIRSET, **4**(7) (2015)
14. Patel, M.S., Reddy, S.L., Naik, A.J.: An efficient way of handwritten English word recognition. In: Satapathy, S.C., Biswal, B.N., Udgata, S.K., Mandal, J.K. (eds.) Proceedings of the 3rd International Conference on Frontiers of Intelligent Computing: Theory and Applications (FICTA) 2014. AISC, vol. 328, pp. 563–571. Springer, Cham (2015). https://doi.org/10.1007/978-3-319-12012-6_62
15. Ansari, S., Patil, P.M.: A research survey of Devnagari handwritten word recognition. Int. J. Eng. Res. Technol. **2**(10), 1010–1015 (2013)
16. Plamondon, R., Srihari, S.N.: Online and off-line handwriting recognition: a comprehensive survey. IEEE Trans. Pattern Anal. Mach. Intell. **22**(1), 63–84 (2000)
17. Shaikh, M.A., Dagade, M.R.: Offline recognition of handwritten devanagari words using hidden markov model. IJIRST, **1**(11) (2015)
18. Shaw, B., Parui, S.K., Shridhar, M.: Offline handwritten Devanagari word recognition: a segmentation based approach. In: 2008 19th International Conference on Pattern Recognition, ICPR 2008, pp. 1–4. IEEE (2008)
19. Singh, B., Mittal, A., Ansari, M.A., Ghosh, D.: Handwritten Devanagari word recognition: a curvelet transform based approach. Int. J. Comput. Sci. Eng. **3**(4), 1658–1665 (2011)
20. Student, R.V.: Off-line handwritten Kannada text recognition using support vector machine using zernike moments. IJCSNS **11**(7), 128 (2011)
21. Zinjore, R.S., Ramteke, R.J., Pathak, V.M.: Segmentation of merged lines and script identification in handwritten bilingual documents. In: Proceedings of the 9th Annual Meeting of the Forum for Information Retrieval Evaluation, pp. 29–32. ACM (2017)

Recognition of Marathi Numerals Using MFCC and DTW Features

Siddheshwar S. Gangonda[1(✉)], Prashant P. Patavardhan[2], and Kailash J. Karande[1]

[1] SKN Sinhgad COE, Pandharpur, Maharashtra, India
sgangonda@gmail.com, kailashkarande@yahoo.co.in
[2] KLS Gogte Institute of Technology, Belagavi, Karnataka, India
prashantgemini73@gmail.com

Abstract. Numeral recognition is one amongst the foremost important problems in pattern recognition. Its numerous uses like reading communication postal code, worker code, bank cheque method etc. To the simplest of our information, very less work has been wiped out Marathi language as compared with other Indian and non-Indian languages. It has mentioned a unique technique for recognition of isolated Marathi numerals. It introduces Marathi numerals and identification technique using MFCC and DTW as attributes. The accuracy of the pre-recorded samples is greater than that of online testing samples. We have got additionally seen that the accuracy of the speaker dependent samples is over that of the speaker independent samples. Another technique known as HMM is additionally discussed. By experimentation, it's ascertained that identification exactness is higher for HMM than DTW, but the training method in DTW is extremely straightforward and quick, as compared to HMM. The time needed for recognition of numerals using HMM is additional as compared to DTW, because it should bear the various states, iterations and lots of additional mathematical modeling, thus DTW is most well-liked for the real-time applications.

Keywords: Hidden Markov Model (HMM) · Mel-Frequency Cepstral Coefficient (MFCC) · Distance Time Warping (DTW)

1 Introduction

Speech identification techniques are employed in totally different fields in our existence. Owing to the fast advancement in this field, we are able to observe several machines with speech input [3]. Speech Synthesis and identification combinedly creates a speech interface. We tend to need such software's to be on the market for Indian languages. Speech recognition in computer domain involves several steps with problems connected with them [1, 2].

K. C. Santosh and R. S. Hegadi (Eds.): RTIP2R 2018, CCIS 1037, pp. 183–193, 2019.
https://doi.org/10.1007/978-981-13-9187-3_17

2 Problem Definition

The goal is to create a speech identification device for Marathi language, which uses MFCC for Attribute Extraction and DTW for Attribute Matching.

3 Marathi Numeral Recognition Using MFCC and DTW Features

The most widely used techniques to examine the form to identify their resemblance are the MFCC and DTW. The MATLAB is used for the implementation of MFCC and DTW attributes.

Feature Extraction (MFCC)
The MFCCs are used for feature extraction. The recognition accuracy for MFCC attribute is taken into consideration as a result of it mimics the human ear perception [4].

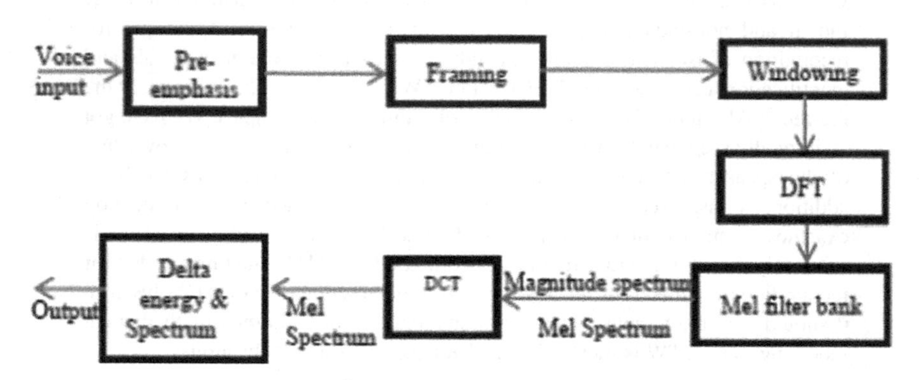

Fig. 1. Block diagram of MFCC [4].

The complete procedure of the MFCC is shown in Fig. 1 which consists of seven computational steps as follows.

Step 1: Pre–emphasis
Step 2: Framing.
Step 3: Windowing
Step 4: Fast Fourier Transformation (FFT)
Step 5: Mel Filter Bank Processing
Step 6: Discrete Cosine Transformation (DCT)
Step 7: Delta Energy and Delta Spectrum

The attributes associated to the variation in cepstral attributes over time are shown by13 delta attributes (12 cepstral attributes and one energy attribute), and thirteen acceleration attributes. The Mel filter bank [3] created is shown in Fig. 2.

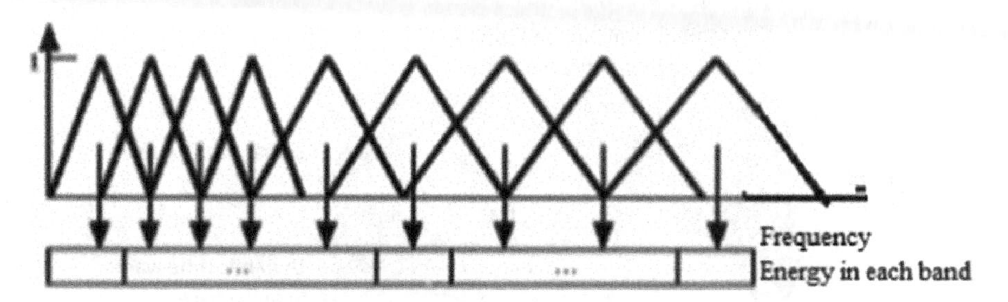

Fig. 2. Mel scale filter bank [2].

4 Features Matching (DTW)

4.1 Outline

In this form of speech identification method, the test information is transformed to templates. To understand the concept of DTW [5], we require to know this parameters:

- Features: The data in every signal has to be exhibited in some fashion.
- Distances: Any type of metric has to be utilized so as to obtain a matched path.

4.2 DTW Algorithm

Speech may be a time-dependent technique. So, the articulations of identical word can have variant durations, and articulations of identical word with identical period can take issue within the middle, because the completely different components of the words are being spoken at variant rates.

DTW depends on Dynamic Programming ways. This technique is for estimating similarity between 2 statistics that will vary in time or speed. Figure 3 shows the instance of however one times series is 'warped' to a different [6].

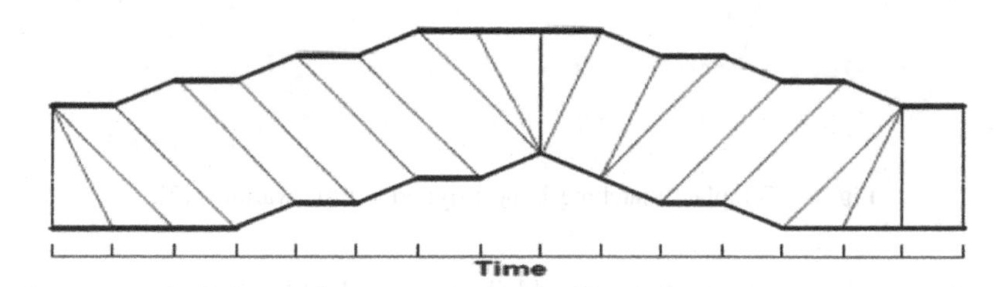

Fig. 3. Warping between two time series [6].

The aim of DTW is to match 2dynamic patterns and calculate its similarity by finding a minimum distance between them (Fig. 4).

Searching for the simplest path that matches 2 time-series signals is that the main task for several researchers, because of its importance in these applications. Dynamic Time-Warping (DTW) is one among the most techniques to accomplish this task,

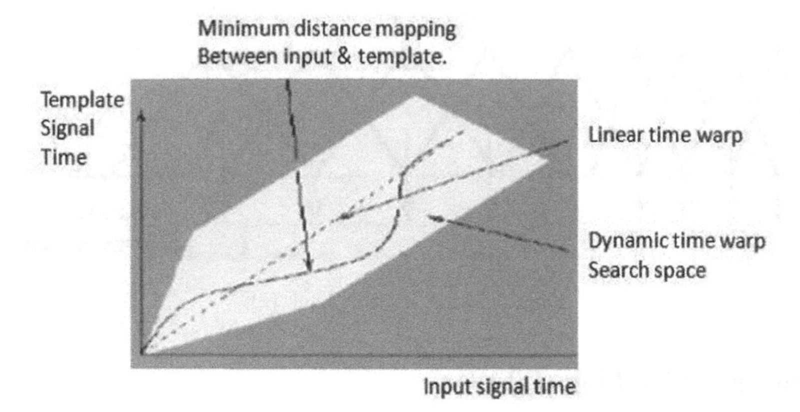

Fig. 4. Example dynamic time warping (DTW) [6].

particularly in speech recognition systems to touch upon totally different speaking speeds. The model with nearest match outlined in a very manner chosen as recognized word [8] (Fig. 5).

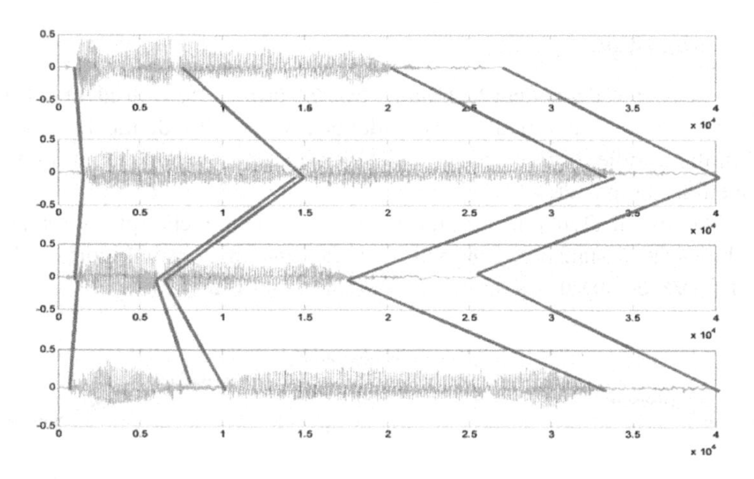

Fig. 5. Template matching issues: dynamic time warping [7].

DTW is mainly used in the embedded speech identification devices like cell phones, speech recognition, data mining, and movement recognition [5].

5 Hidden Markov Model Training and Recognition

The HMMs are widely used statistical tools in recognition system and is given by the formula, $\lambda = (\pi, A, B)$.

π = initial state distribution vector.
A = State transition probability matrix.
B = continuous observation probability density function matrix (Fig. 6).

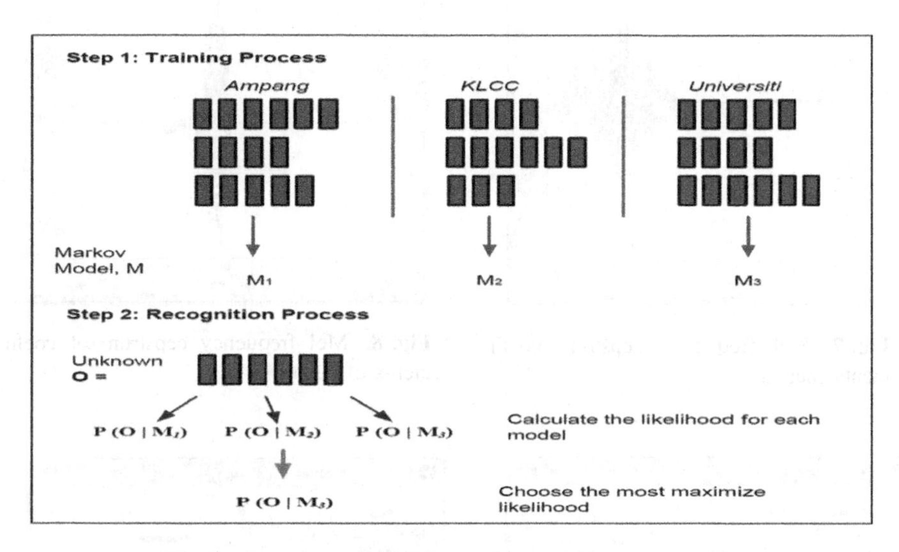

Fig. 6. Overview of HMM speech recognition process [9].

6 Results and Discussions

We recorded Marathi database of numerals zero to nine. In this we have intended to implement a password system with numerals and many other such applications in everyday life. The twenty samples for each numeral were recorded from different people and these samples were then normalized. Then they were decomposed using Dynamic Time Warping. From the twenty samples recorded, 16 are used to train the DTW and the unused 4 are used for test purpose.

In this project, speech identification software is developed using MFCC & DTW techniques. The software would display exact numeral if input signal would be compared to the pre-recorded & online signals.

The Results of some of the extracted features of recorded database of numerals zero to nine in Marathi are shown in the figures below (Figs. 7, 8).

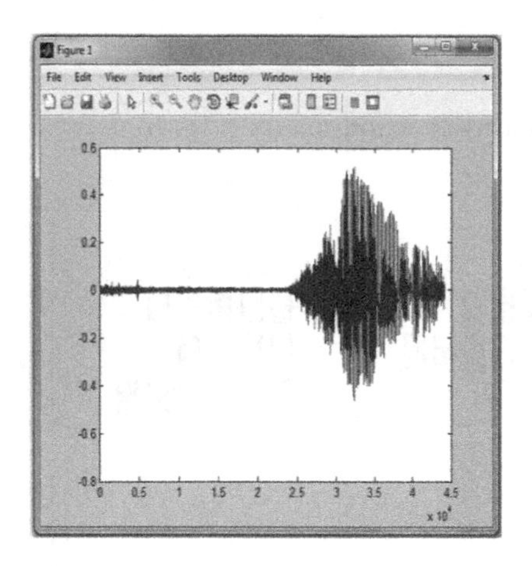

Fig. 7. Mel frequency cepstrum coefficients shunya.

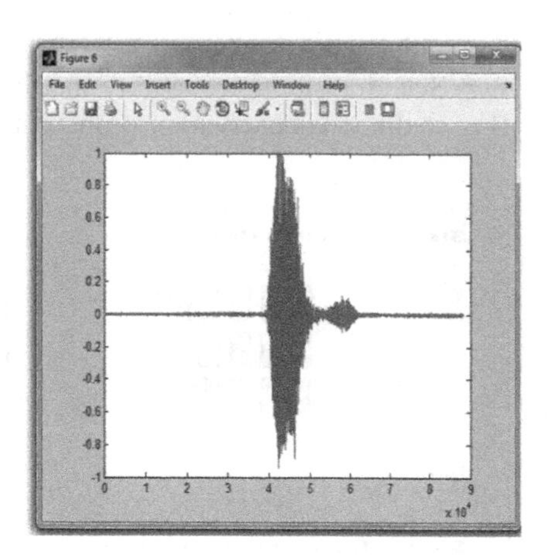

Fig. 8. Mel frequency cepstrum of coefficients of pach.

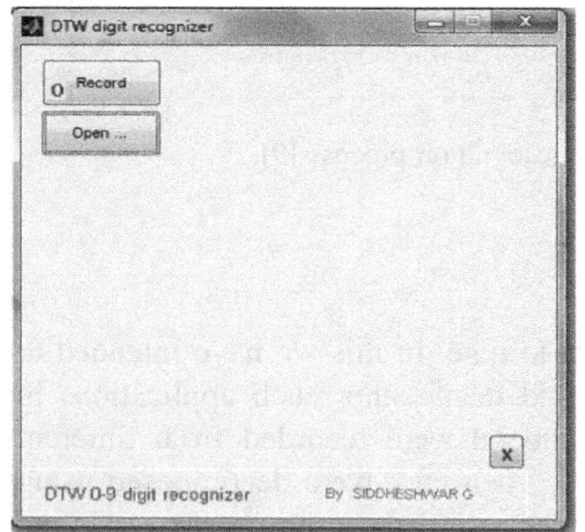

Fig. 9. GUI of DTW digit recognizer.

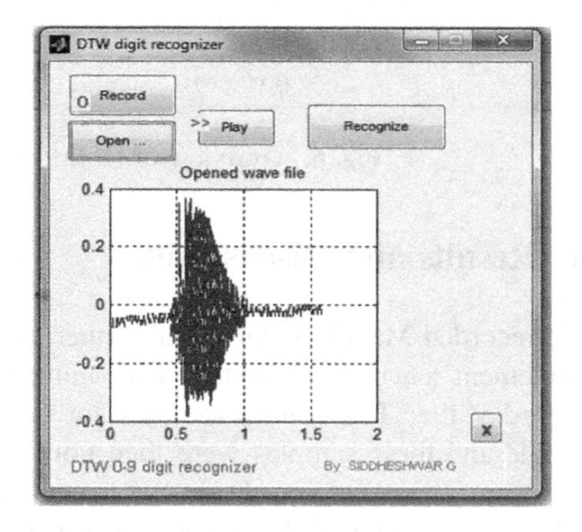

Fig. 10. GUI of opened wavefile.

6.1 Graphical User Interface (GUI) of the System

We have created the GUI of the system for the recognition of the numerals. The DTW 0–9 Digit Recognizer has the various command buttons like record, open, play, recognize etc. It shows the opened wave file.

In this project, we have designed a DTW digit recognizer, in which the command button open reads the pre-recorded numerals and the command button record the online numeral spoken by the speaker. We can play the pre-recorded & online numeral spoken by the speaker, and then we can recognize the numeral using the DTW for feature matching. It matches the template by taking into account the minimum warping distance between the various numerals. The Template with closest match defined in manner chosen as recognized numeral & it is displayed on GUI display (Figs. 9, 10, 11, 12, 13, 14, 15, 16, 17, 18).

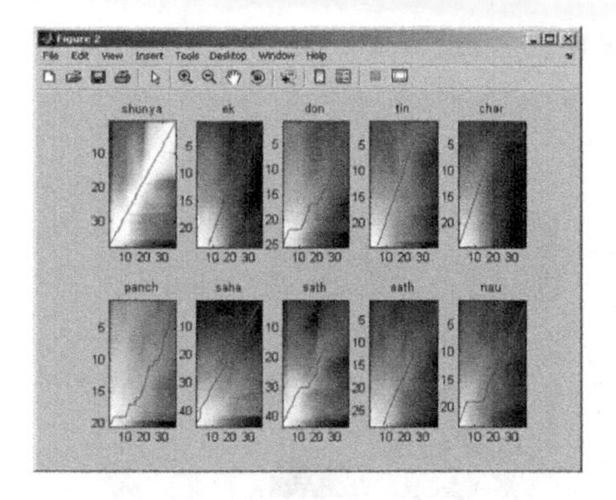

Fig. 11. GUI of pattern matching of shunya.

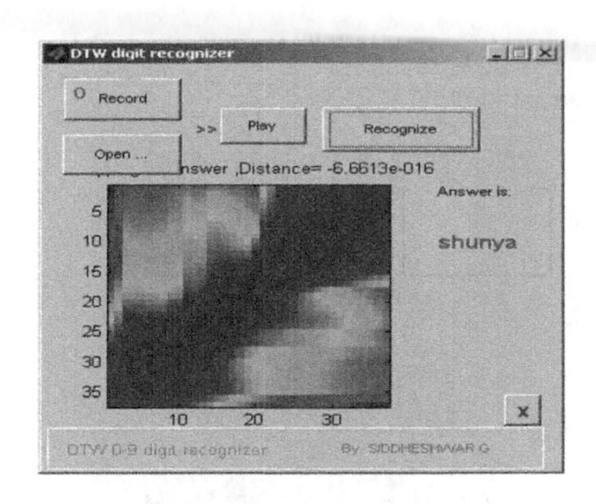

Fig. 12. GUI of recognized numeral shunya.

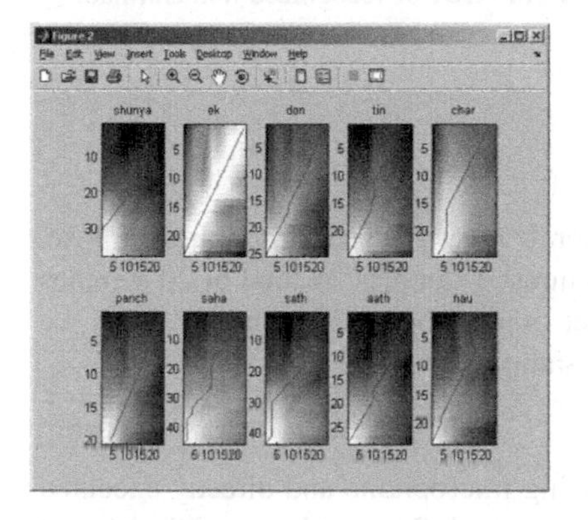

Fig. 13. GUI of pattern matching of ek.

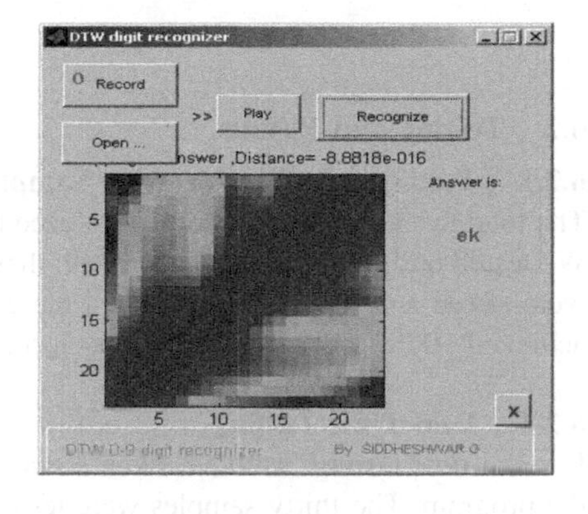

Fig. 14. GUI of recognized numeral ek.

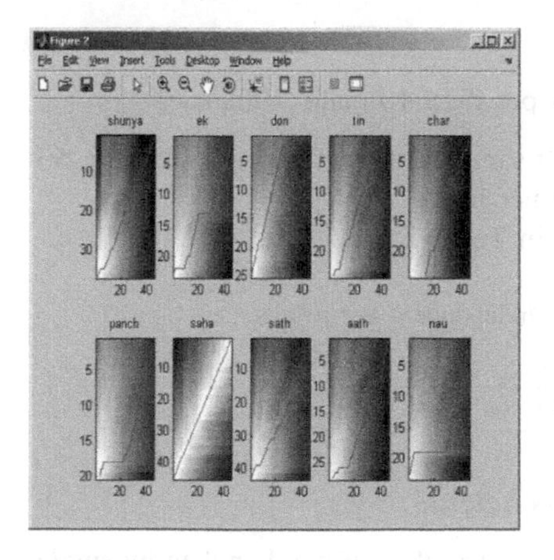

Fig. 15. GUI of pattern matching of saha.

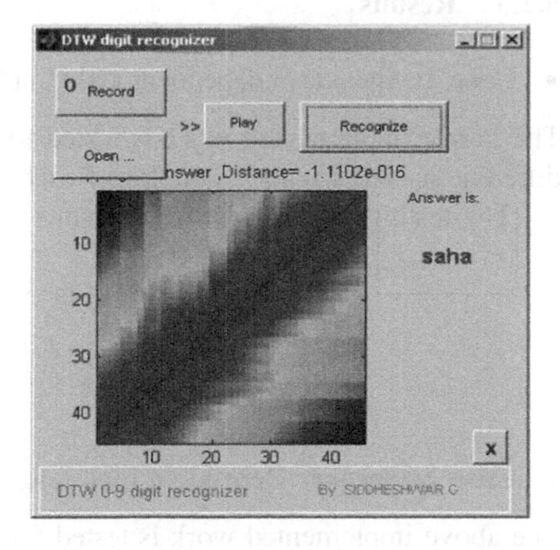

Fig. 16. GUI of recognized numeral saha.

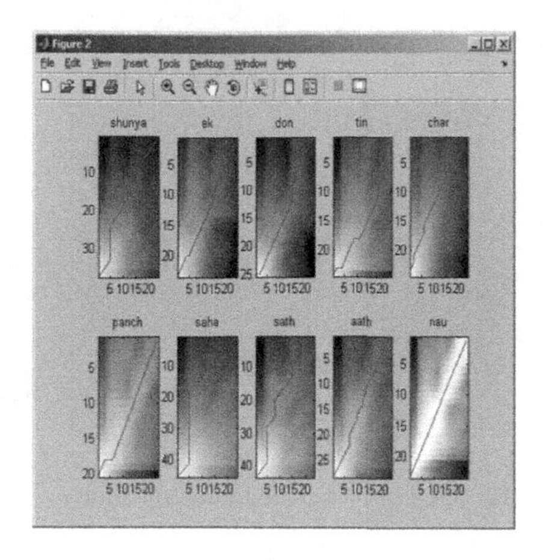

Fig. 17. GUI of pattern matching of nau.

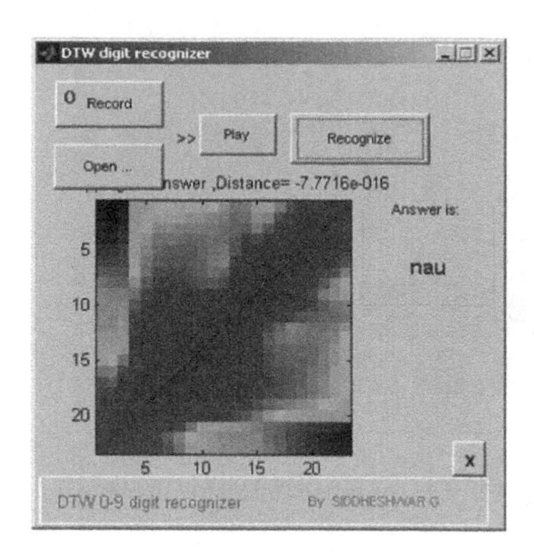

Fig. 18. GUI of recognized numeral nau.

6.2 Testing and Results

6.2.1 Testing with Pre-recorded Samples

The total twenty samples recorded for each numeral, 15 were used for training purpose. We tested our program's precision with these 5 unused samples. A total of 20 samples were tested and the program yielded the correct output for all 20 samples. Thus, we achieved 100% precision for the pre- recorded samples.

6.2.2 Real-Time Testing

For real-time testing, we have taken a sample using microphone and directly executed the program. The thirty samples were tested, from which 24 samples gave the correct output. This obtains an precision of about 80% with real-time samples.

6.2.3 Results

- **Case 1:** Speaker independent (20 templates per digit 10 male, 10female)

The above implemented work is tested for 100 samples of each word spoken by 50 different speakers with 2 samples of each digit per head.

The testing work leads to the results given in Table 1.

Table 1. Accuracy of the speaker independent test results.

Digit	0	1	2	3	4	5	6	7	8	9
% Accuracy	87	88	82	78	79	84	85	81	78	87

- **Case 2:** Speaker Dependent (one template per digit).

The above implemented work is tested for 10 samples of each word spoken by single speaker. The results are given in Table 2.

Table 2. Accuracy of the speaker dependent test results.

Digit	0	1	2	3	4	5	6	7	8	9
% Accuracy	90	91	84	90	87	88	92	84	86	92

It is observed that the accuracy of the pre-recorded samples is more than that of the real-time testing samples. We have also observed that the accuracy of the speaker dependent samples is more than that of the speaker independent samples (Fig. 19) (Table 3, 4, 5).

Table 3. Confusion matrix of the MFCC & DTW recognition.

	ek	don	teen	char	pach	saha	sat	aath	nau	shunya	Avg. %
ek	1	1	1	4	1	1	1	1	1	0	80
don	2	2	2	2	3	2	2	2	2	2	90
teen	3	3	3	3	9	3	3	2	2	2	80
char	4	4	5	4	4	4	4	6	4	4	80
pach	5	5	5	5	5	5	5	5	5	3	90
Saha	6	6	6	6	1	6	6	4	6	6	80
Sat	7	7	8	7	7	7	7	7	7	7	90
Aath	2	8	8	8	8	7	8	8	8	8	80
nau	9	9	4	9	9	5	9	9	9	9	80
shunya	0	0	0	0	5	0	0	2	0	0	80

Table 4. Confusion matrix of the MFCC & HMM recognition.

	ek	don	teen	char	pach	saha	sat	aath	nau	shunya	Avg. %
ek	1	1	1	1	1	1	1	3	1	1	90
don	2	2	2	2	2	2	2	2	2	5	90
teen	3	3	3	3	3	3	1	3	3	3	90
char	4	3	4	4	4	4	4	4	8	4	80
pach	5	5	5	5	5	5	5	5	5	5	100
Saha	6	6	6	8	6	6	6	6	6	6	90
Sat	7	7	7	7	7	7	7	5	7	7	90
Aath	8	8	8	8	8	8	8	8	5	8	90
nau	9	9	9	9	9	7	9	9	9	9	90
shunya	0	0	0	7	0	0	0	5	0	0	80

Table 5. Comparison digit recognition accuracy test results.

Numeral	DTW accuracy	HMM accuracy
ek	80	90
don	90	90
teen	80	90
char	80	80
pach	90	100
Saha	80	90
Sat	90	90
Aath	80	90
nau	80	90
shunya	80	80
Average %	83%	89%

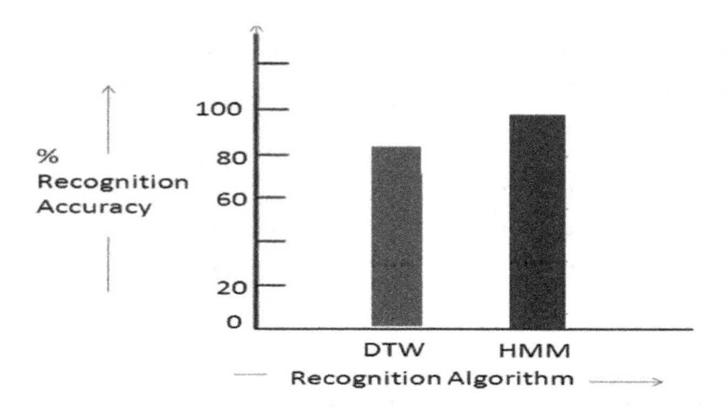

Fig. 19 Recognition accuracy of the DTW & HMM.

The time required for recognition of numerals using HMM is more as compared to DTW, as it has to go through the many states, iterations & many more mathematical modeling, so DTW is preferred for the real-time applications as compared with the HMM.

7 Conclusions and Future Scopes

7.1 Conclusions

- Automatic speech recognition (ASR) continues to be a difficult and troublesome task [1].
- The MFCC is employed as extraction techniques. The nonlinear sequence alignment known as DTW is used as features matching techniques.
- This paper has shown that greater identification rates can be obtained using MFCC attributes with DTW.

- The time required for recognition of numerals using HMM is more as compared to DTW, as it has to go through the many states, iterations & many more mathematical modeling, so DTW is preferred for the real- time applications than the HMM.
- The exactness of the pre-recorded samples is more than that of the real- time testing samples. We have also observed that the accuracy of the speaker dependent samples is more than that of the speaker independent samples.

7.2 Future Scopes

- One of the key areas where future work can be concentrated is the large vocabulary generation & to improve robustness of speech recognition performance [2].
- The preferable interpreting of human abilities and to use it to enhance machine identification execution.
- The future work could be towards Online Speech Summarization.
- The future work could be towards minimizing the time required for recognition of numerals using HMM.

References

1. Saon, G., Picheny, M.: Recent advances in conversational speech recognition using convolutional and recurrent neural networks. IBM J. Res. Dev. **61**(4/5), 1 (2017)
2. O'Shaughnessy, D.: Automatic speech recognition. In: CHILEAN Conference on Electrical, Electronics Engineering, Information and Communication Technologies (CHILECON) (2015). https://doi.org/10.1109/chilecon.2015.7400411
3. Rabiner, L.R., Schafer, R.W.: Digital Processing Of Speech Signals, Low Price Edition. Pearson, London (2007)
4. Anusuya, M.A., Katti, S.K.: Speech recognition by machine: a review. (IJCSIS) Int. J. Comput. Sci. Inf. Secur. **6**(3), 181–205 (2009)
5. Gawali, B.W., Gaikwad, S., Yannawar, P., Mehrotra, S.C.: Marathi isolated word recognition system using MFCC and DTW features. In: Proceedings of International Conference of Advances in Computer Science (2010)
6. Muda, L., Begam, M., Elamvazuth, I.: Voice recognition algorithms using Mel Frequency Cepstral Coefficient (MFCC) and Dynamic Time Warping (DTW) techniques. J. Comput. **2**(3) (2010). ISSN 2151-9617
7. Vimala, C., Radhab, V.: Speaker independent isolated speech recognition system for Tamil language using HMM. In: International Conference on Communication Technology and System Design (2011)
8. Bala, A., Kumar, A., Birla, N.: Voice command recognition system based on MFCC and DTW. Int. J. Eng. Sci. Technol. **2**(12), 7335–7342 (2010)
9. Jiang, H., Li, X., Liu, C.: Large margin hidden Markov models for speech recognition. IEEE Trans. Audio Speech Lang. Process. **14**(5) (2006)

A Filter Based Feature Selection for Imbalanced Text Classification

K. Swarnalatha[1(✉)], D. S. Guru[2], Basavaraj S. Anami[3], and N. Vinay Kumar[2]

[1] Department of Information Science and Engineering, Maharaja Institute of Technology Thandavapura, Mysuru, India
swarnapradyu@gmail.com

[2] Department of Studies in Computer Science, University of Mysore, Manasagangotri, Mysuru 570 006, India
dsg@compsci.uni-mysore.ac.in, vinaykumar.natraj@gmail.com

[3] KLE Institute of Technology, Hubli, India
anami_basu@hotmail.com

Abstract. In this work, a text classification method through a filter type feature selection for imbalanced data is addressed. The model initially clusters the documents associated with a class through a hierarchical clustering there by accomplishing a balanced or near balanced class. Later, a filter type feature selection is recommended to choose the most discriminative features for text classification. Subsequently, the documents are stored in the form of interval valued data. For classification purpose, a suitable symbolic classifier is recommended. The experimentation is done with two standard benchmarking datasets viz., Reuters 21578 and TDT2. The experimental results obtained from the proposed model are better in terms of f-measure when compared to the available models.

Keywords: Imbalance text · Clustering · Feature selection · Symbolic representation · Text classification

1 Introduction

Nowadays, it is very difficult to imagine our life without web information. The main source for the generation of such information is web data. The web data majorly categorized in to four types viz., text, image, video and audio and hence termed as multimedia data. Due to the rapid increase in the multimedia data over the web it is very challenging to handle such data effectively and efficiently. Hence, there is a scope in the development of machine learning algorithms for handling such data effectively and efficiently. Classification and clustering algorithms have been part of the most machine learning algorithms. In this work, the text data is considered for designing an effective text categorization system. In machine learning, the concept of automatic text classification has gained significant attention among the researchers across the globe. Even though plenty of

K. C. Santosh and R. S. Hegadi (Eds.): RTIP2R 2018, CCIS 1037, pp. 194–205, 2019.
https://doi.org/10.1007/978-981-13-9187-3_18

methods [7] available for text categorization, designing an automatic text categorization system is still a challenging task due to the ever increase in the size of text corpus [11]. But, the efficiency of the existing text classification techniques [12,14,22,25] deteriorates due to the class imbalanceness of text documents and is one of the most important problems the researchers are tackling in the field of text categorization. The class imbalanceness occurs due to the improper distribution of samples across the different groups and also, many works found in the available works [23], also made a negative impact on the performance of the system due to the fact of curse of dimensionality which occurs due to vector form representation of text data. Though, this is widely accepted text representation scheme globally, the generic nature of it in representing the terms of a document made it sensitive to curse of dimensionality. Hence, the problem of class imbalanceness for text classification with a proper feature selection strategy is addressed in this paper.

Some of the existing works found on text categorization models which work fine for the balanced datasets and fails to work for the imbalanced data sets [12,14,25]. Though, always a balance data is not expected from the real world environment. The classifier has to be train to handle such imbalance type data for effective text classification. Some of the works which focus on imbalance text classification are [10,17,18]. These works converts the imbalance classes into balance classes by applying class wise clustering. Once the clusters are obtained each cluster has been treated as class and the number of classes increased to number of clusters. In these methods the feature selection method is working better compare to the feature transformation method but the features used in feature selection method was very high. This motivated us to go for a better feature selection method rather than with the feature transformation method to represent documents for efficient classification. The main objective of feature selection is to reduce the dimension of the text document feature vector without compromising the performance of the text classification [16], few works which focus mainly on feature selection methods available in literature are terms-based discriminative information space [9], fisher ratio [21,24], Chi-square [23], Information gain [1], term weightage using term frequency [13,15,20,25], term frequency and document frequency based feature Selection [2], Ontology based [3], global feature selection methods [11,12] etc.

Recently a feature selection frame work proposed by [5] showing significance improvement in text classification in which the features are ranked in group instead of ranking individual features. Hence the same frame work has been used for selecting the features in the proposed method. In the proposed work initially the classes are converted in to clusters and later the clusters are treated as classes. The total number of classes increased to number of clusters. The features are ranked in groups instead of individual ranking and the top ranked one feature is selected from each cluster as a representative. Hence, top features with higher ranks are selected with the optimal number of features to represent a feature vector.

The next sections of this paper are as follows: The Sect. 2 presents new text classification model in which imbalance to balance class conversion, feature selection, symbolic representation and symbolic classification has been presented. The complete experimental setup and dataset is given in Sect. 3 with comparative analysis of various methods for imbalance text classification and in Sect. 4 conclusion of the proposed work is given.

2 Proposed Model

The high level design of the proposed model is given in Fig. 1. The different steps involved in the proposed are Class Balancing, Feature selection, Symbolic representation followed symbolic classification. The detailed explanations of these steps are explained in subsequent sections.

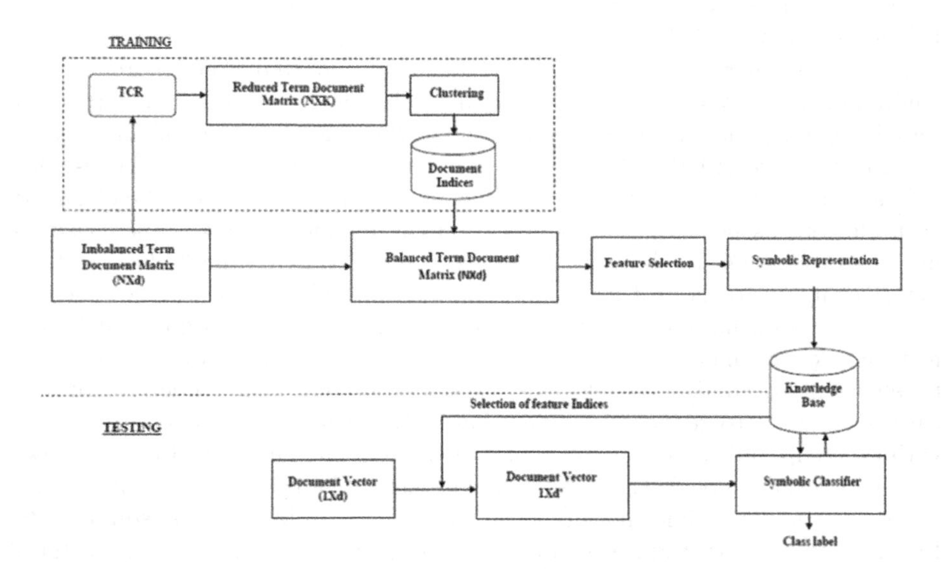

Fig. 1. General architecture of the proposed model.

2.1 Class Balancing Through Clustering

In this section, the procedure for balancing a class in terms of number of documents is explained. For balancing the documents across classes, a clustering technique is recommended. An illustration of the same is shown in Fig. 2.

Let us consider a term-document feature matrix FM of dimension $\mathrm{N} \times \mathrm{d}$, where N represents total number of documents spread across K classes and 'd' represents total number of terms present in the corpus. Let FM(i, j) represents a term frequency of a jth term associated with i^{th} document. The feature matrix

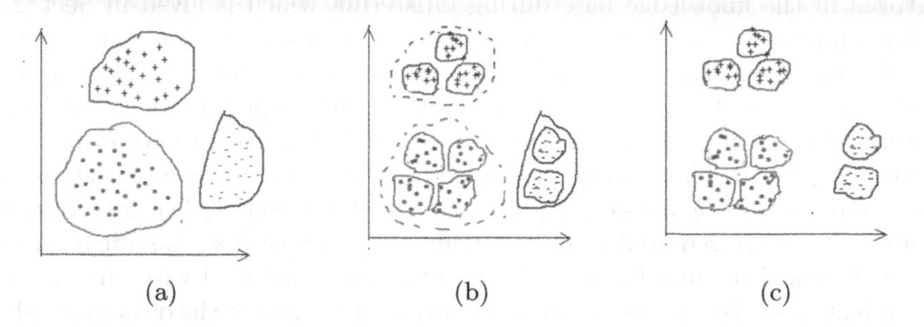

Fig. 2. Illustration of imbalance to balance set of samples through clustering, (a) Samples spread across three groups before clustering, (b) Samples spread across groups after clustering, (c) Samples spread across nine groups after clustering considered as nine balanced classes.

FM is a kind of vector space model (VSM) representation and is sparse in nature due to the presence of more generic features in the text corpus. Basically, the clustering techniques have some limitations to handle sparse feature matrix [8]. Hence, the high dimensional sparse feature matrix has to be represented in the lower dimensional feature space before performing clustering. Hence the documents are represented in lower dimensional space using [6], where the dimension of a document is equal to the number of classes. Hence the dimension of the document reduces from d to K (K $\ll$ d) and it is also equal to the total classes considered in the corpus. The transformed feature matrix is called as term class relevance feature matrix (FM_{TCR}). For each entry of the matrix, $FM_{TCR}(i,j)$ represents the weight of each term t_j with respect to class C_i. Once the features are reduced to only K, the hierarchical clustering technique is applied on class C_i with n_i documents and z number of clusters are obtained. Where n_i is the total number of documents and it is represented by $Q_i = q_1, q_2, q_3, ..., q_z$. The inconsistency co-efficient i_c is used to decide the number of clusters in each class and which is decided through empirically. Henceforth, the clusters correspond to a class C_i are called as a subclasses.

Let $Q_1, Q_2, Q_3, ..., Q_i, ..., Q_z$ are the subclasses obtained for a class and similarly for K different classes, the total number of subclasses be $Q = z * K$. For each class the number of subclasses (z) is different from that of remaining classes and it also depends on number of documents present in each class. Later each subclass is considered as class and after clustering of all the classes, K number of classes thus becomes Q number of classes. This results with documents spread in each class is balanced and thus the problem of imbalance text classification is more or less becomes the problem of balance text classification system.

2.2 Feature Selection

After clustering the documents of each subclass which is considered as class is represented using term document matrix with the help of document indices which are stored in the knowledge base during clustering which is given in Sect. 2.1. Now the dimension of the feature matrix is N × d which is very large and sparse. Hence feature selection is an essential component which is required for text classification. Hence in this work features are ranked with respect to each class using an alternative feature selection method of [5]. Once the features are ranked the top ranked Q features in a group are selected and redundancy is removed among the selected features by considering distinct features among Q features. Further if more features are required to increase the efficiency of text classification the next level ranked distinct features are selected and combined with the already selected features. Like this the process is continued till we get the desired number of features so that the classification is done effectively. The proposed method is very effective in selecting group of more optimal and non redundant features at a time for the classification purpose.

For given collection of N documents with Q balanced classes and d features Chi-square (χ^2) feature evaluation criterion is applied. Thus, upon applying statistic on each pair of term and class, the weight matrix WM is constructed of size Q × d. Where $WM(C_i, t_j) = (\chi^2)(C_i, t_j)$ is the weight of a term with respect to class C_i. In weight matrix each row represents the term weights which are associated with a class for separating other classes from each other. All the rows are sorted based on their weight and the term which is having highest weight gets the highest rank similarly the lowest weight gets the lowest rank. After sorting the weight respective ranked term indices are stored in the term Rank Matrix. Hence the ith row of term Rank Matrix (RM) contain the term indices which are ranked based on their weight. Now each column of a term RM are highly discriminating power with every other classes and each column features are more discriminative in nature than column i + 1.

$$FinalSubsetFeatures = \cup_{i=1}^{r'} \cup_{j=1}^{Q} RM(i,j) \tag{1}$$

where Q and r′ are respectively the total number of classes and columns selected. Hence, at a time Q top ranked features are selected from all the classes and final subset consists of top ranked d′ features which is very less compared to d (d′ ≪ d).

2.3 Symbolic Representation of Text Documents

From recent years the symbolic representation of data is getting more importance in various research fields including text due to its effective representation. Hence, in this work we have used symbolic representation of class samples using interval valued data. It captures the internal variations of a class using a single vector and also it simplifies the task of classification of unknown document to its predefined class.

Let C_i be a class with N_i number of documents represented by $D^i = [D_1^i, D_2^i, D_1^i, ..., D_{N_i}^i]$ where each document is a d′ dimensional feature vector. The method of representation of class C_i symbolically using SV_i is given below.

Consider a sth feature of documents of a class C_i, these features are aggregated in the form of interval valued feature vector $[\mu_s - \sigma_s, \mu_s + \sigma_s]$ where μ and σ are mean and standard deviation of sth feature values of documents N_i of class C_i. Hence, a representative interval vector of d' values are created for class C_i and it is represented as $SV_i = [SV_1^i, SV_2^i, SV_1^i, ..., SV_d^i]$. Here $SV_s^i = [\mu_s - \sigma_s, \mu_s + \sigma_s]$ is the interval vector for the s_{th} feature of the class C_i. The above method is applied for the creation of symbolic vectors for all the Q classes to obtain Q representative vectors for all the classes and stored in the knowledge base for the text classification.

The above method is applied for the creation of symbolic vectors for all the Q classes to obtain Q representative vectors for all the classes and stored in the knowledge base for the text classification.

2.4 Symbolic Classification

The proposed method uses classifier of [4] for symbolic classification of text data. This symbolic classifier well suits for interval valued feature vector which uses nearest neighbour approach in classifying test sample to its predefined class.

During testing, the symbolic classifier is used to classify an unknown test document to a predefined class using only d features. The test document features are matched against knowledge base which is already stored earlier to get the corresponding label of the test document.

3 Proposed Model

3.1 Datasets

To evaluate the efficiency of the proposed model we have conducted experiments on two benchmark imbalanced text datasets. The first one benchmark dataset is Reuters-21578 which is collected from Reuters newswire and we have considered a total 7285 documents from top 10 classes out of 135 classes with features 18221 dimensions. The second dataset which is considered for our experiments is TDT2. A total of 8741 documents have been considered from the top 20 classes out of the 96 classes with 36771 dimensions. The class information of the two datasets before and after clustering is described in Table 1. Similarly, the documents distribution of the datasets before and after clustering is shown in Fig. 3.

Table 1. Class information of the datasets before and after clustering

Dataset	No. of classes before clustering	No. of classes after clustering
Reuters-21578	10	102
TDT2	20	182

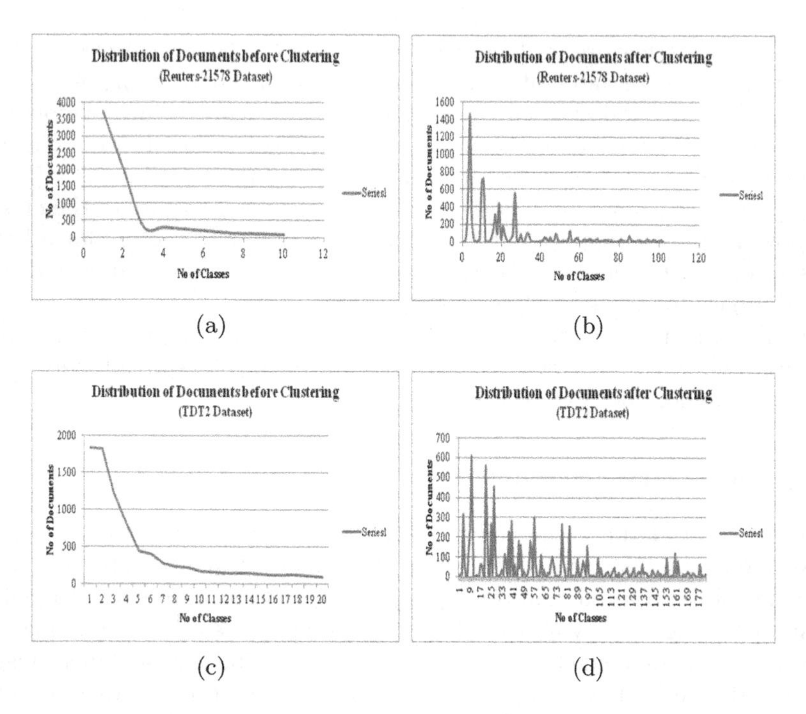

Fig. 3. Documents distribution of two different datasets, (a) Reuter-21578 before clustering, (b) Reuter-21578 after clustering, (c) TDT2 before clustering, (d) TDT2 after clustering.

3.2 Experimental Setup

The experiments are conducted to evaluate the method and to measure the efficiency of the developed model by considering different training and testing sets. Experimentation is conducted on each dataset by considering the 70% of training and 30% of testing samples of the datasets. The efficiency is measured and evaluated in terms of Precision, Recall and F-measure using both micro and macro averaging. The next sections presents results of the imbalanced datasets considered for experimentation.

3.3 Results and Analysis

In this section, we present the results of the proposed method on both the datasets. The experiments were carried out by considering different number of clusters with different inconsistency coefficient value and an optimal number of clusters is decided which produces best results and later these clusters considered as classes. More importantly to evaluate the goodness of this model a comparative analysis with the available models on these imbalance datasets is performed.

Tables 2 and 3 show the results of the proposed method on Reuters-21578 and TDT2 datasets respectively in terms of macro precision, macro recall, macro f-measure and micro f-measure. From these results we can ob serve that the performance is increasing gradually with less number of features to higher number of features.

Table 2. Performance of the proposed model on Reuters-21578-10 dataset on 102 clusters with 70% training and 30% testing (5100 and 2185 documents)

No. of feature groups	Total features	Macro-P	Macro-R	Macro-F	Micro-F
1	88	0.7207	0.7300	0.7253	0.8001
2	175	0.7661	0.8015	0.7834	0.8179
3	264	0.7730	0.8205	0.7960	0.8152
4	355	0.7896	0.8244	0.8066	0.8179
5	441	0.7990	0.8389	0.8184	0.8253
6	528	0.7858	0.8417	0.8128	0.8230
7	622	0.8243	0.8538	0.8388	0.8285
8	714	0.8286	0.8643	0.8461	0.8454
9	803	0.8215	0.8627	0.8416	0.8468
10	895	0.8262	0.8622	0.8438	0.8527

Table 3. Performance of the proposed model on TDT2-20 dataset on 182 clusters with 70% training and 30% testing (6118 and 2623 documents)

No. of feature groups	Total features	Macro-P	Macro-R	Macro-F	Micro-F
1	145	0.9365	0.8738	0.9041	0.8852
2	283	0.9318	0.9075	0.9195	0.9073
3	418	0.9499	0.9200	0.9347	0.9336
4	551	0.9574	0.9282	0.9426	0.9393
5	686	0.9552	0.9272	0.9410	0.9386
6	822	0.9587	0.9159	0.9368	0.9344
7	969	0.9650	0.9189	0.9414	0.9378
8	1109	0.9718	0.9297	0.9503	0.9458
9	1250	0.9748	0.9337	0.9538	0.9489
10	1384	0.9714	0.9336	0.9521	0.9466

To compare the performance of the proposed method with that of the available methods, we have selected three methods which try to handle the class

Table 4. Comparison of the results of the proposed method with the other existing techniques on Reuters - 21578

Method	Macro-F (No. of features)	Micro-F (No. of features)
ACO [1]	0.7842 (>=3600)	0.8908 (>=3600)
DFS + SVM [20]	0.6655 (200)	0.8633 (500)
ALOFT + MOR [12]	0.6213 (135)	0.8047 (135)
MFD + BNS [11]	0.6400 (254)	0.8151 (254)
DFS, RDC + SVM [14]	0.6347 (500)	0.8198 (500)
IG + IGFSS + SVM [19]	0.6753 (500)	0.8647 (300)
Proposed method clustering + Feature selection + symbolic representation	**0.7253 (88)**	**0.8000 (88)**

imbalance by performing class wise clustering. The first method [17] uses class wise clustering for removing class imbalance and χ^2 for feature selection. The second method [10] uses class wise clustering for handling class imbalance and TCR for representation. The third method [18] uses class wise clustering for handling class imbalance and TCR for representation. In [10] each document is represented as feature vector of dimension equal to the number of classes originally present in the dataset whereas in the [18] method each document is represented by a feature vector of dimension equal to the number of clusters identified after class wise clustering.

Table 4 presents the results of existing methods with the proposed method without balancing the classes in terms of macro-F and micro-F for both the datasets with number of features considered. It is showing that the proposed method is better than the [11,12,14] as it considers very less features and the Micro-F is almost same but the Macro-F is better than these method and comparing with the [1,20], the performance is little poor but the features considered in these method is very high which is difficult to process compare to the proposed method. The proposed method also selects at a time group of highly relevant and non redundant features. In [5] K number of features are selected at a time but in this method at a time Q top ranked features are selected Q ($\gg K$). Because in this proposed method K class problem is considered as Q class problem. Hence the proposed method selects the features to the number of clusters so that quickly we can select optimal group of features for text classification.

Table 5 shows the results of the proposed method with that imbalance to balance methods conversion of [10,17,18] in terms of macro-F and micro-F for both the datasets. The number of features used and the total number of clusters formed are also shown. It can be observed from the Table 5 that the proposed method work is better than the model of [10] in terms of both macro-F and micro-F is equal for Reuters-21578 dataset and for TDT2 dataset the performance in terms of both macro-F and micro-F is better than the both the methods of [10,18]. When it comes to the model of [17], the proposed model has less performance. But the number of features used by [17] is very high which compared to the number of features used by the proposed method. Thus the model proposed by [17] is very complex as it involves handling of very high dimensional feature vectors.

Table 5. Comparison of the proposed method against other existing imbalanced methods [10,17,18] with 70% training and 30% testing for Reuters-21578 and TDT2 datasets.

Text corpus	Method	No. of features	Maximum No. of clusters formed	Macro-F	Micro-F
Reuters-21578	Class wise cluster+ Chi-square [17]	5000	453	0.8967	0.9351
	Cluster+ TCR [10]	10	453	0.6843	0.8000
	Method [18]	102	102	0.7471	0.8693
	Proposed method	**88**	**102**	**0.7253**	**0.8000**
TDT2	Class wise cluster+ Chi-square [17]	1000	339	0.9570	0.9631
	Cluster+ TCR [10]	20	339	0.7600	0.8353
	Method [18]	182	182	0.79400	0.8628
	Proposed Method	**145**	**182**	**0.9041**	**0.8852**

4 Conclusion

In this paper, we have proposed a class wise cluster based symbolic representation for imbalanced text classification using feature selection method. To validate our results we have conducted experiments with two different datasets viz. Reuters-21578 and TDT2. The experimental results show that the proposed method works better with the feature transformation methods. Hence the classifier trained on a d' dimensional feature vector can be used to capture the variations across different classes. In future, the text classification can be conducted by this method using different feature selection techniques and clustering documents by considering various parameters like number of clusters and clustering techniques.

Acknowledgement. The author N Vinay Kumar acknowledges the Department of Science and Technology, Govt. of India for their financial support rendered through DST-INSPIRE fellowship.

References

1. Aghdam, M.H., Aghaee, N.G., Basiri, M.E.: Text feature selection using ant colony optimization. Expert Syst. Appl. **36**(3)-2, 6843–6853 (2009)
2. Azam, N., Yao, J.: Comparison of term frequency and document frequency based feature selection metrics in text categorization. Expert Syst. Appl. **39**, 4760–4768 (2012)
3. Elhadad, M.K., Khaled, M., Badran, K.M., Salama, G.: A novel approach for ontology-based dimensionality reduction for web text document classification. In International Conference on Information Systems (ICIS) - 2017, vol. 978, pp. 5090–5507. IEEE (2017)
4. Guru, D.S., Nagendraswamy, H.S.: Symbolic representation of two-dimensional shapes. Pattern Recognit. Lett. **28**, 144–155 (2006)
5. Guru, D.S., Suhil, M., Guru, D.S., Lavanya, N.R., Vinay Kumar, N.: An alternative framework for univariate filter based feature selection for text categorization. Pattern Recognit. Lett. **103**, 23–31 (2018)
6. Guru, D.S., Suhil, M.: A novel term class relevance measure for text categorization. Procedia Comput. Sci. **45**, 13–22 (2015)
7. Harish, B.S., Guru, D.S., Manjunath, S.: Representation and classification of text documents: a brief review. IJCA Spec. Issue Recent. Trends Image Process. Pattern Recognit. (RTIPPR) 110–119 (2010)
8. Jiang, S., Pang, S., Wu, M., Kuang, L.: An improved K-nearest-neighbor algorithm for text categorization. Expert Syst. Appl. **39**, 1503–1509 (2012)
9. Junejo, K.A., Karim, A., Tahir, M.H., Jeon, M.: Terms-based discriminative Information space for robust text classification. Inf. Sci. **372**, 518–538 (2016)
10. Raju, L.N., Suhil, M., Guru, D.S., Gowda, H.S.: Cluster based symbolic representation for skewed text categorization. In: Santosh, K.C., Hangarge, M., Bevilacqua, V., Negi, A. (eds.) RTIP2R 2016. CCIS, vol. 709, pp. 202–216. Springer, Singapore (2017). https://doi.org/10.1007/978-981-10-4859-3_19
11. Pinheiro, R.H.W., Cavalcanti, G.D.C., Ren, T.I.: Data-driven global-ranking local feature selection methods for text categorization. Expert Syst. Appl. **42**, 1941–1949 (2015)
12. Pinheiro, R.H.W., Cavalcanti, G.D.C., Correa, R.F., Ren, T.I.: A global-ranking local feature selection method for text categorization. Expert Syst. Appl. **39**, 12851–12857 (2012)
13. Rehman, A., Javed, K., Babri, H.A., Saeed, M.: Relative discrimination criterion - a novel feature ranking method for text data. Expert Syst. Appl. **42**, 3670–3681 (2012)
14. Rehman, A., Javed, K., Babri, H.A.: Feature selection based on a normalized difference measure for text classification. Inf. Process. Manag. **53**, 473–489 (2017)
15. Sabbaha, T., Selamat, A., Selamat, M.H., Fawaz, S., Viedmae, A.E.H., Krejcarg, O.: Modified frequency-based term weighting schemes for text classification. Appl. Soft Comput. **58**, 193–206 (2017)
16. Sebastiani, F.: Machine learning in automated text categorization. ACM Comput. Surv. **34**(1), 1–47 (2002)
17. Suhil, M., Guru, D.S., Lavanya, N.R., Harsha, S.G.: Simple yet effective classification model for skewed text categorization. In: International Conference on Computing, Communications and Informatics (ICACCI)-2016. IEEE, pp. 904–910 (2016)

18. Swarnalatha, K., Guru, D.S., Anami, B.S., Suhil, M.: Classwise clustering for classification of imbalanced text data. In: Sridhar, V., Padma, M.C., Rao, K.A.R. (eds.) Emerging Research in Electronics, Computer Science and Technology. LNEE, vol. 545, pp. 83–94. Springer, Singapore (2019). https://doi.org/10.1007/978-981-13-5802-9_8. Text categorization. Expert Syst. Appl. **49**, 31–47 (2016)
19. Uysal, A.K.: An improved global feature selection scheme for text classification. Expert Syst. Appl. **43**, 82–92 (2016)
20. Uysal, A.K., Gunal, S.: A novel probabilistic feature selection method for text classification. Knowl.-Based Syst. **36**, 226–235 (2012)
21. Wang, D., Zhang, H., Li, R., Lv, W., Wang, D.: t-Test feature selection approach based on term frequency for text categorization. Pattern Recognit. Lett. **45**, 1–10 (2011)
22. Yang, J., Liu, Y., Zhu, X., Liu, Z., Zhang, X.: A new feature selection based on comprehensive measurement both in inter-category and intra-category for text categorization. Inf. Process. Manag. **48**, 741–754 (2012)
23. Yang, Y., Pedersen, J.O.: A comparative study on feature selection in text categorization. In: Proceedings of the 14th International Conference on Machine Learning, vol. 97, pp. 412–420 (1997)
24. Zeina, D.A., Fawaz, S., Anzi, A.: Employing fisher discriminant analysis for Arabic text classification. Comput. Electr. Eng. **000**, 1–13 (2017)
25. Zong, W., Wu, F., Chu, L.K., Sculli, D.: A discriminative and semantic feature selection method for text categorization. Int J. Prod. Econ. **165**, 215–222 (2015)

A Combined Architecture Based on Artificial Neural Network to Recognize Kannada Vowel Modifiers

Siddhaling Urolagin(✉)

Department of Computer Science, Birla Institute of Technology and Science, Pilani-Dubai, Dubai, UAE
siddhaling@dubai.bits-pilani.ac.in

Abstract. The document image analysis for Indian scripts such as Kannada poses many challenges due to particular characteristics of the script. The Kannada script has huge set of characters consists of vowels, consonants, consonant conjuncts. The character set also includes compound characters which are formed using the basic symbols. A typical procedure to perform Kannada character recognition is to segment words and characters from the document then carry out recognition. But Kannada has the larger character set, and such an approach will have many classes to recognize. Another method is to segment the character into basic symbols and then perform recognition of the basic symbols. The glyph corresponding to a Kannada character has mainly two parts: consonant and vowel modifiers. This paper, a combined architecture is proposed to perform recognition of Kannada vowel modifiers. Gabor filters are employed to carry out precise segmentation of character into basic symbols. A combined architecture using K-Mean clustering and Artificial Neural Network is developed to recognize segmented vowel modifiers. The 10-fold cross validation is performed and an overall recognition rate of 95.04% is observed.

Keywords: Combined classifier · Vowel recognition · Kannada document image analysis

1 Introduction

The Optical Character Recognition (OCR) is concerned with recognizing script specific characters. OCR has found many potential applications such as license plate recognition [1], retrieving book information [2], engineering drawings recognition [3], translation and speech recognition [4], etc. In recent years researchers have focused on developing OCR system for scripts such as Hindi, Kannada, Tamil, Gujarati, etc., [5]. A survey OCR system for Indian languages is presented in [6]. Indian languages have a large number of character. As noted by authors in [6] compared to European languages OCR system for Indian languages are more intricate. Kannada the official language of Karnataka,

K. C. Santosh and R. S. Hegadi (Eds.): RTIP2R 2018, CCIS 1037, pp. 206–215, 2019.
https://doi.org/10.1007/978-981-13-9187-3_19

which has 19090 characters or *aksharas* (characters). Nature of Kannada characters is described in [7]. Many of Kannada *aksharas* have higher similarity in shape [8] causing automatic recognition a challenging task. In [9] OCR system using brick structure is developed. A Radial Basis Function (RBF) network using subspace projection is presented in [8]. Artificial Neural Network (ANN) with wavelet features is proposed in [10]. Kannada vowels (16) and numbers (10) recognition is carried out using modifier invariant moments and nearest neighbor classifier [11]. A two-stage multi neural network on wavelet features is presented in [12]. In [13], a document analysis system for multilingual such as South Indian and English documents is discussed. This multilingual system uses Fourier transform and principal components as features. Support Vector Machines (SVM) based recognition system is described in [7]. In [14] a database approach is used to perform OCR on scanned Kannada documents. A complete OCR for Kannada is discussed in [15]. Authors have developed preprocessing, segmenting, character recognition. They employed a dictionary based post-processing to increase OCR accuracy. A review on Kannada OCR is presented in [16].

The recognition of Kannada characters is complicated because of the character set has many characters and several of them are composed using the basic symbols. Therefore an effective approach is to perform character level segmentation of Kannada *aksharas* and then carry recognition. This paper address the recognition of vowel modifiers which are combined with consonants to form *aksharas*. Initially, glyph corresponding to vowel modifiers is separated using independent component isolation and character level segmentation. A combine architecture is developed using ANN to perform Kannada vowel modifier recognition. In Sect. 2 Kannada script is described and the proposed combined architecture is discussed in Sect. 3. In Sect. 4 Gabor filters based feature extraction is elaborated. The Sect. 5 describes experimental results. Finally in Sect. 6 the conclusion is covered.

2 Kannada Script

The Kannada script has 16 vowels and 34 consonants as basic characters. These are also considered as *aksharas*. *Aksharas* are also composed from the combination of vowels and consonants. The Kannada *aksharas* can be: (i) A basic vowel or a consonant, (ii) A consonant in combination with a vowel and (iii) A consonant in combination with consonants and a vowel. The Kannada vowels are in Fig. 1(a) and consonants are in Fig. 1(d) illustrated. When a vowel forms an *akshara* then it normally appears at the beginning of a word. *Aksharas* composed using the consonant along with a vowel modifiers are depicted in Fig. 1(c). When vowel modifiers combined with a consonant they may appear separately or attached with the consonant. The Fig. 1(b) shows vowel modifiers corresponding to the vowels. The glyph of the vowel modifier is attached to all consonants mostly in a similar way, although, in a few cases the glyphs of the vowel modifier may change depending on the consonant. When a vowel is attached to a consonant, it glyph may be combined to top, right, or bottom positions of the consonant. The *aksharas* are formed by combining several consonants (at most three consonants) with a vowel as depicted in Fig. 1(f). A base consonant (shown in Fig. 1(d)) is combined with consonant conjuncts (*vatthus*) illustrated in Fig. 1(e) and a vowel modifier (shown in Fig. 1(b)). The *vatthus*, when combined with a base consonant, will appear below it.

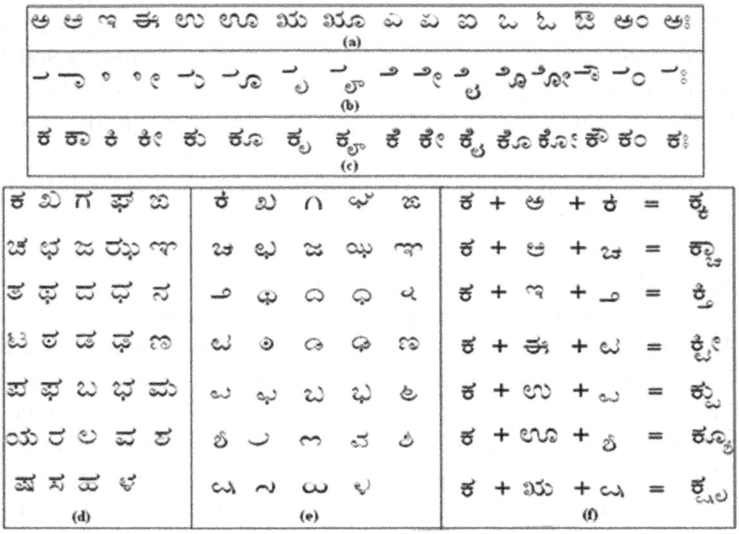

Fig. 1. Kannada Script (a). Vowels. (b). Vowel modifiers. (c) Consonant with a vowel. (d) Consonants. (e) Consonant conjunct or *vatthus.* (f). Consonant-conjunct and a vowel.

In Kannada script, there are 16 vowels, 34 consonants in its alphabet. These 16 vowels can be combined with 34 consonants resulting in 544 *aksharas.* Moreover base consonant along with consonant conjunct and vowel modifier make 18496 possible *aksharas.* Thus in Kannada, there are 19090 *aksharas.*

3 Combined Architecture to Recognize Vowel Modifiers

The combined architecture for the recognition of vowel modifiers is developed, which is shown in Fig. 2. There are 16 vowels and corresponding to them there are 16 vowel modifiers (Fig. 1(b)) which are attached to consonants to form compound *aksharas.* The developed system takes input as a compound *akshara* and then its size is normalized 25 rows and columns accordingly taking into account aspect ratio to perform size independent recognition. The segmentation and independent component identification on *aksharas* carried out and various portions are shown in Fig. 3. The independent components that are present in the *aksharas* are identified and isolated as shown in Fig. 3(a) using method similar to [17].

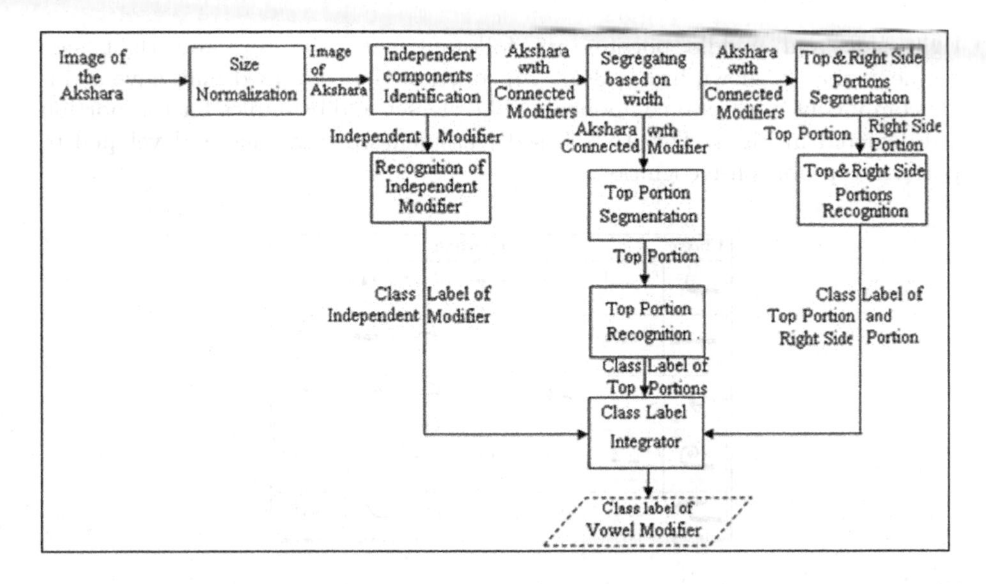

Fig. 2. Architecture of developed system for vowel modifiers recognition.

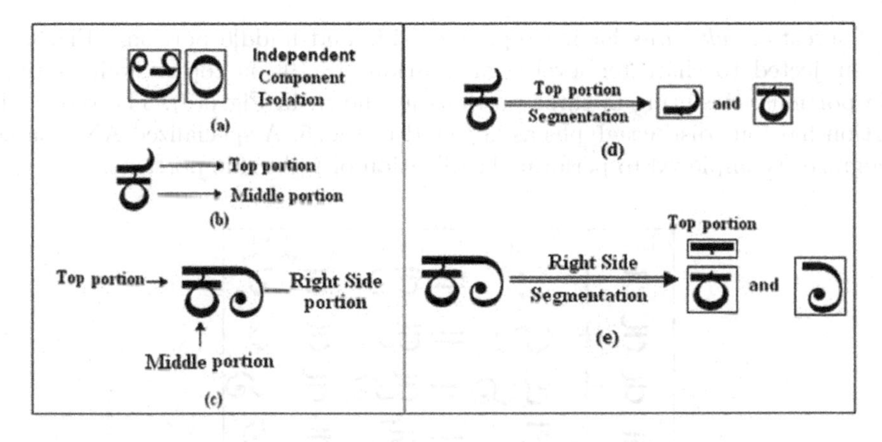

Fig. 3. (a) Isolating independent components of an *akshara*. (b) Top and middle portions. (c) Top, right side and middle portions. (d) Top segmentation. (e) Top and right side segmentation.

The isolated independent vowel modifiers are {ೆ, ು, ೌ, ೃ, ಂ, ಃ}. A specialized classifier based on ANN is developed to recognize these vowel modifiers.

The *akshara* may have top and right side vowel modifier attached to it. Based on the width of the *aksharas*, they are segregated into *aksharas* having only top and middle portion (Fig. 3(b)) or top, right side and middle portion (Fig. 3(c)). The *aksharas* of having top and middle portion are subjected to character level segmentation [18,19] as shown in Fig. 3(d). Aksharas which have the top, right side and middle portion are followed both the top and right side segmentation as shown in Fig. 3(e). Top portion thus obtained may have a single component or several components. The classes and examples for top portion are depicted in Fig. 4. The ANN based classification technique is developed to perform top portion recognition.

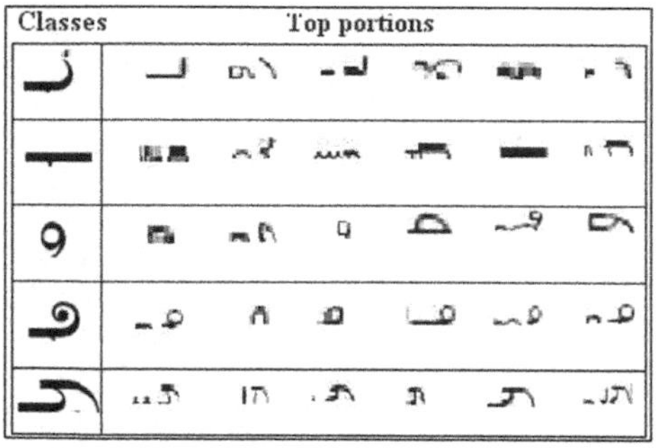

Fig. 4. Five classes of top portions are shown along with different shapes present in each class.

Thereafter, *aksharas* having top, right side and middle portions (Fig. 3(c)) are subjected to character level segmentation [18,19] on top as well as right side portions. Resulting in three portions as shown in Fig. 3(e). The right side portion has four distinct glyphs as depicted in Fig. 5. A specialized ANN based recognizer is employed to perform classification of right side portions.

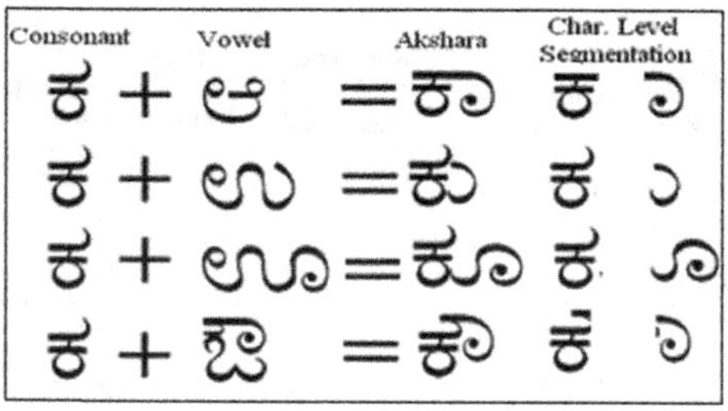

Fig. 5. Four distinct right side portions.

A class label integrator is devised, which collects class labels predicted from recognition of independent modifier, a top portion, and right side portion and subsequently decides the class label of the vowel modifier as shown in Fig. 2.

4 Gabor Filter Based Feature Extraction

Gabor filters based feature extraction scheme is utilized to collect directional features from a segmented portion of *aksharas*. Gabor filters are used to extract features for segmented portions. The 2D Gabor filters have been applied for edge and line detection as in [20,21], features extraction as in [22,23], texture classification [24,25], etc. The Gabor filter $\psi(x,y)$ is two dimensional as given in [26],

$$\psi(x,y) = \frac{1}{2\Pi\alpha\beta}e^{-\frac{1}{2}[(\frac{x'}{\alpha})^2+(\frac{y'}{\beta})^2]}e^{2\Pi jf_0x'} \quad (1)$$

Where $x' = xcos\theta + ysin\theta$ and $y' = ycos\theta - xsin\theta$, (x,y) are 2-dimensional coordinates, f_0 is the fundamental frequency for a filter, θ defines Gaussian envelope used to create a filter. The α and β are the sharpness of the Gaussian along the major axis and minor axis respectively. The 2-dimensional convolution operation is performed on an image $\xi(x,y)$ by the Gabor filter $\psi(x,y)$ to obtain the response $r_\xi(x,y)$.

$$r_\xi(x,y) = \psi(x,y)_{f_0\alpha\beta\theta} * \xi(x,y) = \int\int \psi(x - x_T, y - y_T)\xi(x_T, y_T)dx_T dy_T \quad (2)$$

Where x_T and y_T are integration variables. A general procedure is followed to extract features from segmented portions such as independent vowel modifier, top portions, and a right side portion.

Algorithm 1. Procedure for Feature Extraction

i. Segment and isolate the portion of an *akshara*.
ii. Create a bank of Gabor filters with the angle set as $\theta = \{\theta_1, \theta_2....\theta_n\}$ and frequencies as $f = \{f_1, f_2, ...f_n\}$, $\alpha = 4$ and $\beta = 4$.
iii. Perform the convolution on segmented *akshara* portion using a bank of Gabor filters.
iv. Directional features are accumulated using method similar to [27] from filter responses.

The features are extracted by accumulating the Gabor jets in a spatially localized manner using a grid structure [27]. Directional features accumulating function $fe(r_\xi, \theta)$ is defined, which takes the parameters as the response r_ξ of size mxn and the θ and accumulates the η number of directional features along angle θ in the width δ using a grid structure as shown in Fig. 6.

The angles and frequencies are selected based on the shape of the glyph to be recognized. There are five different independent vowel modifiers can be associated with *aksharas*. Five glyphs corresponding to these independent modifiers are {ೆ, ು, ೂ, ೃ, ಂ, ಃ}. There are five distinct glyphs corresponding to top portions, these are {[illegible]}. Four distinct glyphs are associated with *akshara* on right side. The glyphs are {[illegible]}. In Table 1, the selected angles and frequencies to create Gabor banks specific to portions of *aksharas* are given.

In this research, the procedure of [28] is used to select the parameters of the Gabor filters. The procedure presented in [28] extends the concept of "Information Diagram" (ID) of [23]. The ID represents the magnitude of the Gabor response at a certain interest point (x,y) as a function of θ, α and β while keeping f_0 constant.

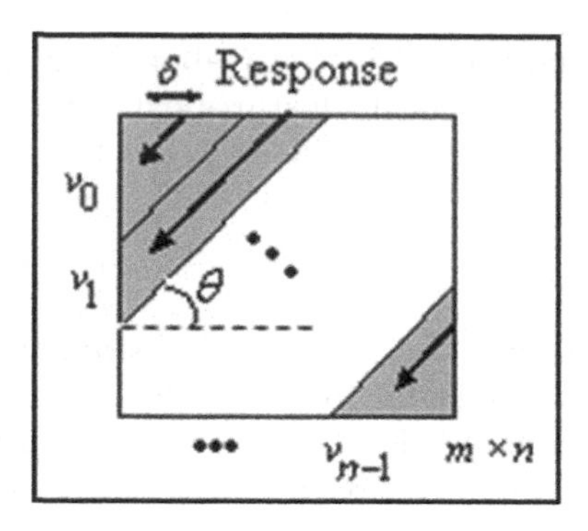

Fig. 6. Directional features accumulating function $fe(r_\xi, \theta)$.

Table 1. Selected angles and frequencies to form Gabor bank.

Sl. Num.	Portion	Shapes	Angles	Frequencies
1.0	Independent modifier	{ೆ, ು, ೂ, ೃ, ಂ, ಃ}	0, 30, 45, 90, 120, 120	0.3, 0.2, 0.19, 0.29, 0.19, 0.22
2.0	Top Portion	[illegible]	0, 80, 90, 100, 142	0.16, 0.19, 0.225, 0.21
3.0	Right Side Portion	[illegible]	0, 45, 90, 135	0.12, 0.225

5 Experimental Results

Experiments are conducted to recognize vowel modifiers on the developed architecture. The database for Kannada script in this work has been prepared from fonts of Baraha multilingual editor. The fonts used include BRH Amerikannada, BRH Bengaluru, BRH Kannada, BRH Kasturi, BRH Sirigannada and BRH Vijay. Multiple pages of Kannada text having different font type and different font size are prepared and scanned through HP Laserjet 3055 scanner. The database thus created includes a total of 10771 characters. Out of these, for experiments on vowel modifiers recognition, 4200 *aksharas* are collected which are composed of the consonants with vowel modifiers. The size of *akshara* is normalized such that it has 25 rows and number of columns is adjusted proportionally taking into consideration its aspect ratio. The ANNs along with Gabor features are used to recognize various portions of the vowel modifiers. Here feedforward ANN with backpropagation learning algorithm is used. Several experiments have been conducted on ANN with varying topologies and each ANN is test using the 10- fold cross-validation. In Table 2 the recognition results are summarized for each ANN.

The recognition results for portions such as independent modifiers, top, right side portions are tabulated in Table 2. The recognition of independent modifiers is performed on ANN with four layers architecture 55-40-20-6, 55 input layer, two hidden layer of 40 and 20 nodes respectively followed by 6 nodes in output layer. The recognition of 82.22% is observed in cross-validation. The top portion

Table 2. Recognition rate in percentage for ANN.

Portion of *akshara*	Topology of ANN	Number layers in ANN	Input nodes	Output nodes	Recognition rate in percentage
Independent modifiers	55-40-20-6	4	55	6	82.22
Top portions	60-40-20-4	4	60	4	90.95
Right side portions	64-30-15-3	4	64	3	99.12

recognition rate of 90.95% is obtained on ANN with topology 60-40-20-4. The right side portion recognition is carried out using the ANN of 64-30-15-3 and observed recognition rate of 99.12%. Vowel modifiers and their class are represented as: {ಾ, ಿ, ೀ, ು, ೂ, ೃ, ೆ, ೇ, ೈ, ೊ, ೋ, ೌ, ಂ, ಃ}. After a given *akshara* size normalized, its different components corresponding to vowel modifiers are isolated, segregated and segmented. The specialized ANN with Gabor features are employed to perform recognition of separated portions. The recognition results obtained at various stages are combined using a class label integrator to predict class label of a vowel modifier. The class integrator uses simple rules to decide the class label of vowel modifier.

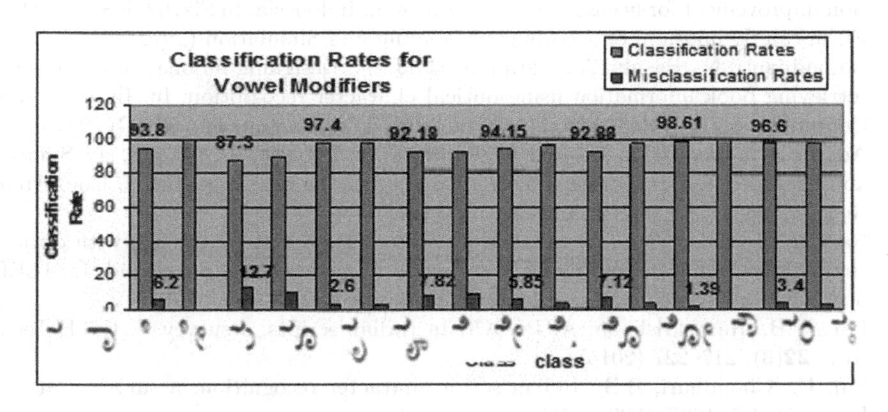

Fig. 7. Classification rates in % for each vowel modifier.

For example, if the class label of independent modifier is ಂ then the class of vowel modifier has to be ೊಂ. On the other hand, if independent modifier recognized is ೀ, then there are 3 possible classes: ೀ, ೇ, or ೋ. The class of vowel modifier is found by segregating *akshara* based on the width and predicted class label of the top portion. With such inferences rules, the recognition rates of vowel modifiers are higher than recognition rates of each portion. The Fig. 7 shows classification rates obtained for each class of vowel modifier on 10-fold cross-validation. An overall classification rate of 95.04% is observed.

6 Conclusion

Optical Character Recognition and Document Image Analysis are gaining popularity due to their potential applications. Indian languages such as Kannada, Hindi, and Gujarati etc., have a large collection of character and making the task of recognition difficult. Moreover, the Kannada language has many characters formed from several basic characters. A better approach for recognition is to perform character level segmentation and separate into various portions. Next, carry out recognition of segmented portions. In this paper, vowel modifier recognition architecture is developed for the Kannada. The Kannada *akshara* is subjected to independent component isolation and character level segmentation to obtain independent modifiers, top and right side portions. Artificial Neural Network along with Gabor features are employed to perform recognition of various portions separated from *aksharas*. Recognition performed by individual ANNs are combined to using a class integrator to predict vowel modifiers. The developed system is tested using 10-fold cross-validation and overall recognition rate of 95.04% is obtained.

References

1. Hidayatullah, P., Syakrani, N., Suhartini, I., Muhlis, W.: Optical character recognition improvement for license plate recognition in Indonesia. In Sixth UKSim/AMSS European Symposium on Computer Modeling and Simulation (2012)
2. Ramadijanti, N., Basuki, A., Agrippina, G.J.W.: Designing mobile application for retrieving book information using optical character recognition. In: International Conference on Knowledge Creation and Intelligent Computing (KCIC) (2016)
3. Adam, S., Ogier, J.M., Cariou, C., Mullot, R., Labiche, J., Gardes, J.: Symbol and character recognition: application to engineering drawings. Int. J. Doc. Anal. Recognit. **3**(2), 89–101 (2000)
4. Ramiah, S., Liong, T.Y., Jayabalan, M.: Detecting text based image with optical character recognition for English translation and speech using Android. In: IEEE Student Conference on Research and Development (SCOReD) (2015)
5. Pati, P.B., Ramakrishnan, A.G.: OCR in Indian scripts: a survey. J. IETE Tech. Rev. **22**(3), 217–227 (2015)
6. Pal, U., Choudhuri, B.B.: Indian script character recognition: a survey. Pattern Recognit. **37**, 1887–1899 (2004)
7. Ashwin, T.V., Sastry, P.S.: A font and size independent OCR system for printed Kannada documents using support vector machines. J. Sadhana **27**(1), 35–58 (2002)
8. VijayKumar, B., Ramakrishnan, A.G.: Radial basis function and subspace approach for printed Kannada text recognition. In: Proceedings of IEEE International Conference on Acoustics, Speech, and Signal Processing, vol. 5, pp. 321–324 (2004)
9. Nagabhushan, P., Pai, R.M.: Modified region decomposition method and optimal depth decision tree in the recognition of non-uniform sized characters - an experimentation with Kannada characters. Pattern Recognit. Lett. **37**, 1467–1475 (1999)
10. Kunte, R.S., Samuel, R.S.: Wavelet descriptors for recognition of basic symbols in printed Kannada text. Int. J. Wavelets Multiresolution Inf. Process. **5**(2), 351–367 (2006)

11. Hangarge, M., Patil, S., Dhandra, B.V.: Multi-font/size Kannada vowels and numerals recognition based on modified invariant moments. IJCA Spec. Issue Recent. Trends Image Process. Pattern Recognit. Part **2**, 126–130 (2010)
12. Kunte, R.S. and Samuel, R.S.: An OCR system for printed Kannada text using two-stage multi-network classification approach employing wavelet features. In: Proceedings of International Conference on Computational Intelligence and Multimedia Applications, pp. 349–353. IEEE Computer Society Press (2007)
13. Aradhya, V.N.M., Kumar, G.H., Noushath, S.: Multilingual OCR system for south Indian scripts and English documents: an approach based on Fourier transform and principal component analysis. Eng. Appl. Artif. Intell. **21**(4), 658–668 (2008)
14. Sagar, B.M., Shobha, G., Ramakanth Kumar, P.: OCR for printed Kannada text to machine editable format using database approach. J. WSEAS Trans. Comput. Arch. **7**(6), 766–769 (2008)
15. Sagar, B.M., Shobha, G., Kumar, P.R.: Complete Kannada optical character recognition with syntactical analysis of the script. In: 2008 International Conference on Computing, Communication and Networking (2008)
16. Indira, K., Sethu Selvi, S.: Kannada character recognition system: a review. InterJRI Sci. Technol. **1**(2), 30–42 (2009)
17. Tonazzini, A., Bedini, L., Salerno, E.: IJDAR **7**, 17 (2004). https://doi.org/10.1007/s10032-004-0121-8
18. Urolagin, S., Prema, K.V., Krishna, R.J., Reddy, N.S.: Segmentation of inflected top portions of kannada characters using gabor filters. In: IEEE Second International Conference on Emerging Applications of Information Technology, Kolkata, India, 18–20 Feb 2011, pp. 110–113 (2011). https://doi.org/10.1109/EAIT.2011.66
19. Urolagin, S., Prema, K.V., Reddy, N.S.: A Gabor filters based method for segmenting inflected characters of kannada script. In: IEEE proceeding of International Conference on Industrial and Information System, pp. 108–113 (2010)
20. Mehrotra, R., Namuduri, K.R., Ranganathan, N.: Gabor filter-based edge detection. Pattern Recognit. **25**(12), 1479–1494 (1992)
21. Chen, J., Sato, Y., Tamura, S.: Orientation space filtering for multiple orientation line segmentation. IEEE Trans. Pattern Anal. Mach. Intell. (PAMI) **22**(5), 417–429 (2000)
22. Kamarainen, J.K., Kyrki, V., Kalviainen, H.: Noise tolerant object recognition using Gabor filtering. In: Proceeding of 14th International Conference on Digital Signal Processing, vol. 2, pp. 1349–1352 (2002)
23. Kamarainen, J.K., Kyrki, V., Kalviainen, H.: Fundamental frequency Gabor filters for object recognition. In: Proceeding of 16th International Conference on Pattern Recognition (ICPR), Quebec, Canada, vol. 1, pp. 628–631 (2002)
24. Bovik, A.C., Clark, M., Geisler, W.S.: Multichannel texture analysis using localized spatial filters. IEEE Trans. Pattern Anal. Mach. Intell. (PAMI) **12**(1), 55–73 (1990)
25. Jain, K., Farrokhnia, F.: Unsupervised texture segmentation using Gabor filters. Pattern Recognit. **24**(12), 1167–1186 (1991)
26. Yoshimura, H., Etoh, M., Kondo, K., Yokoya, N.: Gray-scale character recognition by Gabor jets projection. In: Proceedings of 15th International Conference on Pattern Recognition, vol. 2, pp. 335–338 (2000)
27. Urolagin, S., Prema, K.V., Subba Reddy, N.V.: Robust object recognition using binarized Gabor features under noise and illumination changes. Int. J. Inf. Process. **2**(2), 170–182 (2008)
28. Moreno, P., Bernardino, A., Santos-Victor, J.: Gabor parameter selection for local feature detection. In: Proceeding of 2nd Iberian Conference on Pattern Recognition and Image Analysis, pp. 11–19. IBPRIA, Estoril, Portugal (2005)

English Character Recognition Using Robust Back Propagation Neural Network

Shrinivas R. Zanwar[1](✉), Abbhilasha S. Narote[2], and Sandipan P. Narote[3]

[1] CSMSS, Chh. Shahu College of Engineering, Aurangabad, India
shrinivas.zanwar@gmail.com

[2] S.K.N.ś College of Engineering, Pune, India
a.narote@rediffmail.com

[3] Governmentś Residence Women's Polytechnic, Tasgaon, Sangli, India
snarote@rediffmail.com

Abstract. OCR deals with the handwritten or printed character recognition with the help of digital computers and soft computing. The scanned images of characters and numbers are used as input for system which is analyzed and transformed it into character codes, normally in ASCII format, which is taken for the data processing. Presently, there are lots of issues in recognition of characters and numbers, which can degrade the performance of the system in various ways. Mainly, the rate of recognition is not improved due to distributed neighborhood pixels of an image. Also, there are some techniques used for the OCR are having lack of contrast levels which is well known by fading of the image. So in this paper, the most important concern is to take such measures to enhance the performance of the system for automatic recognition of characters. Here, the operations are performed on the handwritten English alphabets. The dataset is selected from the Chars74K with different shapes and preprocessed it which deals with filtering and edge detection. Then, in the feature extraction process, features are extracted by using independent component analysis and swarm intelligence is used for feature vector selection. Classification of images are done with the back propagation neural network which gives an effective learning approach. The precise contribution which is evaluated in this research work is the uniqueness of classifications using a combination of the feature extraction and feature optimization (instance selection) using extraction of feature vectors. The performance of the developed system is measured in terms of recognition rate, sensitivity and specificity compared with the benchmark.

Keywords: Edge detection · Feature extraction (Independent component analysis and Swarm intelligence) · Backpropagation neural network

1 Introduction

Optical character recognition deals with the mechanical conversion of descriptions electronically, handwritten or published transcript into the machine

K. C. Santosh and R. S. Hegadi (Eds.): RTIP2R 2018, CCIS 1037, pp. 216–227, 2019.
https://doi.org/10.1007/978-981-13-9187-3_20

encoded transcript, which can be from a skim through the manuscript, a picture of a manuscript or from caption text covered on an appearance. OCR is extensively used as a procedure of data entry from published paper data proceedings which includes documents like passport, invoices, statements of bank, receipts generated through computer machines, business cards, printouts of the information, or every suitable documents which is a shared routine of digitizing published manuscripts so that they can be edited electronically, examined, stored more efficiently and also recycled in machine procedures like cognitive work out, machine conversion, text-to-speech scenarios, key information and text analysis [12,13]. OCR is a research field in artificial intelligence, recognition of patterns and computer vision. In the real world, there is a huge interest to modify over the published records into automated records for observing the security of applied information for the clients. Alongside these appearances, the necessity is to create personality acknowledgment software design agenda to achieve image analysis in terms of documents which perform deviations of records to the arrangement of electronic readings [14,15]. Generalized real time block diagram is represented in Fig. 1.

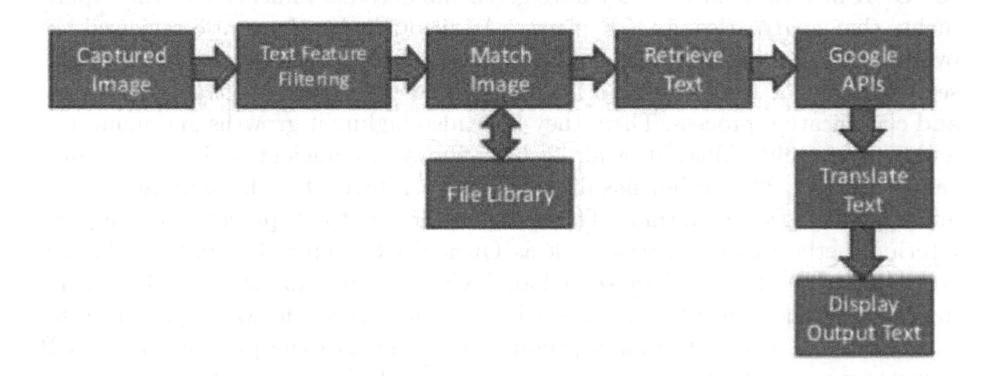

Fig. 1. Generalized real-time block diagram

For this process, there are diverse systems presented. In the middle of each approaches, Optical Character Recognition is available as the main crucial technique to perceive typescripts [1]. The vital goal is to quicken the technique of character response in record arrangements. Consequently, the agenda can prepare an incredible number of records within fewer periods and thus replaces the intervals. Currently, the major OCR construction which was shaped to change above the information is obtainable on papers into computers to get ready for the fit records through which the chronicles can be reusable and editable. The current outline or the past procedure of OCR is only OCR without framework practicality. That is the present outline accomplishes the homogeneous personality response or character recognition of single languages.

The paper is organized as, the related work is shortly expounded in Sect. 2. The proposed method is explained in Sect. 3, which consists of data set preparation, preprocessing, feature extraction and selection, and classifier. In Sect. 4, experimental results are shown in detail. Finally, conclusion is drawn in last Sect. 5.

2 Related Works

This section gives various researches are done for the classification of optical character recognition systems. Chaudhary, Garg, Behera et al. [1] shows the utilization of the optical character recognition concept for character recognition so as to achieve document analysis in terms of image, in which they have arranged the data in the grid form. They deal with the product agenda known as character recognition system. They have found that the need for OCR is to create the atmosphere in terms of programming framework to accomplish Document Image Analysis which shows the deviations of the records in their document organization to the electronic procedure. Hamad, Kaya et al. [6] investigated the OCR in various ways. They have given the detailed indication of the experiments that might arise in OCR phases. Additional, they have also reviewed the overall stages of an OCR scheme like preprocessing of the document or image, segmentation of the image, normalization process, feature extraction process and classification process. Then, they have also highlight growths and main uses and uses of OCR. Ahlawat et al. [2] have shown an efficient review of the processes of OCR. The author has discussed the strategies text from an image into arrangement readable format. The machines accept the typescripts through an altering method which makes work as Optical Character Recognition. Afroge, Ahmed, Mahmud et al. [7] presented an ANN-based method for the acknowledgment of characters in English among FNN named as feedforward system. Noise is measured as one of the main product that damages the presentation of CR (character recognition) system. Their net (network) has several layers named as input, hidden, output layers. The entire system is separated into two segments like training and acknowledgment segment. It comprises of acquisition of image, pre-processing and extraction of features. Lee, Osindero et al. proposed recursive RNN with consideration modeling of character recognition [8]. The main compensations of their systems deals with the usage of recursive CNN which deals with the efficient and actual extraction, learned language model, recurrent neural network, the usage of attention mechanism, image features and end-to-end preparation inside a normal back propagation structure. There are various method for omnidirectional processing of 2D images including recognizable characters [11]. Santosh et al. proposed dynamic time wrapping random based shape descriptor for pattern recognition used to match random features and avoided missing features [18–21].

3 Proposed Method

This is one of the crucial sections for the proposed implementation of the system which is based on optical character recognition system. This system deals with the automatic learning system based on training of the images. The proposed approach is mainly divided into two phases. The very first is the training phase in which the normalization of the image using filtration process and the feature extraction process is done and the last process of training deals with the feature optimization process using particle swarm optimization. The second phase deals with the classification approach which is done using back propagation neural network. The main consider which is to be kept in mind is the proper training of the system which is evaluated in terms of the mean square error rates with respect to the number of epochs. Figure 2 explains the block level flow diagram of the proposed approach. The very first block is the building of the graphical user interface for the human man made interactions.

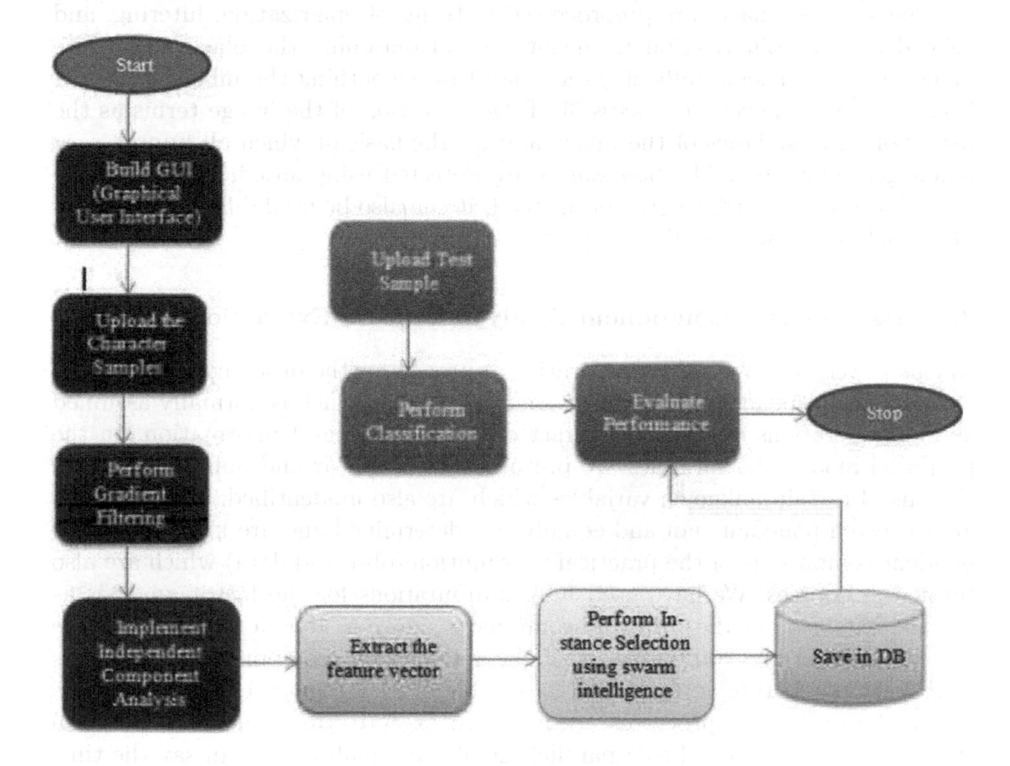

Fig. 2. Flow diagram

3.1 Dataset and Preprocessing

The dataset 3410 samples are selected from the Chars74K with different shapes drown using pc and tablet. This dataset is selected because of consideration of scanned images with the solved problem in constrained situations, like common character fonts and uniform background. Also, images are taken with popular cameras and hand held devices to overcome challenges in character recognition. The dataset consists of both English characters and numbers. We have considered the letters which are written in different styles for the samples. The English alphabets having total 62 classes of different styles of writing for each alphabet. Each Image consists of 900×1200 pixels by size which is high in dimensions and our proposed system is able to operate these high dimension images. In the training phase, 70% of the image data sets are trained and 30% of image dataset samples are taken in the testing phase. It will tell the flexibility of our proposed system to operate on large datasets in the training phase and classification in the testing phase.

The scanned images are preprocessed in terms of binarization, filtering, and edge detection. Binarization is the process of obtaining the black and white image (0 or 1). Then, gradient filter is used for smoothing the image so that it is easy for edge detection process [3]. Edge detection of the image terms as the detection of boundaries of the image and on the basis of which all unnecessary pixels get eliminated. The boundaries are detected using an efficient technique named as the canny edge detector, instead, it can also be used different operator like sobel, log, pewit etc. [5].

3.2 Independent Component Analysis Feature Extraction

In this paper, we have used a generative approach for the observed multivariate data using Independent Component Analysis (ICA), which is normally assumed as huge operations on data to extract out the meaningful information. In the proposed model, the variables are presumed to be linear and nonlinear combinations of certain unknown variables which are also unidentified. The variables are expected non-Gaussian and equally self determined and are known as independent components of the practical information (observed data) which are also termed as sources. We have used ICA computations for the factor analysis as it's the extension of the principle component analysis. It is a power computation process for the computations of the independent components as features to extract the feature vector which is capable of evaluating the factors analysis where the traditional processes are failed to execute. Here, the dimensions of the images are considered as a parallel set of the signals or we can say the time series. The blind source separation process is recycled in it.

The linear mixtures in terms of observing variables T_1, ..., T_n having total P independent components and A is the matrix of the element a_{pn} such that

$$T_p = a_{p1}s_1 + a_{p2}s_2 + a_{p3}s_3 + a_{p4}s_4 + ... + a_{in}s_n \mid for\ all\ p \tag{1}$$

So the notation of the vector matrix is given by

$$T = A * S \tag{2}$$

The S_i matrix is statistically independent. In this work we have used the non-Gaussian distribution which will be used to estimate the A matrix. After the estimation process of A matrix, we will perform its inverse process, say W and will obtain the independent feature vector using

$$s = W * T \tag{3}$$

Where, 's' will be the extracted independent feature values and is organized in the form of a feature vector which is closely related to the blind source separation process. Then the output of the feature vector will be the input to the optimization for the instance selection which is done using particle swarm optimization.

3.3 Particle Swarm Optimization for Selection of Features

PSO algorithm is a global algorithm, which the generalized process used to enhance the problem using iterative scenario which will offer the best explanations and clarifications from various solutions. The algorithm is developed by observing behavior of birds and fishes. In PSO, population is termed as swarm and individuals are particles. These particles will try to get nearest matching, which is termed as personal best (pbest) and global best (gbest) obtained by analyzing closeness of all particles of swarm. The gbest will gives overall best value and its location obtained by any particle in the population. It consists of changing the velocity of each particle as per values of pbest and gbest in each step. It is weighted by separate random numbers being generated for speeding up near pbest and gbest. PSO optimize random solutions particle from the population in D-dimensional space. So that, in PSO neither mutation calculation nor overlapping will be occurred. The collection of procedure limitations can put the huge effect on optimization consequences. In this paper (our research work) we are optimizing the feature set extracted from ICA which is basically known as the instance selection. It is the subset collection rate which is the process of selecting a subsection of appropriate features used in the construction of training model. The feature vector which is extracted from ICA makes new feature vector using operation on feature vector extracted using ICA. Choosing Selection of the input parameters for the PSO is one of the crucial tasks in instance selection process which deals with high performance in the problematic condition in performing optimization. The steps are as follows [9,10].

1. For every particle $j = 1, \ldots$ swarm do
   Swarm will be the total number of rows and columns generated using feature vector
   Set the particle's location with a consistently dispersed random vector X_i
   Set the particle's best recognized location to its initial location P_i

2. If $f(GP) < f(GB)$ then
   Get the swarm updated position which is the selection of the instance or the feature value
   Make the best particle speed V_i
3. Do until all swarms do
4. Evaluate the fitness function and generate the new solution for the next iteration T_i
5. Make updates on the particle's speed and location X_i and V_i
6. Generate the global best solution which is the total set of instance selections and optimize feature vector which is the global best solution G_b.
   Where G_b is the subsequent global best enhanced solution until all iterations are completed.

3.4 Back Propagation Neural Networks

The backpropagation neural network algorithm is proposed in two phases. The very first phase is to set up the input pattern in terms of the layers of the network. The network is repetitive in terms of the number of iterations until it stops training. The network layer consists of the hidden layer which is connected in the form of synaptic weights for the link stability.

This algorithm has been proposed by Sharma et al. [17], to identify the text with respect to the scanned images of the documents which is the form of handwritten printed format. For training and testing purpose this algorithm is useful. After completion of the training we will move to the training phase which deals with the uploading of the test sample comprise of English character which system will automatic classifies and on the bases of which the performance will be evaluated. We have taken the training set as the number of features covered in the feature vector on which the neural perform the back propagation training for the performance evaluations to achieve low updations of weights to make the connections stable.

4 Results and Discussions

This section deals with the valuable simulation of our proposed approach which is taken place in the MATLAB environment.

The Fig. 3 shows the graphical user interface panel using MATLAB toolbox in which human machine interaction panel is taken place and the user friendly environment is created. The user will click on the buttons and the user is able to perform actions and generate some events on it.

4.1 Training and Preprocessing

The Fig. 4 shows the training sample of the image of English Alphabet (D) which is to be processed for the further edge detections and filtration for the normalization process in part (a), where as part (b) shows the edge detection of the image which finds some edges of the image and on the basis of which all unnecessary pixels get eliminated. The Canny edge detection technique is performed.

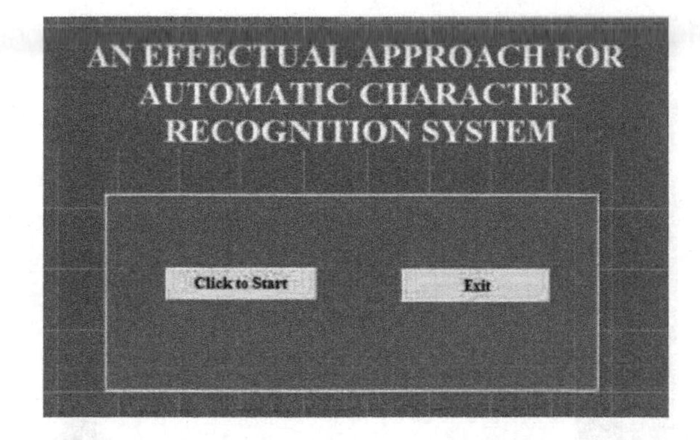

Fig. 3. Main GUI panel

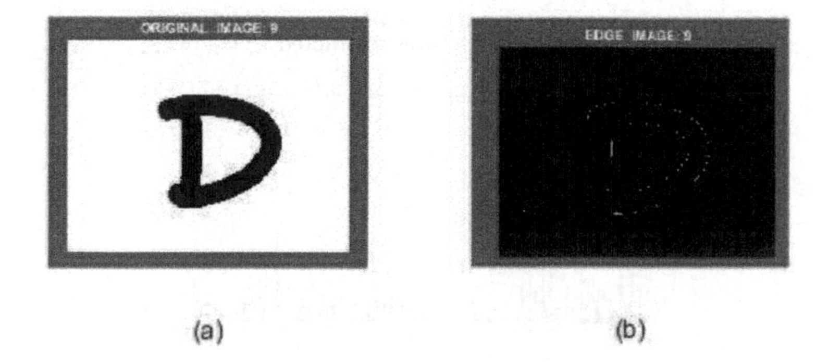

Fig. 4. (a). Training sample and (b). Edge detection

4.2 Feature Extraction

The feature extraction process shows the uploaded sample image is extracted in the content of the image which is done using ICA (Independent Component Analysis). ICA is one of the efficient techniques which uses blind source separation concept to find out the independent components which doesn't disturb the neighborhood pixels and find the characteristics which doesn't affect the other intensities of the image. As shown in Fig. 5, extracted features for sample 9, indicates independent values per bit.

4.3 Feature Vector Selection

As particle swarm optimization is used, it optimizes feature vector which deals with the relevant instances used for the classification of the sample in the testing phase and also acts as input which is directly fed to the neural network to train the whole system and build network layered model using sigmoid function as an activation function. Figure 6 shows optimized feature values.

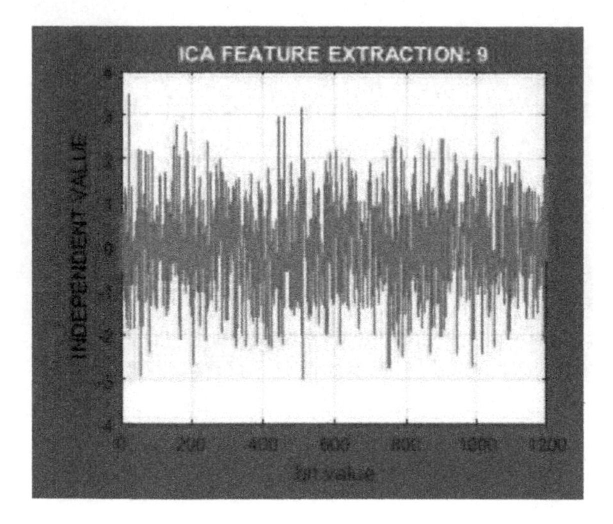

Fig. 5. Features extracted

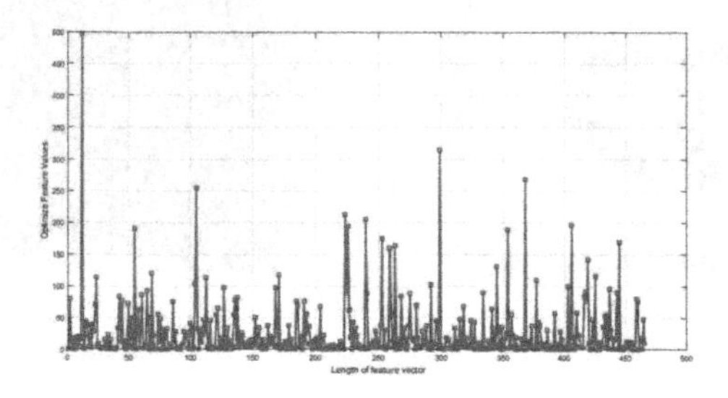

Fig. 6. Optimize feature vector

4.4 Neural Network Training

As shown in Fig. 7, the training of the system using backpropagation neural network which deals with the number of iterations. It signifies the proposed system is taking total 20 iterations out of max limit of iterations which are 1000 to train the whole system which shows the robustness and also the fast response and reaction time. It also deals with the gradient decedent optimization model which decreases the loss function to achieve low mean square error rate with respect to the number of trained samples.

4.5 Performance Evaluation

The Fig. 8 shows the classification output which is done using neural network that the uploaded sample deals with the image number 10 having the alphabet (D) in the training dataset which is automatically classified by the machine or system which shows that our proposed approach is able to perform automatic classification for optical character recognition of the English alphabets [4].

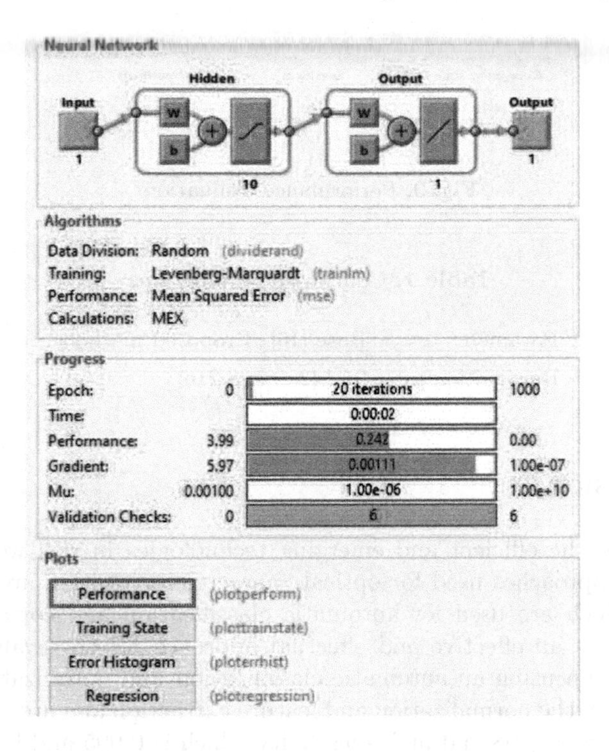

Fig. 7. Neural network training

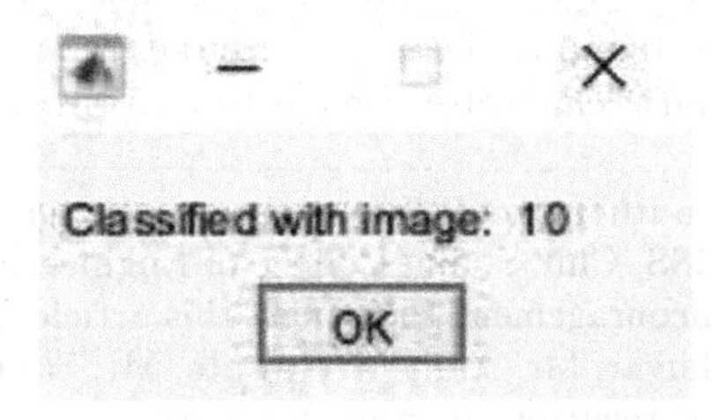

Fig. 8. Classified outputs

Figure 9 shows The performance evaluation in terms of sensitivity, specificity and recognition rate of the proposed system and classification results which signifies that our proposed approach is able to achieve high recognition rate probability which is 0.98216, high sensitivity which signifies the high true positive rates and high specificity which signifies high true negative rates. The recognition, sensitivity and specificity must be high for the low error rate probabilities and less classification error rates.

As recognition rate is focused, Table 1 shows 98.21 % characters are matched using this algorithm.

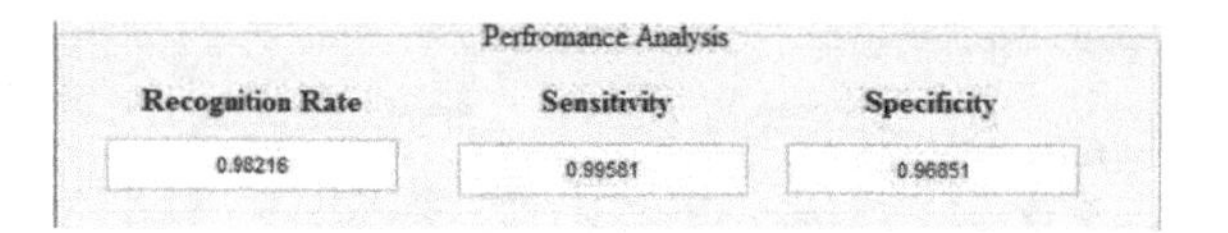

Fig. 9. Performance evaluation

Table 1. Comparison evaluations

Parameter	Base [16]	Proposed method
Recognition rate	96.142	98.216

5 Conclusion

OCR is one of the efficient and emerging technologies in real world scenario. A variety of approaches used for optical character recognition are done by the researchers which are used for automatic classification and correlations. This paper deals with an effective and effectual approach for the evaluation of the proposed solution using an automatic classification and optimization approach which deals with the normalization and feature extraction and are achieving high recognition rates of 0.98 and high sensitivity which is 0.995 and high specificity which is 0.96 and are able to achieve low error rate and classification probabilities. From the results and discussion, it can be noticed that the neural network (backpropagation) provides high reaction and response time to perform high rate of classification based on training data and simulate the network in the testing phase to perform automatic character recognition approach.

Acknowledgment. The authors would like to express sincere gratitude to Dr. Ulhas B. Shinde, Principal, CSMSS, Chh. Shahu College of Engineering, Aurangabad, for his continuous support and encouragement to publish this article. They would also like to thank Mr. Devendra L. Bhuyar, Mr. Amit M. Rawate, Mr. Sanket R. Zanwar, and Mr. Ajit G. Deshmukh for their recurrent help in this work.

References

1. Chaudhary, S., Garg, S., Sathyaraj, R., Behera, A.: An approach for optical character recognition on grid infrastructure using Kohonen neural network. Int. J. Adv. Res. Comput. Sci. **8**(3) (2017). https://doi.org/10.26483/ijarcs.v8i3.3039
2. Ahlawat, D.: A review on character recognition using OCR algorithm. J. Netw. Commun. Emerg. Technol. (JNCET) **7**(5), 56–61 (2017)
3. Gail, H.R., Hantler, S.L.: Method and apparatus for automatic detection of spelling errors in one or more documents. U.S. Patent 9,465,791, issued 11 October 2016
4. Schultz, S.: Method for the automatic material classification and texture simulation for 3D models, U.S. Patent 9,330,494, issued 3 May 2016

5. Zhu, Y., Yao, C., Bai, X.: Scene text detection and recognition: recent advances and future trends. Front. Comput. Sci. **10**(1), 19–36 (2016)
6. Hamad, K.A., Kaya, M.: A detailed analysis of optical character recognition technology. Int. J. Appl. Math. Electron. Comput. **4**(Special Issue–1), 244–249 (2016). https://doi.org/10.18100/ijamec.270374
7. Afroge, S., Raihan, M.A.: Optical character recognition using back propagation neural network. J. Image Process. Pattern Recognit. Prog. 11–18 (2016). https://doi.org/10.1109/icecte.2016.7879615
8. Lee, C.Y., Osindero, S.: Recursive recurrent nets with attention modeling for OCR in the wild. In: Proceedings of the IEEE Conference on CVPR, pp. 2231–2239 (2016). https://doi.org/10.1109/cvpr.2016.245
9. Burry, A.M., Kozitsky, V., Paul, P.: License plate optical character recognition method and system, U.S. Patent 8,644,561, issued 4 February 2014. https://books.google.co.in/books?isbn=1118971647
10. Naz, S., Hayat, K., Razzak, M.I., Anwar, M.W., et al.: The optical character recognition of Urdu like cursive scripts. Pattern Recognit. **47**(3), 1229–1248 (2014). https://doi.org/10.1016/j.patcog.2013.09.037
11. Longacre, J.A.: Method for omnidirectional processing of 2D images including recognizable characters: U.S. Patent 8,682,077, issued 25 March 2014. https://patents.justia.com/inventor/andrew-longacre-jr
12. Bhatia, E.N.: Optical character recognition techniques: a review. Int. J. Adv. Res. Comput. Sci. Softw. Eng. **4**(5) (2014). https://pdfs.semanticscholar.org/7ca5/584576423b366ad7bdf03d5fc136a2e958e6.pdf
13. Mohammad, F., Anarase, J., Shingote, M., Ghanwat, P.: Optical character recognition implementation using pattern matching. Int. J. Comput. Sci. Inf. Technol. **5**(2), 2088–2090 (2014). https://doi.org/10.1.1.661.1089
14. Grimmer, J., Stewart, B.M.: Text as data: the promise and pitfalls of automatic content analysis methods for political texts. Polit. Anal. **21**(3), 267–297 (2013). https://doi.org/10.1093/pan/mps028
15. Mithe, R., Indalkar, S., Divekar, N.: Optical character recognition. Int. J. Recent Technol. Eng. (IJRTE) **2**(1), 72–75 (2013). https://doi.org/10.1093/pan/mps028
16. Kumar B., Kumar N., Palai C., et al.: Optical character recognition using ant miner algorithm: a case study on oriya character recognition. IJCA **61**(3) (2013). https://doi.org/10.5120/9908-4500
17. Sharma, N., Kumar, B., Singh, V.: Recognition of off-line hand printed English characters, numerals and special symbols. In: IEEE International Conference Confluence The Next Generation Information Technology Summit, pp. 640–645 (2014). https://doi.org/10.1109/confluence.2014.6949270
18. Santosh, K.C., Lamiroy, B., Wendling, L.: DTW-Radon-based shape descriptor for pattern recognition. Int. J. Pattern Recognit. Artif. Intell. (2013). https://doi.org/10.1142/S0218001413500080
19. Santosh, K.C., Wendling, L.: Character recognition based on non-linear multi-projection profiles measure. Front. Comput. Sci. **9**(5), 678–690 (2014). https://doi.org/10.1007/s11704-015-3400-2
20. Santosh K.C.: Character recognition based on DTW-Radon. In: 11th International Conference on Document Analysis and Recognition - ICDAR 2011, September 2011, Beijing, China, pp. 264–268. IEEE Computer Society (2011). https://doi.org/10.1109/ICDAR.2011.61
21. Ukil, S., Ghosh, S., Obaidullah, S.M., Santosh, K.C., Roy, K., Das, N.: Deep learning for word-level handwritten Indic script identification. https://doi.org/10.1109/ReTIS.2015.7232880

Recognition of Signature Using Neural Network and Euclidean Distance for Bank Cheque Automation

S. P. Raghavendra[1(✉)] and Ajit Danti[2]

[1] NES Research Foundation, Department of MCA, JNNCE, Shimoga, Karnataka, India
raghusp.bdvt@gmail.com

[2] Department of Computer Science and Engineering, Christ (Deemed to be university), Bangalore, Karnataka, India
ajit.danti@christuniversity.in

Abstract. Handwritten signature recognition plays significant role in automatic document verification system in particularly bank cheque authorization. The proposed method focuses on A novel technique for offline signature recognition approach for bank cheque based on zonal features and regional features. These combined features are used to find genuinety of signature using Euclidean distance as a metric. Extensive experiments are carried out to exhibit the success of the recommended approach.

Keywords: Euclidean distance measure · Zonal features · Neural network · Signature recognition

1 Introduction

While performing financial transactions traditional bank cheques, bank challans, Demand drafts, credit-debit cards and other legal documents plays the remarkable role in current economy. Bank cheques are one of the crucial and authorized means by which individuals and financial organizations transfer money and pay bills. While performing such kinds of transactions authorization of signature is mandatory and this authentication is vulnerable to fraud and forgery. Accordingly, the requirement for research in systematic automated solutions for signature identification and validation has grown in present years to prevent being vulnerable to fraud.

Recognition of signature has been categorized into two major varieties, based on the mechanism of data set collected:

(a) On-line signature identification. (b) Off-line signature identification.

The On-line signature identification is a technique where signature is acquired using an electronic tablet, stylus and further equipment's. In this technique, one can easily fetch information like pressure points, writing speed, acceleration,

K. C. Santosh and R. S. Hegadi (Eds.): RTIP2R 2018, CCIS 1037, pp. 228–243, 2019.
https://doi.org/10.1007/978-981-13-9187-3_21

strokes and fixed attributes of signature, utilizing which the signature can be identified. Conversely, in processing of off-line signature, the signature will be obtained on a paper document which will be later converted into digital image format using a optical high definition scanners at the standard resolution of 300 dpi. The individual signatures vary in different parameters like stokes, curvy style of writing and with physical parameters like pressure, entropy and orientation, hence recognition of signatures poses a huge challenge, in this regards there is a huge demand for a novel and sophisticated methodology in recognition of handwritten signatures in particularly in bank cheques. This paper explains recognition systems for handwritten bank cheque signatures. In this paper, feature extraction based on local and global geometric features of the signature are accomplished. In the proposed methodology our own dataset is constructed with 420 bank cheques from different people in which 350 cheques are genuine and 70 cheques are forged.

Extensive literature survey regarding signature identification and validation has been carried out. Numerous researchers have pursued different techniques to attain recognition of signature viz -In (Boukharouba and Bennia 2017) presented a technique for an effective handwritten digit recognition system consisting of support vector machines (SVM) using Freeman chain code (Nasser and Dogru 2017) present a SIFT and a SURF algorithm which is focuses on enhanced offline signature recognition (Jagtap and Hegadi 2017 and Jagtap et al. 2016) presented and executed a novel approach using upper and lower envelope and Eigen values approach.

In (Pansare and Bhatia 2012) presented methods consists of image preposessing, geometric feature extraction, neural network training with extracted features and verification of handwritten offline signatures.

In (Suryani et al. 2017) propose a system for offline signature recognition and verification which utilizes an efficient fuzzy Kohonen clustering networks (EFKCN) algorithm. The various stages involved in signature verification system consists of five stages viz data acquisition, image processing, data normalization, clustering, and evaluation.

In (Anand and Bhombe 2014) represents a brief review on various approaches based on different datasets, features and training techniques used for verification. (Barbantan et al. 2009) presents offline signature verification system, which considers a new combination of previously used features and introduces two new distance-based ones with the Naive Bayes classifier.

(Putz-Leszczynska 2015) narrates research to restore the template signatures with the hidden signature—an artificial signature is generated by diminishing the mean misalignment connecting itself and the signatures from the enrollment set. (Santosh et al. 2016), presents a new template-free, geometric signature-based technique detects arrow annotations on biomedical images also proposes a method for isolated handwritten or hand printed character recognition using dynamic programming for matching the nonlinear multi projection profiles. (Santosh and Iwata 2012) explains the importance of on-line handwriting recognition over offline for cursive character recognition. (Santosh 2011) presents a method for isolated of-line character recognition using radon features using DTW algorithm. (Santosh et al. 2010) present an innovative approach to integrate spatial relations in stroke clustering for handwritten devanagari character recognition. In (Radhika and Gopika 2014) proposes a technique for both online and offline features of handwritten signatures and to combine their results to identify the signature. Signatures are gathered for both online and offline, also perform the results of online, offline and combined approach.

(Kumar and Singhal 2017) presents a flow chart which represents how the verification is done in previous years, also introduce the SVM. In (Raghavendra et al. 2018) proposed segmentation and recognition of bank cheque number using optical character recognition and statistical correlation function, (Raghavendra and Danti 2018) proposed recognition of bank cheque name using binary patterns and feed forward neural network, (Jayadevan et al. 2011) proposed comprehensive bibliography of many references as support for researchers working in the field of automatic cheque processing.

(Ukil et al. 2018) propose a novel method that uses convolutional neural networks (CNNs) for feature extraction with multilevel 2D discrete haar wavelet transform for handwritten scripts and achieved maximum script identification rate of 94.73% using multi-layer perceptron (MLP).

The proposed paper has been structures as follows, In Sect. 2 proposed methodology for bank cheque signature recognition has been discussed. Section 3, contains the methodology to extract features. Section 4 contains information about Neural network classifier and recognition technique. Section 5 depicts Algorithm proposed and Sect. 6 provides details of the comparision along with the experimental result. Finally in Sect. 7 gives conclusion.

2 Proposed Methodology

The proposed methodology is efficient in terms of speed and robustness to detect the fake or forged bank cheques which are discussed in detail. The Stages of cheque signature recognition process are depicted in the following block diagram.

This paper presents an effective technique for recognizing Indian bank cheque signature using Artificial neural network and Authenticity of the cheque signature will be measured by Euclidean distance to declare the cheque under consideration is genuine or forged signature. After segmentation of the potential cheque signature region, zonal and regional features are extracted for recognition deploying Feed forward artificial Neural network with Euclidean distance (Figs. 2 and 3).

The Proposed methodology comprises two major phases viz training phase and testing phase as follows.

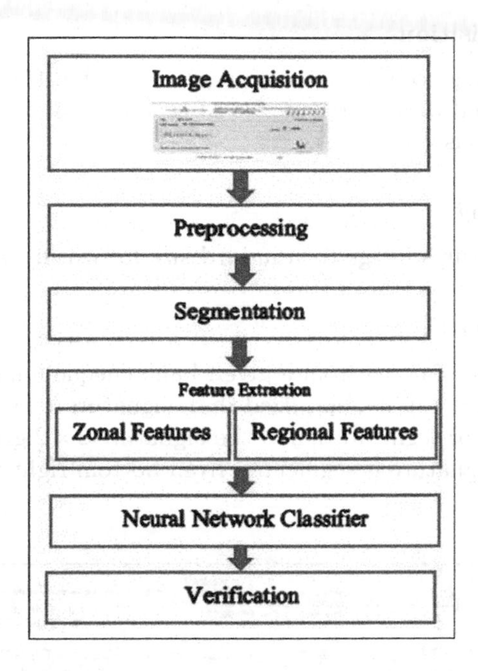

Fig. 1. Block diagram of bank cheque signature recognition.

Fig. 2. Sample experimental results with genuine cheque signature recognition.

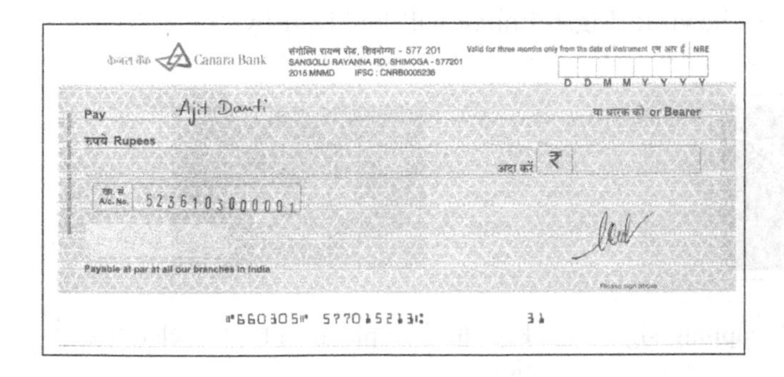

Fig. 3. Sample experimental results with forged cheque signature recognition.

2.1 Image Acquisition

The bank cheque image considered as $f(x, y)$ is acquired from 7 signers with 50 genuine and 10 forged totally 60 samples of each class and then grid is applied to get 4 (2×2) regions.

2.2 Preprocessing

Normalize the signatures image to standard size for feature extraction.

2.3 Segmentation

Signature region is segmented from a given bank cheque image using grid technique. In CTS cheques, it is standardized that, signature is located in the bottom right of the cheque, for segmentation of the signature 2×2 grid is applied to the cheque image and signature is segmented from bottom right portion (Region 4) as shown in Fig. 4.

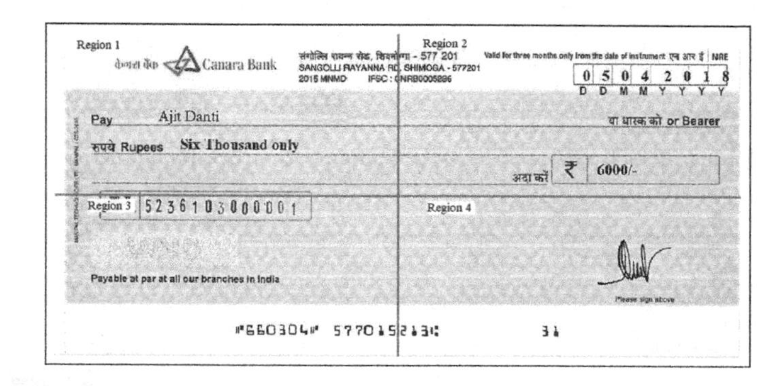

Fig. 4. Sample grid results on four regions of cheque image.

The signature is cropped from the Region 4 as shown in Fig. 5 using following equation

$$Region4 : t(x, y) = g\left(\sum_{i=\frac{x}{2}}^{x} i, \sum_{j=\frac{y}{2}}^{y} j \right) \tag{1}$$

and skeletonized image is shown in Fig. 6 image will be given by:

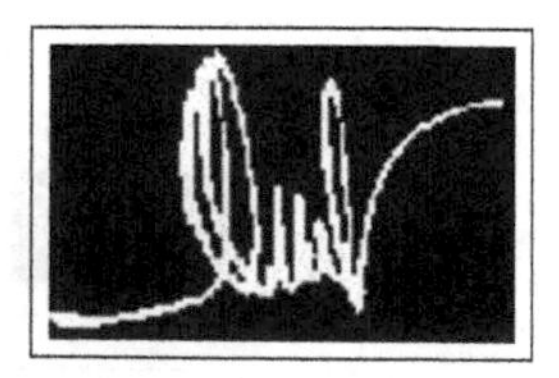

Fig. 5. Sample result of cropping signature.

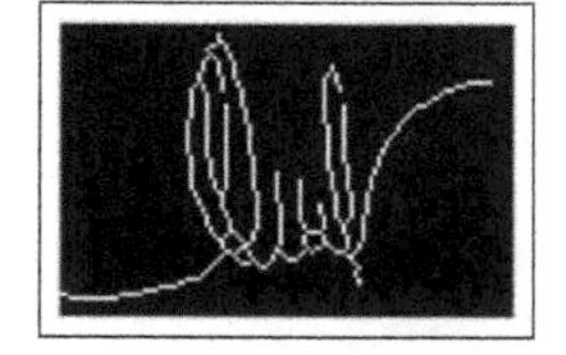

Fig. 6. Sample result of skeletonization.

3 Feature Extraction

In the proposed system, two types of features extracted i.e. zonal features and regional features were used. The signature image is zoned into 9 equal sized segments using the following equations.

$$Zone11 : t(x, y) = \sum_{i=1}^{zh} \sum_{j=1}^{zw} t(x, y) \tag{2}$$

$$Zone12 : t(x, y) = \sum_{i=1}^{zh} \sum_{j=zw+1}^{2+zw} t(x, y) \tag{3}$$

$$Zone13 : t(x, y) = \sum_{i=1}^{zh} \sum_{j=2+zw+1}^{y} t(x, y) \tag{4}$$

$$Zone21 : t(x, y) = \sum_{i=zh+1}^{2+zh} \sum_{j=1}^{zw} t(x, y) \tag{5}$$

$$Zone22 : t(x, y) = \sum_{i=zh+1}^{2+zh} \sum_{j=zw+1}^{zw} t(x, y) \tag{6}$$

$$Zone23 : t(x, y) = \sum_{i=zh+1}^{2+zh} \sum_{j=2+zw+1}^{y} t(x, y) \tag{7}$$

$$Zone31 : t(x, y) = \sum_{i=2+zh+1}^{x} \sum_{j=1}^{zw} t(x, y) \tag{8}$$

$$Zone32 : t(x, y) = \sum_{i=2+zh+1}^{x} \sum_{j=zw+1}^{2+zw} t(x, y) \tag{9}$$

$$Zone33 : t(x, y) = \sum_{i=1}^{zh} \sum_{j=2+zw+1}^{zw} t(x, y) \tag{10}$$

Where $zh = \frac{x}{3}$, $zw = \frac{y}{3}$, $t(x, y)$ is zoned signature image as shown in Fig. 7. The 9 zones will be organized as a grid for feature extraction and represented in the following diagram:

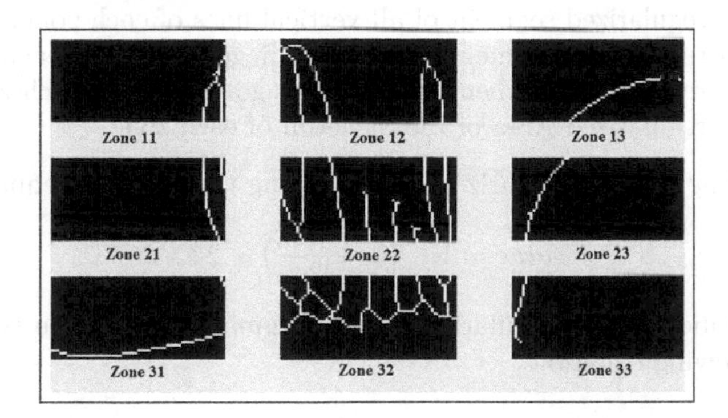

Fig. 7. Zone representations of signature image $t(x, y)$.

Feature extraction is applied to each zone to obtain positions of different types of line segments and later grouped into the following categories viz Vertical line, Horizontal line, Left diagonal line and right diagonal line.

A 3×3 mask is applied on each zone in order to evaluate the direction vector for determining different variants of line types by positioning the mask on each pixel with respect to their surrounding pixels as shown below.

$$\begin{bmatrix} 4 & 5 & 6 \\ 3 & c & 7 \\ 2 & 1 & 8 \end{bmatrix}$$

In the mask, 'C' indicates the center pixel. The surrounding pixels are numbered in a clockwise direction which starts from pixel below the central pixel. In order to evaluate direction vector from the given line segment under consideration, this technique traverses entirely through all the surrounding pixels in each line segments in the same order in which they form the line segment. The direction vector is extracted from the following principles to find new line segments used for classifying each direction vector (dv) evaluated by the following equation.

$$\text{if } \max(\text{dv}) = \begin{cases} 2 \ or \ 6 & lt = rightdiagonal \\ 4 \ or \ 8 & lt = leftdiagonal \\ 1 \ or \ 5 & lt = vertical \\ 3 \ or \ 7 & lt = horizontal \end{cases} \tag{11}$$

where dv: direction vector lt: line type

Every zone has a 9-feature vector given below:

- Sum of horizontal lines of each zone.
- Sum of vertical lines of each zone.
- Sum of Right diagonal lines of each zone.
- Sum of Left diagonal lines of each zone.
- Measure of regularized segment of all horizontal lines of each zone.
- Measure of regularized segment of all vertical lines of each zone.
- Measure of regularized segment of all right diagonal lines of each zone.
- Measure of regularized segment of all left diagonal lines of each zone.
- Measure of regularized Area of the Skeleton of each zone.

The specific line type is regularized by employing the following equation

$$value = 1 - \left(\left(\frac{\sum_{i=1}^{n} ln}{10} \right) * 2 \right) \tag{12}$$

where ln: number of lines regularized length segment of each line type is found using the following equation

$$length\, of\, segment = \frac{\left(\sum_{i=1}^{p} lt \right) * \left(\sum_{j=1}^{q} tp \right)}{\sum_{i=1}^{r} tz} \tag{13}$$

Where lt: linetype, tp: total pixel, p: number of line type q: number of pixels in line type lt tz: Total summation of zonal pixels r: number of zonal pixels Subsequently once zonal feature extraction is completed regional features are extracted using region properties for the entire signature image on the basis of regional descriptors namely:

- **Euler Number:** Interpreted as the dissimilarity between total quantity of components and quantity of holes in the given signature image.

- **Regional Area:** Explained as the proportion of the total quantity of the pixels in the skeleton image to the total quantity of pixels in the signature image.
- **Eccentricity:** Considered as the ratio of the distance between the foci and its major axis length of the signature image.
- **Orientation:** Produces a scalar value that describes the angle between the x-axis and the major axis of the ellipse that has the similar second-moments as the region under consideration.

Combined feature vector consists of both zonal features and regional features are used for signature recognition.

4 Signature Verification

Because of its supremacy and ease of use in object classification and recognition neural network is extensively employed in pattern recognition. The proposed methodology contains two major phases. Firstly, extract the features set exhibiting the zonal and regional features from different bank cheque and create a knowledgebase. The second phase is for the neural network is to train the relationship between a test signature and knowledgebase and classify the class for which the query signature belongs. Third phase is to evaluate the Euclidean distance between the query image feature vectors with identified class signature image vectors to recognize the authenticity class as either "genuine" or "forgery". In this research, the classification of 7 bank customers with 50 samples of each class based on 85 distinct feature vectors to classify them effectively. In order to improve the efficiency of the performance, trained all Neurons (85) in the Neural Network, with 85 invariant feature vectors of signature of the cheque will be given as input to the neural network for the purpose of classification of signatures at later stage. To recognize individual signatures three-layer neural network is designed. The input layer of the network carry neurons in order to encode the input feature vector values. As discussed in the upcoming section, the training data for the network will be consisting of 9 by 9 zonal features and 4 regional features of signatures, and so the input layer contains $85 = 9 * 9 + 4$ neurons. The hidden layer is the second layer of the network. 'n' is the amount of neurons in the hidden layer, and extensive experiments are conducted with different values for 'n'. In the given context hidden layer, with $n = 10$ neurons are considered. 07 neurons are there in the output layer. In case of first neuron activates, i.e., has an output ≈ 1, which specify that the network designates the signature belongs to class 1. In case second neuron activation, will specify that the network designates the signature belongs to class 2. And so on.

Figure 9 represents the creation and design of Neural Network.

Figures 8 and 9 provides the pictorial interpretation of "sample neural network architecture". The output of the neural network classifier estimates the particular signature type of a bank cheque signers in the classification level using Eq. (14), For a Multilayer Feed forward neural network given n input units(where n represents size of the province) the test input 'x' with the network weights considered for H sigmoid nodes and one hidden layer for which a direct output unit is

$$y = \sum_{i=1}^{H} v_i \sigma(z_i) \tag{14}$$

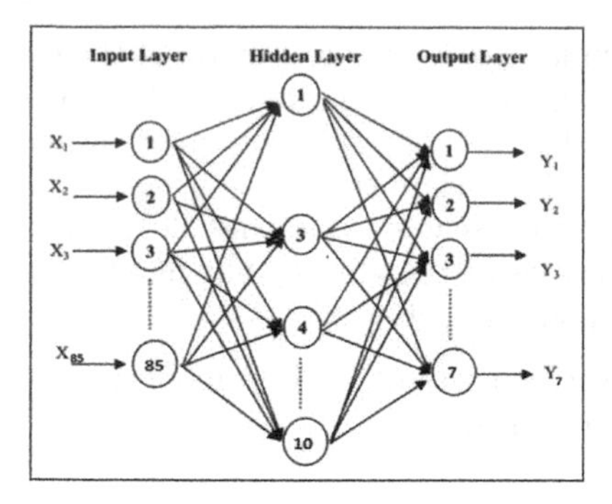

Fig. 8. Architecture of a neural network.

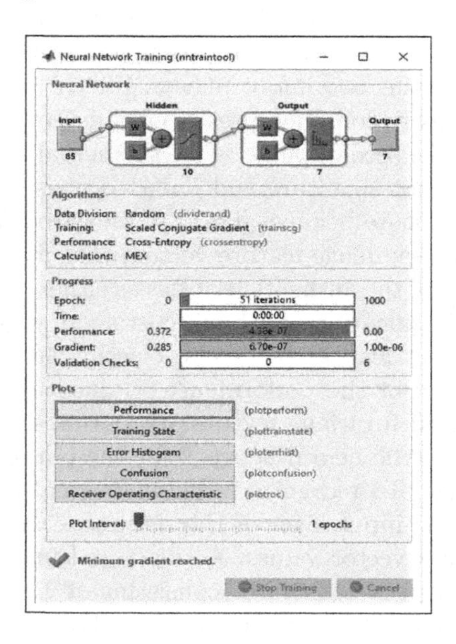

Fig. 9. Formation and exercising of a neural network.

Where v_i: Weight associated in connection from the hidden unit i to the output unit, $\sigma(z_i)$: Matrix of test signature y: signature class anticipated.
The Learning function is given by

$$z_i = \sum_{j=1}^{n} w_{ij}x_j + b_i \tag{15}$$

Where w_{ij}: Weight interconnecting from hidden unit i to input unit j, b_i: Bias of the hidden unit i The Signatures of a bank cheque are classified by employing Neural Network. With knowledgebase $v(i)$ and test signature image the training of the neural network will be accomplished by using the equation (14). Among different classes of signature the query signature will be identified by using the

the value of y with highest value of index to be considered as the class the query signature image belongs by applying the Eq. (16)

$$y_i = max \int_{i=1}^{c_n} (y, i) \tag{16}$$

where c_n: Defines numerous supervised classes (in our case $c_n = 7$) Once the class has been identified for the given query signature image the next phase is to find the Euclidean distance between the query signature feature vector and predicted class of signature feature vector from the Neural network classifier using the Eq. (17).

$$d(p, q) = \left| \sqrt{\sum_{i=1}^{n} (p_i - q_i)^2} \right| \tag{17}$$

The Euclidean distance thus evaluated will be used as a measuring condition in order to authenticate a query signature provided that if the distance lies between the scope of subjective threshold. Nevertheless, the scalar value Euclidean distance thus estimates the matching score which is used in order to compare the the test signature with the other associated signature classes to determine the authenticity of the signature as genuine or forged (Armand et al. 2006).

5 Proposed Algorithm

Proposed algorithm for cheque signature recognition using Neural network and Euclidean distance as given below:
Input: Query cheque image
Output: Recognize bank cheque signature for its authenticity

Algorithm 1. Signature Recognition

1: To Train, input binary segmented cheque signatures of all signers, extract robust invariant feature vector and create a knowledgebase
2: Input the Test Query cheque image, segment the potential cheque signature region using

$$t(x, y) = \sum_{i=topmost}^{lowermost} \sum_{i=leftmost}^{rightmost} t(x, y) \tag{18}$$

as shown in figure 5
3: Apply grid to divide the potential signature regions into 9 uniform zones using the following equations and compute 9 feature from each zone using the equations (2) to (10)
4: Compute euler number, area, eccentricity and orientation as regional properties for signature image.
5: Construct a feature vector of all the features obtained in step 3 and 4 and apply neural network for classification using equation (14) and (15)
6: Determine the Signature whose index is maximum using equation (16)
7: Compute the Euclidean distance between classified signature and knowledgebase and find the classified signature having minimum distance as Genuine or else forged using equation (17)

6 Experimental Results

A manual signature database has been created by gathering different varieties of bank cheques for accomplishing bank cheque signature recognition. For demonstration and experimentation purpose 50 genuine and 10 forged signatures of seven different customers is taken into account for testing. The signature database is constructed from collecting different bank cheques manually for signature recognition. In which 50 genuine and 10 forgery signature samples from 7 people is considered for testing. The signatures which are there inside the database contain skilled, random and simple variants (Angadi et al. 2013). Furthermore the performance and analysis of the proposed methodology will be estimated by using the standard terminologies like False Rejection Rate (FRR) and False Acceptance Rate (FAR). For false rejection cheques, genuine signature are accepted as forgery where as in case of False acceptance, it occurs due to acceptance of forged signature as genuine (Angadi et al. 2013).

$$FAR = \frac{Number\ of forgeries\ accepted}{Number\ of\ forgery\ tested} * 100 \tag{19}$$

$$FRR = \frac{Number\ of genuine\ rejected}{Number\ of\ genuine\ tested} * 100 \tag{20}$$

Sample Experimental results for different cheque signers are shown in the Figs. 10 and 11

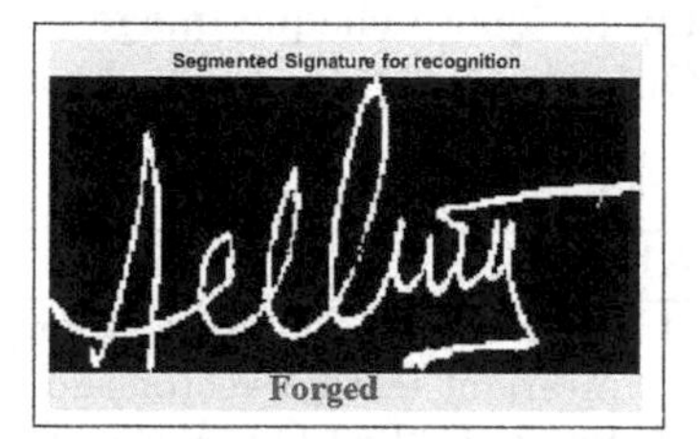

Fig. 10. (a) Genuine signature of user: amj (b) Forged signature of user: amj.

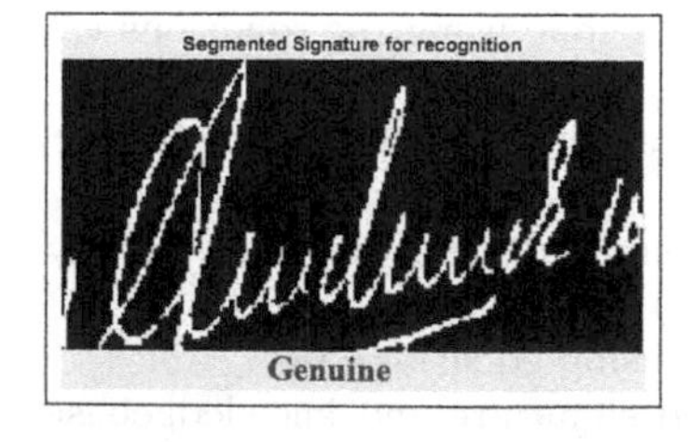

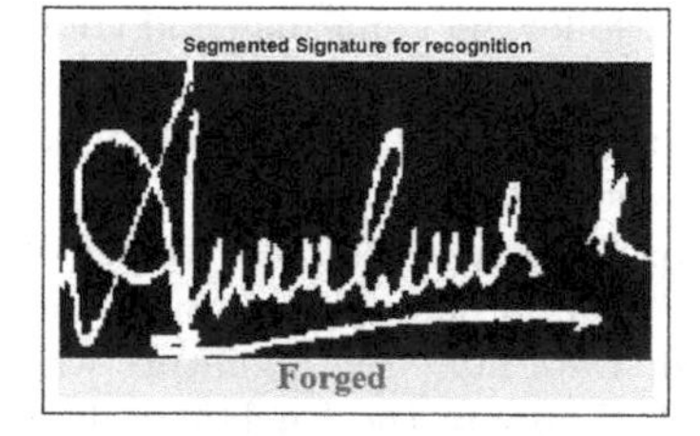

Fig. 11. (a) Genuine signature of user: akl (b) Forged signature of user: akl.

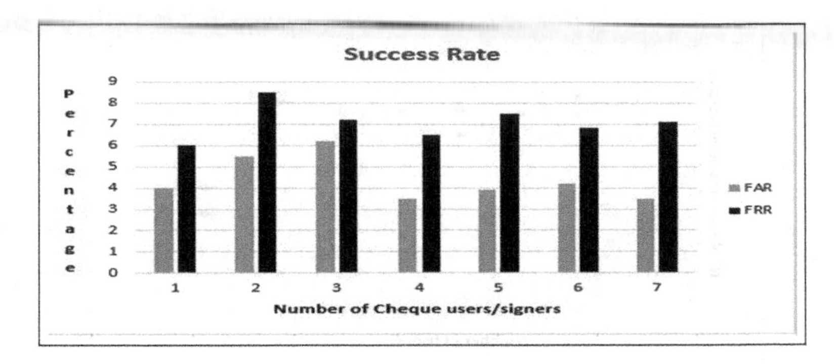

Fig. 12. Bar chart for FAR and FRR of each user.

Table 1. Comparative analysis.

Methodology	No. of cheque signatures	TRR	TAR	FAR	FRR	Accuracy
(Chandra and Maheskar 2016)	540	89.09	89.38	10.62	10.91	89.24
(Sisodia and Anand 2009)	240	95.83	92.72	4.16	7.29	94.22
(Justino et al. 2001)	1600	90.00	78.00	22.00	10.00	91.12
(Armand et al. 2006)	2106	93.7	91.8	8.20	6.30	91.80
(Oliveira et al. 2007)	1700	95.17	94.7	5.30	4.83	91.90
Proposed method	420	92.92	95.60	4.4	7.08	94.26

The complete precision analysis demonstrates the measure of accuracy with respect to the proposed system in which the average accuracy lies between amount of genuine signature authenticated as genuine and amount of forgery signature declared as forged one.

$$Accuracy = \frac{(100-FAR) + (100-FRR)}{2} * 100 \tag{21}$$

Extensive experiments were carried on 7 different bank cheque users. Each class having 50 genuine and 10 forgery cheque signatures. Total 420 bank cheque signatures are taken at 200 dpi resolution for genuine and forged. As shown in Fig. 1. First, train the signature data in knowledgebase through feed forward neural network. Among them one bank cheque signature will be selected and classified as a test from the cheque database. A feed forward neural network is designed and trained with cheque signature database having 280 signatures. The experimental result is performed on a database of 7 users each having 60 signatures with simulation for testing 140 cheque images are considered for which, FAR is 4.4 and FRR is 7.08. The TRR (True Rejection Rate) and TAR (True Acceptance Rate) are evaluated by reducing FRR and FAR by 100.The experimental results are shown in Figs. 9 and 10. Table 1 shows the comparison results

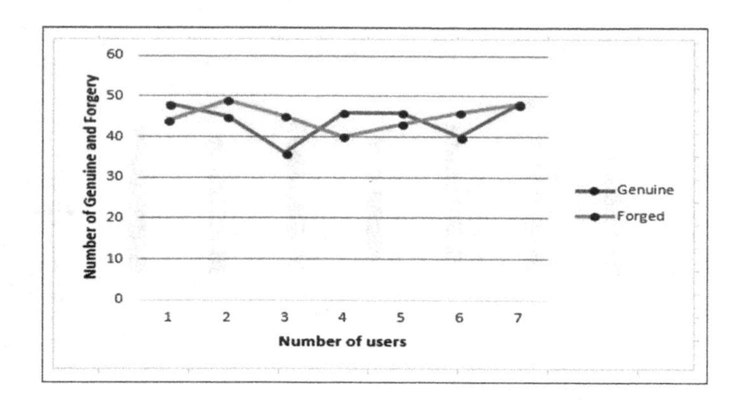

Fig. 13. Number of genuine and forgery signatures recognized.

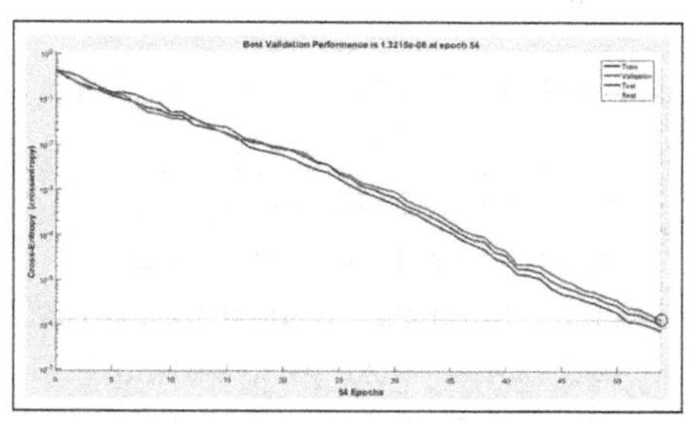

(a) Best validation after 54 epochs for signature recognized.

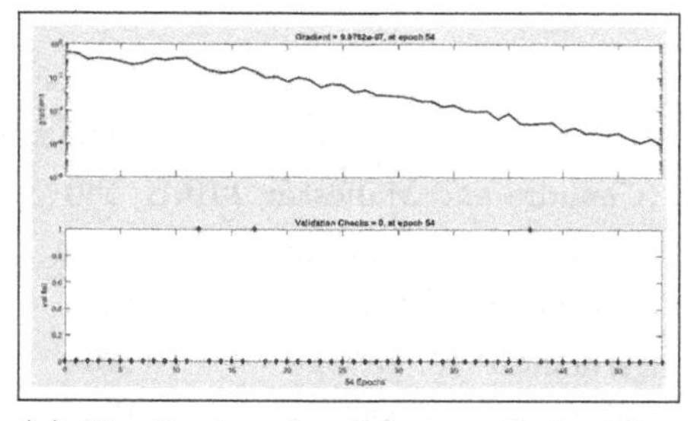

(b) Gradient and validation checks after 54 epochs.

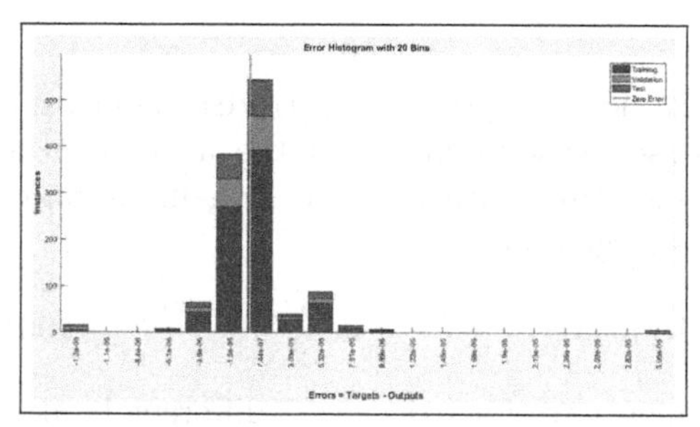

(c) Error Histogram for signature recognition.

Fig. 14. Performance analysis.

with proposed technique and other state of the art methods. Hence the overall comprehensive accuracy of the proposed methodology will be evaluated as a measure of the precision and accuracy value evaluated will be 94.26%. FAR and FRR values of all seven users will be depicted as shown in Fig. 12.

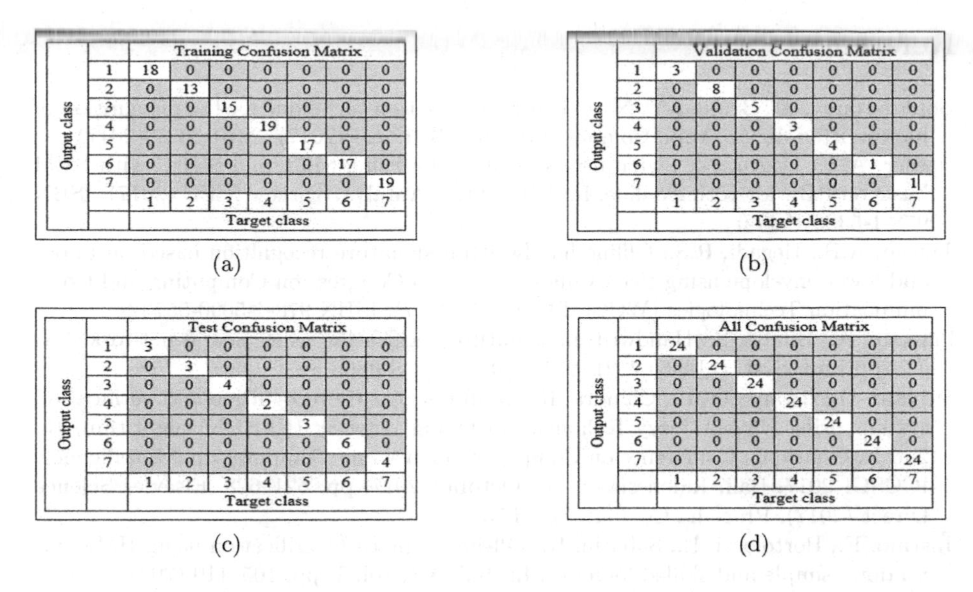

Training Confusion Matrix

Output class \ Target class	1	2	3	4	5	6	7
1	18	0	0	0	0	0	0
2	0	13	0	0	0	0	0
3	0	0	15	0	0	0	0
4	0	0	0	19	0	0	0
5	0	0	0	0	17	0	0
6	0	0	0	0	0	17	0
7	0	0	0	0	0	0	19

(a)

Validation Confusion Matrix

Output class \ Target class	1	2	3	4	5	6	7
1	3	0	0	0	0	0	0
2	0	8	0	0	0	0	0
3	0	0	5	0	0	0	0
4	0	0	0	3	0	0	0
5	0	0	0	0	4	0	0
6	0	0	0	0	0	1	0
7	0	0	0	0	0	0	1

(b)

Test Confusion Matrix

Output class \ Target class	1	2	3	4	5	6	7
1	3	0	0	0	0	0	0
2	0	3	0	0	0	0	0
3	0	0	4	0	0	0	0
4	0	0	0	2	0	0	0
5	0	0	0	0	3	0	0
6	0	0	0	0	0	6	0
7	0	0	0	0	0	0	4

(c)

All Confusion Matrix

Output class \ Target class	1	2	3	4	5	6	7
1	24	0	0	0	0	0	0
2	0	24	0	0	0	0	0
3	0	0	24	0	0	0	0
4	0	0	0	24	0	0	0
5	0	0	0	0	24	0	0
6	0	0	0	0	0	24	0
7	0	0	0	0	0	0	24

(d)

Fig. 15. Performance analysis for classification of 7 users using confusion matrix for signature recognition.

The various analysis undertaken on the proposed methodology viz performance analysis for best validation, Gradient and validation checks at training stage, Error histogram with minimum gradient and confusion matrix for best classification are depicted below in (Figs. 14(a), (b) and (c) and 15)

7 Conclusion

In this paper, zonal and regional based features are extracted for offline signature verification of bank cheque. The proposed method will preprocess and segment the potential signature from the cheque image and is fed to the feed forward neural networks with 7 group of discrete users. Experiments were demonstrated on manual dataset created. By analyzing the graphs in Fig. 13, the best validation match for both test and training will occur at 54 epochs and the validation performance value evaluated will be 1.3215e-06, and more than one hidden layer is used in training deep back propagation neural network the gradient value at this stage will be 9.076e-07 and error histogram graph shows the difference between target and predicted values as mean square error for 20 bins. In the Fig. 14 Confusion matrix drawn which analyses the output class obtained versus targets fixed during training, validation, testing and by adding all of them the final confusion matrix can be generated. The overall accuracy of the proposed method is 94.26% which is very much associative to the varieties of state of the art methods and the accuracy can be enhanced by using higher computational classifiers like support vector machines.

References

Boukharouba, A., Bennia, A.: Novel feature extraction technique for the recognition of handwritten digits. Appl. Comput. Inform. **13**, 19–26 (2017). 2210–8327 (2017)

Nasser, A.T., Dogru, N.: Signature recognition by using SIFT and SURF with SVM basic on RBF for voting online. In: ICET2017, Antalya, Turkey. IEEE (2017). ISBN 978-1-5386-1949-0

Jagtap, A.B., Hegadi, R.S.: Offline handwritten signature recognition based on upper and lower envelope using eigen values. In: World Congress on Computing and Communication Technologies (WCCCT). IEEE (2017). ISBN 978-1-5090-5573-9

Pansare, A., Bhatia, S.: Handwritten signature verification using neural network. Int. J. Appl. Inf. Syst. (IJAIS) **1**(2) (2012). ISSN 2249-0868

Suryani, D., Irwansyah, E., Chindra, R.: Offline signature recognition and verification system using effcient fuzzy Kohonen clustering network (EFKCN) algorithm. In: 2nd International Conference on Computer Science and Computational Intelligence, ICCSCI 2017, Bali, Indonesia, 13–14 October 2017, pp. 621–628. Elsevier Science Direct (2017). Procedia Comput. Sci. **116**

Justino, E., Bortolozzi, E., Saburin, R.: Off-line signature verification using HMM for random, simple and skilled forgeries. In: ICDAR, vol. 1, pp. 105–110 (2001)

Anand, H., Bhombe, D.L.: Relative study on signature verification and recognition system. Int. J. Innov. Res. Adv. Eng. (IJIRAE) **1**(5) (2014). ISSN 2349–2163

Barbantan, I., Vidrighin, C., Borca, R.: An offline system for handwritten signature recognition. IEEE (2009). ISBN 978-1-4244-5007-7

Putz-Leszczynska, J.: Signature verification: a comprehensive study of the hidden signature method. Int. J. Appl. Math. Comput. Sci. **25**(3), 659–674 (2015)

Santosh, K.C., Wendling, L., Antani, S., Thoma, G.R.: Overlaid arrow detection for labelling regions of interest in biomedical images. Pattern Recognition, Part 2. IEEE (2016). ISSN 1541–1672

Santosh, K.C., Iwata, E.: Stroke-based cursive character recognition. Advances in Character Recognition (2012). https://doi.org/10.5772/51471

Santosh, K.C.: Character recognition based on DTW-Radon. In: International Conference on Document Analysis and Recognition, pp. 1520–5363. IEEE (2011)

Santosh, K.C., Nattee, C., Lamiroy, B.: Spatial similarity based stroke number and order free clustering. In: 12th International Conference on Frontiers in Handwriting Recognition. IEEE (2010). ISBN 978-0-7695-4221-8

Sisodia, K., Anand, S.M.: Off-line handwritten signature verification using artificial neural network classifier. Int. J. Recent Trends Eng. **2**(2), 205–207 (2009)

Radhika, K.S., Gopika, S.: Online and offline signature verification: a combined approach. Procedia Comput. Sci. **46**, 1593–1600 (2015). International Conference on Information and Communication Technologies (ICICT 2014)

Oliveira, L.S., Justino, E., Sabourin, R.: Off-line signature verification using writer-independent approach. In: IJCNN-2007, pp. 2539–2544 (2007)

Kumar, R., Singhal, P.: Review on offline signature verification by SVM. Int. Res. J. Eng. Technol. (IRJET) **4**(6) (2017). e-ISSN 2395-0056

Raghavendra, S.P., Danti, A., Suresha, M.: Correlation based template matching for recognition of bank cheque number. Int. J. Comput. Eng. Appl. **XII**(III) (2018). www.ijcea.com. ISSN 2321–3469

Raghavendra, S.P., Danti, A.: A novel recognition of Indian bank cheque names using binary pattern and feed forward neural network. IOSR J. Comput. Eng. (IOSR JCE) **20**(3), 44–59 (2018). e-ISSN: 2278–0661, p-ISSN: 2278–8727

Jayadevan, R., Kolhe, S.R., Patil, P.M., Pal, U.: Automatic processing of handwritten bank cheque images: a survey. IJDAR **15**, 267–296 (2011). https://doi.org/10.1007/s10032-011-0170-8

Angadi, S.A., Gour, S., Bajantri, G.: Offline signature recognition system using radon transform. In: Fifth International Conference on Signals and Image Processing. IEEE (2013). ISBN 978-0-7695-5100-5

Armand, S., Blumenstein, M., Muthukkumarasamy, V.: Off-line signature verification based on the modified direction feature. In: ICPR-2006, pp. 509–512 (2006)

Ukil, S., Ghosh, S., Obaidullah, S.M., Santosh, K.C., Roy, K., Das, N.: Deep learning for word-level handwritten Indic script identification. arXiv preprint arXiv:1801.01627 (2018)

Chandra, S., Maheskar, S.: Offline signature verification based on geometric feature extraction using artificial neural network. In: 3rd International Conference on Recent Advances in Information Technology—RAIT-2016. IEEE (2016). ISBN 978-1-4799-8579-1

Jagtap, A.B., Hegadi, R.S.: Eigen value based features for offline handwritten signature verification using neural network approach. In: Santosh, K.C., Hangarge, M., Bevilacqua, V., Negi, A. (eds.) RTIP2R 2016. CCIS, vol. 709, pp. 39–48. Springer, Singapore (2017). https://doi.org/10.1007/978-981-10-4859-3_4

A Semi-automatic Methodology for Recognition of Printed Kannada Character Primitives Useful in Character Construction

Basavaraj S. Anami and Deepa S. Garag(✉)

K.L.E. Institute of Technology, Hubli 580030, Karnataka, India
anami_basu@hotmail.com, deepagarag@rediffmail.com

Abstract. Every character of the language having script is written using basic units called primitives. One or more number of primitives connected appropriately result in construction of a character. In the process of character construction identification of primitives is considered as high priority. Through this paper we propose a semi-automatic method for extracting features from primitives for their recognition and further Kannada characters' construction. The primitives are recognized automatically by adopting the zone features and neighbor classifier. The feature vectors are obtained for all the primitives of Kannada character set and a knowledge base is created. We have used Euclidean distance measure to establish similarity between test input primitives and existing primitives present in the knowledge base for identifying the primitives in Kannada characters. The suggested methodology is tested for 11520 manually extracted primitive images. Average recognition accuracies observed is in the range of 75% to 100% for printed primitives. Application spreads in various verticals of automating literature like calligraphy, digitizing old manuscripts, multimedia teaching, Robot based assistance in handwriting, animation etc.

Keywords: Kannada printed characters · Primitives extraction · Classification · Zone based features · Feature extraction · Nearest neighbor · K-fold cross validation

1 Introduction

India is rich with its literature having 22 scheduled languages and many such languages are based on the ancient Brahmi script. Kannada is one amongst them. The history takes back us almost more than 2000 years and conform the existence of the Kannada language. The Kannada script is believed to be evolved primarily from stone carving. This is concluded by literature researchers by studying the existence of symmetry in characters and most of the characters are round in representation. Figure 1 shows the typical Kannada text where we can see Kannada word, namely "udaharanegalu" and its primitives which are

K. C. Santosh and R. S. Hegadi (Eds.): RTIP2R 2018, CCIS 1037, pp. 244–260, 2019.
https://doi.org/10.1007/978-981-13-9187-3_22

sub-components of the characters. In other words, primitives can be called as atoms of the characters. It is observed from Fig. 1 that the pattern of characters is different from other scripts and the primitives look more cursive in nature.

The earlier research efforts consider one character as a single unit. The primitives that form character are not focused. But while writing the characters the primitives' recognition becomes important. The primitives are essential in Kannada character construction and such identified primitives are to be retrieved from static database of language [16]. The research scope is vast and with leverage of technology for automation of script writing and learning. Minimal work is noticed in research towards construction of Kannada language character through primitives. The paper deals on recognition of Kannada language primitives images. Consonants and vowels are considered in the work. Work is useful in digitizing old manuscripts, calligraphy, developing multimedia applications and for novice learners etc.

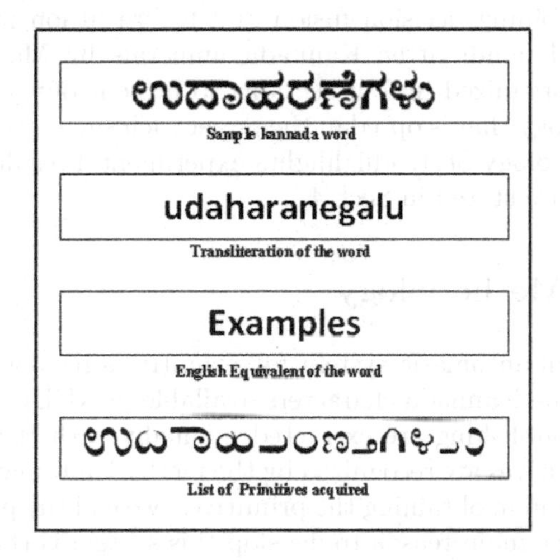

Fig. 1. Sample Kannada word and its primitives.

The paper on recognition of south Indian Kannada printed or hand written character by Pal et al. [1] has discussed various feature extraction and classification techniques for offline hand writing recognition in Indian regional scripts. Sharma et al. [2] have presented various feature extraction techniques to classify handwritten alphabets. Santosh [3,4] paper uses Radon features and DTW algorithm and proposes a method for isolated character recognition. The paper based on Kannada OCR by Sheshadri et al. [5] has used segmentation technique to decompose the character into components from three base classes. The merits are identified and discussed for probabilistic and geometric seeding in k-means technique used. Nithya and Babu [6] have proposed Kannada OCR system, Uses the histogram and connected component methods for segmentation. A Page of Kannada text image is converted in to the editable form for properly non converted characters.

Ragha and Sasikumar [7] proposed and concentrated on the moment's features of Kannada handwritten basic character set. The comparison is carried out between of moment's features of four directional images with original images. Multi layer perceptron with back propagation neural network shows better result. Sangame et al. [8] used chain code features and kNN classifier for an unconstrained handwritten Kannada OCR using invariant moments. The paper by Rajput, Horakeri and Angadi et al. [9–11] uses structural and statistical feature. The hand written Kannada characters are attempted to recognize by SVM classifier. Zone based Dhandra et al. [12] have proposed zone based features to recognize mixer of both printed and handwritten digits of Kannada. The authors used the kNN and SVM technique to classify. Kumar [13] proposes the method to recognize Kannada numerals. Here each numeral is segmented and trying to achieve better response time with minimum number of extracted features. The different feature extraction and classifications techniques are reviewed by Ramappa and Krishnamurthy [14] for recognition of hand written Kannada numerals. The framework by combining decision fusion and future fusion is applied for classifying the isolated handwritten Kannada numerals by Mamatha et al. [15]. Remaining paper organized in four sections to present our approach. Section 2 proposes methodology that is opted in the paper each sub section presents details of stages in methodology Sect. 3 highlights experimental results. The conclusion and Future works are stated in Sect. 4.

2 Proposed Methodology

We proposed a semi-automatic system for primitive's recognition in this work. The primitives of the Kannada characters available in NUDI, Kannada software used for Desktop publishing are extracted manually with the help of Mspaint. The extracted primitives are recognized by the method automatically. Since there is human intervention in obtaining the primitives, we call the proposed system as semi-automatic. The main reason to develop this system is that, the primitives' selection or extraction differs from one script to another and it is dependent on the nature of characters. In addition, different writing styles of individuals make the problem still complex. As a result, it is hard to find common base or fact to develop a method to identify primitives automatically. Therefore, we used human intervention to extract primitives. Next, we extract features based on zones of each primitive image for recognizing primitives.

We have explored zone based features for Kannada character primitives' recognition in this work. The method converts the image to binary image before extracting features, the input image size is changed to a standard size and finally, single pixel width is obtained by applying thinning algorithm. Further, the extracted features are compared with the features of the primitives in the knowledge database using the nearest neighbor classifier to identify the primitives. In summary, the proposed methodology includes of four steps, the first step describes extracting the primitives from printed Kannada characters, second step explains the preprocessing steps, third step deals with feature extraction and fourth step explains the recognition of primitives as shown in Fig. 2.

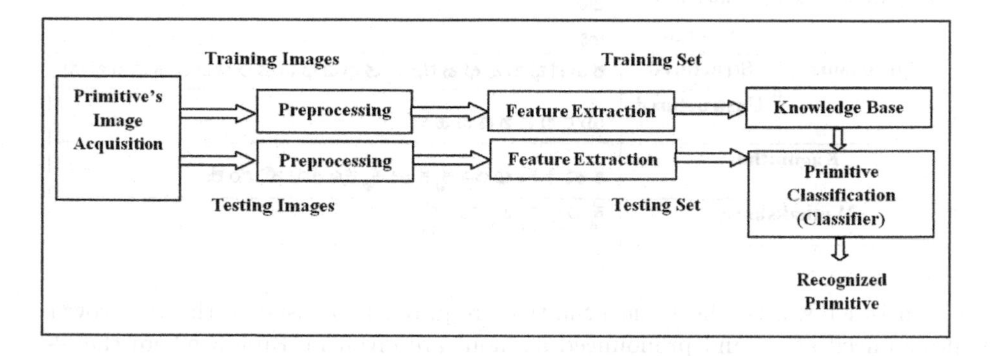

Fig. 2. Phases of semi-automatic system for primitive recognition.

2.1 Primitives Extraction

Generally, the character of any scripts is divided into basic characters called vowels and consonants. Compound characters are formed by combining two or more basic characters. Basis for defining and extracting primitives is the way that is practiced in writing characters during school days. The segment which starts from pen down and pen up is called primitive of the character. The way these primitives are written in order to complete a character forms the sequence of the primitives and the same sequence needs to be followed for recognition. Proper systematic combination of primitives leads to generation of character. Character construction is basic to any medium of learning. At the time of reading, the character is read as whole and while writing, character needs writing primitives in proper sequence. Since extracting primitives in this way for all scripts automatically is hard, we propose experts intervention to extract the primitives in this work.

The classification levels of Kannada script are shown in Table 1 where we can see samples of Vowels, Consonants and Kagunitha. The Kannada language characters are classified into Swaras, Yogavaahakas, Vyanjanas, Kagunitha and Vathakshara as given in Table 1 [16]. Interesting fact is that the primitives of characters can be used to recognize all the characters of Kannada script.

Based on experts' suggestions, to construct all the characters of Kannada script, we have identified 72 primitives as shown in Table 2. These primitives are further classified as: (i) Basic Primitives (BP), which are sub-parts of the characters and (ii) Character Cum Primitives (CcP), which are primitives without sub-parts (complete character). Table 3 shows the order and sequence of BP for two characters.

Table 1. Kannada vowels, consonants and sample Kagunitha.

Vowels	Swaras	ಅ ಆ ಇ ಈ ಉ ಊ ಋ ಎ ಏ ಐ ಒ ಓ ಔ
Yogavaahaka's	Anusvara	ಅಂ
	Visarga	ಅಃ
Consonants	Structured	ಕ ಖ ಗ ಘ ಙ ಚ ಛ ಜ ಝ ಞ ಟ ಠ ಡ ಢ ಣ ತ ಥ ದ ಧ ನ ಪ ಫ ಬ ಭ ಮ
	Unstructured	ಯ ರ ಲ ವ ಶ ಷ ಸ ಹ ಳ
Kagunitha		ಕ ಕಾ ಕಿ ಕೀ ಕು ಕೂ ಕೃ ಕೆ ಕೇ ಕೈ ಕೊ ಕೋ ಕೌ ಕಂ ಕಃ
Wothakshara		ಕ್ಕ ಖ್ಖ ಗ್ಗ ಬ್ಯ ಭ್ರ ಷ್ಟ್ರ

In other words, the basic primitives required to construct the two vowels pronounced as 'e' and pronounced as 'aou' are given in Table 3 where the '+' symbol represents the connection of primitives at appropriate positions to construct a character. We store the features in the same sequence to recognize the primitives. We have listed primitives for all printed characters of Kannada script in Table 4 where the primitives numbered P1-P28 are Character Cum Primitives (CcP) and primitives numbered P29-P72 are Basic Primitives (BP). Therefore, we need to identify 72 primitives to recognize the Kannada characters.

Table 2. Primitives of Kannada characters.

ಅ	ಆ	ಇ	ಉ	ಊ	ಋ	ಎ	ಏ	ಐ	ಒ
ಟ	ಣ	ಬ	ಲ	[illegible]	[illegible]	[illegible]	[illegible]	[illegible]	[illegible]
[illegible]	[illegible]	[illegible]	[illegible]	[illegible]	[illegible]	[illegible]	[illegible]	○	[illegible]
[illegible]	[illegible]	[illegible]	[illegible]	[illegible]	[illegible]	[illegible]	[illegible]	[illegible]	[illegible]
[illegible]	•	[illegible]	[illegible]	[illegible]	[illegible]	[illegible]	[illegible]	[illegible]	[illegible]
[illegible]	-	[illegible]	[illegible]	[illegible]	[illegible]	[illegible]	[illegible]	[illegible]	[illegible]
[illegible]	[illegible]	[illegible]	[illegible]	[illegible]	[illegible]	[illegible]	[illegible]	[illegible]	[illegible]
[illegible]	[illegible]								

Table 3. Example of BP for the characters.

Character	Primitives
ಈ	O + [illegible] + [illegible]
ಔ	[illegible] + [illegible]

2.2 Zone Based Feature Extraction

As discussed in the proposed methodology section, zone based features are considered stable and help in studying the unique structures of the primitives. Before feature extraction, the method applies preprocessing steps to the input image as shown in Fig. 3. Binarization converts an input image to binary image using Otsu thresholding technique. The binarized image is normalized to standard size to make it scale invariant. Negate the normalized binary image. On the negated binary image, we have used thinning algorithm to reduce stroke width to single pixel wide to have invariant features to stroke thickness of the primitives. Each image is normalized to the size of 28×28 of pixel dimension to arrive uniformity among the primitive images. Sample preprocessing steps applied for the primitive P45 are shown in Fig. 4.

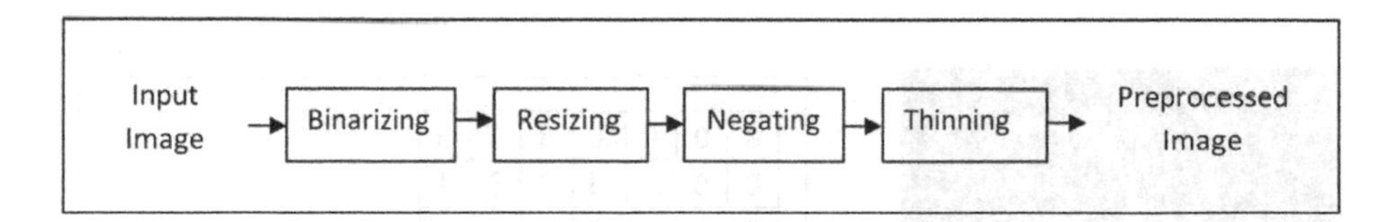

Fig. 3. Preprocessing steps of the primitive images.

In this method the whole primitive image is divided into 7×7 zones of size 4×4 as shown in Fig. 5(a). As a result we obtain 49 blocks of size 4×4 pixels. For each zone, the method extracts stroke strengths by counting number of white pixel as shown in Fig. 5(b), where the edge strength of each block is given. The concatenation of edge strength of all blocks is considered as a single feature vector for the input primitive image. i.e. for the feature vector denoted by $Fi = z1, z2, z3 \ldots z49$, where zi is ith zone value. Adapting the literature [10,12],

Table 4. List of primitives for printed characters.

Primitive Number	Printed Primitive	Primitive Number	Printed Primitive	Primitive Number	Printed Primitive	Primitive Number	Printed Primitive	Primitive Number	Printed Primitive
P1		P16		P31		P46		P61	
P2		P17		P32		P47		P62	
P3		P18		P33		P48		P63	
P4		P19		P34		P49		P64	
P5		P20		P35		P50		P65	
P6		P21		P36		P51		P66	
P7		P22		P37		P52		P67	
P8		P23		P38		P53		P68	
P9		P24		P39		P54		P69	
P10		P25		P40		P55		P70	
P11		P26		P41		P56		P71	
P12		P27		P42		P57		P72	
P13		P28		P43		P58			
P14		P29		P44		P59			
P15		P30		P45		P60			

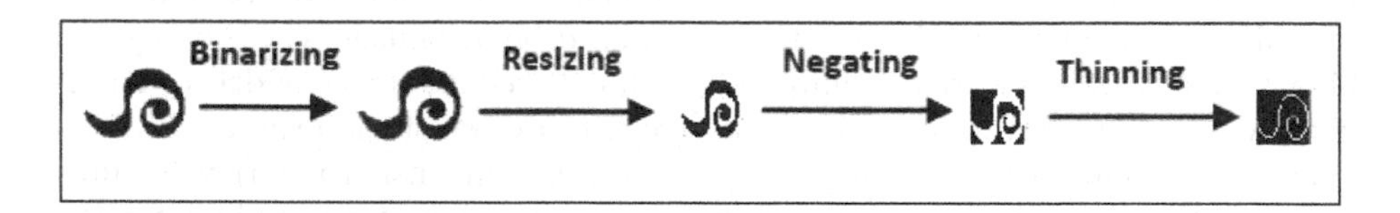

Fig. 4. Preprocessing steps applied for the P45 primitive image.

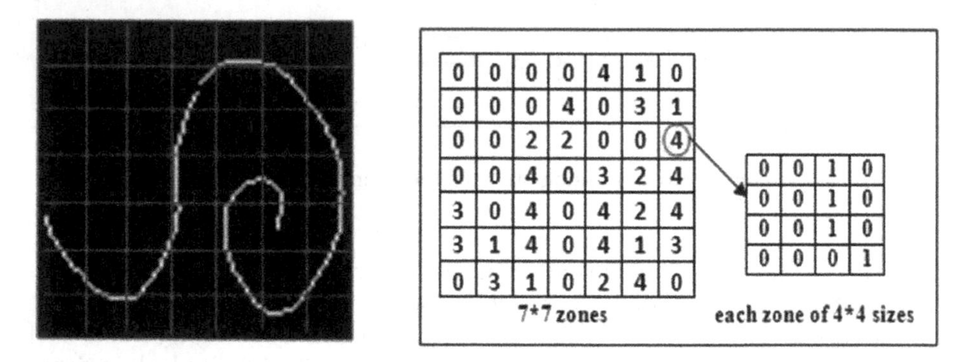

Fig. 5. Intermediate steps for feature extraction (a) Primitive for P45 of 7×7 Zones (b) Sample zone feature values for P45.

the zone based features are extracted for the whole character image either printed or handwriting, while the proposed method extracts features at primitive level. This makes the proposed method computationally effective and stable. In this way, the method extracts feature vectors for all 72 primitives, where zero denotes there are no white pixels. The objective of the paper is to develop semi-automatic system for primitives' recognition, using simple features which require less computational time and give good results.

2.3 Classification Using Distance Measure

For classification, we have used popular and computationally inexpensive, Euclidean distance measure. The Euclidean distance measure is given in Eq. 1. For the unknown input primitive image, the method extracts a feature vector and compares this with the primitives stored in the knowledge database to find minimum distance for classification.

$$d(p,q) = [\sqrt{\sum_{i=1}^{49} (p_k - q_k)^2}] \tag{1}$$

Where 49 is the number of features and pk and qk are respectively the kth feature values of the images p and q.

3 Experimental Results

Since this is the first work on recognition of printed Kannada primitives, there is no standard database available publicly. In consultation with language experts, the primitive database for printed Kannada characters is developed. The 60 font styles and 100 font sizes are considered for the primitives. The various combinations are tried and are as given in Table 5. There are four combinations of font styles and sizes. The bit 0 indicates fixed and 1 indicates varying. The experiment is conducted on 11520 (72 × 160) images of printed Kannada primitives. For each image 72 different font styles are considered to 160 such images. The font size and styles used are given in Table 6. To evaluate success of the proposed method the image set is partitioned as training set and test set. To make the classification, the popular validation technique, K-fold cross validation is adopted.

3.1 Combination 1 Observation

A total of 72 primitives out of which 28 are character cum primitives and 43 are basic primitives are considered to carry out the experiment. The total data set obtained is 2880 (72 × 40). For example, experiment considered 40 images of font size 60 and style - Nudi 0.1 here both font size and style is fixed, the result obtained is 100% accuracy for recognizing both types, the zones based features and nearest neighbor classifier are applied on the primitive testing data set. Table 7 tabulates the results obtained for both Character cum Primitives and Basic primitives.

Table 5. Font Sizes(FS) and Font Styles(FSt) combinations.

Combinations	Font Size(S)	Font Styles(F)	Remarks
1	0	0	Both Font Size and Style are Fixed
2	0	1	Font Size is Fixed and Style is Varied
3	1	0	Font Size is Varied and Style is Fixed
4	1	1	Both Font Size and Style are Varied

Table 6. Used Font Sizes(FS) and Font Styles(FSt).

Sl No	Font Size	Font Style
1.	12,14,16,......,110 (100 sizes)	Baraha 01
....	12,14,16,......,110 (100 sizes)	.....
....	12,14,16,......,110 (100 sizes)	Baraha 30
....	12,14,16,......,110 (100 sizes)	Nudi 01
....	12,14,16,......,110 (100 sizes)	.....
60.	12,14,16,......,110 (100 sizes)	Nudi 30

3.2 Combination 2 Observation

The total 4320 (72×60) images are considered for the experimentation for each primitive, 60 images of 60 fixed font size '60'and 60 varied font styles as presented in Table 6. The 2840 primitive images are taken for training and 1420 primitive images are considered for testing by considering all images with 3-fold cross validation classifier. Tables 8 and 9 give recognition accuracy for this combination of CcP and BP primitives using 3-fold cross validation method. The average of recognition accuracies obtained for both, character cum primitives and basic primitives are given in Table 15.

Table 7. CcP's and BP's recognition results for Fixed FS(SFS) and FSt(SFSt).

Fixed Size and Fixed Font using Euclidean Distance							
Image size=28x28 Zone size=4x4 No of Samples=72x40=2880							
Character cum Primitives				Basic primitives			
Primitives	1_Fold	2_Fold	Average	Primitives	1_Fold	2_Fold	Average
P1-P28	100	100	**100**	P29-P72	100	100	**100**

3.3 Combination 3 Observation

An experiment is carried out on 2880(72×40) images. We have considered 40 images of different font sizes, as given in Table 4, font style of Nudi 01 for each primitive. Tables 10 and 11 give results of classification for varying font sizes and fixed font style using 2-fold cross validation using Euclidean distance. The average accuracies obtained for this combination in case of both character cum primitives and basic primitives are given in Table 15.

Table 8. 3-Fold cross validation of CcP's.

3_Fold Cross Validation				
Same Size and Different Font using Euclidean Distance				
Image size=28x28 Zone size=4x4 No of Samples=72x60=4320				
Primitive	**1_Fold**	**2_Fold**	**3_Fold**	**Average**
P1	80	75	70	75
P2	45	70	55	56.6
P3	80	100	55	78.33
P4	100	95	100	98.33
P5	100	100	95	98.33
P6	85	60	70	71.67
P7	100	85	50	78.33
P8	85	65	70	73.33
P9	85	75	70	76.67
P10	95	70	75	80
P11	65	70	80	71.67
P12	90	85	95	90
P13	60	75	70	68.33
P14	85	85	80	83.33
P15	65	70	80	71.67
P16	55	40	55	50
P17	95	85	80	86.67
P18	75	75	85	78.33
P19	75	55	65	65
P20	55	55	35	48.33
P21	65	50	55	56.67
P22	80	65	80	75
P23	85	80	85	83.33
P24	100	95	80	91.67
P25	75	90	70	78.33
P26	65	85	85	78.33
P27	80	45	85	70
P28	90	70	70	76.67
Average	**79.1**	**73.92**	**73.03**	**75.35**

Table 9. 3-Fold cross validation of BP's.

3_Fold Cross Validation				
Same Size and Different Font using Euclidean Distance				
Image size=28x28 Zone size=4x4 No of Samples=72x60=4320				
Primitive	**1_Fold**	**2_Fold**	**3_Fold**	**Average**
P29	80	85	90	85
P30	85	85	65	78.33
P31	90	80	65	78.33
P32	80	70	80	76.67
P33	85	70	95	83.33
P34	75	75	80	76.67
P35	95	90	85	90
P36	80	70	75	75
P37	80	95	80	85
P38	85	95	85	88.33
P39	95	95	70	86.67
P40	65	100	70	78.33
P41	85	85	95	88.33
P42	100	100	100	100
P43	100	95	100	98.33
P44	80	75	55	70
P45	85	90	85	86.67
P46	80	80	65	75
P47	95	100	100	98.33
P48	70	85	60	71.67
P49	90	85	75	83.33
P50	70	50	70	63.33
P51	75	70	50	65
P52	95	100	100	98.33
P53	95	85	85	88.33
P54	70	85	85	80
P55	80	100	85	88.33
P56	65	80	80	75
P57	95	90	90	91.67
P58	85	100	100	95
P59	75	95	80	83.33
P60	95	95	90	93.33
P61	75	95	95	88.33
P62	85	65	65	71.67
P63	90	80	80	83.33
P64	60	55	60	58.33
P65	80	75	66.667	73.89
P66	65	45	70	60
P67	55	70	85	70
P68	60	40	60	53.33
P69	90	75	100	88.33
P70	100	95	100	98.33
P71	100	100	95	98.33
P72	100	95	100	98.33
Average	**82.72**	**82.84**	**81.06**	**82.2**

3.4 Combination 4 Observation

An experiment is carried out on 7200(72×100) images. We have considered 60 images of different font sizes, as given in the Table 4, of Nudi 01 font style and 40 images of 40 different font sizes for each primitive. Tables 12 and 13 show the results for both varying font sizes and styles using 4-fold cross validation using Euclidean distance. The average accuracy obtained for basic primitives and for character cum primitives is given in Table 15.

We have conducted quantitative experiments on the proposed method on different combination as shown in Table 15 to test its effectiveness. The average accuracy of k-fold cross validation on printed primitives' data for all the combinations are reported in Table 15. It is observed from Table 15 that the combinations 1, 3 and 4 in the proposed method gives good results since the

Table 10. 2-Fold cross validation of CcP's.

2_Fold Cross Validation			
Different Size and Same Font using Euclidean Distance			
Image size=28x28 Zone size=4x4 No of Samples=72x40=2880			
Primitive	**1_Fold**	**2_Fold**	**Average**
P1	100	100	100
P2	100	100	100
P3	100	100	100
P4	100	100	100
P5	100	100	100
P6	100	100	100
P7	100	100	100
P8	100	100	100
P9	100	100	100
P10	100	100	100
P11	100	100	100
P12	100	100	100
P13	100	100	100
P14	100	100	100
P15	100	100	100
P16	95.23	100	97.61
P17	100	100	100
P18	100	100	100
P19	100	100	100
P20	100	100	100
P21	100	100	100
P22	100	100	100
P23	100	100	100
P24	100	100	100
P25	100	100	100
P26	100	100	100
P27	100	100	100
P28	100	100	100
Average	99.82	100	99.91

Table 11. 2-Fold cross validation of BP's.

2_Fold Cross Validation			
Different Size and Same Font using Euclidean Distance			
Image size=28x28 Zone size=4x4 No of Samples=72x40=2880			
Primitive	**1_Fold**	**2_Fold**	**Average**
P29	100	100	100
P30	95	95	95
P31	100	100	100
P32	100	100	100
P33	100	100	100
P34	100	100	100
P35	100	100	100
P36	100	100	100
P37	95	100	97.5
P38	100	100	100
P39	100	95.45	97.72
P40	100	95	97.5
P41	100	100	100
P42	100	100	100
P43	100	100	100
P44	100	100	100
P45	100	100	100
P46	100	100	100
P47	100	95	97.5
P48	95	100	97.5
P49	100	100	100
P50	100	100	100
P51	100	85	92.5
P52	100	100	100
P53	100	95	97.5
P54	100	95	97.5
P55	100	100	100
P56	95	100	97.5
P57	100	100	100
P58	100	100	100
P59	100	95	97.5
P60	90	100	95
P61	100	100	100
P62	100	100	100
P63	100	100	100
P64	100	95	97.5
P65	95	100	97.5
P66	100	90	95
P67	100	100	100
P68	100	95	97.5
P69	100	100	100
P70	100	100	100
P71	100	100	100
P72	100	100	100
Average	**99.2**	**98.41**	**98.81**

Table 12. 4-Fold cross validation of OcP's.

4_Fold Cross Validation					
Different Size and Different Font using Euclidean Distance					
Image size=28x28 Zone size=4x4 No of Samples=38x100=3800					
Primitive	**1_Fold**	**2_Fold**	**3_Fold**	**4_Fold**	**Average**
P1	96	100	84	100	95
P2	88	80	96	100	91
P3	96	96	88	100	95
P4	96	92	100	100	97
P5	100	96	100	100	99
P6	92	96	92	92	93
P7	100	96	100	92	97
P8	96	96	100	92	96
P9	84	92	100	92	92
P10	100	92	100	100	98
P11	88	92	92	100	93
P12	92	100	100	100	98
P13	92	84	84	88	87
P14	92	84	100	100	94
P15	100	88	92	84	91
P16	84	72	84	80	80
P17	88	100	92	100	95
P18	92	100	100	92	96
P19	88	92	84	92	89
P20	72	80	80	84	79
P21	92	92	96	96	94
P22	88	92	100	96	94
P23	96	88	96	92	93
P24	92	100	100	100	98
P25	88	92	100	92	93
P26	100	92	84	76	88
P27	80	84	96	92	88
P28	92	80	100	96	92
Average	**91.57**	**91**	**94.28**	**93.85**	**92.67**

input images are resized to standard size to make them scale invariant before feature extraction. The low accuracy is reported for fixed font size and varying style due to mismatch with the primitives who are similar in shape with slight modifications as given in Table 14. However, Table 15 shows that the proposed method gives promising results for all the combinations and is consistent in recognition for printed primitives' and hence it is considered as the main advantage of the proposed method. Thus, the proposed method provides a good platform for recognition of primitives. However, the method is not compared with existing methods because, as per our knowledge, there are no methods available in the literature for recognition of Kannada primitives. Thus, the proposed method works well for recognition of printed primitives and is considered as a new attempt towards Kannada character construction.

Table 13. 4-Fold cross validation of BP's.

4_Fold Cross Validation					
Different Size and Different Font using Euclidean Distance					
Image size=28*28 Zone size=4*4 No of Samples=44*100=4400					
Primitive	1_Fold	2_Fold	3_Fold	4_Fold	Average
P29	96	100	92	92	95
P30	96	100	100	92	97
P31	100	100	92	92	96
P32	96	80	88	80	86
P33	96	88	96	96	94
P34	100	100	88	92	95
P35	92	100	100	100	98
P36	92	100	92	100	96
P37	92	96	100	100	97
P38	96	96	92	84	92
P39	96	92	100	100	97
P40	88	96	96	100	95
P41	92	84	92	100	92
P42	100	100	100	100	100
P43	100	100	100	100	100
P44	100	88	92	92	93
P45	92	100	100	100	98
P46	92	100	92	96	95
P47	100	100	96	100	99
P48	88	88	100	96	93
P49	80	92	100	100	93
P50	100	96	92	96	96
P51	96	96	100	100	98
P52	84	96	84	92	89
P53	100	100	100	100	100
P54	96	96	96	100	97
P55	100	96	100	92	97
P56	84	92	100	100	94
P57	96	100	100	100	99
P58	100	100	100	100	100
P59	100	100	100	96	99
P60	96	100	100	100	99
P61	96	100	96	100	98
P62	88	80	92	96	89
P63	100	92	92	100	96
P64	76	80	84	84	81
P65	88	92	100	88	92
P66	84	84	84	88	85
P67	88	84	100	96	92
P68	76	84	88	88	84
P69	92	92	92	88	91
P70	100	100	100	96	99
P71	100	100	100	100	100
P72	96	96	96	100	97
Average	93.63	94.45	95.54	95.72	94.84

Table 14. Confusing primitives.

ಅ ಆ ಲ	ಐ ಟ ಣ ಟಿ	[illegible]
ಖ ಷ ಬ	ಒ ಬ ಬಿ ಢಿ	[illegible]
ಎ ಏ ಏ ಬ	ಲ ಲಿ [illegible] [illegible]	[illegible]

Table 15. Recognition rate for the printed primitives of different fonts and font sizes (in %).

Sl No	Combinations	Basic Primitives	Character Cum Primitives
1.	Both Font Size and Style are Fixed	100%	100%
2.	Font Size is Fixed and Style is Varied	82%	75%
3.	Font Size is Varied and Style is Fixed	98%	99%
4.	Both Font Size and Style are Varied	94%	92%

4 Conclusion and Future Work

In this work, we have proposed semi-automatic system for recognizing printed primitives of Kannada characters. Method uses human intervention to extract primitives and then the method recognizes the primitives automatically. The method explores zone based features for recognizing primitives of the Kannada characters. The method has been tested on different font style of primitives to show the effectiveness of the proposed system. To evaluate the performance of the proposed method we have used three-fold cross validation technique. Experimental results reveal that the proposed method gives promising results for printed primitives' recognition. The future work can be extended to recognize handwritten characters of Kannada, construct the characters by connecting primitives at the appropriate connection points using the proposed method. The extended work may easy the work of digitizing old manuscripts of the Kannada literature. Same with the help of archeological department can be extended to read halegannada (old Kannada).

References

1. Pal, U., Jayadevan, R., Sharma, N.: Handwriting recognition in indian regional scripts: a survey of offline techniques. ACM Trans. Asian Lang. Inf. Process. **11**, 1 (2012)

2. Sharma, O.P., Ghose, M.K., Shah, K.B., Thakur, B.K.: Recent trends and tools for feature extraction in OCR technology. Int. J. Soft Comput. Eng. **2**, 220–223 (2013)
3. Santosh, K.C.: Character recognition based on DTW-Radon. In: IAPR, International Conference on Document Analysis and Recognition (ICDAR), pp. 264–268, IEEE, September 2011
4. Santosh, K.C., Wendling, L.: Character recognition based on non-linear multi-projection profiles measure. Front. Comput. Sci. **9**(5), 678–690 (2015)
5. Sheshadri, K., Ambekar, P.K.T., Prasad, D.P., Kumar, R.P.: An OCR system for Printed Kannada using k-means clustering (2010)
6. Nithya, E., Babu, R.: OCR system for complex Printed Kannada characters. Int. J. Adv. Res. Comput. Sci. Softw. Eng. **3**, 102–105 (2013)
7. Ragha, L.R., Sasikumar, M.: Using moments features from Gabor directional images for Kannada handwriting character recognition. In: International Conference and Workshop on Emerging Trends in Technology (ICWET 2010) (2010)
8. Sangame, S.K., Ramteke, R.J., Yogesh, V.G.: Recognition of isolated handwritten Kannada characters using invariant moments and chain code. World J. Sci. Technol. **1**, 115–120 (2011)
9. Rajput, G.G., Horakeri, R.: Shape descriptors based handwritten character recognition engine with application to Kannada characters. In: International Conference on Computer & Communication Technology (ICCCT) (2011)
10. Rajput, G.G., Horakeri, R.: Zone based handwritten Kannada character recognition using crack code and SVM. In: International Conference on Advances in Computing, Communications and Informatics (ICACCI) (2013)
11. Angadi, S.A., Angadi, S.H.: Structural features for recognition of hand written Kannada character based on SVM. Int. J. Comput. Sci. Eng. Inf. Technol. (IJCSEIT) **5**(2), 25–32 (2015)
12. Dhandra, B.V., Mukarambi, G., Hangarge, M.: Zone based features for handwritten and printed mixed Kannada digits recognition. In: 2011 Proceedings of International Conference on VLSI, Communication & Instrumentation (ICVCI). Int. J. Comput. Appl. (2011)
13. Kumar, K.S.P.: Optical Character Recognition (OCR) for Kannada numerals using left bottom 1/4th segment minimum features extraction. Int. J. Comput. Techol. Appl. **3**, 221–225 (2012)
14. Ramappa, M.H., Krishnamurthy, S.: A comparative study of different feature extraction and classification methods for recognition of handwritten Kannada numerals. Int. J. Database Theory Appl. **6**, 71–90 (2013)
15. Mamatha, H.R., Srirangaprasad, S., Srikantamurthy, K.: Data fusion based framework for the recognition of Isolated Handwritten Kannada Numerals. Int. J. Adv. Comput. Sci. Appl. (2013)
16. Anami, B.S., Garag, D.S.: Zonal-features based nearest neighbor classification of images of Kannada printed and handwritten vowel and consonant primitives. Glob. J. Comput. Sci. Technol. GJCST **14-F**(4) (2014)

Distance Based Edge Linking (DEL) for Character Recognition

Parshuram M. Kamble[1(✉)], Ravindra S. Hegadi[1(✉)], and Rajendra S. Hegadi[2]

[1] Solapur University, Solapur 413255, MH, India
parshu1983@gmail.com, rshegadi@gmail.com
[2] Institute of Information Technology, Dharwad, Hubli, India
rajendra.hegadi@gmail.com

Abstract. This article proposes an minimum distance based edge linking algorithms for handwritten character images. Improvement of performance for machine recognition is challenging task due to noise and degraded input images. In the proposed system we enhance the recognition rate of object reconstruction for broken edges by using edge linking. Such edges of objects are reconstructed by using novel Distance based Edge Linking (DEL) approach. Developed new benchmark approach is fill the gaps between nearest edge segment of Binary image map (BIM). We obtain state-of-art performance of proposed system on character recognition (CR) using two datasets MNIST and ISI.

Keywords: Feature extraction · Edge map · Binary image map · Edge segment

1 Introduction

In machine learning automatic object detection and recognition is challenging in the filed of automated computing system. It is commonly used in face recognition, image retrieval, palm-print recognition, fingerprint recognition and optical character recognition (OCR). OCR belongs to the family of machine recognition techniques which perform automatic identification. Automatic identification is the process where recognition system identifies objects automatically. It extracts feature data from object as input and enters data directly into computer systems without human involvement. The OCR are commonly used in office automation, bank and post-office. OCR is mainly used in the field of text document image analysis and processing. The OCR divided into two categories depending upon printing and input method: hand or machine printed and offline or online OCR. In literature we observe, the accuracy of OCR is less due to the noise and broken characters [13]. The sources of noise in text document images are optical device, quality of printing device (pen, paper and printer) and pre-processing techniques [6]. In OCR system, generally various pre-processing techniques are

K. C. Santosh and R. S. Hegadi (Eds.): RTIP2R 2018, CCIS 1037, pp. 261–268, 2019.
https://doi.org/10.1007/978-981-13-9187-3_23

applied on text document image, finally we got characters in good or few characters in bad shape. Character have gaps in the edges segment due to some pixels are missing during the pre-processing and image acquisition. Character recognition based on non-linear multi-projection profiles measure method for isolated handwritten or hand-printed character recognition using dynamic programming for matching the non-linear multi-projection profiles that has produced from the Radon transform [14,15].

Literature we found that many researcher have worked on different areas in pattern detection and recognition. In their work they mentioned accuracy of result was poor due to the broken edge segment of the object [10,17]. Researchers also worked on to minimize the gap between two components of the same image object. Generally gap or broken edge of the object are minimized by using average and morphological dilation operation [8,20]. Average or morphological filter when applied on image it is change the shape of an image object. In traditional method we have used different types size mask on images with various size kernel function because in databases there are different shapes and types of images. Holistic average and morphological filter are applied on entire image to fill the gaps between two different edge segments.

We developed distance oriented edge linking novel approach for connecting gaps and broken edge segment of binary image which is obtained using canny edge detector. Proposed model has tested on offline handwritten and machine printed characters of ISI (INDIAN SCRIPT CHARACTER DATABASES) [10], MNIST(database of handwritten digits) [11] datasets. The experimental results have prepared using five-fold cross validation and recognition results are enhanced.

2 Related Work

In Computer vision and Image processing application edge detection and linking is the step of feature extraction. A traditional edge detection algorithm takes a gray scale image as input and produces a Binary Edge Image Map (BEIM) as output. In BEIM edge pixels are marked an value of edge pixel 1's and non-edge pixel 0's.

In BEIM found noise, Broken edge segment and wide width edges, such edge map are shown in Fig. 1. The BEIM is produced by using traditional canny edge detection algorithms. This edge segment map was calculated in Matlab using widely used Canny edge detection algorithms.

Many researchers worked on broken edge linking (BEL) techniques they proposed various solution to compensate the edges that were not robust connect by the traditional edge detectors [1]. Whichello et al. connected character border using conventional border following method has only deal with connected pixels [19]. Ant System (AS) has proposed swarm based algorithms which exploits the self organizing nature of real ant colonies and their foraging behavior to solve discrete optimization problem [4]. Lu et al. was improved the traditional Ant Colony Optimization (ACO) based system for BEL to reduce the computation cost [9]. Jiang et al. proposed morphological techniques to detect and enhance thin edge features in the low contrast region of an scanned image [5]. Wei et al. had developed sequential edge linking algorithm that enhance the connectivity of the edges but for a rather simplified two region edge detection problem [18]. Shih et al. proposed adaptive structuring elements (ASE) to dilate the broken edges with their slope directions [16].

In this paper, we proposed method to reduce the gaps of broken edge segment by using broken edge linking techniques. In Sect. 3 described the proposed method for broken edge linking. The novel robust algorithm is defined, along with the important initial parameters of the heterogeneous environment in which linking gaps of edge segment pixels. Experimental setup and results discussion on the influence of an method parameters are described in Sect. 4. Finally, in Sect. 5 conclusions and future direction are described.

3 Distance Based Edge Linking (DEL)

We describe here in details of our Distance based Edge Linking algorithms (DEL), which takes as input only the binary edge image map (BEIM) produced by a edge Skeltanization [3] and output is the filling in minimum distance edge segment.

In DEL Algorithm 1 consist of three steps: (1) Finding end tip of edge segments. (2) Calculating the distance between source edge segment and target edge segment. (3) Filling pixels between source and target edge segment minimum distance. If any edge less than the minimum edge segment which are connected to source edge segment. Source edge segment pixel and target edge segment pixel is connected with help of Bresenham algorithms [2]. Following we explained in details of each steps.

Algorithm 1. Distance based edge linking (DEL)

BEIM *Binary edge Image Map.*
CE *Connected Edges.*

DEL(BEIM,MIN_DIST_WIN).
tipMap=findEndTip(BEIM)
ds=disBetSeg(tipmap,BEIM)
CE=FillGaps(ds,BEIM)
Return CE

In Fig. 1 is the Binary edge image Map (BEIM) of handwritten Marathi characters sample result by Canny with low and high threshold values of 15 and 45 respectively. The image first smoothed by using a Gaussian kernel [12] with $sigma = 1.8$. Each binary edge segment have begin and end tip which are shown in green stars.

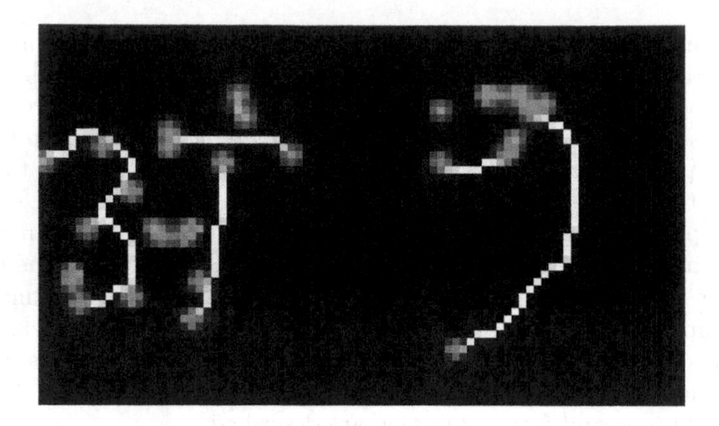

Fig. 1. Green stars are the begin and end tip of each edge segments of sample handwritten Marathi characters (Color figure online)

3.1 Find End Tips of Edge Segment

In BEIM we found their are number of edge segment which disconnected from each other. In proposed method the first stage is to find the tip of such edge segment. Before extracting tip point we assigned the labels of each segment $(L = 1, 2, 3 \ldots n)$. There are three binary edge segment L_1, L_2 *and* L_3 the dark pixel is the end tip and gray pixel is neighbors of each tip pixel of edge segment. In Fig. 2(a) to (h) mask are applied on binary edge map for finding the end tip of binary edge segments, extracted end tips are shown in Fig. 1.

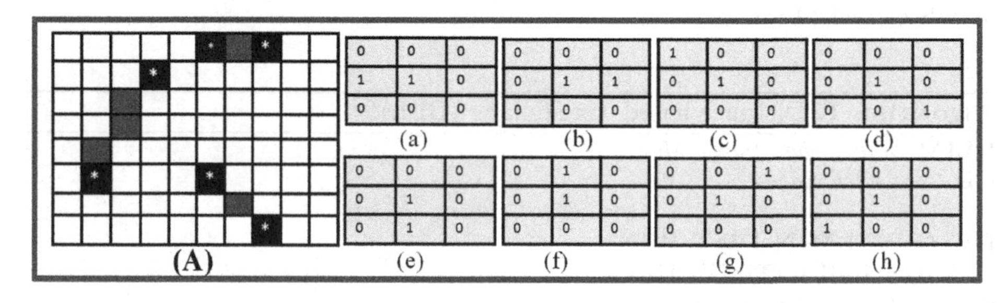

Fig. 2. Figure (A): shows the three binary edge segment the dark pixel is the end tip and gray pixel is neighbors of tip pixel. Figure (a) to (h) mask are used for finding the end tip of binary edge segments.

3.2 Finding Minimum Distance Between Binary Edge Segment

After extracting the end tips of each edge segment the next stage is finding the minimum distance between two different edge tip points. In this algorithms first step is assign label for each an edge segment. After this stage fix the source edge segment L_1, from L_1 segment set the source point $L_1(x, y)$ and marked the 5×5 windows for searching nearest neighbors an edge segment. The $L_1(x, y)$ compared with remaining edge segment in 5×5 windows, if found any edge then marked them target edge segment denoted by $Lt_i(x, y), i = 1, 2, 3, \ldots n$. Finally calculate the distance between $L_1(x, y)$ and $Lt_i(x, y)$ by using euclidean distance algorithm. After the calculating the distance between two different edge segment by using voting schema we finalize the source and target pixels. This source and target edge segment point are passed into next step fill the gap between this segment (Fig. 3).

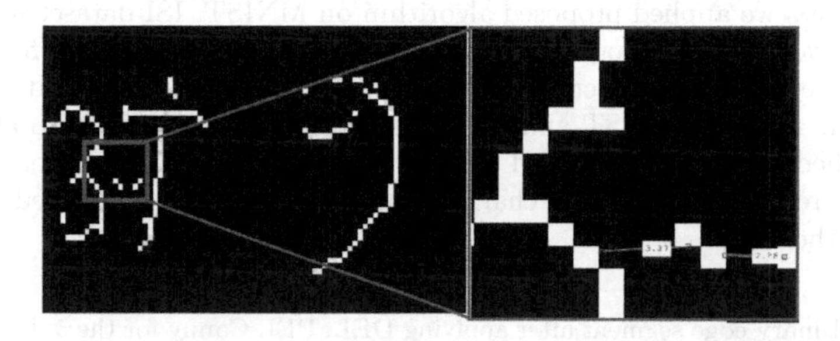

Fig. 3. Shows the calculated minimum distance between corresponding edge segments.

3.3 Filling the Gaps Between Two Edge Segment

After the finalizing source and target pixels next stage is the fill gaps between the them. Here we applied the Bresenham algorithm [2] for filling the begin and end point of each edge segment (Fig. 4).

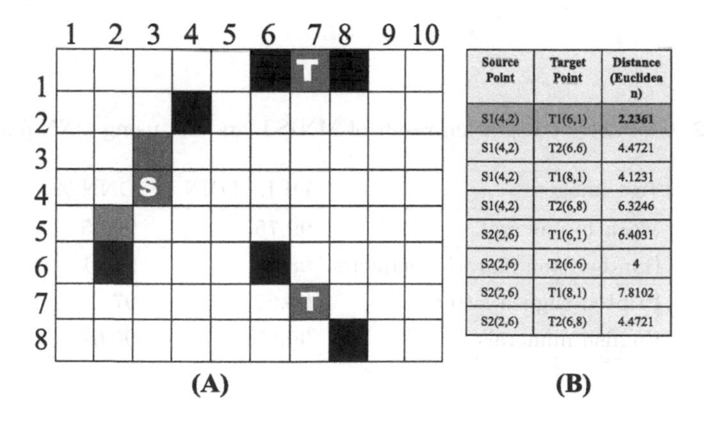

Source Point	Target Point	Distance (Euclidean)
S1(4,2)	T1(6,1)	2.2361
S1(4,2)	T2(6.6)	4.4721
S1(4,2)	T1(8,1)	4.1231
S1(4,2)	T2(6,8)	6.3246
S2(2,6)	T1(6,1)	6.4031
S2(2,6)	T2(6.6)	4
S2(2,6)	T1(8,1)	7.8102
S2(2,6)	T2(6,8)	4.4721

Fig. 4. Figure (A): Sample edge segment map 'S' denote source edge segment and 'T' denotes Target segment. In Figure (B) Minimum distance of source 'S1', 'S2' and target 'T1', 'T2' edge segment.

4 Experimental Setup and Result Discussion

In this section we evaluate the performance of Distance based Edge linking (DEL) algorithms. We compare an contrast DEL performance to that PEL (Predictive Edge Linking) algorithms. In this experiments canny edge maps were obtained by using Sobel kernel with 3×3 size. Before this step image were smoothed by a Gaussian kernel with $\sigma = 1.8$ and gradient threshold value of 20 was used during edge detection. In binary edge segment image map the 1's pixels were replaced by using 255 value. Primary algorithm was applied on MNIST and then total edge segment were calculated as shown in Table 1.

After this we applied proposed algorithm on MNIST, ISI dataset with Deep neural network [7] for recognition of characters. In this experiments firstly both datasets are used for character recognition with DNN algorithm and result are recorded in Table 2. MNIST and ISI datasets are filtered using DEL algorithm then applied DNN algorithm and result recorded in Table 2. In experiment, we found the recognition result of character more enhanced as compared to traditional method.

Table 1. Binary edge segment after applying DEL, PEL Canny for the 5 different test images from MNIST.

Test images	DEL	PEL [1]	Canny
0	7	1	4
1	2	0	3
2	2	2	3
3	2	0	3
4	2	2	3
5	1	0	2

Table 2. Character recognition result of MNIST and ISI using DNN and DEL

Test images	DEL+DNN %	DNN %
Marathi numeral	99.75	99.05
Handwritten marathi character	98.12	94.31
Printed english letter	98.62	97
English numerals	99.85	96.67

5 Conclusion and Future Scope

In this paper we propose an edge linking algorithm based on minimum distance and compared with existing system. DEL algorithm that takes in a binary edge image map (BEIM) generated by any traditional canny edge detection algorithms and convert it to a set of binary edge map (BEM), it links nearest two edge segment and to make it single. Proposed algorithm will be effectively useful on object detection and recognition.

References

1. Akinlar, C., Chome, E.: PEL: a predictive edge linking algorithm. J. Vis. Commun. Image Represent. **36**, 159–171 (2016)
2. Bresenham, J.: A linear algorithm for incremental digital display of circular arcs. Commun. ACM **20**(2), 100–106 (1977)
3. Canny, J.: A computational approach to edge detection. IEEE Trans. Pattern Anal. Mach. Intell. **6**, 679–698 (1986)
4. Dorigo, M., Birattari, M., Stutzle, T.: Ant colony optimization. IEEE Comput. Intell. Mag. **1**(4), 28–39 (2006)
5. Jiang, J.-A., Chuang, C.-L., Lu, Y.-L., Fahn, C.-S.: Mathematical-morphology-based edge detectors for detection of thin edges in low-contrast regions. IET Image Proc. **1**(3), 269–277 (2007)
6. Kamble, P.M., Hegadi, R.S.: Handwritten Marathi basic character recognition using statistical method. In: Emerging Research in Computing, Information, Communication and Applications, vol. 3, pp. 28–33. Elsevier (2014)
7. Kamble, P.M., Hegadi, R.S.: Deep neural network for handwritten Marathi character recognition. Int. J. Imaging Robot.^{TM} **17**(1), 95–107 (2017)
8. Liu, H., Jezek, K.C.: Automated extraction of coastline from satellite imagery by integrating canny edge detection and locally adaptive thresholding methods. Int. J. Remote Sens. **25**(5), 937–958 (2004)
9. Lu, D.-S., Chen, C.-C.: Edge detection improvement by ant colony optimization. Pattern Recogn. Lett. **29**(4), 416–425 (2008)
10. Pal, U., Chaudhuri, B.B.: Indian script character recognition: a survey. Pattern Recogn. **37**(9), 1887–1899 (2004)
11. Qiao, Y.: Modified national institute of standards and technology database (MNIST) (2007). http://www.gavo.t.u-tokyo.ac.jp/~qiao/database.html
12. Ramsay, J.O.: Kernel smoothing approaches to nonparametric item characteristic curve estimation. Psychometrika **56**(4), 611–630 (1991)
13. Santosh, K.C., Nattee, C., Lamiroy, B.: Spatial similarity based stroke number and order free clustering. In: 2010 12th International Conference on Frontiers in Handwriting Recognition, pp. 652–657. IEEE (2010)
14. Santosh, K.C., Nattee, C., Lamiroy, B.: Relative positioning of stroke-based clustering: a new approach to online handwritten Devanagari character recognition. Int. J. Image Graph. **12**(02), 1250016 (2012)
15. Santosh, K.C., Wendling, L.: Character recognition based on non-linear multi-projection profiles measure. Front. Comput. Sci. **9**(5), 678–690 (2015)
16. Shih, F.Y., Cheng, S.: Adaptive mathematical morphology for edge linking. Inf. Sci. **167**(1), 9–21 (2004)

17. Tsang, P.W.M.: A genetic algorithm for affine invariant recognition of object shapes from broken boundaries. Pattern Recogn. Lett. **18**(7), 631–639 (1997)
18. Wei, L., Peng, D., et al.: An improved sequential edge linking model for contour detection in medical images. In: 4th IEEE Conference on Industrial Electronics and Applications, ICIEA 2009, pp. 3757–3760. IEEE (2009)
19. Whichello, A.P., Yan, H.: Linking broken character borders with variable sized masks to improve recognition. Pattern Recogn. **29**(8), 1429–1435 (1996)
20. Zhou, Y., Starkey, J., Mansinha, L.: Segmentation of petrographic images by integrating edge detection and region growing. Comput. Geosci. **30**(8), 817–831 (2004)

Image Analysis in Agriculture

Analysis of Segmentation and Identification of Square-Hexa-Round-Holed Nuts Using Sobel and Canny Edge Detector

Dayanand G. Savakar[1], Ravi Hosur[2(✉)], and Deepa Pawar[2]

[1] Rani Chennamma University, Torvi, Vijayapur, Karnataka, India
dgsavakar@gmail.com
[2] BLDEA's V. P. Dr. P. G. Halakatti College of Engineering and Technology, Vijayapur, Karnataka, India
mca.hosur@bldeacet.ac.in, deepadasupawar20@gmail.com

Abstract. In the existing real-time automobile shop, it is difficult to trace an object and identify its presence. The failure may happen due to its absence or improper match of shape as its identity. So to overcome this, we propose a method which can be used for the automatic identification of vehicular nuts based on the input image that contains a nut with square, hexa, rounded-head and pinned-bolt shapes. The application even works with the nut having clear view or any entity added like mud, noise, colour, etc. on the surface of nut. For the identification process the database has been designed to store different shapes for selected number of nut-shapes. By applying median filter during pre-processing stage, the Sobel-edge-detector and Canny-edge-detector; segmented and identified the captured images by identifying the edges to ascertain shape of the input. With the experimentations carried the method results with an accuracy of 86.1875%

Keywords: Shape-based · Sobel · Canny edge detector · Segmentation · Identification · Vehicular

1 Introduction

Identification of an object is one of the demanding needs in the industry by a machine, too automatically after being captured. In particular is a simple scenario of an industry perspective is to identify the spare parts of the requirement on demand. The earlier image retrieval techniques use of keywords in the text for the retrieval process, in the content based approach work on the observable characters of the image and the characters are colour, shape and texture, etc. Among all the three contents important character for the retrieval process is shape, hence in the proposed work the shape is considered as the one of the important feature for the retrieval of the vehicular objects. Here, the automatic

K. C. Santosh and R. S. Hegadi (Eds.): RTIP2R 2018, CCIS 1037, pp. 271–283, 2019.
https://doi.org/10.1007/978-981-13-9187-3_24

detection of the vehicular spare parts in the automobile industry is one of the important needs, because searching of the specific part from the group of the objects consumes more time and energy. The smart system in the automobile industry will be used to separate the different spare parts of the vehicle. The shape based segmentation and identification of the vehicular nuts is an application used for automatic detection of vehicular spare parts. We can specify the shape based object retrieval technique as the system for fetching the images from the database of the automobile industry or shop, the database consisting of the different images of the vehicular spare parts of different shapes. For the pre-processing step median filter is used depending upon the noise present in the image, then edge based segmentation is performed by using detection algorithms like Sobel-edge and Canny edge. As compared to other edge identification techniques, this approach provides better edges including both strong and weak edges. After applying the Sobel and Canny edge mechanism it provides the distinct edges which are already exist in the image and then by applying centroid method the shape is obtained from the image. Images which already exist in the database are retrieved when shape of the input and database image are matched. The automatic detection of the vehicular spare parts in the automobile industry or shop is one of the important needs. Manual searching is the drawback so to overcome that, this application has been designed by considering one of the important content of the image, which is shape. The retrieval process is carried out by considering the shape, block diagram and algorithm discussed in further sections of the proposal.

2 Related Work

A technique to identify the shapes of automobile nuts like square and hexa shaped using edge detection methods proposed in [1]. A content based image retrieval system with a sequential classifier to identifying a region of interest, are then checked by using BLSTM classifier for the presence of arrows; with the help of Npen++ features proposed in [2]. A realistic 3D model constructed for face using Intel RealSense technology by scanning a face captured from different views combining a 3D face detection, background segmentation and 3D mesh, with texture export [3]. A separation technique to panels that are stitched biomedical research articles in multipanel figures by applying segment detection when the gray-level pixel changes proposed in [4]. Connection of prominent broken lines are done using Vectorization process by eliminating insignificant line segments. The lung regions affected by a foreign agent using digital chest radiographs by computing edge map using edge detection algorithms, and circular Hough transform (CHT) confirms the selection of morphological operations proposed in [4]. Region growing-based [31] and connected component-based [30] segmentation methods are proposed to segment and label the fractured bones from CT images. A technique that identifies circular foreign objects, in particular buttons, within chest X-ray (CXR) images by enhancing the CXRs that helps to find the edge images, confirms the selection of morphological operations in the chest region [5]. The method exhibits thoracic edge maps by studying histograms of any deformation for all possible orientations of gradients by using two CXR benchmark collections using five different regions-of-interest selection [6]. A detection method for lungs using radiographs of chest for quality control with a measure of rib-orientation with the help of line histogram technique for quality control, and therefore augmenting automated abnormality detection. The technique considers normal and abnormal images in the chest radiographs as an aid in quality control [7]. Further in [32] contrast limited histogram equalization (CLAHE) used to enhance synovial cavity region in knee X-ray images. A RSLIC descriptor that allows matching of objects/scenes in terms of their key-lines, line-region properties, and line spatial arrangements proposed in [8]. By considering the face characteristics, method is tested for RSILC performance descriptors for a face matching [9]. A method for constructing 3D using 2D by using a technique centroid clustering process by combining all the depth images in the cluster proposed in [10]. Multi-Depth Generic Elastic Models (MDGEMs) for the construction of 3D face modeling by using varying multiple depth maps of an image proposed in [11]. An algorithm for reconstructing a 3D view using a wavelet transform and Support Vector Machine (SVM) models to get the image focus quality and Mean shift algorithm for the depth scene is proposed in [12]. The results prove to be improving the efficiency better compared to previous techniques. A method for detecting the edge by generating a decision tree [13], a solution 25 times better to a conventional method existing by training the decision tree using supervised learning method in [14] recovers the object boundaries of captured depth images with sharp that refine the use of adaptive block by reallocation of the same position and expand to increase the depth accuracy by avoiding false depth boundary refinement. A technique of Steganography using JPEG images based on 2D block DCT that transform cover image blocks from spatial domain to frequency domain proposed in [15]. The method brought up with 54 bits secret block embedding as information bits and hence a cover image of 417 × 417 pixels can embed 146718 secret bits into it. The experimentation resulted in acceptable image quality and a large message capacity. A model for 3D reconstruction of an indoor scene using a movable standard Kinect system proposed in [16]. The scenes are captured with the help of low-cost scanning, and geometry augmentation and related based interactions. Using the techniques point-plane data association TSDF integration access the scenes are projected with assumed threshold and then truncated with maximum and minimum thresholds. The extraction is done using raycasting and the overlapping is tested using a touch-map technique. A method for JPEG based on 3D formation of original image and the application of 3D discrete-cosine-transformation proposed in [17]. In the process 3D image is represented in 1D as a spiral ladder for a better efficiency in coding [18]. The technique on experimental results proved to be better over JPEG compression in varying bit rates on background images with the deployment in a cube of $8 \times 8 \times 8$. A method for reconstructing 3D scenes from a set of images using 2D photo-editing tools proposed in [19]. The results shown that the Cell Carving algorithm computed the 3D model with maximum volume, and is faster in using 2D intersection operations for computing a polygon module. The user interfaced experiments were carried in visual basic and the algorithm in C++, rendered the scenes from six viewpoints using 3D StudioMax. A 3D watermarking-scheme for the triangle meshes for perturbing the distance between the vertices and the centre of the model that preserves the visuality of the models and intern the strength of the embedded watermark signal proposed in [20]. The mesh has been checked with all parameters like weight, controlling features like extraction and embedding, addition of noise to check for resistance, and also the combination of attacks like simplification, noise and cropping attacks to check the scheme for robustness. In experimentation due to the use of weighting scheme, the watermarking scheme is able to withstand common attacks on 3D models with all possible attacks. A image retrieval system based on its shape using Generic Fourier Descriptor, where image retrieval can be performed by using different contents of the image like colour, texture and shape are used [21]. By observing these contents shape is one of the important content for retrieving the images. In the proposed method image retrieval is done by applying 2-D Fourier transform the Fourier descriptor (FD) is used on the polar shape image. A summarization of curvature scale shape and Fourier Descriptors for shape-based image retrieval proposed in [22]. This technique which works on the content of the image treats shape as one of the feature among the other features which are used for retrieval. In the proposed method two important shape descriptors are used for retrieving the shape of the image, the descriptors are region based and contour based. By using these procedures retrieval is performed. A survey on, where the different types of applications like remote sensing, crime prevention, for searching browsing and retrieving the images the different devices are required to retrieve the details depending on the content of the image presented in [23]. Because of this the various approaches are introduced. In the earlier days the text based approach is developed for the image retrieval afterwards the content-based approach has been developed. If the characteristics are used like keywords and text descriptors then the human effort is required for understanding the images and for the similarity measurement. But is we use the low level image character then they are automatically extracted the human effort is not required. A new class of adaptive filter operators for salt and pepper impulse corrupted images, the proposed work involves two Adaptive filtering methods for removing the salt and pepper kind of noise proposed in [24]. The algorithms used the interactive Adaptive switching median filter (IASMF), it works on the images that are corrupted at the low rate and Adaptive-threshold based median filter (ATMF) algorithm works on images that are highly corrupted for giving the better quality output. Experimental results show that as compared to the other filtering technique it gives better result. A shape descriptors and context, to show the characteristics of the image the visual descriptor is used, which is located close to sector and appearance of the curved lines [25]. In the proposed descriptor two component feature vectors are used. In the first component the local section is partitioned into Zone and their orientation values, incline magnitude are selected. In the second component the local shape features are selected using contour lines. An effective image retrieval scheme using color, texture and shape features, in proposed image retrieval technique all the three characters colour, shape and texture are combined to execute retrieval procedure [26]. To obtain the colour quantization algorithm is used then the steerable filter decomposition is used to select the texture feature. Lastly the pseudo Zernike moments are used for the shape descriptor. A image retrieval based on its content using exact legendre moments and SVM, The technique which is situated upon the content of image uses the in-variant image moments proposed in [27]. Invariant and Zernike moments are good for representing shape but they are not efficient for orthogonal moment, therefore Legendre Moments are used they are orthogonal and efficiently faster. In the proposed work Exact Legendre Moments (ELM) are utilized for the retrieval mechanism. CBIR system based on Polar Raster Edge Sampling Signature proposed in [28]. In the proposed work the shape feature are selected from the image by using the Polar raster Sampling Signature algorithm. By using Euclidean distance similarity whether the images are similar or not is checked, the images present in the database are retrieved when it is appropriate to the input image. The groups of shape descriptors for retrieval and its classification, In the proposed work group of approaches are used by using the weighted sum rule proposed in [29]. This rule is based on the extensively used shape descriptor, for example inner distance shape context, height function. By converting shape descriptor into matrix the features are selected, from which a set of text descriptors are selected. The literature review summarizes that in the earlier days and the new techniques for image retrieval is performed by using text based approach which consumes time and human effort. In the later days the content based approach is came into existence. From the different research work it shows that many techniques used shape for the retrieval of the images by using different shape representation algorithms. We can say that shape is major element used for the retrieval process, Hence in this application it is used.

2.1 Contribution of This Paper

The automatic retrieval of the vehicular spare parts in the automobile industry/shop is one of the important needs. Manual searching is the drawback so to overcome that, this application has been designed by considering one of the important content of the image, which is shape. The retrieval process is carried out by considering the shape, block diagram, algorithm and the working procedure of the proposed work is discussed in the next section.

3 Proposed Work

The below Fig. 1 shows the intended scheme which is shape based object retrieval technique, the flow of this block diagram are explained here first it starts with reading the images from the database and the input image is pre-processed by using median filtering techniques, the pre-processed image is then passed for the segmentation here edge based segmentation is performed by employing the Sobel edge detection procedure. The shape feature is extracted in the feature extraction stage by using convex hull method which is best suited for extracting feasible regions of expected image. If input image shape and images which already exist inside the database are similar at that instant images are retrieved.

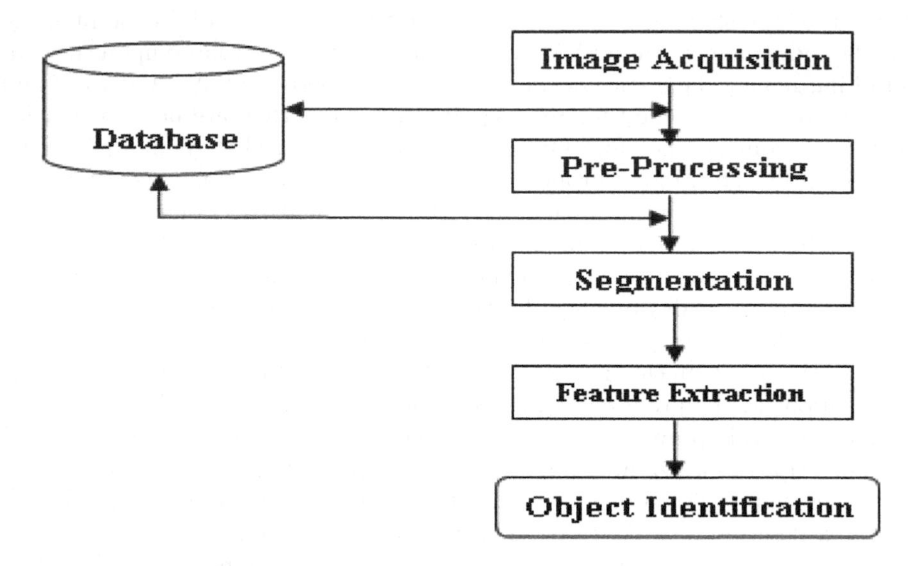

Fig. 1. Block diagram of proposed method

3.1 Image Acquisition

The image acquisition process is defined as the action of obtaining the picture from some source normally the hardware based source, so that any next process can follow afterwards. Image acquisition should perform in the first step because images are very important for the next process, if images are not there next process will not start. In this application the image acquisition is done by collecting the different images of spare parts from camera and from search engines and stored in the database.

3.2 Preprocessing

After collecting the images in the image acquisition step the next step is pre-processing in the pre-processing step the images that are collected and stored in the database are passed as input. Depending on the noise, filtering techniques used are median filter (Figs. 2 and 3).

Depending on the noise, filtering techniques used are median filter.

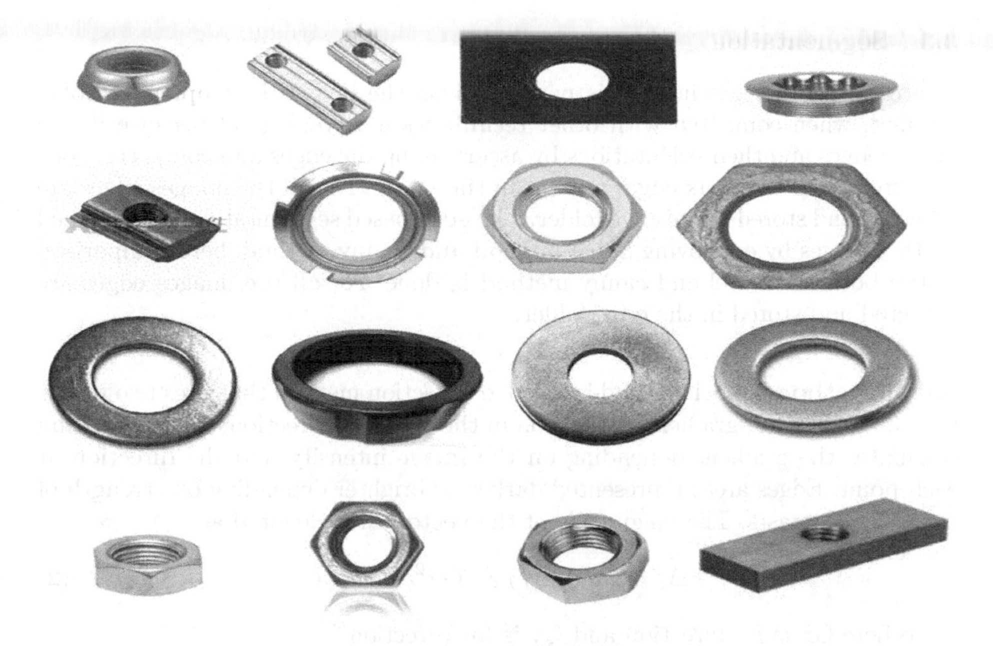

Fig. 2. Nuts of different shapes considered for experimentation

Fig. 3. Converted of captured image into a gray scale image

Selecting Suitable Filter for Reducing Noise. The MSE is abbreviated as the Mean Square Error this is used to compare between original image and compressed image, if we get lower MSE value then the error rate is low

$$MSE = \frac{1}{mn} \sum_{i=0}^{m-1} \sum_{j=0}^{n-1} [I(i,j) - K(i,j)]^2$$

.........(A)

The PSNR is estimated in decibels (DB), this is utilized for the quality estimation among original image and the regenerated image, if we get large PSNR value then quality of the re-established or reconstructed image is better.

$$PSNR = 10 \log 10 \frac{MAX^2}{MSE} \tag{1}$$

3.3 Segmentation

The edge based segmentation is performed on the images by employing Sobel method, when compared with other techniques it is very good because it can detect edges and their orientations by ascertaining the edges and comprises both the strong and the weak edges present in the image. For all the images edges are detected and stored in the edge folder. The edge based segmentation is performed on the images by employing Sobel method and Canny method, here comparison of the both the Sobel and canny method is done. For all the images edges are detected and stored in the edge folder.

Sobel Method. Sobel method is an edge detection method that uses two masks with 3×3 sizes, for gradient estimation in the X and Y directions. The algorithm calculates the gradient depending on the image intensity and the direction at each point. Edges areas represented darker or brighter depending on strength of intensity contrasts. The magnitude of the vector Δf is denoted as

$$\Delta f = mag(\Delta f) = [Gx^2 + Gx^2]^{\frac{1}{2}} \tag{2}$$

Where Gx is for direction and Gy is for direction

Canny Method. The Canny edge detector with Gaussian filter the method, any noise present in an image can be removed; that is:

a. Smooth the image with a two dimensional Gaussian
b. Take the gradient of the image
c. Non-maximal suppression
d. Edge thresholding (Fig. 4).

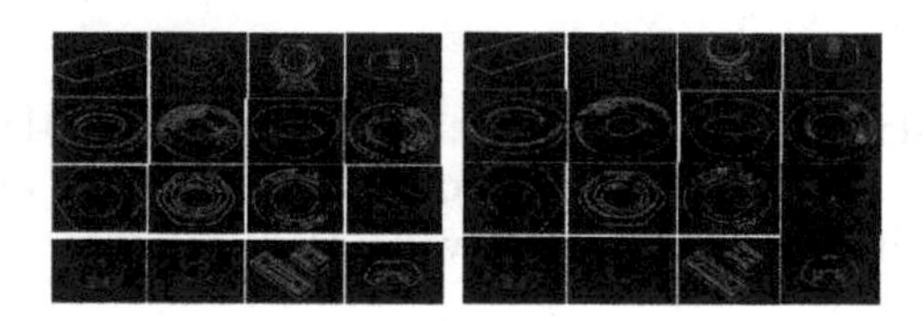

Fig. 4. Detection of the edge by using Sobel method and Canny method

3.4 Feature Extraction

Feature extraction is done by using convex hull method, which is good for describing the shape of the image. In the proposed method retrieval technique is situated on image shape. System generates the shape characters for both query and database images. In this step the extraction of the image can be done by click on the Feature Extraction option within the output, which going to identify the shape of the object from the taken input and it will specify the outer line of the image in the output, as shown in the below Fig. 5, and retrieval of the image also done here.

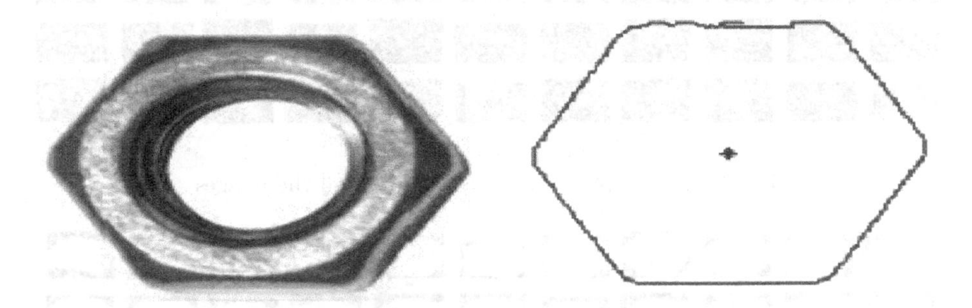

Fig. 5. Shows the feature extraction of the image (a) Input Image (b) Feature extracted Image

4 Results and Discussions

The below Figs. 6, 7, and 8 shows the detection of the edge of the image by using the canny and Sobel edge detection methods. Here the comparison of the images can be take place by using these two methods.

Fig. 6. Database includes collection of hexagon and square nuts

During the experimentation, the samples considered were having the edges better for identification, and also the samples that were not smooth and fine with edges (like ruptured/damaged) when considered; the system was able to identify the sample (nuts/bolts) with a considerable average especially in case of squared and hexa shaped (Table 1).

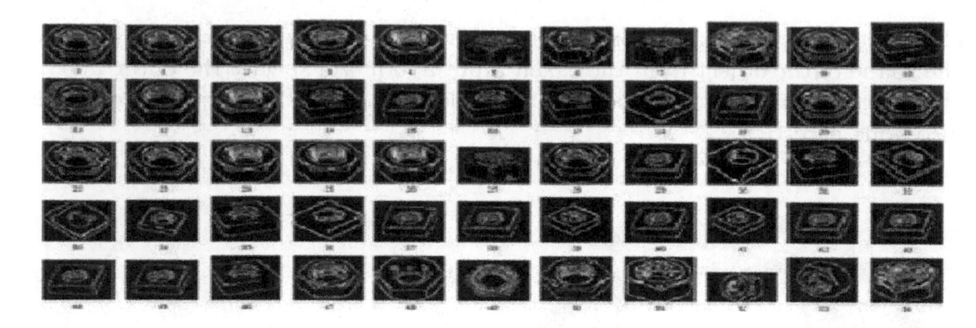

Fig. 7. These are the detected edges of all the images

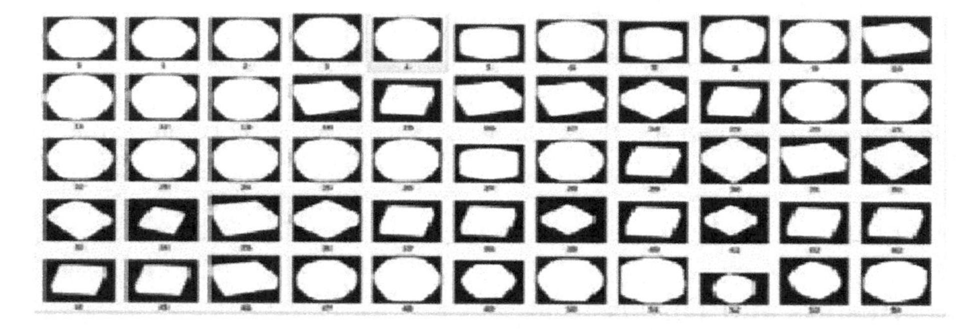

Fig. 8. These are detected shapes of all the images

5 Conclusion

In this work the image segmentation and identification is done for the different shapes of the vehicular spare parts. In the automobile industry the identification of the vehicular parts is one of the important needs. Through the image acquisition stage different images of nuts are collected and they are stored in the database. Shape of the image is identified by using centroid method, segmentation and identification of the objects is done based on the image and the system is successful in segmentation and identification process. Here in this work it involves the detection of the edge which can be done by using the Sobel and

Table 1. Accuracy vs no of images

Sl. No	Type of nut	Number of images	Average accuracy (%)
1.	Square	50	85.5167
2.	Hexa	50	81.8667
3.	Round	50	96.1359
4.	Holed	50	79.2307

Canny edge detection method, and comparison between them will take place. The system is successful in retrieval process with the accuracy of 86.1875%.

5.1 Future Enhancement

In future the proposed system for identifying vehicular/automobile spare parts can still be enhanced and improved for the nuts/bolts whose shapes are ruptured by considering different modes along with shape as multilingual method for identification.

References

1. Hosur, R., Savakar, D.G., Madabhavi, S.: Shape based object retrieval technique for vehicular spare parts. Int. J. Eng. Technol. (UAE) **7**(4.5), 355–359 (2018)
2. Santosh, K.C., Roy, P.P.: Arrow detection in biomedical images using sequential classifier. Int. J. Mach. Learn. Cybern. **9**(6), 993–1006 (2018)
3. Savakar, D.G., Hosur, R.: A relative 3D scan and construction for face using meshing algorithm. Multimedia Tools Appl. **77**(19), 25253–25273 (2018)
4. Santosh, K.C., Aafaque, A., Antani, S., Thoma, G.R.: Line segment-based stitched multipanel figure separation for effective biomedical CBIR. Int. J. Pattern Recogn. Artif. Intell. (IJPRAI) **31**(6), 1–18 (2017)
5. Zohora, F.T., Santosh, K.C.: Foreign circular element detection in chest X-rays for effective automated pulmonary abnormality screening. Int. J. Comput. Vis. Image Process. (IJCVIP) **7**(2), 36–49 (2017)
6. Zohora, F.T., Santosh, K.C.: Circular foreign object detection in chest X-ray images. In: Santosh, K.C., Hangarge, M., Bevilacqua, V., Negi, A. (eds.) RTIP2R 2016. CCIS, vol. 709, pp. 391–401. Springer, Singapore (2017). https://doi.org/10.1007/978-981-10-4859-3_35
7. Santosh, K.C., Vajda, S., Antani, S., Thoma, G.R.: Edge map analysis in chest X-rays for automatic abnormality screening. Int. J. Comput. Assist. Radiol. Surg. (IJCARS) **11**(9), 1637–1646 (2016)
8. Santosh, K.C., Candemir, S., Jaeger, S., Karargyris, A., Antani, S., Thoma, G.: Automatically detecting rotation in chest radiographs using principal rib-orientation measure for quality control. Int. J. Pattern Recogn. Artif. Intell. (IJPRAI) **29**(2), 1557001 (2015)
9. Candemir, S., Borovikov, E., Santosh, K.C., Antani, S., Thoma, G.: RSILC: Rotation- and Scale-Invariant, Line-based Color-aware descriptor. Image Vis. Comput. **42**, 1–12 (2015)
10. Herrera, J.L., del-Blanco Narciso Garcia, C.R.: Edge based depth gradient refinement for 2D to 3D learnt prior conversion. In: IEEE 3DTV-Conference: The True Vision-Capture, Transmission and Display of 3D Video (3DTV-CON) (2015)
11. Wu, Z., Li, J., Hu, J., Deng, W: Pose-invariant face recognition using 3D multi-depth generic elastic models. IEEE (2015)
12. Zhou, Y., Guo, H., Fu, R., Liang, G., Wang, C., Wu, X.: 3D reconstruction based on light field information. In: Proceeding of the 2015 IEEE International Conference on Information and Automation Lijiang, Held on August 2015

13. Kaneko, M., Hasegawa, T., Yamauchi, Y., Yamashita, T., Fujiyoshi, H., Murase, H: Fast 3D edge detection by using decision tree from depth image. In: 2015 IEEE/RSJ International Conference on Intelligent Robots and Systems (IROS) Congress Center Hamburg, Hamburg, Germany, 28 September–2 October 2015, pp. 1314–1319 (2015)
14. Xu, X., et al.: Adaptive block truncation filter for MVA depth image enhancement. In: 2014 IEEE International Conference on Acoustic, Speech and Signal Processing (ICASSP) (2014)
15. Singh, M., Sharma, R., Garg, D.: A new proposed issue for secure image steganography technique based on 2D block DCT and DCT. Int. J. Adv. Res. Comput. Sci. Softw. Eng. **2**, 29–33 (2012)
16. Izadi, S., et al.: KinectFusion: real-time 3D reconstruction and interaction using a moving depth camera. In: UIST 2011 (2011)
17. Alptekin Engin, M., Cavsoglu, B.: New approach in image compression: 3D spiral JPEG. IEEE Commun. Lett. **15**(11), 1234–1236 (2011)
18. Khare, A., Kumari, M., Khare, P.: Efficient algorithm for digital image steganography. J. Inf. Knowl. Res. Comput. Sci. Appl. **1**(1), 1–5 (2010)
19. Bariya, P., Nishino, K.: Scale-hierarchical 3D object recognition in cluttered scenes. In: IEEE Conference on Computer Vision and Pattern Recognition (CVPR), pp. 1657–1664 (2010)
20. Yu, Z., Ip, H.H.S., Kwok, L.F.: A robust watermarking scheme for 3D triangular mesh models. J. Pattern Recogn. Soc. **36**(11), 2603–2614 (2003)
21. Zhang, D., Lu, G.: Shape based image retrieval using Generic Fourier Descriptor. J. Sig. Process. Image Commun. **17**(10), 825–848 (2002)
22. Zhang, D., Lu, G.: A comparative study of curvature scale shape and Fourier descriptors for shape-based image retrieval. J. Vis. Commun. Image Represent. **14**(1), 39–57 (2003)
23. Liu, Y., Zhang, D., Lu, G., Ma, W.-Y.: A survey of content-based image retrieval with high-level semantics. J. Pattern Recogn. **40**(1), 262–282 (2007)
24. Krishnan, N., Varghese, J., Saudia, S., Mathew, S.P., et al.: A new adaptive class of filter operators for salt and pepper impluse corrupted images. Int. J. Imaging Sci. Eng. (IJISE) **1**(2), 44–51 (2007)
25. Peng, S.-H., Kim, D.-H., Lee, S.-L., Chumg, C.-W.: A visual shape descriptor using sectors and shape context of contour lines. J. Inf. Sci. **180**(16), 2925–2939 (2010)
26. Wang, X.-Y., Yu, Y.-J., Yang, H.-Y.: An effective images retrieval scheme using color, texture and shape features. J. Comput. Stand. Interfaces **33**, 59–68 (2010)
27. Rao, S., Srinivas Kumar, S., Chandra Mohan, B.: Content-based image retrieval using exact legendre moment and support vector machine. Int. J. Multimedia Appl. **2**(2), 69–79 (2010)
28. Mathew, S.P., Balas, V.E., Zachariah, K.P., Samuel, P.: A content-based image retrieval system based on polar raster edge sampling signature. Acta Polytech. **11**(3), 25–36 (2014)
29. Nanni, L., Lumini, A., Brahnam, S.: Ensemble of shape descriptors for shape retrieval and classification. Int. J. Adv. Intell. Paradigms (IJAIP) **6**(2), 136–156 (2014)
30. Ruikar, D.D., Santosh, K.C., Hegadi, R.S.: Automated fractured bone segmentation and labeling from CT images. J. Med. Syst. (2019). https://doi.org/10.1007/s10916-019-1176-x
31. Ruikar, D.D., Santosh, K.C., Hegadi, R.S.: Segmentation and analysis of CT images for bone fracture detection and labeling. In: Medical Imaging: Artificial Intelli-

gence, Image Recognition, and Machine Learning Techniques, Chap. 7. CRC Press (2019). ISBN 9780367139612
32. Hegadi, R.S., Navale, D.I., Pawar, T.D., Ruikar, D.D.: Multi feature-based classification of osteoarthritis in knee joint X-ray images. In: Medical Imaging: Artificial Intelligence, Image Recognition, and Machine Learning Techniques, Chap. 5. CRC Press (2019). ISBN 9780367139612

3D Reconstruction of Plants Under Outdoor Conditions Using Image-Based Computer Vision

Abhipray Paturkar(✉), Gaurab Sen Gupta, and Donald Bailey

School of Engineering and Advanced Technology, Massey University, Palmerston North, New Zealand
A.Paturkar@massey.ac.nz

Abstract. 3D reconstruction of plants under outdoor conditions is a challenging task, for applications such as plant phenotyping which needs non-invasive methods. With the availability of new sensors and reconstructions techniques, 3D reconstruction is improving rapidly. However, sensors are still expensive for researchers. In this paper, we propose a cost-effective image-based 3D reconstruction approach which can be achieved by off-the-shelf cameras. This approach is based on the structure-from-motion method. We implemented this approach in MATLAB and Meshlab is used for further processing to achieve an exact 3D model. We also investigated the effect of different adverse outdoor scenarios which affect quality of 3D model such as movement of plants because of strong wind, drastic change in light condition while capturing the images. We have decreased the appropriate number of images needed to get precise 3D model. This method gives accurate results and it is a fast platform for non-invasive plant phenotyping.

Keywords: 3D reconstruction · Structure-from-motion · Feature extraction · Feature matching · Plant phenotying

1 Introduction

In precision agriculture, plant phenotyping is an important aspect. It helps scientists and researchers to collect valuable information regarding plant structure, which is inevitably a basic requirement to enhance plant discrimination and plant selection [1]. 3D reconstruction models of the plant through phenotyping operations are helpful for evaluating plant growth and yield over time. This permits management of the plant to be more extensive [2]. This 3D reconstructed plant models could be used to describe leaf features, discriminate between weed and crop, estimate the biomass of the plant, and classify fruits. Conventionally all these elements have been evaluated by experts in this field depending on the visual score, which was responsible for creating dissimilarity between expert judgements. In addition, this process is tedious.

K. C. Santosh and R. S. Hegadi (Eds.): RTIP2R 2018, CCIS 1037, pp. 284–297, 2019.
https://doi.org/10.1007/978-981-13-9187-3_25

Primarily, the aim of plant phenotyping is to calculate plant features precisely without subjective biases. Nonetheless, developing expertise and knowledge still lack technical advancements in processing and sensing technologies. Most of the modern sensing technologies are primarily only two dimensional, e.g. thermal or hyperspectral imaging. Inferring 3D information from such sensors is greatly reliant on distance and angle to the plants. In contrast, 3D reconstruction is being suggested for morphological classification of plants. 3D reconstruction is developing rapidly and getting tremendous attention. Structured light (Kinect sensor) [3], ToF cameras [4] and LiDAR [5] are active sensing techniques used for 3D reconstruction which basically use their own source of illumination. However, these state-of-the-art systems are costly. On the other hand, image-based passive 3D reconstruction techniques which use radiation present in the scene, which includes, structure-from-motion [6], stereo vision [7] and space carving system [8], only need one or two cameras which results in a very cost-effective system.

ToF cameras have excellent performance and appeared to be a suitable sensor for evaluating plants. ToF cameras are commonly combined with an RGB camera. Kazmi et al. [4] analysed the performance of ToF cameras for close range imaging in different illumination conditions. They found that ToF cameras deliver high frame rates as well as accurate depth data under suitable conditions. However resolutions of depth images are often low; the sensors are sensitive to ambient sunlight, which usually leads to poor performance while working outdoors; the quality of depth values depends on the color of objects, and some sensors have blurring problems while sensing moving objects. Because of these limitations, it is difficult to use TOF cameras for 3D reconstruction under outdoor conditions.

LiDAR is an expansion of the principles applied in radar technology. It estimates the distance between the target and the scanner by illuminating the target using a laser and calculating the time taken for the reflected light to come back [9]. Kaminuma et al. [10] presented an application of a laser range finder for 3D reconstruction which represents the leaves and as a polygonal meshes and then measured the morphological features from those models. Paulus et al. [11] determined that LiDAR is an appropriate sensor for obtaining precise 3D point clouds of plants but it does not give any information on the surface area. In addition, it had poor resolution and a long warm-up time. In contrast, LiDAR has given excellent results under outdoor conditions having a drawback of being very costly. Other disadvantages of the LiDAR sensor are that it needs calibration and multiple captures are required to overcome issues with occlusion. The data from the sensors cannot detect leaves overlapping efficiently and depth and images are not of high quality.

An alternative approach for depth estimation is the use of structured light. In this approach, the light source (either near-infrared or visible) is offset a familiar distance from an imaging device. The luminous from the emitter is reflected into the camera by the target object. Information about the light pattern allows the depth to be derived through triangulation [9]. Baumberg et al. [12] presented a 3D plant analysis based on the technique they called mesh processing. In this work, the authors made a 3D reconstructed model of a cotton using Kinect sensor which performed well under indoor conditions yet struggled under outdoor conditions. Chéné et al. [13] used a depth camera to segment plant leaves and reconstructed the plant in 3D.

As mentioned above, stereo vision and structure-from-motion use passive illumination, which allows these techniques to work efficiently under outdoor conditions. A off-the-shelf digital camera could be used for capturing overlapped images which are processed by a computer to estimate the depth or 3D reconstructed model. Stereo vision has comparatively lower cost than active sensing techniques and has provided excellent 3D reconstructed models. Nevertheless, the camera alignment and spacing between the cameras should be precise. As an illustration, the distance between the plant and camera, which is calculated with the help of the focal length of the camera, there should be an overlapping between the images and the rotation of the plant in different images. Ivanov et al. [14] described maize plants under outdoor conditions by using images captured from various angles to characterize plant structure. Takizawa et al. [15] reconstructed a 3D model of a plant and derived plant height and shape information.

The combination of images and cameras in structure-from-motion generally create a sparse 3D point cloud. Structure-from-motion consists of calculating a set of points from position of cameras, from these set of points, a dense point cloud is created. Jay et al. [16] proposed a method which builds a 3D reconstructed model of a crop row to get the plant structural parameters. This 3D model is acquired using structure-from-motion with the help of colour images captured by translating a single camera along the row. Quan et al. [17,18] proposed a semi-automatic method for modelling plants for application like plant phenotyping, yield estimation based on structure-from-motion which performed well under outdoor conditions but it is computationally expensive.

In summary, each sensing technique has some merits and demerits [9]. The need of current sensors and systems is to reduce the need for manual extraction of phenotypic data. Their performance stays, to a lesser or greater extent, restricted by the dynamic morphological complications of plants [19]. Currently, there is not a 3D system and method which solves all necessities, but one should select depending on the budget and requirements. Moreover, plant structure is generally complex which includes a large amount of self-occlusion (leaves blocking one another). Hence, reconstructing plants in 3D in non-invasive manner stays a serious challenge in phenotyping.

Focusing at contributing a cost-effective solution to above challenge, we present an image-based 3D reconstruction system under outdoor conditions. Our contributions include:

1. An easy and cost-effective system (using just a mobile phone camera)
2. Investigation of effects of adverse outdoor scenarios and possible solutions (movement of plants because of wind and change in light condition because of the movement of clouds while capturing the images)
3. A precise 3D model obtained from a limited number of images

The rest of the paper is organized as follows; Sect. 2 discusses the step by step results of the method we used in this paper. The effect of adverse outdoor scenarios on 3D model along with the possible solution discussed in Sect. 3.

2 Materials and Methods

We selected a chilli plant *Capsicum annum L.* on a commercial field (Palmerston North, New Zealand) for testing our image processing. This chilli plant is selected for its demand over the year and its high value. Images were acquired during December 2017 when plant height was between 15 cm to 20 cm. A crop

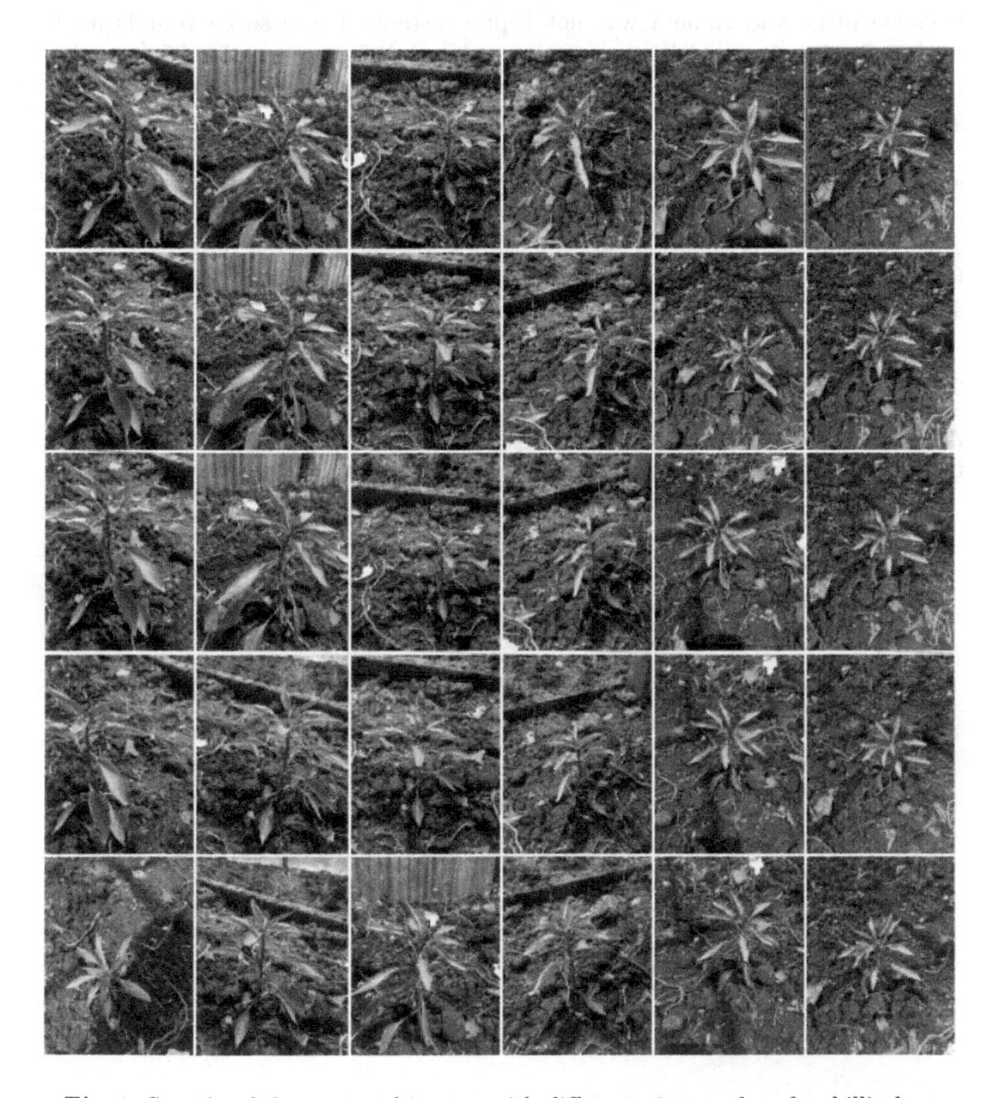

Fig. 1. Sample of the captured images with different view angles of a chilli plant

was planted in lines 90 cm apart, our experiment aimed at modelling individual plants. As a result, other plants did not hinder in the model and only one plant was monitored at a time.

2.1 Image Acquisition

The images were captured sequentially following a circular path with respect to the plant axis. Seven different rounds were taken at various angles, heights and distance. At least 15 images were captured at each path by revolving around the plant with a mobile phone's rear camera (Apple iPhone 6s+ with 12MP rear camera, f/2.2), capturing at every 10° to 15° of the perimeter. The distance between plant and camera was not kept constant. These seven rounds made during the image acquisition process produced 105 images with 95% overlap between successive images. Images were taken under outdoor conditions. This gives us variety of images to work with. The camera positions were chosen to ensure that the plant was entirely in the field of view, and the images were of a good quality (not blurred etc.). Structure-from-motion calculates the intrinsic camera parameters by itself, so camera positions do not have to be calibrated during the image acquisition process. Samples of the captured images with different view angles of chilli plant and image acquisition scheme is shown in Figs 1 and 2 respectively.

2.2 Plant-Soil Segmentation

As we are conducting our experiment under outdoor conditions, plant-soil segmentation has to be robust. This step is to distinguish plant-pixels from soil-

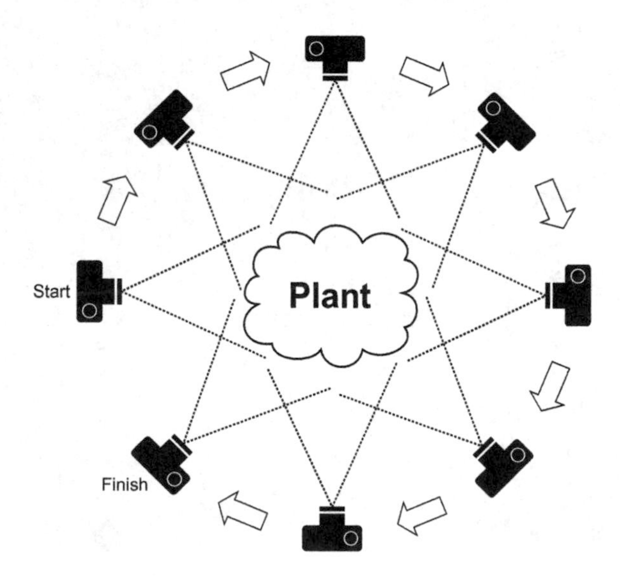

Fig. 2. Image acquisition scheme

pixels. As this process is applied to every image, this segmentation has to be autonomous. The improved vegetation index, excess green (ExG) [20] has been used, which is defined as:

$$ExG = \frac{2G - R - B}{R + G + B} \tag{1}$$

where, R, G, and B are the red, green, and blue pixel components respectively. With ExG, pixels associated with the plant class generally have high ExG values. This makes the discrimination between plant and soil easier. Figure 3 depicts plant-soil segmentation of one of the views.

Fig. 3. Plant-soil segmentation based on improved vegetation index

2.3 Keypoint Detection and Matching

After segmentation between plant and soil, the next task is to find the common keypoints (features) between a pair of images. For this process, we implemented the scale-invariant feature transform (SIFT) [21]. We converted an image into a huge set of keypoint vector, all of them is invariant to image scaling, rotation and translation. The standard steps in SIFT are.

1. **Formation of a scale space:** A basic step of calculation explores over each and every scales and image locations. It is achieved decisively using difference-of-Gaussian (DoG) function to determine potential keypoints that are scale and orientation invariant.

2. **Locating keypoints:** As we located the possible keypoints, a structured model is fit to identify scale and location. These keypoints are chosen hinged on their stability.
3. **Assignment of orientation:** Depend on local image gradient directions, one or additional orientations are elected to the location of every keypoint. Each and every operations are executed on image data which has been transformed corresponding to the elected scale, location, and orientation for each keypoint, by that giving invariance to all these transformations.
4. **Keypoint descriptor:** In the preferred scale in the region near to every keypoint, the local image gradients are calculated. There gradients are transformed into a delineation that permits for considerable levels of change in illumination and local shape distortion. Figure 4 illustrate the keypoints detected in two images.
5. **Matching of keypoints:** Keypoints are matched between pair of images of an object or a scene captured from different view points and angles. Matching is based on finding similar keypoint feature vectors between the two images. Figure 5 shows matching keypoints between two images. The matches are then filtered to remove outliers, and bundle adjustment is used to create a sparse 3D point cloud of matching object or scene and to retrieve camera calibration intrinsic, extrinsic parameters and positions at the same time. Pyramid-like symbol in Fig. 6 represents the positions and angles of the camera and green dots represent the plant structure.

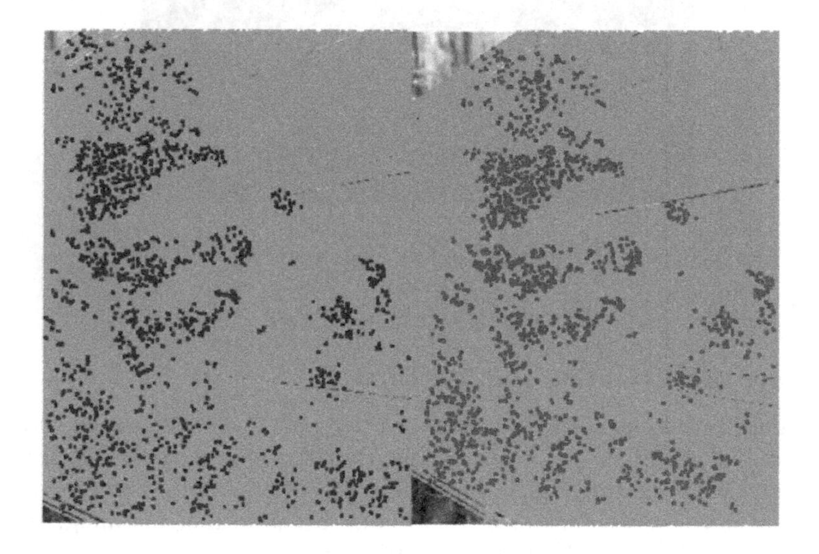

Fig. 4. Keypoint detection in an image pair

Fig. 5. Keypoint matching in an image pair

Fig. 6. Sparse 3D point cloud of the scene

2.4 3D Reconstruction

Finally, the calculated camera positions, parameters, and orientations are used to create a dense 3D point cloud. We implemented a cross-correlation matching method. For a pair of overlapped images, a pixel in the first image is corresponded with the pixel corresponding in the second image on the epipolar line [7]. This process is iterated for each pair of images keeping in mind that the calculated position of a given keypoint to be less noisy. The derived dense 3D point cloud is shown in Fig. 7, because of the page limitations we have added just two views as a resultant 3D model.

Fig. 7. 3D reconstructed model of chilli plant

2.5 Post Processing

The dense 3D point cloud is post processed off-line in an open source software named Meshlab [22]. This software is used to process unstructured dense 3D models using filters and remeshing tools, which helps to clean, smooth and manage our dense 3D model, which helps us to solve the quantization issue. Figure 8 shows the cleaned entire 3D model.

2.6 Selection of Appropriate Number of Images

It is very tricky to decide the number of images needed for plant 3D reconstruction, and hence it is an important factor. In general, a larger number of images will give additional information about the plant. At the same time, it will hold redundant data because of the overlapping regions of same scene, and it will take extra computation time to process more images. Moreover, it was noticed during our experiment that a large number of images caused feature matching error

Fig. 8. Clean 3D reconstructed model with overlapping leaves

which inevitably affects the accuracy. In contrast, with few images, the output 3D model will lack necessary data about the plant. We determined during our experimentation that it is quite difficult to reconstruct the plant in 3D using just 3–4 images, which cover only a limited range of viewpoints.

So based on our above investigation, hypothesis around the connection between multi-view information capturing and the trait of interpreted virtual view were tested to find an appropriate balance between multi-view information capturing and the quality of the 3D reconstructed model [23].

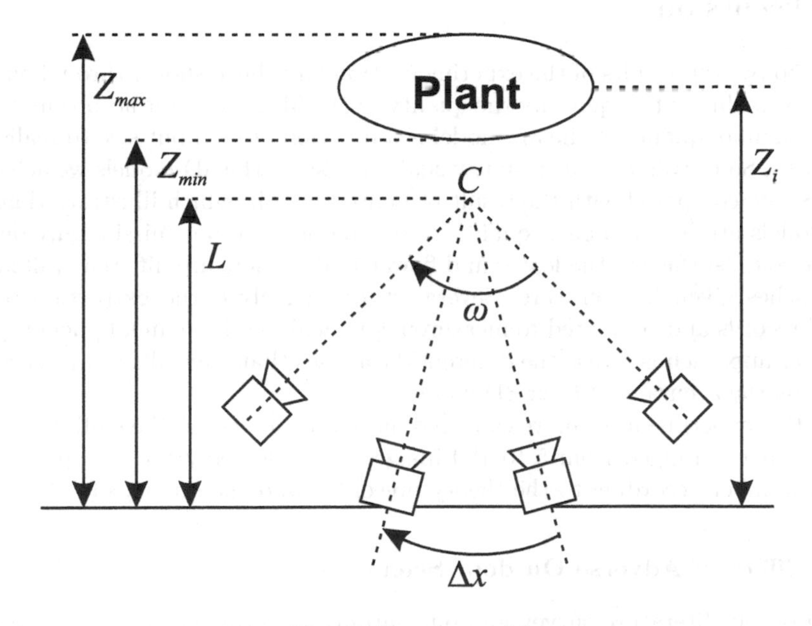

Fig. 9. Camera model

Figure 9 illustrates the camera model used in this experiment. Zi is the distance between plant and camera. ω is an arc which shows the space between each view from the camera with radius L in same pitch which is $= \Delta x$. f_l is a focal length of the camera. Z_{max} is the maximum depth of the plant and Z_{min} is the minimum depth of the plant.

Based on these assumptions and model, an appropriate number of images for 3D reconstruction can be calculated based on the below formulas:

$$\Delta x_{max} = \frac{1}{f_l}\frac{Z_{max}}{Z_{min}} \tag{2}$$

$$f_{nyq} = \frac{L}{2\Delta x_{max}} \tag{3}$$

where

$$L = \frac{2 \times Z_{max} Z_{min}}{Z_{max} + Z_{min}} \tag{4}$$

$$N_A = \frac{\omega L}{\Delta x_{max}} \tag{5}$$

We selected 30 as an appropriate image number for 3D reconstruction based on the above theory and formulas. Due to the page limitations we are not presenting the step by step calculation of the aforementioned formulas but as it is straight forward theory, it is easy to estimate the appropriate number of images.

3 Discussion

The step by step results of the experimentation have been shown throughout the paper. It is difficult to quantify the quality of the 3D model but according to the rule of thumb, quality of the 3D model is a function of its input size to realism it produces. So to validate our result, visual analysis of the 3D models we achieved (Fig. 8) are compared with the result presented in [24], which illustrate that our 3D models are having better quality as our models are not missing any details like petioles, surface of the leaves and flower buds. There are different validation approaches given in literature. Several studies involved the extraction of 2D visual records and compared to measurements achieved by manual phenotyping. Another approach is to use the different databases that have allowed researchers to assess the accuracy of their 3D model [8,25].

In this experimentation, we captured numerous images of the chilli plant and the number is ranged from 5 to 100 images. We selected 30 as an appropriate image number according to the theory presented and the quality of 3D model.

3.1 Effect of Adverse Outdoor Scenarios:

Based on our literature survey and our outdoor experimentations, we analysed that there are still some scenarios which cause problems and need more attention. In general, we know that sometimes under outdoor conditions it can be windy. We acquired another set of images of plants in windy condition, where plants were moving. In another scenario we captured the images when there was change in light conditions because of movement of the clouds. Here, we tried to investigate the effect of these scenarios under outdoor conditions and the effects of these on the resulting 3D models.

Fig. 10. Poor feature matching and 3D reconstruction due to heavy wind

Movement of Plant: In this scenario, we noticed that, because of the displacement of the plant due to wind, there were many feature matching errors resulting in poor 3D model. The resulting 3D model was missing important details in the stem area of the plant with some half reconstructed leaves (see Fig. 10). One possible solution for this scenario is to detect the inconsistent matches between the images because of the wind and filter out those images from the database.

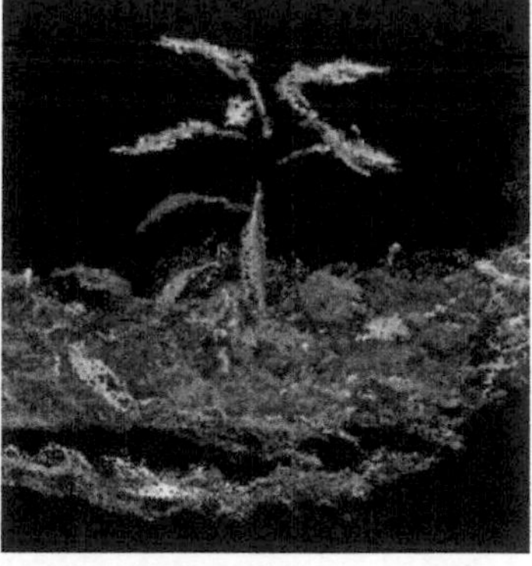

Fig. 11. Poor feature matching and 3D reconstruction due to drastic change in illumination

Change in Illumination: In our study, we try to reduce the error caused by change in illumination, with good results. However, in certain scenarios, there could be a drastic change in the illumination while capturing the plant images. We studied that, in this scenario, the resulting 3D model was missing necessary information about the plant such as plant surface and leaves resulting in blank patches in the 3D model, shown in Fig. 11. One possible solution for this scenario is to pre-process and normalise the acquired images first to reduce the effect of change in illumination in the database.

4 Conclusion

Plant phenotyping is achievable using our approach. Arguably, results of our experiments demonstrated that chilli plant 3D reconstruction is feasible with a low budget and could be used in different scenarios, even under outdoor conditions. Our contribution contains: (1) Easy and cost-effective system operated under outdoor conditions and achieved good results. (2) Investigation of adverse outdoor scenarios and effect on 3D model (3) The appropriate number of images were selected and used for reconstruction. (3) An entire 3D model with limited images. (4) Automatic plant-soil segmentation is implemented. This 3D reconstruction system is gives a cost-effective and efficient platform for non-invasive plant phenotyping, containing informations such as, fruit volume, leaf angles, leaf area index, which are important for assessing the stress and growth on plant features.

References

1. Mishra, K.B., Mishra, A., Klem, K., Govindjee: Plant phenotyping: a perspective. Indian J. Plant Physiol. **21**(4), 514–527 (2016)
2. Li, L., Zhang, Q., Huang, D.: A review of imaging techniques for plant phenotyping. Sensors **14**(11), 20078–20111 (2014)
3. Zhang, Z.: Microsoft Kinect sensor and its effect. IEEE Multimedia **19**(2), 4–10 (2012)
4. Kazmi, W., Foix, S., Alenyà, G., Andersen, H.J.: Indoor and outdoor depth imaging of leaves with time-of-flight and stereo vision sensors: analysis and comparison. ISPRS J. Photogramm. Remote Sens. **88**, 128–146 (2014)
5. Guo, Q., et al.: Crop 3D—a LiDAR based platform for 3D high-throughput crop phenotyping. Sci. China Life Sci. **61**(3), 328–339 (2018)
6. Jebara, T., Azarbayejani, A., Pentland, A.: 3D structure from 2D motion. IEEE Signal Process. Mag. **16**(3), 66–84 (1999)
7. Scharstein, D., Szeliski, R.: A taxonomy and evaluation of dense two-frame stereo correspondence algorithms. Int. J. Comput. Vision **47**(1–3), 7–42 (2002)
8. Cremers, D., Kolev, K.: Multiview stereo and silhouette consistency via convex functionals over convex domains. IEEE Trans. Pattern Anal. Mach. Intell. **33**(6), 1161–1174 (2011)
9. Paturkar, A., Gupta, G.S., Bailey, D.: Overview of image-based 3D vision systems for agricultural applications. In: 2017 International Conference on Image and Vision Computing New Zealand (IVCNZ), pp. 1–6, December 2017

10. Kaminuma, E., et al.: Automatic quantification of morphological traits via three-dimensional measurement of Arabidopsis. Plant J. **38**(2), 358–365 (2004)
11. Paulus, S., Dupuis, J., Riedel, S., Kuhlmann, H.: Automated analysis of barley organs using 3D laser scanning: an approach for high throughput phenotyping. Sensors **14**(7), 12670–12686 (2014)
12. Baumberg, A., Lyons, A., Taylor, R.: 3D S.O.M.—a commercial software solution to 3D scanning. Graph. Models **67**(6), 476–495 (2005)
13. Chéné, Y., et al.: On the use of depth camera for 3D phenotyping of entire plants. Comput. Electron. Agric. **82**, 122–127 (2012)
14. Ivanov, N., Boissard, P., Chapron, M., Andrieu, B.: Computer stereo plotting for 3-D reconstruction of a maize canopy. Agric. For. Meteorol. **75**(1), 85–102 (1995)
15. Takizawa, H., Yamamoto, S., Ezaki, N., Mizuno, S.: Plant recognition by integrating color and range data obtained through stereo vision. J. Adv. Comput. Intell. Intell. Inform. **9**(6), 630–636 (2005)
16. Jay, S., Rabatel, G., Hadoux, X., Moura, D., Gorretta, N.: In-field crop row phenotyping from 3D modeling performed using structure from motion. Comput. Electron. Agric. **110**, 70–77 (2015)
17. Quan, L., Tan, P., Zeng, G., Yuan, L., Wang, J., Kang, S.B.: Image-based plant modeling. ACM Trans. Graph. **25**(3), 599–604 (2006)
18. Tan, P., Zeng, G., Wang, J., Kang, S.B., Quan, L.: Image-based tree modeling. ACM Trans. Graph. **26**(3), 87 (2007)
19. Paproki, A., Sirault, X., Berry, S., Furbank, R., Fripp, J.: A novel mesh processing based technique for 3D plant analysis. BMC Plant Biol. **12**(1), 63 (2012)
20. Meyer, G., Camargo Neto, J.: Verification of color vegetation indices for automated crop imaging applications. Comput. Electron. Agric. **63**, 282–293 (2008)
21. Lowe, D.G.: Distinctive image features from scale-invariant keypoints. Int. J. Comput. Vision **60**(2), 91–110 (2004)
22. Cignoni, P., Callieri, M., Corsini, M., Dellepiane, M., Ganovelli, F., Ranzuglia, G.: Meshlab: an open-source mesh processing tool. In: Scarano, V., Chiara, R.D., Erra, U. (eds.) Eurographics Italian Chapter Conference. The Eurographics Association (2008)
23. Liu, S.-X., An, P., Zhang, Z.-Y., Zhang, Q., Shen, L.-Q., Jiang, G.-Y.: On the relationship between multi-view data capturing and quality of rendered virtual view. Imaging Sci. J. **57**(5), 250–259 (2009)
24. Ni, Z., Burks, T., Lee, W.: 3D reconstruction of plant/tree canopy using monocular and binocular vision. J. Imaging **2**(4), 28 (2016)
25. Pound, M.P., French, A.P., Murchie, E.H., Pridmore, T.P.: Automated recovery of three-dimensional models of plant shoots from multiple color images. Plant Physiol. **166**(4), 1688–1698 (2014)

Automated Soil Tester

P. Kovelan, T. Kartheeswaran(✉), and N. Thisenthira

Department of Physical Science, Vavuniya Campus of the University of Jaffna, Vavuniya, Sri Lanka
karthees@vau.jfn.ac.lk

Abstract. This study proposes an automated soil tester to test the soil for agricultural purposes. The contribution of the computer science to agriculture is essential for the sustainability of human being in the world by keeping the food production up to a satisfactory level. The automation technique is a suitable and efficient solution to overcome difficulties in agriculture. This technique will increase the productivity and reduces the hardness of human effort in the field. The traditional soil testing mechanism has many difficulties and drawbacks such as time-consuming, poor knowledge of sample collection and variation in laboratory results compared to field results. The proposed automated soil tester will be the solid solution to overcome the problems of the traditional soil testing mechanism. The device has a temperature, moisture and pH sensors to measure the soil parameters such as temperature, water level, electro conductivity and pH. The automated soil tester will be able navigate in the given area of the field with the guidance of GPS and it is capable of avoiding obstacles in the filed. The data sensed will be sent to a website to get visualized and will stored in a database.

Keywords: Automation · Soil parameters · Rover · pH sensor · Agriculture

1 Introduction

Automation is a process which performed without the human intervention. Automation is the way to reduce the human workload and make the job easier and effective. Many industries adopt the variety of automation techniques and deliver the zero defect quality product efficiently. Only a limited number of contributions from computer science is given to agriculture. The agriculture is the base of the human survival. Mankind must ensure the food production at a satisfactory level always. However, there is a potential problem that the human involvement in agriculture is continually decreasing over the years [1]. Automation is one of the solutions to overcome this problem.

There are many phases in agriculture, this work especially on how the soil testing can be done without the human involvement in the field. Soil plays a vital role in agriculture by making the crop healthy with more yields. "Soil contain many kinds of nutrients, such as water, air and living organisms that help to

K. C. Santosh and R. S. Hegadi (Eds.): RTIP2R 2018, CCIS 1037, pp. 298–311, 2019.
https://doi.org/10.1007/978-981-13-9187-3_26

create healthy and sustainable gardens and landscapes" [2]. We can improve these qualities by assessing the soil through soil testing and it will give us the soil's pH, acidity, temperature, electrical conductivity (EC) and soil moisture.

There are variety of soil types spread all over the island such as chalky, peaty, sandy, clay, silty and loamy. The soil is made out of 45% of minerals, 25% of water, 25% of air and 5% of organic matters. The suitable soil types are varying crop to crop. Selection of perfect soil according to the crop is an important process in agriculture. The soil from different field must be tested with random samples to determine the type and to select perfect crop according to the parameters learned [3]. It is important to do soil testing to increase the quality of cultivation and to make more yield. The automated soil testing plays a vital role in agriculture rather than the traditional one which solves these kinds of problems. The computer scientist must ensure the contribution of computer science in the agriculture like they do in production, service and other commercial fields. Soil testing is one of the big tasks in agriculture which can be managed using automation techniques with the help of computational methods.

The soil properties can be measured with a guided automated soil tester assembled with needed sensors and actuators. The automated soil testing mechanism may increase the accuracy of the test results and also make the task easy. There are many challenges in developing an automated soil tester to test the soil. The work planned is to do a cost-effective automated guided vehicle to test the soil. We planned to use the Global Positioning System (GPS) sensor to guide the automated soil tester to navigate and test on the desired points (coordinates) of a given filed. Further, we decided to sense the pH using a custom designed mechanism. Recently, there are some research being carried out which explores the possibility for automated soil testing system, to achieve higher accuracy with least cost. But, there is a problem of interference of GPS signal by the bushes and pits.

The purpose of this system is to make the results accurately meanwhile avoiding the physical barriers in the path. Once the device set with everything then it is automatically tests and reports back without human intervention. This can be done with the help of an Arduino Microcontroller board and with some sensors to sense the soil parameters and some communication modules to facilitate the proposed automated soil tester. The automated soil tester will rove and collect samples within the given boundary, and after completing the process it will update the collected results to the computer with the help of Wi-Fi module connected to the microcontroller.

The main objective is to speed up the working process through the device developed. Also, making more accurate measurements is another task of this system. Therefore, we can produce better results faster. Automation of soil testing will provide an easy mechanism than the existing testing methods.

In this modern world, people prefer to do their work using machines because of the facilitated world of work. Hence, this soil testing with hardware device will reduce the need for human resources.

Background. Usually, the soil test is performed manually by the experts with a lot of expensive and complex devices or equipment. The manual soil testing mechanism is a challenging task as the processes need to handle much equipment in an awkward and uncomfortable agricultural field. The soil tester (Human) must pick different random locations and record the parameters for further analysis. Automated soil testing device makes the soil testing task easy and gives more number of samples with accurate measurements. The automated soil tester may reach even the places that can't be efficiently reached by human [4].

Obviously, people who involved in the soil testing process are facing some drawbacks in traditional methods. The traditional soil testing method takes long time to calculate the results. Because, the soil samples are brought to the laboratory for testing purposes. The farmers should wait to plant until they get the final results from the laboratory. Only the experts can handle the traditional system with the aid of complicated equipment. There are lots of paper works, and also it is difficult to keep all the records manually for future references. These types of traditional testing may give less accurate and less efficient results. Sometimes, some kind of traditional testing system is tough to take to a rural area to test the agricultural land. Some experts test the soil by touching with their thumb that means they do the visible testing according to the soil color and also tries to feel it through their bare thumb. So they need more experience to predict the results and which may have more chances to give standard errors.

Traditionally, the extracted samples were analyzed using different methods for different nutrients and minerals. Many laboratories are using sophisticated instruments that can analyze many nutrients simultaneously. Proving ring, sieve shaker, test sieves, hot plate are some of the equipment traditionally used in a soil testing laboratories.

An automated soil tester is made to test the soil in a faster and more natural way to solve these problems. Many hardware and software resources are used to build the system which is easy to handle, and anyone can use it without much effort. Further, collectively it reduces the human resources and gives accurate results compared to traditional testing. The automated soil tester will help to give better results and to make better and prosperous agriculture for the nation.

Literature Review. A research of "Wireless Monitoring of Soil Moisture, Temperature & Humidity Using ZigBee in Agriculture" is done by Chavan and others in 2014 [3]. The research objective is "Monitoring agricultural environment for various factors such as soil moisture, temperature, and humidity along with other factors can be of significance". However, they used LM35 temperature sensor for their prediction. But, in our system, the DS18B20 temperature sensor is used with $\pm 0.5\,^{\circ}$C accuracy and waterproof which is better and accurate than LM35.

Boopathy and others has done another research on "Implementation of Automatic Fertigation System by Measuring the Plant Parameters" [4]. The objective of the research is to monitor the necessary parameters of agriculture in the horticulture field such as pH, temperature, moisture of the soil. Here, they test the pH by making a liquid soil mixed solution, which is a complicated mechanism when its comes to automation. The proposed system uses a unique customized mechanism to measure the pH in the field.

The other study is monitoring the soil using wireless system [5]. The objective of the study is to monitor the level of the soil water content. However, the study says that the researchers are using the EC-5 moisture sensor for moisture analysis and low powered nRF24L01 wireless transceiver with MPC82G516A microcontroller. They test only the moisture of the soil which is not enough to make decision in crop selection and other aspects. We can use low-cost sensor rather than the above mentioned sensors. Our device is capable of reading all the needed parameters of the soil on the field which immensely helps to make better decision.

There is a paper which discusses about controlling the temperature in tea leaves preparation using Arduino Uno and Android App [6]. The objective of the work is to control the temperature of the chamber in different stages of drying. The researchers are using an Arduino Uno board, HC05 Bluetooth module, LM35 temperature sensor and android based terminal application for this purposes.

The monitoring of humidity is one of the other research that have been done in 2011 [7]. The objective of the work is to obtain environmental humidity and temperature information. A combination of humidity and the temperature sensor is used to sense the humidity and temperature information of the environment, and also PCI bus based data acquisition card is used for data collection.

There was another study made by Bugai et al. in guiding an automated soil tester using GPS [8]. The aim of the study is to develop and implement a GPS (Global Positioning System) guided automated soil testing device. They have used comprised of a Perspex base, battery pack, servo motors, Unbox GPS module and an ATMega328 Arduino microcontroller. Here they concluded with a drawback of time-consuming to move the Automated Soil Tester between the given points. The velocity is around 0.8 m/s. However, in the proposed project the time consumption problem has been solved with effective pathfinding algorithm. The digital compass and an efficient GPS module have been used in some GPS guided Automated Soil Tester according to [9–12]. Therefore, the digital compass is used in the proposed system for further enhancement of navigation. The soil sampling techniques were learned from the study [13] which discuss about characterizing and recognizing soil sampling strategies. The algorithm for path planning for the movement of robots is published by Yuksel and Sezgin in 2005 [14]. Their objective is to find the shortest and the low cost path from the given map. Those studies focus on Breadth-first, Dijkstra, and A* algorithms and examine them to use in automated navigation of robot. The above mentioned mechanisms has been taken as basic guideline for finding the path in the proposed system to guide the automated soil tester in the field. The web page on the hosting server can do some operations like read and update data in database. So, the Wi-Fi module has been used for this purpose which has been discussed in the smart home project [15].

Finally, based on the techniques discussed in the literature, it has been decided to use Arduino UNO microcontroller with a DS18B20 temperature, pH, and moisture sensors to measure the soil parameters. Further, the GPS module and digital compass are used to navigate the automated soil tester to the given points using a dedicated pathfinding algorithm. Wi-Fi module is fixed with the system to transfer the data from the automated soil tester to the local web server [15]. The local web server will export the read parameters into a centralized database.

2 System Architecture

The system architecture of the automated soil tester is shown in Fig. 1. The architecture is explained using a simplified diagram that describes the interaction of the technology involved in the proposed project. The connectivity system is working as a transceiver to receive the selected GPS coordinates of the particular agricultural field from the web server. Then, the web server will send the data to the master microcontroller and the master controller will send the final processed data to the web server back. ESP8266 v12F is used to do the transmissions under the TCP/IP protocol [15]. Two Arduino microcontrollers are used to control the whole functionality of the automated soil tester. The master microcontroller is used to interconnect with all sensors and modules.

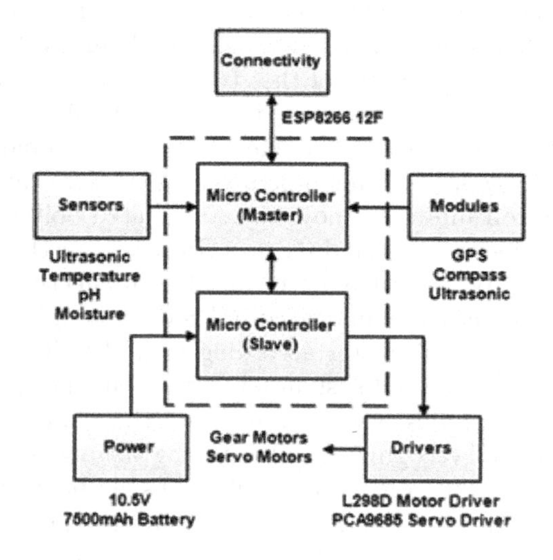

Fig. 1. System architecture

The slave microcontroller is used to interconnect with the motor drivers to control the main four gear motors and other servo motors. The motors are controlled by master microcontroller using I2C protocol through slave. The circuit diagram of the soil tester is described in Fig. 2. Detailed functions of the Sensors, Modules are described below in Sect. 3 "Planning and Execution of the Automated Soil Tester Operation" (PEASO).

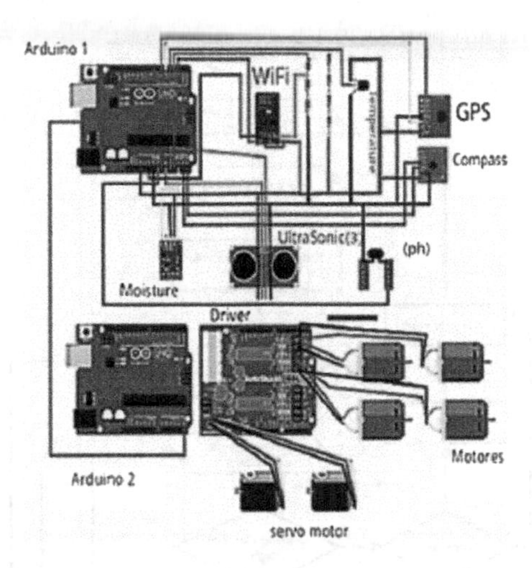

Fig. 2. Circuit diagram

3 Planning and Execution for Automated Soil Tester Operation

Pathfinding. The automated soil tester has to visit all over the location of a selected agricultural land and need to collect many numbers of samples to make the accurate result. The visiting points (as a form of latitude and longitude) are given by the user with the help of Google map embedded in the web interface and sent to the automated soil tester through the Wi-Fi module according to the shortest path computed by the web server.

As a first step automated soil tester finds the current position where it's stationed. The GPS module will pitch the latitude and longitude of the station. According to this first reading, the soil tester finds the distance to the waypoint and the angle of the waypoint from clockwise and true north. After finding out the angle of the waypoint and the current angle of the soil tester must be learn to start the navigation. HMC5883L compass module is used to read the current angle of the soil tester. The soil tester is turned right or left according to the angle difference between the angle of the waypoint and the current angle of the soil tester until the difference reduces to zero.

The NEO6M GPS module is used in the initial state for the GPS positioning. However this module is not capable enough to pitch the points accurately. The NEO6M sensor's accuracy is 10 to 15 m. For the short distances in the waypoints, it is difficult to automated soil tester to pitch the expected point and it may stop moving. The GPS positions have to be very accurate to increase the applicability of the soil tester to all kind of lands. The Fastrax-up501 GPS module is better than NEO6M [16] in various forms such as a number of channels and the accuracy. The pathfinding strategy discussed above is described in Fig. 3.

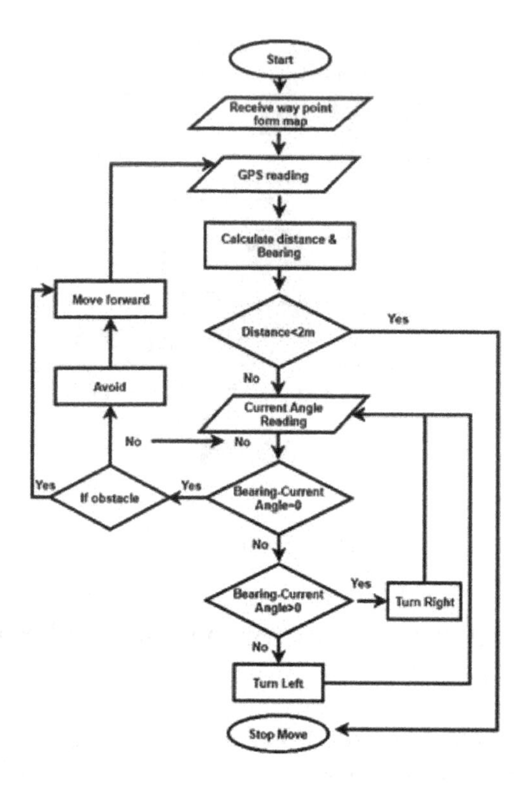

Fig. 3. Pathfinding algorithm

Obstacles Avoidance. There may be possibilities for obstacles in the land such as stones, blocks or plants while the automated soil tester navigates. The soil tester must be able to avoid those obstacles on the path to protect the automated soil tester as well as the plants as it can be used for pre and post-plantation. Ultrasonic Distance Measuring Transducer Sensor HC-SR04 is used in this project to find the obstacle. There are three rotatable ultrasonic sensors are used to avoid the time delay and to perform effective avoidance mechanism. Ultrasonic sensors are fixed in left, right and the middle positions of the soil tester's frontend. It's good enough to cover the soil tester skeleton, and after finding the obstacle, the soil tester chose the best strategy according to the width of the obstacle. The strategy of the obstacles avoidance is described in Fig. 4.

Sensors of the Soil Parameters. After reaching each given point, the soil tester's arm pulls and fixes the sensors into the soil (land). The arm is designed with four metal servomotors MG996. All the sensor readings are taken one by one and send it to the web server through the Wi-Fi module. The selected temperature sensor is a pre-wired and waterproofed version of the DS18B20 sensor. This sensor can be easily used to measure the temperature bit more

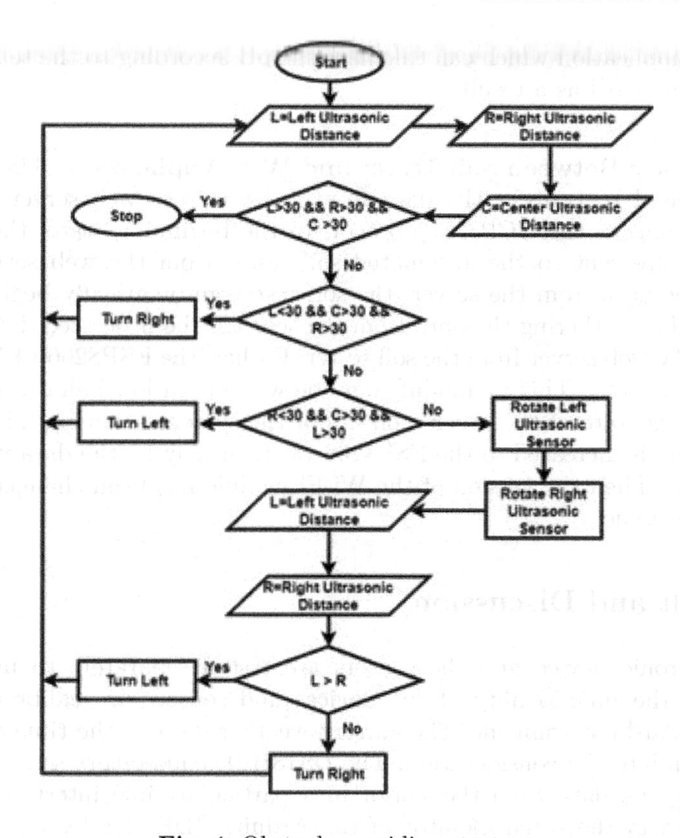

Fig. 4. Obstacles avoiding strategy

deep into the soil, or in wet conditions. We can measure the temperature of the air, liquids like water, oil and the temperature of the soil using this sensor. This sensor measures temperature from −55 °C to +125 °C and the accuracy is ±0.5 °C for reading from −10 °C to +85 °C. Thermometer resolution is user-selectable from 9 to 12 bits. It converts the temperature to 12-bit digital word with in 750 ms. It can be powered from the data line, and the power supply range is 3.0 V to 5.5 V. DS18B20 gives an accurate result compared to an LM35 temperature sensor.

Moisture sensor SEN 13322 is used in this project. The mechanism behind the moisture sensor is that the two probes are acting as a variable resistor and measuring the electrical conductivity. More water in the soil means better conductivity and results in lower resistance and a higher voltage reading is vice-versa. The reading of the pH is taken by a soil pH meter. Electrical conductivity between the two metallic probes is tested with a range of standard pH solutions. The method and the testing result are discussed in the result and discussion section. The derived relationship between the voltage reading and the standard pH solution pH level is used to determine the pH range of the soil sample. However, this can be varying at different temperatures. A customized method applied in the web application which can calculate the pH according to the temperature to give the best pH as a result.

Data Transfer Between Soil Tester and Web Application. The ESP8266 module is used to transfer the data from soil tester to web server and vise-versa according to the TCP/IP protocol. In the beginning state, the selected coordinates are sent to the automated soil tester from the web server. After receiving the data from the server, the soil tester automatically begins to find the path. After gathering the data from the sensors, the processed data need to be sent to the web server from the soil tester. Earlier, the ESP8266 v1 is used for these transmissions. This v1 module may be working in low baud rate in serial read, but baud rate of the other modules of the system are high. Therefore, if the baud rate is increased to the ESP8266 v1, then may be the data will lost in transmission. Then the version of the Wi-Fi module has been changed to v12F to avoid this issue.

4 Result and Discussion

Every electronic device and the sensors are tested separately to understand and ensure the functionality of the devices and sensors are calibrated based on the standard environment. The current weather data at the time of calibration is taken into the consideration. The DS18B20 temperature sensor is tested by receiving the data from the sensor in a particular time interval. The output is shown in the serial monitor of the Arduino IDE (for testing purposes) which shows the response from the sensor is rapid and the temperature changes also measured as efficiently by the DS18B20 sensor. Responses to temperature changes are tested by testing the sensor in different temperatures such as in ice, hot water, and fire flame. Figure 5 shows the temperature sensor reading of the human body temperature. The response speed is plotted.

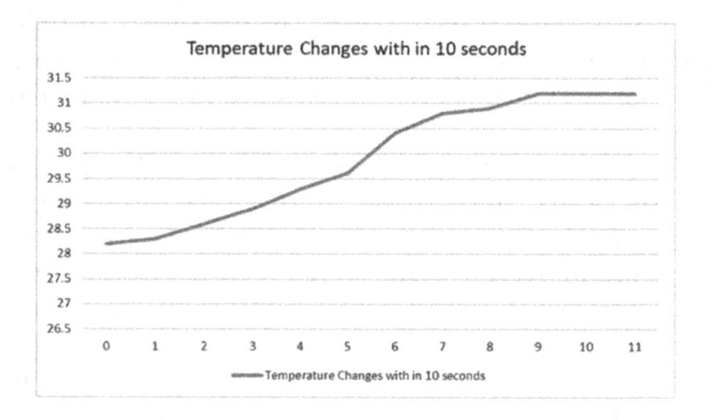

Fig. 5. Temperature changes

Moisture sensor SEN 13322 is tested in different environments such as in the dry, wet soil and the water. The sensor's analog outputs are shown in the serial monitor of the Arduino IDE. Calibration of the moisture sensor is done in the air. The calibration details are shown in Table 1.

Table 1. Results of the moisture sensor readings in wet and dry zones

Time (Sec)	Reading	Remark
1	653	Soil is wet
2	653	Soil is wet
3	653	Soil is wet
4	652	Soil is wet
5	652	Soil is wet
6	651	Soil is wet
7	652	Soil is wet
8	652	Soil is wet
9	652	Soil is wet
10	652	Soil is wet
1	0	Soil is dry
2	0	Soil is dry
3	0	Soil is dry
4	0	Soil is dry
5	0	Soil is dry
6	0	Soil is dry
7	0	Soil is dry
8	0	Soil is dry
9	0	Soil is dry
10	0	Soil is dry

The pH sensor is tested with the various sample pH solutions in chemical laboratory with the help of chemistry expert. We compare the pH reading between the laboratory pH reading with the pH sensor and also voltage of the pH sensor is measured. After that a graph has been plotted against the obtained voltage and the actual pH value as shown in Fig. 6 to retrieve the formula to code the microcontroller (customized method).

AT commands are used to test the basic functions of the Wi-Fi module such as Attention, Version information and Reset. Available Wi-Fi access points are search through the AT+CWLAP commands by setup the module as Wi-Fi receiver. Connections are checked by connecting to the Wi-Fi access point. The Wi-Fi module's IP address is taken when the module changed to a Wi-Fi transfer device. The capability of working as both receiver and transmitter is

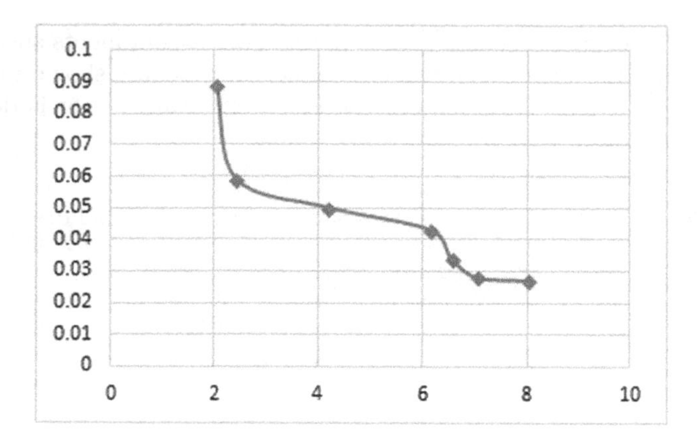

Fig. 6. The test result of the pH sensor

tested through the AT commands. The temperature sensor data has been sent to a website to test the communication of Wi-Fi module. In this test, the API key of ThingSpeak.com is given to the Wi-Fi module and the Wi-Fi module is connected with the internet to transmit the data. HMC5883L compass module is tested to find the angle of the module with the north direction. The analog data of the compass module is read in the serial monitor. The resulting angle also changes according to the rotation when the compass is rotated. The Compass module has been tested to ensure the functionality as shown in Table 2.

Table 2. Testing result of the compass module

Time (Sec)	Heading	Degrees
1	3.43	196.77
2	3.26	186.54
3	3.06	175.14
4	2.92	167.17
5	2.78	159.54
6	2.65	151.75
7	2.50	143.40
8	2.39	137.19
9	2.26	129.54
10	2.13	122.18

5 Final Testing

The automated soil tester is tested in the Agriculture Research Center, Killinochi, Northern province of Sri Lanka. The automated soil tester visited seven points according to the given GPS coordinates shown in Fig. 7 and the final results are recorded and described in Table 3. The **Point No** given in the Table 3 is according to the shortest path algorithm. The sample soil of each point is taken to the laboratory to test the pH and moisture is described in Table 4. The traditional soil testing method is used to test the soil sample taken. The temperature of the sample soil is tested with a digital thermometer which is commonly used in wet laboratories.

Table 3. Final readings are taken by the automated soil tester

Point No	Time took-seconds	Latitude	Longitude	Temperature	Moisture	pH
1	–.–	9.419696	80.406465	28.2	12.50%	6.4
2	32.26	9.419637	80.40653	28.3	12.90%	6.5
3	40.22	9.419537	80.406487	28.1	12.60%	6.4
4	54.12	9.419556	80.40634	28.5	13.20%	6.7
5	76.88	9.419352	80.406379	28.4	13.10%	6.6
6	129.22	9.419534	80.406681	28.4	12.30%	6.3
7	59,86	9.419690	80.406638	28.6	12.80%	6.7

Fig. 7. Satellite map of the GPS coordinates points

Table 4. Final readings are taken by applying manual method

Point No	Temperature	Moisture	pH
1	28.3	13.55%	6.8
2	28.5	12.82%	7.1
3	28.1	13.24%	6.8
4	28.6	12.92%	7.2
5	28.5	13.24%	7.1
6	28.6	12.62%	6.6
7	28.2	12.83%	6.9

6 Conclusion and Recommendations

The temperature resulted from the automated soil tester and the manual thermometer reading are relatively the same, and there are no big differences. The moisture reading is slightly different from the laboratory results. This problem may be due to the time delay to take the sample to the laboratory and some other environmental factors may cause the changes in manual reading. The pH readings are little bit deviated. However, both the results are shown same increment and decrement pattern. Our pH probe can read the pH value from 2.0 to 8.0 which is good enough for the soil testing purposes. Because most probably the pH value of the soil is in the range of 2.0 to 8.0. The pH 7.0 is required for a better plantation, so it's enough to give the soil pH result as less than 7.0 or higher than 7.0 to make a decision. There is a accurate result of pH range in between 6.0 to 8.0 by the customized sensor which is tested from a range of solutions in pH 6 to 8 in the laboratory.

The instant soil testing is crucial. The instant results can be accurate in the point of moisture and temperature testing. Moisture percentage and the pH is needed from the beginning of the plantation for cultivation. This automated soil tester is the best solution to overcome the problems of above discussed bottlenecks of agriculture in the background section.

References

1. Roser, M.: Employment in agriculture (2018). ourworldindata.org. https://ourworldindata.org/employment-in-agriculture
2. College of Agriculture and Natural Resources (2017). Home and Garden Information Center. http://extension.umd.edu/hgic/soils/soil-testing
3. Chavan, C.H., Karande, P.V.: Wireless monitoring of soil moisture, temperature & humidity using ZigBee in agriculture. Int. J. Eng. Trends Technol. **11**(10), 493–497 (2014)
4. Boopathy, S., et al.: Implementation of automatic fertigation system by measuring the plant parameters. Int. J. Eng. Res. Technol. **3**(10), 583–586 (2014)

5. llawlor (2014). Thingspeak. https://github.com/iobridge/thingspeak/blob/master/README.textile
6. Boopathy, S., Ramkumar, N.: Controlling the boiler temperature of tea leaves by using Android application and Arduino UNO. Int. J. Mod. Comput. Sci. (IJMCS) **4**(3), 57–60 (2016)
7. Li, J., Zhou, H., Li, H., Li, Y.: Design of humidity monitoring system based on virtual instrument. In: International Conference on Advances in Engineering, pp. 759–763 (2011)
8. Bugai, T., Salmon, K., Chiba, M.: GPS guided autonomous rover. School of Electrical & Information Engineering, University of the Witwatersrand, 24 May 2013
9. Chandramohan, J., et al.: Intelligent smart home automation and security system using Arduino and Wi-Fi. Department of Electrical and Electronics Engineering, Gnanamani College of Technology, Namakkal, India, March 2017, vol. 6. Issue 3, pp. 20694–20698. Index Copernicus value (2015). 58.10. https://doi.org/10.18535/ijecs/v6i3.53
10. Al-Faiz, M.Z., Mahameda, G.E.: GPS-based navigated autonomous robot. Al-Nahrain University, vol. 3, no. 4, April 2015
11. Srividyadevi, P., Pusphalatha, D.V., Sharma, P.M.: Measurement of power and energy using Arduino. Gokaraju Rangaraju Institute of Engineering and Technology, vol. 2(10), pp. 10–15, October 2013
12. Paradkar, A.D.: GPS guided autonomous robot. California State University, Long Beach (2016)
13. Chung, W.-Y., et al.: Wireless sensor network based soil moisture monitoring system design. In: Federated Conference on Computer Science and Information Systems, pp. 79–82 (2013)
14. Yuksel, T., Sezgin, A.: An implementation of path planning algorithms for mobile robots on a grid based map. Ondokuz Mayis University, Electrical & Electronics Engineering Department, Technical report (2005)
15. Espressif Smart Connectivity Platform: ESP8266. Espressif Systems, 12 October 2013
16. Yen, K.S., Lasky, Ty.A., Adamu, A., Ravani, B.: Application of high-sensitivity GPS for a highly-integrated automated longitudinal travel behavior diary. University of California, January 2007

Crop Discrimination Based on Reflectance Spectroscopy Using Spectral Vegetation Indices (SVI)

Rupali R. Surase[1(✉)], Karbhari V. Kale[1], Mahesh M. Solankar[1], Amarsinh B. Varpe[1], Hanumant R. Gite[1], and Amol D. Vibhute[2]

[1] Department of Computer Science and Information Technology, Dr. Babasaheb Ambedkar Marathwada University, Aurangabad, MH, India
rupalisurase13@gmail.com, kvkale91@gmail.com, mmsolankar13@gmail.com, varpeamarsinh@gmail.com, hanumantgitecsit@gmail.com

[2] Solapur University, Solapur, India
amolvibhute2011@gmail.com

Abstract. This paper represents three main objectives of research, including (1) development of crop spectral library for diverse crops, (2) combination of two varying spectral responses for crop benchmarking, (3) interpretation of spectral features using Spectral Vegetation Indices (SVI). Hyperspectral sensors were used for spectral development including Maize, Cotton, Sorghum, Bajara, Wheat and Sugarcane crops with Analytical Spectral Device (ASD) Spectroradiometer and Earth Observing (EO)-1 Hyperion dataset positioned at Aurangabad region by Latitude 19.897827 and Longitude 75.308666. In precision agriculture, the Spectral Vegetation Indices (SVI) delivers valuable information for crop discrimination and growth monitoring; the present research elaborates about five SVI. The spectral responses were collected at the ripening stage of crops at standard darkroom environment in the laboratory. It was found that there was a progressive correlation 0.92 with squared residual value 4.69 amongst ASD and EO-1 Hyperion. The significant spectral features were recognized inAnthrocyanin Reflectance Index 1 (ARI_1) with R_{550}, R_{700}, for Moisture Stress Index (MSI) R_{1599}, R_{819} wavelength respectively. The experimental analysis was performed using ENVI and python open source software and it was concluded that crops types were successfully discriminated based on spectral parameters with different band combinations.

Keywords: Hyperspectral remote sensing · Precision agriculture · Spectral signature · Spectral vegetation indices · Regression model

1 Introduction

Discrimination of crop types is an important requirement for crop mapping and monitoring. Remote sensing tool is one of the emerging technologies for agricultural research. The imaging and non-imaging sensors were detected as useful for crop type's discrimination using electromagnetic spectrum range. An Aurangabad region was

K. C. Santosh and R. S. Hegadi (Eds.): RTIP2R 2018, CCIS 1037, pp. 312–322, 2019.
https://doi.org/10.1007/978-981-13-9187-3_27

selected for crop study because of consisting varying crop pattern including cotton, maize, Bajara, Jawar, sugarcane and wheat; it's one of the challenging issues to overcome the mixing pixel problem based on spectral features.

Previous research activities have developed approaches to estimate vegetation chlorophyll content from remotely sensed data using spectral indices. The literature reveals that, less research has been touched the crop discrimination using hyperspectral sensor combinations for accurate benchmarking of spectral characteristics [1]. The focused crops were selected for spectral feature estimation including regional crops of Marathwada region.

The SVI is widely used as a prominent indicator of finding spectral variations in vegetation and biophysical parameters [2]. The significant information on rice with phonological phases was extracted using leaf area index from multitemporal seasonal data [3]. The invasive and non-invasive shrubs were discriminated based on ASD with LIDAR dataset with spectral based classification model [4]. The literature reveals that most of the researchers worked on hyperspectral airborne imaging data but only imaging data doesn't provide single crops information for the identification. The pixel resolution of imaging data gives mixed crop spectra. To cover such issue the asd spectroradiometer was applied towards Hyperion for the purpose of increasing spectrum information [5]. The spectral indices were compared including and NDVI and Brightness Index (BI), for cotton, alfalfa and sugar beet crops, but the results signify less mapping of using two indices [6]. The NDVI has also been experienced for yield prediction especially for the wheat crop in many research studies [7–12].

The paper reports on a combined approach to discriminate crop types based on leaf spectral features with photosynthetic contents. This paper contains four sections including an introduction with background knowledge of the existing study. Section 2 comprises specifics about the study area along with dataset. The proposed methodology is described in Sect. 3; the findings of the research are described in Sect. 4 along with estimated output. Finally, conclusions with the future scope are précised in the last section.

2 Study Site

The study area is a meager lengthened strip of Hyperion image located on the Aurangabad region of Maharashtra, India having dense vegetation cover represented in Fig. 1. Dataset used in the research study are categorized into three patches including hyperspectral data, i.e. (1) ASD field spectroradiometer reflectance of 600 samples of six type crops collected from the field (2) ancillary data containing various environmental maps prepared by My GPS Coordinates system and (3) Earth Observing-1 Hyperion. The focused crops were Jawar, Bajara, Sugarcane, Cotton, Maize and Wheat. The major data source is EO1- Hyperion imagery was assimilated on 25 December 2015 collected from USGS site. It contains 220 spectral channels having the 30 m spatial resolution within 400 nm to 2500 nm wavelength range. The swath width of the Hyperion is 7.5 km, whereas characteristic scene size is 7.5 × 100 km [13].

Fig. 1. Study site of experiment with False Color Composite of Hyperion (Aurangabad Region, Maharashtra, India)

3 Experimental Approach

The complete framework of the research study is shown in Fig. 2. The hyperspectral data contains spectral responses of crop leaves in the ripening stage for extraction of chlorophyll, carotenoid and xanthophyll contents. The standard dark environment was followed for leaf spectral measurements. Each sample was placed under a bunch of high-intensity lights facing directly towards the leaves. Moreover, spectral measurements were accumulated through EO-1 Hyperion based on ground control points of crops.

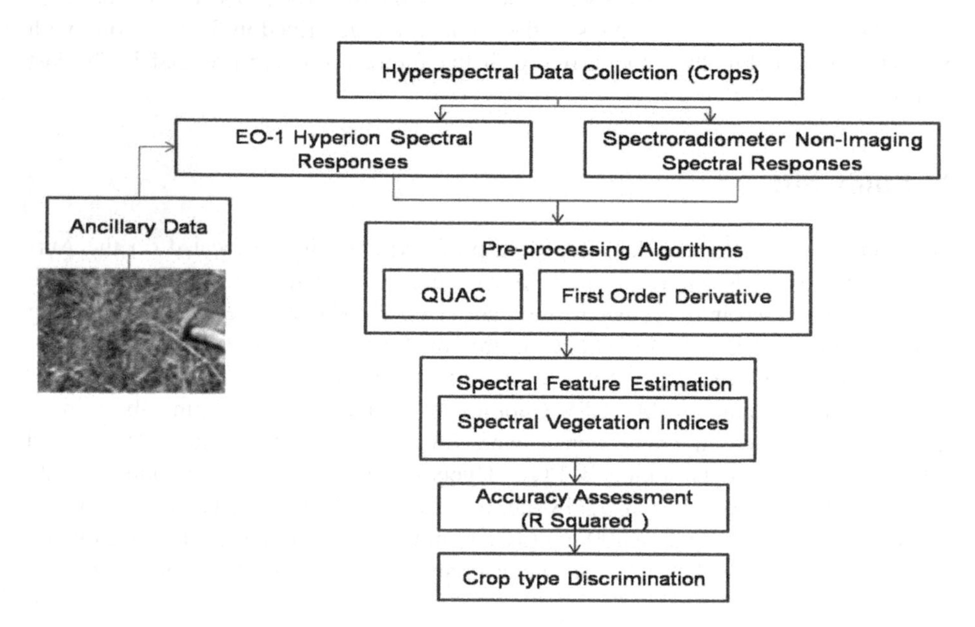

Fig. 2. The proposed framework of research conducted for crop discrimination

Hyperspectral datasets of two different sensors were collected, processed and based on spectral features; the respecting crop types were discriminated.

3.1 Vegetation Sample Assortment

The six crops varieties were collected from different sites, carried out in the airtight bag. All leaves were fully expanded and were selected subjectively to ensure that representative leaves were chosen. Leaves spectral reflectance measurements were made on dark acclimated and illuminated leaves using spectroradiometer (ASD Field Spec 4) by attaching a fiber optic probe onto the tip of the sample. The fiber optic cable reflects towards white panel for calibration and furthermore, the spectral measurements were collected. The spectral sampling range of spectroradiometer was 350 nm–2500 nm, with a sampling interval of 1.4 nm [13]. The leaf samples were placed with the 8-degree field of view in front of tungsten light, the height of the light intensity was fixed at 12. 5 cm from the object placed shown in Fig. 3.

Fig. 3. Spectral measurement setup of spectroradiometer at darkroom laboratory environment

Ten spectral measurements were made per leaf sample to ensure that the spatial homogeneity of each leaf region was captured. The spectral signatures were converted from radiance to reflectance by dividing each leaf measurement's using white standard measurement. The narrowband spectral indices were significantly proven effective

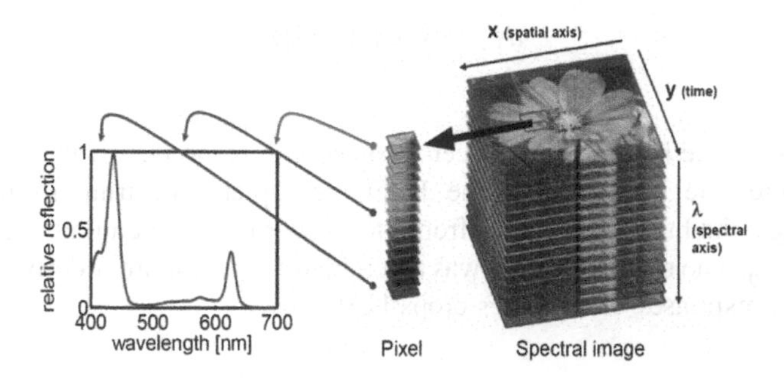

Fig. 4. Hypercube formation of crop samples in hyperspectral remote sensing

outcomes for crop pigments analysis [14]. The wavelength range (400 nm–800 nm) covering the visible region, which was used for crop health analysis. The hypercube of the image was generated as shown in Fig. 4 for crop samples.

3.2 First Derivative Spectral Analysis

The first derivative was used to remove unwanted noise and enhance the quality of spectral features shown in Fig. 5. The splice error correction method was applied at 1100 nm and 1830 nm spectral range. The band 350 nm to 400 nm contains no any information for crop detection and identification, for that those 50 bands were eliminated from a spectrum.

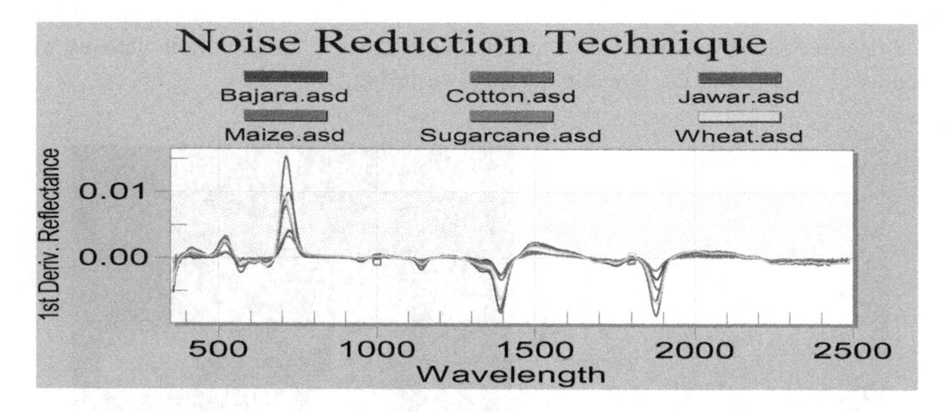

Fig. 5. First order derivative of crops for noise reduction

The X-axis represents wavelength with the sampling interval of 1 nm and Y-axis represents reflectance captured in visible, near infrared and shortwave infrared channels.

The processed spectral responses were assembled and the spectral library was engendered show in Fig. 6. The spectral analysis using the first derivative was done by dividing the difference between successive spectral values by its wavelength interval (bandwidth). The estimation for the first derivative is shown in Eq. (1).

$$\frac{ds}{dy}l_i \approx \frac{s(\lambda_i) - s(\lambda_j)}{\Delta\lambda} \tag{1}$$

Where, $\Delta\lambda$ is the bandwidth between two spectral responses of the measured crop spectrum. Here, the bandwidth is the 1 nm incomplete spectrum from 340 nm to 2500 nm range, the bandwidth varies from i to j according to the number of samples as ij, $\Delta\lambda = \lambda_i - \lambda_j$. The spectral library was assembled from first order derivative acquired using spectral responses of six types crops [15].

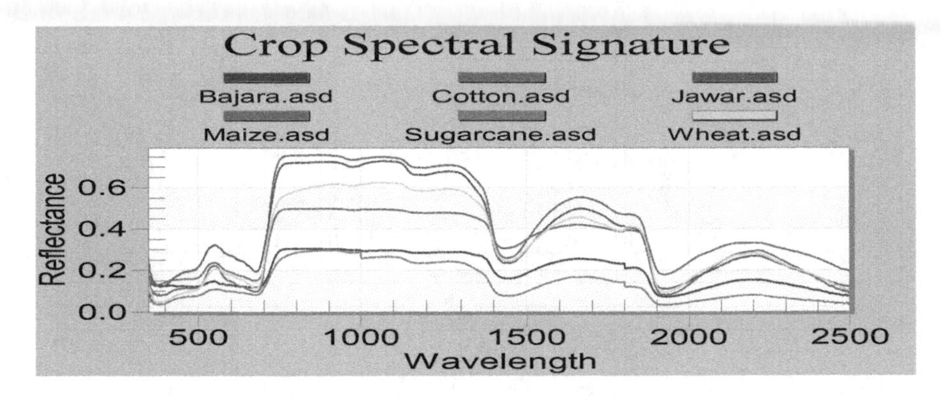

Fig. 6. Spectral responses of crops after noise reduction technique

3.3 Quick Atmospheric Correction (QUAC)

QUAC code performs atmospheric correction on hyperspectral imagery spanning the visible and near-infrared–shortwave infrared (VNIR) spectral range (0.4–2.5 μm) based on the observed judgment that the usual reflectance of an assortment of various objects spectrum is effectively scened independently. QUAC needs given steps to perform atmospheric corrections based on radiance and reflectance values.

The radiance to reflectance is possible using Eq. (2)

$$\text{Rad to Reflectance} = \frac{\pi L\lambda\, d^2}{\cos\theta * ESUN_\lambda} \tag{2}$$

Here, L_λ is the spectral radiance sensor ($W.m^{-2}.sr^{-1}.\mu m^{-1}$), and d is Earth to Sun distance, θ_s is astral zenith angle, $ESUN_\lambda$ is mean astral atmospheric irradiance. QUAC approach is based on radiance values for the atmospheric correction. It has a pragmatic approach based on the mean reflectance values of an end member spectrum based field of view [16]. The QUAC approach based on three steps including the selection of pure spectral responses from imagery, estimation of baseline and reflectance can be solved using given Eq. (3).

$$\text{QUAC} = (.n)/n^{*}1\,2, (\rho = \rho + \rho + \rho) \tag{3}$$

Where, n specifies the number of End-members. The pre-processed end members of crops were estimated based on GCP shown in Fig. 7. The X-axis represents wavelength number and the Y-axis represents spectral reflectance of the object.

Table 1 describes, a number of spectral responses collected through ASD and extracted through the Hyperion sensor for spectral feature estimation.

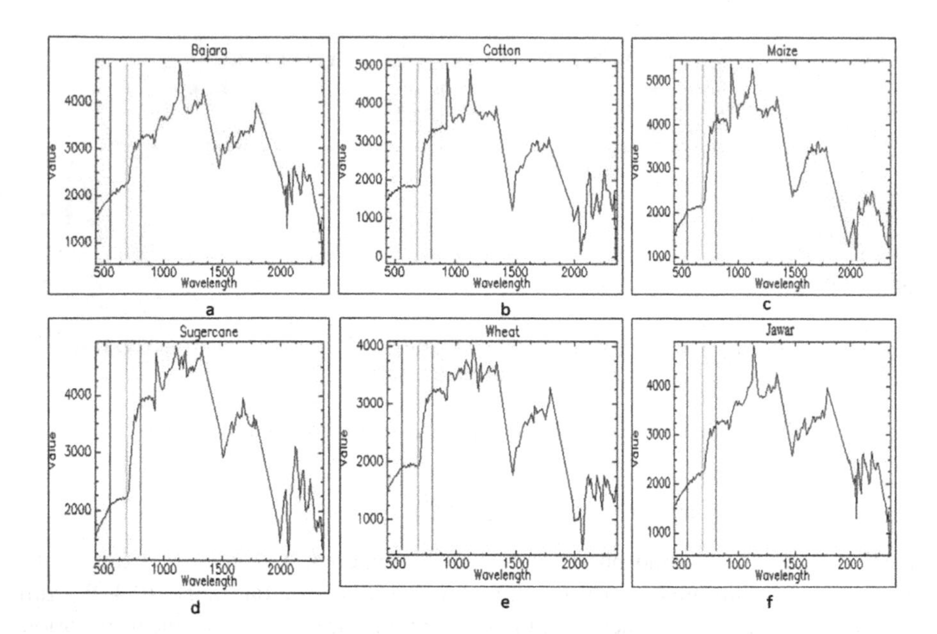

Fig. 7. Spectral responses of crops extracted through EO-1 Hyperion Imagery

Table 1. The crops spectral responses collected at different growth stages

Sensor collection	Crop types	Growth stage	Date of spectrum	No. of spectrum
ASD& EO-1 Hyperion	Cotton	Growing	Nov 2017	600
	Bajara	Flowering	Jan 2017	
	Jawar	Grain setting	Dec 2017	
	Maize	Growing	Jan 2017	
	Sugarcane	Growing	Dec 2017	
	Wheat	Ripening	Dec 2017	

3.4 Spectral Reflectance: Vegetation Indices

In the era of hyperspectral remote sensing, inventors have established spectral vegetation indices (VI) for qualitatively and quantitatively appraising homogenous shields using spectral dimensions. The spectrum of vegetated regions represents a composite combination of vegetation, soil brightness, environmental effects, shadow, soil color and moisture [17]. Herewith we had applied some vegetation indices for the comparison of Hyperion dataset with the spectral dataset. Normalized Difference Vegetation Index (NDVI) varies between +1 to −1, it measures the vegetation cover in Near Infrared (NIR) and Red Edge. Anthrocyanin Reflectance Index (ARI1) is a modified index for Anthocyanin pigments identification. It works to detect the stress level of crops. Moisture Stress Index (MSI) measures the total moisture stress in leaves through reflectance. Transformed Chlorophyll Absorption Reflectance Index (TCARI), TCARI represents the effect of soil towards crop growth. Vogelmann Red Edge Index1 (VogREI1) is modified index which provides details of sensitive information about crops stress in red edge band as given in Table 2.

Table 2. The list of narrow-band hyperspectral vegetation indices along with applicable formulas and key references

SVI	Relevant formulas	Key citations
NDVI	$(R_{800} - R_{670})/(R_{800} + R_{670})$	Blackburn et al. [18]
ARI_1	$R_{900}(1/R_{550} - 1/R_{700})$	Daughtry et al. [19]
MSI	R_{1599}/R_{819}	Cohen et al. [20]
TCARI	$3[R_{700} - R_{670} - 0.2(R_{700} - R_{550})(R_{700}/R_{670})]$	Haboudane et al. [21]
Vogel REI_1	R_{740}/R_{720}	Kim et al. [22]

4 Result and Discussion

This section describes the empirical approach of spectral methods followed by analysis of research work. The SVI estimate results addressed below, it is clear that the NDVI, MSI, Volg1 and TCARI clearly showed positive performance from research. The red edge parameters are more sensitive to capture environmental changes, so Jawar, Sugarcane, Maize and Wheat are significantly correlated through spectral parameters represented in Fig. 8.

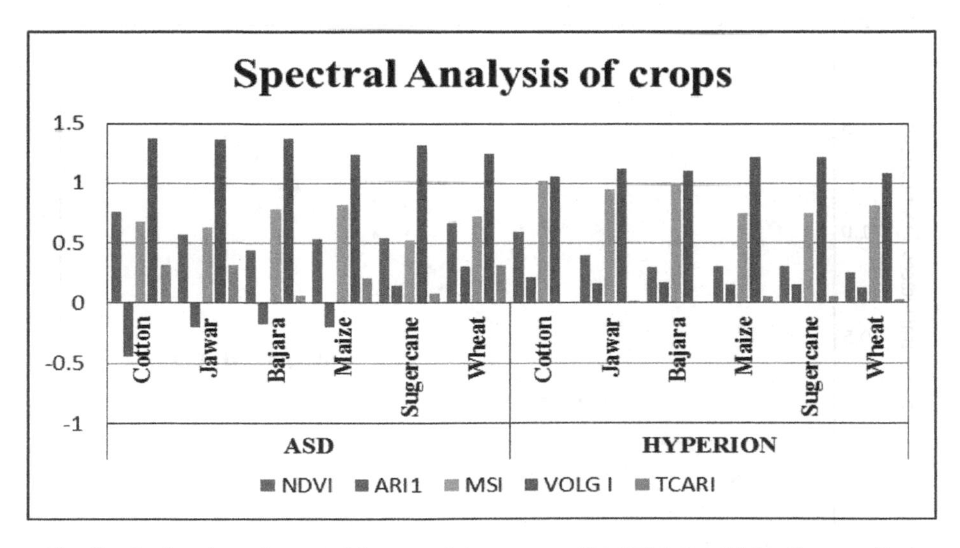

Fig. 8. Exploration of spectral features for crop type discrimination (Color figure online)

The X and Y intercept shows the individual intercept values of spectral features. The sum of residual values is 0.33, it is significant for the analysis of results and squared residual is 4.96. The R-squared values show that spectral features between both sensors give a positive correlation in maize, Jawar, sugarcane and wheat. The less correlation was found in cotton and Bajara because of mixing pixels with grain setting stage of Jawar. The 0.92 is considerable accuracy for crop discrimination based on hyperspectral responses given in Table 3.

Table 3. R-Squared analysis of spectral features

A slope of intercept	x	0.93290512
	y	−0.00245078
Sum of residual	0.338301725776	
Squared residual values	4.698495	
R squared values	0.9279978	

The polyfit correlation between two spectra with six types of crops is plotted in Fig. 9. The maize, sugarcane and wheat clusters are grouped together because of identical vegetation population of spectral features. The X-axis represents the spectral correlation of spectral features of spectroradiometer and Y-axis represents spectral features of Hyperion imagery.

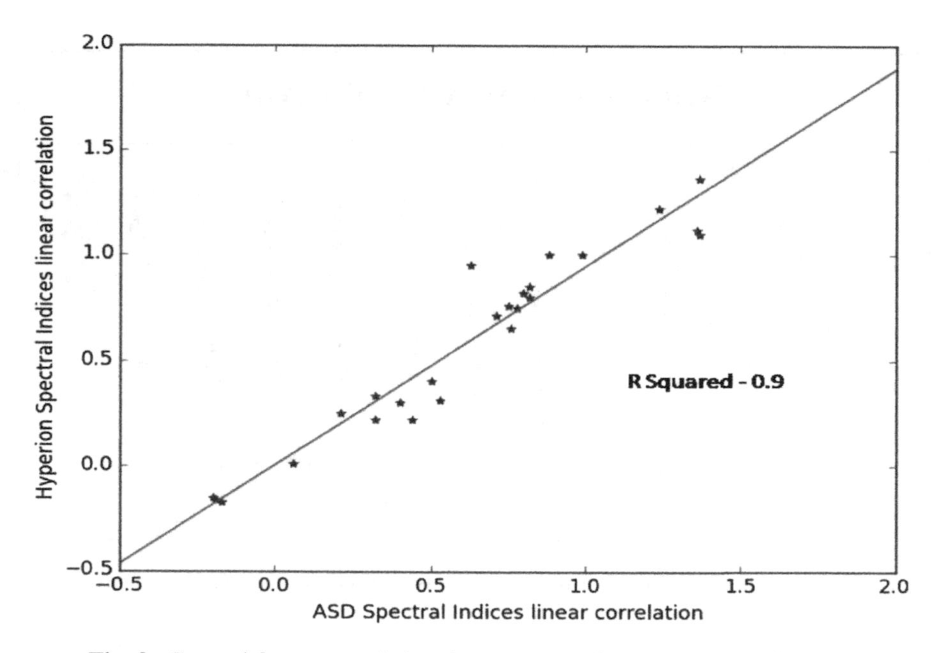

Fig. 9. Spectral feature correlations between Hyperion and Spectroradiometer

5 Conclusion and Future Scope

The crop type discrimination was successfully achieved using hyperspectral data. This paper highlights a positive correlation in spectral responses of spectroradiometer and EO-1 Hyperion. It is found that there was a progressive correlation 0.92 with squared residual value 4.69 amongst ASD and EO-1 Hyperion spectral features for ARI1 with R_{550}, R_{700}, for MSI R_{1599}, R_{819} respectively. The experimental analysis was performed using ENVI and python software. In the conclusion, crops types are successfully discriminated based on spectral parameters with different band combinations which will be useful for crop growth monitoring and precision farming. Also, future directions for current research will be to implement the PLSR method for complete fitting analysis of spectral parameters.

Acknowledgements. Authors would like to acknowledge for providing partial technical support under UGC SAP (II) DRS Phase-II, DST-FIST and NISA to Department of Computer Science & IT, Dr Babasaheb Ambedkar Marathwada University, Aurangabad, Maharashtra, India and also thanks for financial assistance under UGC-BSR research fellowship for this research work.

References

1. Haboudane, D., Tremblay, N., et al.: Estimation of plant chlorophyll using hyperspectral observations and radiative transfer models: spectral indices sensitivity and crop type effects. In: IGARSS. IEEE (2008). 978-1-4244-2808-03
2. Hatfield, J.L., Gitelson, A.A., Scherpers, J.S., et al.: Application of remote sensing for agronomic decisions. Agron. J. **100**, S-117–S-131 (2006)
3. Holecz, F., Barbieri, M., Collivignarelli, F., Gatt, L.: An operational remote sensing based service for rice production estimation at a national scale. In: Proceeding of ESA Living Planet Symposium, pp. 1–11 (2013)
4. Lehmann, J.R.K., Oldeland, J., Romer, M.: Field spectroscopy in the VNIR-SWIR region to discriminate between Mediterranean native plants and exotic-invasive shrubs based on leaf tannin content. Remote Sens. **7**, 1225–1241 (2015). https://doi.org/10.3390/rs70201225
5. Ling, C., Liu, H., Ju, H., Zhang, H., You, J., Li, W.: A study on spectral signature analysis of wetland vegetation based on ground imaging spectrum data. J. Phys. **910**, 012045 (2017)
6. Silleos, N., Misopolinos, N., Perkis, K.: Relationship between remote sensing spectral indices and crops discrimination. Geocarto Int. **7**(2), 41–51 (1992)
7. Verma, K.S., Saxena, R.K., Hajare, T.N., Ramesh-Kumar, S.C.: Gram yield estimation through SVI under soil and management conditions. Int. J. Remote Sens. **19**, 2469–2476 (1998)
8. Stanhill, G., Kafkafi, U., Fuchs, M., Kagan, Y.: The effect of fertiliser application on solar reflectance from a wheat crop. Israel J. Agric. **22**, 109–118 (1972)
9. Sridhar, V.N., Dadhwal, V.K., Chaudhari, K.N., Sharma, R., Bairagi, G.D., Sharma, A.K.: Wheat production forecasting for a predominantly unirrigated region in Madhya Pradesh. Int. J. Remote Sens. **15**, 1307–1316 (1994)
10. Reynolds, C.A., Yitayew, M., Slack, D.C., Hutchinson, C.F., Huete, A., Petersen, M.S.: Estimating crop yields and production by integrating the FAO crop specific water balance model with real-time satellite data and ground-based ancillary data. Int. J. Remote Sens. **21**, 3487–3508 (2002)

11. Dadhwall, V.K., Sridhar, V.N.: A non-linear regression for vegetation indexcrop yield relation incorporating acquisition date normalization. Int. J. Remote Sens. **18**, 1403–1408 (1997)
12. Maselli, F., Romanelli, S., Bottai, L., Andmaracchi, G.: NDVI data for yield forecasting in the Sahelian region. Int. J. Remote Sens. **21**, 3509–3523 (2000)
13. Surase, R.R., Varpe, A., Solankar, M., Gite, H., Kale, K.: Development of non-imaging spectral library via Field Spec4 spectroradiometer. Int. J. Res. Eng. Appl. Manage. (IJREAM). ISSN 2454-9150 Special Issue - NCCT – 2018
14. Magney, T.S., Griffin, K.L., Eitel, J., Vierling, L.A., et al.: Spectral determination of concentrations of functionality diverse pigments in increasing complex arctic tundra canopies. Oecologia **182**(1), 85–97 (2016)
15. Arun Prasad, K., Gnanappazham, L.: Species discrimination of mangroves using derivative spectral analysis. ISPRS Ann. Photogramm. Remote Sens. Spat. Inf. Sci. **II-8**, 45 (2014)
16. Mazer, A.S., Lee, M., et al.: Image processing software for imaging spectrometry analysis. Remote Sens. Environ. **24**(1), 201–210 (1988)
17. Bannari, A., Morin, D., Bonn, F., Huete, A.R.: A review of vegetation indices. Remote Sens. Rev. **13**, 95–120 (1995)
18. Blackburn, G.A.: Spectral indices for estimating photosynthetic pigment concentrations: a test using senescent tree leaves. Int. J. Remote Sens. **19**(4), 657–675 (1998)
19. Daughtry, C.S.T., et al.: Discriminating crop residues from soil by short-wave infrared reflectance. Agron. J. **93**, 125–131 (2001)
20. Cohen, W.B.: Response of vegetation indices to changes in three measures of leaf water stress. Photogramm. Eng. Remote Sens. **57**(2), 195–202 (1991)
21. Haboudane, D., Miller, J.R., Tremblay, N., et al.: Integrated narrow-band vegetation indices for prediction of crop chlorophyll content for application to precision agriculture. Remote Sens. Environ. **81**(2–3), 416–426 (2002)
22. Kim, Y., Michael Glenn, D., Park, J., Lehman, B.L.: Hyperspectral image analysis for plant stress detection. An ASABE Meeting presentation, Paper Number-1009114 (2010)

Use of Spectral Reflectance for Sensitive Waveband Determination for Soil Contents

Chitra M. Gaikwad[1](✉) and Sangeeta N. Kakarwal[2](✉)

[1] Government College of Engineering, Aurangabad, India
gaikwadchitra@gmail.com

[2] PES Engineering College, Aurangabad, India
s_kakarwal@yahoo.com

Abstract. In this paper we present the study of soil reflectance for organic matter content in soil based on their spectral signatures. We present the study of soil reflectance obtained from ASD Field spec spectrometer in the wavelength range 350–2500 nm. These values of reflectance are used to find the organic matter content in soil. Spectral curves of 8 soil samples are studied which are collected from Maharashtra state of India. Correlation between the spectral reflectance values of soil and the values obtained from chemical analysis in laboratory of soil contents is carried out. The predictions are carried out using the correlation coefficient. The content of soil organic matter in the soil samples is predicted for the wavelengths from 1801 to 1872 nm.

Keywords: Spectral signature · ASD Field spec spectrometer · Reflectance · Correlation · Soil analysis

1 Introduction

Soil contains various nutrients which are responsible for the growth of plants and vegetation. Extraction of the information of soil nutrients from soil is gaining importance for the healthy growth of vegetation in agriculture [1]. Traditional methods are time consuming, costly and its disposal is not easy [2]. The use of remote sensing using reflectance spectroscopy is one such technique which does not affect or harm the nutrients in the soil [3]. It is cost effective, non destructive and can accurately measure the various properties of soil [4,5]. Spectroscopy contains the wide bands of wavelength from visible (400–700) and infra red region (700–2500). Studies have shown that spectroscopy using laboratory spectrometer is used for predicting the correlation between spectral and chemical properties of soil [6–8]. Soil organic matter (SOM) is one of the important content in soil which gives the information of essential nutrients. The correlation between SOM content and composition with soil reflectance is very strong [9].

The soil properties are spread in the VIS, IR and SWIR region of electromagnetic spectrum. VIS (400–700 nm) range gives the information about colour of

K. C. Santosh and R. S. Hegadi (Eds.): RTIP2R 2018, CCIS 1037, pp. 323–328, 2019.
https://doi.org/10.1007/978-981-13-9187-3_28

soil. Wavelength Bands near 500–700 nm represent iron oxides, oxy hydroxides, hydroxides, narrow bands near 1400–1900 nm for hydroxyl and water molecules. Clay minerals, organic constituents, carbonates, salt minerals are represented beyond 2000 nm wavelength [10].

The aim of this paper is to determine the correlation between soil spectral reflectance with soil organic matter (SOM). Soil provides nutrients to plants and soil organic matter is the main source of these nutrients. In the global C cycle it has a huge amount of carbon [11]. The work is carried out on 8 different types of soil samples from different locations of Maharashtra districts in India.

2 Data Preparation

The soil samples were collected from different locations in the state of Maharashtra, in India. In all 8 topsoil samples were collected. The soil material was first air dried and sieved with the 2 mm mesh. The reflectance from soil material was measured in the laboratory using ASD FieldSpec field spectrometer. For Spectral measurements we used a fiber optic probe with 10 degree angle. Before recording the spectral reflectance of soil measurements it was calibrated with a white reference Spectralon panel. The wavelength covered by ASD Field spec Pro Analytical Spectral device ranges from 350 nm–2500 nm. 20 scans of each soil samples were carried out. The mean of these 20 scans was then calculated. The values of reflectance in ASD format were obtained and then they were exported to spreadsheet software using Viewspec pro software.

3 Soil Organic Matter Content

The air dried soil samples were sieved using a 2 mm sieve for laboratory analysis. The soil chemical properties were determined using the standard laboratory methods. Soil organic matter for all the soil samples was estimated. These values of soil organic matter were used to find the correlation with the spectral reflectance.

4 Correlations and Covariance

The correlation and covariance analysis between 8 soil samples' reflectance and the values of SOM contents in them obtained from chemical laboratory analysis was carried out with the following formula: Covariance - The Covariance Matrix represents covariance between elements.

$$cov = \frac{\sum_{i=1}^{N}(r_i - r_i')(s_i - s')}{N-1} \tag{1}$$

Correlation - Correlation is a technique used for finding the relation between two variables.

$$corr = \frac{\sum_{i=1}^{N}(r_i - r_i')(s_i - s')}{\sqrt{\sum_{i=1}^{N}(r_i - r_i')^2 \sum_{i=1}^{N}(s_i - s')^2}} \tag{2}$$

The variables used to find the values of correlation and covariance coefficients are as follows:

1. r is the reflectance for corresponding wavelength and r' is its mean
2. s is the organic matter content for corresponding soil sample and s' is its mean
3. N is the sample size

Soil properties are linearly correlated with the soil contents' chemical properties [12].

5 Results

The 8 soil samples were taken from Bhivandi, Kasara, Chikalthana, Igatpuri, Naregaon, Kumbhephal, Shendra, University located in the state of Maharashtra. The soil organic matter was found out by the chemical analysis of soil for the 8 soil samples. The covariance and correlation between spectral reflectance and SOM actual values was calculated using Eqs. 1 and 2. Table 1 gives statistical values for values obtained by performing the chemical analysis of soil to find the organic matter.

Table 1. Descriptive statistics

Statistics	Min	Max	Mean
SOM Content	2.94	6.96	3.96

The spectral reflectance of soil was measured in the wavelength range 350–2500 nm. Figure 1 shows the spectral reflectance curves of 8 soil samples.

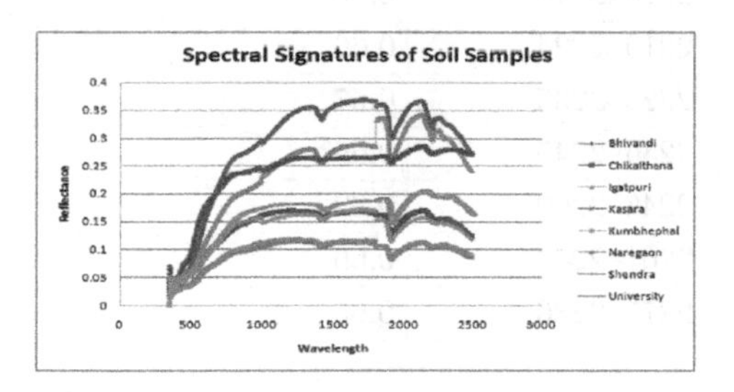

Fig. 1. Spectral reflectance curves

The maximum value of correlation coefficient in organic matter content and reflectance was obtained to be 0.71 in the wavelength range 1801–1872 nm. The maximum covariance value was 0.00027. Table 2 shows the range of wavelengths

Table 2. Wavelength range with correlation coefficient values

Wavelength range	Correlation coefficient
1801–1872	0.71
1873–1882	0.7
1883–1887	0.69
1888–1890	0.68
1891–1894	0.67
1895–1896	0.66
1897–1899	0.65
1900–1901	0.64
1902–1904	0.63
1905–1908	0.62
1909–1921	0.61
1922–1929	0.62
1930–1939	0.63
1940–1951	0.64
1952–1967	0.65
1968–1982	0.66
1983–1998	0.67
1999–2022	0.68
2023–2141	0.69
2142–2154	0.68
2155–2168	0.67
2169–2192	0.66
2193–2200	0.65
2201–2214	0.64
2215–2218	0.65
2219–2224	0.66
2225–2235	0.67
2236–2245	0.66
2246–2260	0.69
2261–2274	0.66
2275–2286	0.65

for which the correlation coefficients values are having higher values. The values in Table 2 range from 0.65–0.71. From Table 1 we get the maximum value of SOM content as 6.96. This shows that the soil samples contain organic matter from both the chemical analysis and the spectral analysis of soil samples.

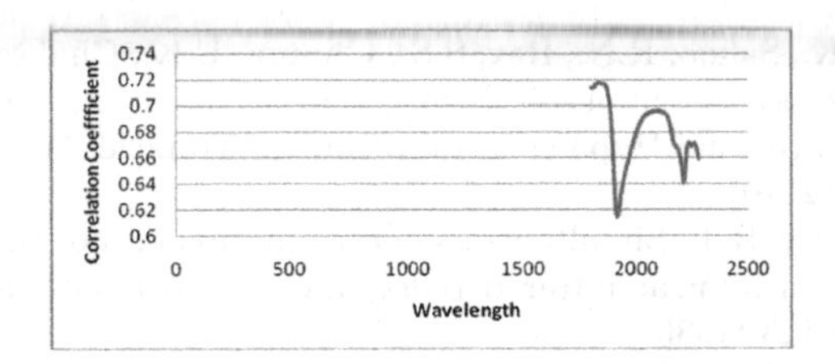

Fig. 2. Correlation Coefficient vs. Wavelength (nm)

Figure 2 shows the graphical representation of wavelength versus the correlation coefficients. The peak value is at 1801 nm.

6 Conclusion

In this study, the correlation between soil reflectance and chemical values of soil contents is found out. The peak for correlation between actual soil organic matter and reflectance is 1801 nm in the range 350–2500 nm of wavelength. The correlation for SOM values is higher for wavelengths from 1801 to 1872 nm and then they start decreasing gradually. We can conclude that for the soil samples collected from different locations the reflectance in the wavelength range 1801–1872 nm gives us the soil organic matter content in soil. Thus we can conclude that 1801 to 1872 nm wavelength can be used for predicting soil organic matter in soil. In future the correlation between various transforms of soil reflectance can be used for further predictions.

References

1. Rozeinstein, O., Kagan, T.P., Salbach, C., Karneili, A.: Comparing the effect of preprocessing transformations on methods of land use classification derived from spectral soil measurements. IEEE J. Sel. Top. Appl. Earth Observations Remote Sens. **8**(6), 2393–2404 (2015)
2. Dinakaran, J., Bidalia, A., Kumar, A., Hanief, M., Meena, A., Rao, K.S.: Near infrared-spectroscopy for determination of carbon and nitrogen in Indian soils. Commun. Soil Sci. Plant Anal. **47**, 1503–1516 (2016). ISSN: 0010–3624 (Print) 1532–2416 (Online)
3. Cheng, C.-W., Laird, D.A., Mausbach, M.J., Hurburgh, C.R.: Near-infrared reflectance spectroscopy-principal components regression analyses of soil properties. Soil Sci. Soc. Am. J. **65**, 480–490 (2001)
4. Gmur, S., Vogt, D., Zabowski, D., Monika Moskal, L.: Hyperspectral analysis of soil nitrogen, carbon, carbonate, and organic matter using regression trees. Sensors **12**, 10639–10658 (2012)
5. Ben-Dor, E., Patkin, K., Banin, A., Karnieli, A.: Mapping of several soil properties using DAIS-7915 hyperspectral scanner data-a case study over clayey soils in Israel. Int. J. Remote Sens. **23**, 1043–1062 (2002). ISSN 0143–1161 print/ISSN 1366–590 1 online 2002 Taylor and Francis Ltd

6. Kadupitiya, H.K., Sahoo, R.N., Ray, S.S., Chopra, U.K., Chakraborty, D., Ahmed, N.: Quantitative assessment of soil chemical properties using visible (VIS) and Near Infrared (NIR) Proximal Hyperspectral data. In: Tropical Agriculturist 2010, vol. 158, pp. 41–60 (2010)
7. Dalal, R.C., Henry, R.J.: Simultaneous determination of moisture, organic carbon, and total nitrogen by near infrared reflectance spectrophotometry. Soil Sci. Soc. Am. J. **50**, 120–123 (1986)
8. Brown, D.J., Shepherd, K.D., Walsh, M.G., Dewayne Mays, M., Reinsch, T.G.: Global soil characterization with VNIR diffuse reflectance spectroscopy. Geoderma **132**, 273–290 (2006)
9. He, T., Wang, J., Lin, Z., Cheng, Y.: Study on spectral features of soil organic matter. Int. Arch. Photogrammetry Remote Sens. Spatial Inf. Sci. J. XXXVII, Part B7 Beijing (2008)
10. Csorba, A., et al.: Identification of soil classification units from VIS-NIR spectral signature. In: 20th World Congress of Soil Science (2014)
11. Li, R., Kono, Y., Liu, J., Peng, M., Raghavan, V., Song, X.: Soil organic matter mapping with fuzzy logic and GIS. In: International Geoscience and Remote Sensing Symposium. IEEE (2012)
12. Pinheiro, E.F.M., Ceddia, M.B., Clingensmith, C.M., Grunwald, S., Vasques, G.M.: Prediction of soil physical and chemical properties by visible and near-infrared diffuse reflectance spectroscopy in the central amazon. Remote Sens. **9**, 293 (2017)

Image Processing Based Vegetation Cover Monitoring and Its Categorization Using Differential Satellite Imageries for Urmodi River Watershed in Satara District, Maharashtra, India

Wasim A. Bagwan[1] and Ravindra S. Gavali[2(✉)]

[1] Department of Environmental Science, School of Earth Sciences, Solapur University, Solapur 413255, India
wasim.bagwan16@gmail.com

[2] National Institute of Rural Development and Panchayati Raj, Hyderabad 500030, India
rsgavali@gmail.com

Abstract. Rapid monitoring of vegetation cover with precision has always been a challenge for maintaining accuracy over a large area. Remote Sensing (RS) based satellite imagery has significantly contributed in monitoring vegetation and land cover categorization. As the vegetation has a close relationship with detachment of soil and its sedimentation, regular monitoring of vegetation is essential especially in the catchment area of dams and reservoirs. In this study, vegetation maps were prepared through imaging processing of satellite imageries. With the help of Vegetation Index (VI) based maps, we were able to study the vegetation phenology in the watershed. The Normalized Difference Vegetation Index (NDVI), Soil Adjusted Vegetation Index (SAVI) and Enhanced Vegetation Index (EVI) were obtained using the spectral bands of Landsat 8 and Sentinel 2 A satellite data. The classes were made in accordance to no vegetation cover (<0.1), low vegetation cover (0.1–0.3), moderate vegetation cover (0.3–0.4), high vegetation cover (>0.4). The area under each category was calculated with vector files. Further, the relationship between pixel values of Landsat 8 and Sentinel 2 was analyzed by downscaling the spatial resolution of Landsat 8 maps. The pixel value of two satellite based NDVI and SAVI shows same R^2 value, that is 85.12 and EVI 83.15 respectively. Basically, the low vegetation cover depicted by the two imageries shows enormous difference which is quite huge for assessing the land/soil degradation. It was also revealed from the study that, Sentinel 2 imagery was very useful in computing EVI where the high density vegetation cover is present as compared to Landsat 8.

Keywords: Land degradation · Vegetation Index (VI) · NDVI · SAVI · EVI

K. C. Santosh and R. S. Hegadi (Eds.): RTIP2R 2018, CCIS 1037, pp. 329–341, 2019.
https://doi.org/10.1007/978-981-13-9187-3_29

1 Introduction

Land use simply refers to utilization of land for various human activities and land cover refers to factors on soil surface [1]. Land Use Land Cover (LULC) has many classes such as agriculture, water body, barren land, settlement and forest. Vegetation is an inevitable part of LULC. Natural vegetation cover and cropland are forms a normal interest of Remote Sensing (RS) investigations. The data collected through these techniques is a source of knowledge which can be used to monitor and examine the Earth's vegetation resources [2]. Vegetation cover is probably one of the most crucial factors in soil erosional activity [3]. For the monitoring of this, spectral vegetation indices are used to show the vegetation dynamics. The vegetation index (VI) is an arithmetic calculation on satellite spectral bands that focuses on vegetation greenness over soil signals and indicator of vegetation greenness [4]. The improvement in superior resolution sensors has enhanced the applicability of remote sensing data for the precise study of earth resources. Further, narrow spectral reflectance bands provide the chance to monitor physical, chemical and biological features of earth's resources [5]. The suitable study of vegetation dynamics helps to monitor changes in vegetation due to various factors and also could be utilized to predict its impact on ecosystem [6]. Water erosion in tropical countries is very profuse and hence conservation measure need to be taken and one of that is to establish a natural vegetation cover which has great potential to protect the soils [20]. According to Arnoldus and Riquier [19] in case of soil erosion by water, devoid of vegetation could lead to soil degradation. For monitoring of vegetation, Normalized Difference Vegetation Index (NDVI) has been used and reported by Rouse et al. [7]. Geographical Information Science (GIS) and Remote Sensing (RS) with integration to algorithms allows to monitor changes in Earth's surface processes at various spatio-temporal scales and was found to be more feasible through ground census techniques [8]. Therefore, the aim of this study was to assess the performance to the vegetation indices and its categorization for the current study area by two different satellite imagery within the same week.

2 Study Area

The Urmodi River Watershed (Fig. 1) covers approximately 412 km^2, and is spread across two talukas namely Satara and Jaoli in Satara district of Maharashtra, India. Spatial extent is 73°47′38.432″E to 74°6′58.267″E and 17°28′36.893″N to 17° 44′35.072″N. The study region is located on the Deccan peninsula. The elevation range of the watershed is 685 m derived from Survey of India toposheets. The maximum elevation at Kaas plateau having height 1265 m and minimum elevation at 580 m at Urmodi River mouth meeting to Krishna River. The Urmodi River originates at Kaas pond and on the flow path there is a waterfall named Vajrai. During the monsoon the Kaas plateau acts as touristic place as a 'Valley of flowers' The Urmodi River is located on the right bank of Krishna River which is the major river origination in the Western Ghats At the middle of the watershed Urmodi dam is present which is an irrigational

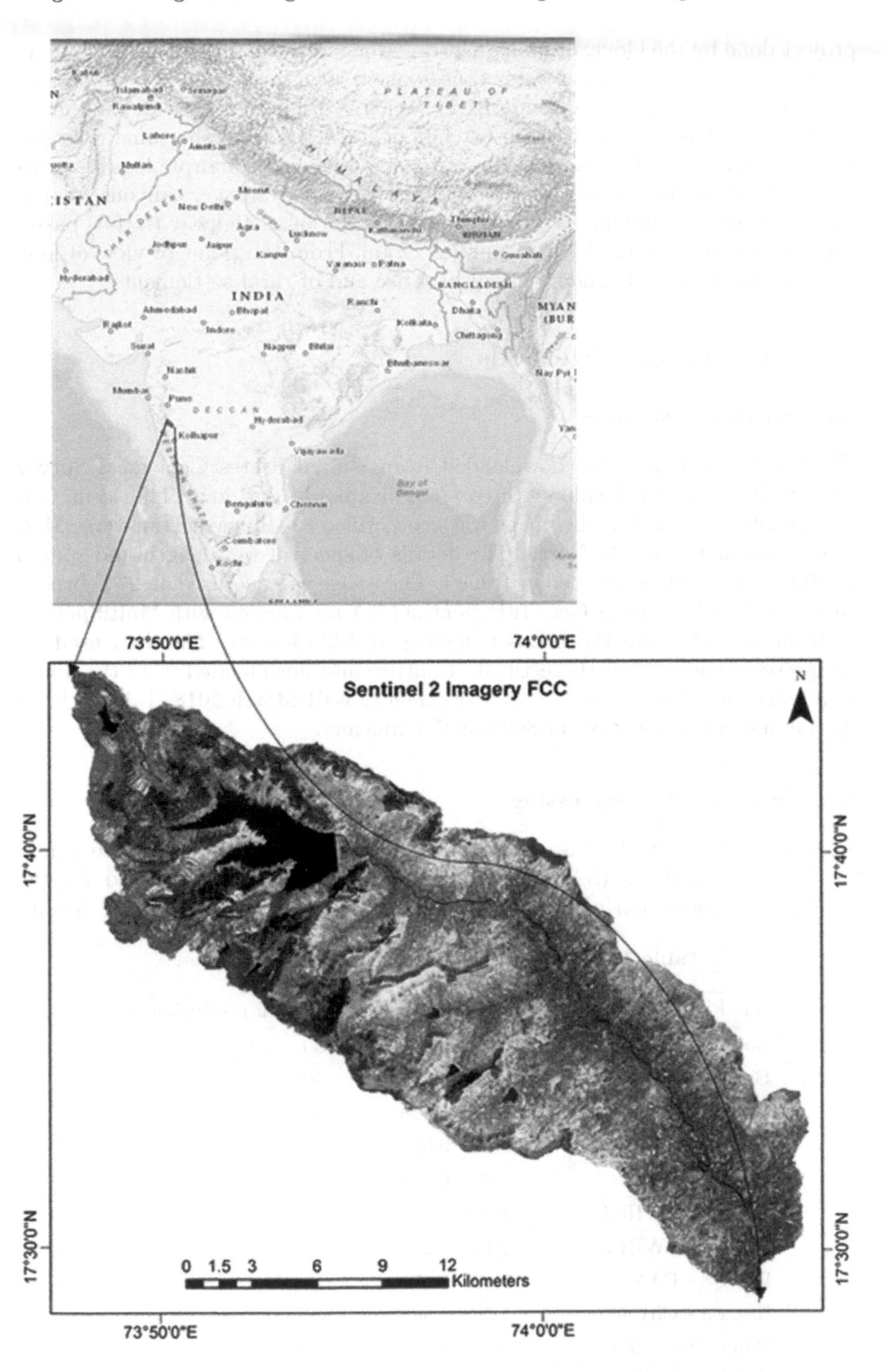

Fig. 1. Location of the Urmodi River watershed

project done by the Govt. of Maharashtra. Along with the main reservoir catchment also contain five small water conservation structures. On left side of river Kanher dam is situated on Venna River which is originated at Mahabaleshwar. On the right hand of the Urmodi catchment Tarali dam on the same river has been constructed. The study region is having various geomorphological structures like plateaus, mesas and buttes, hills, upland drainage, canyons, plains, mountain tops, deeply incised streams etc. The National Highway (NH-4) passes through the lower part of basin running 22 km. From the point of view of land use most of the land is under agricultural use and of rural settlements.

3 Materials and Methods

3.1 Satellite Datasets

The Landsat 8 image was downloaded from United States Geological Survey (USGS) portal Earth Explorer (http://earthexplorer.usgs.gov). The scene falls in path 147 and row 48. The image was georectified to Universe Transverse Mercator coordinate zone 43 North. The details of spectral wavelength and spatial resolution has been mentioned in Table 1. The image was acquired on 27 February 2018. The band format is GeoTIFF. Sentinel 2 A is equipped with MultiSpectral Instrument (MSI). All the bands is having JP2 file format. The data used for the present study was orthorectified in nature and downloaded from the above mentioned site. The acquisition date of imagery is 01 March 2018. Table 2 shows spatial and spectral details of Sentinel 2 A imagery.

3.2 Imagery Pre-processing

The General Public License (GPL) package named System for Automated Geoscientific Analyses (SAGA) version 4.1.0 GIS software was used for the atmospheric correction of Landsat 8 bands. The software has inbuilt facility

Table 1. Band information about Landsat 8 dataset.

Band number	Wavelength (nm)	Spatial resolution (m)
Band 1 - Aerosol/coastal	0.435–0.451	30
Band 2 - Blue	0.452–0.512	30
Band 3 - Green	0.533–0.590	30
Band 4 - Red	0.633–0.673	30
Band 5 - NIR	0.851–0.879	30
Band 6 - SWIR 1	1.566–1.651	30
Band 7 - SWIR 2	2.107–2.294	30
Band 8 - PAN	0.503–0.676	15
Band 9 - Cirrus	1.363–1.384	30
Band 10 - TIR 1	10.60–11.19	100
Band 11 - TIR 2	11.50–12.51	100

Table 2. Band information about Sentinel 2 dataset.

Band number	Wavelength (nm)	Bandwidth (nm)	Spatial resolution (m)
B1 - Aerosol	443	20	60
B2 - Blue	490	65	10
B3 - Green	560	35	10
B4 - Red	665	30	10
B5 - Red edge I	705	15	20
B6 - Red edge II	740	15	20
B7 - Red edge III	783	20	20
B8 - NIR	842	115	10
B8b - Red edge IV	865	20	20
B9 - Water vapor	945	20	60
B10 - Cirrus	1380	30	60
B11 - SWIR I	1610	90	20
B12 - SWIR II	2190	180	20

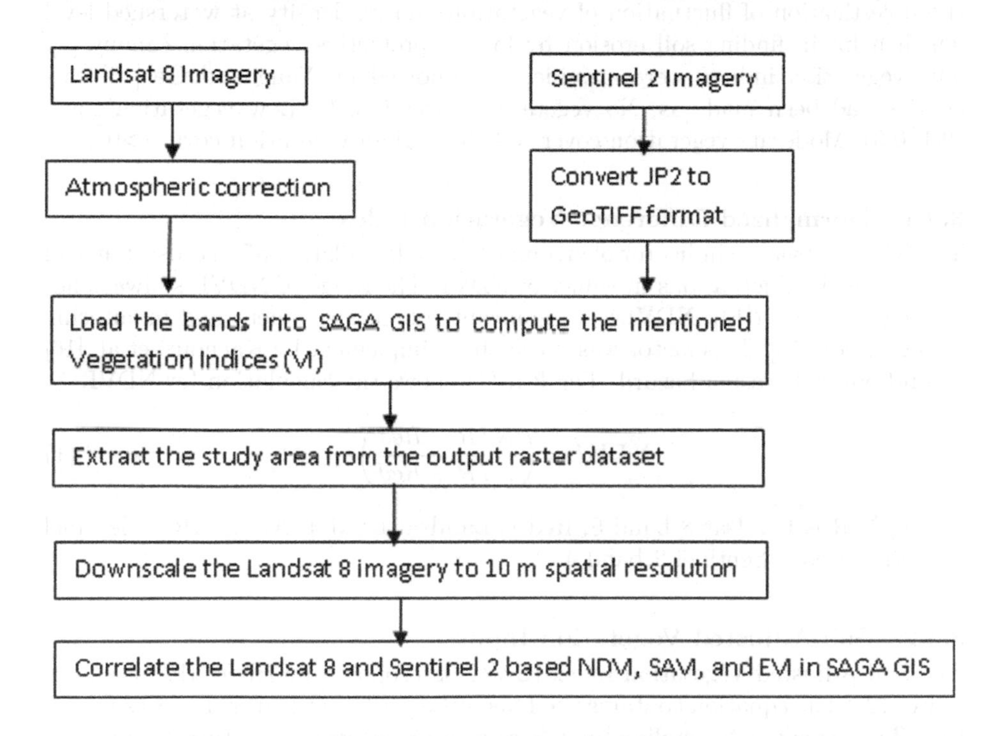

Fig. 2. Execution of methodology

of Top of Atmospheric Reflectance which helps in finishing the raw bands like haze removal. There is no need to process the for Sentinel 2 imagery as it is of level 1C. Both of the imagery was cloud free, hence we could get clear ground reflectance. Figure 2 shows the adopted methodology for the current study.

3.3 Vegetation Monitoring

Vegetation cover has an important role in safeguarding the soil from being loss due to water induced soil erosion. Distribution of vegetation cover is strongly influenced by altitude, slope of land, topographic factor, and solar input. Along with these conditions soil development and type of vegetation and its distribution also matters as they are the natural drivers of vegetation [9]. The Vegetation Index (VI) compared with actual measurement has benefits like saving time, frequency data, and saving expenses on labor [10]. The classic vegetation indices based on Red and NIR bands, because of their characteristic of reflectance, which has great help in monitoring it. There are two types of indices namely ratio and orthogonal. The ratio indices are independent on soil reflectance characteristics on the other side orthogonal indices considers the baseline specific to soil background [11]. Generally, protecting the top soil against the raindrop energy enhances the infiltration into the soil and reduces the swift of rainfall runoff. Cover factor like vegetation is an important environmental aspect which is a functioning part in decreasing, describing and overall assessment of water led soil erosion [12]. It is also a dynamic indicator in the investigation of fluctuation of vegetation and its density at watershed level which helps in finding soil erosion by loss of protective vegetation canopy [3]. The vegetation indices were classified as reported by Yang and Ge [13] categories had been made as, No vegetation cover (<0.1), Low vegetation cover (0.1–0.3), Moderate vegetation cover (0.3–0.4), High vegetation cover (>0.4).

3.3.1 Normalized Difference Vegetation Index

The NDVI acts as an indicator of greenness. The degradation of an ecosystem can be assessed by fluctuations in values of NDVI. The range of NDVI is always lies between −1 to +1[14]. NDVI has significant role in assessment and monitoring of vegetation [15]. This factor was successfully implement by Kachouri et al. [16] to find out soil erosion hazard. The Eq. (1) shows the calculation for NDVI.

$$NDVI = \left(\frac{NIR - Red}{NIR + Red}\right) \tag{1}$$

where, NIR is Landsat 8 band 5, Red is Landsat band 4. Also, NIR is Sentinel 2 band 8, Red is Sentinel 2 band 4.

3.3.2 Soil Adjusted Vegetation Index

The Soil Adjusted Vegetation Index (SAVI) is proposed by Huete [17] as shown in Eq. (2). The equation contains the L as soil-adjustment factor. L varies from 0 to 1. The value can be applied to wide range of vegetation densities. And for the intermediate vegetation (L = 0.5). Also, it was proposed by Huete [17] that, the problems associated with soil in NDVI analysis can be reduced by using SAVI.

$$SAVI = \left(\frac{NIR - Red}{NIR + Red + L}\right) \times (1 + L) \tag{2}$$

where, L is soil adjustment factor which is 0.5, Red is Landsat 8 band 4, NIR is Landsat 8 band 5. Also, for Sentinel 2 imagery, Red is Sentinel 2 band 4, NIR is Sentinel 2 band 8.

3.3.3 Enhanced Vegetation Index

The aim to develop Enhanced Vegetation Index to optimize the vegetation signal with improved sensibility in biomass and vegetation monitoring with de-coupling the canopy background signal and minimizing the atmospheric influence [18]. For its calculation Near Infra Red, Red and Blue bands are used. The equation is as shown below:

$$EVI = \left(G \times \frac{NIR - Red}{NIR + C1 \times Red - C2 \times Blue + L} \right) \quad (3)$$

where, Red is Landsat 8 band 4, NIR is Landsat 8 band 5. Also, for Sentinel 2 imagery, Red is Sentinel 2 band 4, NIR is Sentinel 2 band 8. C1 and C2 are coefficients with the value of 6, 7.5 respectively, L is 1 indicates the canopy background adjustment, and G (gain factor) taking value as 2.5.

4 Results and Discussion

The current research focuses on RS based vegetation cover monitoring using two different multispectral instruments that is Landsat 8 and Sentinel 2 A satellite imagery. These two satellites have different spatial resolution. Figure 3 depicts the three vegetation indices of the Urmodi River watershed using two different satellite sensor acquired imagery. Table 3 describes the descriptive statistics of various raster layers used for the assessment.

Table 3. Descriptive statistics of the Landsat 8 and Sentinel 2 derived vegetation indices

Satellite dataset	Raster layer	Minimum	Maximum	Average	Standard deviation
Landsat 8	NDVI	−0.32	0.76	0.29	0.15
	SAVI	−0.48	1.15	0.44	0.22
	EVI	−0.04	0.37	0.09	0.04
Sentinel 2	NDVI	−0.2	0.71	0.21	0.12
	SAVI	−0.3	1.07	0.32	0.18
	EVI	−0.05	0.52	0.09	0.06

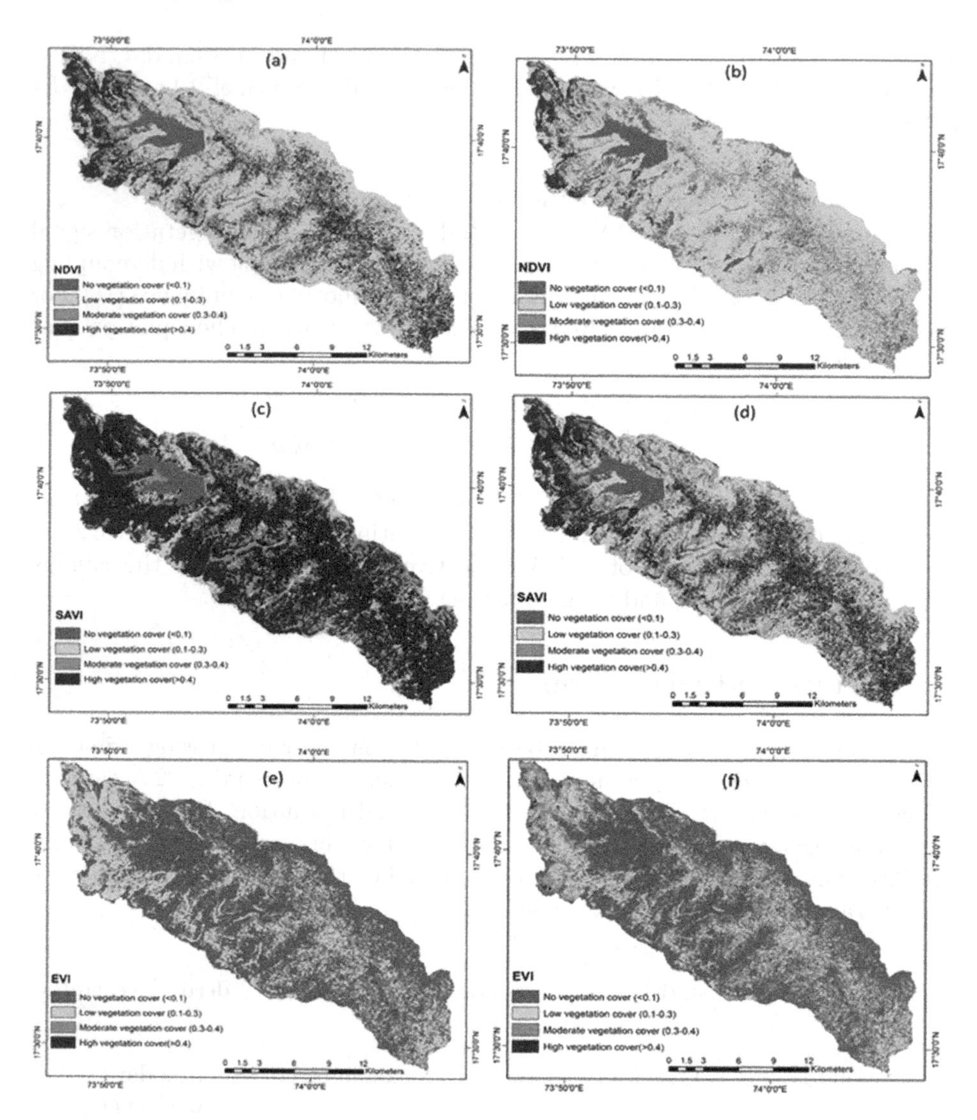

Fig. 3. (a) NDVI using Landsat 8, (b) NDVI using Sentinel 2, (c) SAVI using Landsat 8, (d) SAVI using Sentinel 2, (e) EVI using Landsat 8, (f)EVI using Sentinel 2

4.1 Normalized Difference Vegetation Index

From the prepared slope based vegetation indices like NDVI, we found that, the water body area like reservoir, small catchment structures and perfectly bare land shows negative value. The mountain slope shows less value as there is no vegetation due to steepness. The valley area, basically behind the reservoir due the presence of protected forest shows high values for this parameter. It also indicates that, there is a good correlation between the pixels values after the

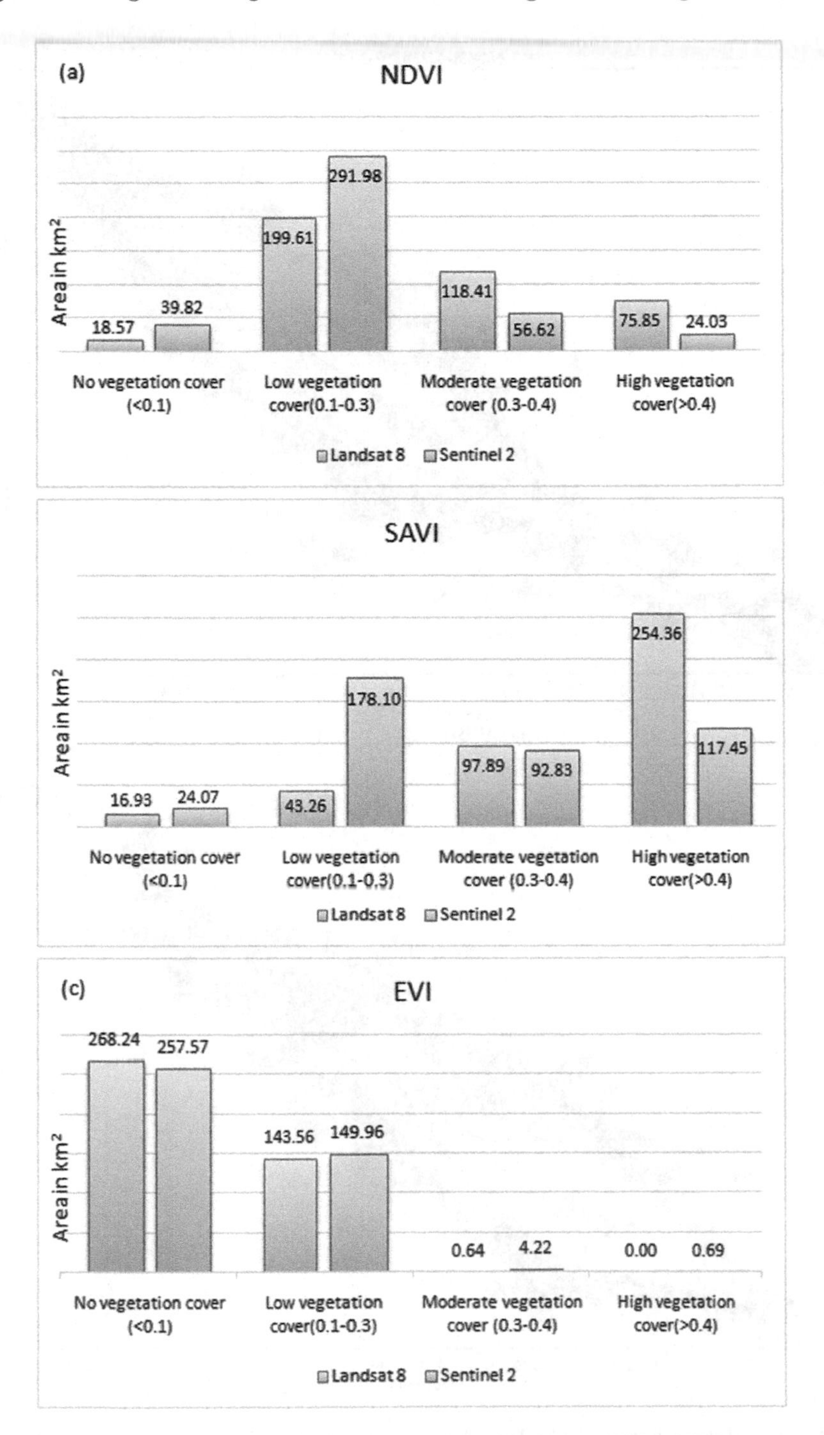

Fig. 4. Comparison of land area (km^2) covered by vegetation classes of Landsat 8 and Sentinel 2; (a)NDVI of Landsat 8 and Sentinel 2, (b) SAVI of Landsat 8 and Sentinel 2, (c) EVI of Landsat 8 and Sentinel 2

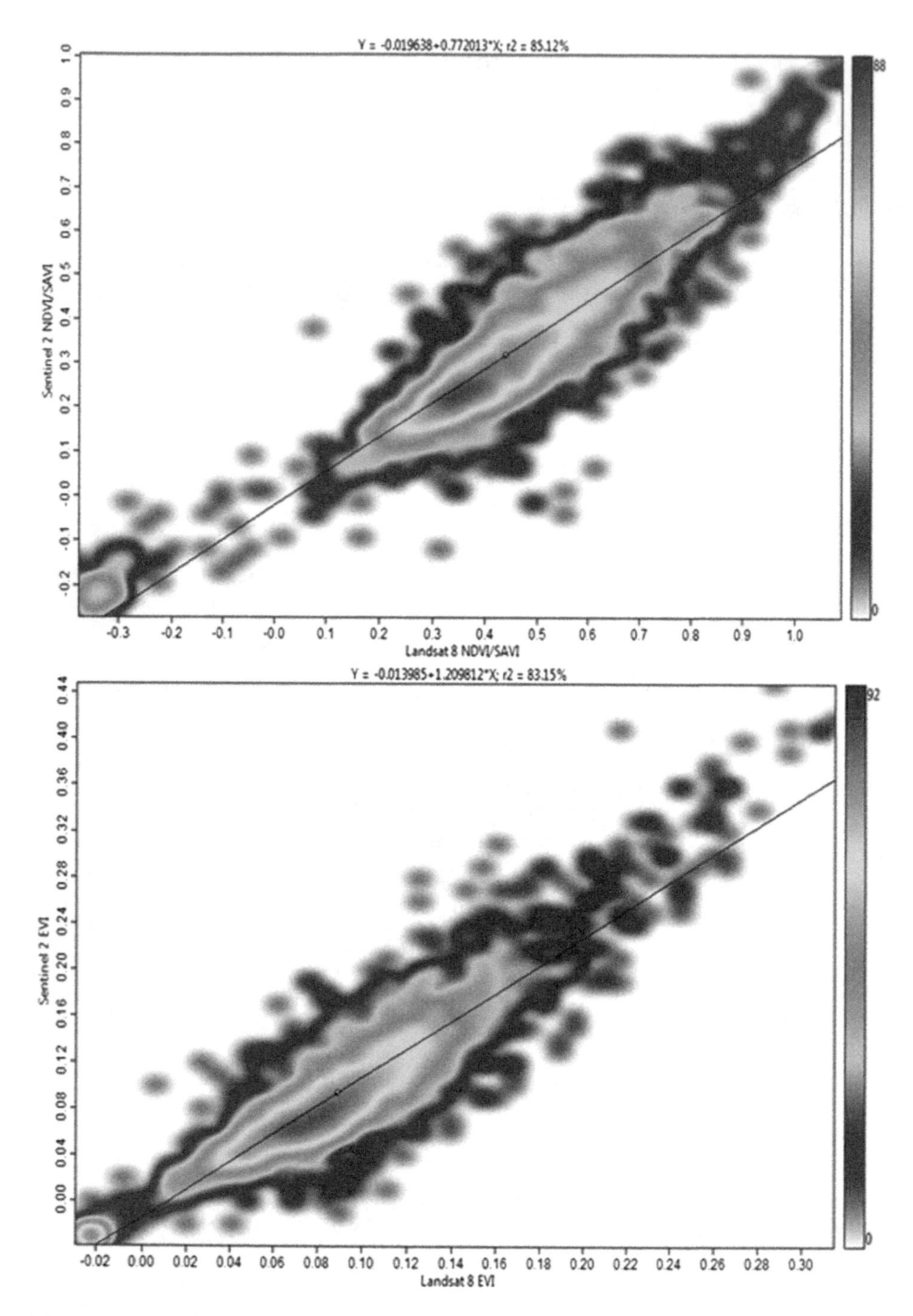

Fig. 5. Correlation between pixels of Landsat 8 and Sentinel 2 imagery for the vegetation indices (a) NDVI/SAVI, and (b) EVI

downscaling of Landsat 8 imagery pixel to 10 m. It was also found that, the low vegetation density (0.1–0.3) zone shows high fluctuation in its area showing difference of 92.36 km^2. So, in such case the huge area lost with proper vegetation density mapping which could lead to threat of land degradation.

4.2 Soil Adjusted Vegetation Index

It is a vegetation index which uses the Near Infra Red and Red band along with the soil adjustment factor L = 0.5. From the categorization it was clear that, on the mountain slopes it is showing the moderate and high vegetation cover. The dense vegetation canopy having value greater than 0.4 includes the land cover of dense forest, open forest and agriculture land. On the plateau area and steep slope it shows the low (0.1–0.3) to moderate (0.3–0.4) vegetation densities. From Fig. 4, it is clear that, the low vegetation cover category suddenly rises in its area coverage and as it was found with the NDVI categorization. Here, we found the difference of 134.84 km^2. In case of high vegetation density (>0.4) it shows difference of 136.91 km^2. This accounted an area which is nearly the 33% of study area and can mislead data for use planning.

4.3 Enhanced Vegetation Index

For the current study, EVI show very minute variation in case of No vegetation cover zone mapping using Landsat 8 and Sentinel 2 imagery that is 10.67 km^2 other vegetation density has been shown in Table 4. But the main difference was found that, we are unable to find the fourth category of vegetation density that is, high vegetation cover (>0.4) using the Landsat 8 imagery for the present band reflectance. However, Sentinel 2 imagery was found to be very helpful in uncovering the same.

Table 4. Distribution of vegetation cover densities derived from different indices in Urmodi River watershed

Vegetation category	Distribution of vegetation covers (km^2)					
	NDVI		SAVI		EVI	
	Landsat 8	Sentinel 2	Landsat 8	Sentinel 2	Landsat 8	Sentinel 2
No vegetation cover (<0.1)	18.57	39.82	16.93	24.07	268.24	257.57
Low vegetation cover(0.1–0.3)	199.61	291.98	43.26	178.10	143.56	149.96
Moderate vegetation cover (0.3–0.4)	118.41	56.62	97.89	92.83	0.64	4.22
High vegetation cover(>0.4)	75.85	24.03	254.36	117.45	0.00	0.69

4.4 Relationship Between NDVI, SAVI, and EVI

The analysis was carried out using SAGA GIS software. Before the correlation, the pixel size was equalized of Landsat 8 imagery for the purpose to retain the pixel information. From the Fig. 5(a), it is clear that, the relationship between Landsat 8 and Sentinel 2 NDVI and SAVI maps shows positive correlation of R2 = 85.12 and for its computation it uses the same bands namely NIR and Red. Interestingly, Fig. 5(b), the EVI shows nearly the same correlation between the pixels of Landsat 8 and Sentinel 2 imagery that is, $R^2 = 83.15$.

5 Conclusion

The present study depicts the information about the vegetation indices (VI) based on RS techniques, yet differentiating the vegetation density cover based variation. The synoptic view of whole watershed can be easily handled using GIS. The study has a crucial role in understanding the slope based vegetation indices. Moreover, it was found to be very useful in preparation of vegetation maps. The index like NDVI has a potential role in suggesting the management practices for soil conservation. A wide difference in area categorization based on vegetation density indicates that, how Sentinel 2 imagery is efficient to solve the ambiguity in canopy measurement. In case of EVI, it shows less difference in the vegetation density value which also clears that, Landsat 8 imagery can be efficiently used to monitor the same parameter. Moreover, high resolution cell size gives better understanding of canopy density as in case of Sentinel 2 imagery. And thus, the discrimination of VI classes made clear representation of vegetation over the terrain.

Acknowledgments. We are very thankful to the European Space Agency (ESA) for provide Sentinel- 2 data through portal- Earth Explorer and also for providing Landsat 8 imagery. The first author is thankful to the School of Earth Sciences, Solapur University, Solapur for their financial support in the form of Departmental Research Fellowship (DRF).

References

1. Durigon, V.L., Carvalho, D.F., Antunes, M.A.H., Oliveira, P.T.S., Fernandes, M.M.: International journal of remote NDVI time series for monitoring RUSLE cover management factor in a tropical watershed. Int. J. Remote Sens. **35**, 441–453 (2014)
2. Mróz, M., Sobieraj, A.: Comparison of several vegetation indices calculated on the basis of a seasonal SPOT XS time series, and their suitability for land cover and agricultural crop identification. Tech. Sci. **7**, 39–66 (2004)
3. Sharma, A.: Integrating terrain and vegetation indices for identifying potential soil erosion risk area. Geo-Spatial Inf. Sci. **13**, 201–209 (2010)
4. Lamchin, M., Park, T., Lee, J.Y., Lee, W.K.: Monitoring of vegetation dynamics in the mongolia using MODIS NDVIs and their relationship to rainfall by natural zone. J. Indian Soc. Remote Sens. **43**, 325–337 (2015)
5. Patel, N.K., Saxena, R.K., Shiwalkar, A.: Study of fractional vegetation cover using high spectral resolution data. J. Indian Soc. Remote Sens. **35**, 73–79 (2007)
6. Verma, R., Dutta, S.: Vegetation dynamics from denoised NDVI using empirical mode decomposition. J. Indian Soc. Remote Sens. **41**, 555–566 (2013)

7. Rouse, J.W., Haas, R.H., Schell, J.A.: Monitoring the vernal advancement and retrogradation (greenwave effect) of natural vegetation (1974)
8. McFarland, T.M., van Riper, C.I.: Use of normalized difference vegetation index (NDVI) habitat models to predict breeding birds on the San Pedro River, Arizona, p. 42 (2013)
9. Shifaw, E., Sha, J., Li, X., Bao, Z., Ji, J., Chen, B.: Spatiotemporal analysis of vegetation cover (1984–2017) and modelling of its change drivers, the case of Pingtan Island. China. Model. Earth Syst. Environ. **0**, 0 (2018)
10. Mokarram, M., Hojjati, M., Roshan, G., Negahban, S.: Modeling the behavior of vegetation indices in the salt dome of Korsia in North-East of Darab, Fars, Iran. Model. Earth Syst. Environ. **1**, 9 (2015)
11. Jaishanker, R., Senthivel, T., Sridhar, V.N.: Comparison of vegetation indices for practicable homology. J. Indian Soc. Remote Sens. **33**, 395–404 (2005)
12. Maimouni, S., EL-Harti, A., Bannari, A., Bachaoui, E.-M.: Water erosion risk mapping using derived parameters from digital elevation model and remotely sensed data. Geo-spatial Inf. Sci. **15**, 157–169 (2012)
13. Yang, Z., Ge, Y.U.: Spatio-temporal distribution of vegetation index and its influencing factors–a case study of the Jiaozhou Bay, China*. Chinese J. Oceanol, Limnol. **35**(6), 1398–1408 (2017)
14. Meneses-Tovar, C.L.: NDVI as indicator of degradation. Unasylva. **62**, 39–46 (2011)
15. Alphan, H., Derse, M.A.: Change detection in Southern Turkey using normalized difference vegetation index (NDVI). J. Environ. Eng. Landsc. Manag. **21**, 12–18 (2013)
16. Kachouri, S., Achour, H., Abida, H., Bouaziz, S.: Soil erosion hazard mapping using analytic hierarchy process and logistic regression: a case study of Haffouz watershed, central Tunisia. Arab. J. Geosci. **8**, 4257–4268 (2015)
17. Huete, A.R.: A soil-adjusted vegetation index (SAVI). Remote Sens. Environ. **25**, 295–309 (1988)
18. Huete, A., Didan, K., Miura, H., Rodriguez, E.P., Gao, X., Ferreira, L.F.: Overview of the radiometric and biopyhsical performance of the MODIS vegetation indices. Remote Sens. Environ. **83**, 195–213 (2002)
19. Arnoldus, H.M.J., Riquier, J.: World assessment of soil degradation - Phase I. In: FAO Soils Bulletin Assessing Soil Degradation. Food and Agriculture Organization of the United Nations, Rome (1977)
20. Roose, E.: Land use and soil degradation. In: FAO Soils Bulletin Assessing Soil Degradation. Food and Agriculture Organization of the United Nations, Rome (1977)

Discrimination Between Healthy and Diseased Cotton Plant by Using Hyperspectral Reflectance Data

Priyanka Uttamrao Randive(✉), Ratnadeep R. Deshmukh(✉), Pooja V. Janse(✉), and Rohit S. Gupta(✉)

Department of Computer Science and Information Technology, Dr. Babasaheb Ambedkar Marathawada University, Aurangabad, India
priyankarandive7@gmail.com, rrdeshmukh.csit@bamu.ac.in, puja.janse@hotmail.com, rohitgupta8844@gmail.com

Abstract. Cotton is major cash crop in India. Whenever disease occurs on the plant it causes reduction in production and also it effects on economy. Traditional way of monitoring disease is very hectic and time consuming. Healthy and Diseased leaves of cotton plant are collected from Harsul Sawangi regions of Aurangabad region. In this study ASD FieldSpec4 Spectroradiometer device is used for collection of hyperspectral data of cotton plant. This paper aims to examine the effect of disease on cotton plant. Spectral data is compared statistically. Discrimination is done among the healthy and diseased leaves for different regions of electromagnetic radiation. Ranges of Region: Blue (400 nm–525 nm), Green (525 nm–605 nm), Yellow (605 nm–655 nm), Red (655 nm–750 nm), and NIR (750 nm–1800 nm). Found higher reflectance in healthy leaves of than the diseased leaves of cotton plant.

Keywords: Cotton crop · ASD FieldSpec4 · Hyperspectral data · Discrimination · Remote sensing

1 Introduction

The Indian economy is based on agriculture. Among all the major crops that produced in India Cotton (Gossypium sp) is a main crop. In about 70 countries for its valuable fiber, around 20 million tons of cotton produces in each year. China, United State, India, Pakistan, Uzbekistan and West Africa these countries account for over 75% of global cotton Production. India ranked first in production of cotton [1]. That's why it also called as 'white Gold' by Indian farmers. The disease on cotton can be bacterial, fungal, viral or can be damaged by the insects. Any disease on plants that badly effect on economy and production. Manually examine disease on plant it gets difficult due to lack of experience or the eye vision limitations. Diseased caused by microorganism which can't be seen by the human eyes. This traditional method fails down many times to identify disease correctly [2]. After seeing and observing the symptoms on the plant to verify the disease by using disease detection technique. Laboratory method is also time consuming and costly.

K. C. Santosh and R. S. Hegadi (Eds.): RTIP2R 2018, CCIS 1037, pp. 342–351, 2019.
https://doi.org/10.1007/978-981-13-9187-3_30

1.1 Spectroscopy Method for Plant Disease Estimation

Remote sensing is wide area of research which is useful for many purposes for example data mining, in precision agriculture, soil content determination, geological application etc. In Hyperspectral Remote sensing spectroscopy is non-destructive method that examines physical and chemical properties of any material which is also helpful for plant disease assessment closely that includes individual leaf – level and plant –level estimation under controlled condition in laboratory to spectroscopic measurement on the field [3]. Remote sensing helps to early detection of disease and the infection caused by the insects within the crop which helps to reduce potential production losses and also decrease the cost of the farming [4].

Plant leaves contains water, amino acid, protein, nucleic acid and other material which are the reason for the photosynthesis capacity and the final production [5].

1.2 Internal Structure of Leaf

Figure 1 shows internal structure of the leaf. Leaf is important part of the Plant. Leaves plays vital role in photosynthesis. There are three major classes of pigments found in plant [6]. They are chlorophylls, Carotenoids and anthocyanin. Photosynthesis potential and primary production can directly determine by the Chlorophyll [7,8]. Carotenoid shows strong light absorption in the blue region of

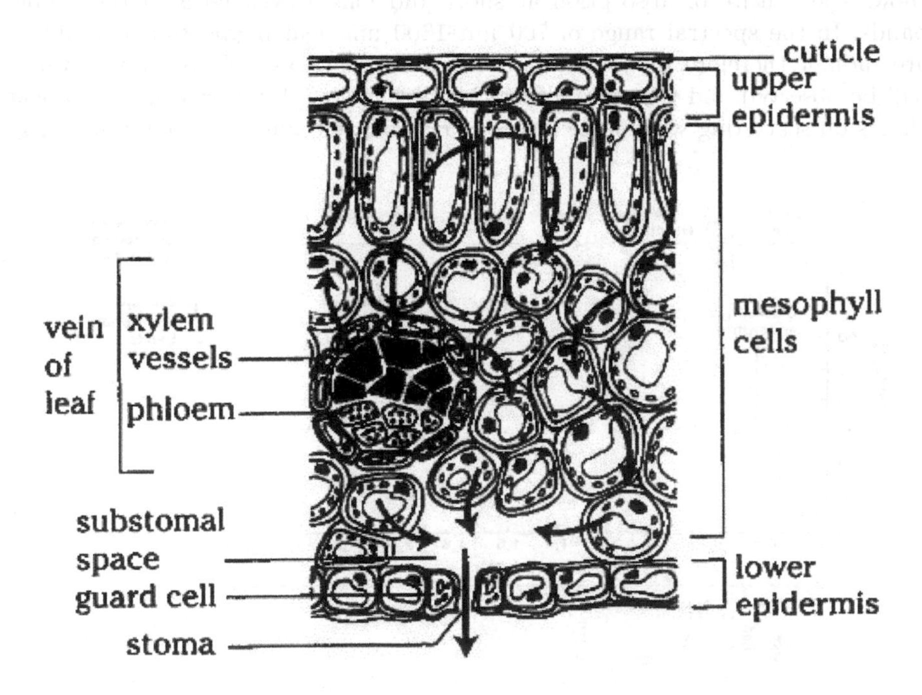

Fig. 1. Internal structure of leaf.

electromagnetic spectrum and it non-uniformly distributed in photosystem [9]. The Anthocyanin is pigment responsible for red colorization of leaves [10].

1.3 Spectral Characteristics of Leaf Reflectance

Figure 2 shows general spectral signature of leaf. Effect on visible region (Blue, Green, and Red) of electromagnetic radiation is due to leaf pigments present in the leaf likely effect on cell structure can be analyzed by observing Near Infrared (NIR) region and Short Wave Infrared (SWIR) region effects due to the water content present in leaf. Absorption by leaf pigments is the best significant process which mainly focused to low reflectance and transmittance values as well. The strong reason behind that leaf pigments such as chlorophyll a and chlorophyll b, carotenoids, polyphenols, xanthophyll and entire pigments have overlying absorption characteristics. Around 65% of the overall pigments are Chlorophyll a (Chl a) and chlorophyll b (Chl b). Chl a is the fundamental component of developed plants species. Chl a shows very high absorption of light in the 410 nm–430 nm and 600 nm–690 nm region, whereas in 450 nm–470 nm range Chl b shows very high absorption. In these absorption bands get a reflectance high value at nearby 550 nm in the green space [11].

Carotenoid shows absorption of light in the range of 440 nm and 480 nm. Polyphenols shows absorption of light with falling amount of intensity from the blue region to the red region and also active when the leaf is dry [12]. Chl b shows the feature of absorption at short and long wavelengths in the visible bands. In the spectral range of 700 nm–1300 nm, leaf pigments and cellulose are radiant therefore maximum values of reflectance as well as transmittance will be observed and therefore absorption is very low. Interior structure of leaf effects on scattering within the leaves at the air, cell, and water interfaces [13].

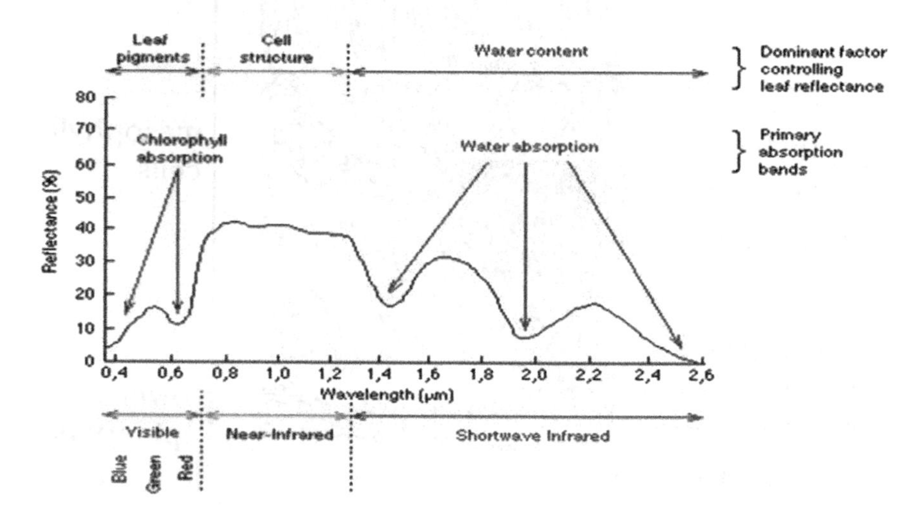

Fig. 2. Spectral signature of leaf. (Color figure online)

The reflectance level increases in NIR region with increasing inter-cell spaces, cell layers and cell size. Scattering of light happens due to multiple refractions and reflections in the air spaces and hydrated cellular walls. In the short wave infrared (SWIR) region, leaf spectral properties are mainly affected by water and supplementary foliar constituents. The 1450 nm, 1940 nm and 2700 nm these are major water absorption bands and subordinate features occur at 960 nm, 1120 nm, 1540 nm, 1670 nm and 2200 nm [14].

2 Materials and Methodology

2.1 Study Area

Cotton leaf Samples are collected from Harsul Sawangi regions (19.95412, 75.36038) of Aurangabad City which is located in Maharashtra. The annual mean temperature of Aurangabad region ranges from 17 to 33 °C and average annual rainfall is 710 mm.

2.2 Spectral Reflectance Measurement

Twenty leaf samples were randomly selected such as 10 healthy and 10 disease leaves for the spectral discrimination. Samples were collected in 14th October 2017 at 12.20 pm. The leaves kept fresh in plastic bag for less than 2 h before spectral reflectance were measured. The spectral reflectance of leaf samples collected using a Fieldspec4 in the range of 350 nm to 2500 nm. With sampling interval 3 nm (from 350 to 1000 nm) to 10 nm (from 1000 to 2500 nm). 50 W quartz halogen lamp is worked as light source for illumination purpose set over the samples. The leaf samples were put at the distance 40 cm and height of the lamp from the source kept 42 cm. Samples kept at distance 11 cm from Spectral gun. To optimize the signal and calibrate the accuracy and detector response the white reflectance panel was used. Takes 10 spectral signature of each sample and considered mean spectra of these 10 spectral signatures for further analysis.

2.3 Spectral Data Analysis

For Data Analysis ViewSpecPro Software is used. Spectral reflectance captured by spectral gun of Fieldspec4 spectroradiometer that is stores on laptop with the help of RS3 software. For Data Analysis and visualization ViewSpecPro Software is used.

3 Result and Discussion

After taking spectral signature of healthy and diseased leaves mean spectral signature of each sample is evaluated using ViewspecPro software. 10 spectral signature of each sample is taken in consideration for mean spectral signature. Figure 3 shows the mean spectral signature of all healthy leaf samples and Fig. 4 shows mean spectral signature of all diseased leaf samples. Figure 5 shows comparative signature of healthy and diseased leaf. In this figure red line shows the diseased leaf signature and green line shows the healthy leaf signature.

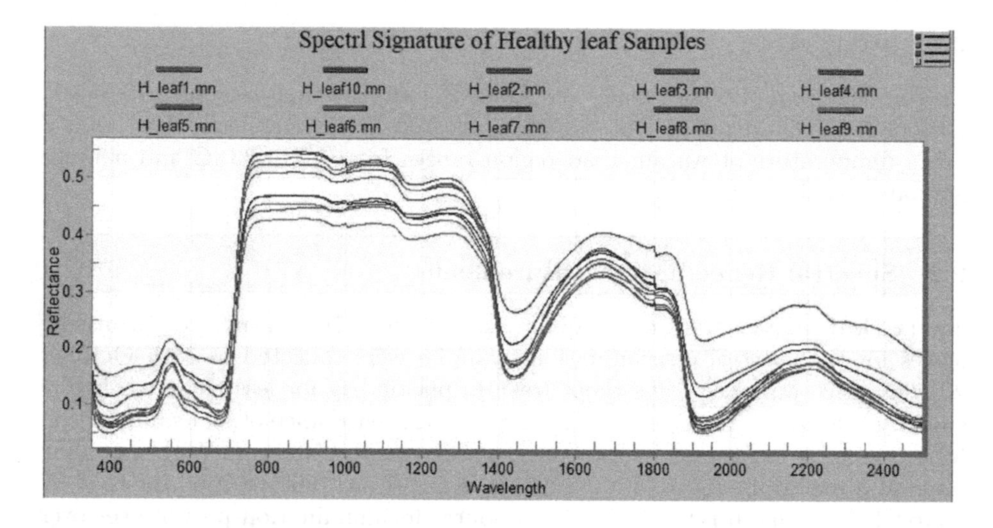

Fig. 3. Spectral signature of healthy leaf samples.

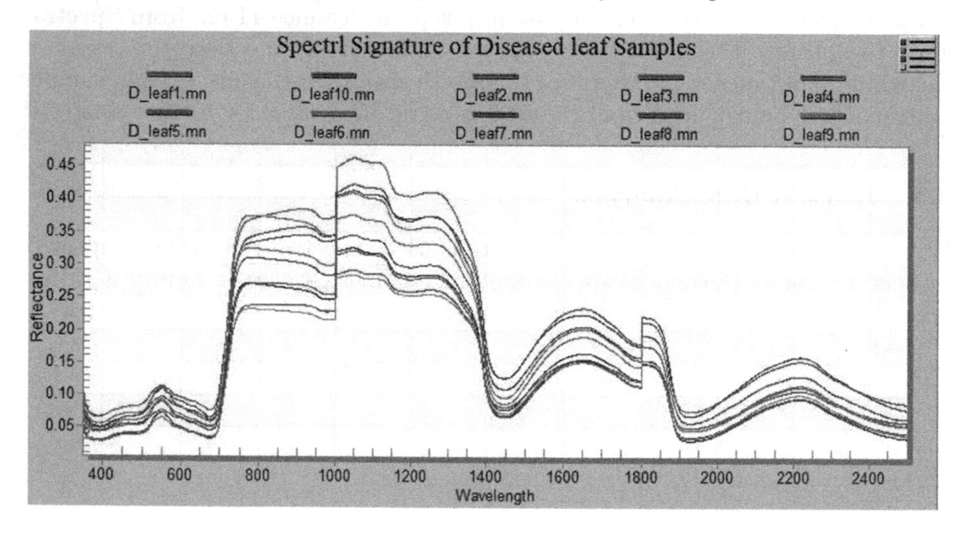

Fig. 4. Spectral signature of diseased leaf samples.

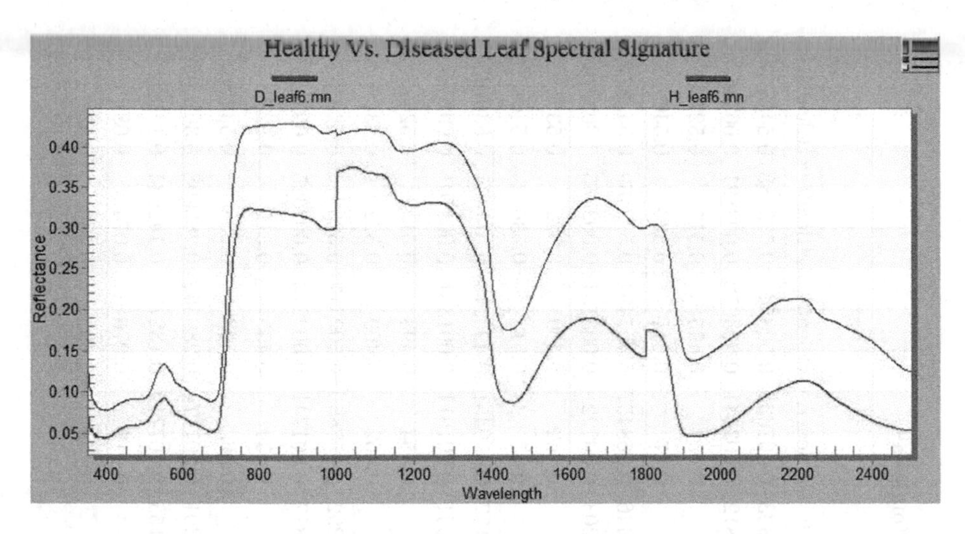

Fig. 5. Healthy Vs. Diseased leaf. (Color figure online)

3.1 Statistical Analysis

Two group of cotton leaf samples are taken in consideration for the statistical Analysis. 10 samples of healthy and 10 samples of diseased leaf are selected. Discrimination between healthy and diseased leaf samples are done by studying the effect on change in spectral signature in all regions of electromagnetic spectrum. Ranges of Region: Blue (400 nm–525 nm), Green (525 nm–605 nm), Yellow (605 nm–655 nm), Red (655 nm–750 nm), and NIR (750 nm–1800 nm). Found higher reflectance in healthy group than the diseased group. Mean, Standard Deviation (Std.Dev), Min, Max these statistical calculations is done for Blue, Green, Yellow, Red and NIR regions. Found high Mean in all regions of healthy group is greater than diseased group. Std.Dev in healthy group is less as compare to diseased group. Min value is found higher in healthy group than the Diseased Group. Max value found higher in healthy group as compare to diseased group. Tables 1 and 2 shows the statistical calculation of two groups of cotton leaves namely Healthy and Disease.

Table 1. Statistical analysis of healthy group of cotton leaves

Group name		Healthy									
Region name	Sample name	H_leaf1	H_leaf2	H_leaf3	H_leaf4	H_leaf5	H_leaf6	H_leaf7	H_leaf8	H_leaf9	H_leaf10
Blue	Mean	0.443238	0.542865	0.523667	0.468087	0.453413	0.425333	0.467635	0.50854	0.533183	0.542873
	Std Dev	0.003171	0.002821	0.002766	0.003483	0.003143	0.003812	0.003141	0.003072	0.003821	0.00386
	Min	0.432	0.529	0.511	0.452	0.438	0.409	0.454	0.495	0.517	0.527
	Max	0.448	0.544	0.525	0.47	0.455	0.429	0.47	0.511	0.536	0.546
Green	Mean	0.448222	0.541062	0.52242	0.467568	0.45284	0.427346	0.468457	0.507494	0.532469	0.543877
	Std Dev	0.000949	0.003075	0.003134	0.002674	0.002477	0.002203	0.00222	0.003131	0.003755	0.003014
	Min	0.445	0.533	0.513	0.46	0.446	0.421	0.462	0.499	0.522	0.535
	Max	0.449	0.544	0.525	0.47	0.455	0.429	0.47	0.51	0.536	0.546
Yellow	Mean	0.444431	0.527588	0.51002	0.456843	0.443176	0.418157	0.459451	0.495118	0.517451	0.530765
	Std Dev	0.001616	0.005108	0.002102	0.000987	0.001212	0.001541	0.001911	0.001351	0.001701	0.00238
	Min	0.443	0.51	0.508	0.456	0.442	0.415	0.454	0.493	0.515	0.524
	Max	0.449	0.533	0.515	0.46	0.446	0.421	0.462	0.499	0.522	0.535
Red	Mean	0.453156	0.51475	0.522042	0.462719	0.449771	0.419292	0.457865	0.500417	0.525844	0.529802
	Std Dev	0.001743	0.001563	0.002699	0.002242	0.001738	0.001812	0.001741	0.002213	0.002625	0.002121
	Min	0.449	0.511	0.515	0.457	0.445	0.415	0.454	0.495	0.519	0.525
	Max	0.455	0.517	0.525	0.465	0.452	0.421	0.46	0.503	0.528	0.532
NIR	Mean	0.348582	0.296644	0.28286	0.265461	0.260534	0.280301	0.301174	0.289552	0.285098	0.302663
	Std Dev	0.078757	0.142128	0.144623	0.12558	0.124217	0.091153	0.107719	0.132725	0.142634	0.141666
	Min	0.217	0.061	0.063	0.068	0.054	0.137	0.119	0.076	0.074	0.081
	Max	0.455	0.515	0.523	0.464	0.45	0.42	0.459	0.501	0.526	0.53

Table 2. Statistical analysis of disease group of cotton leaves

Group name		Disease									
Region name	Sample name	D_leaf1	D_leaf2	D_leaf3	D_leaf4	D_leaf5	D_leaf6	D_leaf7	D_leaf8	D_leaf9	D_leaf10
Blue	Mean	0.37181	0.335365	0.236873	0.306992	0.371317	0.321127	0.267556	0.269452	0.229841	0.344881
	Std Dev	0.007516	0.005345	0.002135	0.002207	0.002643	0.002202	0.002197	0.002477	0.001394	0.007854
	Min	0.348	0.318	0.227	0.297	0.36	0.313	0.263	0.257	0.224	0.319
	Max	0.381	0.342	0.238	0.309	0.374	0.324	0.271	0.271	0.232	0.354
Green	Mean	0.382148	0.343074	0.23642	0.299765	0.362864	0.313333	0.259284	0.266778	0.224951	0.354568
	Std Dev	0.002351	0.00143	0.001942	0.004016	0.004606	0.003691	0.003385	0.002894	0.002464	0.002208
	Min	0.374	0.338	0.231	0.289	0.351	0.304	0.251	0.259	0.219	0.347
	Max	0.384	0.345	0.238	0.305	0.368	0.318	0.264	0.27	0.228	0.356
Yellow	Mean	0.376706	0.341137	0.232824	0.289	0.350059	0.305275	0.252588	0.259647	0.220922	0.34551
	Std Dev	0.021277	0.0187663	0.015393	0.0145	0.015473	0.018972	0.019829	0.016057	0.017439	0.018016
	Min	0.369	0.335	0.227	0.284	0.344	0.299	0.246	0.254	0.215	0.342
	Max	0.449	0.405	0.285	0.338	0.403	0.37	0.32	0.314	0.28	0.41
Red	Mean	0.457615	0.408594	0.287823	0.340958	0.405156	0.371333	0.321229	0.317375	0.280771	0.717833
	Std Dev	0.003413	0.002767	0.00222	0.002615	0.002792	0.002767	0.002759	0.002363	0.002408	0.003171
	Min	0.449	0.405	0.285	0.337	0.4	0.367	0.317	0.314	0.277	0.41
	Max	0.462	0.413	0.291	0.345	0.409	0.375	0.325	0.321	0.284	0.422
NIR	Mean	0.223111	0.213113	0.146782	0.154021	0.186556	0.175342	0.146061	0.143789	0.13936	0.198254
	Std Dev	0.115977	0.101686	0.072366	0.09162	0.108226	0.099709	0.087962	0.085982	0.07463	0.108586
	Min	0.076	0.068	0.05	0.036	0.046	0.046	0.031	0.034	0.034	0.058
	Max	0.456	0.406	0.285	0.337	0.4	0.367	0.317	0.314	0.277	0.417

4 Conclusion

Higher reflectance found in healthy group of cotton leaves than the reflectance found in diseased group of cotton. NIR region is mostly affected in diseased leaves shows less reflectance. That due to disease effect on cell structure of leaf. We can say that cell structure is damaged in diseased leaves due to the disease. Whereas in healthy leaves shows higher reflectance in all regions of electromagnetic spectrum. This information is helpful for studying disease effect on the plants body.

Acknowledgement. DST-FIST has supported this work with sanction number-SR/FST/ETI340/2013. Authors are thankful to DST-FIST and Department of Computer Science and Information Technology of Dr. Babasaheb Ambedkar Marathwada University, Aurangabad, Maharashtra, India. For providing necessary infrastructure and support.

References

1. Prabhakar, M., et al.: Hyperspectral indices for assessing damage by the solenopsis mealybug (Hemiptera: Pseudococcidae) in cotton. Comput. Electron. Agriculture **97**, 61–70 (2013)
2. Randive, P.U., Deshmukh, R.R., Janse, P.V., Kayte, J.N.: Study of detecting plant diseases using non-destructive methods: a review. Int. J. Emerg. Trends Technol. Comput. Sci. (IJETTCS) **7**(1), 66–71 (2018)
3. Slonecker, E.T.: Analysis of the effects of heavy metals on vegetation hyperspectral reflectance properties. In: Thenkabail, P.S., Layon, J.G., Huete, A. (eds.) Hyperspectral Remote Sensing of Vegetation, pp. 561–578 (2012)
4. Atherton, D., Choudhary, R., Watson, D.: Advanced detection of early blight (Alternaria solani) disease in potato (Solanum tuberosum) plants prior to visual disease symptomonology. Int. J. Agric. Environ. Res. **03**(03) (2017)
5. Li, J., Li, C., Zhao, D., Gang, C.: Hyperspectral narrowbands and their indices on assessing nitrogen contents of cotton crop applications. In: Thenkabail, P.S., Layon, J.G., Huete, A. (eds.) Hyperspectral Remote Sensing of Vegetation, pp. 579–589 (2012)
6. Gitelson, A.A.: Nondestructive estimation of foliar pigment (chlorophylls, carotenoids, and anthocyanins) contents: evaluating a semianalytical three-band model. In: Thenkabail, P.S., Layon, J.G., Huete, A. (eds.) Hyperspectral Remote Sensing of Vegetation, pp. 141–165 (2012)
7. Curran, P.J., Dungan, J.L., Gholz, H.L.: Exploring the relationship between reflectance red edge and chlorophyll content in slash pine. Tree Physiol. **7**, 33–48 (1990)
8. Filella, I., Serrano, L., Serra, J., Penuelas, J.: Evaluating wheat nitrogen status with canopy reflectance indices and discriminant analysis. Crop Sci. **35**, 1400–1405 (1995)
9. Blackburn, G.A.: Quantifying chlorophylls and caroteniods at leaf and canopy scales: an evaluation of some hyperspectral approaches. Remote Sens. Environ. **66**, 273–285 (1998)

10. Gloud, K., Kevin, D., Winefield, C. (eds.) Anthocyanins: Biosynthesis, Functions and Applications, p. 330. Springer, NewYork (2008). https://doi.org/10.1007/978-0-387-77335-3
11. Janse, P.V., Deshmukh, R.R.: Hyperspectral remote sensing for agriculture: a review. Int. J. Comput. Appl. (0975–8887) **172**(7) (2017)
12. Jensen, J.R.: Remote Sensing of the Environment: An Erath Resource Perspective. Prentice-Hall, Upper Saddle River (2000)
13. Hunt, J., Ramond, E., Rock, B.N.: Detection in changes in leaf water content using near and mid-infrared reflectance. Remote Sens. Environ. **30**, 45–54 (1989)
14. Ustin, S.L., Roberts, D.A., Green, R.O., Zomer, R.J., Garcia, M.: Remote sensing methods monitor natural resources. Photon. Spectra **33**(N10), 108–113 (1999)

An Automated Model to Measure the Water Content of Leaves

I. A. Wagachchi[1], B. B. D. S. Abeykoon[1], R. D. K. Rassagala[1], D. P. Jayathunga[1], and T. Kartheeswaran[2](✉)

[1] Department of Computer Science and Technology, Uva Wellassa University, Badulla, Sri Lanka
[2] Department of Physical Science, Vavuniya Campus of the University of Jaffna, Vavuniya, Sri Lanka
karthees@vau.jfn.ac.lk

Abstract. Leafy product industries like tea, tobacco, Palmyra, green vegetables, Ayurveda productions are playing significant role to uplift the Sri Lankan economy. The current water content in the leaves is an essential factor for leafy productions to maintain their quality. Naked eye observation of an expert is the general method to identify the water content. The objective of this study is to introduce a novel and easy method to measure the water content of the detached plant leaves using digital image processing. As a result, a simple computational water content prediction method has been built using image processing techniques to obtain a quality output at the end of production processes. The findings of this study help to identify the water content without an expert in an efficient manner. First, the color images were captured in a controlled environment from the drying leaves and simultaneously the weight was measured traditionally to find the water loss. Several features of the images have been analyzed to find the best features which show a better correlation with the changes of the water content in the leaves. The textural and statistical features were extracted. The green matrix of the RGB image is taken for feature extraction to get the better results. The best features among the selected features are chosen through a correlation test. The classification was done with the K-Nearest Neighbor algorithm by training with the selected best features of the training set of images. Finally, a simple model was built using the significant features which have a relationship with the water content measurement. Accuracy has been achieved at a satisfactory level, and the model derived can be used to predict the water content of a particular green leaf (dried leaf also) through images. This model will be a turning point for measuring the water content of the leaves in the industries.

Keywords: Image processing · Water content · KNN

1 Introduction

Digital Image Processing is a subfield of computer science which focuses mainly on images and agriculture is the basis for the survival of the human. When

K. C. Santosh and R. S. Hegadi (Eds.): RTIP2R 2018, CCIS 1037, pp. 352–362, 2019.
https://doi.org/10.1007/978-981-13-9187-3_31

digital image processing techniques combined with the agriculture, it contributes profoundly to the economic growth of the world by adding value.

The leafy product quality is based on the moisture content in a production system [1] Therefore, measuring the water content of leaves is extremely important in leafy productions like tea (Camellia sinensis), tobacco (Nicotiana), Palmyra (Borassus flabellifer), ayurvedic products, spice products, leafy vegetables, etc. Hence, the relationship between water content and the visual changes of the leaves are identified in this study by measuring the weight of the plant leaves and by analyzing the features using digital image processing simultaneously.

Sri Lanka is famous for its high-quality tea products, and this high quality is directly related to a well-controlled withering process. During the process of withering, a series of physical and chemical changes take place when the moisture content in the fresh tea leaves disappears gradually [2]. Therefore, controlling the required water content is essential to gain the best quality. And also, in the Tobacco industry, the Tobacco can be cured through several methods, including Sun-cured mode. It is done by uncovering the leaves to the sun. Thus, it has to be controlled very carefully until it comes to the necessary moisture as the quality of the Tobacco leaves is highly considerable in the export market.

Spice productions are some export commodities which can be made especially using leaves such as curry leaves (Murraya koenigii), Pandan leaves (Pandanus amaryllifolius), etc. Drying of spices has been used for disinfestations, microbial decontamination, and long-term preservation [3]. But the leaves should not get dry entirely as there should be particular moisture content to have the necessary flavors and the fragrances which lead to a quality product. And also, small-scale industries like Palmyra industry gives value to Sri Lanka by their uniqueness in the products such as hats, mats, etc. Therefore, the leaves which use for the production purposes need to dry accordingly to the requirement of the final product. More drying leads to the brittleness of the products as they are not much flexible in production processes while less drying leads to fungal infections as well as less rigidity.

In Ayurveda treatments, leaves like Adathoda vasica (pawatta leaves), Vitex negundo (Nika leaves), Azadirachta indica (Neem leaves), etc. are used as medicinal herbs after drying to some extent. However, drying must be performed carefully to preserve the aroma, appearance and nutritional characteristics of the raw herbs as much as possible [4]. Also, there is a high value for the fresh fruits and leafy vegetables in the Sri Lankan market. If the freshness is high, the demand of the customers for them also considerably high.

Currently, in the industry, the required water content is identified by the experts. People may distinguish the small differences between similar colors or appearances, but they can hardly identify the specific levels [2]. So the correct amount of water percent in the leaves identified by the several experts can differ. Available methods of measuring the water content of the plant leaves do not have high accuracy, and they are not cost effective. Therefore, to find a solution, we focused on the similarities between the changes of the basic features and water content of the detached Pepper leaves (Capsicum) with the continuous drying under sunlight. The proposed method is using Digital Image Processing techniques. A software is used to analyze the pixel values of the images and gain the values for the basic features, and the classification is done using K-Nearest Neighbor algorithm.

Methodology. This research proposes a model to identify the water content of the leaves using digital image processing techniques. Therefore, anyone can identify the water level within the leaves without any expert and produce best quality products made from leaves.

The linear attenuation coefficient of fresh leaves is higher than dry leaves and the difference between fresh leaves and dry leaves indicates that they contain water in leaves [5]. Therefore, the Pepper leaves were selected to find the relationship between the weight reduction with the water loss and some visual features of the leaves for this research study as they show high variations with time while drying continuously after detaching from the wines. Further, as this is a first trial work the Pepper is selected as is widely available with zero cost. Therefore, as the first step randomly selected 30 pepper leaves were detached from the wines and the leaves were kept under sunlight from 10 am to 2 pm hung parallel to get the optimum sunlight. RGB images were captured while drying under sunlight using an image box which provides the same condition for all the leaves in the image capturing process. The previous analysis performed showed that both blue and red components present lack of information [1]. Therefore, for further analysis green components were taken from the RGB images. Other than that, the weight of the leaves was measured using an analytical balance at each instance the image was captured. After 2 pm to until 10 am on next day, the leaves were kept in the refrigerator (4 °C) as a controlled environment. After taking images for ten instances simultaneously as shown in Fig. 1. And Fig. 2., the feature extraction was done and the values for basic features were taken from the images using the software.

The image processing technique involves different processes such as image acquisition, image pre-processing, feature extraction and classification [6]. So the software was used to perform the preprocessing, image enhancement and feature extraction. Statistical software was used for the statistical analysis. Relationships were found between weight measurements got using the analytical balance equipment, and feature values got through software. Moreover, classification learner app was used to train the model. The values for the basic features described in Table 1 such as Mean, Median, Homogeneity, Skewness, Contrast, Energy, Entropy, Correlation, Min, Max, Standard Deviation, Variation, Kurtosis, IQR, and Range were taken for the images.

The plots were obtained using software for each separate reducing weight values of 30 leaves by taking the time as ten instances as shown in Fig. 3. Then plots were taken for the basic features separately for the 30 leaves. As it was difficult to compare the graphs keeping them in different plots, they were obtained taking the weight values and feature values together for one plot as shown in Fig. 4.

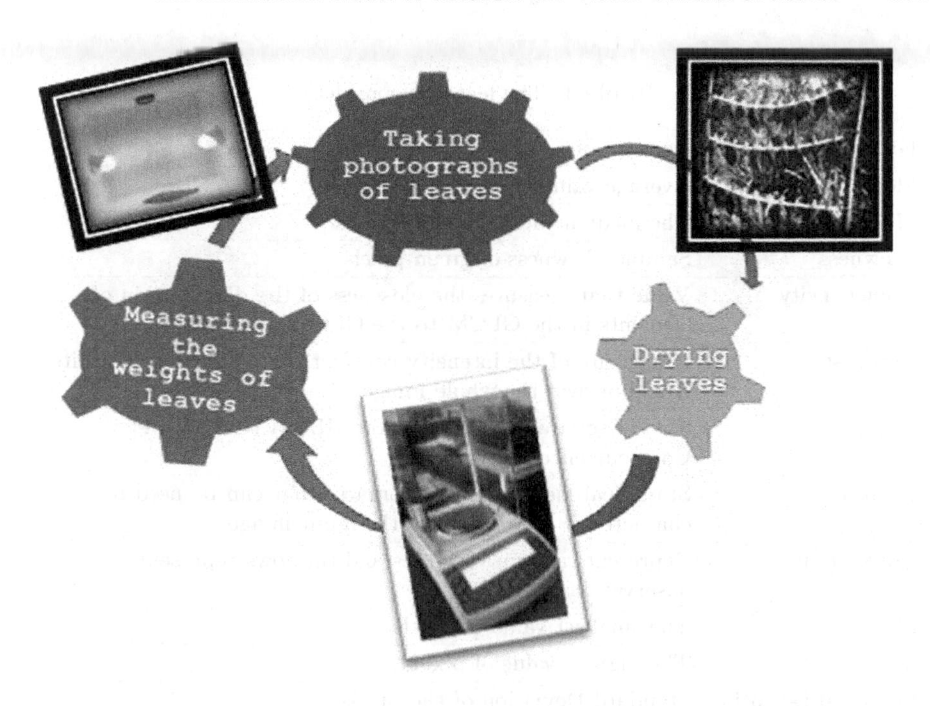

Fig. 1. The procedure continued for leaves

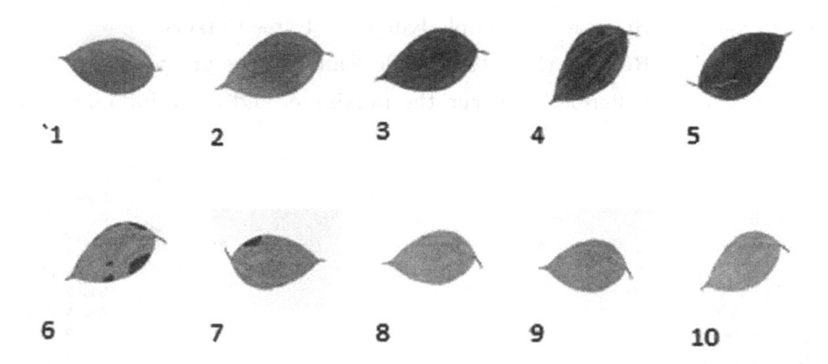

Fig. 2. Drying stages of a leave

Then plots were taken by separating for each leaf to get visually clear graphs so that each graph contain one leaf with weight and one feature for the ten instances in one plot. Randomly graphs of six leaves (1, 5, 10, 15, 20, and 30) were selected, as all the leaves have similar patterns. Other than that, graphs were obtained containing one leaf with weight and feature sets. After taking a visual understanding from some points of the graphs which show some similarity, statistical analysis was done using the data sets.

Table 1. The features considered

Feature	Description
Mean	Average value of green pixels
Median	The median value of green pixels
Skewness	Sample Skewness of green pixels
Homogeneity	Value that measures the closeness of the distribution of elements in the GLCM to the GLCM diagonal
Contrast	A measure of the intensity contrast between a pixel and its neighbor over the whole image
Energy	Sum of squared elements in the GLCM (Gray-Level Co-occurrence Matrix)
Entropy	Statistical measure of randomness that can be used to characterize the texture of the input image
Correlation	Represent random variables and the rows represent observations
Min	The smallest value of pixels
Max	The highest value of pixels
Standard Deviation	Standard Deviation of the pixels
Variance	The Variance of the elements
Kurtosis	Returns the sample Kurtosis of green pixels
IQR	Returns the Interquartile Range of the green values
Range	Difference between the maximum and the minimum of values

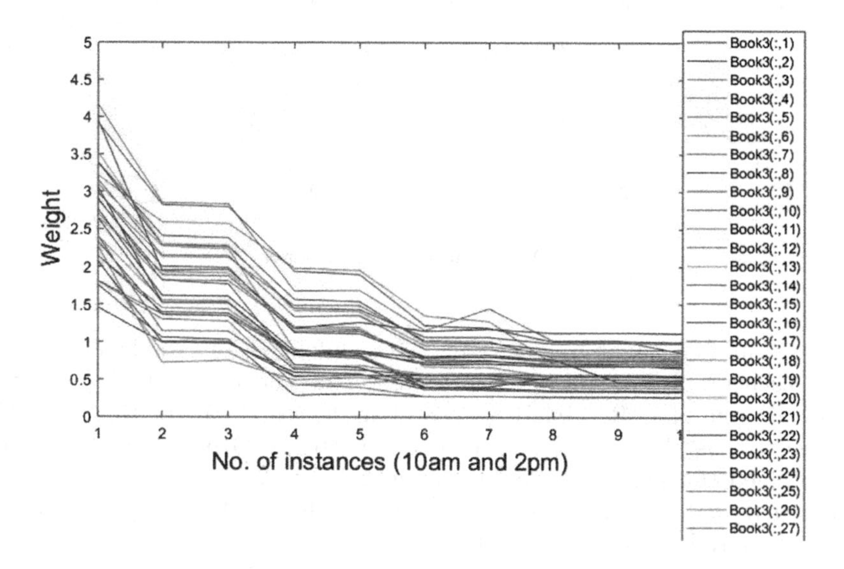

Fig. 3. The graph for changes in variation of leaves weights

Then the selected features were combined with each other to check whether there are relationships and hidden patterns exist. A model is built accordingly, using the relationships gained by the values obtained from the statistical analysis.

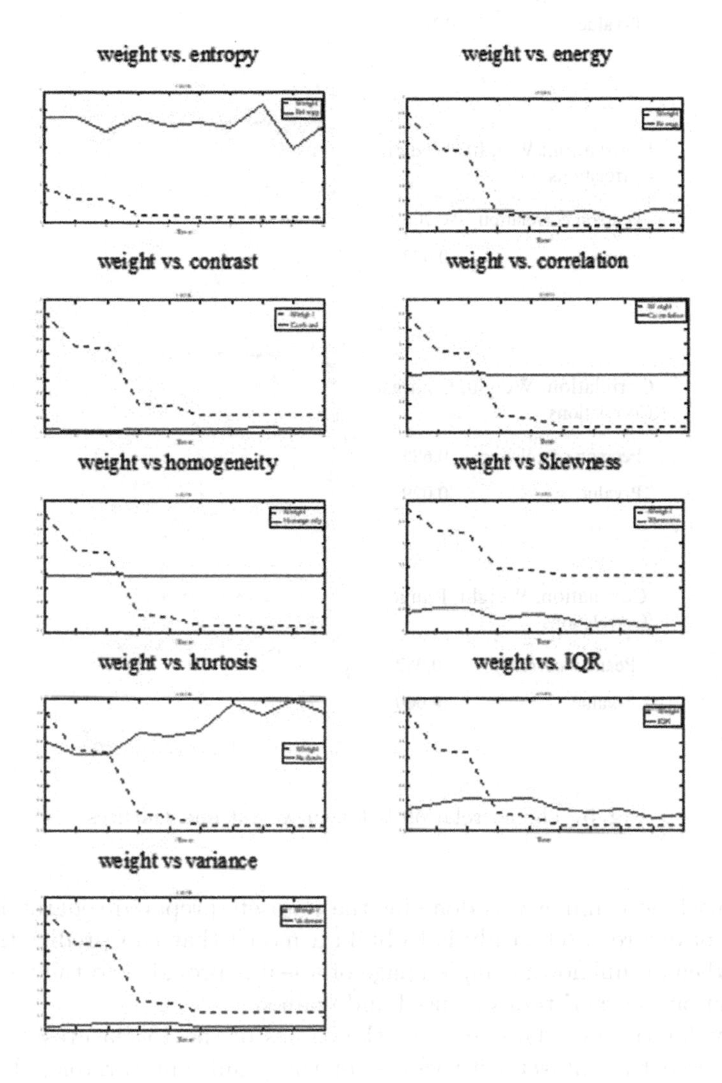

Fig. 4. The plots obtained for weight vs. features

The data collected using software and the data collected using the analytical balance for the leaf weights were taken into one data set. Then the regression analysis was done to find the relationship between weight and features. The most significant features were selected by considering Pearson Coefficient values calculated as mentioned in Fig. 5. The selected features were used to make the model by combining the features.

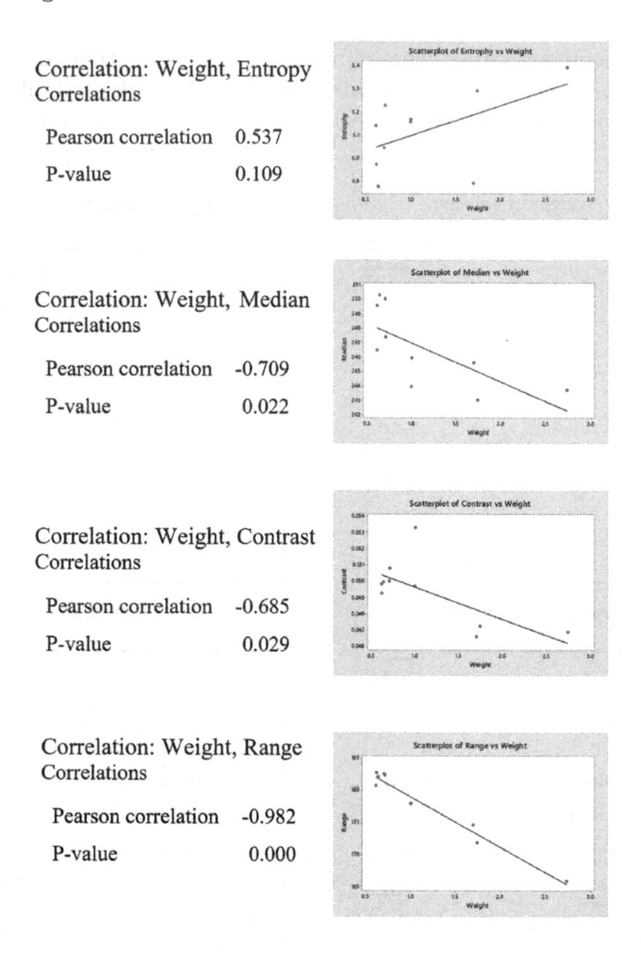

Fig. 5. The correlation between weight and features

The machine training was done for the datasets prepared separately as the objective of our research study is to build a model that can predict the water content when an unknown sample image of a leaf is provided to the model; The 'Classification Learner' tool was used and trained.

Firstly, for the raw data set with the values of the 15 features were taken and then used the dataset with values for the significant variables. Finally, it is trained with the dataset consisted of the values of the combinations of the significant variables. In here, the dataset is imported to the tool and selected the response variable and the predictor variable. The response variable is the water content and the predictor variables are the selected features. The algorithms such as Tree, Linear Discriminant, Quadratic Discriminant, KNN, and Ensemble were used to train the machine and to check the accuracy.

The training dataset of known data against the testing data set is given to the model, and the validation is done. It was done while training the dataset to remove the unusable values. In the machine training, the numbers of folds were selected to 75% of the dataset goes to train the model and the rest portion of 25% used to test the model.

2 Analysis

The dataset assessed by taking the ten instances, correlation analysis was done separately for each feature with weight. Then, the regression analysis was done to find the relationship between weight and features. The most significant features were selected as indicated in Fig. 6., by considering Pearson Coefficient values calculated. The selected features were used to make the model by combining the features as shown in Fig. 7. Sequential Sum of Squares in statistical software was used to calculate the R2 statistics. Adjusted R2 was used to compare models with different numbers of predictors. R2 always increased when we added a predictor to the model, even when there is no real improvement to the model and adjusted R2 value incorporated to determine the number of predictors in the model to choose the correct model.

General Linear Model: response versus time

Factor	Type	Levels	Values
time	fixed	10	1, 2, 3, 4, 5, 6, 7, 8, 9, 10

Analysis of Variance for response, using Adjusted SS for Tests

Source	Model DF	Reduced DF	Seq SS
time	9	9	133.2759
Median	1	1	2.6386
Contrast	1	1	0.3600
Correlation	1	1	0.2028
Energy	1	1	0.0058
Homogeneity	1	1	0.0105
Range	1	1	0.0550
Skewness	1	1	0.1664
IQR	1	1	0.0279
Kurtosis	1	1	0.2125
Median*Contrast	1	1	0.1032
Median*Correlation	1	1	0.2187
Median*Energy	1	1	0.0001
Median*Homogeneity	1	1	0.0871
Median*Range	1	1	0.4190
Median*Skewness	1	1	0.6937
Median*IQR	1	1	0.1667
Median*Kurtosis	1	1	1.4055
Contrast*Correlation	1	1	0.5701
Contrast*Energy	1	1	0.0067
Contrast*Homogeneity	1	1	0.0807
Contrast*Range	1	1	0.5020
Contrast*Skewness	1	1	1.5808
Contrast*IQR	1	1	0.5671
Contrast*Kurtosis	1	1	0.0812
Correlation*Energy	1	1	0.4851

Fig. 6. The seq SS values retrieved

```
S = 0.00620919   R-Sq = 100.00%   R-Sq(adj) = 99.99%

1. time
2. Median
3. Median*Kurtosis
4. Contrast*Skewness
5. Energy*Range
6. Median*Contrast*Kurtosis
7. Median*Energy*Kurtosis
8. Median*Homogeneity*Kurtosis
9. Contrast*Correlation*Kurtosis
10. Homogeneity*Range*IQR Median*Contrast*Correlation*IQR
11. Contrast*Energy*Homogeneity*Kurtosis
12. Energy*Range*Skewness*IQR
13. Median*Contrast*Homogeneity*IQR*Kurtosis
14. Median*Contrast*Homogeneity*Range*Kurtosis
15. Median*Contrast*Skewness
16. Correlation*Homogeneity*Range
```

Fig. 7. The summary of statistical analysis

3 Discussion

In this research, while drying the leaves under sunlight, the weight of the pepper leaves was reduced. Therefore, by the weight variation with drying of leaves, the green matrix of the images was selected to identify the variation of basic features of the images as the green component shows a high variation than red and blue components.

The amount of water content was obtained as follows: The reduction of the weight is stated in the dataset obtained while drying the leaves and ten readings can be seen when considering about a single leave. Then, it is assumed that the water content of the leave in the time of detaching it from the plant is 100% and it is 0% when the constant weight is gained when drying continuously. The constant weight can be assumed as the weight of the Carbonic compounds without water content. Finally, for further processing of the leave at the constant weight is deducted from the initial weight.

The accuracy of the dataset with water content values as the response variables and the feature values as the predictor variables was obtained as the first step and that accuracy was less than the dataset with water content as the response variable and the combination of feature values as the predictors. The accuracy of maximum 65.3% is obtained from the K-Nearest Neighbor algorithm as shown in Fig. 8 and the other algorithms gave an accuracy less than that for the dataset of 300 values which 270 values used for the training and 30 values used for the testing.

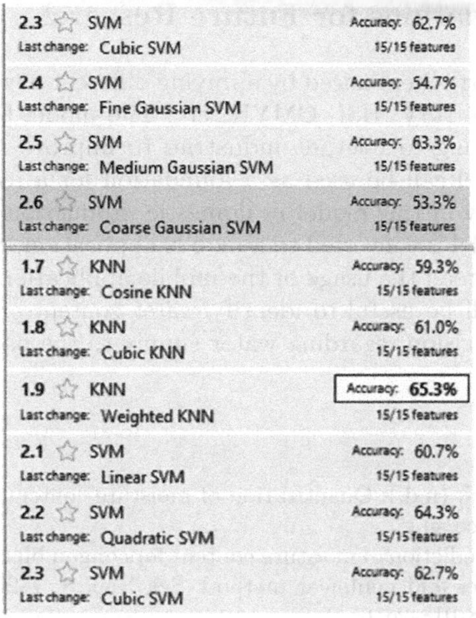

2.3 SVM Accuracy: 62.7%
Last change: Cubic SVM 15/15 features
2.4 SVM Accuracy: 54.7%
Last change: Fine Gaussian SVM 15/15 features
2.5 SVM Accuracy: 63.3%
Last change: Medium Gaussian SVM 15/15 features
2.6 SVM Accuracy: 53.3%
Last change: Coarse Gaussian SVM 15/15 features
1.7 KNN Accuracy: 59.3%
Last change: Cosine KNN 15/15 features
1.8 KNN Accuracy: 61.0%
Last change: Cubic KNN 15/15 features
1.9 KNN Accuracy: 65.3%
Last change: Weighted KNN 15/15 features
2.1 SVM Accuracy: 60.7%
Last change: Linear SVM 15/15 features
2.2 SVM Accuracy: 64.3%
Last change: Quadratic SVM 15/15 features
2.3 SVM Accuracy: 62.7%
Last change: Cubic SVM 15/15 features

Fig. 8. The testing is done by classification learner

4 Conclusion

The objective of this research study is to identify a relationship between the water content and the visual changes of leaves and to build a model. Thus, the model can predict the water content when an unknown sample image of a leaf is provided to the model.

The images of the plant leave dried in the sunlight were undergone into an image processing process, and finally, the data obtained through green pixels of them were used continuously for statistical analysis and the machine learning. Finally, the model obtained through the feature combinations was trained, and testing was done for several algorithms in software.

The model was trained, and the accuracy of maximum 65.3% was obtained from the K-Nearest Neighbour algorithm, and the other algorithms gave an accuracy less than that for the data set of 300 values which 270 values used for the training and 30 values used for the testing. This accuracy can be increased by providing a larger dataset with huge number of values.

Although this model was built using the software on a computer, it is not practicable to use in the industry as it is difficult to process each photograph using a computer in which the dedicated software is installed. Thus, as a further improvement, a mobile application can be build using the model build. Therefore, any person in the industry can use the facility to identify the water content in the leaves just by taking a photograph using their mobile phones to predict the amount of water.

5 Recommendations for Future Research

The model can be further enhanced by applying different new features with other color models such as HSV, HSI, CMYK, etc. The model build can be further enhanced by modifying to relevant industries to improve the accuracy. Other than that, this model can be used as a foundation for a mobile application to improve the usability of this model in domestic productions, small-scale industries, etc. If this model is converted to a mobile application, it can be added with localized GUIs to spread the usage of the mobile application in rural areas. The model build also will be useful to identify water content of live leave with the plant and to take decision regarding water supply to the particular plant.

References

1. Leiva, L., Acosta, N. (n.d.): Quantization of moisture content in Yerba Mate leaves through image processing
2. Liang, G., et al.: Prediction of moisture content for congou black tea withering leaves using image features and nonlinear method. Sci. Rep. **8**, 1–22 (2018). https://doi.org/10.1038/s41598-018-26165-2
3. Schweiggert, U., Carle, R., Schieber, A.: Conventional and alternative processes for spice production-a review. Trends Food Sci. Technol. **18**, 260–268 (2007)
4. Crivelli, G., Nani, R.C., Di Cesare, L.F.: Influence of processing on the quality of dried herbs. In: Atti VI Giornatescientifiche SOI, Spoleto, 23–25 April 2002, vol. II, pp. 463–464 (2002)
5. Pattanashetti, I.I., Galagali, P.M.N.: Attenuation coefficient and water content determination of almond leaves using beta radiation. Int. J. Innovative Sci. Eng. Tech. **3**(3), 569–573 (2016)
6. Robert, B.R.: Methods for measuring water status and reducing transpirational water loss in trees. J. Arboric. **13**, 56–61 (1987)

Estimation of Crop Chlorophyll Content by Spectral Indices Using Hyperspectral Non-imaging Data

Pooja Vinod Janse, Ratnadeep R. Deshmukh(✉), Jaypalsing N. Kayte(✉), and Priyanka U. Randive(✉)

Department of CS and IT, Dr. B. A. M. University, Aurangabad, India
puja.janse@hotmail.com, rrdeshmukh.csit@bamu.ac.in, jaypalsing@gmail.com, priyankarandive7@gmail.com

Abstract. Leaf chlorophyll plays very important role in photosynthesis process which enables conversion of light energy into biochemical energy which is directly related with plant stress, relationship between environment and plants, plants nutritional stress which is important for agriculture management. Conventional methods for estimating leaf chlorophyll content are destructive, time consuming and difficult for taking repeated measurement. So we have estimated leaf chlorophyll content using hyperspectral non-imaging data. Leaf reflectance has been captured by using FieldSpec 4 Spectroradiometer. Different spectral indices were applied for estimation of chlorophyll content and to develop non-destructive model. Spectral indices which have been applied over spectral reflectance have been reported as sensitive to chlorophyll content present in leaf and the correlation between chlorophyll content and indices shows medium to good R^2 values.

Keywords: Chlorophyll content · Spectral indices · Hyperspectral remote sensing · Spectral reflectance

1 Introduction

In study of vegetation, it have been seen that leaf chlorophyll content plays very important role in photosynthesis process. Photosynthesis process provides energy in plants and also it is source of carbon for all organic compounds in plant. Plants having lower chlorophyll concentration will face problem of non-optimal photosynthesis. This will not only effects on plant growth but also reduces carbon fixation.

In this paper we focused on remote measurement of leaf chlorophyll content. High Performance Liquid Chromatography (HPLC) or leaf content extraction using organic material using wet chemistry method are traditional methods for measurement of leaf chlorophyll content. But recently, monitoring plant's physiological status with the help of spectral reflectance measurement offers number of

K. C. Santosh and R. S. Hegadi (Eds.): RTIP2R 2018, CCIS 1037, pp. 363–371, 2019.
https://doi.org/10.1007/978-981-13-9187-3_32

advantages over traditional methods. The most important advantages are reliability, sensitivity, simplicity etc. this method is non-destructive, inexpensive, quick and saves great deal of manual labor. Recent developments in hyperspectral remote sensing makes us able for crop monitoring applications.

The reflectance of variety of crops specifies crop characteristics like physical structure, biochemical compositions, water contents, crop yield, chlorophyll content, nitrogen content, plant stress, moisture present in plant, carotenoid pigments etc. Solar light is used as source of energy by plants for photosynthesis process, with reference to electromagnetic spectrum visible region provides reflectance information of chlorophyll content, Near Infrared (NIR) region is suitable for estimating biomass or Leaf Area Index (LAI), and other biochemical properties. Shortwave Infrared (SWIR) region provides us reflectance information of water content. Therefore some spectral indices have been used for estimating leaf chlorophyll content [1–3].

2 Spectral Features of Leaf Reflectance

Leaves represents main surface of plants where energy and gases are exchanged. The general spectral reflectance of leaf is shown in following Fig. 1.

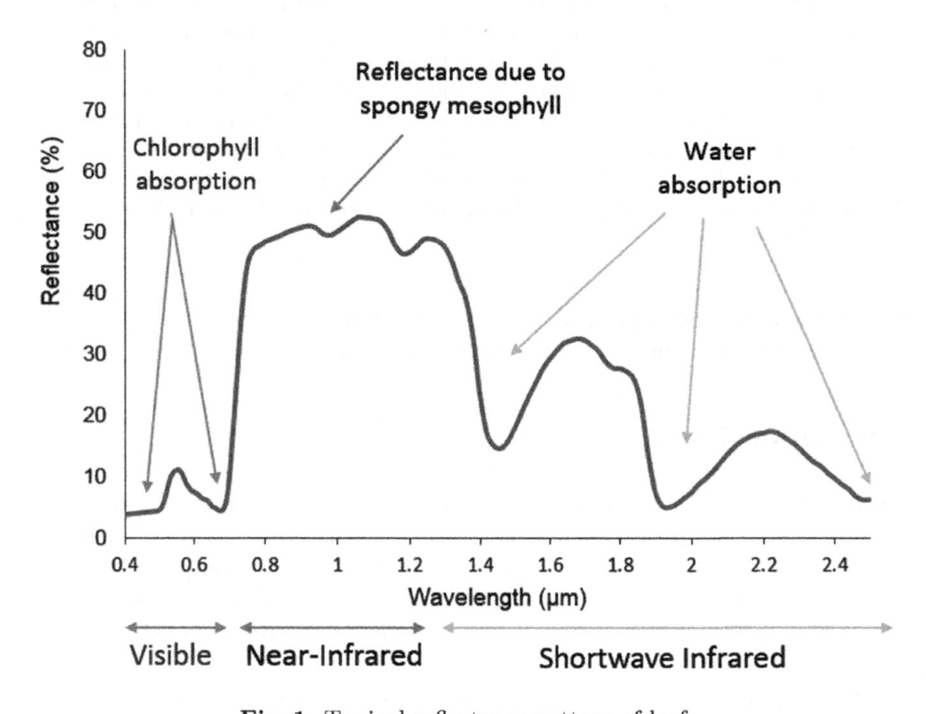

Fig. 1. Typical reflectance pattern of leaf.

Chlorophyll a and b, carotenoids, polyphenols and xanthophyll are the main light absorbing pigments in leaf. Maximum energy absorption by chl a observed in 410–430 nm and 600–690 nm, whereas chl b shows absorption in 450–470 nm. Between 440 and 480 nm carotenoid absorbs most efficiently. Other leaf pigments and cellulose are transparent in nature so they absorbs very less light energy therefore reflectance and transmittance reaches to higher value in Near Infrared (NIR) region because air-cell-water interfaces are responsible for it. In Shortwave-Infrared (SWIR) region, water and other foliar constituents affects leaf optical properties. Maximum absorption by water occurs at 1450 and 1940 nm and secondary features at 960, 1120, 1540, 1670 and 2200 nm [4,5].

3 Data Collection

We have collected spectral reflectance of Wheat and Jowar in the spectral range of 350–2500 nm with the help of ASD FieldSpec 4 Spectroradiometer (Analytical Spectral Device) in a laboratory in controlled condition. The Field of View (FOV) was set to both 1o and 8o, and the distance between head of the Spectroradiometer and leaf sample was kept 10 cm. For optimizing the ASD FieldSpec 4 Spectroradiometer, a Spectralon white reference panel have been used to get reference signal preceding to measure leaf reflectance. 10 spectral signatures of each single leaf have been recorded and mean spectra is used for further analysis. We have collected spectral signatures of Wheat, Bajara, Jowar and Maize. Viewspec pro was used for converting reflectance values in ASCII format. Spectral signatures of all four crops are shown in following Fig. 2.

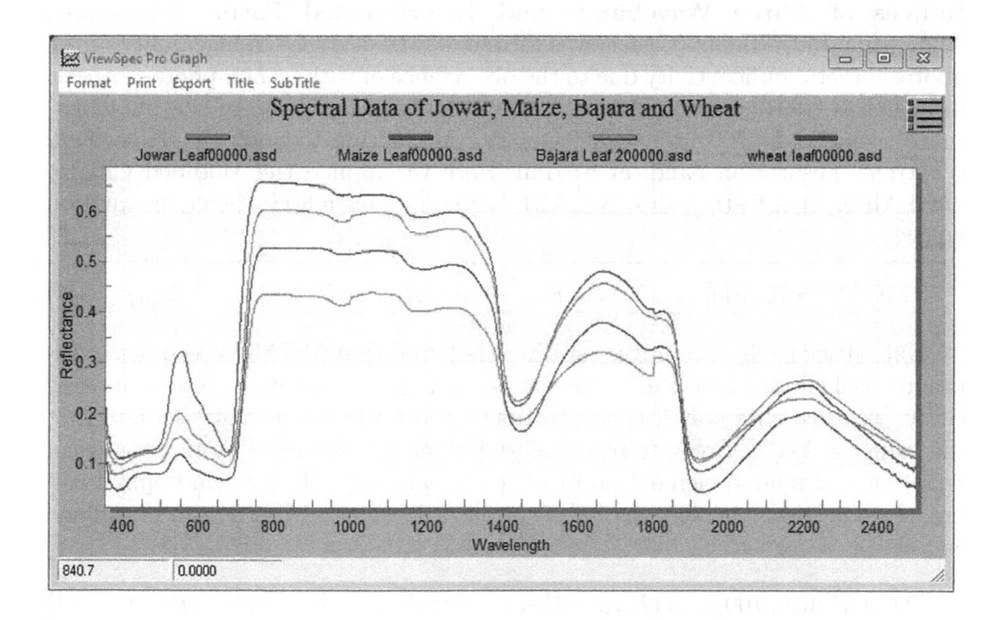

Fig. 2. Spectral signatures of Jowar, Maize, Bajara and Wheat.

4 Materials and Methods

4.1 Spectral Vegetation Indices for Estimation of Chlorophyll Content

Normalized Difference Vegetation Index. The best extensively used and recognized vegetation index is the normalized difference vegetation index (NDVI). It shows difference between chlorophyll's extreme absorption in red region and leaf cellular structure causes the extreme reflection in the IR region [6]. The formula for NDVI is shown in following Eq. 1:

$$\text{NDVI}\ [670, 800] = (R_{800} - R_{670})/(R_{800} + R_{670}) \tag{1}$$

where R = reflectance on the given wavelength.

Simple Ratio Indices. Indices of simple ration are used to compare among the peak of reflection and peak of absorption by chlorophyll content which states that they are subtle to variations in chlorophyll content. Simple ratio indices are subjective to environmental influences, like soil and cloud. Therefore Modified Simple Ratio (MSR) have been calculated to remove this drawback.

$$\text{MSR}\ [670, 800] = [(R_{800}/R_{670}) - 1]/[(R_{800}/R_{670}) + 1]^{\frac{1}{2}} \tag{2}$$

Indices of Three Wavebands and Incorporated Form. Chlorophyll Absorption Ratio Index (CARI) was formulated to decrease changeability of the photosynthetic radioactivity due to the occurrence of various non photosynthetic ingredients. CARI uses bands which shows lowest absorption of the photosynthetic elements at 700 nm and 550 nm, in combination through the chlorophyll a extreme absorption band, at 670 nm. Here we applied the Modified Chlorophyll Absorption Ratio Index (MCARI) defined by Daughtry, shown in equation below:

$$\text{MCARI}\ [670, 700] = [(R_{700} - R_{670}) - 0.2(R_{700} - R_{550})](R_{700}/R_{670}) \tag{3}$$

But on the basis of literature study, it is found that MCARI was quiet sensitive to background reflectance properties, non-photosynthetic components specially with low chlorophyll concentrations and it was problematic to interpret the values at low LAI. So, to recompense for the differences of characteristics of reflectance of related element and to rise the sensitivity by low chlorophyll values, the transformed chlorophyll absorption ratio index (TCARI) can be defined as follows:

$$\text{TCARI}\ [670, 700] = 3[(R_{700} - R_{670}) - 0.2(R_{700} - R_{550})](R_{700}/R_{670}) \tag{4}$$

Daughtry found that when MCARI in combination with a soil line vegetation index like optimized soil-adjusted vegetation index, the sensitivity to the primary reflectance properties of soil can be decreased. OSAVI is member of the soil-adjusted vegetation index family and is well-defined by equation shown below:

$$\mathrm{OSAVI}\,[670, 800] = [(1 + 0.16)(R_{800} - R_{670})]/(R_{800} + R_{670} + 0.16) \quad (5)$$

So, the two combined formulae of three reflectance wavebands can be well-defined as

$$\frac{\mathrm{TCARI}}{\mathrm{OSAVI}} = \frac{3\{[(R_{700} - R_{670}) - 0.2(R_{700} - R_{550})]\,(R_{700} / R_{670})\}}{[(1+0.16)(R_{800} - R_{670})] / (R_{800} + R_{670} + 0.16)}$$

$$\frac{\mathrm{MCARI}}{\mathrm{OSAVI}} = \frac{[(R_{700} - R_{670}) - 0.2\,(R_{700} - R_{550})]\,(R_{700}/ R_{670})}{[(1+0.16)(R_{800} - R_{670})] / (R_{800} + R_{670} + 0.16)}$$

Fig. 3. Two combined formulae of reflectance wavebands.

5 Results and Discussion

5.1 Preprocessing of Spectral Data

First and Second derivative transforms are most commonly used methods for preprocessing of spectral data. Derivative transforms reduces or eliminates background interference and atmospheric scattering effects. The first derivative transform can partially removes linear and quadratic background noise and second derivative can fully eliminates linear background noise. The first and second derivative spectra of spectral signature derived from ViewSpec Pro is shown in following Figs. 4 and 5.

5.2 Spectral Indices

The spectral indices which were applied on spectral reflectance of various crops are sensitive to leaf chlorophyll content. We have applied both original and revised spectral indices on data. It have been observed that reflectance at 750 nm and 705 nm are best suited for chlorophyll estimation than that of 670 nm and 800 nm.

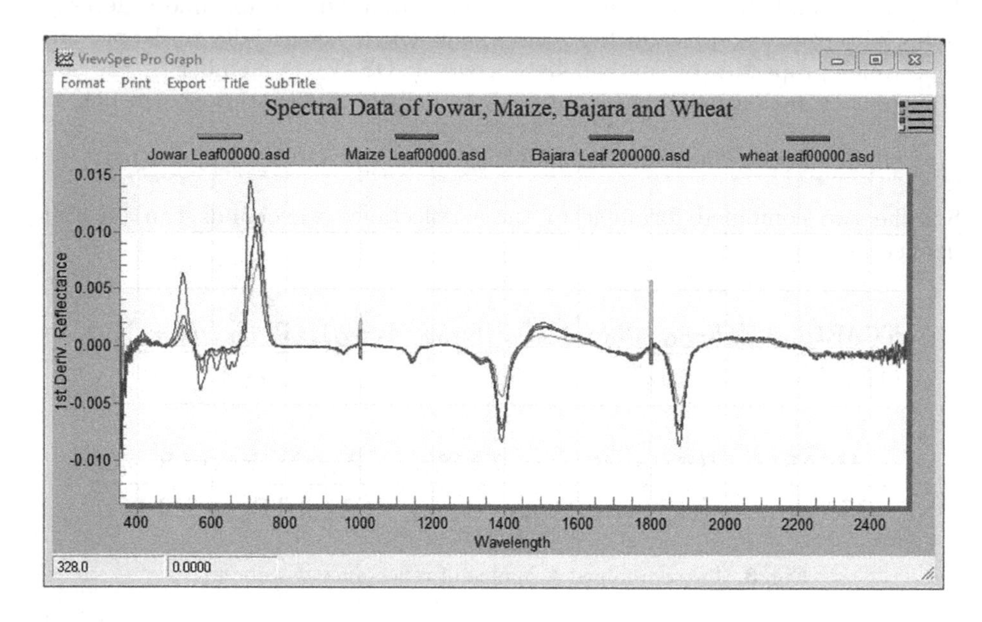

Fig. 4. First derivative of spectral signatures

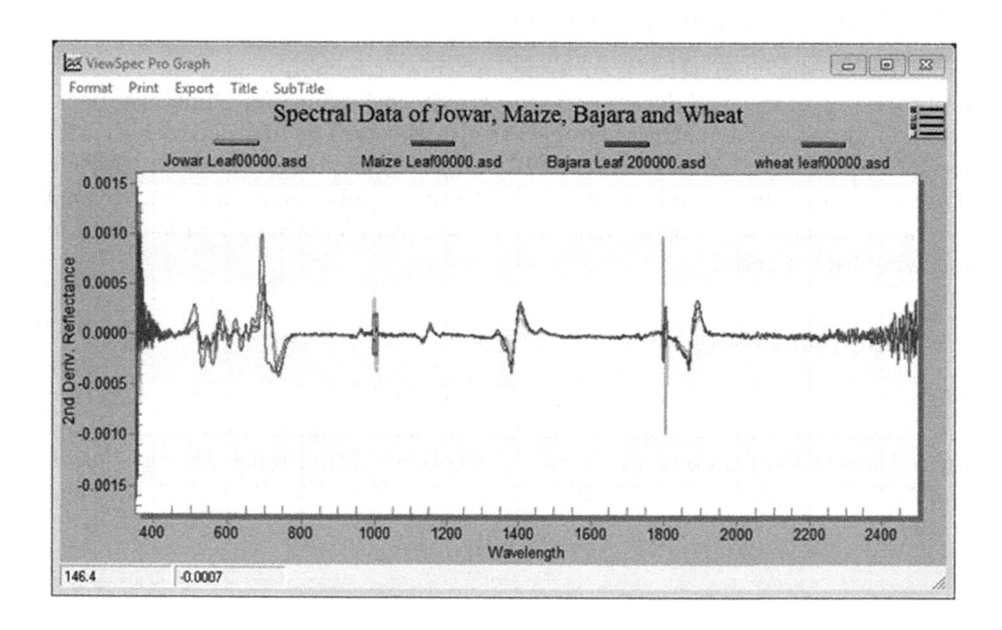

Fig. 5. Second derivative of spectral signatures

Table 1. Original Spectral Indices value

Indices	Bajara	Wheat	Maize	Jowar
NDVI	0.6774252	0.768859871	0.666336596	0.648702196
MSR	1.6868366	2.261767622	1.636424672	1.547836477
MCARI	0.092907	0.070866413	0.286475784	0.078591781
TCARI	0.278721	0.212599238	0.859427353	0.235775343
OSAVI	0.6430446	0.703547306	0.637853903	0.577700863
TCARI/OSAVI	0.4334306	0.302171262	1.347491188	0.408121998
MCARI/OSAVI	0.1444769	0.100723754	0.449163729	0.136040666

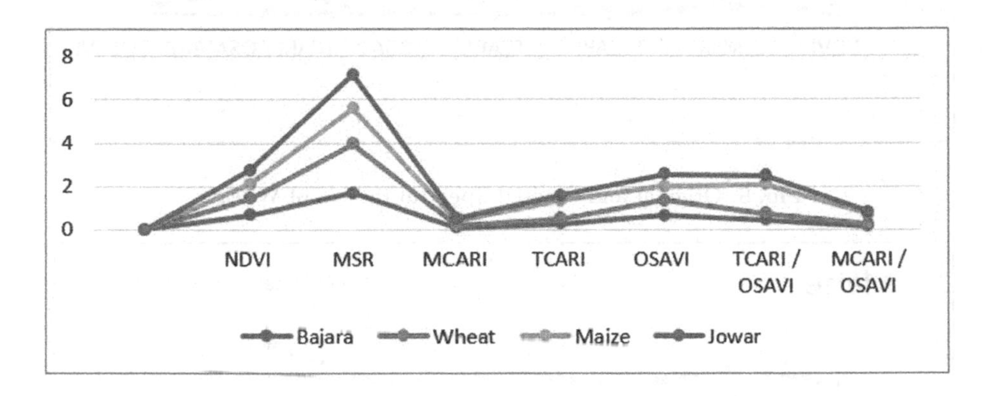

Fig. 6. Graphical representation of Original Spectral Indices

Table 2. Revised Spectral Indices values

Indices	Bajara	Wheat	Maize	Jowar
NDVI	0.46982	0.546243	0.316630416	0.430448
MSR	0.91257	1.146813	0.541852601	0.806624212
MCARI	0.82406	0.956283	0.461782908	0.499685046
TCARI	2.47218	2.868848	1.385348723	1.499055
OSAVI	0.45363	0.50935	0.314	0.392172
TCARI/OSAVI	5.44925	5.63234	4.411844208	3.822419
MCARI/OSAVI	1.81642	1.877447	1.470614736	1.27414

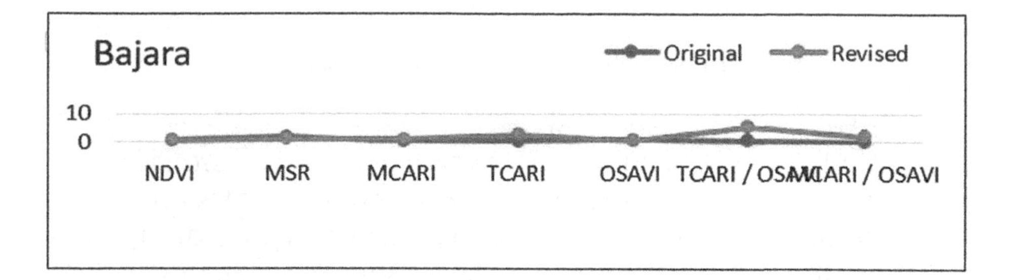

Fig. 7. Original and Revised Spectral Indices of Bajara

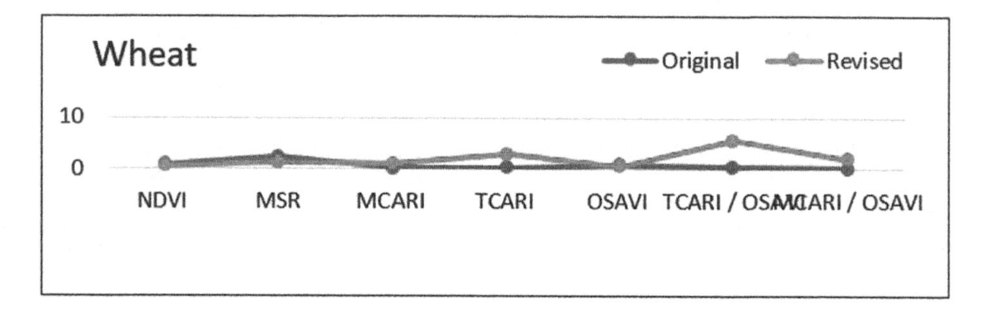

Fig. 8. Original and Revised Spectral Indices of Wheat

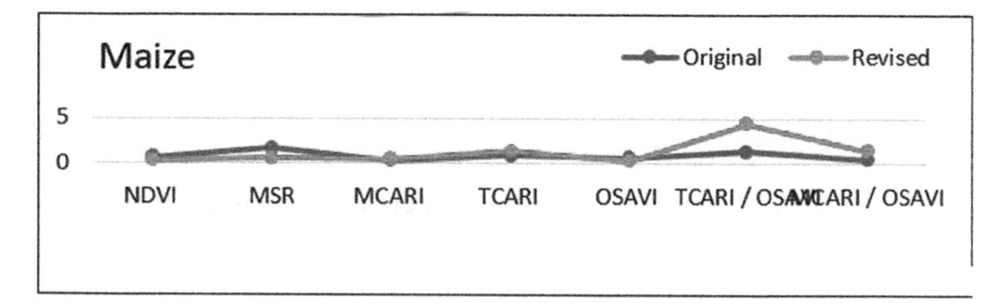

Fig. 9. Original and Revised Spectral Indices of Maize

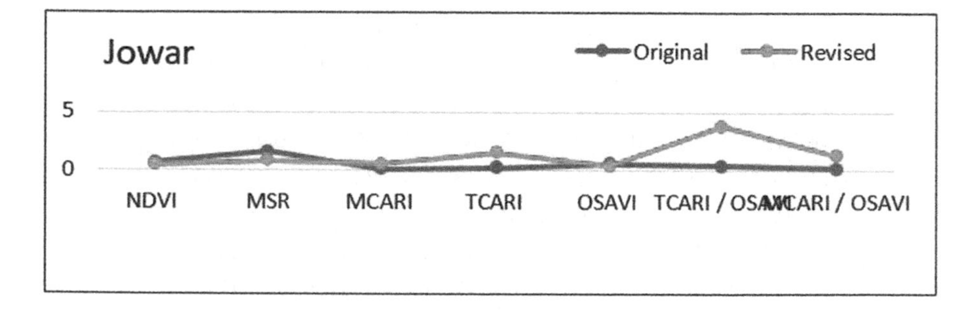

Fig. 10. Original and Revised Spectral Indices of Jowar

6 Conclusion

In this paper we have calculated seven different vegetation indices for estimation of content of chlorophyll. These vegetation indices are revised for getting better results. We have used 705 nm and 750 nm reflectance values which have proven that these both bands are best for estimation of chlorophyll content than 670 nm and 700 nm. It have also been observed that spectral indices which have been applied over spectral reflectance have been reported as sensitive to content of leaf chlorophyll and the correlation between content of chlorophyll and indices shows medium to good R^2 values.

Acknowledgement. The authors are thankful to UGC for providing BSR fellowship as a financial support for this research work and also for formation of UGC SAP (II) DRS Phase-I and Phase-II, the authors also extends our deepest thanks to DST-FIST for their support for this work with consent no. SR/FST/ETI-340/2013 to Dept. of CS and IT, Dr. B. A. M. University, Aurangabad. The authors are also thankful to Authorities of Department and University for providing the setup and necessary backing to the research.

References

1. Janse, P.V., Deshmukh, R.R.: Hyperspectral remote sensing for agriculture: a review. IJCA **172(7) (2017)**
2. Sahoo, R.N., Ray, S.S., Manjunath, K.R.: Hyperspectral remote sensing of agriculture. Curr. Sci. **108**(5), 848–859 (2015)
3. Thenkabail, P.S., Smith, R.B., Pauw, E.D.: Hyperspectral vegetation indices and their relationships with agricultural crop characteristics. Remote Sens. Environ. **71**, 158–182 (2000)
4. Randive, P.U., Deshmukh, R.R., Janse, P.V., Kayte, J.N.: Study of detecting plant diseases using non-destructive methods: a review. Int. J. Emerg. Trends Technol. Comput. Sci. **7**(1), 66–71 (2018)
5. Martinez, J., García, C.A.R.: Estimation of chlorophyll concentration in maize using spectral reflectance. Int. Arch. Photogrammetry, Remote Sens. Spat. Inf. Sci. XL-7/W3 (2015)
6. Perez-Patricio, M., et al.: Optical method for estimating the chlorophyll contents in plant leaves. Sensors **18**, 650 (2018)
7. Wu, C., Niu, Z., Tang, Q., Huang, W.: Estimating chlorophyll content from hyperspectral vegetation indices: modelling and validation. Elsevier, pp. 1230–1241 (2008)

MFDS-m Red Edge Position Detection Algorithm for Discrimination Between Healthy and Unhealthy Vegetable Plants

Anjana Ghule[1(✉)], R. R. Deshmukh[2], and Chitra Gaikwad[1]

[1] IT Department, GEC, Aurangabad, India
anjanaghule@gmail.com, gaikwadchitra@gmail.com
[2] CS and IT Department, Dr.B.A.M. University, Aurangabad, India
rrdeshmukh.csit@bamu.ac.in

Abstract. Spectral Reflectance of crop shows very distinguished sensitivity in spectral regions according to biophysical and biochemical parameters. Red Edge Position is the inflection point in the red edge region of electromagnetic spectrum which is between 680–780 nm. This is sensitive indicator of crop health. Red Edge Position is used to discriminate between healthy and unhealthy plants. Analytical Spectral Devices (ASD) Fieldspec spectroradiometer instrument having spectral range from 350 nm to 2500 nm, was used to collect lab spectra of vegetable plants. An algorithm is proposed based on a Maximum First Derivative Spread – mean and its reflectance magnitude in Red Edge Region. Maximum First Derivative Spread-mean (MFDS-m) algorithm is proposed to detect Red Edge Position which will be further used to discriminate healthy and unhealthy plants. Results are compared with Four Point Linear Interpolation, Extrapolation and Maximum First Derivative Techniques.

Keywords: MFDS-m · Spectral Reflectance · Hyperspectral remote Sensing · Red Edge Position

1 Introduction

Major part of a plant is covered with leaves. Leaf of a plant is involved in photosynthesis. We can say that leaf is representative of crop canopy. Leaves are prime indicators of plant health and direct impact on eco system health can be noticed.

Hyperspectral remote Sensing is collection of reflectance of object surface over various narrow bands of electromagnetic spectrum. One of the important areas of application of Hyperspectral remote Sensing is monitoring of vegetation and vegetation health across the globe. Characterization and quantification of vegetation parameters has gained importance. Multispectral remote Sensing is used for monitoring and study of vegetation, but there are limited broad bands. To study vegetation characteristics in detail large number of narrow bands is required. Hyperspectral remote sensing provides a good number of narrow bands, so it is used for study of crop characteristics. For the experimentation Lab Spectra of healthy and infected plant leaves was collected.

Spectral properties depend upon reflectance and transmittance value. Spectral characteristics of vegetation can be described in major four regions, visible region is

K. C. Santosh and R. S. Hegadi (Eds.): RTIP2R 2018, CCIS 1037, pp. 372–379, 2019.
https://doi.org/10.1007/978-981-13-9187-3_33

having spectral range between 400–700 nm. Major light absorbing pigments are Chlorophyll a (*Chl a*) and Chlorophyll b (*Chl b*). Maximum absorption of *Chl a* is observed over blue band region between 410–430 nm and Green Band region of 600–690 nm range. *Chl b* shows maximum absorption in 450–470 nm of Blue Band region. Cartonoids shows absorption between 440–480 nm. Brown pigments are observed when the leaf is losing moisture drying to dead. Intensity decreases from blue region to red region.

In NIR and FNIR region between 700–1300 nm, leaf pigment and cellulose is almost transparent, so absorption is very low. SWIR region 1300–2500 nm largely influenced by water contents and biochemical properties. Protein, cellulose, lignin and starch also influence SWIR region [1].

Sudden change in leaf reflectance is observed between 680 and 780 nm. This region of inflection is known as Red Edge Position. This region is affected by bio-physical and biochemical properties of a plant [2, 3].

2 Experimental Design and Data Sets

2.1 Experimental Design

Field data is collected from a farm where different vegetables are grown. Experiment farm is near to Aurangabad city, Maharashtra, India. It is located at Lat 19.85 and Long 75.85. Total area of farm is around 100 Acres and vegetables grown, are further supplied to the city. Four vegetables Brinjal, Cluster Beans, Okra and Long Beans were planted in the farm. Sample collection was done in the month of August 2016. Leaves of Brinjal and Long Beans were collected for experimentation. Total 25 samples of both healthy and unhealthy plants were collected from five different locations of the farm. Hand-picked and clipped leaves were kept in polythene bags. Leaf spectra were measured within 3 h after collection [4].

2.2 Leaf Spectra Measurements

Leaf spectral reflectances were acquired with an ASD FieldSpecPro spectroradiometer, in the Lab of Computer Science and Information Technology Department of Dr. Babasaheb Ambedkar Marathwada University, Aurangbad. ASD FieldSpecPro provides measurements in the 350–2500 nm spectral range and 1 nm sampling step. Spectral measurements were taken in the dark room using a fibre optic with a 8° field of view. The sensor was kept at about 15 cm above the leaf blade which was covering the complete leaf. To illuminate the target A 50 W halogen lamp used above the sensor. White reference scans were taken and calibration was done. Five iterations of spectral measurements of each healthy and unhealthy leaves were collected using RS3 software. ASD files were converted to data files using ViewSpec-Pro software for further experimentations.

3 Algorithms for Red Edge Position

3.1 Maximum First Derivative Spectrum

Sudden rise in the red region is observed in the vegetation spectra. This region with maximum slope shows maximum first derivative and this is defined as Red Edge Position. Using first-difference transformation of the reflectance spectrum, First Derivative is calculated. Here λ and $R_{\lambda i}$ represents wavelength and reflectance at i^{th} wavelength [3].

Figure 1.(a) Shows spectral reflectance of Healthy and Unhealthy leaves of Brinjal and Cluster Beans in the range of 350–2500 nm. Figure 1.(b) shows First Derivative of the same samples in the range of 680–780 nm.

Maximum of FDR in the Red Edge Region is considered as REP.

$$MFDR_{(\lambda i)} = \left(R_{\lambda(j+1)} - R\right)/\Delta\lambda \tag{1}$$

This method finds REP by considering total spread of First Derivative Curve.

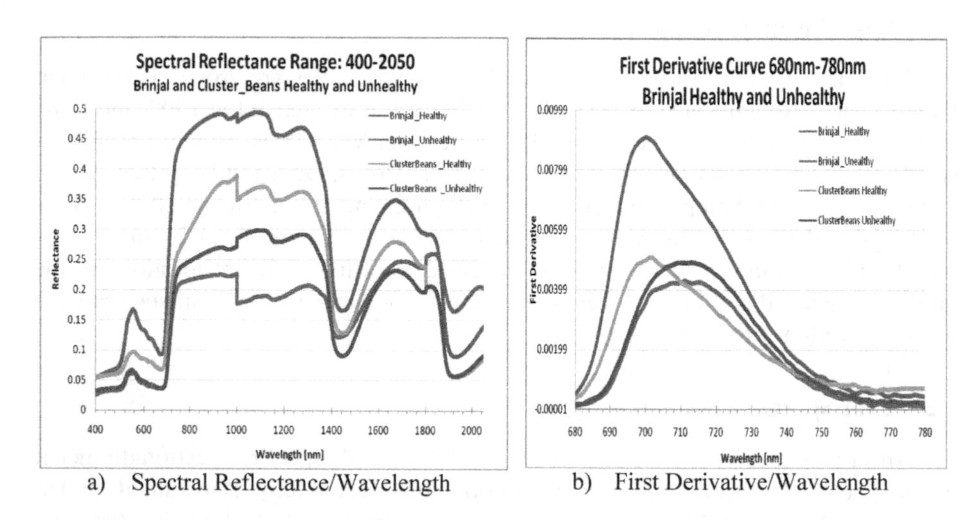

a) Spectral Reflectance/Wavelength b) First Derivative/Wavelength

Fig. 1. (a) Spectral Reflectance of healthy and unhealthy plants of Brinjal and Cluster Beans in range of 350 nm–2500 nm X-axis- Wavelength in nm and Y-axis- Reflectance (b) First Derivative curve of healthy and unhealthy plants of Brinjal and Cluster Beans in range of 680 nm–780 nm X-axis- Wavelength in nm and Y-axis- Magnitude of First Derivative.

3.2 Linear Interpolation Technique

It is observed that maximum reflectance is in the NIR region at around 780 nm and minimum reflectance in the Chlorophyll absorption which is around 680 nm. Considering these two as important regions, four points are identified. These four wavebands are 673 nm, 700 nm, 740 nm and 780 nm. This is a two step computation [5].

1. REP computation of reflectance using linear interpolation

$$R_{REP} = \frac{R_{780} + R_{673}}{2} + R_{673} \tag{2}$$

R_{780} and R_{673}: Reflectance at 780 nm and 673 nm wavelengths respectively

2. Calculation of the Red Edge Position (REP):

$$REP = 700 + (740 - 700) * \left[\frac{R_{REP} - R_{700}}{R_{740} - R_{700}}\right] \tag{3}$$

700 and 740 are constants of interpolation.

This method is based on fixed four pre-defined points. But it is observed that First Derivative curve of does not always show same characteristics. From Fig. 2.(c) and (d) we can observe that start and end of the curve do not show unique characteristics. Curve shift is observed in unhealthy plants with respect to healthy plants.

3.3 Linear Extrapolation Technique

A modified approach of Linear Interpolation is Linear Extrapolation technique [6]. Two regions are selected Far Red region i.e. (680–700 nm) and second one is NIR region (725–780 nm). First Derivative of the reflectance between 680 nm and 780 nm is taken. Two points on the far-red (680–700 nm) and two points on the NIR (725–780 nm) are selected. Two straight lines are drawn on edges of the First Derivative curve. The REP is calculated at the point of intersection of two straight lines. Calculations are done in two steps:

$$FarRedLineFDR = (m_1\lambda + c_1) \tag{4}$$

$$NIRLineFDR = (m_2\lambda + c_2) \tag{5}$$

Here *m* represents slop and *c* represents the intersection point. The wavelength at the point of intersection is Red Edge Position which is as below

$$REP = \frac{-(c_1 - c_2)}{(m_1 - m_2)} \tag{6}$$

REP is calculated using four coordinates two from 680–700 and two from 725–780 region. m1 and c1 are calculated from two coordinates of Far Red region and m2 and c2 are calculated from NIR region.

3.4 Proposed Algorithm: Maximum First Derivative Spread-Mean (MFDS-Mean)

It is observed that we do not get double peak in the First Derivative of every crop. So Interpolation and Extrapolation techniques may not be required in every case for finding REP. Figure 2 we can observe, only one peak in the First Derivative curve.

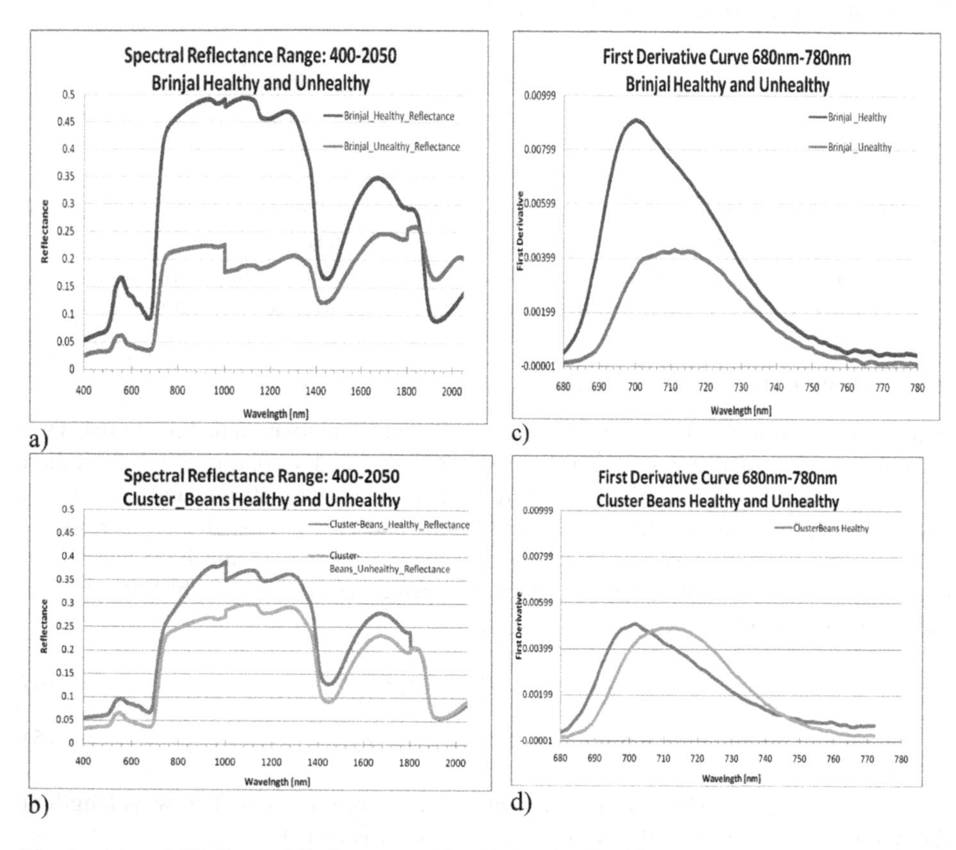

Fig. 2. (a) and (b) Spectral Reflectance of healthy and unhealthy plants of Brinjal and Cluster Beans in range of 350 nm–2500 nm X-axis- Wavelength in nm and Y-axis- Reflectance (c) and (d) First Derivative curve of healthy and unhealthy plants of Brinjal and Cluster Beans in range of 680 nm–780 nm X- axis- Wavelength in nm and Y-axis- Magnitude of First Derivative

This algorithm is based on Maximum First Derivative method. It is observed that we do not get a single point of maximum derivative. We get maximum value of First Derivative, over a range of wavelengths. This algorithm finds the REP from the mean wavelength of Maximum First Derivative Spread. Further REP and Reflectance at the REP is compared to discriminate healthy and infected vegetables.

MFDS-m - Red Edge Position Detection Algorithm:

1. Mean of spectral measurements is calculated
2. First Derivative transform of 680nm to 780 nm : FD $_{\lambda i = 680\text{-}780nm}$
3. Mean of First Derivative was calculated separately healthy and unhealthy vegetable

Here λ represents wavelength and $R_{\lambda i}$ Represents Reflectance at ith waveband

Input : 1) Feature Set FD $_{\lambda i}$ = 680nm-780nm2) Set of samples of healthy and unhealthy vegetables

Output: 1) MFDS-mean$_{\lambda}$: Red Edge Position 2) REP R_{λ} : Reflectance at Red Edge Position 3) Maximum First Derivative 4) FD $_{\lambda min}$: Wavelength where First Derivative curve starts 5) FD $_{\lambda max}$: Wavelength where First Derivative curve ends

For each sample in the set, from 680nm to 780nm:

i. FD $_{\lambda min}$ = (λ)min where FD$_{\lambda i}$ >0 (Minimum wavelength from where FD curve starts)
ii. FD$_{\lambda max}$ = (λ)max where FD$_{\lambda i}$ >0(Maximum wavelength where FD curve ends)
iii. MFD= FD $_{max}$ (Find maximum magnitude of FD)
iv. MFD $_{\lambda min}$= (λ)min where FD = MFD(Minimum wavelength from where MFD Spread starts)
v. MFD $_{\lambda max}$: = (λ)max where FD = MFD(Maximum wavelength where MFD Spread ends)
vi. MFDS-mean$_{\lambda}$ = mean$_{\lambda}$ (MFD$_{\lambda max}$, MFD $_{\lambda min}$)
vii. REP $_{\lambda}$ = MFDS-mean$_{\lambda}$
viii. REP R_{λ} : Reflectance at Red Edge Position

Algorithm 1: MFDS-m Red Edge Position Algorithm

4 Result and Discussion

Calculation of correct inflection point is very crucial for determining REP. We have discussed algorithms for finding REP in previous section. REP was calculated using maximum first derivative, linear interpolation and linear extrapolation methods. It is observed that, every spectral signature do not give double peak in the First Derivative Curve.

Maximum first derivative method considers complete curve and finds spectral band having maximum magnitude of first derivative. In practice we do not get only one spectral band having maximum magnitude. We may extract the first band only which gives maximum magnitude ignoring rest all bands having same magnitude. This will lead to incorrect REP.

Linear Interpolation is applicable when there is double peak in the curve. This is based on fixed four points. It is observed that spectral bands of starting and ending of the curve cannot be fixed [7]. There is curve shift in case of healthy and unhealthy plants. REP calculation is not precise. In Linear Extrapolation two lines are drawn in Far Red and NIR region. It is difficult to locate perfect points for finding c1, c2 and m1, m2. Table 1 shows REP calculated by four methods.

Table 1. Red Edge position calculated by different methods for healthy and unhealthy vegetables

REP calculation method	Brinjal healthy	Brinjal unhealthy	Custer Beans healthy	Custer Beans unhealthy
MFDS-mean	701	712	703	712
Maximum FD	725	720	741	722
Linear interpolation	714	719	717	719
Extrapolation	705	721	696	708

All above methods are only used for REP calculations, but could not be used for directly discriminating healthy and infected crops.

Table 1 shows Red Edge Position calculated for Healthy and Unhealthy plants of Brinjal and Cluster Beans. REP shift is observed when we compare FD curve of Healthy and Unhealthy plants. Table 2 depicts reflectance difference between Healthy and Unhealthy plants. Reflectance magnitude is greater in case of healthy plants. Table 3 shows features selected for discrimination between healthy and unhealthy vegetable plants. REP shift can be observed. Maximum First Derivative magnitude and spread is calculated. We observed that MFDS (Maximum First Derivative Spread) of unhealthy plant is more than healthy plant. And Reflectance of healthy plant at REP is greater than Reflectance of unhealthy plant.

Table 2. Reflectance values at Red Edge Position calculated by different methods for Healthy and Unhealthy Vegetables

REP calculation method	Brinjal healthy	Brinjal unhealthy	Custer Beans healthy	Custer beans unhealthy
MFDS-mean	0.37	0.14	0.25	0.17
Maximum FD	0.2	0.11	0.14	0.12
Linear interpolation	0.3	0.14	0.2	0.15
Extrapolation	0.23	0.15	0.11	0.1

Table 3. Features selected for discrimination between healthy and unhealthy vegetables using MDFS-mean algorithm

Features selected	Brinjal healthy	Brinjal unhealthy	Custer Beans healthy	Custer Beans unhealthy
MFD	0.009	0.004	0.005	0.005
MFD $_{\lambda max}$ − MFD $_{\lambda min}$	8	24	11	16
REP_{λ}	701	712	703	712
$REP_{R\lambda}$	0.37	0.11	0.25	0.17
$FD_{\lambda min}$	680	688	682	687
$FD_{\lambda max}$	769	752	799	756

5 Conclusion

In this experimentation and study we propose an algorithm for determining more precise inflection point for finding REP. This method is applicable for the First Derivative curves of spectral signature, which do not give double peak. We have found different features which will be used to discriminate healthy and unhealthy vegetables. In this study only spectral reflectance and first derivative of 680 nm–780 nm was used. Optical and SWIR regions can be used further for finding disease infections at different stages. Biochemical and biophysical pathological analysis along with spectral analysis will be very useful for finding sensitive bands regarding plant health.

References

1. Sahoo, R.N., Ray, S.S., Manjunath, K.R.: Hyperspectral remote sensing of agriculture special Section. Hyperspectral Remote Sens. Curr. Sci. **108**(5), 10 (2015)
2. Horler, D.N.H., Dockray, M., Barber, J.: The red-edge of plant leaf reflectance. Int. J. Remote Sens. **4**, 273–288 (1983)
3. Dawson, T.P., Curran, P.J.: A new technique for interpolating red edge position. Int. J. Remote Sens. **19**(11), 2133–2139 (1998)
4. Lamb, D.W., Steyn-Ross, M., Schaare, P., Hanna, M.M., Silvester, W., Steyn-Ross, A.: Estimating leaf nitrogen concentration in ryegrass (Lolium spp.) pasture using the chlorophyll red-edge: theoretical modelling and experimental observations. Int. J. Remote Sens. **23**(18), 3619–3648 (2002)
5. Guyot, G., Baret, F.: Utilisation de la haute résolution spectrale pour suivre l'état des couverts végétaux. In: Proceedings of the 4th International Colloquium on Spectral Signatures of Objects in Remote Sensing, ESA SP-287, Assois, France, pp. 279−286 (1988)
6. Cho, M.A., Skidmore, A.K.: A new technique for extracting the red edge position from hyperspectral data: the linear extrapolation method. Remote Sens. Environ. **101**, 181–193 (2006)
7. Das, P.K., Choudhary, K.K., Laxman, B., Kameswara Rao, S.V.C., Seshasai, M.V.R.: A modified linear extrapolation approach towards red edge position detection and stress monitoring of wheat crop using hyperspectral data. Int. J. Remote Sens. **35**(4), 1432–1449 (2014)

Evaluation of Pretreatment Methods for Prediction of Soil Micronutrients from Hyperspectral Data

Shruti U. Hiwale[1(✉)], Amol D. Vibhute[2], and Karbhari V. Kale[1]

[1] Department of Computer Science and IT, Dr. Babasaheb Ambedkar Marathwada University, Aurangabad 431004, MS, India
shruti.hiwale@gmail.com
[2] School of Computational Sciences, Solapur University, Solapur 413255, MS, India

Abstract. An assessment of soil quality is vital for monitoring of crop growth and agricultural practices along with its management. Moreover, soil quality is essential to fulfill the requirements of agricultural as well natural resource planning. However, the conventional methods do not suffice to identify the Soil Macro Nutrients (SMN) which is useful for soil quality evaluation. Recently, visible near infrared (VNIR) reflectance spectroscopy is widely acceptable technology for detecting and estimating the soil attributes in effective and rapid manner. Nevertheless, the acquired reflectance spectra by spectroscopy are affected by sensor error or illumination errors. Though, the affected errors can be diminished by the VNIR pretreatment methods. In this study, efforts made to identify the SMN from VNIR spectroscopy. The important data has been extracted by using data mining techniques and algorithm such as the various pretreatment methods: Standard Normal Variate (SNV), First Derivative (FD) and Maximum Normalization Continuum Removal (MNCR) were computed for obtaining pure spectra. The Partial Least Squares Regression (PLSR) algorithm was used for estimating the SMN from thirty soil samples collected from agricultural sectors. The experimental results depict that, the SMN was identified and estimated better after implementing the said pre treatment methods on VNIR spectra. The R^2 value was 0.87 for raw spectra and it was 0.93, 0.95 and 0.94 for SNV, FD and MNCR respectively. Whereas, Root mean square error (RMSE) was 0.037, 0.006, 0.049 and 0.028 for raw spectra, SNV, FD and MNCR spectra respectively. In conclusion, the FD method provided betters results than other tested methods. The present research is beneficial for farmers and decision makers to detect and determine SMN from soil samples in better way.

Keywords: Soil micro nutrients · Standard Normal Variate · First derivative · Maximum normalization · Continuum removal · Partial Least Squares Regression · Visible near infrared

K. C. Santosh and R. S. Hegadi (Eds.): RTIP2R 2018, CCIS 1037, pp. 380–390, 2019.
https://doi.org/10.1007/978-981-13-9187-3_34

1 Introduction

Soil is essential natural asset for production of a nutriment in agriculture. It not only controls the motion of water from the landscape also works as a filter for metals and different contents which may draw into specific sphere of the environment. The soil ability is used to manage any of these functions which recite on its physiochemical properties. However, these transient properties vary everywhere. Traditionally, soil analysis is done by hazards chemical treatment to make the interrelationship between soil physicochemical properties as well particular attribute of soil. Though the chemically the soil analysis is complicated due to the availability of soil components. Moreover, an interpretation of results is also difficult [1]. Consequently, the management and planning of natural resources like soil is more vital for food production. Recently, VNIR spectroscopy is giving choice to improve or replace ordinary laboratory methods for assessment of soil; these methods are non-disastrous, speedy and low rate which fulfills the spatiotemporal variability [2].

Visible and Near Infrared Spectroscopy

Visible and near-infrared (400–2500 nm) spectroscopy is a physical constructive, speedy, reportable and low-prize method that discriminate samples from their reflectance with wavelength range from 400 to 2500 nm. VNIR spectroscopy was come to be good analysis option for costly physiochemical laboratory soil analysis which is needed for differentiate soil types and prediction of a massive range and significant scope of soil characteristics [3, 4].

1.1 Work Done so Far

An author Xu, Xie and Fan [5] has worked on the soil spectra pretreatment properties using the impact of spectral pretreatment and selection of wavelength by using prediction method with total kalian (TK) for analysis. The first order derivative has the maximum result among all pretreatments, including Savitzky-Golay smoothing (S-G) method. Linear regression (MLR) method with first derivative used. This method combined with PC transformation given good TK predictive method, with maximal regression result R^2: 0.774 and less RMSE: 1.109. In the study [6], extended multiplicative signal correction method was derived for estimation of soil moisture. The EMSC method was used and compared with first derivate, second derivate, MSC and SNV and combine of PLSR to get strong method with good prediction and reduce data set. The good classification results were given by EMSC with first derivates. The author Yixing et al. [7] has worked on two algorithms which were as a successive projections algorithm and with a support vector machine regression. SPA-SVMR model gives results to enhance the precision of soil organic carbon (SOC) contents from soil reflectance. On the basis selection of wavelength the impact of spectra treatment methods such as Log (1/R), Log (1/R) combine with Savitzky-Golay smoothing, first derivative (FD), second derivative with SG smoothing (SD), SG, mean center (MC) and multiplicative scatter correction (MSC), standard normal variate (SNV) were analyzed on spectra. Results were affected by various preprocessing methods by SG preprocessing and support vector machine for 28 samples given a good result R^2 = 0.73, RMSE = 2.78, RPDV3 = 1.89. For grass quality which estimated by N, P, K, Ca, and Mg absorption is complex. In this study three methods were used band depth and band depth ratio continuum removed derivative. It is described from continuum remove to estimate the content N, P, K, Ca, Mg by using stepwise linear regression for selection of wavelengths from absorption feature of spectra. With training data set, three methods were estimated as results were high with continuum removal and depth ratio data as R^2 for N: 0.70, P: 0.80, K: 0.64, Ca: 0.50 and Mg: 0.68 [8].

2 Materials and Methods

Surface Soil samples were collected from site near area of Tahsil, Jalna District Maharashtra, India. The soil samples were collected from farming sector of region geologically placed at 75°87′38.0227″N latitude and 19°84′22.7844″ E longitude. Soil type was black farm soil with sandy loam texture and average temperature is 31 °C to 33 °C. The surface soil (0–20 cm) was collected from 30 km^2 farming in air tight bag with sunny atmosphere. The surface crop residue was removed while collecting soil samples. The collected soil samples were air dried and went through 2 mm strainer.

2.1 Spectral Measurements Using the ASD

An ASD Field Spec 4 spectroradiometer have spectral range from 350–2500 nm. Thirty soil samples collected with the distance 5 cm from farm were air dried and measured using the ASD field Spect 4. The instrument is with resolution 3 nm and 10 nm for 350–1000 nm and 1000–2500 nm respectively. Reading interval is 1.4 nm and 2 nm. Soil samples placed in petri dish under high intensity source probe. It has a 75 W tungsten quartz halogen filament lamp for illumination of sample. The lamp placed as 45 cm at distance 46 cm above soil sample. The reflected light was collected with 1 nm bandwidths is from 350–2500 nm and 8° (FOV) field of view with cable of field spec gun which placed at a distance of 15 cm above the soil samples. The device firstly optimized. Afterward it calibrated by white reference, obtaining reflectance with lab readings before recording of soil samples.

2.2 Proposed Methodology

There were total 30 soil samples were taken and 75% of data was used for training for calibration, while 25% was used for validate. The quantitative relationship between physiochemical soil contents and soil spectrum evaluated using three preprocessing methods. Moreover, two approaches (A and B) were performed. Where an approach A is based on spectral feature extraction coupled with preprocessing methods whereas PLSR as a multivariate statistical method is implemented as an approach B.

Approach A
Spectral dataset were preprocessed with the three methods (SNV), (FD) and Spectral absorption features were analyzed by maximum normalized continuum removal (CR) and spectral curve.

Approach B
For comparison a calibration model made upon spectral dataset by using PLSR algorithm, which is well established regression method [11]. The proposed methodology is depicted in Fig. 1.

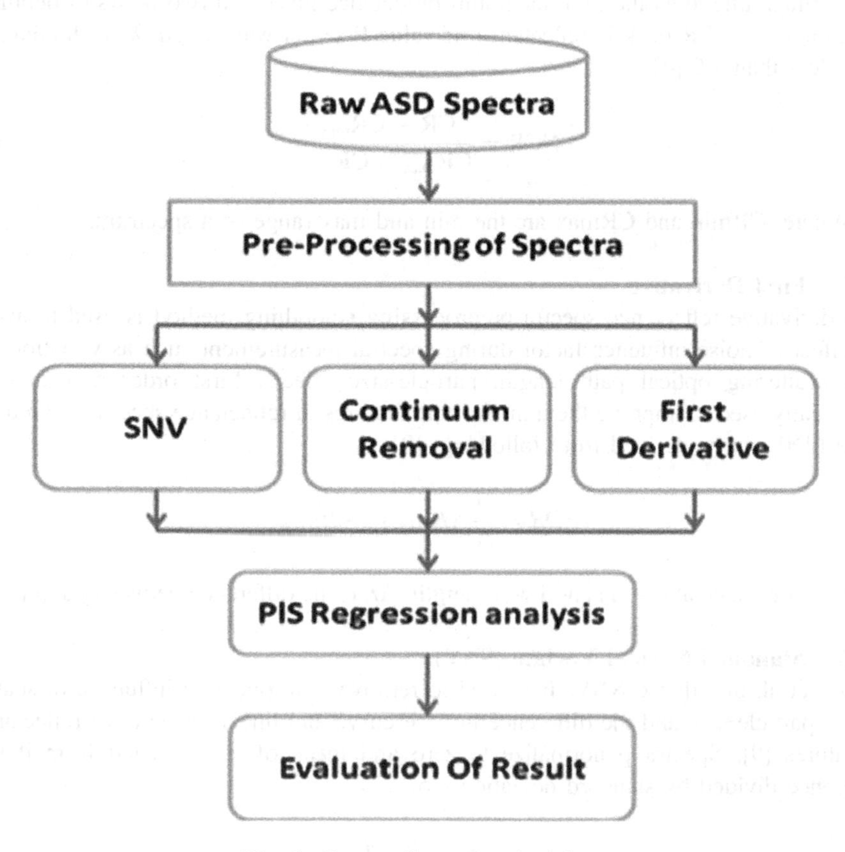

Fig. 1. Process flow of methodology

2.3 Importance of Spectral Data Preprocess

Spectral properties of soil nutrients with dried and sieved soil samples were helpful to reduce the effect of humidity on reflectance spectra of soil. Different soil sample and properties was complex to understand, and soil spectral properties were influenced by scattered parts which affect to the related information. It is hidden between tested work and chemical analysis. The spectrum during collection can be influenced by unknown noises, the surrounding influence, path difference etc. So, it was necessary to apply pretreatment methods to reduce impact of noise of spectrum and other disturbance, so it is required to choose range of wavelength to enhance the precession and prediction [2, 5]. Importance of pretreatment method is to select required a spectral pretreatment method and the band range, which is used to enhance the precision.

2.3.1 Continuum Removal

Continuum removal is specifically used to notice the absorption of content at particular distinctive wavelength predicting that other content features should not strong absorption feature near place of this distinctive wavelength. With continuum is approximate line joins the two maximum reflectance places on two peaks of depth of wavelength (λ). CR peak is reflectance of value R (λ) at wavelength λ, with contains value less than 1.0 [4].

$$\mathrm{NCR} = \frac{\mathrm{CR} - \mathrm{CR_{min}}}{\mathrm{CR_{max}} - \mathrm{CR_{min}}} \tag{1}$$

Where, CRmin and CRmax are the min and max range of a spectrum.

2.3.2 First Derivative

First derivative reflectance spectra preprocessing smoothing method is used to avoid the effect of noise influence factor during spectral measurement such as variations in light scattering optical path length, particle-size effects. First order derivative is specifically used to improve the unnoticeable features of reflectance spectra and reduce noise FDR can be derived from following: [9]

$$FDR = \frac{1}{\Delta\lambda}\left(R_{\lambda(j+1)} - R_{\lambda(j)}\right) \tag{2}$$

$R\lambda$ reflectance at $(j + 1)$ and j wavelength, Δλ is the difference between j and j + 1.

2.3.3 Standard Normal Variate (SNV)

Barnes et al. introduced SNV. It is used to remove multiplicative influence of scatter due to particle size and the difference in base curve and linear property of reflectance signatures [9]. Spectra ρ normalize to zero and mean of ρ subtracted from it this difference divided by standard deviation σρ.

$$SNV = \frac{\rho - \bar{\rho}}{\sigma_p} \tag{3}$$

2.4 Partial Least Squares Regression

This method (PLSR) uses two combine methods: partial least squares (PLS) analysis and multiple linear regression (MLR), and a latent variable which depends on multi variable method. It predict response variables **Y** from correlated variables **X**. [10]

It has the advantages of reducing dimensionality, noise and calculative time, which avoids multiple correlations and the loss of data. The plsr equation as:

$$\mathbf{Y} = \mathbf{XB}\text{PLS} + \mathbf{F}.$$

To estimate **BPLS** is regression coefficient F is residue

After calibrated data plsr were applied for test data. Performance was evaluated for dataset and method which were compared and estimated with correlation coefficient R^2 and RMSE as predicted and measured data [10, 11].

2.5 R-Square

It is used to measure the percentage variability in any of the data matrix, which is accounted for by the built model. The value closer to 1 of R^2 shows a better prediction.

$$R^2 = \frac{SSR}{SST} \qquad \text{SST} = \text{SSR} + \text{SSE}$$

$$SSR = \left(\sum\nolimits_{i=1}^{n} (\hat{y} - \tilde{y})^2\right) \quad SSE = \left(\sum\nolimits_{i=1}^{n} (y - \tilde{y})^2\right) \tag{4}$$

$\hat{y}$ is mean of y, $\tilde{y}$ is total of y

2.6 Mean Square Error

MSE is the most significant equation for comparing and determining different data mining techniques. MSE measures the difference between actual and predicted test outputs. Smaller MSE is good and large values shows poor prediction. The MSE of the predictions is the mean of squared difference between observed and measured values.

$$RMSE = \sqrt{\frac{\sum_{i=1}^{n} (\tilde{y} - y)^{2i}}{n}} \tag{5}$$

$\hat{y}$ is predicted y is observed

3 Results and Discussion

Reflectance signature for thirty soil samples is shown in Fig. 2. In order to detect the zinc content absorption band firstly for maximum normalization continuum removal is applied.

Specifically, this transform was used to notice the specific particular places of the absorptions properties with wavelength absorption depth and characteristics. Hence, the absorption characteristics were noticed with each reflectance spectra and continuum removal spectral signature as shown in Fig. 3.

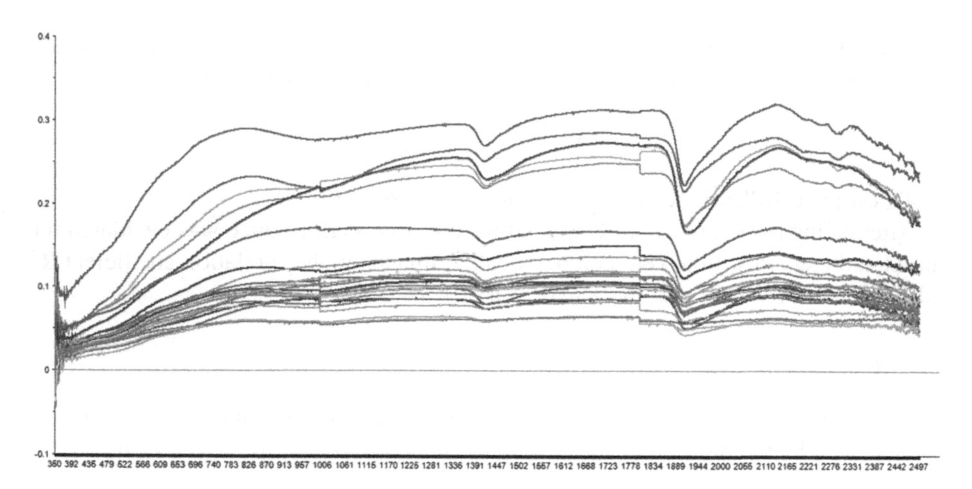

Fig. 2. Raw reflected spectra

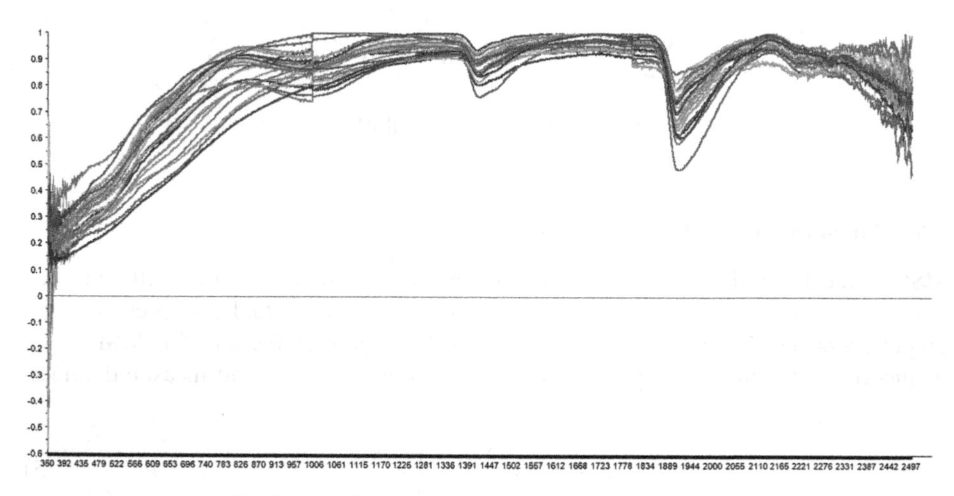

Fig. 3. Maximum normalization continuum removal

SNV importantly used to remove difference in curvilinearity and baseline shift, with intrusion of scatter due to partial size which affects absorbance band, as Fig. 4 shows scattering corrected spectra by using SNV

FDT specifically for border removal of spectra and also smoothed also remove complex surrounding interference by excluding coincided spectral wavelength. It also reduces reflectance having fewer signals to noise ratio was excluded by FD method and transformation spectra was used for next analysis processing. Figure 5 shows the noise removed spectra of 30 soil samples which conserve spectral properties [12].

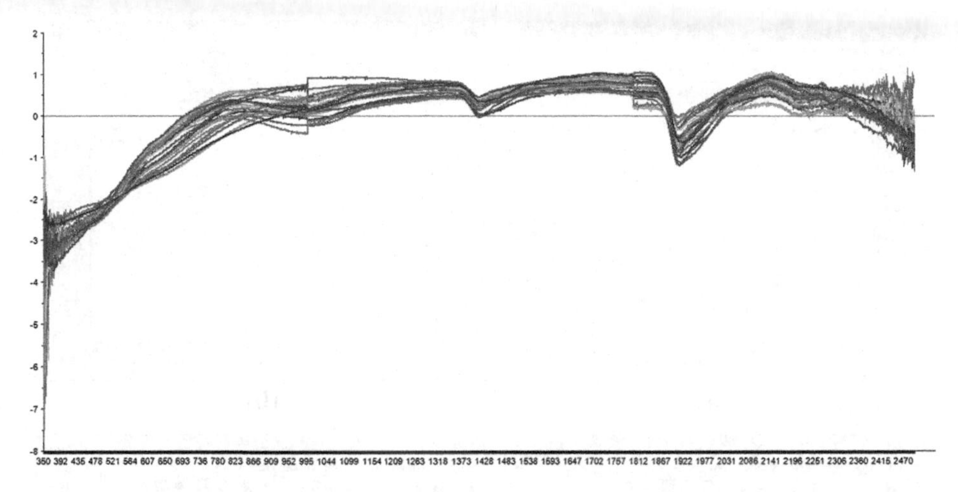

Fig. 4. SNV

The calibration performances for Zn resulting from the different pretreatments were all comparably good. The effect of different pretreatment on spectral wavelength can be analyzed before and after preprocessing on spectra [13] as shown in Figs. 3, 4 and 5 with three different pretreatment methods. Specifying spectral absorption by using pretreatment on spectra will determine the soil contents in soil with the calibration and validation method which affects the accuracy of results methods and satisfactory result obtained in terms of R^2 and RMSE as shown in Table 1.

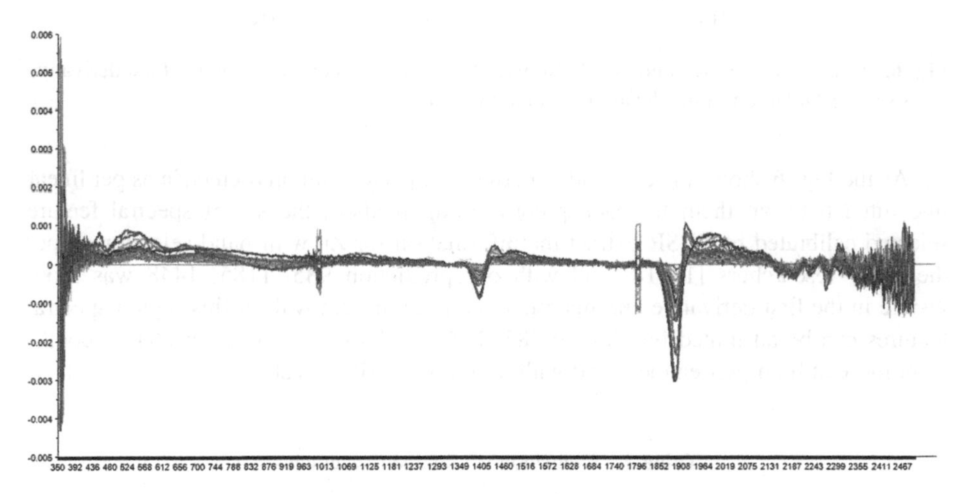

Fig. 5. 1st order derivative transform

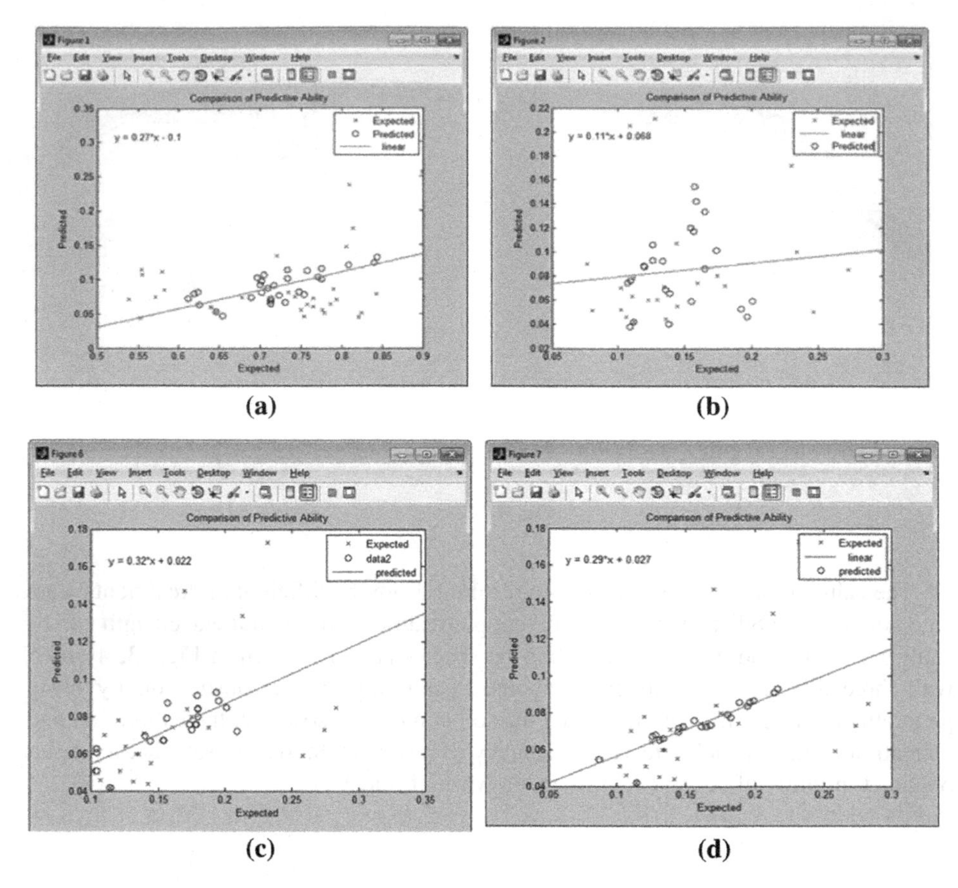

Fig. 6. Scatter plot for prediction of Zn from (a) Data without preprocessing. (b) First derivative (c) SNV (d) Maximum normalization continuum removal.

As the Fig. 6 shows the correlation between expected and predicted Zn as per linear line fitted between them for each preprocessing method, the set of spectral feature selected calibrated by PLSR extract the information for Zn with band selection as per the study researchers [14–18] and with our prediction 985, 1185, 1448 was more visible in the first derivative and maximum continuum removal. In this region spectral features can be attributed for Zinc to 985, 1185, and 1448 showed correlation coefficient more at both pretreatment and with minimum RMSE as shown in above Table 1.

Table 1. Evaluation of obtained accuracies

Dataset	Calibration		Validation	
	R^2	RMSE	R^2	RMSE
Original data without preprocessing	0.893	0.045	0.871	0.037
SNV	0.933	0.002	0.934	0.0062
1st order derivative	0.910	0.002	0.955	0.049
Maximum normalization continuum removal	0.938	6.7544e−04	0.9458	0.0283

4 Conclusions

Pretreatment methods of spectra are important and needed for estimation of soil contents from hyperspectral data due to complexity of soil characteristics. This study signifies the application of three transformation methods for spectral characteristics of 30 soil samples collected from the vegetable farm of District Jalna, Maharashtra. Each sample of spectra includes different patterns. When preprocessing methods applied on it will extract important features of soil spectra which are helpful for band selection and plsr regression model for estimation of soil attribute such as for Zn, sensitive band 985 nm, 1185 nm and 1448 nm with good prediction accuracies are ($r^2 = 0.955$, rmse = 0.049) shown in table no.1 with FD and other SNV and CR so it conclude that pretreatment methods are imperative for developing plsr regression with good accuracy results for estimation of content using prediction and analysis model.

Acknowledgment. The above study is supported by Department of Computer Science and Information Technology. Authors are thankful for technical supports under UGC SAP (II) and partial financial funds for DST-FIST to Dr. Babasaheb Ambedkar Marathwada University Aurangabad,

References

1. Gholizadeh, A., Borůvka, L., Saberioon, M., Vašát, R.: Visible, near-infrared, and mid-infrared spectroscopy applications for soil assessment with emphasis on soil organic matter content and quality: state-of-the-art and key issues. Appl. Spectrosc. **67**(12), 1349–1362 (2013)
2. Gholizadeh, A., Amin, M.S.M., Saberioon, M.M.: Potential of visible and near infrared spectroscopy for prediction of paddy soil physical properties. J. Appl. Spectrosc. **81**(3), 534–540 (2014)
3. Bellon-Maurel, V., McBratney, A.: Near-infrared (NIR) and mid-infrared (MIR) spectroscopic techniques for assessing the amount of carbon stock in soils–Critical review and research perspectives. Soil Biol. Biochem. **43**(7), 1398–1410 (2011)
4. Gomez, C., Lagacherie, P., Coulouma, G.: Continuum removal versus PLSR method for clay and calcium carbonate content estimation from laboratory and airborne hyperspectral measurements. Geoderma **148**(2), 141–148 (2008)
5. Xu, L., Xie, D., Fan, F.: Effects of pretreatment methods and bands selection on soil nutrient hyperspectral evaluation. Procedia Environ. Sci. **10**, 2420–2425 (2011)
6. dos Santos Panero, P., dos Santos Panero, F., dos Santos Panero, J., da Silva, H.E.B.: Application of extended multiplicative signal correction to short-wavelength near infrared spectra of moisture in marzipan. J. Data Anal. Inf. Process. **1**(03), 30 (2013)
7. Peng, X., Shi, T., Song, A., Chen, Y., Gao, W.: Estimating soil organic carbon using VIS/NIR spectroscopy with SVMR and SPA methods. Remote Sens. **6**(4), 2699–2717 (2014)
8. Mutanga, O., Skidmore, A.K.: Continuum-removed absorption features estimate tropical savanna grass quality in situ. In: Earsel Workshop on Imaging Spectroscopy, vol. 3, pp. 13–16, May 2003

9. Buddenbaum, H., Steffens, M.: The effects of spectral pretreatments on chemometric analyses of soil profiles using laboratory imaging spectroscopy. Appl. Environ. Soil Sci. **2012** (2012)
10. Bayer, A., Bachmann, M., Müller, A., Kaufmann, H.: A comparison of feature-based MLR and PLS regression techniques for the prediction of three soil constituents in a degraded South African ecosystem. Appl. Environ. Soil Sci. **2012** (2012)
11. Bayer, A., Bachmann, M., Müller, A., Kaufmann, H.: A comparison of feature-based MLR and PLS regression techniques for the prediction of three soil constituents in a degraded South African ecosystem (2012)
12. Vibhute, A.D., et al.: Assessment of soil organic matter through hyperspectral remote sensing data (VNIR spectroscopy) using PLSR method. In: 2017 2nd International Conference on Man and Machine Interfacing (MAMI), pp. 1–6. IEEE, December 2017
13. Wang, J., et al.: Desert soil clay content estimation using reflectance spectroscopy preprocessed by fractional derivative. PLoS ONE **12**(9), e0184836 (2017)
14. Rossel, R.V., Walvoort, D.J.J., McBratney, A.B., Janik, L.J., Skjemstad, J.O.: Visible, near infrared, mid infrared or combined diffuse reflectance spectroscopy for simultaneous assessment of various soil properties. Geoderma **131**(1–2), 59–75 (2006)
15. Guo, Y.B., Feng, H., Chen, C., Jia, C.J., Xiong, F., Lu, Y.: Heavy metal concentrations in soil and agricultural products near an industrial district. Pol. J. Environ. Stud. **22**(5) (2013)
16. Omran, E.S.E.: Inference model to predict heavy metals of Bahr El Baqar soils, Egypt using spectroscopy and chemometrics technique. Model. Earth Syst. Environ. **2**(4), 200 (2016)
17. Li, W., Lu, J., Dong, M.: Quantitative analysis of calorific value of coal based on spectral preprocessing by laser-induced breakdown spectroscopy (LIBS) (2018)
18. Yang, H., Mouazen, A.M.: Vis/near-and mid-infrared spectroscopy for predicting soil N and C at a farm scale. In: Infrared Spectroscopy-Life and Biomedical Sciences. InTech (2012)

Role of Lens Position and Illumination Source for Acquiring Non-imaging Hyperspectral Data to Estimate Biophysical Characteristics of Leaves

Amarsinh Bhimrao Varpe[1,2(✉)], Rupali R. Surase[1,2], Amol D. Vibhute[3], Dhananjay B. Nalawade[1,2], and Karbhari V. Kale[2]

[1] Geospatial Technology Research Laboratory, Dr. Babasaheb Ambedkar Marathwada University, Aurangabad 431004, MS, India
varpeamarsinh@gmail.com, rupalisurase13@gmail.com, dhananjay.bamu@gmail.com

[2] Department of C.S and IT, Dr. Babasaheb Ambedkar Marathwada University, Aurangabad 431004, MS, India
kvkale91@gmail.com

[3] School of Computational Sciences, Solapur University, Solapur 413255, MS, India
amolvibhute2011@gmail.com

Abstract. The procedure of recognition of plant species is very important in various application such as, classification of plant species along with yield estimation and present status. In the current study, we have carry out the analysis of anthocyanin (ARI) and xanthophyll (X) content of different types of plants with healthy leaves. The collected healthy leaves were scanned through ASD FieldSpec4 spectroradiometer with two positions such as nadir and off-nadir. The recorded data was used for further processing. The variations of ARI and X contents of healthy leaves were examined using spectral indices. The statistical analysis has been carried using open source environment. The experimental observations shows the highest mean value for ARI was (0.084) for off-nadir position. Whereas for the nadir position it was (1.833). Similary, for X the mean value was (−0.509) for off-nadir and for nadir it was (−0.845). It was possible to estimate content of ARI and X in plant leaves using the hyperspectral non imaging data.

Keywords: Spectral reflectance · Correlation of off-nadir and nadir position · Leaves analysis · Spectral indices

1 Introduction

The monitoring of ARI and X contents dynamically plays important role in growth analysis of plant species. [1,2]. The core photosynthetic pigment is directly related to leave status [3,4]. For the analysis of leaf content within

K. C. Santosh and R. S. Hegadi (Eds.): RTIP2R 2018, CCIS 1037, pp. 391–396, 2019.
https://doi.org/10.1007/978-981-13-9187-3_35

canopy vertical distribution plays crucial role. Various studies has been carried to estimate the plant contents. However, the fewer amount of leave also useful for detection of the senescence of the leaves [5]. Recently, hyperspectral remote sensing provides the appropriate datasets to estimate the contents of plants without using any traditional ways. Consequently, numerous studies have been worked on the hyperspectral non imaging data for identification and estimation of plant leaves. On the other hand, spectral indices have been used for the analysis of spectral characteristics of hyperspectral data. To maximize the sensitivity of leave content spectral indices are estimated using two or more spectral wavebands. Currently researchers applying new remote sensing methods for quantifying vertical leaves. Whereas the nadir position captures more information of leaves [6] and off-nadir provides less or varied information [7]. In spite of some research studies which shows sensitivity of these spectral indices for the estimation of leave contents, same spectral index finds useful for different plant species [8]. The current study highlights the advantages of remote sensing technique for analyzing the off-nadir and nadir positioning dataset analysis of plant leaves for Aurangabad region of Maharashtra, India. We have analyzed two types of content such as anthocyanin and xanthophyll. The present paper is organized in five sections like introduction, study region along with used data, proposed methodology, result and discussion.

2 Study Area

The study region covers the part of the botanical garden of Dr.B.A.M.University Aurangabad, Maharashtra, India. The geographical location of the study area is 19°54′3.7944 N latitude and 75°21′8.9208 E longitude. Annual average rainfall is 970 mm with 32 °C to 45 °C annual temperatures. The plant leaves samples were collected from the botany garden on 21 March 2016.

2.1 Hyperspectral Non-imaging Datasets

Many researchers have defined their own dataset and very few are freely available online. However, in the present work, the Analytical Spectral Device (ASD) is utilized for storing the spectral response of plant leaves. The data scanning was done immediately after the sample collection. The perfect dark room and black background environment has been generated for acquisition process of reluctance spectra. White reference panel was used for optimization and calibration before samples recording. Each sample has been stored for 30 times to receive spectra and then averaged for getting pure spectrum. The RS3 tool was utilized for collecting the reflected spectra of leaves. The remaining details are given in (Table 1) along with their spectral wavelength.

3 Applied Methods

3.1 Spectral Indices for Evaluation

The obtained spectral signatures of said plants are indicated in (Fig. 1) collected with off-nadir and nadir position. The details of wavelength versus reflectance

Table 1. Hyperspectral non imaging data parameters used for current research study

Instrument	ASD FieldSpec4	
Position of lamp and FOV	Off-Nadir	Nadir
Angle of FOV	45°	90°
Distance from the sample	15 cm	10 cm
Product type	Healthy leaves	
Acquistion date	21 March 2016	
Wavelength	350–2500 nm	
Sampling interval	1.4 nm to 2 nm	
Spectral resolution	3 nm to 10 nm	
Halogen lamp	75 w	
FOV	8°	
Number of samples	30	

of given leave having 350–2500 nm spectral range with visible, near infrared and short wave infrared of electromagnetic spectrum.

The used hyperspectral indices are listed in (Table 2). With its formulation and the indices are normalized. The indices were implemented on collected and recorded data by off-nadir and nadir position.

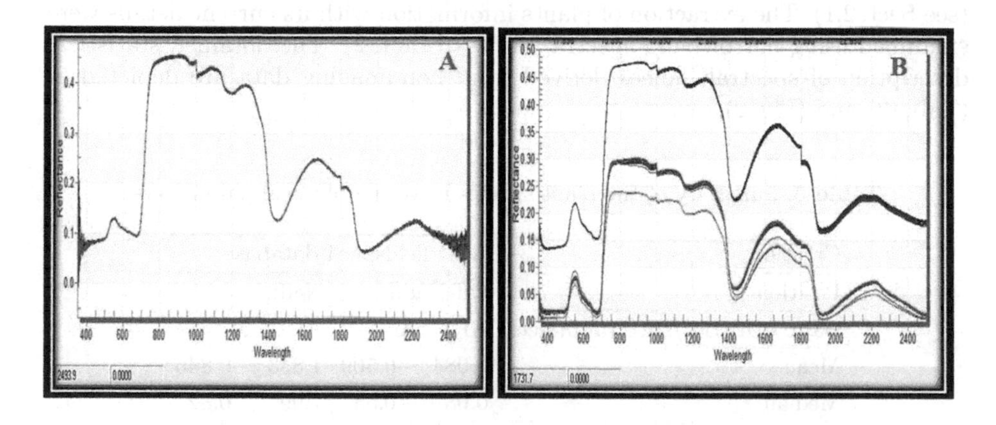

Fig. 1. Healthy leaves using (A) Off-Nadir and (B) Nadir position

3.2 Anthocyanin (ARI) Index

The ARI is a reflectance measurement that is sensitive to anthocyanins in plant leaves. Increases in ARI indicate canopy changes in leave via new growth or death. The ARI index detects higher concentrations of anthocyanins in vegetation or crops [9,10]. The formula of ARI index is given in Eq. 1.

$$ARI = 1/R550 - 1/R700 \quad (1)$$

3.3 Xanthophyll (X) Index

The xanthophyll content was used for reduction of samples weight which were measured as the amount of X. Xanthophyll was extracted for measuring the quantity of other compound from it [11]. The mathematical formulation of X is shown in Eq. 2.

$$X = R528 - R567/R528 + R567 \tag{2}$$

Table 2. Summary of the used vegetation indices

Index	Algorithm	Reference
Anthocyanin (ARI)	1\R550 - 1\R700	Gitelson et al. [10]
Xanthophyll (X)	R528 - R567/R528 + R567	Blackburn et al. [12]

4 Results and Its Discussion

After the field work the plant samples were recorded instantly in the controlled laboratory conditions to extract the features of leaf & its evaluation (see Sect. 2.1). The extraction of plants information with its current details were examined using narrowband spectral indices (Table 2). The obtained statistical description of spectral indices derived from non-imaging data are depicted in (Table 3) and (Fig. 2).

Table 3. Summary of the plant species leaves biophysical parameters

Creation	ASD fieldspec4 database			
Position	Off-Nadir		Nadir	
Basic statistics/spectral indices	ARI	X	ARI	X
Mean	0.084	−0.509	1.833	−0.845
Median	0.08	−0.53	1.99	−0.82
Standard deviation	0.011	0.035	0.653	0.051
Minimum	0.07	−0.54	0.22	−0.99
Maximum	0.11	−0.46	2.65	−0.82
Sum	2.52	−15.27	55.01	−26.2
Count	30	30	30	30

The (Table 3) shows the results of off-nadir and nadir position data along with its corresponding mean, median and standard deviation for plant leaves of healthy conditions.

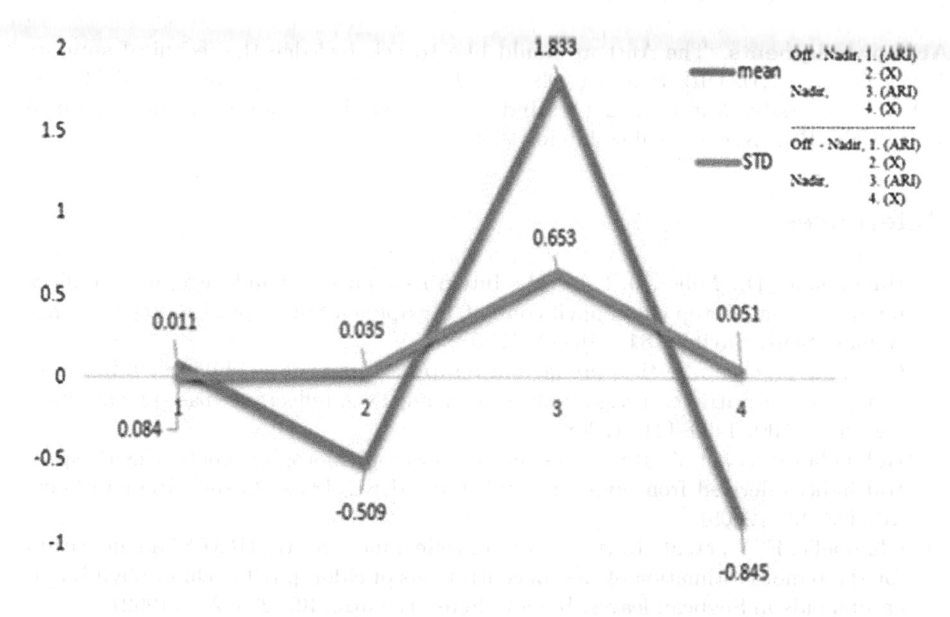

Fig. 2. Standard deviation and mean of ARI and X contents

The min and max values for off-nadir were observed (0.07) and (0.11) for ARI and for nadir it was (0.22) and (2.65) respectively. The computed mean of healthy leaves with nadir position has provided highest value i.e. (1.833) and cor-responding value for off-nadir position was (0.084). Similarly, the X index was used and resulted the values of minimum (−0.54) and (−0.99), whereas maximum value was (−0.46) and (−0.82) on nadir and off-nadir position respectively. The X indices of mean largest value of nadir position was (−0.845) and lowest value of off nadir was (−0.509).

5 Conclusion and Future Scope

The current study has explored the use of spectral indices applied on plant spectral response to recognize the anthocyanin and xanthophyll content. The plant species leaves contents were calculated using hyperspectral datasets. The spectral indices are helpful in plant analysis with proper results. The experimental analysis shows for analyzing hyper spectral non-imaging datasets based on narrow bands spectral indices founds useful. Most of the wave bands are frequently appeared in the NIR portion of the spectrum at the 700 nm to 1200 nm wavelengths. Hence, red edge region was the most useful for the prediction of biophysical variables. In the green region a peak was observed at 550 nm to 700 nm. The wavelength characteristics of anthocyanin absorption and xanthophyll was found at 528 nm and 567 nm associated with biomass variables. The present study will be useful for the identification of plant leaves content within time which is useful for agricultural applications. In future, the study will be focused on correlation of multiple samples and more chemical content estimation of plant leaves.

Acknowledgments. The Authors would like to acknowledge the technical support from UGC SAP(II), DRS Phase-II, DST-FIST and NISA to Dept. of CS and IT, Dr. B.A.M. University, Aurangabad (MS)India and also thanks for financial assistance under UGC-BSR research fellowship for this work.

References

1. Haboudane, D., Miller, J.R., et al.: Integrated narrow band vegetation indices for prediction of crop chlorophyll content for application to precision agriculture. Remote Sens. Environ. **81**, 416–426 (2002)
2. Ciganda, V., et al.: Vertical profile and temporal variation of chlorophyll in maize canopy: quantitative crop vigor indicator by means of reflectance based techniques. Agron. J. **100**, 1409–1417 (2008)
3. Haboudance, D., et al.: Remote estimation of crop chlorophyll content using spectral indices derived from hyperspectral data. IEEE Trans. Geosci. Remote Sens. **46**, 423–437 (2008)
4. Chappelle, E.W., et. al.: Ratio analysis of reflectance spectra (RARS) an algorithm for the remote estimation of the concentrations of chlorophyll a, chlorophyll b and carotenoids in Soybean leaves. Remote Sens. Environ. **39**, 239–247 (1992)
5. Curran, P.J., et al.: Reflectance spectroscopy of fresh whole leaves for the estimation of chemical concentration. Remote Sens. Environ. **39**, 153–166 (1992)
6. Wang, Q., Li, P.H.: Canopy vertical heterogeneity plays a critical role in reflectance simulation. Agric. For. Meteorol. **169**, 111–121 (2013)
7. Hung, W., et al.: Estimation of vertical distribution of chlorophyll concentration by bidirectional canopy reflectance spectra in water wheat. Precis. Agric. **12**, 165–178 (2011)
8. Edward, J., et al.: Progress in field spectroscopy. Remote Sens. Environ. **113**, S92–S109 (2009)
9. Arafat, S.M., et al.: Internet based spectral for different land covers in egypt. Adv. Remote Sens. **2**, 85–92 (2013)
10. Gitelson, A.A., et al.: Optical properties and nondestructive estimation of anthocyanin content in plant leaves. Photochem. Photobiol. **74**, 38–45 (2001)
11. Blackburn, G.A.: Quantifying chlorophylls and carotenoids at leave and canopy scale an evaluation of some hyperspectral approaches. Remote Sens. Environ. 657–675 (1998)
12. Kannadasan, T., et al. Extraction of natural dye xanthophyll from marigold flower. J. Adv. Sci. Res. 48–50 (2013)

Effect of Time on Aluminium Oxide FESEM Nanopore Images Using Fuzzy Inference System

Parashuram Bannigidad[1], Jalaja Udoshi[1(✉)], and C. C. Vidyasagar[2]

[1] Department of Computer Science, Rani Channamma University, Belagavi 591156, Karnataka, India
parashurambannigidad@gmail.com, prof.jalaja@gmail.com
[2] Department of Chemistry, Rani Channamma University, Belagavi 591156, Karnataka, India
vidya.891@gmail.com

Abstract. The applications in nanotechnology require customized nanopore membrane. The structure and number of nanopore on the oxidized metal template rely upon the anodizing parameters used in the electro-chemical cell during the nanopore synthesis. The fundamental idea of this paper is to develop an automated system to quantify the effect of time on aluminum nanopore through advanced minuscule FESEM images. The test results foresee that, the increase in anodization time results in gradual increment in porosity and pore size estimating from 0.234% to 2.034% and 32 nm to 78 nm respectively and shrinking in nanopore wall thickness from 58 nm to 41 nm. The anticipated after effects of the following conceivable development of aluminum nanopore size and wall thickness are processed by applying factual investigation (statistical analysis) and building the principles of fuzzy inference system. The manual and test results are compared, analyzed and deciphered to exhibit the competence of the proposed technique.

Keywords: AAO · Nanopore · Nanomaterial · Digital image analysis · Fuzzy · FIS · FESEM

1 Introduction

The customised nano templates have gained high appreciation in nano applications like chemical and biochemical sensors [1–4], nano filtration [5], drug-delivery [6], super hydrophobic surfaces [7] and nano generators [8–10]. Anodized aluminium oxide (AAO) has ended up being a profoundly flexible material framework in such applications [11–13]. It is requisite to acquire pores with monotonous width and form, to orchestrate and install them in a superstructure. Fabrication of nanopores with different pore size, wall thickness and porosity is the need of an industry to meet different applications like filtering solutes of various molecular size, vacillating from small (urea, creatinine) to middle (vancomycin, inulin) to retain large weight molecules (albumin) [20]. Contribution in

K. C. Santosh and R. S. Hegadi (Eds.): RTIP2R 2018, CCIS 1037, pp. 397–405, 2019.
https://doi.org/10.1007/978-981-13-9187-3_36

this field include; S. Singh analyzed optimization on nanoparticles [14]. Parashuram detected the effect of interim on AAO [15]. Muneesawang measured pore size [16]. Soham extracted pores in noisy image [17]. The influence of anodizing time on the nanopore was presented by Vidyasagar et al. [18]. FIS was used by Parashuram to analyze bacterial cell growth and division [19]. Fuzzy tool for detecting arrow annotations has been introduced by santosh et al. [21]. Work on sequential classifier was proposed by santosh et al. [22].

The FESEM images of Al_2O_3 films caught at customary interim of time and constant in concentration, temperature and voltage are shown in the Fig. 1.

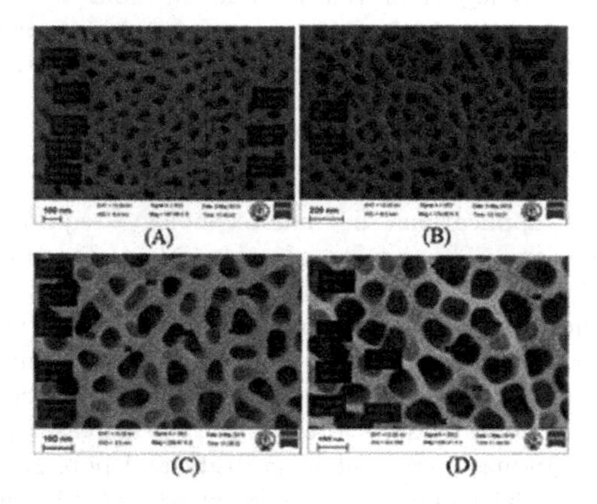

Fig. 1. FESEM images of Al_2O_3 films caught at customary time (A–5 min, B–9 min, C–20 min and D–30 min) and constant in concentration, temperature and voltage.

The programmed minuscule image analysis system provides an efficient tool for subjective analysis in present day material science and biological studies. The statistical technique and fuzzy inference system is applied for anticipating the progressions in nanopore size and wall thickness.

2 Proposed Method

The present investigation has intended to build an programmed device to determine the impact of time on Al_2O_3 nanopore structures using FESEM images. The physical properties of aluminium nanopores; pore size, wall thickness, porosity, pore diameter depends on four major anodization characteristics, namely; anodization time (min), voltage (V) applied, reaction temperature (°C) and concentration (%) of electrolyte. In this paper, an algorithmic method is suggested to measure and predict the effect of time on nanopore images using the statistical analysis and fuzzy logic. The geometrical features are used to extract the properties of aluminium nanopores and are defined below:

- Nanopore diameter (D_p): Mean of major and minor axis of the nanopore.
- Interpore distance (D_i): The average distance from the nanopore centroid to its neighbouring nanopore centroid.

– Porosity(α): Porosity is defined using Eq. (1).

$$\alpha = \frac{\pi}{2\sqrt{3}}\left(\frac{D_p}{D_i}\right)^2 \quad (1)$$

The proposed algorithm and procedures for predictions is given below:

1. Input nanopore FESEM image.
2. Perform pre-processing operations on input image.
   i Image enhancement
   ii Binarization using global thresholding.
   iii Morphological operations to reduce noise.
3. Extract pores.
4. Compute interpore distance, diameter, wall thickness and porosity.
5. Analysis, interpretation of the results and prediction for further growth of the Al_2O_3 nanopore size and wall thickness using statistical analysis and building the fuzzy inference system (FIS).

Methodology for predicating the pore size and wall thickness based on statistical analysis:

1. Construct the correlation graph depicting the growth of pore size and declination of wall thickness.
2. Extend the graph to required anodization time depicted on x-axis.
3. Record the values on y-axis showing the pore size and wall thickness with respect to time on x-axis.

Methodology for predicating the pore size and wall thickness based on fuzzy inference system:

1. Construct the fuzzy inference system using time (min) as input variable and nanopore size and wall thickness (nm) as output variable.
2. Obtain the predictions of nanopore area and wall thickness by considering the anodization time (min) as input variable to the FIS and the nanopore size (nm) and wall thickness (nm) as output variable by building the fuzzy rules:
   i If the time (min) is between 1 to 5 then the range of pore size is 5 nm to 32 nm and wall thickness is 5 nm to 58 nm.
   ii If the time is between 4 to 9 then the range of pore size is 5 nm to 35 nm and wall thickness is 5 nm to 56 nm.
   iii If the time is between 4 to 20 then the range of pore size is 6 nm to 61 nm and wall thickness is 4 nm to 48 nm.
   iv If the time is between 4 to 30 then the range of pore size is 6 nm to 78 nm and wall thickness is 4 nm to 41 nm.
   v If the time is between 4 to 40 then the range of pore size is 7 nm to 98 nm and wall thickness is 3 nm to 36 nm.
   vi If the time is between 4 to 50 then the range of pore size is 8 nm to 118 nm and wall thickness is 2 nm to 27 nm.
   vii If the time is between 4 to 60 then the range of pore size is 9 nm to 138 nm and wall thickness is 1 nm to 9 nm.
3. Interpretation of predicated results.

3 Experimental Results and Discussion

The experiment is implemented for 350 nanopores on Intel Core i3-6100 CPU @ 3.70 GHz using MATLAB R2016 software.

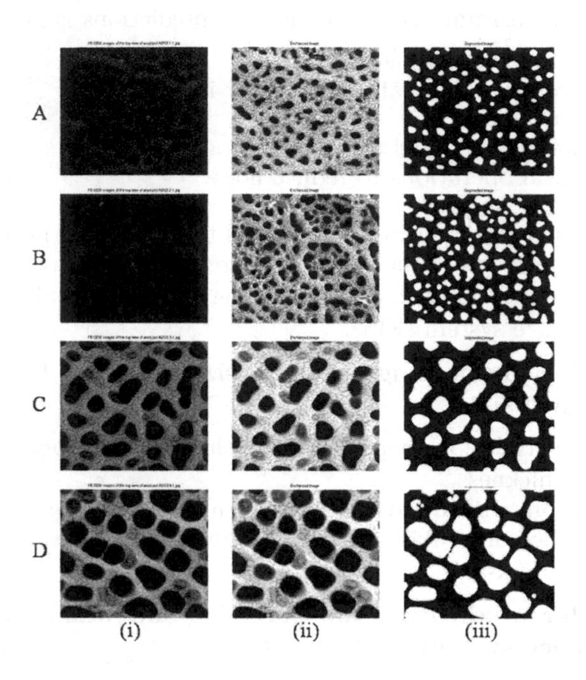

Fig. 2. (i) FESEM images, (ii) Grayscale images (iii) Segmented images

Every Al_2O_3 FESEM image utilized in the investigation are captured at customary interim of time (min) and keeping constant in concentration (%), temperature (°C) and voltage (V) (Fig. 2(i)). The input image is converted into grayscale image (Fig. 2(ii)) and global threshold segmentation is applied (Fig. 2(iii)) to isolate the nanopores (foreground) from background. Pore size, interpore distance and porosity has been extricated for each labelled segment. Later, the results are compared, analyzed and deciphered. The interpreted results are publicized in the Table 1.

It is observed that, the pore size is increased and the wall thickness is decreased, when the time (min) increases (i.e. 5, 9, 20, 30 min) and concentration (%), temperature (°C) and voltage (V) remains constant. The experimental results also predict that, the increase in anodization time results in gradual increase in porosity and nanopore size from 0.234% to 2.034% and 32 nm to 78 nm respectively and decrease in nanopore wall thickness from 58 nm to 41 nm. The manual outcomes acquired by synthetic specialists and processed aftereffects of time versus wall thickness are portrayed in the Fig. 3. Similarly, the consequences of time versus nanopore measure are appeared in the Fig. 4. The effect of time on wall thickness, pore diameter

Table 1. Geometric feature values of Aluminium Oxide nanopore Images of Fig. 2 and predicted results using statistical and fuzzy inference system approach.

Image	Anodization time (min)	Wall thickness (nm)				Pore size (nm)			
		Manual	Proposed methods			Manual	Proposed methods		
			GA	SA	FA		GA	SA	FA
A	5	58	58	58	57.24	32	32	32	33.57
B	9	56	57	57	56.14	34	34.74	34	37.01
C	20	48	49	49	48.53	58	61.25	61	60.72
D	30	26	41	41	41.66	81	78.62	78	78.96
PV	40	-	-	35	36	-	-	98	98.71
	45	-	-	32	32.51	-	-	108	108.98
	50	-	-	28	27.87	-	-	118	118.73

GA: Geometrical Analysis; SA: Statistical Analysis; FA: Fuzzy Analysis; PV: Predicted Values

and porosity of the Al_2O_3 nanopores are shown in the Fig. 5. The correlation graph plotted based on the proposed results is depicted in the Fig. 6, shows, relationship between the anodization time and geometrical features like nanopore size and wall thickness. Assuming that the facts will continue to remain true, the graph can be extended to record the pore size and wall thickness on y axis against the predicted anodization time (min) on x axis, these results are shown in Table 1. Further, the mamdani model is used to build the fuzzy inference system with time (min) as input variable and pore size and wall thickness as output variables. The Gaussian membership functions are utilized as information guidelines (input rules) to construct the fuzzy inference system and is portrayed in Fig. 7. The anticipated fuzzification results are mentioned in Table 1. The manual outcomes are compared with the three proposed methods namely; geometrical analysis method, statistical analysis method and fuzzy inference method. In case of geometrical analysis method

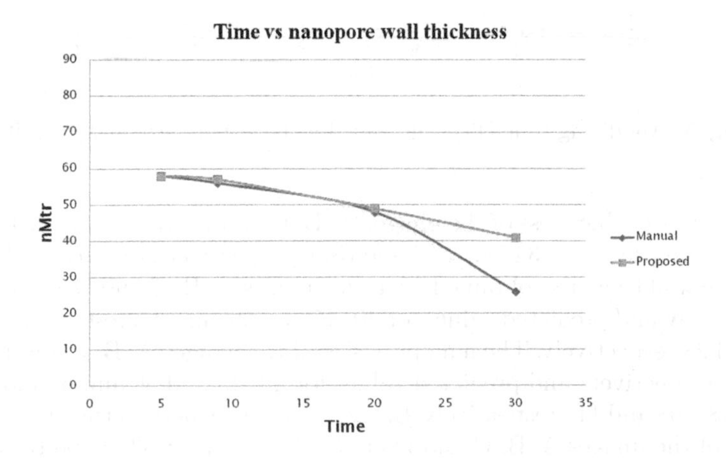

Fig. 3. Time versus wall thickness

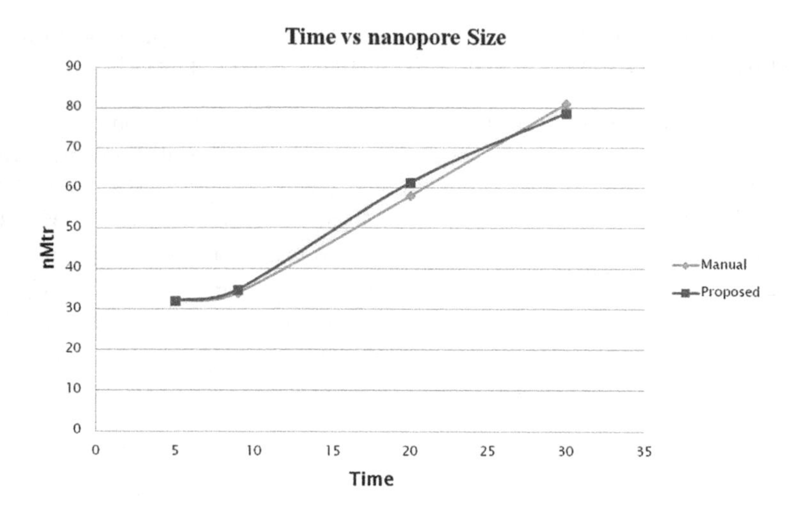

Fig. 4. Time versus pore size

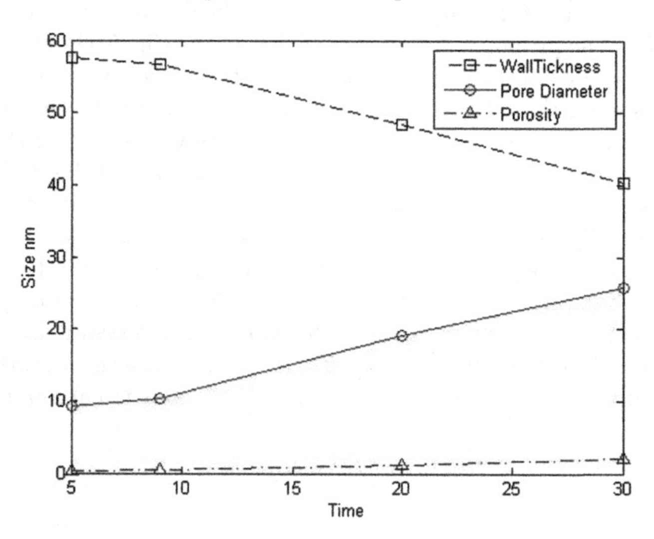

Fig. 5. Anodizing time effect on wall thickness, pore size and porosity

the nanopore wall thickness of the images A, B, C and D are 58, 57, 49, 41 respectively and pore size are 32, 34.74, 61.25 and 78.62 respectively. In statistical analysis method the wall thickness obtained from the images A, B, C and D are 58, 57, 49, 41 respectively and predicted values for 40, 45 and 50 min of anodization time are 35, 32 and 28 respectively. The nanopore size of the images A, B, C and D are 32, 34, 61, 78 respectively and predicted values for 40, 45 and 50 min of anodization time are 98, 108 and 118 respectively. In fuzzy inference method the predicted wall thickness of the images A, B, C and D are 57.24, 56.14, 48.53, 41.66 respectively and predicted values for 40, 45 and 50 min of anodization time are 36, 32.51, 27.87

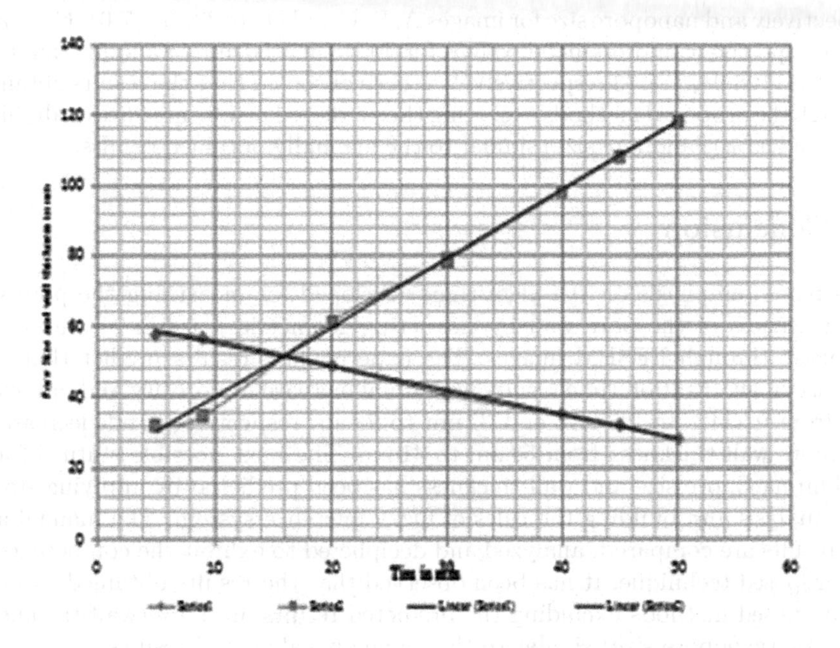

Fig. 6. Statistical analysis of Al_2O_3 nanopore images of Fig. 2

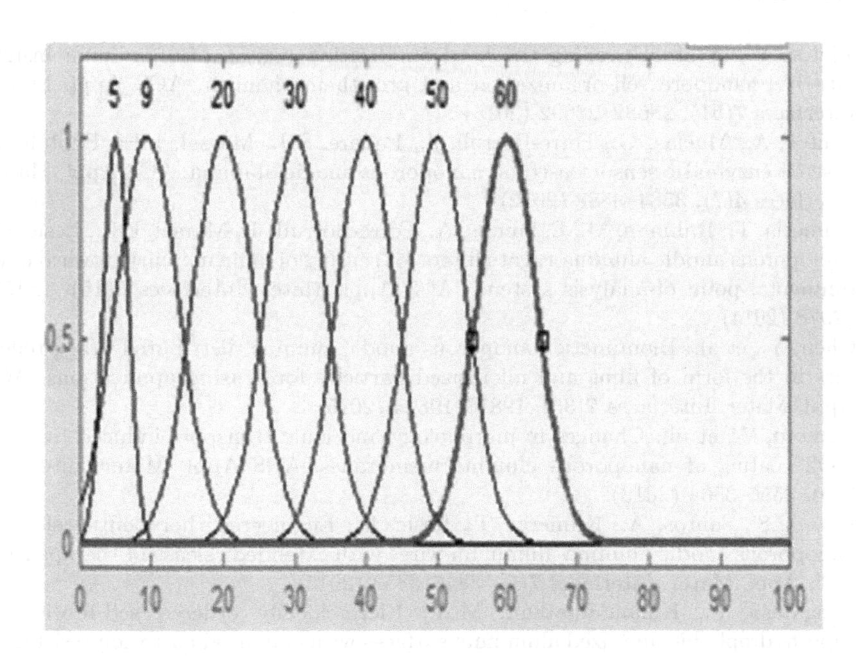

Fig. 7. Fuzzy inference system based on time (min) as input

respectively and nanopore size for images A, B, C and D are 33.57, 37.01, 60.72 and 78.96 respectively and predicted values for 40, 45 and 50 min of anodization time are 98.71, 108.98, 118.73 respectively. It is been observed that the results obtained through the proposed methods excluding the predicted results are 96% (wall thickness) and 99% (nanopore size) similar to the manually obtained results.

4 Conclusion

This paper, emphasis on the algorithm, developed for measuring the pore size and thickness of the pore wall by extracting geometric features of aluminium nanopore through FESEM images. The experimental results predict that, the increase in anodization time results in gradual increase in porosity and nanopore size from 0.234% to 2.034% and 32 nm to 78 nm respectively and decrease in nanopore wall thickness from 58 nm to 41 nm. The next possible status of aluminium nanopore size and wall thickness has been predicted by applying statistical analysis and building the rules of fuzzy inference system. The manual and test results are compared, analyzed and deciphered to exhibit the competence of the proposed technique. It has been observed that the results obtained through the proposed methods excluding the predicted results are 96% (wall thickness) and 99% (nanopore size) similar to the manually calculated results.

References

1. Victor, V., et al.: Unveiling the hard anodization regime of aluminum: insight into the nanopore self-organization and growth mechanism. ACS Appl. Mater. Interfaces **7**(51), 28682–28692 (2015)
2. Santos, A., Macías, G., Ferre-Borrull, J., Pallare, S.J., Marsal, L.F.: Photoluminescent enzymatic sensor based on nanoporous anodic alumina. ACS Appl. Mater. Interfaces **4**(7), 3584–3588 (2012)
3. Kumeria, T., Rahman, M.M., Santos, A., Ferre-Borrull, J., Marsal, L.F., Losic, D.: Nanoporous anodic alumina rugate filters for sensing of ionic mercury: toward environmental point-of-analysis systems. ACS Appl. Mater. Interfaces **6**(15), 12971–12978 (2014)
4. Chen, Y., et al.: Biomimetic nanoporous anodic alumina distributed bragg reflectors in the form of films and microsized particles for sensing applications. ACS Appl. Mater. Interfaces **7**(35), 19816–19824 (2015)
5. Romero, V., et al.: Changes in morphology and ionic transport induced by ALD SiO2 coating of nanoporous alumina membranes. ACS Appl. Mater. Interfaces **5**(9), 3556–3564 (2013)
6. Law, C.S., Santos, A., Kumeria, T., Losic, D.: Engineered therapeutic-releasing nanoporous anodic alumina-aluminum wires with extended release of therapeutics. ACS Appl. Mater. Interfaces **7**(6), 3846–3853 (2015)
7. Vengatesh, P., Kulandainathan, M.A.: Hierarchically ordered self-lubricating superhydrophobic anodized aluminum surfaces with enhanced corrosion resistance. ACS Appl. Mater. Interfaces **7**(5), 1516–1526 (2015)

8. Munoz, R.M., Grauby, S., Rampnoux, J.M., Caballero-Calero, O., Martín-Gonzalez, M., Dilhaire, S.: Fabrication of Bi_2Te_3 nanowire arrays and thermal conductivity measurement by 3ω-scanning thermal microscopy. J. Appl. Phys. **113**, 054308 (2013). https://doi.org/10.1063/1.4790363
9. Bohnert, T., Vega, V., Michel, A.K., Prida, V.M., Nielsch, K.: Magneto-thermopower and magnetoresistance of single Co-Ni alloy nanowires. Appl. Phys. Lett. **103**, 092407 (2013). https://doi.org/10.1063/1.4819949
10. Dudem, B., Ko, Y.H., Leem, J.W., Lee, S.H., Yu, J.S.: Highly transparent and flexible triboelectric nanogenerators with sub-wavelength-architectured polydimethylsiloxane by a nanoporous anodic aluminum oxide template. ACS Appl. Mater. Interfaces **7**(37), 20520–20529 (2015)
11. Banerjee, P., Perez, I., Henn-Lecordier, L., Lee, S.B., Rubloff, G.W.: Nanotubular metal-insulator-metal capacitor arrays for energy storage. Nat. Nanotechnol. **4**, 292–296 (2009)
12. Liang, Y., et al.: Direct access to metal or metal oxide nanocrystals integrated with one-dimensional nanoporous carbons for electrochemical energy storage. Am. Chem. Soc. **132**(42), 15030–15037 (2010)
13. Min, H.L., et al.: Roll-to-roll anodization and etching of aluminum foils for high-throughput surface nanotexturing. Nano Lett. **11**, 3425–3430 (2011)
14. Shwetabh, S.: Microscopic image analysis of nanoparticles by edge detection using ant colony optimization. J. Comput. Eng. **11**(3), 84–89 (2013)
15. Bannigidad, P., Vidyasagar, C.C.: Effect of time on anodized Al_2O_3 nanopore FESEM images using digital image processing techniques: a study on computational chemistry. IJETTCS **4**(3), 15–22 (2015). ISSN 2278-6856
16. Muneesawang, P., Sirisathitkul, C.: Size measurement of nanoparticle assembly using multilevel segmented TEM images. J. Nanomater. **16**, 58 (2015)
17. De, S., Biswas, N., Sanyal, A., Ray, P., Datta, A.: Detecting subsurface circular objects from low contrast noisy images: applications in microscope image enhancement. World Acad. Sci. Eng. **67**, 1317–1323 (2012)
18. Vidyasagar, C.C., Bannigidad, P., Muralidhara, H.B.: Influence of anodizing time on porosity of nanopore structures grown on flexible TLC aluminium films and analysis of images using MATLAB software. Adv. Mater. Lett. **7**(1), 71–77 (2016)
19. Hiremath, P.S., Bannigidad, P.: Digital microscopic bacterial cell growth analysis and cell division time determination for *escherichia coli* using fuzzy inference system. In: Thilagam, P.S., Pais, A.R., Chandrasekaran, K., Balakrishnan, N. (eds.) ADCONS 2011. LNCS, vol. 7135, pp. 207–215. Springer, Heidelberg (2012). https://doi.org/10.1007/978-3-642-29280-4_23
20. Belwalkar, A., Grasing, E., Van Geertruyden, W., Huang, Z., Misiolek, W.Z.: Effect of processing parameters on pore structure and thickness of anodic aluminum oxide (AAO) tubular membranes. J. Memb. Sci. **319**, 192–198 (2008)
21. Santosh, K.C., Wendling, L., Antani, S., Thoma, G.R.: Overlaid arrow detection for labeling regions of interest in biomedical images. IEEE Intell. Syst. **31**(3), 66–75 (2016)
22. Santosh, K.C., Roy, P.P.: Arrow detection in biomedical images using sequential classifier. Int. J. Mach. Learn. Cybern. **9**(6), 993–1006 (2018)

Infrared Image Pedestrian Detection Techniques with Quantitative Analysis

Rajkumar Soundrapandiyan[1], K. C. Santosh[2](✉), and P. V. S. S. R. Chandra Mouli[3](✉)

[1] School of Computer Science and Engineering, Vellore Institute of Technology, Vellore, India
[2] Department of Computer Science, University of South Dakota, Vermillion, USA
santosh.kc@ieee.org
[3] Department of Computer Applications, NIT Jamshedpur, Jamshedpur, India
mouli.chand@gmail.com

Abstract. Pedestrian detection in infrared (IR) images is important due to widely used IR images in many applications including surveillance, night vision, searching, environmental monitoring, driving assistant system etc. Among these pedestrian detection in defense gained more attention in the infrared images. However, there are still many problems existed in pedestrian detection in infrared images are low signal to noise ratio, low contrast, complex background, pedestrians are prone to occluded by other things and lack of shape. In this paper, Global background subtraction, adaptive filter and local adaptive thresholding based Pedestrian Detection method proposed to overcome these problems. Further, the proposed method tested on the OSU thermal pedestrian database. In addition, proposed method result is compared along with the popular existing traditional methods using quantitative measures. From experimental results deduced that the proposed method earned excellent detection rate when compared to other methods.

Keywords: Pedestrian detection · Infrared images · Thresholding · Mean · Variance · Histogram · Misclassification error · Relative foreground area error

1 Introduction

In recent era, pedestrian detection is a active local research area in infrared images. Infrared image applications such as tracking, night vision, automatic recognition of pedestrian in rescue missions, surveillance, content retrieval etc. Among these extensive applications detection of pedestrian is one of the major field in defense [1]. Although, IR images has its own challenges such as low signal to noise ratio (SNR), low contrast, low gray-level foreground object, capture the heat emitting objects, foreground can occlude with background, lack in shape of object which make the detection process as hard [2].

K. C. Santosh and R. S. Hegadi (Eds.): RTIP2R 2018, CCIS 1037, pp. 406–415, 2019.
https://doi.org/10.1007/978-981-13-9187-3_37

In literature, in the past decades, a good number of research work is done on detection of pedestrian in infrared images. These methods are generally classified as filter based methods, template based methods, threshold methods, saliency based methods and so on [2]. Filter methods comprise maximum and mean filter [3], maximum and median filter, non-linear subtraction filter [4], 2D least mean square filter, Laplacian of Gaussian (LoG) filter [2]. The filters are failed to identify the dim pedestrian when the IR image contains the low contrast. Template based methods include scale invariant and improved template matching methods [5,6]. These methods are failed to extract the shape of the pedestrian. Threshold methods include iterative thresholding [7], local thresholding [8], thresholding using gray-level histogram [9]. These methods were failed to detect the pedestrian due to discontinuity in the pixels and complex background noise. Saliency based detection methods include multi-scale saliency extraction method [10], saliency feature analysis using multi-scale decomposition [11]. These methods are may be time consuming process and processing on the noisy image is the hard.

Further more, Wang et al. [12] proposed the method using high-pass filter and support vector machine. The drawback of this method is This method is unable to detect the exact shape of the pedestrian and failed to detect the pedestrian with the discontinuous pixels. Liu et al. [13] presented the improved template matching method and it is failed in the detection of shape of the image. In addition, An efficient approach proposed by the Li et al. [14] which is the combination of histogram oriented gradient (HOG) features, and two geometrical features – mean contrast and standard deviation are used to detect the human in thermal imagery. Li et al. [4] method failed to sense the pedestrians in hot sunny day. Rajkumar et al. [15] proposed the background subtraction and filter based pedestrian detection (BSFPD) method. Rajkumar et al. [15] method unsuccessful to identify the pedestrian during day time.

From the previous work, it is analyzed that there is an opportunity for upgrade of pedestrian detection in IR image. This problem inspire us to propose a simple pedestrian detection method using global background subtraction filter, adaptive filter and local adaptive threshold. The signal to noise ratio (SNR) of the input image can be increased by making the use of background subtraction filter. The region can be enhanced and the residual clutter along with the noise in image can be removed using the adaptive filter. The local adaptive threshold can be used to partition the pedestrian region from the image.

The proposed method has the following advantages

- Any prior information about the input image is not required.
- Image characteristics are used to compute the parameter values of the background subtraction filter, adaptive filter and local adaptive threshold.
- The exact shape of the pedestrians can be determined.

The paper is organized in the following manner: Sect. 2 contains brief discussion about the proposed system; Sect. 3 contains explanation of quantitative analysis metrics; The experimental results and performance analysis is discussed in Sect. 4; Finally the work is concluded in Sect. 5.

2 Proposed System

The proposed system comprise of three stages: Global Background Subtraction Filtering (GBSF), Adaptive Filtering (AF) and Local Adaptive Thresholding (LAT). Figure 1 shows the block diagram of the proposed method. The following stages are involved in the proposed method:

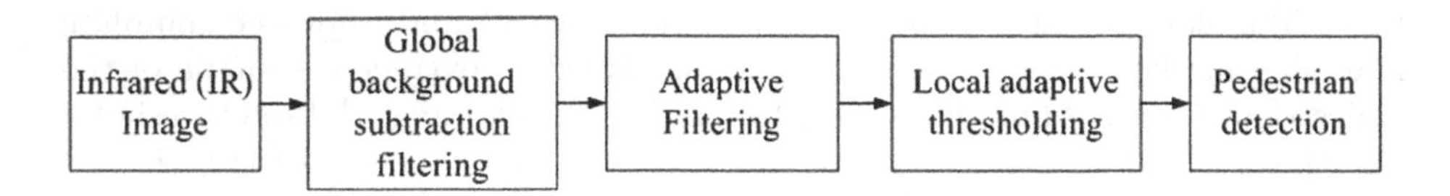

Fig. 1. Block diagram of the proposed method

- Initially, Registered IR image taken as input
- Secondly, IR image background informations are removed using GBSF in Sect. 2.1
- Next, Remove the remaining noise and enhance the background subtracted image using AF in Sect. 2.2
- Finally, Pedestrians are detected using LAT in Sect. 2.3

2.1 Global Background Subtraction Filtering

In IR images imperfect illumination, climatic conditions, optical defocusing etc. are general problems while capturing images. These conditions reflect in IR images and the background visualization becomes difficult. In order to reduce the background information, a simple method using maximum occurrence of intensity value of an image is proposed.

Let $f(k,l)$ be the grayscale input IR image, the formation of which is defined in Eq. 1.

$$f(k,l) = X(k,l) + Y(k,l) \tag{1}$$

where $X(k,l)$ denotes the foreground information of an image and $Y(k,l)$ represents the background information of an image. The background subtraction method works with the pixel properties of the image. Suppressing the background information of the image is the objective of this model. The background suppression of the image is achieved by subtracting the estimated the grayscale background from the image pixel and is defined in Eq. (2).

$$f_X(k,l) = \begin{cases} f(k,l) - P_e, & f(k,l) \geq P_e \\ 0, & otherwise \end{cases} \tag{2}$$

where $f_X(k,l)$ represents the background suppressed image, P_e is the estimated peak obtained from the maximum occurrence of intensity value of the image. The background suppression elevates the warm objects. Statistically, it is observed that there is an increase in SNR value compared to the original image.

2.2 Adaptive Filtering

Elevation of warm objects takes place in background subtraction but it introduces some textural and shape variations. In order to preserve the shape and textural properties, adaptive filtering method is used.

The adaptive filter designed using the two statistical parameters mean and variance. The reason for choosing these two parameters is they are quantities related to the characteristic of the image. The mean gives an average intensity in the region, whereas variance gives a contrast in the region. To enhance the contrast region in the background subtracted image an adaptive filter is defined in Eq. (3)

$$f_{XE}(k,l) = f_X(k,l) - \frac{\sigma_\eta^2}{\sigma_L^2}\left[f_X(k,l) - m_L\right] \tag{3}$$

where $f_{XE}(k,l)$ is the resultant image of adaptive filter, $f_X(k,l)$ is the background subtracted image, m_L indicate the mean of the image, σ_L^2 indicate the local variance of the image.

2.3 Local Adaptive Thresholding

In general, in IR image, objects have higher temperatures than the background. The background is suppressed and foreground objects, enhanced using the above steps. So, the foreground objects have the maximum gray-level intensities in an image. To detect only the true pedestrians, adaptive thresholding is designed as

1. To eliminate false targets, adaptive threshold is defined as

$$T = \mu(f_{XE}(k,l)) + K\sigma(f_{XE}(k,l)) \tag{4}$$

   where K is the scalar constant, which is calculated using entropy of the f(k,l) image.
2. The bright pixels from the image is detected by utilizing Eq. (5)

$$F(k,l) = \begin{cases} 0, & \text{if } f_{XE}(k,l) \geq T \\ 1, & otherwise \end{cases} \tag{5}$$

   where $F(k,l)$ is the resultant image.

3 Quantitative Analysis Measures

This section discuss the quantitative metrics used to analyze the proposed method.

Misclassification Error (ME). It [17] refers to the wrongly identified percentage of the background pixels to be the desired areas and vice-versa. This kind of error can be classified defined in Eq. (6).

$$ME = \left(1 - \frac{|F_i \cap F_T| + |B_i \cap B_T|}{|F_i| + |B_i|}\right) \times 100 \quad (6)$$

In Eq. (6), B_i and F_i indicates number of background pixels and pixels in the desired area of the input image. B_T and F_T denotes to the number background pixels and pixels in the desired area of the resultant image. The common pixels between the two areas is denoted by sign $\cap$. The range of ME lies between zero to one, where the extremities i.e. zero refers to the best classification and one refers to the worst classification.

Relative Foreground Area Error (RAE). It [17] is used to measure the area of feature and is defined in the Eq. (7).

$$RAE = \begin{cases} \frac{A_i - A_T}{A_i} & if A_T < A_i \\ \frac{A_T - A_i}{A_T} & if A_T \geq A_i \end{cases} \quad (7)$$

where A_i is the input image, and A_T is the thresholded image.

Detection Rate. Detection rate (DR) [18] is a measure of true pedestrians and is defined in the Eq. (8)

$$DR = \frac{\text{True pedestrians detected in images}}{\text{True pedestrians existing in images}} \times 100 \quad (8)$$

4 Results and Performance Evaluation

The results and performance analysis of the proposed method is discussed here.

4.1 Input Images

To evaluate the proposed method OSU thermal pedestrian database [16] is used. This database contains 10 groups of 284 images with 984 pedestrian objects and its size as 360 × 240. These images are captured in various environmental conditions such as cloudy, rain, haze and so on. The details of the each collection are shown in Table 1.

4.2 Subjective Results

Figure 2 exhibits the detection results of proposed method for sample images. Figure 2(a) displays the sample input image, Fig. 2(b) presents the result of global background subtraction filter, Fig. 2(c) displays the result of adaptive

Table 1. OSU Thermal pedestrian database details

Group	Image	Pedestrians	Environmental conditions			
			Climates	Humidity	Temperature	Visibility
1	31	91	Light rain	82%	45F	3 miles
2	28	100	Partly cloudy	76%	37F	8 miles
3	23	101	Partly cloudy	38%	53F	10 miles
4	18	109	Fair	45%	41F	10 miles
5	23	101	Partly cloudy	47%	57F	10 miles
6	18	97	Mostly cloudy	77%	53F	8 miles
7	22	94	Light rain	70%	68F	9 miles
8	24	99	Light rain	80%	62F	4 miles
9	73	95	Haze	89%	50F	2.5 miles
10	24	97	Haze	80%	55F	3 miles

filter, Fig. 2(d) demonstrates the result of localthresholding, Fig. 2(e) shows the result of pedestrian detection. In addition, the pedestrian detection result of proposed method compared with the existing methods such as max-median method, BSFPD method, and wang et al. method are expressed in Fig. 3. Figure 3(a) is the sample input images, Fig. 3(b) demonstrate the result of proposed method, Fig. 3(c) demonstrates the result of max-median method, Fig. 3(d) displays the result of BSFPD method and Fig. 3(e) indicates the result of Wang et al. method. From the Fig. 3, noticed that the proposed method detection rate is higher than the existing methods.

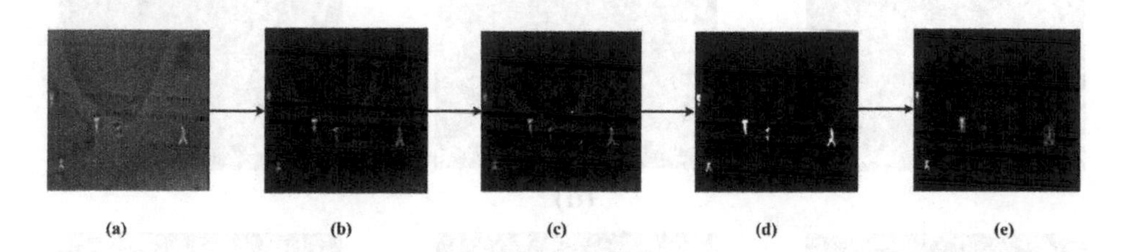

Fig. 2. Sample result of the proposed method

Performance Analysis of Results. In this section, the detection results obtained from proposed method compared with the existing methods using quantitative analysis measures ME, RAE and DR are discussed and it is shown in Table 2. For the comparison max-median method, BSFPD method, and wang et al. method are used. From the Table 2, it is observed that the value of ME close to 0 in proposed method and it is indicates that the proposed method has less error than the existing methods.

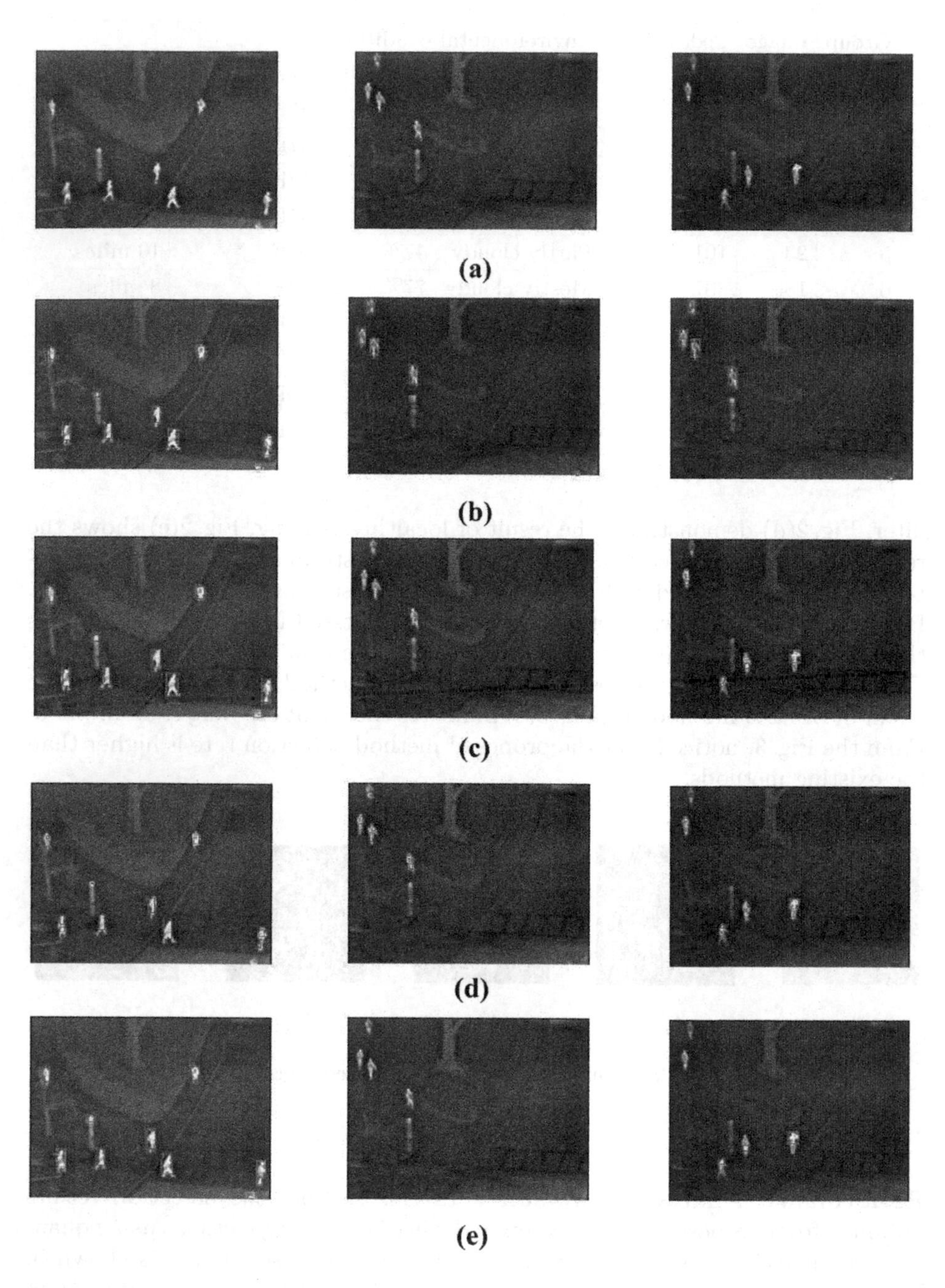

Fig. 3. Subjective comparison of pedestrian detection result

Table 2. Comparison of proposed method with the existing method using ME measure

Class	#Image	#Total pedestrians	Proposed method	Max-median method	BSFPD method	Wang et al. method
1	31	91	0.0259	0.0809	0.0300	0.0717
2	28	100	0.0076	0.0801	0.0290	0.0671
3	23	101	0.0329	0.3754	0.0700	0.1980
4	18	109	0.0229	0.0854	0.0400	0.0168
5	23	101	0.0229	0.1694	0.0100	0.0244
6	18	97	0.0224	0.1663	0.0120	0.0336
7	22	94	0.0183	0.1755	0.0100	0.0320
8	24	99	0.0244	0.1846	0.0160	0.0412
9	73	95	0.0351	0.1190	0.0160	0.0366
10	24	97	0.0214	0.0839	0.0420	0.0214

Table 3 shows the comparison of proposed method with the existing method using RAE measure. From Table 3 it is clear that the value of RAE is lesser than the existing methods for all the classes.

Table 3. Comparison of proposed method with the existing method using RAE measure

Class	#Image	#Total pedestrians	Proposed method	Max-median method	BSFPD method	Wang et al. method
1	31	91	0.0215	0.2655	0.0348	0.0484
2	28	100	0.0072	0.2764	0.0104	0.0446
3	23	101	0.0659	0.2509	0.1036	0.0402
4	18	109	0.0049	0.2810	0.1041	0.0266
5	23	101	0.0131	0.2691	0.0603	0.0324
6	18	97	0.0023	0.2849	0.0700	0.0382
7	22	94	0.0011	0.2840	0.0502	0.0345
8	24	99	0.0106	0.2716	0.0653	0.0453
9	73	95	0.0639	0.2139	0.1434	0.0883
10	24	97	0.0021	0.2730	0.1124	0.0436

Figure 4 display the detection rate of the proposed method and existing methods. From Fig. 4 discern that the proposed method superior than the existing methods in-terms of detection rate.

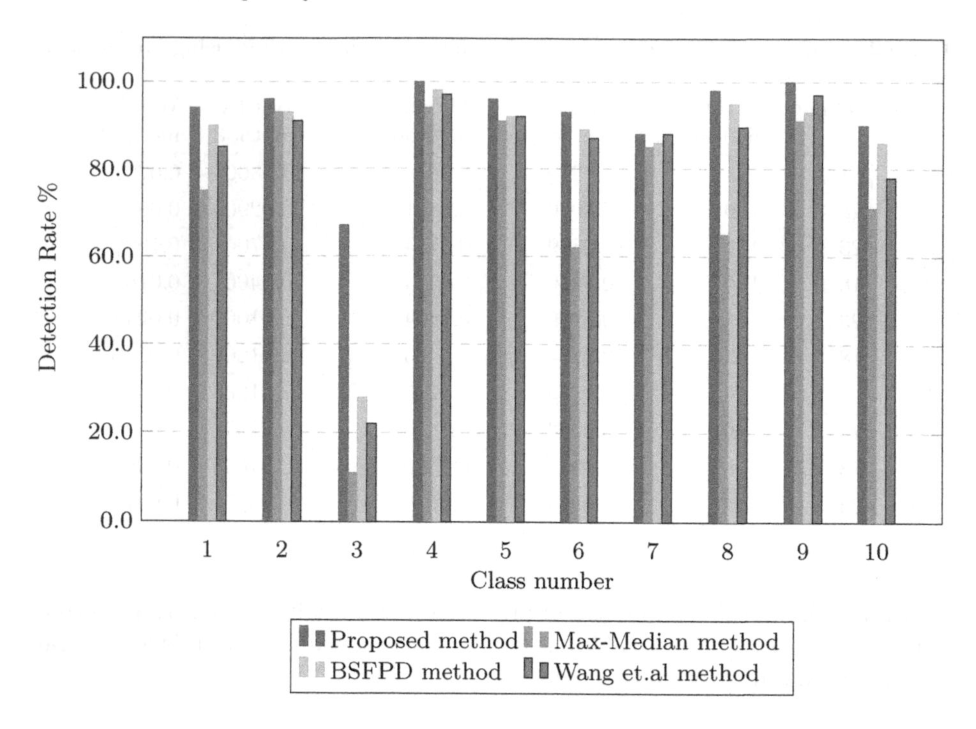

Fig. 4. Comparison of detection rate

5 Conclusion

This paper proposed a simple pedestrian detection method using global background subtraction filter, adaptive filter and local thresholding concepts. This proposed method tested on OSU thermal pedestrian database which contains the 10 classes of data and it is captured in different environment conditions. The experimental results of proposed method compared with both subjectively and objectively. From the experimental results, perceived that the proposed method obtained 92% of detection rate and which is superior than the existing methods.

References

1. Rajkumar, S., Mouli, P.C.: Target detection in infrared images using block-based approach. In: Informatics and Communication Technologies for Societal Development, New Delhi, pp. 9–16 (2015)
2. Soundrapandiyan, R., Mouli, P.C.: Adaptive pedestrian detection in infrared images using fuzzy enhancement and top-hat transform. Int. J. Computat. Vis. Robot. **7**(1–2), 49–67 (2017)
3. Deshpande, S.D., Meng, H.E., Venkateswarlu, R., Chan, P.: Max-mean and max-median filters for detection of small targets. In: Proceedings of the International Society for Optical Engineering, Signal and Data Processing of Small Targets, USA, pp. 74–83 (1999)

4. Barnett, J.: Statistical analysis of median subtraction filtering with application to point target detection in infrared backgrounds. In: Proceedings of the International Society for Optical Engineering, Infrared Systems and Components III, USA, pp. 10–18 (1989)
5. Liu, R., Lu, Y., Gong, C., Liu, Y.: Infrared point target detection with improved template matching. Infrared Phys. Technol. **55**(4), 380–387 (2012)
6. Yoo, J., Hwang, S.S., Kim, S.D., Ki, M.S., Cha, J.: Scale-invariant template matching using histogram of dominant gradients. Pattern Recognit. **47**(9), 3006–3018 (2014)
7. Kapur, J.N., Sahoo, P.K., Wong, A.K.: A new method for gray-level picture thresholding using the entropy of the histogram. Comput. Vision Graphics Image Process. **29**(3), 273–285 (1985)
8. Otsu, N.: A threshold selection method from gray-level histograms. IEEE Trans. Syst. Man Cybern. **9**(1), 62–66 (1979)
9. Sun, S.G., Kwak, D.M.: Automatic detection of targets using center-surround difference and local thresholding. J. Multimedia **1**(1), 16–23 (2006)
10. Qi, S., Ma, J., Tao, C., Yang, C., Tian, J.: A robust directional saliency-based method for infrared small target detection under various complex backgrounds. IEEE Geosci. Remote Sens. Lett. **10**(3), 495–499 (2013)
11. Zhao, J., Feng, H., Xu, Z., Li, Q., Peng, H.: Real-time automatic small target detection using saliency extraction and morphological theory. Opt. Laser Technol. **47**(1), 268–277 (2013)
12. Wang, J.T., Chen, D.B., Chen, H.Y., Yang, J.Y.: On pedestrian detection and tracking in infrared videos. Pattern Recognit. Lett. **33**(6), 775–785 (2012)
13. Liu, Y., Zeng, L., Huang, Y.: An efficient HOG-ALBP feature for pedestrian detection. Sig. Image Video Process. **8**(1), 125–134 (2014)
14. Li, W., Zheng, D., Zhao, T., Yang, M.: An effective approach to pedestrian detection in thermal imagery. In: Proceedings of Eighth International Conference on Natural Computation, China, pp. 325–329 (2012)
15. Soundrapandiyan, R., Mouli, P.C.: Adaptive Pedestrian Detection in Infrared Images Using Background Subtraction and Local Thresholding. Procedia Comput. Sci. **58**(1), 706–713 (2015)
16. http://www.cse.ohio-state.edu/otcbvs-bench. Accessed 01 June 2015
17. Rajkumar, S., Mouli, P.C.: Pedestrian detection in infrared images using local thresholding. In: Proceedings of 2nd International Conference on Electronics and Communication Systems, Coimbatore, pp. 259–263 (2015)
18. Soundrapandiyan, R., Mouli, P.C.: A novel and robust rotation and scale invariant structuring elements based descriptor for pedestrian classification in infrared images. Infrared Phys. Technol. **78**(1), 13–23 (2016)

Classifying Arabic Farmers' Complaints Based on Crops and Diseases Using Machine Learning Approaches

Mostafa Ali(✉), D. S. Guru, and Mahamad Suhil

Department of Studies in Computer Science, University of Mysore, Mysore, India
mam16@fayoum.edu.eg, dsg@compsci.uni-mysore.ac.in, mahamad45@yahoo.co.in

Abstract. In this paper, two models are proposed to automatically categorize the farmers' complaints on various diseases that may affect crops. In the first model, a complaint which is expressed in Arabic free text is classified into its respective crop and then into a particular disease, but in the second model, the complaint is classified directly into diseases. A dataset of farmers' complaints described in Arabic script consisting of complaints from five different crops raised by farmers of different parts of Egypt is created. Separate lexicons are created for crops and their diseases by considering all possible technical terms related to a crop and its diseases including their possible slang synonyms. Each preprocessed complaint is represented in the form of a binary vector using the vector space model (VSM) with the help of crop lexicon so that machine learning techniques can be applied. Experiments are conducted on the dataset by varying the percentage of training with multiple trials using SVM and KNN classifiers. It has been observed from the results that the proposed model is performing on par with the human expert and can be deployable for real-time operations.

Keywords: Arabic text classification · Vector space model · Farmers' complaints · Diseases lexicon · SVM · KNN

1 Introduction

Nowadays, text analysis has become a vital area of research because of the increasing and wide spread demand for internet textual applications and communication network gates. Text Classification (TC) is the process of routing each document to its labeled class. Analyzing Arabic text is more difficult than analyzing English text as there are many additional features in Arabic language that do not exist in any other languages. Complex morphology of Arabic language words makes it difficult to extract features and to apply subsequent text mining tasks. Agriculture is one of the major fields that use information technology in publishing its data for farmers, guides and decision makers.

There are many web portals, blogs and government web sites for agricultural activities such as food and agriculture organization (FAO) [1] and rural agricultural development communication network [2]. They help the farmers through

K. C. Santosh and R. S. Hegadi (Eds.): RTIP2R 2018, CCIS 1037, pp. 416–428, 2019.
https://doi.org/10.1007/978-981-13-9187-3_38

different announcements and recommendations for effective farming. These technologies contain an application module through which farmers can get their agricultural problems resolved. A farmer can describe his problem in Arabic language and he should wait for an expert to response with the answer. Agriculture portals allow farmers to do that and it is noticed that most of farmers describe the complaints in slang Arabic text instead of standard words.

This paper addresses an important issue that helps decision makers in automating the process of complaints classification. In this work, we have processed models to classify a collection of complaints into their respective disease using machine learning techniques.

The major contributions of this work are a proposal of a method to extract features on various diseases from unstructured text complaints of farmers and also to develop a classifications model. To achieve the first contribution, we have built a lexicon for 5 crops which contains the names and synonyms of each crop and another lexicon describing all harmful weeds, fungal diseases, insects and other pests that affect each crop during its planting and growing stage. We have proposed two alternative approaches for classifying the complaints into diseases. The first approach classifies the complaints into their respective crops, then to diseases, where the second approach classifies the complaints directly into their diseases.

This paper is organized as follows; Sect. 2 presents a survey of literature on analyzing and classifying Arabic text. Section 3 gives a view about Arabic Language in Egyptian Dialect. Section 4 refers to the Arabic complaints dataset. Section 5 explains the newly built crops and diseases lexicons. Section 6 processes the proposed classification models and finally, Sect. 7 shows the research results with evaluation and comparison.

2 Literature Survey

In this section, we present a brief overview of the previous work on Arabic text analysis for different approaches. In customers' complaints, there are many works on analyzing complaints and extracting the main features. In the service of online chat to solve customer complaints, the authors in [3] utilized logistic regression, SVM with radial kernel and random forest to classify sessions. Many works can be marked on Arabic text mining. In [4] the authors used ontology to extract features from the text data and used TF-IDF to make indexing. They clustered the complaints and solutions using GCluto mining tool to summarize the problems. Also in [5] the authors presented a multi-faceted object extraction methodology to extract knowledge from the problems and their solution. For Arabic text classification, many researchers have investigated the techniques for classifying Arabic documents. In [6], the authors used TF-IDF for reducing features dimensionality and used ANN as a classification technique to classify Arabic text documents. In [7], the authors used hybrid approaches of machine learning, N-grams approach and Semantic orientation approach to classify Egyptian tweets to get the polarity of each tweet. But for Multi-label classification for Arabic news articles, the authors in [8] used SVM and NB through

WEKA tool. In [9], the author used N-Gram frequency statistics technique employing the Manhattan distance to classify Arabic text documents. Moreover, in [10], the authors compare KNN and SVM in classifying Arabic news articles. They select features using TF-IDF method and CHI statistics as a ranking metric. The authors proved that SVM showed better results and prediction time when compared to KNN. In [11] the authors introduced a new measure called term_class relevance to compute the relevancy of a term in classifying a document into a particular class. The authors in [12] showed that N-gram for indexing is more accurate compared to the traditional way of indexing (bag of words) by classifying Arabic news articles using KNN algorithm, but [13] used CHI square for feature selection and used SVM as a classifier and got less accurate result by other classifiers. Term Class Weight-Inverse Class Frequency (TCW-ICF) has been proposed by [14] as a novel term weighting scheme to classify Arabic farmers' complaints based on crop class. Then, TCW-ICF has been applied on English newsgroup skewed dataset as a novel feature selection technique for news categorization [15]. The term Frequency-Inverse Class Frequency (TF-ICF) has been applied on the Arabic farmers' complaints dataset by [16] for a comparison of different term weighting schemes namely; TF, TF-IDF, and TF-ICF on Arabic text classification.

3 Arabic Language in Egyptian Dialect

In Egypt, Arabic language is spoken in slang way and most of Egyptians are not committed with the formal language in speaking and writing. So, the analysis process becomes hard to apply on text documents especially on those which are written by laymen.

Arabic formal language has many constructs which can be used to negate a word. "ما، ليس، ليست لا، لن، لم" (they can be read as "la, ln, lm, ma, lays, laysat") are the commonly used Arabic formal language negation constructs which mean "No, Not" and negative suffixes and prefixes in English [14]. However, in slang Arabic language, there are new negation constructs such as "شي ش،" which are suffixed to get the opposite meaning of the Arabic verbs. For example, "يستخدمش" or "يستخدمشي" are used to get the negative meaning of the verb "use" where the formal Arabic construct is "لا يستخدم". In addition to this, there are many new slang Arabic words that do not belong to the formal Arabic dictionary like "كويس" whose formal Arabic equivalent is "جيد" (which means "Good"). Because of these issues, the analysis of informal Arabic text is more difficult than the analysis of formal Arabic text. This necessitates the design of a new lexicon to handle the slang Arabic text effectively. A work can also be traced in literature which has shown that the inclusion of new lexicons for slang Egyptian dialects besides a formal dictionary helps in improving the classification results [17].

4 Arabic Farmers' Complaints Dataset

Most of Egyptian farmers are poor and they are not familiar with technology. So, it is difficult to collect Arabic complaints with the Egyptian Dialect. The complaints are written in unstructured way of Arabic text which is not well formatted. All the complaints present different causes that affect the plants, like harmful weed, fungal diseases or insects and other diseases. A dataset has been created by collecting 3700 Arabic text complaints related to five different crops and three main diseases from VERCON website [18]. The number of complaints in diseases' categories is well balanced. Table 1 shows the number of complaints in each crop. Some examples of the complaints and their translation into English are shown in Table 2.

Table 1. Number of complaints in each crop

Crop	Rice	Wheat	Corn	Cotton	Beans
# of Complaints	754	990	585	735	709

Table 2. Farmers' complaints and their translation to English

Translated Complaints in English	Complaints in Arabic Language
Land of wheat has been cultivated in mid-November and appeared grasses broad and high, what is the appropriate control mechanism to handle weeds in sandy land?	ارض قمح تم زراعتها فى منتصف نوفمبر ظهر حشائش عريضة ورفيعة فما هى المكافحة المناسبة مع العلم بان الارض رملية ؟
An area of 1 acre is planted Giza 178 rice by seeding planting way, whât is the proper herbicide and proper amount to handle Donaiba and Agira weeds.	مساحة فدان منزرع ارز جيزة ٨٧١ بطريقة البدار ما هو نوع المبيد و الكمية لحشائش العجيرة و الدنيبة.
There are irregular yellow spots on the top surface of leafs in corn species (Watanyea 4). The plants age is 60 days.	وجود بقع صفراء غير منتظمه على السطح العلوى لاوراق الاذره صنف فردى ٤ وطنيه فى عمر ٠٦ يوم.

The examples shown in Table 2 are the complaints related to different crops which occur during planting and growing stage. The context of the complaints is complex to detect and hence preprocessing is required to prepare the complaints for classification task. The farmers usually write the complaints without following Arabic language grammar rules, so there is difficulty in tokenizing and analyzing the complaints. Also, they used many slang synonyms for each disease name like; ("بقع، مسحوق، صدأ") which in English ("Spots", "Powder", "Rust") and they mean (" ") which in English ("Fungal Disease"). Sometimes, the farmers talk about two types of diseases in the same complaint and that increases the difficulty in classification process like this example: "فى نباتات القمح من تحت التربة واصفرار النبات و تهدلة قرط." which in English means "Erosion in wheat plants from under the soil and yellowing of the plant and sagging". The farmer here mentioned "قرط" (Erosion) which indicate that he is facing an insect's disease and mentioned "اصفرار" (yellowing) to indicate that he is also facing a fungal disease on his plants.

5 Crops and Diseases Lexicons

Most of the researchers use formal language lexicon, dictionary and WordNet to extract features for text analysis. But, it is important to take into consideration the new descriptive words used by users when writing Arabic text for better analysis. For that, we have built 2 lexicons containing formal names of crops and diseases, and their synonyms which are used by the farmers when they describe their complaints. Hence, the lexicons are expected to support in enhancing the accuracy of a classifier.

Based on the recommendations of Agriculture Research Center (ARC) [18] and Virtual Extension and Research Communication Network (VERCON) [19], we have built one lexicon containing all crops' synonyms and keywords and another one containing all varieties of disease causes of crops. The lexicons have been built using XML. Table 3 shows some of the synonyms present in the lexicon with respect to different crops and Table 4 shows examples of causes of different disease with respect to the Wheat crop. Diseases lexicon is presented as three parts representing the disease types: Weeds, Insects and Pests. Each crop might be affected by one of these three types of different diseases. Each type of disease causes has varieties of subtypes. There was a problem in the complaints that they refer to a specific type of disease with different synonyms. So, we collected all the synonyms for each disease by manual view in a sample of the dataset of the complaints. The lexicons have been reviewed and confirmed by an agriculture expert who is working in ARC [18].

Table 3. Examples of synonyms of the crops in crop lexicon

Crop	Examples of the Species of Crops
Wheat	جيزه ١٥٥، جيزه ١٥٦، جيزه ١٥٧، شناب ٧٠، قمح Giza 155, Giza 156, Giza 157, Shnab 70 etc.
Rice	جيزه ١٧٧، جيزه ١٧٨، جيزه ١٨١، جيزه ١٨٢، سخا ١٠١، سخا ١٠٢، سخا ١٠٣، سخا ١٠٤، ارز عطري، مشاتل، ارز، رز مُتث Giza 177, Giza 178, Giza 181, Giza 182, Sahka 101, Sakha 102, Sakha 103, Sakha 104, Aromatic rice
Corn	جيزه١، جيزه بلدي، سبعيني، فيشار، سكري، منغوزه etc Giza 1, Native Giza, Sabeeny, Pop corn (everta), Sweet (saccharata), indentata

Table 4. Shape Examples of the causes of wheat diseases in the lexicon

Disease	Examples of the Types of Causes
Weeds	Annual Broad Leaved weeds: Sonchus oleraceus (جعضيض), Rumex dentatus (حميض), Lathyrus hirsutus (دحريج), Beta vulgaris(سلق), Brassica Kaber (كبر). Annual Narrow Leaved Weeds: Avena spp (الزمير), Polypogen monspeliens (ديل القط), Phalaris sp(الفلارس),Lolium temulentum(الصامه). Perennial Broad Leaved weeds: Conyza discoroides (برنوف),Alhagi spp (عاقول),Convolvulus arvensis (عليق). Perennial Narrow Leaved weeds: Cynodon dactylon(النجيله),Cyperus spp (السعد),Impreta cylindrical (حلفا),Phragmites spp (حجنه).
Fungi	Yellow Rust (الصدأ الأصفر), leaf rust (صدأ الاوراق),stem rust (الصدأ الأسود او صدأ الساق), Ustilago tritice (التفحم السائب), Blumeria graminis (البياض الدقيقى).
Insects and other causes	Aphid (حشرة المن), Birds (العصافير), snails (القواقع), mice (الفئران)

6 Proposed Model

Figure 1 presents the approach that we designed to work on this problem. First, we had started with retrieving the complaints from whole complaints dataset. Second, we prepared and pruned the complaints to be ready for work. After that we initialized the feature selection process by using the lexicons. Then we applied two different classifiers to classify the complaints into diseases. Finally we visualized the outcome results. The following sections explain each step in detail.

6.1 Complaints Preparing

Before representing the data in the form of vectors to apply processing tasks, we should prepare the complaints for data preprocessing by removing stop words. In [20–23], the authors used a list of stop words to remove unwanted words that have high weights like; (in, the, therefore, from) in English language and like; (من، الى، على، لذلك، هذا) in Arabic language. We used a list of more than 600 stop words to eliminate them from the complaints.

The second part of data preparation is the correction step which replaces some wrong written Arabic alphabetic letters with common letters to facilitate features extraction in addition to the removal of unwanted characters. For example, the letters (آ، إ، أ) will be replaced by (ا), the letter (ة) will be replaced by (ه) and the letter (ي) will be replaced by (ى).

Stemming step is not recommended in our model as it has been noticed through the application of different stemmers that they change the meaning of some disease types and affect the feature extraction negatively. We applied stemmer of [20] and it gave a stem "عجر" for the Word "العجيره" that refers to a weed name which is not required. Also we tried stemmer of [24,25] and they did not give a correct stem for the plural words like "حشائش" which means "Weeds" they stems it to the same word.

6.2 Feature Extraction Technique

Feature extraction techniques are used to extract knowledge from text. In [26,27], the authors used a list of Lexicon synonyms and lexicons to extract features from the text, but in [28–30], the authors used ontology. In [22,30], the researchers used POS to get sentences parts as extracted features. Moreover, TF-IDF is used in [23,31]. In our work, we have built a lexicon which contains the crops synonyms and another one which contains disease types of each crop as mentioned before. We can use these lexicons in extracting the crops and disease causes from the complaints. As the complaint consists of at most two sentences and each word item is rarely repeated so, we used binary representation method [32] to represent the items of each complaint vector. $VLC = \{w_i : w_i \in \{\text{Crop Lexicon items}\}, \text{ where } i = 1, 2, 3..n\}$, where n is number of features in crops lexicon. As a preparing for each complaint to be classified, we used VSM to represent the terms of each complaint as a Boolean value $C_{vj} = \{1 \text{ iff } t_i \in VLC,$

0 iff $t_i \notin VLC, i = 1, 2, 3, ..n\}$, where $j = 1, 2, 3, ..m$ and m is the number of complaints.

Similarly, a vector has components related to three types of diseases (Weed, Fungi and Insects) is formed. It is a common vector for all diseases, $VLD = \{w_i : w_i \in \{\text{Disease Lexicon items}\}$ where $i = 1, 2, 3..N$, where N is number of features in disease lexicon. Then we used VSM to represent the complaints as a binary vector model $\in \{0, 1\}$, for each complaint there is one vector to represent the items of the complaint $C_{vj} = \{ 1 \text{ iff } t_i \in VLD, 0 \text{ iff } t_i \notin VLD, i = 1, 2, 3, ..n\}$ where $j = 1, 2, 3, \ldots m$ where m is number of complaints. Each vector contains same number of items in diseases lexicon vector. The item will be 1 if the item is present in the diseases vector and 0 otherwise.

6.3 Disease Classification Models

(Model 1) Disease Classification Through Classification of Crops: This model classifies a given complaint in two levels. In level-1, a classifier is trained to classify an unknown complaint into one of the crops using the crop lexicon. In level-2, another classifier is trained to classify the complaint into one of the diseases of a particular crop. So, given is an unknown complaint, it is first classified into an appropriate crop using the classifier trained at level-1.

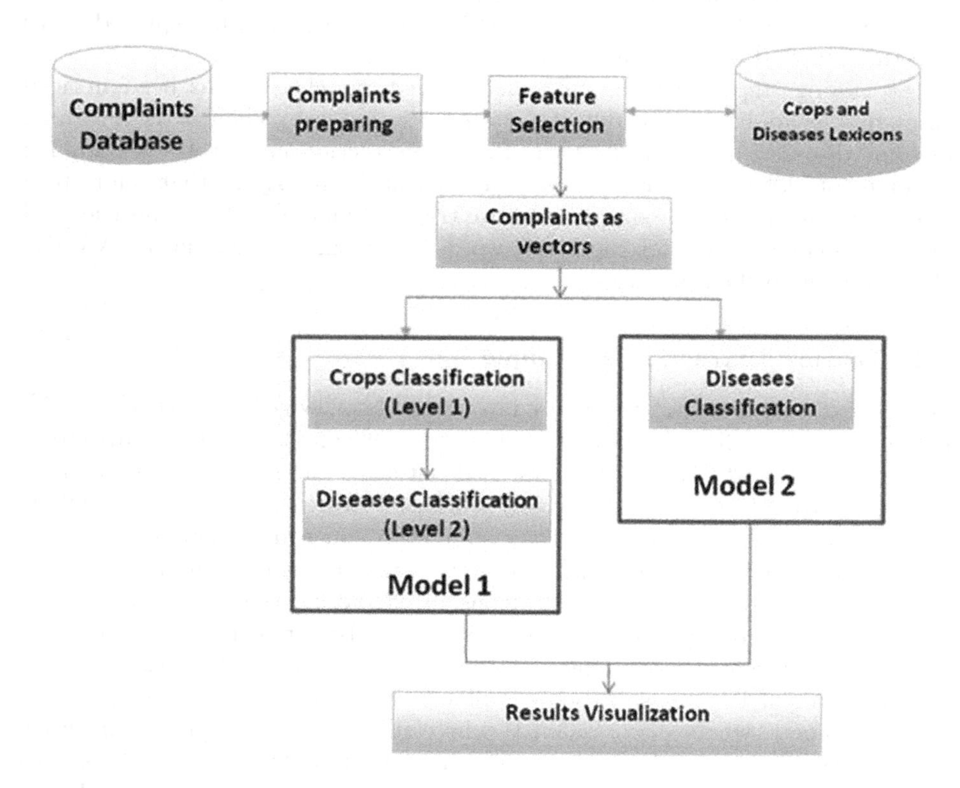

Fig. 1. Architecture of the proposed Farmers' complaints classification model

Then, the complaint is given to the classifier trained at level-2 to classify it as a member of one of the diseases of the crop decided at level-1.

(Model 2) Disease Classification Without Classification of Crops: This model classifies the complaints directly into disease classes. A classifier is trained with a training set consisting of various percentages of samples from the dataset.

7 Experimentation and Results

This section presents the results obtained by the proposed classification model. Because of the simplicity, the fame and the practicality, Support vector machine (SVM) and K nearest neighbor (KNN) classifiers are used to evaluate the proposed models. The classifiers are trained under varying percentages of the dataset to study the behavior of the proposed model. We trained the classifiers with 10%, 20%, 30%, 40% and 50% of the data set and with 10 trials in each.

7.1 Model 1: 2 Levels of Classification

Level-1 (Crops Classification): All the complaints are classified based on crops using SVM and KNN classifiers. Figure 2 shows the average accuracy obtained under varying percentages of the dataset for training for both SVM and KNN classifiers. For KNN, the K is empirically fixed to 5.

In Fig. 2, we noticed that the accuracy increases with the increase in the percentage of training. It can be observed that when 10% of dataset is used for training, the model has attained a lowest average accuracy of 86.68% by SVM and 88.95% by KNN. However, the model has achieved a highest average accuracy of 93.38% by SVM and 93.53% by KNN when 50% of the dataset is

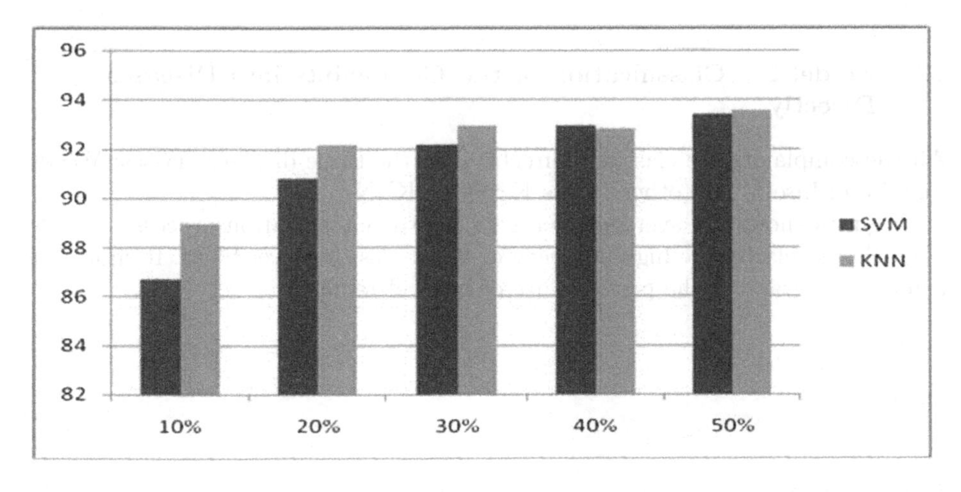

Fig. 2. SVM and KNN Accuracies based on training percentage (Model 1 Leve-1)

used for training. We can see the performance of KNN classifier is relatively high compared to that of SVM classifier when 10%, 20% and 30% of dataset is used for training. However, the performances of both the classifiers appear to be more or less similar when the training percentage is 40% and 50% respectively.

Level-2 (Diseases Classification): After the complaints have been classified based on crop, they are classified into disease, viz., Weeds, Fungi and Insects, and others where K is 3 for KNN.

Figure 3 shows the average accuracies for the classification process of level-2. It is observed that the accuracy is higher in KNN classifier than in SVM for all crops.

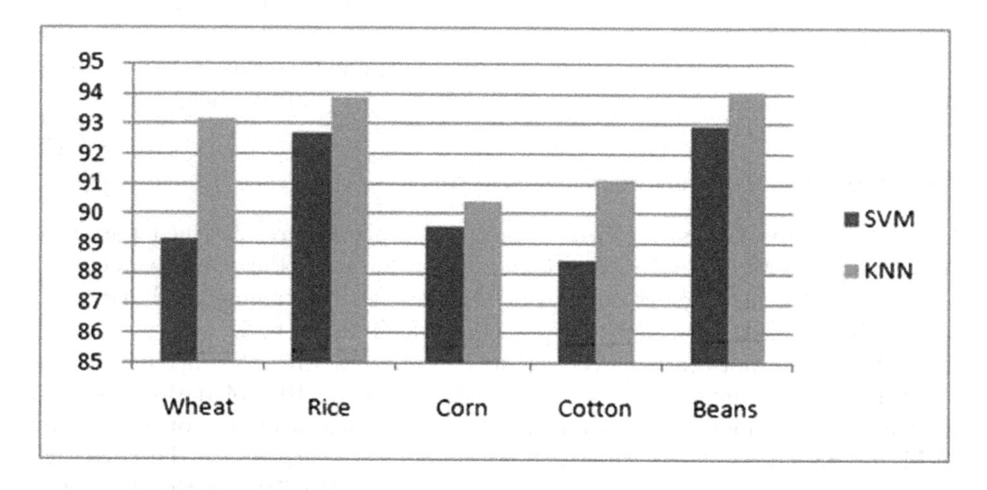

Fig. 3. SVM and KNN Accuracies based on crop class (Model 1 Leve-2)

7.2 Model 2 – Classification of the Complaints into Diseases Directly

All the complaints are classified directly into the three diseases classes; Weeds, Fungi and Insects and others. Here K = 3 for KNN.

Figure 4 shows the average accuracies for the classification process of model 2 which also clarify the high accuracy of KNN classifier over SVM. It shows the comparison graph of the performance of both classifiers.

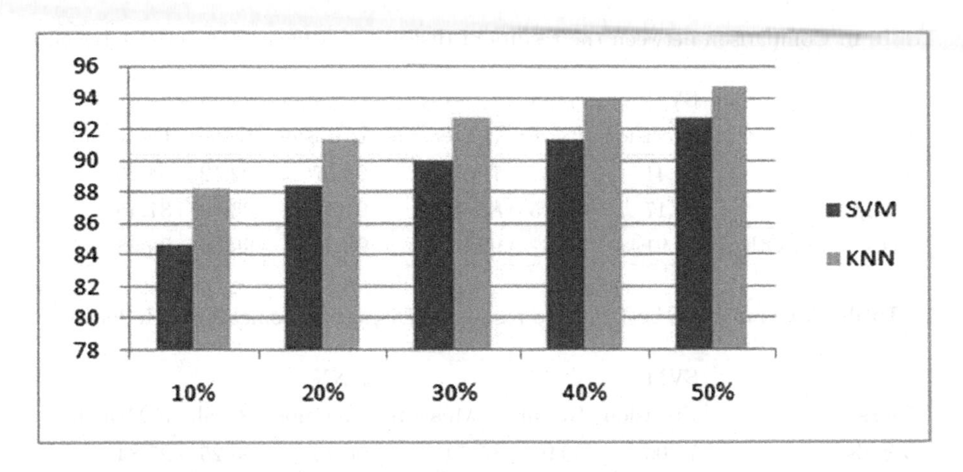

Fig. 4. SVM and KNN Accuracies based on disease training percentage (Model 2)

7.3 Classification Results Comparison

In this part, we provide a numerical and visual comparison between the results of the two models of classifiers with Precision, Recall and F-Measure. As we see in Table 5, F-Measure for Wheat is the highest and then F-Measure for Cotton, but the lowest F-Measure appears for Beans class in SVM classifier. In contrast, the Beans class has the highest F-Measure and Cotton class has got the lowest in KNN. Table 6 shows the results of level-2 classification. It can be seen that the class of Insects and others has the highest F-Measure results which are 92.9 by SVM and 96.8 by KNN. For diseases classification, Table 7 shows that the complaints have been classified directly with the highest F-Measure in Insects class and lowest in Fungi class by SVM and KNN classifiers.

Table 5. Comparison between the results of Crops classification (Model 1-Level1)

	SVM			KNN		
Class	Precision	Recall	F-Measure	Precision	Recall	F-Measure
Wheat	93.68	97.49	95.55	93.96	97.58	95.74
Rice	89.13	97.27	93.02	96.85	89.71	93.14
Corn	80.55	95.43	87.36	94.16	88.92	91.47
Cotton	93.06	95.32	94.18	83.87	93	88.2
Beans	96.82	75.43	84.8	95.24	97.62	96.42

Finally, it can be seen that the classification accuracy has been improved with the inclusion of two new lexicons for crops and diseases. Another observation is that both the models have almost similar performances in terms of their

Table 6. Comparison between the results of diseases classification (Model 1-Level2)

	SVM			KNN		
Class	Precision	Recall	F-Measure	Precision	Recall	F-Measure
Weeds	66.41	99.02	79.5	96.57	92.12	94.29
Fungi	71.17	99.75	83.07	91.57	83.68	87.45
Insects and Others	99.91	86.81	92.9	93.66	99.06	96.28

Table 7. Comparison between the results of Diseases classification (Model 2)

	SVM			KNN		
Class	Precision	Recall	F-Measure	Precision	Recall	F-Measure
Weeds	75.09	99.9	85.74	96.45	95.25	95.84
Fungi	69.13	99.98	81.74	94.37	83.88	88.81
Insects and Others	100	85.49	92.18	94.57	99.02	96.74

classification accuracies. The value of F-measure which obtained for Model-1 is almost similar to that obtained of the Model 2 except for the weed class in case of SVM classifier. It is also noticed that, there are many crops affected with a same disease.

8 Conclusion

Analyzing Arabic farmers' complaints with the Egyptian dialect is not an easy task. In this work, we prepared a farmers' complaints dataset consisting of complaints belonging to five different crops. There are three types of diseases; harmful weed, fungal diseases and insects and other diseases. Two new lexicons have been built describing different diseases and crops respectively. Further, two alternative models for classifying the formers' complaints into different diseases have been proposed. The first model classifies the complaints into diseases based through identification of the crop whereas the second classifies each complaint into different diseases directly. The experimentation conducted on the newly created dataset using SVM and KNN classifiers has demonstrated the applicability of the proposed models for real time operations.

References

1. http://www.fao.org/home/en/
2. http://www.radcon.sci.eg/
3. Park, K., et al.: Mining the minds of customers from online chat logs. In: Proceedings of the 24th ACM International on Conference on Information and Knowledge Management, pp. 1879–1882 (2015)

4. El-Beltagy, S.R., Rafea, A., Mabrouk, S., Rafea, M.: Mining farmers problems in web based textual database application. In: 12th International Conference on Enterprise Information Systems, ICEIS (2010)
5. El-Beltagy, S.R., Rafea, A., Mabrouk, S., Rafea, M.: An approach for mining accumulated crop cultivation problems and their solutions. In: Proceedings of the International Computer Engineering Conference (ICENCO) (2010)
6. Zaghoul, F.A., Al-Dhaheri, S.: Arabic text classification based on features reduction using artificial neural networks. In: UKSim 15th International Conference on Computer Modelling and Simulation (UK-Sim). IEEE (2013)
7. Shoukry, A., Rafea, A.: A hybrid approach for sentiment classification of Egyptian dialect tweets. In: First International Conference on Arabic Computational Linguistics (ACLing), pp. 78–85 (2015)
8. Ahmed, N., Shehab, M., Al-Ayyoub, M., Hmeidi, I.: Scalable multi-label Arabic text classification. In: The International Conference on Information and Communication Systems (ICICS) (2015)
9. Khreisat, L.: Arabic text classification using N-gram frequency statistics: a comparative study. In: Proceedings of the DMIN 2006, Las Vegas, USA (2006)
10. Ismail, H., Bilal, H., Eyas, E.-Q.: Performance of KNN and SVM classifiers on full word Arabic articles. Adv. Eng. Inform. **22**(1), 106–111 (2008)
11. Guru, D.S., Suhil, M.: A novel term_class relevance measure for text categorization. Procedia Comput. Sci. **45**, 13–22 (2015)
12. Al-Shalabiand, R., Obeidat, R.: Improving KNN Arabic text classification with N-grams based document indexing (2008)
13. Mesleh, A.M.: Chi square feature extraction based SVMs Arabic language text categorization system. J. Comput. Sci. **3**(6), 430 (2007)
14. Guru, D.S., Ali, M., Suhil, M.: A novel term weighting scheme and an approach for classification of agricultural Arabic text complaints. In: 2nd IEEE International Workshop on Arabic and derived Script Analysis and Recognition (ASAR), pp. 24–28 (2018)
15. Guru, D.S., Ali, M., Suhil, M.: A novel feature selection technique for text classification. In: Abraham, A., Dutta, P., Mandal, J., Bhattacharya, A., Dutta, S. (eds.) Emerging Technologies in Data Mining and Information Security, vol. 813, pp. 721–733. Springer, Singapore (2019). https://doi.org/10.1007/978-981-13-1498-8_63
16. Guru, D.S., Ali, Mostafa, Suhil, Mahamad, Hazman, Maryam: A Study of Applying Different Term Weighting Schemes on Arabic Text Classification. In: Nagabhushan, P., Guru, D.S., Shekar, B.H., Kumar, Y.H.Sharath (eds.) Data Analytics and Learning. LNNS, vol. 43, pp. 293–305. Springer, Singapore (2019). https://doi.org/10.1007/978-981-13-2514-4_25
17. Hassan, T., Soliman, A., Ali, M.A.: Mining social networks' Arabic slang comments. In: Proceedings of IADIS European Conference on Data Mining (ECDM 2013), pp. 22–24 (2013)
18. http://www.arc.sci.eg/NARIMS_upload/NARIMSdocs/79523/الفنية لمكافحة الحشائش نشر التوصيات2010.pdf
19. http://www.vercon.sci.eg/indexUI/uploaded/wheatinoldsoil/wheatinoldsoil.htm
20. Tutubalina, Elena: Target-Based Topic Model for Problem Phrase Extraction. In: Hanbury, Allan, Kazai, Gabriella, Rauber, Andreas, Fuhr, Norbert (eds.) ECIR 2015. LNCS, vol. 9022, pp. 271–277. Springer, Cham (2015). https://doi.org/10.1007/978-3-319-16354-3_29
21. Mansour, N., Haraty, R.A., Daher, W., Houri, M.: An auto-indexing method for Arabic text. J. Inf. Process. Manag. **44**(4), 1538–1545 (2008)

22. Lee, A.J.T., Yang, F.-C., Chen, C.-H., Wang, C.-S., Sun, C.-Y.: Mining perceptual maps from consumer reviews. J. Decis. Support Syst. **82**, 12–25 (2016)
23. Haddi, E., Liu, X., Shi, Y.: The role of text pre-processing in sentiment analysis. Procedia Comput. Sci. 17, 26–32 (2013). Information Technology and Quantitative Management (ITQM)
24. Khoja, S., Garside, R.: Stemming Arabic text. Lancaster University, Lancaster, UK, Computing Department (1999)
25. Elbeltagy, S.R., Reafea, A.: An accuracy-enhanced light stemmer for Arabic text. ACM Trans. Speech Lang. Process. (TSLP) **7**(2), 1–22 (2011)
26. Dong, S., Wang, Z.: Evaluating service quality in insurance customer complaint handling throught text categorization. In: International Conference on Logistics, Informatics and Service Sciences (LISS), 27–29 July 2015, pp. 1–5 (2015)
27. Georgiou, T., Abbadi, A.E., Yan, X., George, J.: Mining complaints for traffic-jam estimation: a social sensor application. Proceedings of the IEEE/ACM International Conference on Advances in Social Networks Analysis and Mining, ASONAM **2015**, 330–335 (2015)
28. Lee, C.-H., Wang, Y.-H., Trappey, A.J.C.: Ontology-based reasoning for the intelligent handling of customer complaints. J. Comput. Ind. Eng. 84(C), 144–155 (2015)
29. Zirtiloglu, H., Yolum, P.: Ranking semantic information for e-government: complaints management. In: Proceedings of the First International Workshop on Ontology-Supported Business Intelligence, OBI 2008. ACM (2008). Article no. 5
30. Al Zamil, M.G.H., Al-Radaideh, Q.: Automatic extraction of ontological relations from Arabic text. J. King Saud Univ. Comput. Inf. Sci. 28(1), 1–146 (2016)
31. Liang, Y.H.: Integration of data mining technologies to analyze customer value for the automotive maintenance industry. Expert Syst. Appl. **37**, 7489–7496 (2010)
32. Lan, M., Tan, C.L., Su, J., Lu, Y.: Supervised and traditional term weighting methods for automatic text categorization. IEEE Trans. Pattern Anal. Mach. Intell. **31**(4), 721–735 (2009)

Neural Network Modeling and Prediction of Daily Average Concentrations of PM_{10}, NO_2 and SO_2

Sateesh N. Hosamane(✉)

Department of Chemical Engineering, KLE Dr. M.S. Sheshgiri College of Engineering and Technology, Udyambag, Belgaum 590008, Karnataka, India
satishosamane@gmail.com

Abstract. Three-layer principal component based artificial neural network (ANN) model is used to predict PM_{10}, NO_2 and SO_2 concentration. The developed model predictions are compared with the measured pollutant concentrations. The daily average pollutant concentrations and five meteorological variables are used to develop pollution forecast models. The selected monitoring site is a typical residential area with high traffic influence and the air pollution is because of nearby industries. A principal component regression (PCR) model is used for comparing the results obtained by the developed neural network model. The performance of the developed model were assessed using various performance index. Developed models exhibit a decent performance >70–95% for three measured pollutants. The future models performed with good accuracy and the predicted pollutant concentrations were confirmed to be adequate after computing the accuracy using performance indicators.

Keywords: Pollutants · Meteorological parameters · PCA · PCR · ANN

1 Introduction

Air pollution is a state in which the air pollutants originate in the environment at sufficient concentrations, higher than their prescribed ambient levels [1]. The primary sources of air pollutants in most of the cities are mobile, stationary and area sources. Air pollution is expressly significant in urban areas in view of a multitude of people suffering from various health conditions. In many urban areas, pollutant concentration become very serious and is due to predominantly adverse meteorological conditions [2]. Air pollution associated problems have result in increased public awareness of air quality across the world [3]. Lack of statutory guidelines for implementing environmental regulations is contributing to bad air quality in most of the Indian cities. Fossil fuel combustion is one the most significant sources of ambient air pollution since fossil fuels are the main source of energy consumption. These fuels, upon combustion, produce nitrogen

K. C. Santosh and R. S. Hegadi (Eds.): RTIP2R 2018, CCIS 1037, pp. 429–442, 2019.
https://doi.org/10.1007/978-981-13-9187-3_39

dioxides (NO_2), sulphur dioxides (SO_2), dust (flying ash), soot (a black deposit), smoke and other suspended particulate matter (SPM) [4]. The air quality study is carried out at Udyambag of Belagavi city and it was found that the level of SPM is very high and it is due to excessive concentrations of population and manufacturing industries [5]. Central Pollution Control Board (CPCB) has been collecting data for particulate matter that have a diameter less than 10 μm (PM_{10}), NO_2 and SO_2. Air pollution control and management is very crucial to prevent this condition from getting unhealthy in the long period. The high concentrations of these pollutants can be life threatening and cause skin irritation, dizziness, fatigue, headache and difficulty in breathing [6,7]. Conversely, immediate prediction of air quality is necessary to take protective action during incidences of air pollution [8]. Local regulatory authorities of health and environmental agencies are responsible to forecast daily air pollution concentration levels for public advisory and assessment regarding reduction measures of air quality [9]. Therefore, the development of operational models for air pollutant concentrations in urban areas are important. Supervision, control and public forewarning approaches for the pollutant levels requires accurate estimates of the pollutant concentrations. The air pollution predictions are frequently based on statistical interaction between meteorological settings and ambient air pollution concentrations. Such relationships between pollutant concentrations and meteorological conditions have been examined in various studies by using combination of statistical regression [10,11]. A number of methods and their combinations involving multiple linear regression (MLR), principal component regression (PCR) and ANN have been used to form air pollution predictive models [12–16]. However, air pollutant concentrations and weather relationships are very complex and nonlinear in nature and are well modeled by neural networks. These models deliver a better alternative to statistical approach because of their speed and accuracy of computation and simplification capability [17–20]. Many other investigations have shown that the ANN models have performed better in comparison to similar statistical methods and many air pollution predictive models are put into operation in many cities across the world [16,21,22]. The best method to solve the complexity of problem is the principal component analysis (PCA), which has been getting increased consideration as a recognised tool in complex environmental data analysis and pattern recognition [11,23]. In the current study PCA based ANN techniques is used for PM_{10}, NO_2 and SO_2 predictive models at udyambag industrial area at Belagavi City of Karnataka state, India. PCR is used to compare the performance of principal component based ANN (PCANN) models [11,23,24]. PCA is very useful when dealing with a large set of data and when the number of variables involved in the analysis are complex [23,25]. The PCA was applied to decrease the complexity of input variables in the ANN model with reduction of training time and to improve the accuracy of ANN model [26]. The main objective of this paper is to exhibit a methodology for extracting various meteorological parameters on measured pollutant concentrations and to build a strong PC based ANN mode.

2 Materials and Methods

2.1 Study Area and Data

Belagavi is a city in the southwest Indian state of Karnataka located in its northern part along the Western Ghats. Belagavi is located at 15.87 °N 74.5 °E with an altitude of 751 m or 2,464 feet. The city borders the two states, Maharashtra and Goa. On the demographic front, the city has a population of 488,157 as per the Indian census of 2011. The Fig. 1 shows the satellite picture of air pollution monitoring site. Air samples were collected at Udyambag, an industrial area with small and medium scale industries. It is also surrounded by many schools and colleges and witnesses a heavy traffic during the daytime.

2.2 Data Set

Air pollutant PM_{10} (μg/m^3) sampling and analysis was carried out using respirable dust sampler (APM 460 NL-Envirotech) and gas sampler (APM 433-Envirotech) was used for the measurement of NO_2 (μg/m^3) and SO_2 (μg/m^3) concentrations using appropriate chemical reagents. The metrological and pollutant concentrations data sets are measured based on daily average (24-h) for the period 2011 to 2015. The meteorological parameters Temperature (T), Wind Speed (WS), Wind Direction (WD), Relative Humidity (H) and Rainfall (R) along with air pollutant concentrations PM_{10}, NO_2 and SO_2 have been used for investigation. The Table 1 shows the basic statistics of various meteorological parameters and pollutant concentrations collected during the study period. The daily average (24 h) meteorological parameter temperature (T,°C), wind speed (WS, m/s), wind direction (WD, degree), rainfall (R, mm) and relative humidity (H. %) were collected from Sambra Meteorological station near Belagavi city. Annual average temperature is 25.33 °C, annual average wind speed is 2.82 m/s, annual average wind direction is 194.72°, annual monthly average rainfall is 54 mm and annual average relative humidity is 64.58%.

2.3 Computation of Principal Components

PCA is an advanced computational approach that originated from the multivariate statistical method that allows the detection of important information also used to express the data highlighting the differences and similarities. Principal components (PC) were calculated by the covariance of input information matrix using Minitab 7 software. The PC related among the highest eigenvalue, the first principal component (PC1), signifies the linear grouping of the variables that contribute for the highest total changeability in the data. The second PC explicates the highest changeability that is not contributed for the PC1 [27]. The components with eigenvalues greater or equal to 1 were used for the investigation. After generating the PCs, the initial data set is altered into the orthogonal set and is multiplied by the eigenvectors to the preliminary data set. The altered matrix is used to model the process using MLR technique and is represented by the Eq. (1) [28].

$$Y = \beta_0 + \beta_1(PC_1) + \beta_2(PC_2) + ... + \beta_n(PC_n) + \varepsilon \quad (1)$$

Where β_0,β_1,β_2...........β_n are the coefficients of the model equation.

The PCs were derived and used as input for the regression method known as principal component regression (PCR). The Table 2 represents the Pearson correlation matrices of the five meteorological variables and three pollutant concentrations. Statistically significant correlation coefficients ($P < 0.05$) are bolded in the table. PM_{10} concentration is positively correlated with T and negatively correlated with WS and WD. The SO_2 is negatively correlated with WS, WD, R and H. The NO_2 concentrations were positively correlated with T. The correlation between meteorological parameters with SO_2 and NO_2 was very poor ($P > 0.05$) and the overall correlations of pollutant concentrations with meteorological parameters are relatively weak.

Table 3 shows the sum of variance explicated by each component for the monitoring site at udyambag. PCs with an eigenvalue >1 or ~60% of variance explained by the correlation matrix by PCs are used for the analysis. PCA result for Udyambag is summarised in the component Table 3 in which T, WS, WD, R, H and PM_{10} are grouped as the first factor and is explained by 30.8% by PC1 loading. T, WD and NO_2 loading by PC2 are explained by 14.8% of the total variance of the data. Similarly, T, WS, PM_{10}, SO_2 and NO_2 are grouped into the third factor as PC3 and is explained by 13.2% of the sum of variance of the data.

The sum of PC are plotted on x axis and cumulative eigenvalue is plotted on y axis for the monitored data set. The connection (dotted line) in the figure determines the number of PCs used for the study. The first three PCs with eigenvalues >1 is used for PCR which is represented in Fig. 2.

Predicted values PM_{10}, NO_2 and SO_2 obtained by PCR modeled equation (2), (3) and (4) are developed using linear regression. The PCR model is developed by using 2011–2014 data and the process is validated using 2015 data which is not used to build the model.

$$PM_{10} = 16.48 - 0.6746 * vPC1 - 0.5343 * vPC2 + 0.6420 * vPC3 \quad (2)$$

$$SO_2 = 6.816 - 0.0268 * vPC1 - 0.02023 * vPC2 + 0.0328 * vPC3 \quad (3)$$

$$NO_3 = -19.65 + 0.0064 * vPC1 - 0.1872 * vPC2 + 0.7085 * vPC3 \quad (4)$$

2.4 Model Building

The simple neural network structure, unidirectional feed forward network used for the architecture consists of solitary set of neural units are coupled to the input layer. The feed forward neural network (FFNN) is capable of solving a large class of problems [29]. The network architecture used in the model was the multilayer perceptron (MLP) with one hidden layer. The network consists of input, hidden and the output layer. The hidden layer is optimised by conducting numerous tests with different network structure subsequently this produces smaller mean square error (MSE) values and smaller convergence times. The present research work is to use PM_{10}, NO_2 and SO_2 (24-h average concentration) modeling for forecasting. The data set used are daily average concentration of PM_{10}, NO_2 and SO_2, and daily averages of T, WS, WD, H and R since all these factors are proven to be of importance for illustrating the effect of pollutant concentration. The meteorological parameter data sets and pollutant datasets were used via the PCA method discussed in the preceding Sect. 2.2. The ANN models are developed using the MATLAB software. It is used to mimic the human brain's capability to organise patterns or to make prediction or conclusion based on the experience [30]. MATLAB is designed to aid the designer who have a minimum of specialization and comprehension to build neural network Models [28]. MATLAB can acquire patterns from training data and generates its predictions when accessible with new hidden data. Levenberg - Marquardt (trainlm) algorithm is used to train the developed architecture using PCs as input parameters. The method of development of PC based ANN is shown in Fig. 3.

2.5 Data Preparations

The data preparation is based on training and testing sets, the data set for the present study were arbitrarily split into a training set and the testing set is in the proportion of 3:1. The monitored data for each subset was divided into training, validation and testing sets. The 2011 to 2014 datasets are used for the training set, while 2015 data sets are used for testing and validation. The developed model is used to forecast PM_{10}, NO_2 and SO_2 daily concentrations (24-h average).

2.6 Index for the Model Performance

Evaluation of the developed model for the predictable performance is very significant to check the forecast precision. The performance of the model is verified by using measures of error, based on the deviation between predicted values and original deviation [31]. The developed model is evaluated using Mean-Absolute Error (MAE), Mean Square Error (MSE), Root Mean Square Error (RMSE), Mean Absolute Percentage of Error (MAPE) and Correlation Coefficient (R) that are given by Eqs. (5), (6), (7), (8) and (9) respectively.

$$MAE = \frac{\sum_{i=1}^{n} |Y_i - X_i|}{n} \tag{5}$$

$$MSE = \frac{\sum_{i=1}^{n} (Y_i - X_i)^2}{n} \tag{6}$$

$$RMSE = \sqrt{\frac{\sum_{i=1}^{n} (Y_i - X_i)^2}{n}} \tag{7}$$

$$MAPE = \sqrt{\frac{\sum_{i=1}^{n} |(Y_i - X_i)/X_i|}{n}} \tag{8}$$

$$R = \frac{\sum_{i=1}^{n} (Y_i - \overline{Y_i})(X_i - \overline{X_i})}{\left\{\left[\sum_{i=1}^{n} (X_i - \overline{X_i})^2\right]\left[\sum_{i=1}^{n} (Y_i - \overline{Y_i})^2\right]\right\}^{1/2}} \tag{9}$$

Where,n is the number of data, Y_i is the predicted concentration, X_i is the observed concentration. Zero Error indicate that prediction of pollutant concentrations computed by PCR and PC based ANN models were completely matching the observed pollutant concentrations.

3 Results and Discussion

The PCs are selected based on the variance explicated by the eigenvalues of the correlation matrix. The cumulative amount of PCs with eigenvalues discussed in Sect. 2.2 was used to develop ANN models. Only three PCs were used instead of 8 variables. The smallest error was attained for trainlm function and the MSE was used to select the model performance and is optimised by changing in the sum of hidden nodes and maintaining the logsig transfer function [32]. The ANN architecture with back propagation that yielded best results with one hidden layer is trained with 1000 epochs [31]. The performance evaluation with respect to the number of nodes is shown in Table 4. The PM_{10} testing and training findings for all neuron combinations for the data under the study is captured in Table 4. The MAPE values were found to be between 0.065 to 0.104 for testing set with 10 to 25 neurons in the hidden layer. 20 neurons have performed well with MAPE of 0.065 for the training and 0.082 testing of the dataset. Similarly, for SO_2, MAPE values were found to be 0.072 to 0.094 for testing set with 5 to 20 neurons in the hidden layer. 10 neurons have performed well with MAPE of 0.352 for the training and 0.215 testing of the data set. For NO_2 MAPE is found to be 0.093 to 0.155 for testing set with 15 to 30 nodes.

The network architecture employed for prediction of PM_{10}, SO_2 and NO_2 concentration levels and effectiveness of the developed ANN model for udyambag is presented in Table 5. From the results of the optimised network architecture, the following observations can be made. For PM_{10}, the data sets tend to under fit if the number of neurons is <20, over fit was observed by increasing neuron number to >20. It is observed from the Table 5 that the MSE was increasing from 11.544 to 24.617 as the neurons increase in number from 20 to 25. A good conformity was observed among the observed and predicted PM_{10} concentration at udyambag. It may be attributed to the strong influence of selected inputs. The high and low concentrations of PM_{10} was well predicted by the models due to the type of inputs used during the model training. For SO_2 datasets tend to underfit if the number of neurons is <5, overfit was observed by increasing neuron number to >5. It is observed from the Table 5 that the MSE is increasing from 5.513 to 6.955 with an increase in neuron number from 5 to 15. Performance of SO_2 has not been accurately predicted by optimized architecture. The inability of the model to predict the low concentrations of SO_2 is probably because of selection of input information and the negligible concentration of SO_2. Similarly for NO_2, MAPE values were found to be 0.042 to 0.150 for testing set with 15 to 30 neurons at the hidden layer. 25 neurons have performed well with MAPE of 0.038 for training and 0.087 for testing the dataset. Inspecting the results, the model evaluation for PM_{10}, SO_2 and NO_2 is extensive and developed models have performed satisfactorily. The performance of architecture was measured

Table 1. Basic statistics of meteorological parameters and pollutant concentrations form 2011–2015.

Varaible	Mean	Standard deviation	Minimum	Maximum
Temperature, °C	25.33	2.93	19.75	34.09
Wind speed, m/s	2.83	1.07	0.17	12.09
Wind direction, degree	194.52	81.49	22.5	337.5
Rainfall, mm	0.41	0.81	0	9.23
Humidity, %	64.58	20.27	17.5	99.07
PM_{10}, ($\mu g/m^3$)	58.84	20.89	9.87	152.36
SO_2, ($\mu g/m^3$)	8.45	4.08	0.47	28.01
NO_2, ($\mu g/m^3$)	20.02	12.55	3.4	81.78

Table 2. Pearson's correlation matrix for Udyambag

Varaible	T	WS	WD	R	H	PM_{10}	SO_2	NO_2
T,°C	1.000	**−0.421**	**0.101**	**−0.192**	**−0.391**	**0.182**	**0.084**	**0.120**
WS, (m/s)		1.000	**0.261**	**0.223**	**0.500**	**−0.200**	−0.071	−0.041
WD, (degrees)			1.000	**0.212**	**0.439**	**−0.211**	**−0.083**	0.011
Rain, (mm)				1.000	**0.445**	**−0.291**	**−0.110**	0.151
H, (%)					1.000	**−0.290**	**−0.101**	0.056
PM_{10}($\mu g/m^3$)						1.00	0.08	0.153
SO_2($\mu g/m^3$)							1.000	0.050
NO_2($\mu g/m^3$)								1.00

Table 3. PCA results for Udyambag

Variable	PC1	PC2	PC3	PC4	PC5	PC6	PC7	PC8
T	**−0.352**	**0.561**	**0.366**	−0.090	0.017	0.076	−0.459	0.449
WS	**0.452**	−0.111	**−0.301**	−0.034	0.422	−0.034	−0.716	−0.010
WD	**0.324**	**0.576**	0.265	−0.156	0.418	0.060	0.261	−0.471
R	**0.386**	0.116	0.084	0.035	−0.665	0.542	−0.227	−0.203
H	**0.532**	0.164	−0.161	−0.004	0.024	0.068	0.379	0.718
PM_{10}	**−0.332**	0.085	**−0.419**	0.256	0.334	0.720	0.105	0.000
SO_2	−0.143	0.066	**−0.395**	−0.894	−0.120	0.065	0.035	−0.023
NO_2	−0.063	**0.537**	**−0.583**	0.317	−0.279	−0.410	−0.036	−0.139
Eigenvalue	**2.468**	**1.186**	**1.058**	0.956	0.808	0.724	0.469	0.332
Proportion	**0.308**	**0.148**	**0.132**	0.119	0.101	0.09	0.059	0.041
Cumulative	**0.308**	**0.457**	**0.589**	0.709	0.809	0.900	0.959	1.000

Table 4. Evaluation of ANN model performance for PM_{10}, SO_2 and NO_2

NODES	MAE	MSE	RMSE	MAPE	MAE	MSE	RMSE	MAPE
PM_{10}- TRAINING					PM_{10}- TESTING			
10	1.798	24.633	33.567	0.09	3.285	63.105	36.285	0.092
15	4.388	29.287	30.254	0.104	2.796	54.544	35.796	0.087
20	2.753	11.544	27.567	0.065	1.993	42.294	44.993	0.082
25	4.005	24.617	28.356	0.09	3.451	45.324	35.451	0.075
SO_2- TRAINING					SO_2- TESTING			
5	1.856	5.513	6.856	0.36	1.366	2.889	3.366	0.188
10	1.915	5.996	7.315	0.352	1.732	4.598	4.732	0.215
15	2.096	6.955	6.096	0.415	1.67	4.318	6.67	0.231
20	1.842	5.341	7.842	0.37	1.499	3.477	5.499	0.213
NO_2- TRAINING					NO_2- TESTING			
15	0.87	1.381	27.83	0.061	3.597	21.704	23.597	0.108
20	0.583	0.574	30.15	0.042	2.615	11.711	12.615	0.093
25	0.588	0.723	22.45	0.038	2.771	13.121	13.771	0.087
30	2.047	6.855	28.047	0.15	4.525	35.282	42.525	0.155

Table 5. Optimised of PCANN model performance for PM_{10}, SO_2 and NO_2

Input- 3 PCs		Network*	Training	Testing	Validation
PM_{10}	R	3-20-1-1	96.53%	96.01.%	93.970%
	RMSE	3-20-1-1	27.567	44.993	29.523
SO_2	R	3-5-1-1	72.78%	66.65%	61.60%
	RMSE	3-5-1-1	6.856	3.366	6.261
NO_2	R	3-25-1-1	96.40%	89.34%	90.10%
	RMSE	3-25-1-1	22.451	13.771	25.296

for testing, training, validation and overall performance of data sets shown in Table 5. The predicted PM_{10}, SO_2 and NO_2 concentration by the optimised model were compared with observed concentrations as shown in Fig. 4.

Table 6. Comparison of validation results for PCR and PCANN

	PCR				PCANN			
	MAE	MSE	RMSE	R	MAE	MSE	RMSE	R
PM_{10}	12.012	231.25	57.001	0.203	2.526	10.049	12.523	0.937
SO_2	2.924	13.101	21.235	0.416	2.877	12.979	5.261	0.616
NO_2	8.349	100.465	31.26	0.395	1.296	14.829	9.296	0.901

Fig. 1. Satellite picture of monitoring site

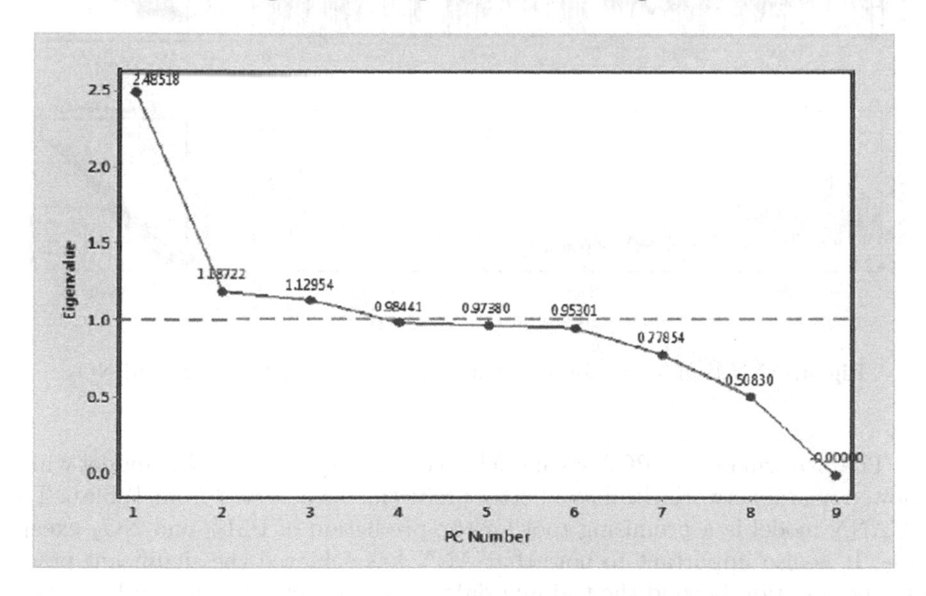

Fig. 2. Selection of PCs for Udyambag dataset

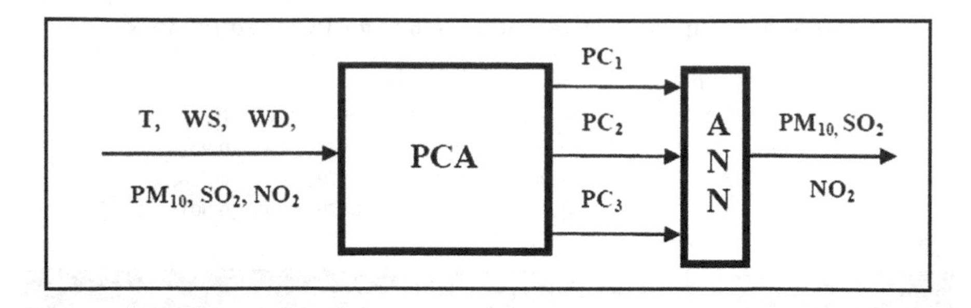

Fig. 3. PC based ANN model

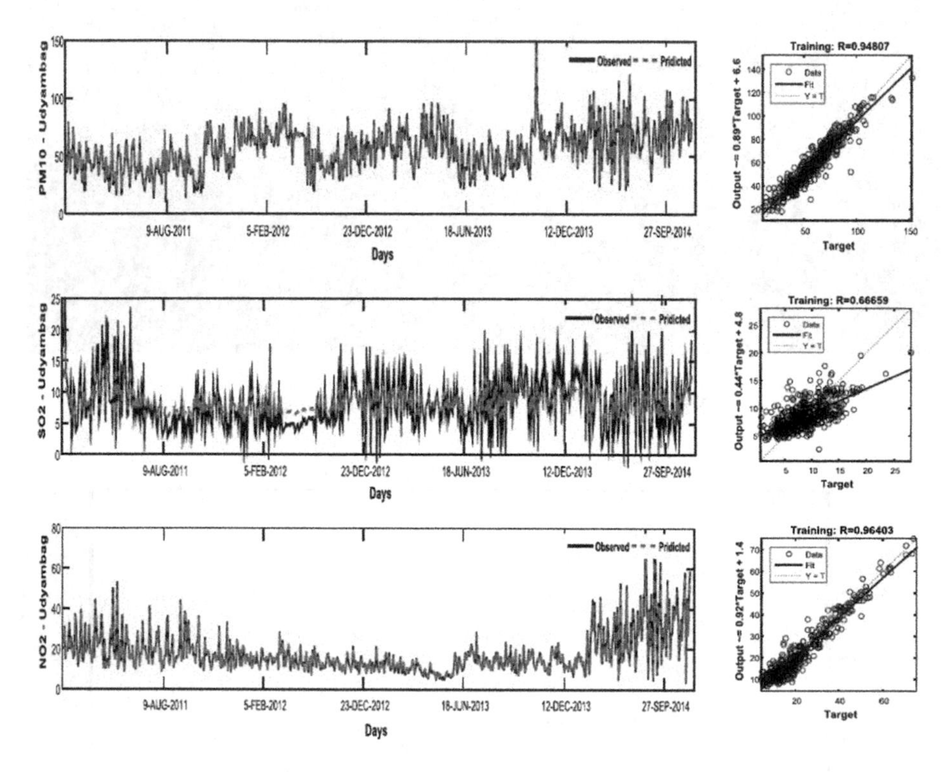

Fig. 4. ANN training performance, and prediction of PM_{10}, SO_2 and NO_2

The performance of PCANN model is compared with the PCR model which showed neural networks lead to a better prediction as observed from Table 6. The PCANN model is a promising tool for the prediction of PM_{10} and NO_2 except SO_2. It is also important to note that ANN has achieved the significant power of generalisation beyond the training data. It was observed from the Fig. 5 that highest measured concentrations were underestimated, and also that the model was unable to predict lower concentrations of SO_2. This was observed mainly

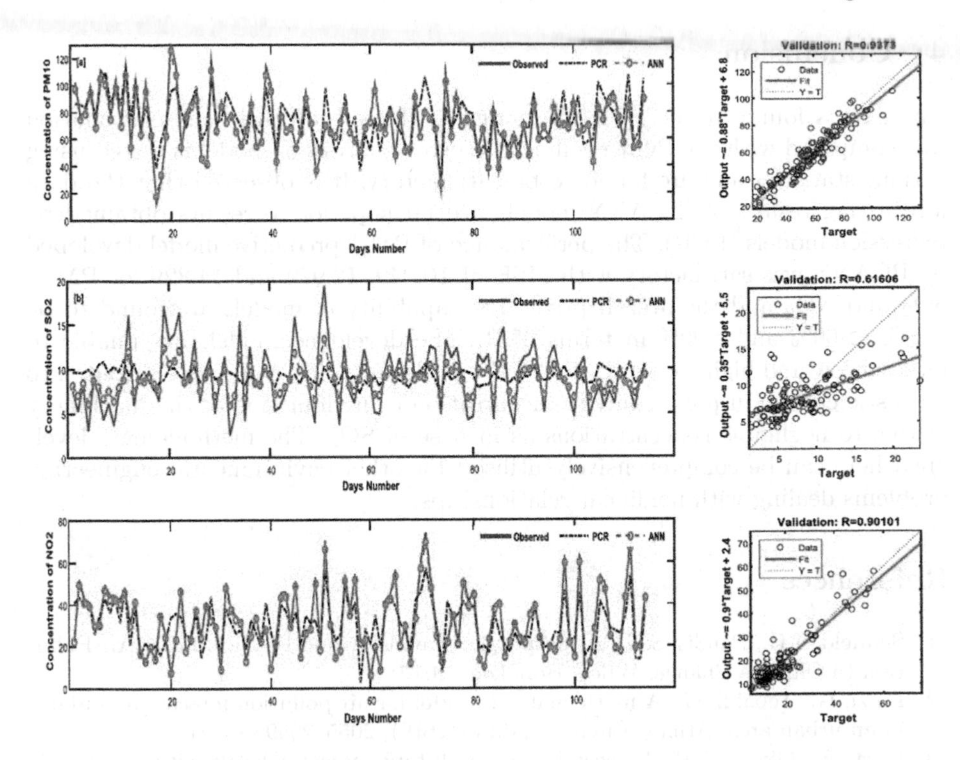

Fig. 5. PCR and ANN model performance comparison and validation of PCANN model [a] PM_{10} [b] SO_2 and [c] NO_2

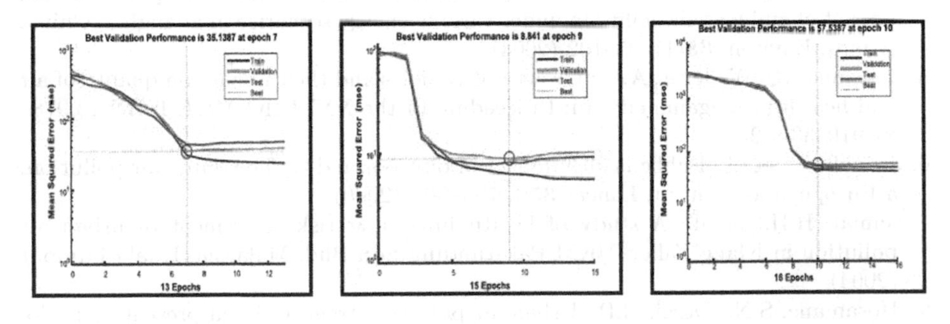

Fig. 6. Validation performance of PCANN Model for PM_{10}, SO_2 and NO_2

because, during the validation period 36% percentage of data points are having concentration less than 1 $\mu g/m^3$ and during the training period it was only 18%. The validation performance is measured using mean square of error and is shown in Figure 6. The performance index, obtained using PCANN was almost the same for training and validation periods.

4 Conclusion

The results found by the PCANN models with feed forward back propagation was compared with the PCR results. The error analysis of model is tested using various statistics and are found to be satisfactory. It is observed that the data forecasting results of the ANN models are far superior to results obtained by regression models (PCR). The performance of PM_{10} predictive model developed by PCANN was satisfactory with MSE of 10.049, 12.979 and 14.829 for PM_{10}, SO_2 and NO_2 and the overall predictive capability of models are found to be >90%, >60% and >90% in terms of R. The developed models are unable to forecast low and high concentration of the pollutants that could be attributed to a drastic change in point source concentration or malfunction of the machinery or due to negligible concentrations as in case of SO_2. The methodology developed here can be comprehensively utilised for other environmental engineering problems dealing with nonlinear relationships.

References

1. Seinfeld, J.H., Pandis, S.N.: Atmospheric Chemistry and Physics: From Air Pollution to Climate Change. Wiley, Hoboken (2012)
2. Finzi, G., Tebaldi, G.: A mathematical model for air pollution forecast and alarm in an urban area. Atmos. Environ. (1967) **16**(9), 2055–2059 (1982)
3. Kurt, A., Oktay, A.B.: Forecasting air pollutant indicator levels with geographic models 3 days in advance using neural networks. Expert Syst. Appl. **37**(12), 7986–7992 (2010)
4. Kan, H., Chen, B., Chen, C., Fu, Q., Chen, M.: An evaluation of public health impact of ambient air pollution under various energy scenarios in Shanghai, China. Atmos. Environ. **38**(1), 95–102 (2004)
5. Basanna, R., Wodeyar, A.K.: Growth of vehicles and their impact on quality of air and health in Belgaum city. In: Proceeding of the XXXI IIG Meet. ISBN-13:978–81-910533-0-2
6. Künzli, N., et al.: Public-health impact of outdoor and traffic related air pollution: a European assessment. Lancet **356**, 795–801 (2000)
7. Jamal, H.H., et al.: A study of health impact & risk assessment of urban air pollution in Klang Valley. UKM Pakarunding Sdn Bhd, Malaysia, Kuala Lumpur (2004)
8. Hosamane, S.N., Desai, G.P.: Urban air pollution trend in India present scenario. Int. J. Innovative Res. Sci. Eng. Technol. **2**(8), 3738–47 (2013)
9. Wang, W., Xu, Z., Weizhen Lu, J.: Three improved neural network models for air quality forecasting. Eng. Comput. **20**(2), 192–210 (2003)
10. Akpinar, E.K., Akpinar, S., Öztop, H.F.: Statistical analysis of meteorological factors and air pollution at winter months in Elaziğ, Turkey. J. Urban Environ. Eng. **3**(1), 7–16 (2009)
11. Comrie, A.C.: Comparing neural networks and regression models for ozone forecasting. J. Air Waste Manag. Assoc. **47**(6), 653–663 (1997)
12. Sousa, S.I.V., Martins, F.G., Alvim-Ferraz, M.C.M., Pereira, M.C.: Multiple linear regression and artificial neural networks based on principal components to predict ozone concentrations. Environ. Model. Softw. **22**(1), 97–103 (2007)

13. Schlink, U., et al.: A rigorous inter-comparison of ground-level ozone predictions. Atmos. Environ. **37**(23), 3237–3253 (2003)
14. Abdul-Wahab, S.A., Bakheit, C.S., Al-Alawi, S.M.: Principal component and multiple regression analysis in modelling of ground-level ozone and factors affecting its concentrations. Environ. Model. Softw. **20**(10), 1263–1271 (2005)
15. Lengyel, A., Héberger, K., Paksy, L., Bánhidi, O., Rajkó, R.: Prediction of ozone concentration in ambient air using multivariate methods. Chemosphere **57**(8), 889–896 (2004)
16. Abdul-Wahab, S.A., Al-Alawi, S.M.: Assessment and prediction of tropospheric ozone concentration levels using artificial neural networks. Environ. Model. Softw. **17**(3), 219–228 (2002)
17. Chelani, A.B.: Predicting chaotic time series of PM_{10} concentration using artificial neural network. Int. J. Environ. Stud. **62**(2), 181–191 (2005)
18. Boznar, M., Lesjak, M., Mlakar, P.: A neural network based method for short term predictions of ambient SO_2 concentrations in highly polluted industrial areas of complex terrain. Atmos. Environ. Part B. Urban Atmos. **27**(2), 221–230 (1993)
19. Cogliani, E.: Air pollution forecast in cities by an air pollution index highly correlated with meteorological variables. Atmos. Environ. **35**(16), 2871–2877 (2001)
20. Mishra, D., Goyal, P.: Development of artificial intelligence based NO2 forecasting models at Taj Mahal, Agra. Atmos. Pollut. Res. **6**(1), 99–106 (2015)
21. Chelani, A.B., Rao, C.C., Phadke, K.M., Hasan, M.Z.: Prediction of sulphur dioxide concentration using artificial neural networks. Environ. Model. Softw. **17**(2), 159–166 (2002)
22. Gardner, M.W., Dorling, S.R.: Neural network modelling and prediction of hourly NOx and NO_2 concentrations in urban air in London. Atmos. Environ. **33**(5), 709–719 (1999)
23. Lu, W.Z., Wang, W.J., Wang, X.K., Xu, Z.B., Leung, A.Y.T.: Using improved neural network model to analyze RSP, NOx and NO_2 levels in urban air in Mong Kok, Hong Kong. Environ. Monit. Assess. **87**(3), 235–254 (2003)
24. Sanchez, M.L., Casanova, J.L., Ramos, M.C., Sanchez, J.L.: Studying the spatial and temporal distribution of SO_2 in an urban area by principal component factor analysis. Atmos. Res. **20**(1), 53–65 (1986)
25. Karatzas, K.D., Kaltsatos, S.: Air pollution modelling with the aid of computational intelligence methods in Thessaloniki, Greece. Simul. Model. Pract. Theory **15**(10), 1310–1319 (2007)
26. Al-Alawi, S.M., Abdul-Wahab, S.A., Bakheit, C.S.: Combining principal component regression and artificial neural networks for more accurate predictions of ground-level ozone. Environ. Model. Softw. **23**(4), 396–403 (2008)
27. Slini, T., Kaprara, A., Karatzas, K., Moussiopoulos, N.: PM10 forecasting for Thessaloniki, Greece. Environ. Model. Softw. **21**(4), 559–565 (2006)
28. Sahin, U., Ucan, O.N., Soyhan, B., Bayat, C.: Modeling of CO distribution in Istanbul using artificial neural networks. Fresenius Environ. Bull. **13**(9), 839–845 (2004)
29. Goyal, P., Kumar, A., Mishra, D.: The impact of air pollutants and meteorological variables on visibility in Delhi. Environ. Model. Assess. **19**(2), 127–138 (2014)
30. Zhang, G., Patuwo, B.E., Hu, M.Y.: Forecasting with artificial neural networks: the state of the art. Int. J. Forecast. **14**(1), 35–62 (1998)
31. Baawain, M.S., Al-Serihi, A.S.: Systematic approach for the prediction of ground-level air pollution (around an industrial port) using an artificial neural network. Aerosol Air Qual. Res. **14**(1), 124–134 (2014)

32. Hosamane, S.N., Desai, G.P.: Air pollution modelling from meteorological parameters using artificial neural network. In: Hemanth, D., Smys, S. (eds.) Computational Vision and Bio Inspired Computing. LNCVB, vol. 28, pp. 466–475. Springer, Cham (2018). https://doi.org/10.1007/978-3-319-71767-8_39

Smart Irrigation and Crop Yield Prediction Using Wireless Sensor Networks and Machine Learning

D. L. Shanthi(✉)

BMS Institute of Technology and Management, Bangalore, Karnataka, India
gopalaiahshanthi@bmsit.in

Abstract. Agriculture decides the monetary development of country and is known to be its backbone. Farmers, specialists, and specialized makers are joining endeavors to discover more effective answers for taking care of different distinctive issues in agriculture to enhance current generation and procedures Precision. The proposed structure for exactness agriculture utilizes ease natural sensors, an Arduino Uno prototyping board and a couple of remote handsets (XBee ZB S2) alongside inciting circuit to give robotized water system and checking of harvests. The proposed model uses XBee convention which depends on ZigBee innovation. The vital attributes of ZigBee innovation ideal for accuracy agriculture are; low information rate, low power utilization and bigger scope region. Along these lines, because of previously mentioned attributes, ZigBee innovation happens to be the main decision for actualizing exactness farming.

The recently developing innovation i.e. Wireless Sensor Networks spread quickly into numerous fields resembles therapeutic, living space observing, bio-Innovation and so forth. Yield forecast is an intricate phenomenon that is affected by agro-climatic information parameters. Agriculture input parameters shifts from field to field and rancher to agriculturist. Gathering such data on a bigger zone is an overwhelming errand. The colossal such informational indexes can be utilized for foreseeing their impact on significant yields of that specific region or place. There are diverse estimating strategies created and assessed by the analysts everywhere throughout the world in the field of agriculture or related sciences. Here we are providing comparative analysis of the results using different models.

Keywords: Agriculture · Precision agriculture · Wireless Sensor Networks · Crop yield prediction

1 Introduction

Agriculture is the foundation of India. Around 70% of India's populace relies upon horticulture (cultivation) [1]. Agriculture is one of the fundamental areas to be affected by various sources like climatic changes, soil traits, occasional changes and so forth. Keeping in mind the end goal to take care of various issues happening in conventional agriculture like poor continuous information securing, little observing scope territory,

K. C. Santosh and R. S. Hegadi (Eds.): RTIP2R 2018, CCIS 1037, pp. 443–452, 2019.
https://doi.org/10.1007/978-981-13-9187-3_40

over the top prerequisite of labor, and so forth. Exactness horticulture is utilized [8]. Conventional farming is gradually changed into computerized horticulture to be specific Precision Agriculture (PA). The customary water system procedures incorporate surface, sprinkler, miniaturized scale sprinkler and dribble frameworks. Conventional water system is performed without exact information about the soil and prompts non-uniform water for the plants, which brings about less yields, tedious, water wastage and then some human endeavors. As of now, nursery is beginning to end up a typical technique in crops business. Past nurseries are furnished with old innovation and not appropriate to be installed in the new exactness agriculture innovation. One of the targets of this undertaking is for moving little scale horticulture into enormous scale farming keeping in mind the end goal to add to national financial development. [9] Internet of Things (IoT) is a system of sensors and availability to empower application like horticulture ideal water system.

Remote sensor network (WSN) [7] and Wireless Moisture Sensor System (WMSN) are parts of IoT. Remote Sensor Network (WSN) can be the answers for an extensive assortment of utilizations. The application illustrations are wellbeing checking, horticulture checking, climate observing, air quality observing and land slide observing. WSN is consolidates different advances, for example, sensor innovation, organizing innovation, control innovation, data stockpiling and data handling innovation. Precision Agriculture (PA) is where the rancher ready to get to parameters identified with his homestead and ready to control the related parameters either physically, in plan or naturally. Perusing and composing the parameters are done carefully. Each plot of territory of the harvests estate would have the capacity to be observed and controlled. At last Precision Agriculture (PA) would empower better generation. Accuracy agriculture that have moved farming utilizing an arrangement of advancements into the automated data-based world and is intended to enable agriculturists to oversee the administration of ranch activities.

Yield forecast is a mind-boggling wonder that is impacted by [5, 11] agro-climatic information parameters. Agriculture input parameters shifts from field to field and agriculturist to rancher. Gathering such data on a bigger zone is an overwhelming errand. Be that as it may, the climatic data gathered in India at each 1 sq.m zone in various parts of the area are organized by Indian Meteorological Division. The tremendous such informational indexes can be utilized for foreseeing their effect on significant yields of that specific locale or place. There are distinctive determining systems created and assessed by the scientists everywhere throughout the world in the field of horticulture or related sciences [10]. A remote detecting system is a spatially appropriated self-sufficient sensor, which co-operatively pass their information through the system to a primary area. The information acquired from WSN is typically spared as numerical information in the focal base station.

2 Literature Survey

Present day cultivating can be accomplished by including new ideas, for example, such Internet of Things (IoT), Wireless Sensor Networks (WSN) and Precision Agriculture (PA). Exactness horticulture is describing as the condition of workmanship and study of receiving propelled innovation to increase the yield development [4]. The [1, 10] proposed system for accuracy agribusiness utilizes low cost environmental sensors, an Arduino Uno prototyping board and a couple wireless transceivers (XBee ZB S2) alongside actuating circuit to give computerized water system and monitoring of yields. The proposed model uses XBee convention [2, 3] which depends on ZigBee innovation furthermore, is worked over IEEE 802.15.4 standard [4]. The model created not just gives monitoring of crops in real time in greenhouse environment by detecting parameters like temperature, dampness and dampness substance of the soil which are basically required for the development of the products yet in addition gives mechanized water system. Automated irrigation optimizes water and fertilizer usage.

The productivity of input control technique in green house water system, [10] a test was led to see the distinctive strategies. The strategies are water system by timetable or input based water system. Water system by plan is to supply water to the plant at particular periods. Criticism based water system is to flood plant when the dampness or level of media wetness came to predefined esteem. [5] A product apparatus named 'Crop Advisor' has been created as an easy to use web page for foreseeing the impact of climatic parameters on the product yields. C4.5 calculation is used to discover the most affecting climatic parameter on the harvest yields of chose trims in chosen locale of Madhya Pradesh. This product gives a sign of relative impact of various climatic parameters on the harvest yield, other agro-input parameters in charge of product yield are not considered in this instrument, and utilization of these info parameters differs with singular fields in space and time.

3 Methods and Materials

- The soil moisture sensor (YL-38) senses the moisture level in the soil and displays the data on the laptop wirelessly through Xbee.
- The temperature sensor (DHT22) senses the room temperature respectively and sends the data to the laptop.
- According to the above factors, water is pumped to the soil at regular intervals through the water pump.
- The vapor pressure (BMP280) sensor is used to sense the atmospheric pressure and it is one of the parameter used to predict the yield of a particular crop. This data is also sent to the laptop.
- The processing program on the laptop will read the data from the serial port and it will save the data in the CSV file.
- It will aggregate the data to get the minimum, maximum and the average temperature along with the vapor pressure.
- This data is then sent to the PHP file and we have to insert the rainfall value.
- The PHP program will call the RAPI and the RAPI will give the predicted output of the crop yield.

4 Hardware

Sensors: The sensors utilized for detecting the field parameters are; a soil moisture sensor (YL-38 + YL-69) and a DHT22 humidity/temperature sensor. The soil moisture sensor is fundamentally an electrical protection sensor. The tests (YL-69) is utilized to gauge the electrical resistance between the two tests. The electrical resistance is a component of measure of dampness (water) show in soil. On the off chance that the soil is dry, at that point the protection is extensive and if soil is moist, at that point the protection between the two tests is little. YL-38 is a comparator circuit utilizing chip LM393. The sensor gives both simple and advanced yields [6]. On the off chance that exact moisture esteems are required, at that point simple yield can be utilized or else advanced yield stick can be utilized to get course dampness levels. Simple yield shifts from 0 to 1023, 0 shows least protection esteem (most elevated dampness) though 1023 demonstrates most elevated protection esteem (most minimal dampness). The benefits of this dampness sensor are its ease, capacity to give simple and computerized yields, low power utilization and high affectability. In spite of the fact that there are other dampness/temperature sensors accessible in the market, we picked DHT22 for its ease, high precision and higher temperature and moistness ranges. We have also incorporated a BMP280 vapor pressure sensor which senses the atmospheric pressure and sends it to the laptop. It is one of the parameters to predict crop yield.

RF Module: XBee ZB (S2) [3, 4] are implanted RF modules which can give remote availability to gadgets in ZigBee work systems while being anything but difficult to utilize. They do have other vital highlights as they don't require arrangement or extra advancement and furthermore the system setup time is low. The principle explanation behind choosing these modules for building up the model is a result of their straightforwardness in programming, effortlessness in equipment and programming outlines and capacity of steering information and being compatible (switch, facilitator and end gadget).

Arduino Uno R3: Arduino Uno R3 [3] advancement board comprises of ATmega328 microcontroller. Arduino happens to be an open-source, prototyping stage which is exceptionally famous in light of its straightforwardness. It has got 14 computerized input/yield pins (6 pins can give Pulse Width Modulation (PWM) yield), 6 simple data sources, a precious stone oscillator timed at 12 MHz, a USB association port, a power jack, an in-circuit serial programming (ICSP) header, and a reset catch. It is a ready-made microcontroller board with information sources and yields and it simply should be associated with PC utilizing USB link or an AC-to-DC connector or battery to begin. Arduino comes in numerous flavors, to give some examples like Uno, Mega 2560, Leonardo, Nano, Lilypad etc.

Relay: A relay is an electrically worked switch. Numerous transfers utilize an electromagnet to mechanically work a switch, however other working standards are likewise utilized, for example, strong state transfers. Transfers are utilized where it is important to control a circuit by a different low-control signal, or where a few circuits must be controlled by one signal.

Water Pump: We use a submergible water pump. It is used to irrigate the soil. It is connected in series with a relay that acts as a switch and a 9 V battery that acts as an input voltage source. The pumping of water is a basic and practical technique, far more practical than scooping it up with one's hands or lifting it in a hand-held bucket.

5 Software

XCTU: XCTU is a free, multi-stage application perfect with Windows, MacOS and Linux XCTU is a free multi-stage application intended to empower designers to collaborate with Digi RF modules through an easy to-utilize graphical interface. It incorporates new apparatuses that make it simple to set-up, design and test XBee RF modules. XCTU incorporates the greater part of the devices a designer needs to rapidly get up and running with XBee. One of a kind highlights like graphical system see, which graphically speaks to the XBee arrange alongside the flag quality of every association, and the XBee API outline manufacturer, which naturally constructs and decipher API outlines for XBees being utilized as a part of API mode, join to make improvement on the XBee stage less demanding than at any other time.

Arduino IDE: It is an open-source Arduino Software which helps in writing code and uploading it to the Arduino board. It supports Windows, Mac OS X, and Linux installations. This software works with any of the aforementioned Arduino boards. The programming language used to write codes is C and C++. The program or the code written in IDE is known as sketch. Once the sketch is ready, it is compiled. If without errors, the code is uploaded to the Arduino board.

RStudio: RStudio is a free and open-source coordinated advancement condition (IDE) for R, a programming dialect for factual processing and illustrations. RStudio is accessible in open source and business versions and keeps running on the work area (Windows, macOS, and Linux) or in a program associated with RStudio Server or RStudio Server Pro (Debian, Ubuntu, Red Hat Linux, CentOS, openSUSE and SLES). RStudio is composed in the C++ programming dialect and utilizations the Qt system for its graphical UI.

PDE: The Processing Development Environment (PDE) makes it simple to compose Processing programs. Projects are composed in the Text Editor and began by pressing the Run option. In Processing, a PC program is known as an outline. Portrayals are put away in the Sketchbook, which is an envelope on your PC. Preparing has distinctive programming modes to make it conceivable to convey draws on various stages and program in various ways. The Java mode is the default. Other programming modes might be downloaded by choosing "Include Mode…" from the menu in the upper-right corner of the PDE (Fig. 1).

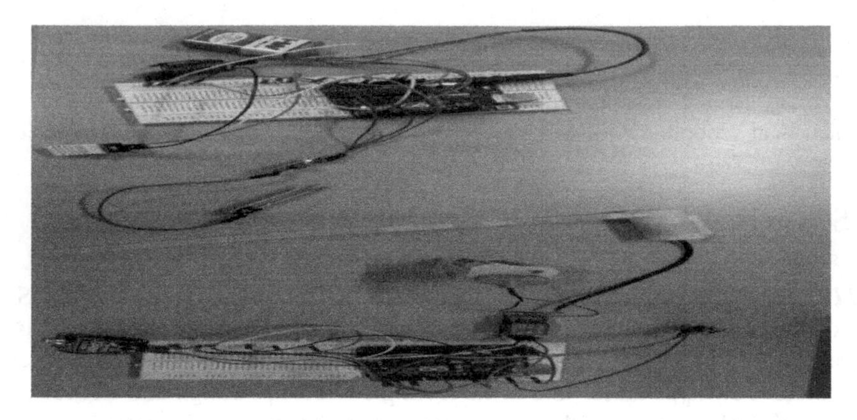

Fig. 1. The device includes sensors, actuators and motor along with XBee ZB (S2) and RF modules.

6 Results and Discussions

We have tested the result for prediction of crop yield with three different algorithms. They are:

- Multiple linear regression model
- Neural net for regression
- K Nearest neighbor regression

The input parameters that we have considered for crop yield prediction are: Rainfall, Min_Temperature, Max_Temperature, Avg_Temperature, Cloud coverage, Vapor pressure (Table 1).

Table 1. Dataset of climatic conditions in various districts of Karnataka

	A	B	C	D	E	F	G
1	DISTRICT	YEAR	MIN TEMP	MAX TEMP	AVG TEMP	CLOUD COVER	VAPOUR PRESSURE
2							
3	BAGALKOT	1998	15.38	37.797	26.18433333	38.4715833333	20.5404166667
4	BAGALKOT	1999	14.792	37.167	25.61183333	40.0906666667	19.8734166667
5	BAGALKOT	2000	15.78	37.244	25.62275	38.1925	19.8018333333
6	BAGALKOT	2001	16.248	36.786	25.881	39.9728333333	20.07675
7	BAGALKOT	2002	15.606	37.468	26.11608333	38.27175	20.1028333333
8	BANAGALORE RURAL	1997	16.383	33.558	25.09416667	48.5915	21.1118333333
9	BANAGALORE RURAL	1998	16.858	35.598	25.38975	48.72975	21.4451666667
10	BANAGALORE RURAL	1999	16.485	34.588	24.75833333	47.1391666667	20.7793333333
11	BANAGALORE RURAL	2000	16.346	34.752	24.72283333	48.2210833333	20.9578333333
12	BANAGALORE RURAL	2001	16.712	34.831	25.11916667	51.2769166667	21.1789166667
13	BANAGALORE RURAL	2002	17.085	35.19	25.22891667	47.6205	21.241

Table 2. Models and accuracy

Model	Deviation (RMS = Root Mean Squared Error)
Multiple linear regression model	13.718
Neural net for regression	14.7
K nearest neighbor regression	16.914

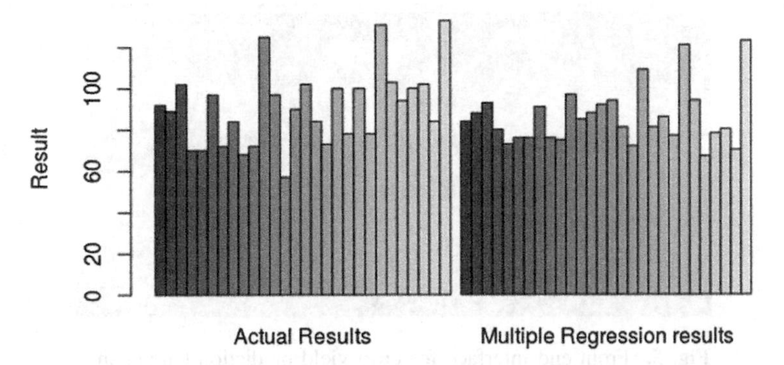

Fig. 2. Results comparison actual with Multiple linear regression model

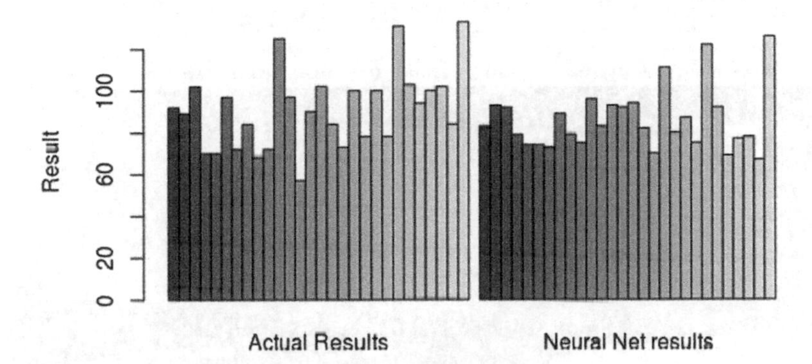

Fig. 3. Results comparison actual with Neural Net

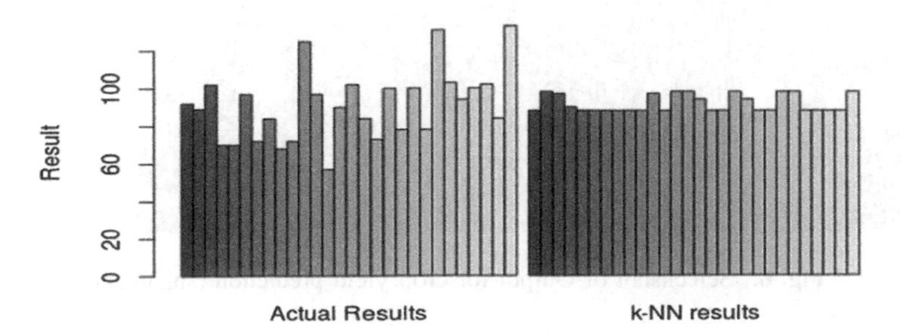

Fig. 4. Results comparison actual with k-Nearest Neighbor

The collected dataset includes climatic data for some districts of Karnataka over a period of years and the data collected from the field with the help of soil moisture sensor, temperature sensor, vapor pressure sensor are given as input to the prediction models mentioned.

Fig. 5. Front end interface for crop yield prediction (sugarcane)

Fig. 6. Screenshot of Output for crop yield prediction (sugarcane)

The web interface is designed to interact with the registered user Fig. 5. The range of values can be varied over the considered threshold values for the selected crop (sugarcane) rainfall r = 768, cloud coverage c = 40, and vapor pressure v = 19 through the interface. These results are tested with respect to the actual results and calculated root mean squared error for the models from Table 2 and refer Figs. 2, 3, and 4.

The vapor pressure sensor data and the temperature sensor data is sent to the respective website and by inputting the values of rainfall and cloud coverage, we get the predicted yield. The following Fig. 6 represents the input value and the predicted value:

7 Conclusion

Precision agriculture (PA), satellite cultivating or site particular product administration (SSCM) is a cultivating administration idea in light of watching, estimating and reacting to entomb and intra-field inconstancy in crops, for entire homestead administration with the objective of enhancing returns on inputs while saving resources. Precision agriculture is an exhaustive framework intended to streamline agriculture production. Utilizing the key components of data, innovation, and administration, accuracy agriculture can be utilized to build generation productivity, enhance item quality, enhance the proficiency of harvest concoction utilize, ration vitality, and secure the environment.

We were for sure ready to indicate how the viability of ZigBee innovation as far as critical qualities like low information rate, low power utilization and bigger region scope can be adequately used to assemble a financially effective answer for precision agriculture. The sensors chose were financially effective as well as having high exactness when contrasted with other accessible sensors in the class. The model could give automation to water system, low asset (water) use and furthermore requires low work association.

Through these outcomes, it can be demonstrated that the remote sensors when utilized alongside the remote innovation like ZigBee, the edge work will help the agriculturists to chop down the cost of resources above all the rare resources, i.e. Water and work included and in turn get profited regarding enhanced generation quality and amount. The crop yield prediction helps the farmer to know his exact outcome of the yield for the year.

References

1. Math, R.K., Dharwadkar, N.V.: A wireless sensor network based low cost and energy efficient framework for precision agriculture. In: International Conference on Nascent Technologies in the Engineering Field (ICNTE-2017), pp. 1–6, 978-1-5090-2794-1/17
2. Veenadhari, S., Misra, B., Singh, C.D.: Machine learning approach for forecasting crop yield based on climatic parameters. In: International Conference on Computer Communication and Informatics (ICCCI-2014), Coimbatore, India, 03–05 January 2014

3. Kumbhar, H.: Wireless sensor network using XBee on Arduino platform an experimental study. In: International Conference on Computing Communication Control and Automation (ICCUBEA) (2016). https://doi.org/10.1109/iccubea.2016.7860081
4. Sahitya, G., Balaji, N., Naidu, C.D., Abinaya, S.: Designing a wireless sensor network for precision agriculture using ZigBee. In: International Advance Computing Conference (IACC) (2017). ISSN 2473-3571. https://doi.org/10.1109/iacc.2017.0069
5. Veenadhari, S., Misra, B., Singh, C.D.: Machine learning approach for forecasting crop yield based on climatic parameters. In: International Conference on Computer Communication and Informatics (ICCCI-2014)
6. https://en.wikipedia.org/wiki/Artificial_neural_network
7. Georgieva, T., Paskova, N., Gaazi, B., Todorov, G., Daskalov, P.: Design of wireless sensor network for monitoring of soil quality parameters. In: 5th International Conference on Agriculture for Life, Life for Agriculture, pp. 431–437 (2016). ScienceDirect. Agricultural and Agricultural Science Procedia 10
8. Hamouda, Y.E.M., Elhabil, B.H.Y.: Precision agriculture for greenhouses using a wireless sensor network. In: Palestinian International Conference on Information and Communication Technology, pp. 78–83 (2017). https://doi.org/10.1109/picict.2017.20
9. Saraf, S.B., Gawali, D.H.: IoT based smart irrigation monitoring and controlling system. In: 2nd IEEE International Conference On Recent Trends in Electronics Information and Communication Technology (RTEICT), India, pp. 815–819, 19–20 May 2017
10. Mat, I., Kassim, M.R.M., Harun, A.N., Yusoff, I.M.: IoT in precision agriculture applications using wireless moisture sensor network. In: IEEE Conference on Open Systems (ICOS), Langkawi, Malaysia, pp. 24–29, 10–12 October 2016
11. Menak, K., Yuvaraj, N.: A survey on crop yield prediction models. Indian J. Innovations Dev. **5**(12), 1–7 (2016)

Evaluation and Analysis of Plant Classification System Based on Feature Level Fusion and Score Level Fusion

Pradip Salve[1], Pravin Yannawar[1(✉)], and Milind Sardesai[2(✉)]

[1] Vision and Intelligence Lab, Department of Computer Science and IT, Dr. Babasaheb Ambedkar Marathwada University, Aurangabad, MS, India
pradipslv@gmail.com, pravinyannawar@gmail.com

[2] Floristic Research Lab, Department of Botany, Savitribai Phule Pune University, Pune, India
sardesaimm@gmail.com

Abstract. This paper describes the automatic leaf recognition based on feature level fusion and score level fusion of vein orientation angles, GLCM, SIFT, SURF as a features. However, to obtain the sophisticated leaf recognition, the system must be undergo through numerous difficulties such as intra and inter-class variations in plants and defining the proper local and global image descriptors which can deal with the color, shape and textual, information for the classification. Selection of the meticulous features plays key role in designing the best classification system. In this paper we proposed multi-modal plant classification where several components are fused together for a more precise classification. The results shows that the proposed system for feature level fusion achieved 93.72% GAR with 6.27% of EER and for score level fusion system achieves 97.13% GAR and 2.86% EER. It is found that the performance of the classification has been increased by 3.79% of EER when score level fusion applied to the system.

Keywords: Leaf veins · Plant recognition · Plant classification · Leaf features

1 Introduction

Plant possesses several contents that can be useful in our day to day life. To utilize and conserve plants automatic plant classification is needed in this era. This task can be done using many characteristics like its leaf shape, leaf colour, wood, roots, flowers, etc. Among all plant characteristics the leaf is consider to be most significant due to is unique properties like vascular structures that contains important information towards automatic leaf recognition. Leaves carries significant information which can be contributed for plant recognition such as texture, color, shape etc. but these are not satisfactory characteristics due to they may get influenced by environmental change and ecological areas. Combining multiple

K. C. Santosh and R. S. Hegadi (Eds.): RTIP2R 2018, CCIS 1037, pp. 453–470, 2019.
https://doi.org/10.1007/978-981-13-9187-3_41

features like leaf vein, gradient patterns with other commonly used features and classifiers can increase the performance and may lead state-of-the-art leaf recognition. Extraction of leaf venation information involves enhancement, data normalization, primary and secondary vein detection, tracing followed by connecting disconnected veins and thresholding, extraction. However, variations may occurs in the venation between same classes of plants due to differences in size, style, orientation, alignment and noise make the problem of automatic leaf venation extraction extremely challenging. Recently [1] proposed leaf venation structures for classification of leaf since leaf venation is an important feature for botanists to recognize the plant species and categories its taxonomic nomenclature. [2–5] proposed leaf descriptors based classification of plants. While comprehensive surveys of related problems such as leaf shape extraction, leaf size extraction, leaf margin features extraction and leaf texture extraction can be found, the problem of leaf venation extraction is not well surveyed. This study leads to do help in plant taxonomy and automatic plant classification using leaf features.

2 Related Work

Numerous methods have been proposed to achieve robust plant classification system. This work provides method to catalogue the plant classification methodologies based on selection of the features. The recent advances in plant classification have been listed in this section mostly based on leaf shape, texture, vein structure, morphological approaches endorsed by most of the researchers.

Rahmadhani et al. [6], used Hough transform and Fourier descriptor to recognize leaf shapes. Sun et al. [7] point cloud has been utilized along with mesh algorithm to extract leaf veins. Clarke et al. [8] scale space analysis and edge detectors were utilized. Siravenha et al. [9] combined multi-resolution techniques by analysing leaf textures, authors have adopted two-dimensional Discrete wavelet Transform (2D-DWT), Grey-Level Co-occurrence Matrices and statistical models to perform classification over Flavia leaf dataset. An overall classification accuracy was achieved 91.85%. Goeau et al. [10] in LifeCLEF 2017 plant identification challenge carried experiments on 10,000 plant species of Pl@ntNet and PlantCLEF datasets. The Mean Reciprocal Rank (MRR) and convolutional neural networks (CNN) were used for evaluation purpose. Wu et al. [11] proposed a Probabilistic Neural Networks (PNN) to classify 32 kinds of plants all experiments done on the Flavia leaf dataset. Total 12 morphological features were extracted and passed to PNN they achieved an accuracy of 90% Prasad et al. [12] proposed shape profile curve transform (Angle View Projection (AVP)) to identify leaf shapes. Chaki et al. [13] proposed Shape Selection Template (FSST) technique in their expedients for shape feature extraction. The Heu color channel was employed to categorise green and non-green leaves. Hamrouni et al. [14] utilized the parallel combination of Naïve Bayes (NB), K-Nearest Neighbour (KNN) and Support Vector Machine (SVM) classifiers. Experiments have been evaluated on Flavia dataset [15]. Lee et al. [16] used Convolutional Neural Networks (CNN) and Deconvolutional Network (DN) for leaf shape extraction. Zhang et al. [16]

proposed two stage approach to classify leaf, Euclidean distance and linear combined all training samples from the candidate neighbour subsets. Murat et al. [17] proposed a hybrid approach that combines four type of leaf descriptor over 45 plant species of myDAUN dataset. They have employed support vector machine (SVM), k-nearest neighbour (k-NN), artificial neural network (ANN), random forest (RF) and linear discriminant analysis (LDA) as a classifier. Features were extracted from proposed technique morphological shape descriptors (MSD) with other commonly used techniques. Pawara et al. [18] employed Convolutional neural networks (CNNs) for classification. Experiments were performed on various datasets viz. AgrilPlant, LeafSnap, and Folio. Bags of visual words and local features were extracted for better classification. adopted for classification purpose. Ghazi et al. [19] deep convolutional neural networks applied on LifeCLEF 2015 datasets to identify the plant species. GoogLeNet, AlexNet, and VGGN utilized as a deep learning architectures. Barré et al. [20] LeafNet, a CNN-based used as a classifier. LeafSnap, Flavia and Foliage datasets were used for evaluation. Goeau et al. [21] utilized convolutional neural networks (CNN) classifier and LifeCLEF 2017 dataset. Lasseck et al. [22] presents deep learning techniques i.e. Deep Convolutional Neural Networks (DCNNs) for classify 10,000 species of LifeCLEF 2017 dataset and used mean reciprocal rank (MRR) for evaluation purpose. Some researchers also addressed the issue of line segmentation and content based image retrieval (CBIR) from biomedical image processing which can be adopted in leaf vein extraction and automatic leaf retrieval systems. Santosh et al. [23] proposed technique applies local line segment detection followed by vectorization process to connect prominent broken lines for eliminating insignificant line segments within the subpanels of the image. Santosh et al. [24] proposed method to detect arrow annotations on biomedical images using template-free geometric signature. Fuzzy binarization was used to extract region of interest (ROI). Candemir et al. [25] Utilized key line-regions and spatial arrangements of objects for object matching. They also proposed Line-based Color-aware descriptor (RSILC) for Rotation and Scale-Invariant.

3 Methodology

In order to design an automatic plant classification mechanism, various experiments have been designed and validated. The identification and verification are two critical steps involved, therefore performance of proposed system are verified using False Rejection Rate (FRR), Equal Error Rate (EER), False Acceptance Rate (FAR) and Genuine Acceptance Rate (GAR). The system is discussed by multiple stages and each stage has dependent hierarchically, which is comprises by the data acquisition, noise removal, data normalization, feature extraction, Matching, Decision and Results Evaluation. Figure 1 depicts the proposed methodology and the overall system flow (Fig. 2).

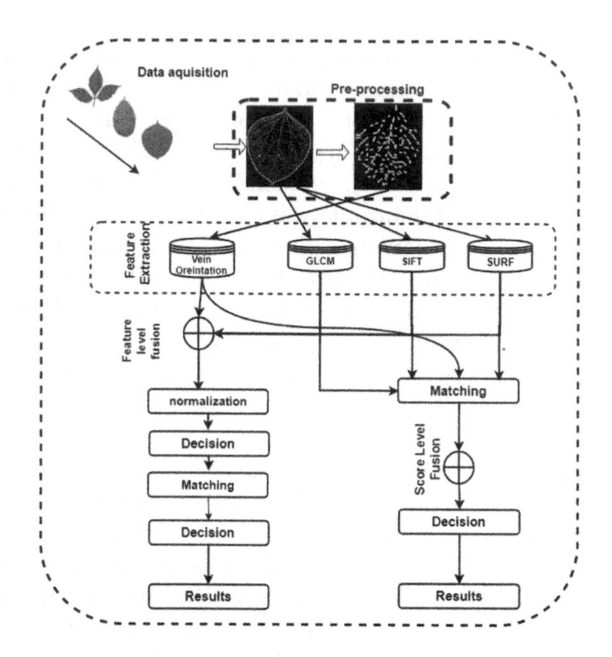

Fig. 1. Block diagram of system flow

3.1 Data Acquisition

The leaf images from VISLeaf data set were used. VISLeaf dataset was exclusively prepared for this research work. The VISLeaf dataset is comprised of sixty types of diverse plants, which was obtained from the Botanical garden and around the of Dr. B. A. M. University campus, Aurangabad (MS) India. Each plant comprised of 10 leaf samples which were scanned under digital scanner. Therefore, size of dataset is 600 i.e. $(60 * 10)$ samples. Fresh leaf samples with all possible variations were grabbed during data collection. The resolution of the digital scanner was set to be 300 dpi during scanning process of the leaves. The datasheet containing scientific names to the proper leaf samples were prepared and verified from taxonomic experts. The features were extracted from each scanned leaf and preserved in the form of matrix on the system, further the feature matrix was divided as training and testing to obtain the results. The list of plants with taxonomic name is provided below.

A. indicum, A. mexicana, A. nervosa, A. roxburghianus, Acalypha indica, Acropogon Bullatus, Amaranthus Roxburghianus, Anisomeles malabarica, Antigonon Leptopus, Azadirachta indica, B. diffusaB. prionitisB. racemosaC. grandis C. procera, Caesalpinia bonduc, Careya arborea, cassia fistula, Chenopodium album, Citrus Limon, Eucalyptus globulus, Eucalyptus polybrachtea, Euphorbia geniculata, Ficus Retusa, Gliricidia sepium, Grewia hirsuta, H. rosa sinensis, Hamelia patens, Hiptage Enghalensis, Holoptelea integrifolia, Ixora coccinea, Jasmineae Jasminum, Jatropha podagrica, Lantana Camara, M. indica, Magnolia Champaca, Millettia Pinnata, Mimusops Elengi, Nerium Oleander, P. dulce, P. granatum,

Parthenium hysterophorus, Perilla frutescens, Polyalthia, Rauvolfia Serpentina, S. Tora, Siamese Cassia, Sicyos waimanaloensis, Spathodea Campanulata, T. diversifolia, T. rhomboidea, T. stans, T. zeylanicum, Tephrosia villosa, Terminalia bellirica, Tinospora Cordifolia, Trigonella foenum graecum, Triumfetta pentandra, Ziziphus mauritiana, Ziziphus rotundifolia.

Fig. 2. Leaf samples from 'VISLeaf' dataset

3.2 Pre-processing

The scanned leaves are composed of (Red + Green + Blue) channels, therefore Green channel is selected due to its supremacy of formation leaf Green channel image is pre-processed using Discrete Wavelet Transform (DTW) for the noise removal and better conservation of edge. Since, the edges are considered to be the most vital information towards extraction of better leaf venation pattern. It was seen that traditional filter excluding wavelet filter makes the image blurry and also degrades significant edge information from the leaf sample. Therefore, to extract the edges of the scanned leaf samples Discrete Wavelet Transform was applied.

$$\psi(x) = \sum_{k=-\infty}^{\infty} (-1)^k a_{N-1-k} \psi(2x - k) \quad (1)$$

Where, N is an even integer. It was seen that only few of the coefficients 'a', 'k' are nonzero which further simplifies the calculations. The leaf venation patterns from leaf sample were generated using binarized image as shown in Fig. 4. The vein is the principle feature and contribute significantly towards generating unique properties therefore its venation angle is captured. Where, Venation angle is the angle between principal vein and ancillary veins associated with it (Fig. 3).

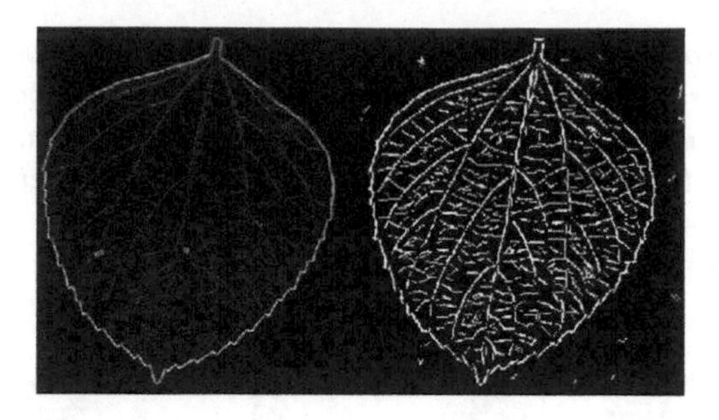

Fig. 3. Depicts the leaf venation produced from original leaf image

3.3 Feature Extraction

The multimodal plant classification and recognition systems accepts four types of features viz. Leaf vein orientation angles, GLCM, SURF, SIFT and used for classification purpose. The feature extraction process in these segments are classified as follows.

Venation Orientation Angle Features Extraction. The Leaf venation structures were obtained from the images produced by Eq. (1). The image produced by DWT was binarized. The venation angle of image represents the directionality of vein. The angle of every vein(edge) is derived from the equation, The angle orientations for each vein (i, j) was produced using Eq. (2):

$$\theta(i,j) = tan^{-1}(\frac{Gy(i,j)}{Gx(i,j)}) \tag{2}$$

Where, Gx, Gy are the derivatives with respect to x and y direction. The image obtained by using Eq. (1) have been used for computing Gx, Gy (Table 1). The length of each veins was calculated by using the Eq. 3 given below.

$$L = \int_a^b \sqrt{(1 + (\frac{dx}{dy})^2} \, dx \tag{3}$$

Table 1. Sample of venation orientation angles features

Sample 1	0.40	0.98	0.66	0.47	0.68	0.45	0.85	0.40	0.34	0.40
Sample 2	0.36	0.38	0.96	0.72	0.51	0.96	0.96	0.98	0.72	0.51
Sample 3	0.94	0.77	0.47	0.94	0.94	0.40	0.62	0.40	0.40	0.34
Sample 4	0.49	0.53	0.57	0.60	0.53	0.55	0.62	0.62	0.60	0.64
Sample 5	0.51	0.83	0.66	0.66	0.87	0.92	0.66	0.62	0.98	0.94
Sample 6	0.81	0.60	0.83	0.60	0.85	0.40	0.43	0.98	0.40	0.70
Sample 7	0.51	0.83	0.66	0.66	0.87	0.92	0.66	0.62	0.98	0.94
Sample 8	0.62	0.83	0.55	0.47	0.94	0.36	0.70	0.72	0.77	0.66
Sample 9	0.62	0.83	0.55	0.47	0.94	0.36	0.70	0.72	0.77	0.66
Sample 10	0.68	0.68	0.62	0.87	0.62	0.51	0.98	0.70	0.79	0.64

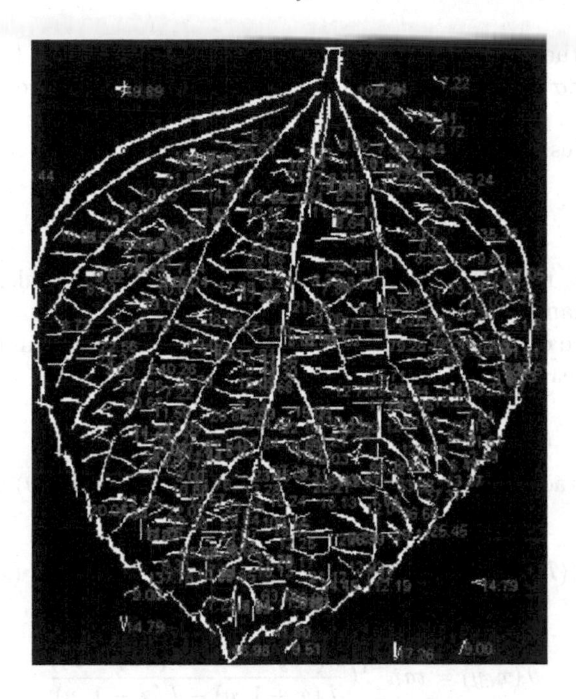

Fig. 4. Venation angle

SIFT Features Extraction. SIFT is a method for detecting significant and constant feature points in an image. The features obtained from SIFT are invariant to rotation and scale. A Single SIFT key point selected from a leaf sample that divide image into 16×16 pixel gradients. And 4 * 4 cells key point descriptor with 8 pixel orientations of each block. The length of a single SIFT key point descriptor is 4 * 4 * 8 (128) element. Following are the steps involved in the calculation of SIFT descriptor [26].

Building the Gaussian pyramid of an image using (x, y, σ), following Eqs. 4, 5 and 6.

$$G\left(x, y, \delta\right) = \frac{1}{2\prod\sigma^2} exp^{\frac{-(x^2+y^2)}{2\sigma^2}} \tag{4}$$

$$L\left(x, y, \sigma\right) = G\left(x, y, \sigma\right) * I(x, y) \tag{5}$$

$$D\left(x, y, \sigma\right) = L\left(x, y, k\sigma\right) - L\left(x, y, \sigma\right) \tag{6}$$

Where σ is the scale parameter, G(x,y,σ) is Gaussian filter, I(x,y) is smoothing filter, L(x,y,σ) is Gaussian pyramid and D(x,y,σ) is difference of Gaussian (DoG)
Compute the Hessian Matrix

$$\begin{bmatrix} I_{xx}(x,\sigma)\ I_{xy}(x,\sigma) \\ I_{xy}(x,\sigma)\ I_{yy}(x,\sigma) \end{bmatrix} \tag{7}$$

Where I_{xx} is the second order Gaussian smoothed image derivatives which detect signal changes in two orthogonal directions.

Calculate the determinant of the Hessian matrix as shown in the Eq. (8) and eliminate the weak key points.

$$(Det\,(H) = I_{xx}\,(x,\sigma)\,I_{yy}\,(x,\sigma) - (I_{xx}(x,\sigma))^2 \tag{8}$$

Calculate the gradient magnitude and orientation as in Eqs. (9) and (10).

$$Mag\,(x,y) = ((I\,(x+1,y) - I\,(x-1,y))^2 + ((I\,(x,y+1) - I\,(x,y-1))^2)^{1/2} \tag{9}$$

$$\theta\,(x,y) = tan^{-1}(\frac{I\,(x,y+1) - I\,(x,y-1))}{I\,(x+1,y) - I\,(x-1,y)} \tag{10}$$

Apply the sparse coding feature based on SIFT descriptors as in Eqs. (11) and (12).

$$min\sum_{i=1}^{M}(\|\,x_i - \sum_{j=1}^{M}a_i^{(j)}\varnothing^{(j)}\|\,^2 + L) \tag{11}$$

$$L = \lambda\sum_{j=1}^{M}|a_i^{(j)}| \tag{12}$$

Where x_i is the SIFT descriptor feature, a^j is sparse (zeroes), ϕ basis of sparse coding, λ is the weights vector (Table 2).

Table 2. Sample of SIFT features

Sample 1	0.0470	0.0470	0.0550	0.0400	0.0440	0.0430	0.0400	0.0400	0.0540	0.0400
Sample 2	0.0400	0.0410	0.0450	0.0400	0.0400	0.0430	0.0380	0.0400	0.0450	0.0400
Sample 3	0.0550	0.0400	0.0400	0.0400	0.0450	0.0400	0.0400	0.0390	0.0430	0.0400
Sample 4	0.0410	0.0580	0.0530	0.0400	0.0470	0.0400	0.0400	0.0400	0.0440	0.0450
Sample 5	0.0410	0.0550	0.0400	0.0400	0.0610	0.0400	0.0400	0.0350	0.0400	0.0400
Sample 6	0.0450	0.0400	0.0540	0.0400	0.0430	0.0400	0.0490	0.0400	0.0450	0.0610
Sample 7	0.0590	0.0670	0.0400	0.0390	0.0400	0.0400	0.0400	0.0400	0.0390	0.0400
Sample 8	0.0550	0.0400	0.0400	0.0400	0.0400	0.0450	0.0430	0.0400	0.0450	0.0480
Sample 9	0.0430	0.0530	0.0400	0.0400	0.0400	0.0410	0.0720	0.0720	0.0400	0.0410
Sample 10	0.0540	0.0410	0.0400	0.0450	0.0430	0.0430	0.0400	0.0440	0.0450	0.0430

SURF Features Extraction. SURF's descriptor is faster in term of computation time. There are two steps involve in SURF feature extraction first is Interest Point Detection and second is Interest Point Description [27,28]. The Interest Point are Detected using following step: SURF concert original image to integral image. Integral Image which summed area tables is an intermediate representation of the image. It is the sum of intensity values of all pixels in input image. the sub regions (rectangular) were formed at any pixel I(x,y) on image with origin (0,0) . It offers fast computation of box type convolution filters using Eq. (13) [29,30].

$$I_{\sum}(X) = \sum_{i=0}^{i\leq x}\sum_{j=0}^{j\leq y} I(x,y) \tag{13}$$

Integral image is convoluted with box filter. Box filter is approximate filter of Gaussian filter. The Hessian matrix $H(X,\sigma)$, (as Eq. 14) where $X = (x, y)$ of an image I, at scale σ is defined as follows:

$$H(X,\sigma) = \begin{bmatrix} L_{xx}(X,\sigma) & L_{xy}(X,\sigma) \\ L_{xy}(X,\sigma) & L_{yy}(X,\sigma) \end{bmatrix} \tag{14}$$

Where L_{xx} (X,σ) (Laplacian of Gaussian) is the convolution of the Gaussian second order derivative $\frac{\delta_{y^2}}{\delta_{x^2}} g(\sigma)$ with the image I in point X and similarly for L_{xy} (X,σ) and $L_{yy}(X,\sigma)$. The Interest Point are Described using following steps:
The reproducible orientations were identified in order to achieve invariant rotation of an image. The Haar wavelet responses were computed by convolving the Haar filter in the x, y direction around the interest point within circular neighbourhood of radius 6. The size of the filter is depends upon the scale which determines the sampling steps, particularly the length of filter is 4s. Then wavelet response are weighted with Gaussian $\sigma = 2s$ cantered at interest points. Subsequently, find maximum the sum of all wavelet responses which is response in each sliding window ($\pi/3$ window orientation). Local orientation vector is fabricated by summing all vertical and horizontal responses. The longest vector within all windows is consider to be the orientation of the interest point. The rectangle of size 20 s is created on interest points to extract the descriptor. The responses of dx and dy from wavelet were summed from each rectangle (subregion) to yield the first set of features in the feature vector. The polarity of intensity changes were found by summing the absolute values of responses. The final descriptor vector was constructed by concatenating the 4 * 4 sub regions $(\Sigma d_x, \Sigma d_y, \Sigma|d_x|, \Sigma|d_y|)$ this results to yield the descriptor vector of length 64 (Table 3).

GLCM Texture Feature Extraction. Gray level Co-occurrence Matrix (GLCM) was employed to calculate gray levels in an image. Gray levels are exactly equal to the number of rows and columns in the image sample [31, 32]. Co-occurrence matrices are constructed using four spatial orientations

Table 3. Sample of SURF features

Sample 1	0.6000	0.6285	0.6000	0.4570	0.6000	0.6285	0.4285	0.6285	0.6000	0.4570
Sample 2	0.5116	0.3954	0.5116	0.5350	0.4884	0.5350	0.3954	0.3489	0.5116	0.3489
Sample 3	0.4048	0.3571	0.5714	0.5238	0.5477	0.3809	0.3809	0.3571	0.3571	0.3571
Sample 4	0.4999	0.4999	0.5228	0.5228	0.3636	0.4999	0.5228	0.5228	0.4772	0.6363
Sample 5	0.5640	0.5640	0.5640	0.5640	0.5898	0.5385	0.5385	0.4102	0.4102	0.4360
Sample 6	0.4894	0.4680	0.3191	0.4468	0.3191	0.4680	0.4680	0.3191	0.6809	0.3191
Sample 7	0.3571	0.5477	0.3571	0.5000	0.4048	0.3809	0.5238	0.5477	0.5238	0.5714
Sample 8	0.4858	0.6285	0.4285	0.4285	0.6285	0.4858	0.4570	0.6572	0.4285	0.6000
Sample 9	0.5142	0.5715	0.6572	0.6000	0.6285	0.6000	0.4570	0.5715	0.5715	0.5715
Sample 10	0.5116	0.5350	0.5582	0.3257	0.3489	0.3489	0.3954	0.3954	0.3721	0.4884

$(0^o, 45^o, 90^o, 135^o)$. Another matrix is constructed as the average of Co-occurrence matrices. Each element of Co-occurrence matrix represents the frequency of each pixel spatially related to gray level to the neighbouring pixel's gray level. Initially, the value of every pixel is zero. The value is updated as per the occurrence of each pixel together. GLCM produces Contrast, Correlation Dissimilarity, Energy, Entropy, Mean, Variance and Standard Deviation. In this study we used Correlation, Homogeneity and Standard Deviation (Table 4). Feature extraction calculated using Eqs. 15–17 given below

$$\sum_{i,j=0}^{N-1} p_{i,j} \left[\frac{(i-\mu_i)(j-\mu_j)}{\sqrt{(\sigma_i^2)(\sigma_j^2)}} \right] \tag{15}$$

$$\sum_{i,j=0}^{N-1} \frac{P_{i,j}}{1+(i-j)^2} \tag{16}$$

$$\sigma_i^2 = \sum_{i,j=0}^{N-1} P_{i,j}(i-\mu_i)^2, \quad \sigma_j^2 = \sum_{i,j=0}^{N-1} P_{i,j}(j-\mu_j)^2, \tag{17}$$

3.4 Feature Normalization.

The features were normalized using feature normalization method. This will provide the most prominent features suitable for classification from feature set. The fabricated features were saved as a matrix in the system to perform all the experiments. The variance between features were normalized to discriminate the leaf samples. The feature normalization was performed using the Eq. (18)

$$NF = \frac{FV - \mu}{\sigma} \tag{18}$$

Where, μ, σ is mean and standard deviation, respectively. FV is the feature vector that represents the each leaf sample.

Table 4. Sample of GLCM features

Sample 1	0.6590	0.6010	0.4680	0.2030	0.2030	0.4680	0.6010	0.6590	0.1890	0.3250
Sample 2	0.4700	0.4240	0.3260	0.1420	0.1420	0.3260	0.4240	0.4700	0.1610	0.2350
Sample 3	0.6340	0.5700	0.4340	0.1860	0.1860	0.4340	0.5700	0.6340	0.1980	0.2840
Sample 4	0.3010	0.2800	0.2300	0.1170	0.1170	0.2300	0.2800	0.3010	0.1380	0.1800
Sample 5	0.3170	0.2900	0.2300	0.1150	0.1150	0.2300	0.2900	0.3170	0.1570	0.1830
Sample 6	0.6420	0.5880	0.4640	0.2100	0.2100	0.4640	0.5880	0.6420	0.1690	0.2960
Sample 7	0.5970	0.5470	0.4290	0.1870	0.1870	0.4290	0.5470	0.5970	0.1580	0.2680
Sample 8	0.6120	0.5580	0.4350	0.1890	0.1890	0.4350	0.5580	0.6120	0.1890	0.2960
Sample 9	0.5840	0.5290	0.4160	0.1860	0.1860	0.4160	0.5290	0.5840	0.2070	0.2600
Sample 10	0.1470	0.1360	0.1100	0.0530	0.0530	0.1100	0.1360	0.1470	0.0460	0.0760

3.5 Identification Method

The classification was performed using feature matrix obtained from each feature extraction technique. The feature matrix from each feature extraction technique further divided as a training and testing set. The ratio for training set retained as 70% and for testing 30% remained data have been utilized. Therefore, the size of testing data was 180 samples and 420 samples for training from each plant class. Consequently, the score matrix was produced using Euclidean distance measure. Equation (19) was used to calculate the scores.

$$S(i,j) = \sqrt{\sum_{i=1}^{n} (q_i - p_j)^2} \tag{19}$$

The score matrix was generated and passed for decision where system determines the species of the leaf which it belong to. To generate the threshold value for each leaf sample, the criteria of deciding threshold value from [33,34] can summarized as shown in Eq. (20):

$$\Delta = \frac{max(FV_i) - min(FV_i)}{\beta} \tag{20}$$

Where, pre-determined constant β was used to divide the threshold values into 'N' parts.

$$\theta i = \min(FV) + \Delta i \tag{21}$$

$\theta_i (i = 1, 2, \ldots, N)$ are selected when the value outreach to the minimal as theoretically FRR or FAR value is observed to be minimal and it is also depending on the specifications required by the system.

3.6 Decision Criteria

Finally, threshold value (T), has been used to make decisions, where the class of the leaf sample accepted by the system if distance score (DS) is less than or equal to '*T*' otherwise if Distance score is greater than '*T*' it simply rejected ($ifDS \leq T$) or rejects ($ifDS > T$). Distance Score (DS) is the measurement of the similarity and dissimilarity associated between each training and testing sample.

3.7 Evaluation Criteria

The performance of the system was obtained from various nations used in the evaluation process viz. False acceptance rate (FAR), False rejection Rate (FRR) and the Equal error rate (EER), Genuine Acceptance Rate (GAR).

The FAR was calculated by obtaining the percentage (fraction) of number of incorrect (inter-class) samples recognised by classification system. FAR was calculated using Eq. (22)

$$FAR = \frac{the\ score\ of inter\ class\ samples\,(Impostor) >\ threshold}{number\ of\ all\ impostor\ score} * 100 \quad (22)$$

The FRR was calculated by obtaining the percentage (fraction) of number of known (intra-class) samples rejected by the classification system. The FRR was calculated using Eq. (23)

$$FRR = \frac{score\ of\ inter\ class\ samples\,(Genuine) < threshold}{number\ of\ all\ genuine\ score} * 100 \quad (23)$$

The GAR was calculated by finding the percentage (fraction) of correct leaf samples recognised by the system. Equation (24) was used to calculate GAR:

$$FAR = 100 - FRR \quad (24)$$

Similarly, EER is a threshold point set to evaluate system performance, it is plotted on the crossing point between the GAR verses FAR over the graph which is mid-point section. The EER used to measure the FAR performance against FRR (The minimum EER considered to be the highest accuracy) by obtaining the similarity (nearest) between False Acceptance Rate (FAR) and False Rejected Rate. EER was calculated using Eq. (25)

$$EER = \frac{FAR - FRR}{2} \quad (25)$$

3.8 Feature Level Fusion of Leaf Samples

In this section the feature level fusion from Vein orientation, SIFT, SURF and GLCM modalities were summed together to re-generate single feature matrix by using following Eq. 26.

$$FM_{(i,j)} = \sum_{f=1}^{4} FV_{(VeinOri(fi))}, FV_{(SIFT(fi))}, FV_{(SURF(fi))}, FV_{(GLCM(fi))} \quad (26)$$

Where, feature vector *(FV)* was grabbed from each technique and *FM* is the newly produced feature matrix. *FM* was used for matching in subsequent stage.

3.9 Score Level Fusion of Leaf Samples

The scores obtained from the threshold values from decision criteria section for Vein orientation, SIFT, SURF and GLCM techniques have been fused by add sum rule, by using Eq. (27) in this study.

$$Score_{(i,j)} = \sum_{f=1}^{4} FV_{(Score(fi))}, FV_{(Score(fi))}, FV_{(Score(fi))}, FV_{(Score(fi))} \quad (27)$$

4 Experimental Results

The proposed system of automatic identification of plants was developed using two modalities that is feature level fusion and score level fusion. For both modalities features were constructed from 60 plants, each plant consists of 10 leaf samples hence 600 samples was stored to perform experiments to construct classification system. All the experiments was accomplished over four types of features, particularly Vein orientation Features, GLCM Features, SIFT and SURF features.

System Evaluation Based on Feature Level Fusion. Features extracted from each techniques Vein orientation, GLCM, SIFT and SURF have been composed together by add sum rule. The concatenated feature matrix were normalized to reduce the difference between each features sets combined together. Consequently the normalized features used for performing experiments. The similar approach used in the previous stages adopted to evaluate the performance of the system. The performance of the system was found the maximum with accuracy of 93.72% when threshold value was 0.001. At the same, minimum EER 6.27% achieved on the fraction of FAR and FRR at threshold value of 12.53% and 0.018% respectively. It was observed that when threshold value is between 0.001812% to 0.001359% the FRR was found to be lowest although the ERR was increased up to 93.72%. Hence, the most appropriate threshold is 0.0013 for feature level fusion method when EER, FAR, FRR are moderates and GAR is appropriate or acceptable to 93.72% which is highest among all single modal classification techniques. In Table 5 indicated few significant system generated performances and Fig. 5 depicts the relation between FAR and FRR. Similarly, Fig. 6 depicts the ROC curve of the classification model.

Table 5. The performance of the classification based on optimal threshold values using feature fusion method

Method	Threshold	FAR (%)	FRR (%)	EER (%)	GAR (%)
Feature fusion	0.001359	12.53968	0.018832	6.279257	93.72074
	0.001812	12.53968	0.091472	6.315577	93.68442
	0.002264	12.53968	0.227334	6.383508	93.61649
	0.002717	12.53968	0.567662	6.553672	93.44633
	0.00317	12.53968	1.068066	6.803874	93.19613
	0.001359	12.53968	0.018832	6.279257	93.72074

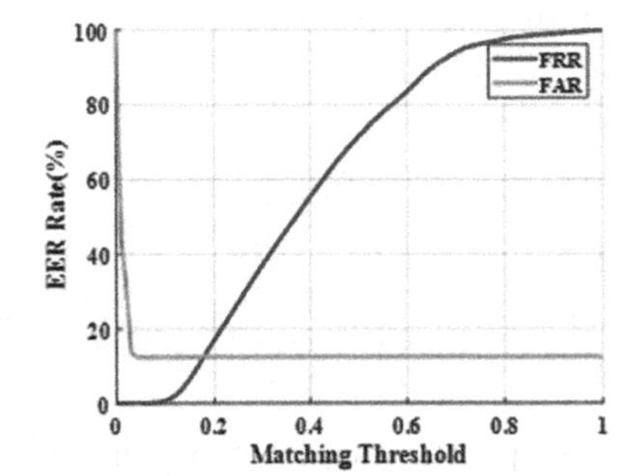

Fig. 5. The performance of using features level fusion based on threshold value

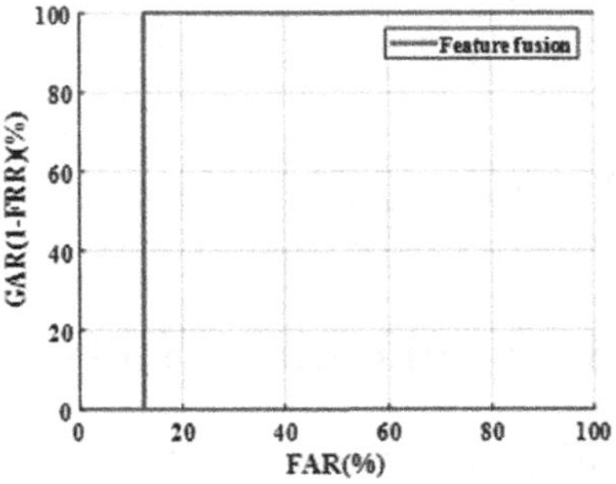

Fig. 6. ROC curve of classification produced using feature level fusion

System Evaluation Based on Score Level Fusion. Features extracted from Vein orientation, GLCM, SIFT and SURF concatenated by sum rule. The concatenated feature matrix normalized reduce the difference between each score sets were combined together. The score matrix normalized passed further for performing the experiments. The similar approach used in the previous stages adopted to evaluate the performance of the system (Table 6). The maximum accuracy obtained by the system was 97.13% when threshold value set to be 0.04235. Similarly at the same, 2.86% EER was achieved significantly minimum on the fraction of FAR and FRR at threshold value of 5.71% and 0.0080% respectively (Figs. 7 and 8).

From Tables 5 and 7 it is found that the score level fusion method outperforms the feature level fusion method. It was seen that when Feature level fusion of Vein Orientation angles, GLCM, SIFT and SURF are utilized 93.72% GAR were reported on "VISLeaf" dataset. Similarly in case of score level fusion 93.13% GAR were reported.

Table 6. Genuine and Impostor scores of Vein orientation, GLCM, SIFT and SURF features

SURF feature Scores		GLCM feature Scores		SIFT Feature Scores		Vein Angle Feature score	
Genuine	impostors	Genuine	impostors	Genuine	impostors	Genuine	impostors
0.020487	0.023739	0.003412	0.026259	0.017265	0.017265	0.051952	0.051952
0.018194	0.021079	0.007606	0.024246	0.035765	0.035765	0.044804	0.044804
0.008248	0.022881	0.04538	0.020528	0.04959	0.04959	0.020681	0.020681
0.011259	0.032779	0.01127	0.065601	0.02736	0.02736	0.042769	0.042769
0.009133	0.021316	0.009922	0.010666	0.029459	0.029459	0.034468	0.034468
0.016712	0.014711	0.0324	0.009148	0.055047	0.055047	0.039389	0.039389
0.021238	0.024696	0.002372	0.022716	0.027689	0.027689	0.051785	0.051785
0.019084	0.006649	0.005182	0.027502	0.02871	0.02871	0.045434	0.045434
0.007859	0.019286	0.043908	0.016135	0.055245	0.055245	0.019443	0.019443
0.025137	0.027507	0.024513	0.043274	0.054309	0.054309	0.05327	0.05327
:	:	:	:	:	:	:	:
:	:	:	:	:	:	:	:
:	:	:	:	:	:	:	:
:	:	:	:	:	:	:	:
:	:	:	:	:	:	:	:
:	:	:	:	:	:	:	:
0.007005	0.014202	0.008933	0.028338	0.020369	0.020369	0.030365	0.030365
Fused Genuines	Fused Impostors						
0.072629	0.242861						
0.088176	0.109919						
0.115651	0.10861						
0.081399	0.278105						
0.073848	0.103407						
0.126836	0.072812						
0.081846	0.1527						
:	:						
:	:						
:	:						
:	:						
:	:						
:	:						
:	:						
0.101691	0.151017						

Table 7. The performance of the system based on the score level fusion method on optimal threshold values

Method	Threshold	FAR (%)	FRR (%)	EER (%)	GAR (%)
Score level fusion	0.04235	5.714286	0.008071	2.861178	97.13882
	0.046585	5.714286	0.028249	2.871267	97.12873
	0.05082	5.714286	0.090126	2.902206	97.09779
	0.055055	5.714286	0.196395	2.95534	97.04466
	0.05929	5.714286	0.417003	3.065644	96.93436

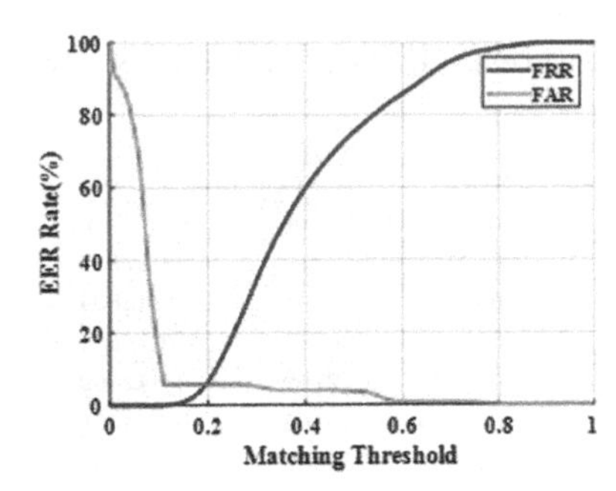

Fig. 7. The performance of all features obtained using score level fusion based on threshold value

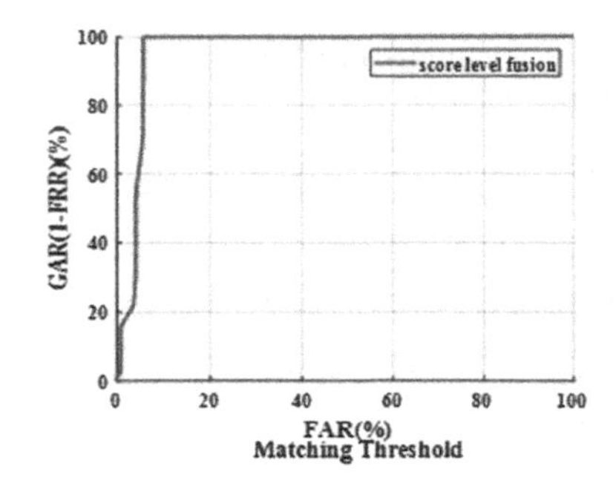

Fig. 8. ROC curve of classification produced using score level fusion

5 Conclusion

The proposed multimodal automatic plant classification system was evaluated on VISLeaf data set. The score level fusion method was performed exceptionally well towards elevating performance of classification up to 97.13% as Genuine Acceptance Rate with corresponding 2.86% EER, 5.71 FAR and 0.0080 FRR respectively. The 70:30 ratio was maintained for training and evaluation purpose for the datasets. The score level fusion method has outperformed over feature level fusion with overall 3.41 improvement in GAR. The work may be extended towards classification of medicinal plants using feature fusion method.

Acknowldgment. The authors would like to acknowledge Department of Computer Science & IT, Dr. Babasaheb Ambedkar Marathwada University, Aurangabad, Maharashtra, India providing support for the infrastructure during the research work and UGC-MANF fellowship for financial support.

References

1. Fern, B.M., et al.: Stratified classification of plant species based on venation state. Biomed. Res. **28**(13), 5660–5663 (2017)
2. Salve, P., Sardesai, M., Manza, R., Yannawar, P.: Identification of the plants based on leaf shape descriptors. In: Satapathy, S.C., Raju, K.S., Mandal, J.K., Bhateja, V. (eds.) Proceedings of the Second International Conference on Computer and Communication Technologies. AISC, vol. 379, pp. 85–101. Springer, New Delhi (2016). https://doi.org/10.1007/978-81-322-2517-1_10
3. Bonnet, P., et al.: Plant identification: experts vs. machines in the era of deep learning. In: Joly, A., Vrochidis, S., Karatzas, K., Karppinen, A., Bonnet, P. (eds.) Multimedia Tools and Applications for Environmental & Biodiversity Informatics. MMSA, vol. 379, pp. 131–149. Springer, Cham (2018). https://doi.org/10.1007/978-3-319-76445-0_8
4. Amlekar, M., Manza, R.R., Yannawar, P., Gaikwad, A.T.: Plant classification based on leaf features. IBMRD's J. Manag. Res. **5**(1), 30–34 (2016)

5. Amlekar, M.M., Ashok T.G.: Plant classification based on leaf Shape features using Neural Network. Int. J. Adv. Res. Sci. Eng. 635–639 (2017)
6. Rahmadhani, M., Herdiyeni, Y.: Shape and vein extraction on plant leaf images using Fourier and B-spline modeling. In: AFITA International Conference, the Quality Information for Competitive Agricultural Based Production System and Commerce, pp. 306–310 (2010)
7. Sun, Z., Lu, S., Guo, X., Tian, Y.: Leaf vein and contour extraction from point cloud data. In: 2011 International Conference on Virtual Reality and Visualization (ICVRV), pp. 11–16. IEEE (2011)
8. Clarke, J., et al.: Venation pattern analysis of leaf images. In: Bebis, G., et al. (eds.) ISVC 2006. LNCS, vol. 4292, pp. 427–436. Springer, Heidelberg (2006). https://doi.org/10.1007/11919629_44
9. Siravenha, A.C., Carvalho, S.R.: Plant classification from leaf textures. In: 2016 International Conference on Digital Image Computing: Techniques and Applications (DICTA), Gold Coast, QLD, pp. 1–8 (2016). https://doi.org/10.1109/DICTA.2016.7797073
10. Goeau, H., Bonnet, P., Joly, A.: Plant identification based on noisy web data: the amazing performance of deep learning. In: CLEF 2017-Conference and Labs of the Evaluation Forum (LifeCLEF 2017) (2017)
11. Wu, S.G., Bao, F.S., Xu, E.Y., Wang, Y., Chang, Y., Xiang, Q.: A leaf recognition algorithm for plant classification using probabilistic neural network. In: IEEE 7th International Symposium on Signal Processing and Information Technology, Cairo, Egypt (2007)
12. Prasad, S., Kumar, P.S., Ghosh, D.: An efficient low vision plant leaf shape identification system for smart phones. Multimed. Tools Appl. **76**(5), 6915–6939 (2017)
13. Chaki, J., Parekh, R., Bhattacharya, S.: Plant leaf classification using multiple descriptors: a hierarchical approach. J. King Saud Univ.-Comput. Inf. Sci. (2018), ISSN 1319-1578. https://doi.org/10.1016/j.jksuci.2018.01.007
14. Hamrouni, L., Bensaci, R., Kherfi, M.L., Khaldi, B., Aiadi, O.: Automatic recognition of plant leaves using parallel combination of classifiers. In: Amine, A., Mouhoub, M., Ait Mohamed, O., Djebbar, B. (eds.) CIIA 2018. IAICT, vol. 522, pp. 597–606. Springer, Cham (2018). https://doi.org/10.1007/978-3-319-89743-1_51
15. Lee, S.H., Chee, S.C., Simon, J.M., Remagnino, P.: How deep learning extracts and learns leaf features for plant classification. Pattern Recogn. **71**, 1–13 (2017)
16. Zhang, S., Wang, H., Huang, W.: Two-stage plant species recognition by local mean clustering and Weighted sparse representation classification. Cluster Comput. **20**(2), 1517–1525 (2017)
17. Murat, M., Chang, S.-W., Abu, A., Yap, H.J., Yong, K.-T.: Automated classification of tropical shrub species: a hybrid of leaf shape and machine learning approach. PeerJ **5**, e3792 (2017)
18. Pawara, P., Okafor, E., Surinta, O., Schomaker, L., Wiering, M.: Comparing local descriptors and bags of visual words to deep convolutional neural networks for plant recognition. In: ICPRAM, pp. 479–486 (2017)
19. Ghazi, M.M., Yanikoglu, B., Aptoula, E.: Plant identification using deep neural networks via optimization of transfer learning parameters. Neurocomputing **235**, 228–235 (2017)
20. Barre, P., Stőver, B.C., Müller, K.F., Steinhage, V.: LeafNet: a computer vision system for automatic plant species identification. Ecol. Inf. **40**, 50–56 (2017)

21. Goeau, H., Bonnet, P., Joly, A.: Plant identification based on noisy web data: the amazing performance of deep learning. In: CLEF 2017-Conference and Labs of the Evaluation Forum (LifeCLEF 2017), pp. 1–13 (2017)
22. Lasseck, M.: Image-based plant species identification with deep convolutional neural networks. In: Working Notes of CLEF 2017 (2017)
23. Santosh, K.C., Antani, S., Thoma, G.: Stitched multipanel biomedical figure separation. In: 2015 IEEE 28th International Symposium on Computer-Based Medical Systems (CBMS), pp. 54–59. IEEE (2015)
24. Santosh, K.C., Wendling, L., Antani, S., Thoma, G.R.: Overlaid arrow detection for labeling regions of interest in biomedical images. IEEE Intell. Syst. **31**(3), 66–75 (2016)
25. Candemir, S., Borovikov, E., Santosh, K.C., Antani, S., Thoma, G.: RSILC: rotation-and scale-invariant, line-based color-aware descriptor. Image Vis. Comput. **42**, 1–12 (2015)
26. Fouad, M.M.M., Zawbaa, H.M., El-Bendary, N., Hassanien, A.E.: Automatic Nile Tilapia fish classification approach using machine learning techniques. In: 2013 13th International Conference on Hybrid Intelligent Systems (HIS), pp. 173–178. IEEE (2013)
27. Mistry, D., Banerjee, A.: Comparison of feature detection and matching approaches: SIFT and SURF. GRD J.- Global Res. Dev. J. Eng. **2**(4), 7–13 (2017), ISSN 2455-5703
28. Herbert, B., Andreas, E., Tinne, T., Luc, V.G.: Speeded up robust feature (SURF). J. Comput. Vis. Image Underst. **110**(3), 346–359 (2008)
29. Utsav, S., Darshana, M., Asim, B.: Image registration of multi-view satellite images using best feature points detection and matching methods from SURF. SIFT PCA-SIFT **1**(1), 8–18 (2014)
30. Bay, H., Ess, A., Tuytelaars, T., Gool, L.V.: SURF: speeded up robust features. Comput. Vis. Image Underst. (CVIU) **110**(3), 346–359 (2008)
31. Vijayakumar, V., Neelanarayanan, V., Veeramuthu, A., Meenakshi, S., PriyaDarsini, V.: Big data, cloud and computing challengesbrain image classification using learning machine approach and brain structure analysis. Proc. Comput. Sci. **50**, 388–394 (2015)
32. Shijin, K.P.S., Dharun, V.S.: Extraction of texture features using GLCM and shape features using connected regions. Int. J. Eng. Technol. (IJET) **8**(6), 2926–2930 (2016)
33. Salve, P., Yannawar, P., Sardesai, M.: Multimodal plant recognition through hybrid feature fusion technique using imaging and non-imaging hyper-spectral data. J. King Saud Univ. - Comput. Inf. Sci. (2018), ISSN 1319-1578. https://doi.org/10.1016/j.jksuci.2018.09.018
34. Salve, P., Sardesai, M., Yannawar, P.: Classification of plants using GIST and LBP score level fusion. In: Thampi, S.M., Marques, O., Krishnan, S., Li, K.-C., Ciuonzo, D., Kolekar, M.H. (eds.) SIRS 2018. CCIS, vol. 968, pp. 15–29. Springer, Singapore (2019). https://doi.org/10.1007/978-981-13-5758-9_2

Assessment of Rainfall Pattern Using ARIMA Technique of Pachmarhi Region, Madhya Pradesh, India

Papri Karmakar[1], Aniket A. Muley[2(✉)], Govind Kulkarni[3], and Parag U. Bhalchandra[3]

[1] Department of General and Applied Geography, Dr. H. G. Central University, Sagar, M.P., India
karmakarpapri@gmail.com

[2] School of Mathematical Sciences, S.R.T.M. University, Nanded, India
aniket.muley@gmail.com

[3] School of Computational Science, S.R.T.M. University, Nanded, India
govindcoolkarni@gmail.com, srtmun.parag@gmail.com

Abstract. Rainfall prediction is a crucial event as large portion in India is depends upon it. Since, agriculture is one of the most important constituent on Indian economy and rainfall has an indirect impact on it. In this paper, an attempt has been made to forecast the rainfall activities in terms of pattern matching data analytics work carried over rain fall time series. The major aspect is to study pattern of rainfall over Pachmarhi region. To forecast rainfall of Pachmarhi region data during the years 2000 to 2017 has been collected and Auto Regressive Integrated Moving Average (ARIMA) method was applied to forecast the rainfall for next five years.

Keywords: Time series analysis · ARIMA · Rainfall · Pachmarhi region

1 Introduction

The rainfall is significant constituent of the hydrological succession. Among a variety of precipitation, rainfall embraces most significant form and commonly apparent in tropical climate. This important facet constitutes the climate type of every region. It is the product of pressure variation and successive formation of clouds, etc. It performs an imperative function in assessment of atmospheric moisture balance. It also deserves a prime position in its realistic value as it controls humidity and aridity of an area and accordingly, the agricultural efficiency as well as ideal condition for human habitation. Due to its practical and climatic inferences, it is rather necessary to place intense interest on its characteristics viz., temporal variation i.e. duration of rainfall; monthly and seasonal variation, percentage, intensity and spatial variability or aerial distribution [1]. Rainfall is usual climatic incident. Apart from its prediction it is also demanding due to its relevance in the planning and management of agricultural scheme as well as water resources systems [19, 20]. An annual rainfall activity in India covers most of the part of city especially in monsoon season. Further, change in rainfall time and its

K. C. Santosh and R. S. Hegadi (Eds.): RTIP2R 2018, CCIS 1037, pp. 471–481, 2019.
https://doi.org/10.1007/978-981-13-9187-3_42

quantity have probable effect on water resource. The rainfall forecasting ensures to supply of rainwater for planning purpose particularly in rain fed areas becomes of prime significant.

In India, various parts of rainfall prediction studies has been completed through different models viz., monthly precipitation in Sylhet city [2], time series analysis model to forecast rainfall for Allahabad region [5], modelling and prediction of rainfall using ANN and ARIMA techniques [4, 6, 8, 10, 16]. Singh [17] proposed new prototype to resolve fuzzy time series forecasting. Artificial neural network (ANN) based algorithm was applied to generate the intervals. Further, fuzzy based time series data set was applied with FTS theory for establishment of fuzzy logical relations [17]. Jain and Mallick [7] introduced an approach to improve the daily temperature data. Major objective of their study was to analyze the affecting parameters. Wang et al. [19, 20] studied the role of cloud coverage on solar energy irradiance and ground based cloud images were analyzed through ARIMA time series model. The ARIMA time series model forecasted the short term cloud coverage. Farajzadeh and Alizadeh [5] proposed a hybrid model for prediction of rain fall time series of Urmia Lake watershed. To forecast the monthly rainfall activities ARIMAX and least squares support vector machine models were used on discrete wavelet transform data set. Dabral and Murry [3] utilized SARIMA model on month wise, week wise and day wise time series data set of monsoon rainfall. Based on the autocorrelation function (ACF), partial autocorrelation function (PACF), and the minimum values of Akaike Information Criterion (AIC) with Schwarz Bayseian Information (SBC) best SARIAM model was identified. Mehdizadeh et al. [10] used two hybrid novel models gene expression programming-autoregressive conditional heteroscedasticity and ANN autoregressive conditional heteroscedasticity for estimation of monthly rainfall forecasting values. Meher et al. [11] used univariate time series ARIMA model to develop and designed for simulation and forecasting of average rainfall with Theissen weights over the Mahanadi River Basin in India ACF and PACF plots were used to observe the trend and seasonality. Uba [18] applied Box-Jenkins methodology on monthly rainfall data of Maiduguri Airport Station with 372 readings of period 1981 to 2011 for forecasting of rainfall activities in study region. Olatayo and Taiwo [13] investigated the rainfall data with modern method fuzzy time series to forecast the future rainfall values. The aim of their study was to identify the behavioral pattern in rainfall phenomena based on past observations. The ARIMA is the emerging fuzzy time series model and the non-parametric method Theil's regression was used to evaluate the prediction efficiency. Pazvakawambwa [14] applied time series technique to forecast the monthly rainfall data with descriptive summary statistics with measures of centrality and dispersion, time series plots, and autocorrelation functions. For the identification of best suited model for the data set, the Box-Jenkins's ARIMA model was applied with estimation and validation techniques. Residual analysis was performed to assess the adequacy of the identified models.

In the next subsequent sections, a detailed study of rainfall prototype is done and by ARIMA for model is identified to predict its forecasting future pattern. Rainfall forecasting for Pachmarhi hasn't been prepared previously. Thus, prediction of future rainfall is ultimately necessary for such a rain fed area. The prime spotlight of this study is on development of a consistent forecasting of rainfall over Pachmarhi to supervise water resource management effectively. The forecasting of rainfall has been completed using data for the time span of 2000–2017.

2 Materials and Methodology

2.1 Study Area

Pachmarhi is the most attractive tourist site in Madhya Pradesh fascinating with its natural aesthetic beauty. Figure 1 represents the Pachmarhi geographical location collected from the Google earth. The Pachmarhi hills and its surroundings area mainly expand in Hoshangabad district. It is lying in the southern part of Hoshangabad district, Madhya Pradesh. Pachmarhi is situated on top of a dissected plateau near left bank of Denwa River in Narmada basin. Small part of area is spread over the other three blocks (Ghoradongri, Junnardeo and Tamia) of two districts i.e. Betul and Chhindwara respectively. There are great disparities in average height of these rain gauge stations above mean sea level from study area. It is maximum rainfall receiving area of the state but completely rely on monsoon rainfall. The area in recent times particularly the Pachmarhi Township and some specific pockets of this territory are facing a newly emerging problem of ground water depletion. Thus, among rain gauge stations only rainfall data of Pachmarhi station is collected to avoid the elevation differences.

Fig. 1. Pachmarhi region

A continuous record of meteorological data on rainfall from 2000 to 2017 for Pachmarhi station has been sourced from the India meteorological Department, Nagpur and Land Record Department, Hoshangabad. In present context we applied time series ARIMA model approach to forecast the rainfall average for next five years. The ARIMA model consists of iterative, exploratory, process which is divided in three different phases as identification, estimation and diagnostic. Further, ARIMA consist of auto-regressive component (AR) denoted as p, integrated component (I) as d, and moving average (MA) as q. The AR part corresponds to persistent belonging to earlier observations. Integrated component (I) represents trends, including seasonality. Also, MA represents lingering effects of previous random shocks or error. In ARIMA model the order of component starts with the small integer values (as 0, 1 or 2). The order of these components is set to identify the best parsimonious model with estimated parameters. Time series operation is based on the mathematical function which states the nature of data. These functions are useful to predict, to monitor or to control the data set in systematic pattern. The ARIMA (p, 1, q) has been obtained from ARIMA (p, d, q) model for d = 1 with difference of Y_t [1, 9, 12, 15]. Further, it is represented in the form of following Eqs. (1–5):

$$\Delta_d Y_t = W_t \text{OR} W_t = Y_t - Y_{t-1} \tag{1}$$

$$\begin{aligned} \text{When} : W_t &= Y_t - Y_{t-1} \\ W_{t-1} &= Y_{t-1} - Y_{t-2} \\ W_{t-2} &= Y_{t-2} - Y_{t-3} \end{aligned}$$

Also, when: $t = t - p$ then $W_{t-p} = Y_{t-p} - Y_{t-p-1}$

$$W_t = \Phi_1 W_{t-1} + \Phi_2 W_{t-2} + \cdots + \Phi_p W_{t-p} + \varepsilon_t - \theta_1 \varepsilon_{t-1} - \theta_2 \varepsilon_{t-2} - \cdots \theta_q \varepsilon_{t-q} \tag{2}$$

The subtraction operation gives the following equations Eqs. (3–4):

$$\begin{aligned} W_t = {} & \Phi_1 (Y_{t-1} - Y_{t-2}) + \Phi_2 (Y_{t-2} - Y_{t-3}) + \cdots + \Phi_p \left(Y_{y-p} - Y_{t-p-1}\right) + \varepsilon_t - \theta_1 \varepsilon_{t-1} \\ & - \cdots - \theta_q \varepsilon_{t-q} \end{aligned} \tag{3}$$

$$\begin{aligned} Y_t - Y_{t-1} = {} & \Phi_1 Y_{t-1} - \Phi_1 Y_{t-2} + \cdots + \Phi_p Y_{t-p} - \Phi_p Y_{t-p-1} + \varepsilon_t - \theta_1 \varepsilon_{t-1} \\ & - \cdots - \theta_q \varepsilon_{t-q} \end{aligned} \tag{4}$$

Hence, it becomes (Eq. 5):

$$\begin{aligned} Y_t = {} & (1 + \Phi_t) Y_{t-1} + (\Phi_2 - \Phi_1) Y_{t-2} + (\Phi_3 - \Phi_2) Y_{t-3} + \cdots + \left(\Phi_p - \Phi_{p-1}\right) Y_{t-p} \\ & + \Phi_p - Y_{t-p-1} + \varepsilon_t - \theta_1 \varepsilon_{t-1} - \theta_2 \varepsilon_{t-2} - \cdots - \theta_q \varepsilon_{t-q} \end{aligned} \tag{5}$$

3 Result and Discussion

3.1 Monthly Rainfall

The average monthly rainfall data collected from the year 2000 to 2017. The average rainfall of each month has been calculated and represented and visualized in Fig. 2.

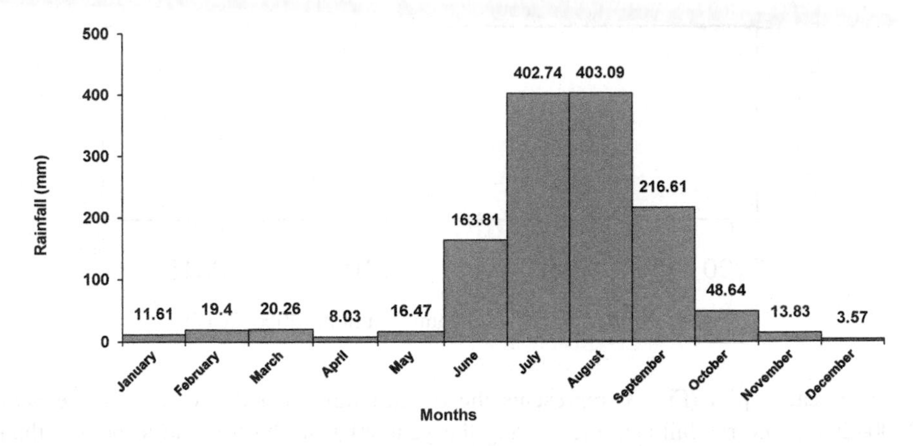

Fig. 2. Average rainfall

July and August are the principal rainy months together making 63.07% of the annual rainfall; followed by September and June in decreasing order, together making about 27.65%. Thus, the rainy season extending from June to September receive about 90.72% of total annual rainfall. Due to intense evaporation a large amount of rainfall in the month of June is lost and only a small part of it reaches to the soil moisture zone. The data reveals that, in the month of July a part of rainfall is utilized for saturating the soil moisture. The rest of it goes mostly as runoff on steep slope and evapotranspiration losses and a very little goes towards the groundwater increments. In the month of August, ground water addition to ground water reservoirs takes place and the soil moisture zone water requirements are fulfilled with sufficient rainfall. Due to infiltration in upper aquifer bodies ground water rise is possible in September.

3.2 Annual Rainfall

The rainfall data available from the year 2000 to 2017 of Pachmarhi station have been collected from the India Meteorological Department, Pune, Nagpur and Land Record Department, Hoshangabad. These data have been shown in the form of graph (Figs. 3 and 4). The results inferred from these data are given in Table 1.

Table 1. Summary of the rainfall data

Sr. No.	Item description	Pachmarhi station
1	Period of rainfall record	2000–2017
2	Number of years	18
3	Average rainfall (mm) of 18 years	110.68
4	Maximum rainfall (mm) with year	203.40 (2013)
5	Minimum rainfall (mm) with year	1.20 (2004)

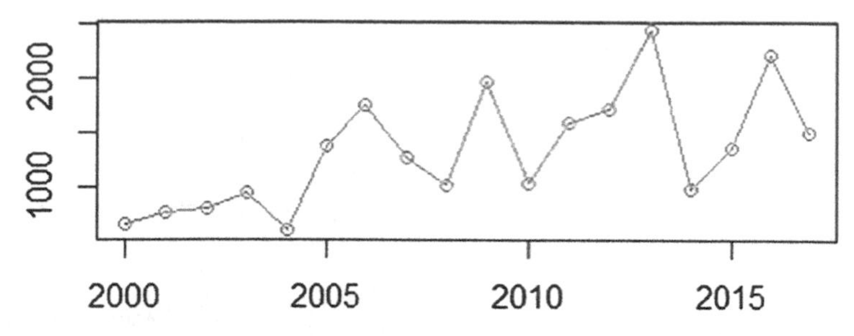

Fig. 3. Scatter plot of total annual rainfall (2000–2017)

The scatter plot (Fig. 3) represents the overall rainfall activities during the year 2000–2017. The rainfall activities during the year 2000 to 2005 found to be less than 1000 mm. Further, the highest rainfall activities were found during the period of 2010 to 2017. During the time period of 2005 to 2015 the rainfall activities were lies in between 1000 to 2000 (mm). Further, the overall average rainfall activities were considered for the time series analysis.

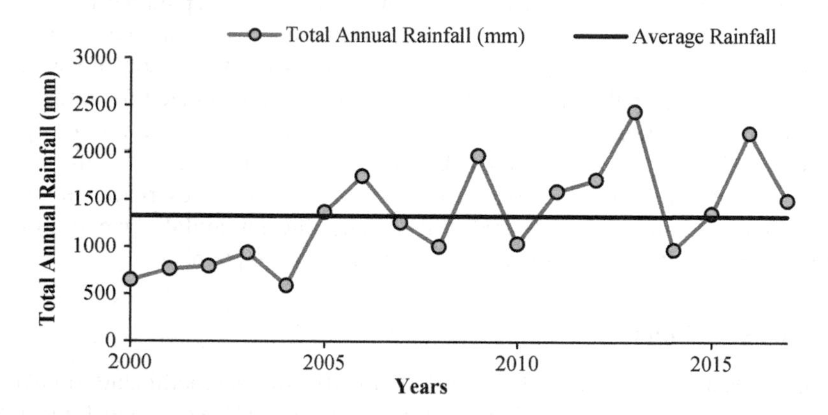

Fig. 4. Total annual rainfall (2000–2017)

3.3 Frequency Analysis

The statistical analysis of the rainfall data is transformed into the frequencies; cumulative frequency and percentage frequency have been determined and has been represented in Table 2.

Table 2. Frequency data of rainfall in mm (2000–2017)

Sr. No.	Range (mm)	Frequency	Cumulative frequency	Percentage
1	500–700	2	2	11.11
2	700–900	2	4	22.22
3	900–1100	4	8	44.44
4	1100–1300	1	9	50.00
5	1300–1500	3	12	66.67
6	1500–1700	1	13	72.22
7	1700–1900	2	15	83.33
8	1900–2100	1	16	88.89
9	2100–2300	1	17	94.44
10	2300–2500	1	18	100.00

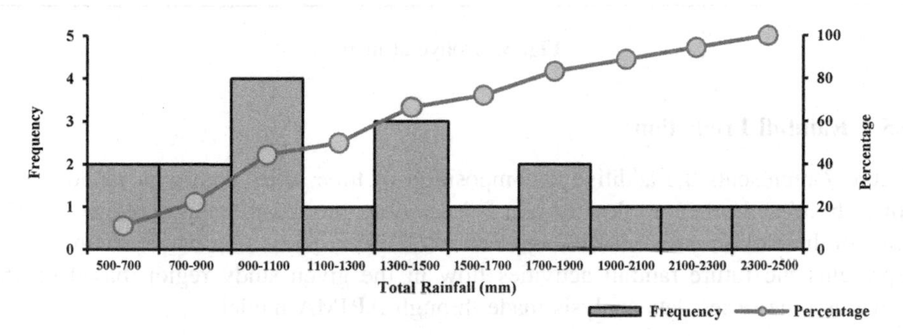

Fig. 5. Frequency graph and histogram of rainfall (2000–2017)

Table 2 represents the percentage frequency graphs and histogram of rainfall has been prepared and represented by histogram (Fig. 5). It illustrates total annual rainfall of the area under investigation usually registered not below 500 mm and normally hovered around 900 to 1100 mm. The extremity of total annual rainfall (2300–2500 mm) is only 5.56%.

3.4 Isohyetal Map

Isohyetal map has been prepared by plotting the depth of rainfall from TRMM (Tropical Rain Measuring Mission) is shown in Fig. 6. It is arranged on a horizontal scale of 1 cm to 2 km along with 25 mm isohyets rainfall. At a glance, the isohyetal configuration indicates that the highest value of isohyets is 1375 mm in the south-east part of the territory while the lowest value of isohyets is 1275 mm in the north-west part of the study area near Tawa reservoir. Amount of rainfall in this province represented an increasing trend from north-west to south-east. The average annual rainfall of the area under investigation is 1351.86 mm.

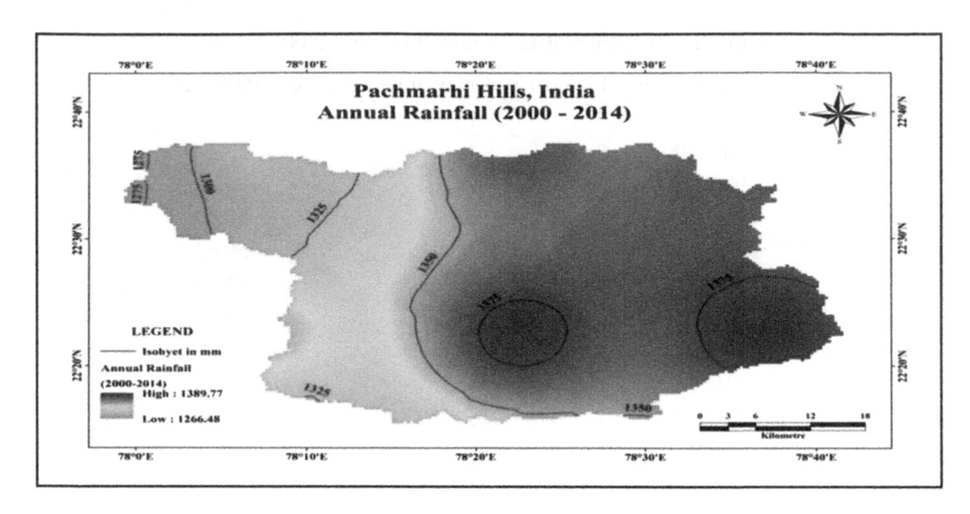

Fig. 6. Isohyetal map

3.5 Rainfall Prediction

Figure 7 represents the additive decomposition of time series data with random, seasonal, trend and observed flow of rain fall activities predicted for next five years. The scale of these observations is different from each other. These four nature of time series represents the future rainfall activities flow in the given study region based on the previous time series data analysis made through ARIMA model.

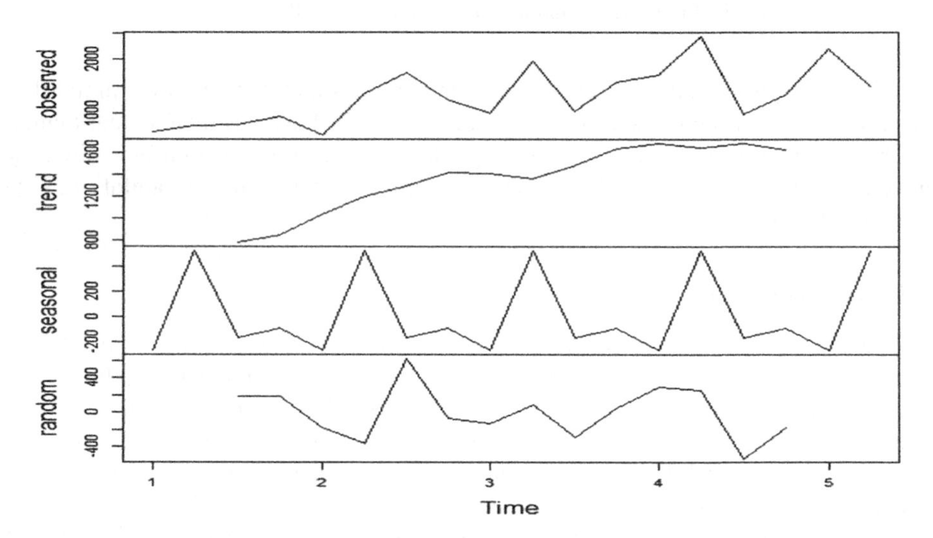

Fig. 7. Decomposition of additive time series

Figure 8 reveals the future level of rainfall activities predicated for next five years i.e. for 2018–2022. The forecasted values are represented with redline with 95% of confidence values. The actual time series data of 2000–2017 is represented with simple line. The error bounds line represents the possible rainfall values in upper and lower bound respectively. The forecasted values are calculated for next five years with average rainfall prediction. Figure 8 reveals the future level of rainfall activities predicated for next five years i.e. for 2018–2022. The forecasted values are represented with redline with 95% of confidence values. The actual time series data of 2000–2017 is represented with simple line. The error bounds line represents the possible rainfall values in upper and lower bound respectively. The forecasted values are calculated for next five years with average rainfall prediction. The R command is used as:

```
arima(x = r1, order = c(1, 0, 0), seasonal = list(order = c
(2, 1, 0), period = 4))
```

Table 3 represents the summarized results obtained through (Eq. 5) of time series model for our data set as:

Table 3. Result of the model.

Coefficients:			S.E			$sigma^2$ estimated	log likelihood	AIC
ar1	sar1	sar2	ar1	sar1	sar2			
0.15	0.08	−0.21	0.29	0.35	0.42	298387	−108.33	224.66

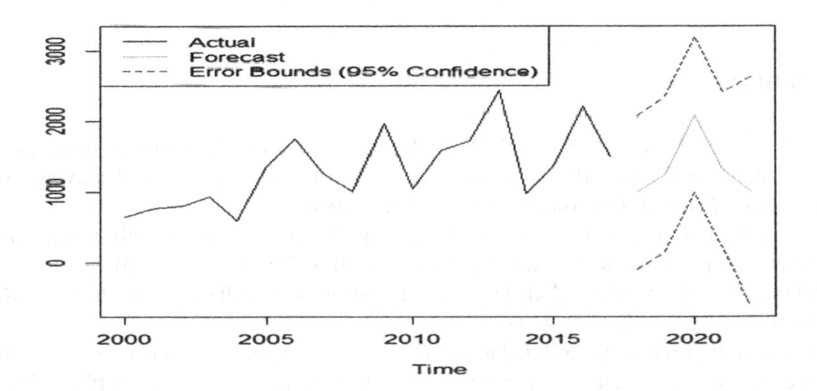

Fig. 8. Forecasted values of rainfall

Table 4. Rainfall prediction results (mm).

Year of prediction	Rain fall prediction for next 5 years (in mm)
2018	−100.1261 to 2084.868
2019	137.1670 to 2348.612
2020	986.5469 to 3198.632
2021	209.0949 to 2421.196
2022	−611.5687 to 2627.084

The predicted values obtained through time series ARIMA model reveals the average rainfall activities in the respective years i.e. 2018–2022 (Table 4). The ARIMA time series model reveals that, for the year 2018, average rainfall activities will lies in between −100.1261 to 2084.868 (mm) of lower and upper bound values. Further, for 2019 the forecasted values will ranges between the 137.1670 to 2348.612 (mm). The rainfall activities in the year 2020 will lies in 986.5469 to 3198.632 (mm) of lower and upper bound values. The average rainfall activities for 2021 are forecasted with the lower and upper bound values of 209.0949 to 2421.196 (mm) respectively. The ARIMA model forecasted the rainfall nature in the study region for 2022 as −611.5687 to 2627.084 (mm).

4 Conclusions

In India agriculture plays an important support to Indian GDP, it is therefore necessary to have awareness regarding possibility of a good rain fall. So, it is essential have an accurate forecast of the rain fall. In present investigation, we applied future forecasting time series based ARIMA model for rainfall activities. Our major interest was to develop a time series model which satisfies normality, independence without any autocorrelation. Based on the model selection criteria, we have revealed average rainfall activities in the respective years i.e. 2018–2022. Hence, we recommended that, this average rainfall gives brief idea about storage capacity of rainwater in the Pachmarhi region. Further, it will be helpful to increase the agriculture yield.

References

1. Ab Razak, N.H., Aris, A.Z., Ramli, M.F., Looi, L.J., Juahir, H.: Temporal flood incidence forecasting for Segamat River (Malaysia) using autoregressive integrated moving average modelling. J. Flood Risk Manage. **11**, 794–804 (2018)
2. Bari, S.H., Rahman, M.T., Hussain, M.M., Ray, S.: Forecasting monthly precipitation in Sylhet City using ARIMA model. Civil Environ. Res. **7**(1), 69–77 (2015)
3. Dabral, P.P., Murry, M.Z.: Modelling and forecasting of rainfall time series using SARIMA. Environ. Process. **4**(2), 399–419 (2017)
4. Duangdai, E., Likasiri, C.: Rainfall model investigation and scenario analyses of the effect of government reforestation policy on seasonal rainfalls: a case study from Northern Thailand. Atmos. Res. **185**, 1–2 (2017)
5. Farajzadeh, J., Alizadeh, F.: A hybrid linear-nonlinear approach to predict the monthly rainfall over the Urmia Lake watershed using wavelet-SARIMAX-LSSVM conjugated model. J. Hydroinformatics **20**(1), 246–262 (2018)
6. Graham, A., Mishra, E.P.: Time series analysis model to forecast rainfall for Allahabad region. J. Pharmacognosy Phytochem. **6**(5), 1418–1421 (2017)
7. Jain, G., Mallick, B.: A study of time series models ARIMA and ETS. I.J. Mod. Educ. Comput. Sci. **4**, 57–63 (2017)
8. Kalanker, N.V., Somvanshi, V.K., Pandey, O.P., Agrawal, P.K., Prakash, M.R., Chand, R.: Modelling and prediction of rainfall using artificial neural network and ARIMA techniques. J. Ind. Geophys. Union **10**(2), 141–151 (2006)

9. Kulkarni, G.E., Muley, A.A., Deshmukh, N.K., Bhalchandra, P.U.: Autoregressive integrated moving average time series model for forecasting air pollution in Nanded city, Maharashtra, India. Model. Earth Syst. Environ. **4**(1), 1–10 (2018)
10. Mehdizadeh, S., Behmanesh, J., Khalili, K.: New approaches for estimation of monthly rainfall based on GEP-ARCH and ANN-ARCH hybrid models. Water Resour. Manage. **32**(2), 527–545 (2018)
11. Meher, J., Jha, R.: Time-series analysis of monthly rainfall data for the Mahanadi River Basin, India. Sci. Cold Arid Reg. (SCAR) **5**(1), 73–84 (2013)
12. Naveen, V., Anu, N.: Time series analysis to forecast air quality indices in Thiruvananthapuram District, Kerala, India. Int. J. Eng. Res. Appl. **7**(6), 66–84 (2017)
13. Olatayo, T.O., Taiwo, A.I.: Statistical modelling and prediction of rainfall time series data. Global J. Comput. Sci. Technol. **14**(1), 1–9 (2014)
14. Pazvakawambwa, G.T.: A time-series forecasting model for Windhoek Rainfall, Namibia. pp. 1–11 (2017). https://digitalcommons.andrews.edu/cgi/viewcontent.cgi?article=1146&context=arc
15. Rojo, J., Rivero, R., Romero-Morte, J., Fernández-González, F., Pérez-Badia, R.: Modeling pollen time series using seasonal-trend decomposition procedure based on LOESS smoothing. Int. J. Biometeorol. **61**(2), 335–348 (2017)
16. Sahai, A.K., Soman, M.K., Satyan, V.: All India summer monsoon rainfall prediction using an artificial neural network. Clim. Dyn. **16**(4), 291–302 (2000)
17. Singh, P.: Rainfall and financial forecasting using fuzzy time series and neural networks based model. Int. J. Mach. Learn. Cybern. **9**(3), 491–506 (2018)
18. Uba, E.S., Bakari, H.R.: An application of time series analysis in modeling monthly rainfall data for Maiduguri, North Eastern Nigeria. Math. Theory Model. **5**(11), 24–33 (2015)
19. Wang, K.W., Deng, C., Li, J.P., Zhang, Y.Y., Li, X.Y., Wu, M.C.: Hybrid methodology for tuberculosis incidence time-series forecasting based on ARIMA and a NAR neural network. Epidemiol. Infect. **145**(6), 1118–1129 (2017)
20. Wang, Y., Wang, C., Shi, C., Xiao, B.: Short-term cloud coverage prediction using the ARIMA time series model. Remote Sens. Lett. **9**(3), 275–284 (2018)

Land Use and Cover Mapping Using SVM and MLC Classifiers: A Case Study of Aurangabad City, Maharashtra, India

Abdulla A. Omeer(✉), Ratnadeep R. Deshmukh, Rohit S. Gupta, and Jaypalsing N. Kayte

Department of Computer Science and Information Technology, Dr. Babasaheb Ambedkar Marathwada University, Aurangabad, MH, India
binomeer1979@yahoo.com, rrdeshkukh.csit@bamu.ac.in, rohitgupta8844@gmail.com, jaypalsing@gmail.com

Abstract. The fast developing cities like Aurangabad need an effective analysis of Land Use Land Cover. In this paper we examine the use of MLC and SVM for mapping the LULC from satellite Imagery. This paper investigates the accuracy of Support Vector Machine SVM and Maximum Likelihood Classification MLC for multi-spectral images of Aurangabad city and Waluj area. The satellite images collected from IRS-1C LISS III and PlanetScope Imagery. Our objective is to produce LULC map for Aurangabad and Waluj and estimate the change that happened to each class of the land cover. The results show that the increase of buildings area was significant from the period of 2008 to 2018. The accuracy of MLC was 94.13% and 85.65% with kappa 0.92, 0.81 in 2008 and 2018 respectively, and for SVM the accuracy was 93% with kappa 0.91, 94.4% with kappa 0.92 in 2008 and 2018 respectively. It was noticed that both SVM and MLC can be used effectively on LULC data analysis.

Keywords: Remote sensing · Land Use Land Cover · Maximum Likelihood Classifier · Support Vector Machine · GIS

1 Introduction

Remote sensing and its related technologies have been used extensively in many applications of land cover change monitoring. This methods and technologies have been developed to meet the requirements and interests of planners and land managers [1]. Classification of Land Used Land Cover involves the use of many classifiers which assign the pixels to particular land cover. Many classification algorithms have been applied and produced different accuracies. The most common classifiers are Maximum Likelihood classifiers, Minimum Distance, Mahalanobis, ANN, and SVM [2]. MLC is used successfully in assessment of land cover change, but it is based on statistical assumption on data. Each pixel assigned to the class with highest likelihood [3]. The MLC requires the training

K. C. Santosh and R. S. Hegadi (Eds.): RTIP2R 2018, CCIS 1037, pp. 482–492, 2019.
https://doi.org/10.1007/978-981-13-9187-3_43

data to be normally distributed on all classes and this is not always possible on remote sensing data [4]. SVM is non-parametric algorithm developed by Vapnik and Chervonenkis (1971). The classifier quality based on how well the process trained [5]. SVM and other kernel algorithms have become standard tools in techniques of machine learning and high dimensional data analysis because of limit number of parameters to tune, easy deployment into application and high performance on large dimensional data [6]. Extracting objects and shapes like buildings and roads from remote sensing data like spaces born sensors would require more better approaches like Fuzzy Binarization [7,17,18]. In this study, we investigate the use of Maximum Likelihood Classifier and Support Vector Machine classifier for Land use Land Cover mapping and estimating the land cover change in Aurangabad city and Waluj area. We aim to (i) Compare the results of MLC and SVM, (ii) Asses the land cover change along two periods 2008 and 2018.

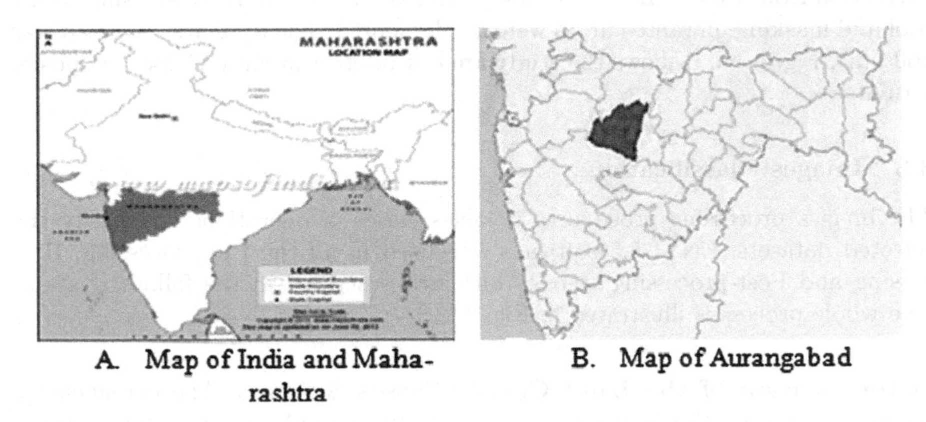

Fig. 1. (A) Map of India and Maharashtra and (B) Map of Aurangabad

2 The Study Area

The study area on this paper is Aurangabad city and Waluj Area in Maharashtra, India (Fig. 1). The geographical co-ordinate of area of interest is: latitude (N19° 53′59″) and longitude (E 75° 22′46″). It is 581 m high on the sea mean level. The total area of interest is 260 km^2. The land cover of this area is classified in to five classes: Built-up Area, Agricultural Area, Barren Land Area, Fallow Land Area, and Water Bodies.

3 Materials and Methods

3.1 Datasets Acquisition

In this study we used datasets acquired by Indian Remote Sensing Satellite ResourceSat LISS III and PlanetScope satellites. LISS III datasets were acquired on 23 October 2008 after rainy season with four Bands and spatial resolution of 23.5 m downloaded from Bhuvan ISRO web service, and Planetscope dataset were downloaded on August 2018 from Planet Imagery Website with four bands (RGB, NIR) and high resolution of 3 m. During the training data selection Google earth Pro Historical Data from 2008 used for ground points checking.

3.2 Datasets Pre-processing

Preparing satellite remote sensing data is fundamental step in GIS processing procedures. Image needs to be corrected atmospherically before further processing step. Dark object subtraction were applied to remove the effect of atmospheric scattering [8] which is common in remote sensing images [9]. Then image mosaicking was applied for the tiles covering the interested area. The image dataset subjected to False Color Composition FCC data format for better data extraction from LULC [10]. Then Subsetting the Region of Interest using ROI's tool and masking unwanted areas were implemented. The total area selected was 260 km^2. Figure 3a, b shows the study area of interest in False Color Composite format.

3.3 Images Classification

The images processing procedures involves series of operations to classify the selected datasets. ENVI 5.5 software was used in all the Pre-processing, Processing and Post-processing steps, which are described in the following steps. The whole process is illustrated in Fig. 2.

Determination of the Land Cover Classes Scheme. The classification scheme determine the land cover classes to be considered in remote sensing data. We determined five classes as our knowledge to the interested area. These classes are Built-up area (B), Agricultural Area (A), Barren Land Area (BA), Fallow Land Area (F), and Water Bodies (W).

Selecting the Training Data. The training data selection is potential step to train the classifier in supervised classification process, and it is time consuming step. The success of classification process depends on selecting appropriate pixels representing the classes under investigation. In this study, we used Google earth pro for investigating the selected pixels from LISS III image, and we used ENVI reference map Link tool for PlanetScope image as an addition tool for checking while its high resolution gives us satisfied clearance to select the pixels correctly. The distribution of classes in each training dataset is shown in Table 1.

Post Classification. Image post classification is necessary to refine the output image. We used the majority/minority analysis to change spurious pixels within a large single class to that class, and we also used Edit Image Classification tool to convert the areas which are outside the interested region to unclassified class.

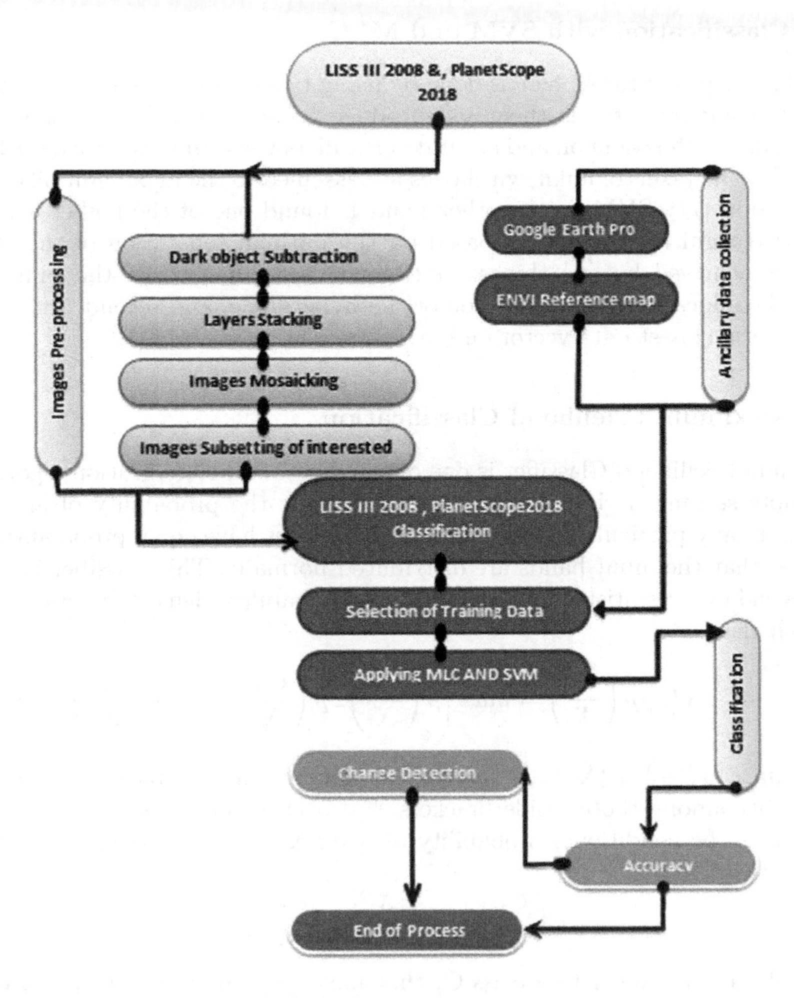

Fig. 2. Chart of the processing steps

Table 1. Distribution of classes in training dataset.

No. of pixels in 2008 dataset			No. of pixels in 2018 dataset	
	SVM	MLC	SVM	MLC
Built-up (B)	894	733	7853	7853
Agricultural (A)	2083	2264	2409	2409
Fallow land (F)	771	129.43	3238	3238
Barren land (BA)	1288	1396	1839	1839
Water bodies (W)	881	1017	1143	1143
Total	5917	5932	16482	16482

3.4 Classification with SVM and MLC

The classification process here is done by using two famous classifiers: SVM and MLC to compare between them and produce accurate land cover classes map. MLC is one of the common and accurate classifiers when the data are distributed normally. The pixels of unknown classes are assigned to the maximum Likelihood membership [11]. SVM on the other hand is found one of the higher accuracy pattern recognition algorithms based on the optimal separation of the classes which is achieved by two things; first, each vector data from the same class should be allocated to its same side of the hyperplane, and second, the margin between the closest data vector on both classes is maximized [12].

3.5 Maximum Likelihood Classification

Maximum Likelihood Classifier is one of the most used classification algorithms in remote sensing. It based on decision rule that the probability of any pixel belongs to any particular class and all of the classes have equal probability. It is assumes that the input bands are distributed normally. This classifier based on the second-order statistics of the Gaussian probability density function model for each class.

$$X \in C_j ifp\left(\frac{C_j}{X}\right) = max\left[p\left(\frac{C_1}{X}\right), p\left(\frac{C_2}{X}\right), \ldots, p\left(\frac{C_m}{X}\right)\right] \tag{1}$$

where $max\left[p\left(\frac{C_1}{X}\right), p\left(\frac{C_2}{X}\right), \ldots, p\left(\frac{C_m}{X}\right)\right]$ is a function that returns the highest probability among those inside brackets. For each pixel X belongs to class C_j, $p(C_j/X)$ is the conditional probability of pixel X. And this could be solved by Bayes's theorem:

$$p\left(\frac{C_j}{X}\right) = p\left(\frac{X}{C_j}\right) * \frac{p(C_j)}{p(X)}. \tag{2}$$

If we allocate a pixel X to a class C_j that maximize the probability expression (1), the results is called maximum likelihood solution [13].

3.6 Support Vector Machine

The development of Support Vector Machine (SVM) was initially released by the exploration of learning machine capacity control. Although the initiation and emerging of SVMs were in the late of 1970, but they have not got important attention until the recent years. Unlike the traditional classifiers, SVM is reducing the classification error. SVM was linear binary classifies, but now it is extended to variety of applications such as, Pattern Recognition, HandWriting, Document Identification, Clustering and Regression. In remote sensing SVM presents an improved results compared to traditional classifiers like Maximum Likelihood [14]. SVM is related to the general category of kernel methods algorithms. And the kernel algorithms are depend on data that comes through the dot-products in some possibly high-dimensional feature space. The kernel function basically having two advantages, First advantage is the ability to generate non-linear decision boundaries using methods designed for linear classifiers. And the second advantage is to use of kernel function to allow the user to apply a classifier to data that having no obvious fixed dimensional vector space representation [5].

4 Results

4.1 Comparison Between MLC and SVM

MLC and SVM were used for classification in this study case. The output maps of these classifiers are shown in Fig. 3. The coverage area from these maps is illustrated in Tables 2 and 3. The following section describes the change between these two classification approaches in term of coverage percentage and coverage area. MLC shows that there is an increase in Built-up area from 66 km^2 2008 to 78.3 km^2 in 2018 which represents an increasing rate of 25.83% to 30.12% of total area 260 km^2. The Agricultural land coverage area increased from 122 km^2 in 2008 to 131.4 km^2 which represents 46.92% in 2008 and 50.54% in 2018, this increases because of the vegetation cover after short time of rainy season, most of the Barren land covered by vegetation and that is explain why the barren land decreased from 68 km^2 in 2008 to 46 km^2 in 2018 which represents 26.15% in 2008 and 17.92% in 2018. Water body also increased from 4 km^2 to 5 km^2 which represents increasing rate from 1.54% in 2008 to 1.92% in 2018. SVM shows that there is increasing

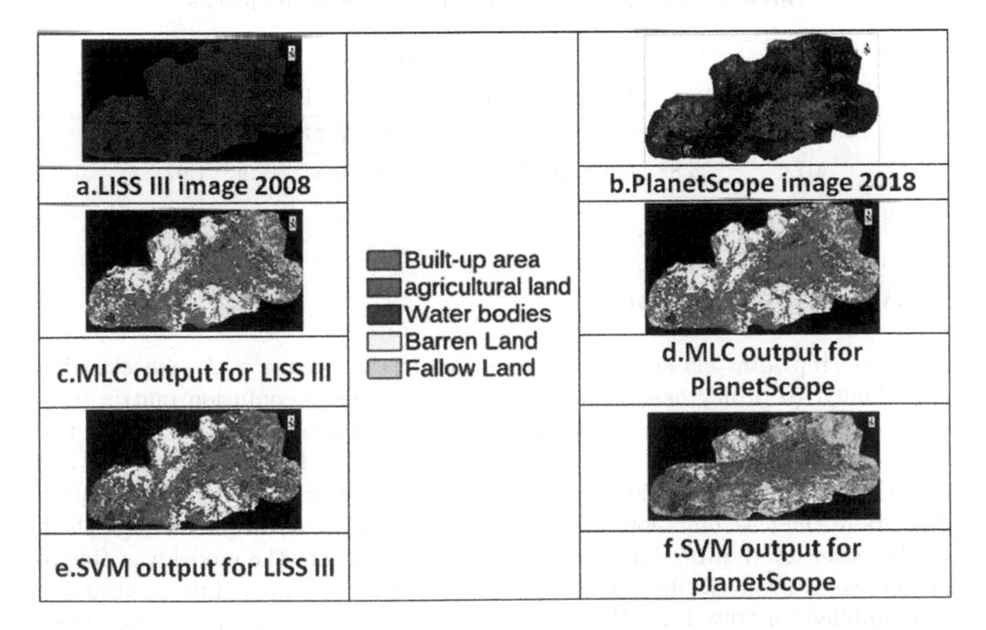

Fig. 3. Input and output images of MLC and SVM processes

rate in Built-up area from 53 km^2 2008 to 72.69 km^2 in 2018 which represents an increasing rate 20.26% in 2008 to 27.96% of total area 260 km^2, and increasing of Agricultural land from 129.43 km^2 in 2008 to 147.6 km^2 which represents 49.34% in 2008 and 46.49% in 2018. The Barren land decreased from 74.12 km^2 in 2008 to 38.15 km^2 in 2018 which represents 28.2% in 2008 and 14.67.59% in 2018. Water body also decreased from 5.61 km^2 to 2.3 km^2 in 2018 which represents decreasing from 2.61% in 2008 to 0.88% in 2018. A relevant survey for LISS III dataset showed a close result for maximum likelihood classifier accuracy around 92.6% [15]. From the results, we can say that the both classifiers show increasing of buildings area at expense of agricultural land which decreases continuously. Areas resulted from the two classifiers which are different in values and this thing is expected because of different working principles of MLC and SVM and different times of images acquisition. Our main objectives was to compare between the SVM and MLC and produce LULC map by using the two classifiers mentioned earlier.

Table 2. Area of each class in km^2

2008					2018			
Classifier	B	A+F	Ba	W	B	A+ F	Ba	W
SVM	53	129.43	74.12	5.61	72.69	147.6	38.15	2.3
MLC	66	122	68	4	78.3	131.4	46.6	5

Table 3. Area coverage in percentage cover for each class

2008					2018			
Classifier	B	A+F	Ba	W	B	A+ F	Ba	W
SVM	20.26	49.34	28.26	2.14	27.96	56.77	14.67	0.88
MLC	25.83	46.92	26.15	1.54	30.12	50.54	17.92	1.92

5 Accuracy Assessment

The next step after classification is to measure the accuracy of each classifier individually. Accuracy assessment was performed using confusion matrix from region of interest. In this study we computed the confusion matrix for the two periods datasets with total of 100 ground control points for 2008 image, and 300 ground control points for 2018 image. User accuracy provides us an indication of errors in the case omission and how well the training data was distinguished. On the other hand, procedure accuracy indicates whether the model was able to predict itself. From Tables 4 and 5, we can see that the procedure accuracy of MLC applied on 2008 LISS III image was high on identifying the barren land, water, and agricultural land, but low in identifying Built-up area and fallow land, its clearly that the 20% of barren land testing data were classified as built up area and this indication to the similarity in the reflectance of barren soil and building in the urban areas. As a result, we can conclude that MLC was unable to distinguish between barren land and built-up area with low resolution dataset. PlaneScope with MLC shows high performance in all the classes and this may return to the high resolution of the image (3 m). SVM in Tables 6 and 7 also shows high accuracy or all the classes expect built-up area in LISS III dataset where barren land training data classified as built-up area, In general, overall accuracy indicate high accuracy for both SVM and MLC classifiers as it is shown in Table 8, where SVM accuracy increased with the high resolution from 93% with LISS III to 94.5% with PlanetScope and MLC accuracy decreased from 94% in LISS III to 85.56% with PlanetScope.

Table 4. Confusion matrix for classification image in 2008 with MLC.

Class	F	BA	W	A	B	Total	User Acc%
Unclassified	0	1	1	0	1		
F	86	0	0	0	1	87	98.85
BA	0	96	0	0	7	103	93.20
W	0	0	99	0	0	99	100
A	14	0	0	100	2	116	86.21
B	0	3	0	0	89	92	96.74
Total	100	100	100	100	100		
Pro Acc%	86	69	99	100	89		

Table 5. Confusion matrix for classification image in 2018 with MLC.

Class	F	BA	W	A	B	Total	User Acc%
Unclassified	0	0	0	0	0	0	0
F	299	32	0	0	0	331	90.33
BA	0	253	0	0	152	405	62.47
W	0	0	150	2	0	152	98.68
A	1	0	0	392	0	393	99.75
B	0	15	0	6	148	169	87.57
Total	300	300	150	400	300		
Pro Acc%	99.67	84.33	100	98.00	49.33		

Table 6. Confusion matrix for classification image in 2008 with SVM

Class	F	BA	W	A	B	Total	User Acc%
Unclassified	0	0	1	0	0	0	0
F	92	0	4	4	3	103	89.32
BA	0	99	0	0	7	106	93.40
W	0	0	95	0	0	95	100
A	7	0	0	96	6	109	88.07
B	1	1	0	0	84	86	97.67
Total	100	100	100	100	100		
Pro Acc%	92	99	95	96	84		

Table 7. Confusion matrix for classification image in 2018 with SVM

Class	F	BA	W	A	B	Total	User Acc%
Unclassified	0	0	0	0	0	0	0
F	298	26	2	0	0	326	91.41
BA	0	252	2	0	26	280	90.00
W	1	0	146	0	0	147	99.32
A	1	0	0	400	0	401	99.75
B	0	22	0	0	274	296	92.57
Total	300	300	150	400	300		
Pro Acc%	99.33	84	97.33	100	91.33		

Table 8. Overall accuracy and kappa for all classifiers.

2008			2018	
Classifier	ACC %	KAPPA	ACC %	KAPPA
SVM	93.2	0.91	94.4	0.92
MLC	94	0.92	85.65	0.81

6 Conclusion

The present study focused on the remote sensing capability to map the cover change on concerned study area with the help of geographical information system (GIS). The analyzing of Land Use Land Cover with two classifier shows an increasing of using the land for human demands on buildings which in return leads to a decrease in the agricultural lands area. The two supervised classification approaches used in the study show higher performance if they trained well. It has been noted that the size of training data affects the accuracy of SVM and

MLC. Further analysis and extensive work on identifying the suitable algorithms and classifications of LULC is required before conducting LULC analysis. Further future work should be used with improved traditional classifiers, such as Fast K-nearest Neighbor classifier [16].

Acknowledgements. I would Like to thank the DST-FIST with sanction no. SR/FST/ETI-340/2013 to the Department of Computer Science and Information Technology, Dr. Babasaheb Ambedkar Marathwada University for supporting and funding this work. Also i would like to thank the university and department authorities for facilitating this work.

References

1. Giri, C.P.: Remote Sensing of Land Use and Land Cover: Principles and Applications. CRC Press, Boca Raton (2012)
2. Jensen, J.R.: Remote Sensing of the Environment: An Earth Resource Perspective, 2nd edn. Pearson Education India, Koramangala (2009)
3. Ahmad, A., Quegan, S.: Analysis of maximum likelihood classification on multispectral data. Appl. Math. Sci. **6**(129), 6425–6436 (2012)
4. Mather, P.M.: Computer Processing of Remotely-Sensed Images: An Introduction. Wiley, New York (2011)
5. Ben-Hur, A., Weston, J. : A user's guide to support vector machines. In: Data Mining Techniques for the Life Sciences, pp. 223–239 (2009)
6. Otukei, J.R., Blaschke, T.: Land cover change assessment using decision trees, support vector machines and maximum likelihood classification algorithms. Int. J. Appl. Earth Obs. Geoinf. **1**(12), S27–S31 (2010)
7. Santosh, K.C., Wendling, L., Antani, S., Thoma, G.R.: Overlaid arrow detection for labeling regions of interest in biomedical images. IEEE Intell. Syst. **31**(3), 66–75 (2016)
8. Ding, H., Shi, J., Wang, Y., Wei, L.: An improved dark-object subtraction technique for atmospheric correction of Landsat 8. In: Proceeding of SPIE 9815, MIPPR 2015: Remote Sensing Image Processing, Geographic Information Systems, and Other Applications, 98150K, 14 December 2015
9. Jensen, J.R.: Introductory Digital Image Processing: A Remote Sensing Perspective. Pearson Education Incorporated, Glenview (2016)
10. Gupta, R.P.: Remote Sensing Geology, 2nd edn. Springer, Heidelberg (2017). https://doi.org/10.1007/978-3-662-55876-8
11. Verma, R.K., Kumari, K.S., Tiwary, R.K.: Application of remote sensing and GIS technique for efficient urban planning in India. In: Geomatrix Conference Proceedings (2009)
12. Schlkopf, B., Smola, A.J.: Learning with Kernels: Support Vector Machines, Regularization, Optimization, and Beyond. MIT Press, Cambridge (2018)
13. Gao, J.: Digital Analysis of Remotely Sensed Imagery. McGraw-Hill Professional, New York (2009)
14. Mather, P., Tso, B.: Classification Methods for Remotely Sensed Data. CRC Press, Boca Raton (2016)
15. Saha, A.K., et al.: Land cover classification using IRS LISS III image and DEM in a rugged terrain: a case study in Himalayas. Geocarto Int. **20**(2), 33–40 (2005)

16. Vajda, S., Santosh, K.C.: A fast k-nearest neighbor classifier using unsupervised clustering. In: Santosh, K.C., Hangarge, M., Bevilacqua, V., Negi, A. (eds.) RTIP2R 2016. CCIS, vol. 709, pp. 185–193. Springer, Singapore (2017). https://doi.org/10.1007/978-981-10-4859-3_17
17. Santosh, K.C., Wendling, L., Antani, S.K., Thoma, G.R.: Scalable arrow detection in biomedical images. In: ICPR 2014, pp. 3257–3262 (2014)
18. Santosh, K.C., Roy, P.P.: Arrow detection in biomedical images using sequential classifier. Int. J. Mach. Learn. Cybern. **9**(6), 993–1006 (2018)

Impact of Dimensionality Reduction Techniques on Endmember Identification in Hyperspectral Imagery

Mahesh M. Solankar(✉), Hanumant R. Gite(✉), Rupali R. Surase(✉), Dhananjay B. Nalawade(✉), and Karbhari V. Kale(✉)

Department of Computer Science and IT, Dr. B. A. M. University, Aurangabad, MS, India
mmsolankar13@gmail.com, hanumantgitecsit@gmail.com, rupalisurase13@gmail.com, dbnalawade@gmail.com, kvkale91@gmail.com

Abstract. The image derived hyperspectral endmembers fulfil the requirement of ground truth information required for supervised classification and spectral unmixing of the hyperspectral scenes. In hyperspectral images endmembers are very difficult to identify by visual inspection, since their population within the scene is significantly low. There are several algorithms (PPI, ATGP, NFINDR, CCA, VCA and SGA) developed for endmember identification. Most of these algorithms begins with MNF transformation based data dimensionality reduction, even though there other dimensionality reduction algorithms are available in the literature. This paper critically evaluates the comparative performances of NFINDR and ATGP endmember finding algorithms using original hyperspectral scene and PCA, MNF and ICA transformed data sets separately. The experimental outcomes are evaluated using two important parameters. First parameters compares the execution time of EM identification algorithms. Second parameter compares the spatial coordinates and their corresponding spectral signatures of NFINDR and ATGP identified endmembers. The comparative experimental analysis showcase that the execution time of NFINDR and ATGP algorithms is significantly improved with ICA transformed principle components.

Keywords: Hyperspectral image · Dimensionality reduction · Endmember extraction · PCA · MNF · ICA · NFINDR · ATGP

1 Introduction

The hyperspectral image encloses high volume of information because of its wide range of spatial and spectral coverage. To discriminate the surface materials based on its chemical properties, the hyperspectral data is been recorded into continuous fashion in hundreds of spectral channels generally covering 400 nm to 2500 nm portion of the electromagnetic spectrum. These continuously recorded

K. C. Santosh and R. S. Hegadi (Eds.): RTIP2R 2018, CCIS 1037, pp. 493–504, 2019.
https://doi.org/10.1007/978-981-13-9187-3_44

spectral channels contains tremendous amount of redundant information leading to both time and space complexity issues during the data exploration and analysis. This data redundancy problem continues in hyperspectral endmember identification too. In hyperspectral imaging, the endmembers are well-defined as the unique spectral signatures representing distinct surface materials within the scene under observation [1]. In hyperspectral image classification and spectral unmixing, endmembers acts as a training data and fulfils the requirement ground truth information [2,3]. There are several endmember identification algorithms designed, developed and used in hyperspectral data exploitation. These algorithms includes are Pixel Purity Index [4], Convex Cone Analysis [5], NFINDR [6], Vertex Component Analysis [7], Simplex Growing Algorithm [8], and Automatic Target Generation Process [9,10]. All of these algorithms performs the dimensionality reduction before proceeding to the endmember identification. To perform the same, Minimum Noise Fraction [11] dimensionality reduction techniques is commonly used by most of the endmember finding algorithms, even though there are other techniques are available for the dimensionality reduction including Principal Component Analysis (PCA) and Independent Component analysis (ICA). The dimensionality reduction and endmember identification techniques used for experimental analysis are highlighted below.

1.1 Dimensionality Reduction Techniques

Principal Component Analysis (PCA): The PCA transformation is a statistical approach which minimized the dimensions of the original image while preserving maximum variations within the data. It proceeds with an assumption that, the adjacent spectral channels are having higher correlation and encloses similar information. The PCA achieves dimensionality reduction by finding the directions, termed as principal components, along which the data variations are maximal. In PCA transformation, higher band correlation is removed and the significant information is represented with few principal components rather than with thousands of variables. The significant amount of information is made available in first few PCs, where the amount of information get decreased with increase in number of PCs [8,12].

Minimum Noise Fraction (MNF): The MNF transformation is a two-step process. In first step, it uses the PCs of noise covariance matrix to de-correlate and rescale the noise within the data. It transforms the data where the noise is having unit variance and no band-to-band correlation. This procedure is also termed as 'Noise Whitening'. In second step, the noise whitened data is used to identify the PCs, further the inherent data dimensions is identified by examining the final eigenvalues and the associated images [11,13].

Independent Component Analysis (ICA): The ICA transformation is a probabilistic and statistical approach to decompose a complex hyperspectral data into independent sub-parts. To minimize the data dimensions, ICA transforms the original data into maximum independent and non-Gaussian components. It proceeds with an assumption that, the hyperspectral data is mixed linearly with two or more independent signal sources. ICA performs unmixing of these signal sources based on their independent statistical features [14].

1.2 Endmember Identification Techniques

The two popular endmember extraction techniques are used for experimental analysis.

NFINDR: The principle of NFINDR endmember finding approach is centered on the assumption that the volume of simplex formed by hyperspectral endmembers is maximum as compared to the simplex volume formed by any other combination of other pixels. In NFINDR, before proceeding to endmember identification, data dimensionality is reduced using MNF. Then the predefined number of pixels are selected in random fashion as an initial endmembers, and their corresponding volumes are calculated iteratively. Finally, the maximum volume is obtained by increasing simplex within the scene. These hyperspectral pixels, producing maximum volume, are considered as final endmembers [6].

Automatic Target Generation Process (ATGP): The ATGP endmember finding algorithm proceeds with the repetitive use of an orthogonal subspace projection to find the hyperspectral data sample vectors of significant interest. The algorithm begins without having any prior information, regardless of what type of information lf enclosed within the data sample vectors. The algorithm gets terminated upon satisfying the stopping rule, which is determined by the total number of endmembers required to identify [9].

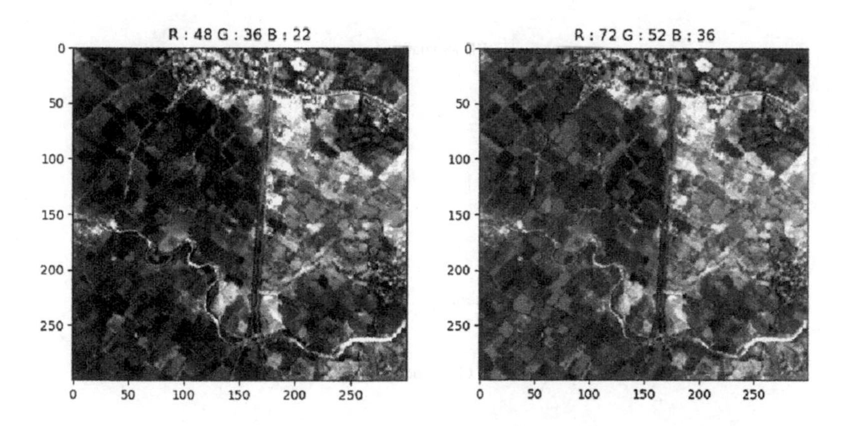

Fig. 1. True Color View (left) and False Color View (right)

1.3 Dataset

The subset of AVIRIS-NG hyperspectral scene covering wavelength from 380 nm to 2500 nm is used for experimental analysis. The original scene is captures over the Jhagadia (Gujrat) site with spatial dimension 5447 × 697 and spectral dimension 425 spectral channels [15]. The scene contains vegetation, soil, water and urban area. Figure 1 Shows the true color view (left) and false color composite view (right) of the scene respectively.

2 Experimental Framework

The experimental work is performed step-by-step as per the proposed experimental framework (Fig. 2). In first step, the original hyperspectral scene is preprocessed spatially as well as for further experiments. Then the notion of Virtual Dimensionality (VD) is used to identify the number of principal components to preserve after dimensionality reduction and the number of endmembers to be identified automatically. In EM identification-1, two well-known endmember identification algorithms (NFINDR and ATGP) are used to identify the hyperspectral endmembers. Then in next stage, the same dataset is transformed using three popular dimensionality reduction techniques (PCA, MNF and ICA) to retain only 10 principal components. These transformed scenes are used for endmember identification using NFINDR and ATGP algorithms independently. Further in result analysis, the comparative outcomes of EM identification 1 and EM identification 2 is performed.

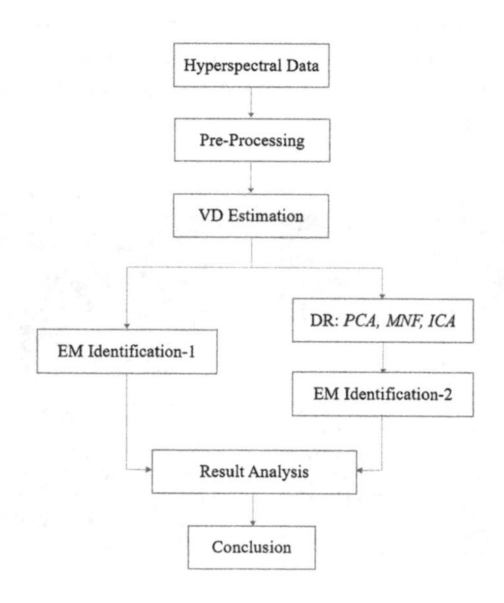

Fig. 2. Methodology

3 Experimental Work

The step-by-step experimental work performed is elaborated below:

3.1 Preprocessing

The original hyperspectral scene is having spatial dimension 5447 × 697 and spectral dimension 425 spectral channels. In preprocessing, original scene is resized both spatially and spectrally. In spatial subset, the 300 × 300 spatial portion of the original scene is selected, where first pixel begins at x = 200 and y = 290 coordinate. In spectral subset, the spectral channels (1–20, 195–214, 286–318 and 414–425) affected by vertical strips are removed. The final subset taken for an experiments is having dimension 300 × 300 × 340.

3.2 VD Estimation

The HySime VD estimation technique is used to automatically select the number of principal components and the number of endmembers to be identified. HySime estimated that there are ten (10) unique materials within the scene, so on the basis of HySime estimated VD count, the total ten (10) number of principal components and eleven (11) number of endmembers are extracted.

3.3 EM Identification-1

In EM Identification-1, the endmembers are identified from whole data without performing any dimensionality reduction using NFINDR and ATGP endmember identification algorithms. The total number of endmembers (i.e. 11) to be identified is defined by the HySime VD estimation technique. The NFINDR and ATGP identified endmember signatures are recorded with respect to their corresponding spatial coordinates and used for comparative analysis.

3.4 Dimensionality Reduction

In PCA, MNF and ICA transformation, the hyperspectral scene with 340 spectral channels is reduced to only ten principal components. The same HySime VD estimation is used to identify the number of principal components to preserve. The separate stacks of PCA, MNF and ICA transformed hyperspectral scenes are formulated and used for EM Identification-2.

3.5 EM Identification-2

In EM Identification-2, the endmembers are identified separately from PCA, MNF and ICA transformed hyperspectral scene using NFINDR and ATGP endmember identification algorithms. The EM identification-2, the number of endmember identified are similar to that of identified from original scene in EM Identification-1 (i.e. 11). The final sets of all the NFINDR and ATGP identified endmember signatures are recorded with respect to their corresponding spatial coordinates and used for comparative analysis.

4 Comparative Result Analysis

This section critically compares the endmembers identified from pre and post dimensionality reduced hyperspectral scenes. The two parameters are taken in account to compare these pure spectral signatures. The first parameter is the spatial location of the pixel recorded to be pure and the spectral signature corresponding to that particular pixel.

Table 1. Endmember coordinates extracted using NFINDR

Original scene		PCA transformed scene		MNF transformed scene		ICA transformed scene	
X,Y	T	X,Y	T	X,Y	T	X,Y	T
9,172	18 S	9,172	10 S	9,172	12 S	225,175	9 S
57,198		23,165		252,22		9,172	
30,130		30,120		23,165		239,290	
145,60		45,255		82,3		57,198	
225,175		57,198		233,17		45,255	
45,255		276,14		82,192		30,120	
276,14		141,59		142,59		145,60	
215,64		224,176		45,256		215,64	
239,290		296,85		80,37		276,14	
36,206		45,16		155,143		76,40	
91,31		215,64		287,21		36,206	

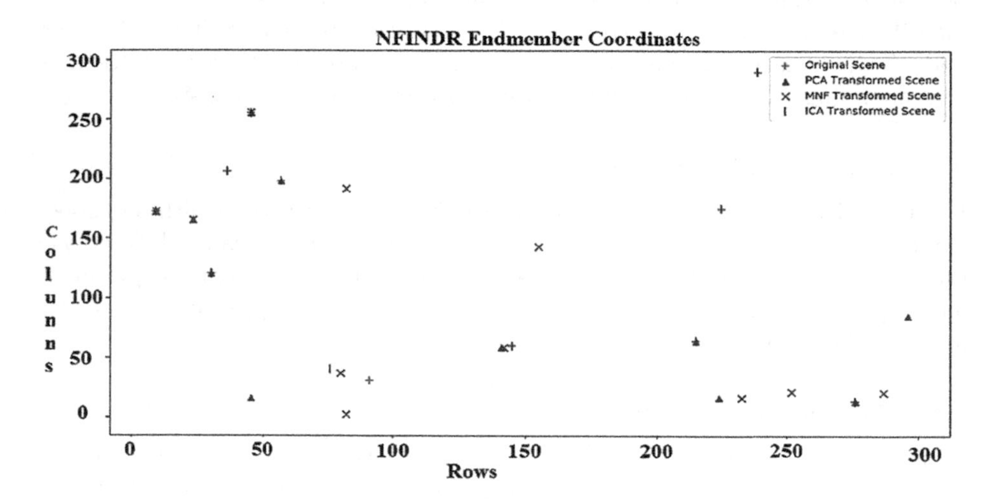

Fig. 3. Comparative spatial coordinates of NFINDR identified endmembers

Table 1 holds the outcomes of NFINDR EM identification algorithm performed separately on original scene, PCA transformed scene, MNF transformed scene and ICA transformed scene along with execution time respectively. In case of execution time, it is observed that the NFINDR algorithm performed a bit faster on ICA transformed scene as compared to PCA and MNF transformed scenes. As far as the spatial locations of the endmembers are considered, majority of the pixel coordinates are overlapping (Fig. 3), and others are found having diverse spatial locations. To analyze the spectral response of these endmembers having diverse spatial coordinates, their corresponding spectral signatures analyzed and found pure. Figures 5, 6, 7 and 8 shows the NFINDR identified spectral signatures of endmember coordinates (See Table 1) from original scene, PCA transformed scene, MNF transformed scene and ICA transformed scene respectively.

Table 2 holds the outcomes of ATGP EM identification algorithm performed separately on original scene, PCA transformed scene, MNF transformed scene and ICA transformed scene along with execution time respectively. In case of execution time, it is observed that the ATGP algorithm performed a bit faster on ICA transformed scene as compared to PCA and MNF transformed scenes. As far as the spatial locations of the endmembers are considered, majority of the pixel coordinates are overlapping (Fig. 4), and others are found having diverse spatial locations. To analyze the spectral response of these endmembers having diverse spatial coordinates, their corresponding spectral signatures analyzed and found pure. Figures 9, 10, 11 and 12 shows the ATGP identified spectral signatures of endmember coordinates (See Table 2) from original scene, PCA transformed scene, MNF transformed scene and ICA transformed scene respectively.

Table 2. Endmember coordinates extracted using ATGP

Original scene		PCA transformed scene		MNF transformed scene		ICA transformed scene	
X,Y	T	X,Y	T	X,Y	T	X,Y	T
9,172	20 S	9,172	4 S	9,172	4 S	9,120	2 S
224,69		224,69		82,192		30,120	
225,175		225,175		23,165		224,176	
30,120		30,120		252,22		144,60	
144,60		145,60		224,69		76,40	
45,256		57,198		225,175		60,198	
57,198		78,39		81,35		239,290	
23,165		45,255		45,256		220,69	
83,34		60,163		166,12		42,255	
110,210		196,74		23,165		162,55	
193,103		93,103		233,17		30,120	

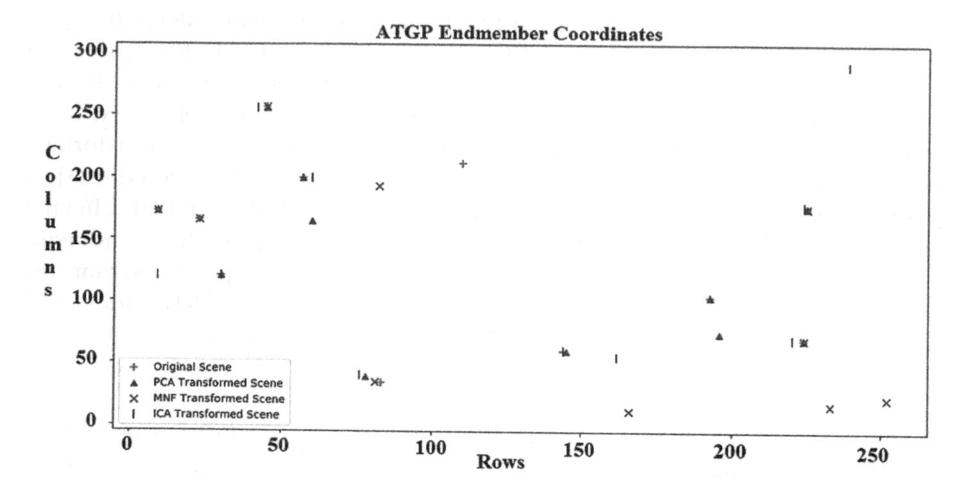

Fig. 4. Comparative spatial coordinates of ATGP identified endmembers

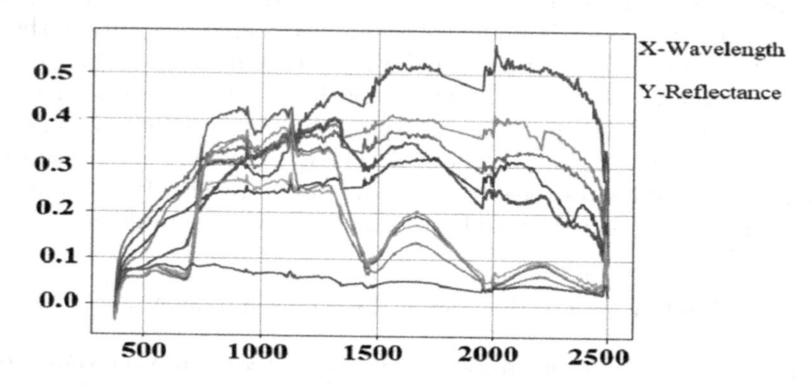

Fig. 5. NFINDR identified endmember signatures from original scene

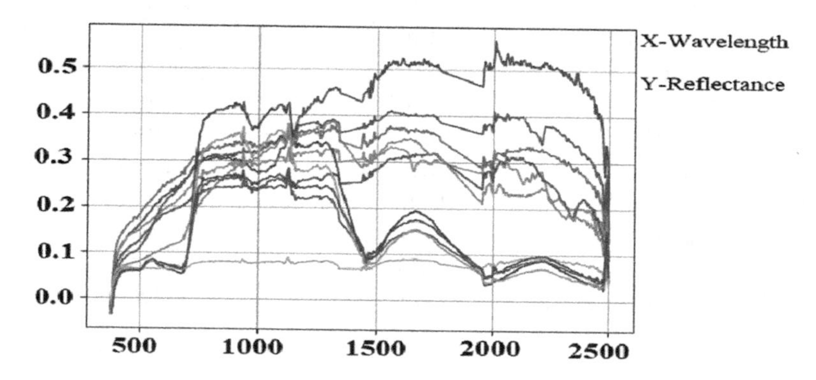

Fig. 6. NFINDR identified endmember signatures from PCA transformed scene

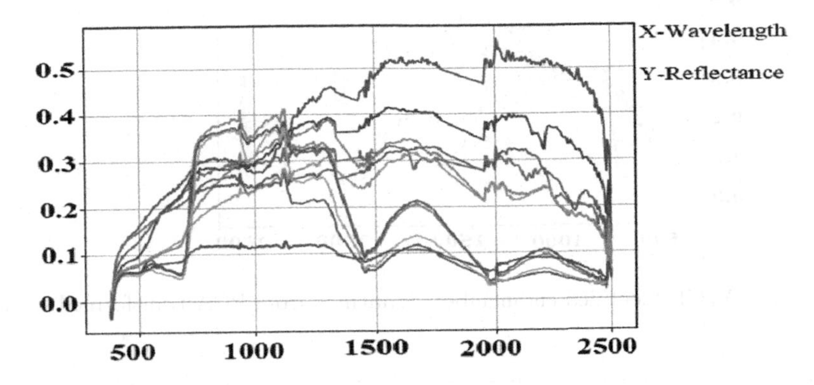

Fig. 7. NFINDR identified endmember signatures from MNF transformed scene

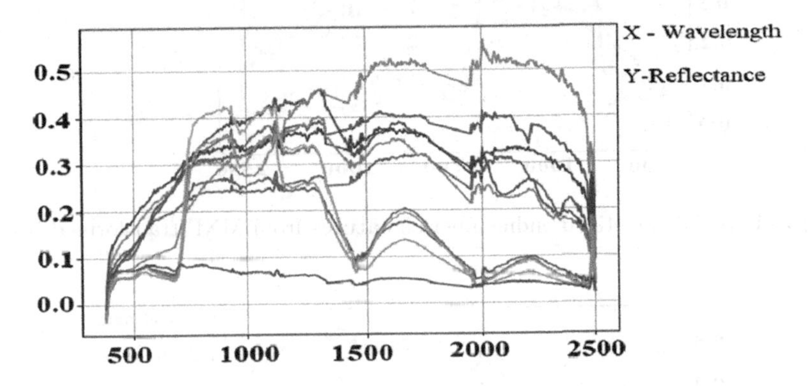

Fig. 8. NFINDR identified endmember signatures from ICA transformed scene

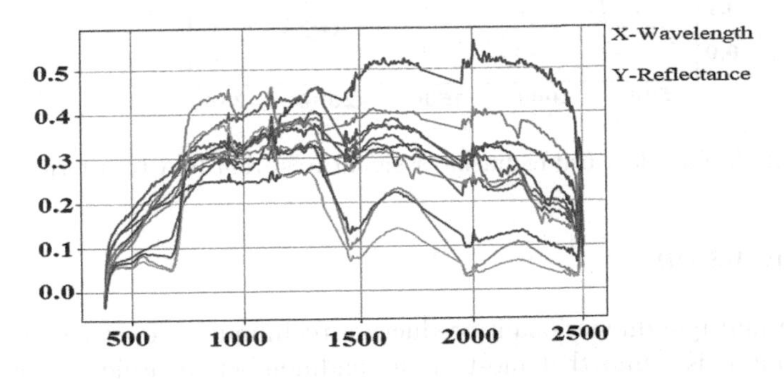

Fig. 9. ATGP identified endmember signatures from original scene

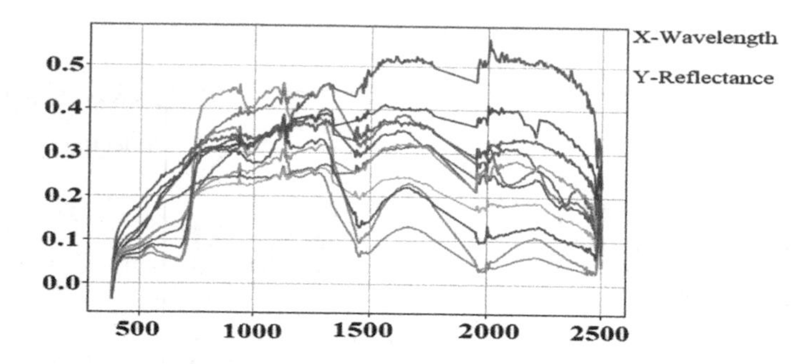

Fig. 10. ATGP identified endmember signatures from PCA transformed scene

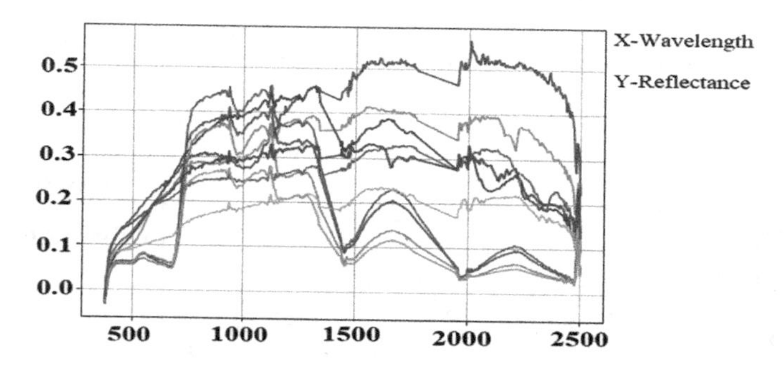

Fig. 11. ATGP identified endmember signatures from MNF transformed scene

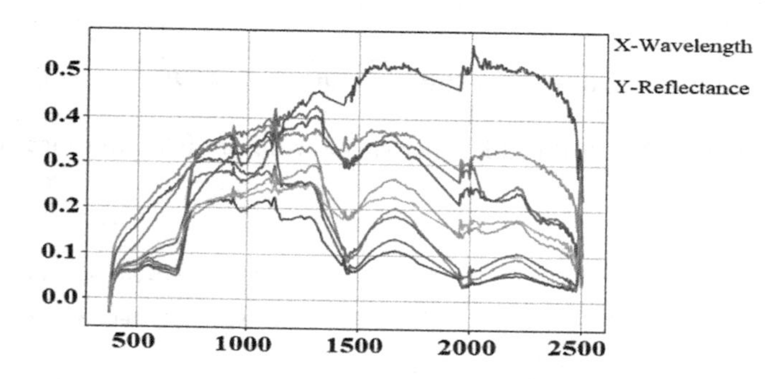

Fig. 12. ATGP identified endmember signatures from ICA transformed scene

5 Conclusion

There are multiple dimensionality reduction techniques are available in the literature, but it is found that most of the endmember identification algorithms prefer to use MNF transformation to perform the data dimensionality reduction. This paper critically compares the impact of three data dimensionality reduction techniques on endmember identification using NFINDR and ATGP algorithms.

The experimental outcomes are compared based on two major parameters, first, the execution time of endmember identification algorithms, and, second, the spatial locations and their corresponding spectral signatures of identified endmembers are compared. The experimental outcomes shows that the execution time of the endmember identification algorithms significantly improved with ICA transformed hyperspectral scene. For endmember spatial locations, it is observed that NFINDR and ATGP identified endmember locations are diverse at large extent, but this resulted because of the random initial endmember selection property of NFINDR and ATGP algorithm, not because of the data transformation techniques. Through the overall experimental observations, it can be concluded that, to improve the execution speed of endmember identification algorithms, the ICA data transformation technique can be used for data dimensionality reduction prior to endmember identification.

Acknowledgment. The Authors acknowledge to DST, GOI, for financial support under major research project (No. BDID/01/23/2014-HSRS/35 (ALG-V)) and for providing AVIRIS-NG data. The authors also extend sincere thanks to UGC SAP for providing lab facilities to the Department of Comp. Science and IT, Dr. B. A. M. University, Aurangabad-(MS), India.

References

1. Kale, K.V., Solankar, M.M., Nalawade, D.B., Dhumal, R.K., Gite, H.R.: A research review on hyperspectral data processing and analysis algorithms. Proc. Natl. Acad. Sci., India, Sect. A Phys. Sci. **87**(4), 541–555 (2017)
2. Nandibewoor, A., Hegadi, R.: A novel SMLR-PSO model to estimate the chlorophyll content in the crops using hyperspectral satellite images. Cluster Comput., 1–8 (2018)
3. Solankar, M.M., Gite, H.R., Dhumal, R.K., Surase, R.R., Nalawade, D., Kale, K.V.: Recent advances and challenges in automatic hyperspectral endmember extraction. In: Krishna, C.R., Dutta, M., Kumar, R. (eds.) Proceedings of 2nd International Conference on Communication, Computing and Networking. LNNS, vol. 46, pp. 445–455. Springer, Singapore (2019). https://doi.org/10.1007/978-981-13-1217-5_44
4. Boardman, J.W., Kruse, F.A., Green, R.O.: Mapping target signatures via partial unmixing of AVIRIS data (1995)
5. Ifarraguerri, A., Chang, C.I.: Multispectral and hyperspectral image analysis with convex cones. IEEE Trans. Geosci. Remote Sens. **37**(2), 756–770 (1999)
6. Winter, M.E.: N-FINDR: an algorithm for fast autonomous spectral end-member determination in hyperspectral data. In: SPIE's International Symposium on Optical Science, Engineering, and Instrumentation. International Society for Optics and Photonics, pp. 266–275 (1999)
7. Nascimento, J.M., Dias, J.M.: Vertex component analysis: a fast algorithm to unmix hyperspectral data. IEEE Trans. Geosci. Remote Sens. **43**(4), 898–910 (2005)
8. Chang, C.I., Wu, C.C., Liu, W., Ouyang, Y.C.: A new growing method for simplex-based endmember extraction algorithm. IEEE Trans. Geosci. Remote Sens. **44**(10), 2804–2819 (2006)

9. Plaza, A., Chang, C.I.: Impact of initialization on design of endmember extraction algorithms. IEEE Trans. Geosci. Remote Sens. **44**(11), 3397–3407 (2006)
10. Chang, C.I., Chen, S.Y., Li, H.C., Chen, H.M., Wen, C.H.: Comparative study and analysis among ATGP, VCA, and SGA for finding endmembers in hyperspectral imagery. IEEE J. Sel. Top. Appl. Earth Obs. Remote Sens. **9**(9), 4280–4306 (2016)
11. Phillips, R.D., Watson, L.T., Blinn, C.E., Wynne, R.H.: An adaptive noise reduction technique for improving the utility of hyperspectral data. In: Proceedings of the 17th William T. Pecora Memorial Remote Sensing Symposium, pp. 16–20 (2008)
12. Rodarmel, C., Shan, J.: Principal component analysis for hyperspectral image classification. Surv. Land Inf. Sci. **62**(2), 115 (2002)
13. Wang, J., Chang, C.I.: Independent component analysis-based dimensionality reduction with applications in hyperspectral image analysis. IEEE Trans. Geosci. Remote Sens. **44**(6), 1586–1600 (2006)
14. Hyvärinen, A., Oja, E.: Independent component analysis: algorithms and applications. Neural Netw. **13**(4), 411–430 (2000)
15. Green, R., Landeen, S., McCubbin, I., Thompson, D., Bue, B.: Airborne visible/infrared imaging spectrometer next generation (AVIRIS-NG), 1st edn. [PDF] JPL, California Institute of Technology (2015). http://vedas.sac.gov.in:8080/aviris/pdf/20150726_AVRISINGDataGuide_v4.pdf. Accessed 1 July 2017

Landslide Susceptibility Zonation (LSZ) Using Machine Learning Approach for DEM Derived Continuous Dataset

Muskan Jhunjhunwalla[1], Sharad Kumar Gupta[2], and Dericks P. Shukla[2](✉)

[1] National Institute of Technology Hamirpur, Hamirpur 177001, Himachal Pradesh, India

[2] School of Engineering, Indian Institute of Technology Mandi, Kamand 175005, Himachal Pradesh, India

dericks@iitmandi.ac.in

Abstract. Landslide-prone areas can be shown by depicting occurrence of landslides; landslide susceptibility zonation map (LSZ), landslide hazard zonation map (LHZ) and landslide risk zonation map (LRZ). However, for the preparation of LRZ map, we need LHZ map and for LHZ map, we need LSZ map. In this work Logistic Regression (LR), Fisher Discriminant Analysis (FDA) associated with the weighted linear combination (WLC) and ANN are used for the preparation of LSZ maps. Seven causative factors that give continuous dataset are taken into consideration i.e. aspect, slope, digital elevation model (DEM), topographic wetness index (TWI), tangential curvature, profile curvature and plan curvature for the part of Mandakini river basin in Garhwal Himalayas. Geology, geomorphology, soil type, thrust/fault buffer, road buffer, drainage buffer etc., which are also important landslide governing factors, were not used as they give a discrete/classified dataset. The study area is spread in 275.60 km^2 area where total 122 landslides occurred between 2004 and 2017. The landslides occurred from 2004 to 2012 (46 landslides with 1203 pixels) have been used for training of the models and from 2013 to 2017 (76 landslides) have been used for testing of the models. The susceptibility maps were classified/categorized into five different zones (very low, low, moderate, high and very high) based on the natural break in data. The landslide locations with the index value greater than 0.55 have been considered for the validation of the maps. The assessment of accuracy is done based on the Heidke Skill Score (HSS). The HSS score for FDA, LR and ANN is obtained as 0.89, 0.98 and 0.96. Based on the HSS score, the LR method can be selected as the best method amongst the three.

Keywords: Landslide susceptibility zonation (LSZ) · Artificial neural network · Fisher discriminant analysis · Logistic regression · Weighted linear combination · Heidke skill score

1 Introduction

The outcomes of landslides are severe with respect to human life and overall economy of the countries across the world. Worldwide, they cause thousands of

K. C. Santosh and R. S. Hegadi (Eds.): RTIP2R 2018, CCIS 1037, pp. 505–519, 2019.
https://doi.org/10.1007/978-981-13-9187-3_45

deaths and billions of bucks in property harm every year [1]. It represents more or less 9% of the natural disasters occurred throughout the 1990s, worldwide [2]. Around 15% of the Indian land surface area are susceptible to landslide hazard [1]. Therefore, landslide susceptibility mapping is necessary. Landslide mapping is used by the engineers, risk managers, etc. for getting accurate information about the occurrence of landslides. There are various causes of slope instability in the Himalayas, which are broadly classified into two types – Geological factors and Anthropogenic factors. Some geological factors are lithology, rock structure, weathering, Geotechnical properties of the soil, natural slopes, vegetation, ground water, land surface temperature, cloud bursts and precipitation, etc. and some of the anthropogenic factors are road and construction activities, improper land use, extension of agriculture on higher altitudes and faulty agricultural practices, etc. [3]. Understanding these causes helps in developing the fundamental principles of landslide hazard zonation, observation and forecasting of landslide hazards for better landslide mitigation and management [3].

For the preparation of LSZ maps, generally there are 4 approaches. These are probabilistic approach, deterministic approach, heuristic approach and machine learning approach [1]. Nowadays, the most commonly used approach is machine learning approach as the model learns the known dependency of nonlinear causative factors and the occurrence of landslides and thus models the unknown dependency of the causative factors and the occurrence of landslide (whether the landslide will occur or not). There are various methods based on this approach that are used frequently i.e. regression analysis, Artificial neural network (ANN), fuzzy logic, discriminant analysis, Support vector machine etc.

Generation of landslide susceptibility maps has been done by various authors previously using different approaches such as probabilistic approach like evidence belief function [4], weight of evidence and certainty factor [5], machine learning approaches like ANN [2,6,7], logistic regression [8,9] and SVM [10,11], and heuristic approach like analytical hierarchical process (AHP) [12,13]. From the literature review, it can not be suggested that which method is better amongst all. However, it can be inferred on basis of different parameters for measuring accuracy. For e.g. if we select causative factors such as slope gradient, altitude, slope aspect, land use, lithology, distance from roads, distance from streams, TWI, distance from faults, stream transport index, plan curvature and stream power index etc., then it is seen that the weight of evidence has higher accuracy than certainty factor [5].

This work aims at using the weights calculated from different methods such as Fisher discriminant analysis, Logistic regression and ANN to generate LSZ maps. Further, HSS has been used to measure the accuracy of these three methods. Based on these scores the best method for generating LSZ maps is selected.

2 Materials

2.1 Study Area

Preparation of the LSZ map has been done for a part of the Mandakini river basin of Garhwal Himalaya in Uttarakhand. It originates from the Chorabari Glacier near Kedarnath in Uttarakhand in India and is a tributary of the Alaknanda river. It covers about 275.60 sq. km area and is situated between 78°49′00″E to 79°21′13″E longitude and 30°19′00″N to 30°49′00″N latitude falling in the Survey of India Toposheet Nos. 53 J and 53 N [1]. This region is extremely prone to landslides during the monsoon season. Landslides occur in this area every year. Therefore, it is necessary to generate LSZ maps for this area. Figure 1 shows the study area.

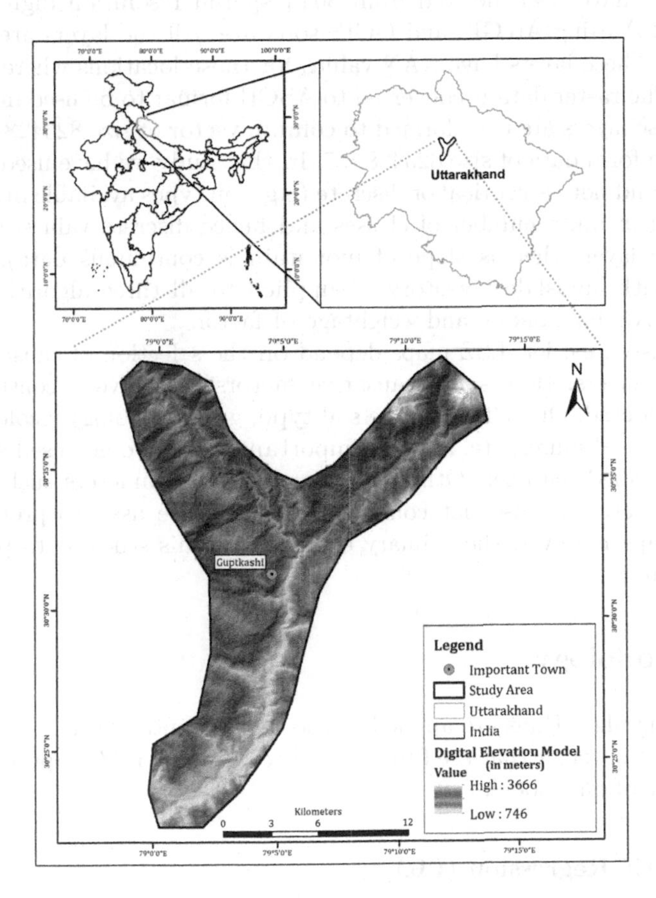

Fig. 1. Location map of the study area showing digital elevation model prepared from SRTM (Data source: boundary of India and Uttarakhand provided by survey of India, DEM provided by USGS)

2.2 Experimental Data Used

The study area comprises of total 122 landslides occurred between 2004 and 2017. The landslides occurred from 2004 to 2012 (46 landslides with 1203 pixels) have been used for training of the models and from 2013 to 2017 (76 landslides) have been used for testing of the models. A landslide inventory is used as a target variable of landslide pixel (1) and non landslide pixel (0). This inventory was prepared manually by looking at the satellite images of different years. However this process could be enhanced by using automatic extraction of feature (landslide) from multi-temporal satellite images. This can be done using active learning approach [14–16].

In this study, seven causative factors/layers i.e. aspect, slope, TWI, elevation, profile curvature and plan curvature are considered for preparation of LSZ maps. These layers have been derived from 30 m spatial resolution digital elevation model (DEM) using ArcGIS and QGIS software. All the layers are of the size 1028×801. These layers have NAN values for those locations where data is not available. The raster data is converted to ASCII format to be used in MATLAB. Each of these layers are transformed to column vector of size 823428×1 and are combined to form data of size 823428×7. In this study we have used continuous data layers and not categorical or discrete (e.g. soil types available in an area can be classified in finite number of classes and hence discrete values are assigned to soil type layer whereas slope of mountain is continuous data). This data combined with landslide inventory is supplied to all three algorithms viz LR, FDA and ANN for ranking and weightage of factors.

As discussed earlier, LSZ maps depend on the selection of causative factors or the data layers. Here seven causative factors that give a continuous data set are considered. Other factors like soil type, geomorphology, geology, thrust/fault buffer, road buffer, etc. are also important, but were not used as they give discrete/classified data set. Other time dependent parameters such as rainfall, earthquakes, etc. are also not considered as they are used to prepare hazard zonation maps however, the primary objective of this study is to prepare susceptibility maps.

3 Methodology

The following algorithms are applied to the experimental data set for the calculation of layer's weight that are further used to generate LSZ maps and calculate the accuracy of our model.

3.1 Logistic Regression (LR)

Logistic Regression is a supervised classification algorithm. It is the most common statistical method that is utilized in landslide assessment [8]. The predicted outcome is binary (0/1; T/F) for a given set of predictor variables. The independent variable can be interval or categorical while the dependent variable can be multinomial or binary [1]. In this study we have used binary outcomes (1- landslide, 0- no landslide). We have modeled the probabilities associated with the outcomes, instead of directly predicting the outcomes.

Logistic regression is a special case of linear regression and predicts the probability of the occurrence of an event by using logit function. Logit is the log of odds of success [1] and is represented as (1):

$$l(x) = \log\left(\frac{h_\theta(x)}{1-h_\theta(x)}\right) \tag{1}$$

The hypothesis of LR is that logit has a linear relation with the predictors(x) as represented in (2) and (3):

$$l(x) = \theta_0 + \theta_1 x_1 + \theta_2 x_2 + ---------- + \theta_n x_n = \log\left(\frac{h_\theta(x)}{1-h_\theta(x)}\right) \tag{2}$$

$$h_\theta(x) = \frac{e^{\theta_0+\theta_1 x_1+\theta_2 x_2+---------+\theta_n x_n}}{1+e^{\theta_0+\theta_1 x_1+\theta_2 x_2+------+\theta_n x_n}} \tag{3}$$

Here n = 7 represents the seven landslide causative factors taken into consideration. $x_1, x_2, x_3....x_n$ are the independent variables, $\theta_1, \theta_2, ...\theta_n$ are the LR coefficients and θ_0 is the intercept or the bias. Equation (3) is known as the LR function. The LR coefficients are actually the weights corresponding to each of these seven layers and are used for the generation of LSZ maps.

3.2 Fisher Discriminant Analysis (FDA)

FDA is a method used in machine learning, statistics and pattern recognition to find a linear combination of features which separates or characterizes two or more classes of events or objects. The resulting combination may be used for dimensionality reduction before classification [17]. In FDA, "within-class variance" is minimized and "between-class variance" is maximized as in Fig. 2.

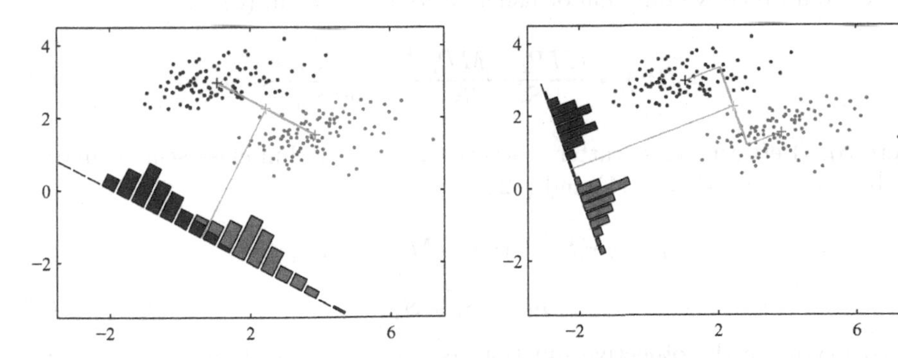

Fig. 2. The left plot shows samples from two classes (depicted in blue and red) along with the histogram ensuing from projection onto the line connecting the class means. The right plot shows the corresponding projection supported the Fisher linear discriminant [18]. (Color figure online)

For two class problem (0-no landslide and 1-landslide), let us assume that we have a set of n-dimensional samples X (here we have 7 dimensions) where $X = x^{(1)}, x^{(2)}, ..., x^{(m)}$ and out of X, D1 samples belong to class L1 (landslide) and D2 belong to class L2 (no landslide). We can obtain y by projecting these samples as (4).

$$y = w^T X \tag{4}$$

Our aim is to maximize the objective function which is the ratio of "between class variance" and "within class variance" and can be defined as Eq. (5)

$$J(w) = \frac{between\ class\ variance}{within\ class\ variance} \tag{5}$$

Let, MS is the sample mean as represented in Eq. (6) and MP is the project mean as represented in (7)

$$MS_i = \frac{1}{D_i} \sum_{x \in L_i} x \tag{6}$$

$$MP_i = \frac{1}{D_i} \sum_{x \in L_i} w^T x \tag{7}$$

Now, a measure of separation between two class is the distance between the projected means [17], so the between class variance is Eq. (8) from Eq. (4)

$$MP_2 - MP_1 = w^T (MS_2 - MS_1) \tag{8}$$

Similarly, we can define within-class variance as (9)

$$S_i^2 = \sum_{y \in L_i} (y - MP_i)^2 \tag{9}$$

Now our objective function or fisher ratio is (10) from (5)

$$J(w) = \frac{(MP_2 - MP_1)^2}{S_1^2 + S_2^2} = \frac{w^T S_B w}{w^T S_W w} \tag{10}$$

where S_B (Between-class scatter matrix) and S_W (within-class scatter matrix) can be calculated [17] as (11) and (12)

$$S_B = (MS_2 - MS_1)(MS_2 - MS_1)^T \tag{11}$$

$$S_W = S_1 + S_2 \tag{12}$$

To maximize the objective function, equate the derivative of the objective function to 0. So, finally we got that J(w) is maximum when $w = S_w^{-1}(MS_2 - MS_1)$. Here w gives us the direction of projection and also the weights for our landslide data layers.

3.3 Artificial Neural Network

It is a computational system inspired by the human brain (Fig. 3a). It consists of neurons, an output layer, input layer and hidden layers. All these layers are interconnected and message from the input layer is passed to output layer by passing through the hidden layers. There can be no, one or many hidden layers depending upon our data complexity. The simplest neural network is known as the perceptron having only a single neuron. ANN is useful where relationships between causative factors and responses are complex like landslides. There are two major types of neural networks in which we can model our problem, i.e., feed forward networks and feed forward back propagation neural nets.

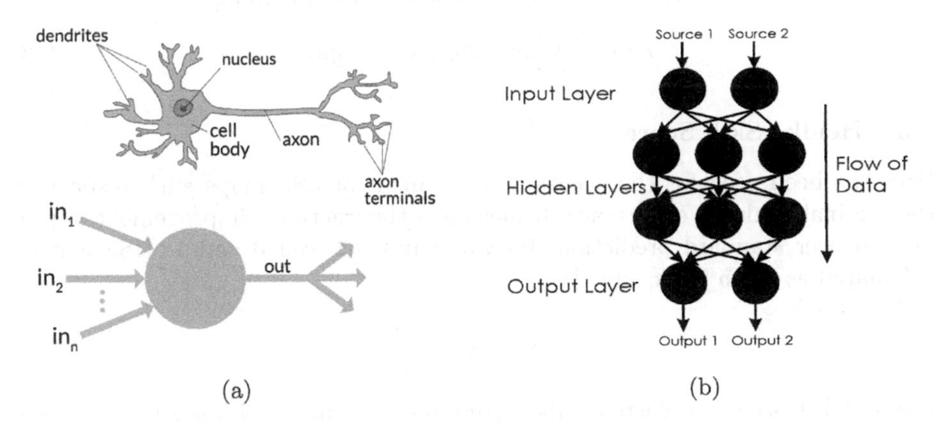

Fig. 3. (a) Human brain neuron and simple perceptron, (b) Feed forward neural network

In this work feed forward neural net has been used (Fig. 3b). It is a type of artificial neural network in which the connection does not form any loop. As the name suggests, the information moves from the input layer to hidden layers and then to the output layer and finally stops. It is primarily used in supervised learning where our data is time independent and not sequential. Feed forward nets can be used for any input to output mapping [19]. Performance of the neural net depends on the number of hidden layers (structure), activation function and the way the connection has been achieved.

There are 1203 landslide pixels and approximately 0.3 million non-landslide pixels in the study area. Due to this irregular distribution of pixels, the model becomes biased towards the class with high number of pixels. Hence, to overcome this problem, we have used $30 \times \textit{No. of Landslide Pixels}$ (i.e. 30×1203 pixels) randomly chosen for non-landslide class. For the training of the model 70% of the input data has been used, however 15% data has been used for testing and 15% for validation of the model. The above mentioned statistics represent the data for computation of the weights. In this study, the number of input neurons is equal to the number of causative factors (i.e. 7). For training "trainlm" function is used. The best results were obtained with 4 hidden layers. These hidden layers have been fixed after rigorous experimentation of the model.

3.4 Weighted Linear Combination

WLC is an analytical method which is useful when more than one feature or attribute is taken into consideration. It allows trade-offs and flexibility amongst all the parameters used [20]. WLC model is one of the most widely used GIS-based decision rules [21]. In this method, the layer weights that are calculated from the above algorithms are multiplied to corresponding layer values and then the products are summed up to give an index. This index is known as Landslide Susceptibility Index [1]. We can calculate LSI using WLC as shown in Eq. (13). LSI can then be classified into five different zones (from very high to very low) based on natural break in the data. This method gives us a composite landslide map layer as its output which is further used in LSZ mapping.

$$LSI = \sum attributes * weights \tag{13}$$

3.5 Heidke Skill Score

HSS has been used for the accuracy assessment of LSZ maps with respect to the original landslide occurrence. It measures the fractional improvement of prediction over standard prediction. Its value lies between 0 and 1. HSS can be calculated as given in Eq. (14) [1].

$$HSS = \frac{CF - EF}{SF - EF} \tag{14}$$

where, CF-Correct forecasted, this represents the number of landslides that are correctly predicted. EF- Error in forecasting, this represents the number of landslides that are incorrectly predicted. SF- Standard forecasted, this represents the total number of landslides.

4 Results and Discussions

Weights are important for making the susceptibility maps. Weights obtained by LR FDA and ANN for all the seven causative factors are shown in Table 1. The LSZ maps that are generated using the weights for LR, FDA and ANN as shown in Figs. 4, 5 and 6 respectively.

The obtained weights are multiplied with the thematic layers to obtain the LSI. The LSI are then normalized between 0 and 1 and are then classified into 5 different zones based on natural break i.e., LSI from 0–0.35 as very low susceptibility zone, from 0.35–0.45 as low susceptibility zone, from 0.45–0.55 as moderate susceptibility zone, from 0.55–0.65 as high susceptibility zone and 0.65–1 as very high susceptibility zone. For validation purpose landslides occurred between 2013 and 2017 are taken into consideration. There are 76 landslides for validation purpose. As mean and median for all the methods (except ANN) is observed to be near 0.55 (refer Table 2) so 0.55 is set as a threshold for classification. The landslides with LSI greater than 0.55 are considered to be correctly classified and below 0.55 are considered to be falsely classified.

Table 1. Weights calculated from different models

Causative factors	FDA	LR	ANN
Aspect	0.31	−1.11	0.66
DEM	−3.69	−3.72	0.38
Plan curvature	4.59	1.00	0.17
Profile curvature	0.80	−2.35	0.05
Slope	3.36	5.04	0.36
Tangential	−3.49	0.51	−0.43
TWI	0.72	2.42	0.72

Table 2. Statistics calculated (mean, median and standard deviation) for all the three methods

Method	Mean	Median	Standard derivation
LR	0.58	0.58	0.11
FDA	0.55	0.56	0.12
ANN	0.43	0.42	0.17

As it is seen from Table 2 that mean and median for LR and FDA is nearly equal to 0.55. The similarity in mean and median for both the methods can be observed in their susceptibility maps also which also looks similar with major area coming under high and very high susceptibility zones. Shift of mean and median towards lower values is observed in ANN indicating more area in low and very low regions. This is clearly seen that the LSZ map of ANN (Fig. 6) has more green area, verifying the statistical results in Table 2.

Table 3. The number of pixels, total area covered and % of total area covered by each of the susceptibility zones for each of these methods.

Susceptibility zones	LR			FDA			ANN		
	No. of pixels	Area covered (sq. km)	% of total area covered	No. of pixels	Area covered (sq. km)	% of total area covered	No. of pixels	Area covered (sq. km)	% of total area covered
Very low ($0 <= LSI <= 0.35$)	7234	6.51	2.36	13545	12.19	4.42	112696	101.43	36.79
Low ($0.35 < LSI <= 0.45$)	26667	24.00	8.71	46382	41.74	15.14	56339	50.71	18.39
Moderate ($0.45 < LSI <= 0.55$)	83475	75.13	27.25	86539	77.88	28.25	56715	51.04	18.51
High ($0.55 < LSI <= 0.65$)	110338	99.30	36.02	95813	86.23	31.28	45676	41.11	14.91
Very high ($0.65 < LSI <= 1$)	78627	70.76	25.67	64062	57.66	20.91	34915	31.42	11.39

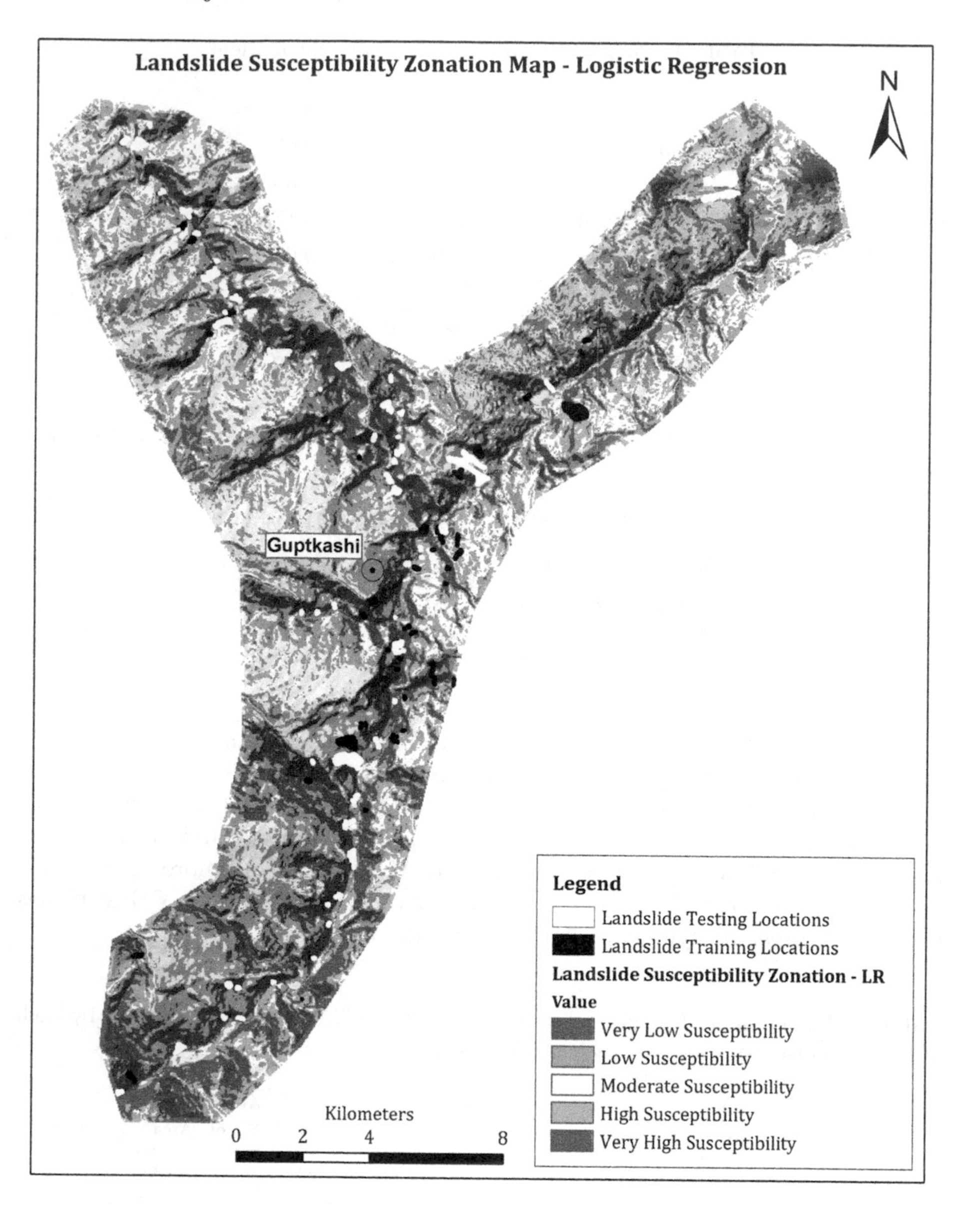

Fig. 4. Landslide susceptibility zonation map obtained using weights from Logistic Regression. (Color figure online)

The number of pixels of all the susceptibility zones can be seen in Table 3. These pixels are further multiplied by 900 to get the area covered by each of these susceptibility zones. Thus, we calculate the percentage of the total study area which is covered by each of these susceptibility zones as shown in Table 3.

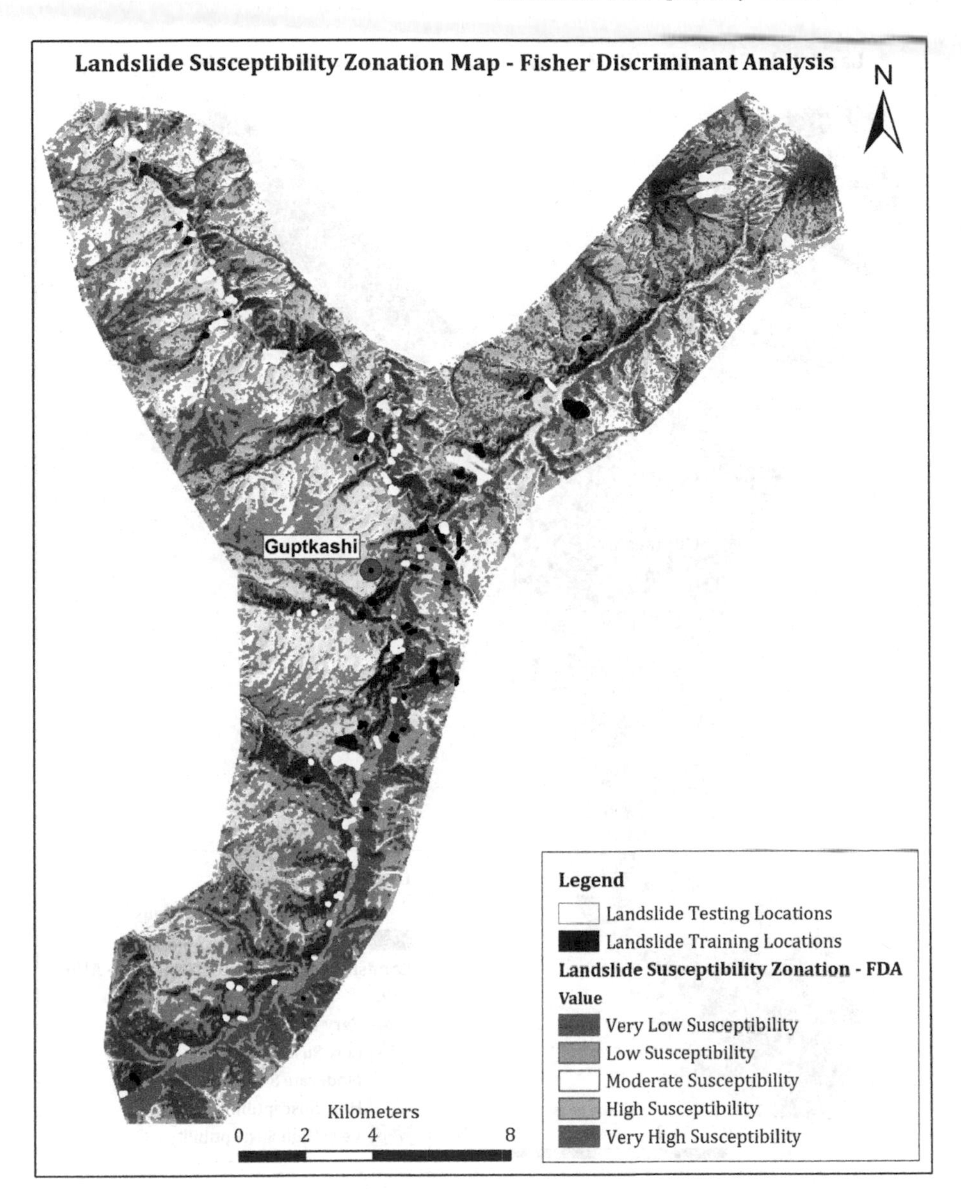

Fig. 5. Landslide susceptibility zonation map obtained using weights from Fisher discriminant analysis. (Color figure online)

As we can see from Table 3 that percentage of area covered by high susceptibility zone and very high susceptibility zone are more as compared to other three zones for LR and FDA (clearly seen from Figs. 4 and 5 that major area of the LSZ map are in red and orange region) but in ANN the percentage of area covered by very low and low susceptibility zone is more as compared to other

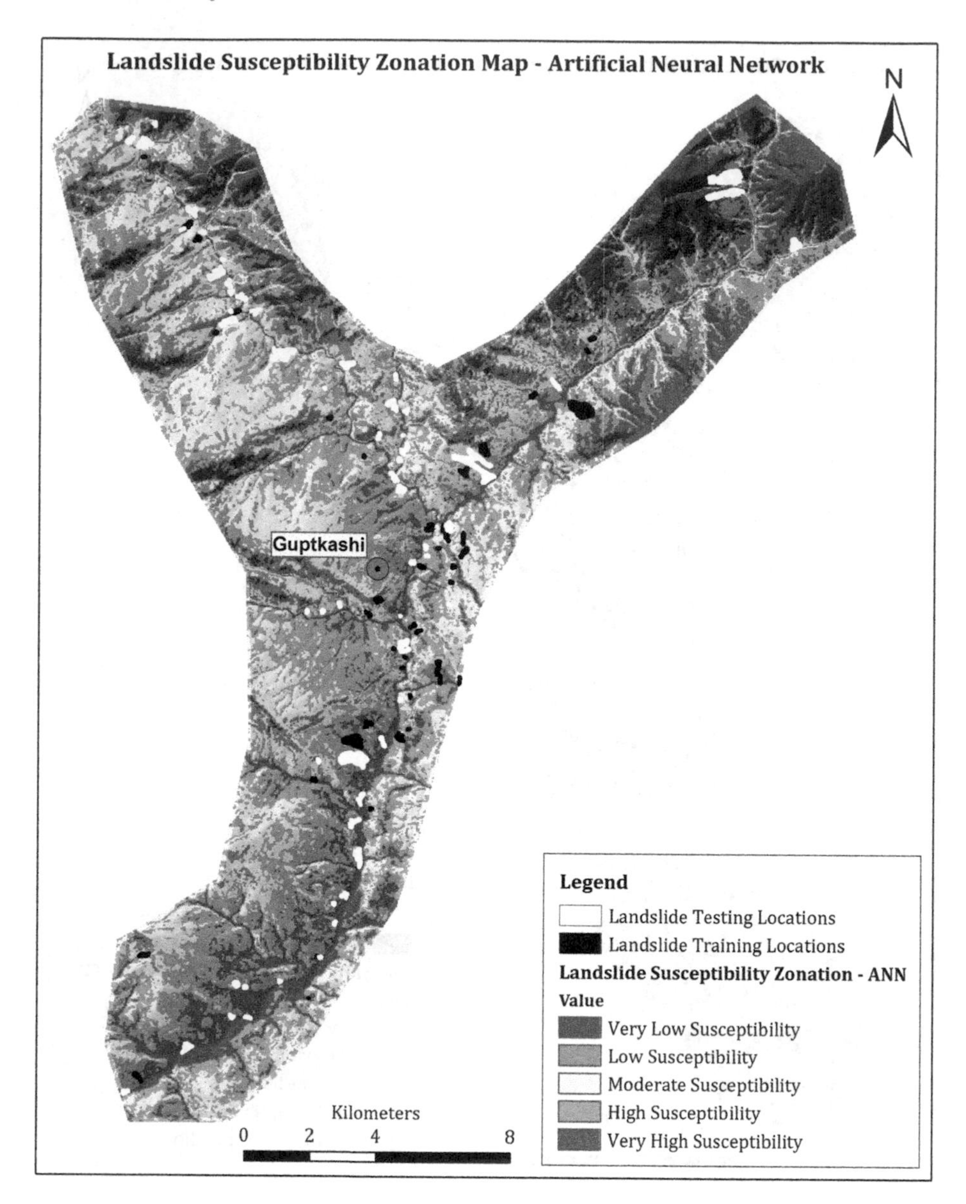

Fig. 6. Landslide susceptibility zonation map obtained using artificial neural network. (Color figure online)

zones(see Fig. 6 that LSZ map contains more green area than that of orange(high susceptibility) red (very high susceptibility) area).

For quantitative assessment of the accuracy of susceptible maps, we've used HSS. Value comes out in the range of 1 and 0, where higher accuracy of prediction is indicated by higher value. Summary for all the three methods which includes the number of landslides correctly classified and misclassified is given in Table 4.

In the Table 4, landslides falling under high and very high susceptibility regions are much more than moderate, very low and low susceptibility regions combined together.

Table 4. Summary of all the three models

Methods	Correctly classified (LSI >= 0.55)	Wrongly classified (LSI < 0.55)	HSS
FDA	69	7	0.89
LR	75	1	0.98
ANN	73	3	0.95

Out of total 76 landslides, 7 landslides are misclassified by FDA, 1 misclassified by LR and 3 landslides are misclassified by ANN. Even though for ANN more area is covered by low and very low susceptible zones but landslides falling under these zones are very much less than that of high and very high zones, resulting in a good HSS score for ANN model. Thus, based on the accuracy of prediction, LR seems better for generation of LSZ maps but we simply cannot rule out ANN as its HSS score is nearly same as LR.

The accuracy of our models can further be enhanced by using active learning approach for selection of landslide inventory which should be used for model training [14–16]. It involves a feedback loop between a sample/human annotator and machine resulting in the tuning of the model. The model begins with some labeled data of landslide occurrences to judge or predict the outcome of the received data. A sample of the output of the machine is labeled by the expert, and then it is feedback over into the model. This process of labeling continues until the model achieves highest accuracy for labeling the feature thus extracting landslide areas.

5 Conclusion

Landslides are natural hazards that cannot be ignored as it is catastrophic and is detrimental to human life and the economy. Thus, landslide zonation maps proved to be a great tool for detection of landslides. In this study we have used three different techniques (LR, FDA and ANN) to obtain weights for each of the data layers and then used WLC to obtain a composite map data layer for each of the method and finally used HSS to obtain the accuracy of each of these methods in terms of score. It is observed that LR has the highest score (0.98), followed by ANN (0.95) and FDA (0.89) respectively. So, LR proved to be the best method among these three. ANN and LR have nearly the same score so we can say ANN is also a good method. Even though ANN is a very effective and powerful method but, it has no close boundary. So, rigorous hit and trial is required to obtain best results. The results of this study can be helpful to the engineers, planners and developers for planning of land-use and proposing new infrastructure projects in landslide prone regions of the Himalayas. Based on the results of the study it can be inferred that restriction may be imposed in areas with higher susceptibility for landslides. However, while utilizing these models for site specific designing and development, proper investigation must be carried out.

References

1. Gupta, S.K., Shukla, D.P., Thakur, M.: Selection of weightages for causative factors used in preparation of landslide susceptibility zonation (lsz). Geomatics Nat. Hazards Risk **9**(1), 471–487 (2018)
2. Zare, M., Pourghasemi, H.R., Vafakhah, M., Pradhan, B.: Landslide susceptibility mapping at Vaz Watershed (Iran) using an artificial neural network model: a comparison between multilayer perceptron (MLP) and radial basic function (RBF) algorithms. Arab. J. Geosci. **6**(8), 2873–2888 (2013)
3. Singh, A.K.: Causes of slope instability in the Himalayas. Disaster Prev. Manag. Int. J. **18**(3), 283–298 (2009)
4. Pradhan, B., Abokharima, M.H., Jebur, M.N., Tehrany, M.S.: Land subsidence susceptibility mapping at Kinta Valley (Malaysia) using the evidential belief function model in GIS. Nat. Hazards **73**(2), 1019–1042 (2014)
5. Pourghasemi, H.R., Pradhan, B., Gokceoglu, C., Mohammadi, M., Moradi, H.R.: Application of weights-of-evidence and certainty factor models and their comparison in landslide susceptibility mapping at Haraz watershed. Iran. Arab. J. Geosci. **6**(7), 2351–2365 (2013)
6. Lee, S., Pradhan, B.: Landslide hazard mapping at Selangor, Malaysia using frequency ratio and logistic regression models. Landslides **4**(1), 33–41 (2007)
7. Melchiorre, C., Matteucci, M., Azzoni, A., Zanchi, A.: Artificial neural networks and cluster analysis in landslide susceptibility zonation. Geomorphology **94**(3–4), 379–400 (2008)
8. Nourani, V., Pradhan, B., Ghaffari, H., Sharifi, S.S.: Landslide susceptibility mapping at Zonouz Plain, Iran using genetic programming and comparison with frequency ratio, logistic regression, and artificial neural network models. Nat. Hazards **71**, 523–547 (2013)
9. Youssef, A.M., Pradhan, B., Jebur, M.N., El-Harbi, H.M.: Landslide susceptibility mapping using ensemble bivariate and multivariate statistical models in Fayfa area. Saudi Arabia. Environ. Earth Sci. **73**(7), 3745–3761 (2015)
10. Bui, D.T., Tuan, T.A., Klempe, H., Pradhan, B., Revhaug, I.: Spatial prediction models for shallow landslide hazards: a comparative assessment of the efficacy of support vector machines, artificial neural networks, kernel logistic regression, and logistic model tree. Landslides **13**(2), 361–378 (2015)
11. Kumar, D., Thakur, M., Dubey, C.S., Shukla, D.P.: Landslide susceptibility mapping and prediction using support vector machine for Mandakini River Basin, Garhwal Himalaya, India. Geomorphology **295**, 115–125 (2017)
12. Boroumandi, M., Khamehchiyan, M., Nikoudel, M.R.: Using of analytic hierarchy process for landslide hazard zonation in Zanjan province. Iran. Eng. Geol. Soc. Territ. **2**, 951–955 (2015)

13. Pourghasemi, H.R., Pradhan, B., Gokceoglu, C.: Application of fuzzy logic and analytical hierarchy process (AHP) to landslide susceptibility mapping at Haraz watershed. Iran. Nat. Hazards **63**(2), 965–996 (2012)
14. Bouguelia, M.R., Nowaczyk, S., Santosh, K.C., Verikas, A.: Agreeing to disagree: active learning with noisy labels without crowdsourcing. Int. J. Mach. Learn. Cybern. **9**(8), 1307–1319 (2018)
15. Stumpf, A., Lachiche, N., Malet, J.P., Kerle, N., Puissant, A.: Active learning in the spatial domain for remote sensing image classification. IEEE Trans. Geosci. Remote Sens. **52**(5), 2492–2507 (2014)
16. Stumpf, A., Lachiche, N., Kerle, N., Malet, J.P., Puissant, A.: Adaptive spatial sampling with active random forest for object-oriented landslide mapping. In: 2012 IEEE International Geoscience and Remote Sensing Symposium, pp. 87–90. IEEE (2012)
17. Li, C., Wang, B.: Fisher linear discriminant analysis (2014)
18. Bishop, C.M.: Fisher's Linear Discriminant. Springer, New York (2006)
19. Beale, M.H., Hagan, M.T., Demuth, H.B.: Neural network toolbox^{TM} user's guide. In: R2017a, The MathWorks Inc., 3 Apple Hill Drive Natick, MA 01760–2098. Citeseer (2017). http://www.mathworks.com
20. Michael, E.A., Samanta, S.: Landslide vulnerability mapping (LVM) using weighted linear combination (WLC) model through remote sensing and GIS techniques. Model. Earth Syst. Environ. **2**(2), 1–15 (2016)
21. Malczewski, J.: On the use of weighted linear combination method in GIS: common and best practice approaches. Trans. GIS **4**(1), 5–22 (2000)

Design and Development of Ground Truth Collection Platform Using Android and Leaflet Library

Sandeep V. Gaikwad[1,2(✉)], Amol D. Vibhute[3], Karbhari V. Kale[2], Dhanajay B. Nalawade[1,2], and Monali B. Jadhav[1,2]

[1] Geospatial Technology Research Laboratory, Dr. Babasaheb Ambedkar Marathwada University, Aurangabad 431004, Maharashtra, India
sandeep.gaikwad22@gmail.com,
dhananjay.bamu@gmail.com, monalij28@gmail.com

[2] Department of Computer Science and IT, Dr. Babasaheb Ambedkar Marathwada University, Aurangabad 431004, Maharashtra, India
kvkale91@gmail.com

[3] School of Computational Sciences, Solapur University, Solapur 413255, MS, India
amolvibhute2011@gmail.com

Abstract. Traditionally, the earth surface features were collected by field visit for collecting the details of earth surface features with coordinates. However, the conventional method is slow tedious and costly. Recently, smartphone applications are providing the details of locations including latitude, longitude which is basically used for ground truth data collection for Remote Sensing (RS) applications. Nevertheless, there are limitations found in existing system as they do not provide offline and online mode, location data management, and visualization into the single system. The objective of our research was thus to overcome the limitations in the existing ground truth collection system by exploring a novel solution that will be based on Smartphone, Cloud server, and WebGIS platform. Consequently, an android application and web-based tool is developed for ground truth data collection and its management. The present research reports, the novel solution for collecting the real-time ground truth and mapping feature for RS research.

Keywords: Leaflet · Real-time ground truth · Android · Geospatial survey · Mapping

1 Introduction

The smartphone technology has made a revolution in the world. Nowadays, it is as much as powerful tool as a desktop computer. The smartphone has a dominant operating system, apps, and support of Global Positioning System (GPS). It is providing low-cost solutions like mapping and surveying related task using application. Therefore, the speed of geospatial survey, real-time mapping of surface object, location accuracy, and integrity has increased significantly [1–5]. The mapping application

K. C. Santosh and R. S. Hegadi (Eds.): RTIP2R 2018, CCIS 1037, pp. 520–528, 2019.
https://doi.org/10.1007/978-981-13-9187-3_46

plays an important role in a mitigation and strategy in a natural and manmade disaster like flood, thunderstorm, tsunami, land sliding, fire, drought, earthquake etc. [5–7]. The post assessment of the disaster can be precisely performed with the help of field observation and high-resolution satellite data [8, 9]. The ground truth data includes GPS coordinates and field observation which is an imperative input to the RS and GIS research [10, 11].

The various researchers have developed a smartphone application to fulfill their specific need of ground truth data collection. Learn-IT Team has developed the Simple GPS Survey app for recording the GPS coordinate with location information. The app uses a SQLite database to store the survey records. The limitation of the app is that, it can not visualize the survey data on the map as well as it is not useful for the team-based real-time survey [12]. The Digrasoft UG has developed My GPS Location app for the collection of ground truth data, it has a various feature like weather, information, location sharing, add placemark facility. But there is a lack of server-side dashboard for handling huge location data [13]. Roya Olyazadeh et al. 2015, has developed a mobile application for Landslide data acquisition and Land cover mapping. The application is known as ROOMA (Rapid Offline–Online Mapping Application) for the rapid ground truth data collection of landslide disaster. The app is integrated with WebGIS application which is deployed on a cloud server [14].

The existing system has a lack of certain features like Offline mode, mapping of spatial objects, and data visualization and management on the server side. Moreover, the existing app does not support to download the location data in a various format like KML (Keyhole Markup Language), CSV (Comma Separated Value), XLS (Microsoft Excel) and PDF (Portable Document Format). The Geospatial Technology Research Laboratory (GTRL) has developed DroughtGIS platform for the Internet of Things (IoT), Weather data acquisition from Open Weather Map services, and Smartphone application for mapping, Meteorological and Agricultural drought analysis. In the current research, we have developed GeoMapper web tool integrated with DroughtGIS platform. The proposed system includes an android app and Web-based tool for ground truth data acquisition and management.

2 Technical Development of GeoMapper

2.1 Android Application

Google has developed Android mobile operating system for smartphone, and tablet computers. Furthermore, Android has extended its support to TV, Auto and wearable device. The interactive android application was developed in android studio 3.0 using Java programming language. Figure 1 shows the architecture of the Android application. The SQLite database was used to store the records in phone memory which is further sent to the server using Volley library. The algorithm has designed for record synchronization between SQLite and MySQL which is based on the application mode and status of internet connectivity.

Algorithm 1. Records synchronization between SQLite and MySQL

```
  // -----------------------------------------------------
  // Retrieve the input data from Intent and store into
the variables
  getInputDataFromIntent();
  // By default save data into the SQLite database
  saveDataToSQLite();
  // default value of SYNC_COLUMN set the  FALSE
  if(APP_MODE=ONLINE){
  // check the mode of app if it is ONLINE then execute
the following code
  while(CurrentRecordNo<=RowCount){
  //  Traverse  through  the  each  records  using  loop
statement
  if(SYNC_COLUMN=FALSE){
  // Check the synchronization status of each record
  if (CHECK_NETWORK_STATUS){
      // check the status of internet connection if it is
TRUE then execute the following function
      syncDataToServer();
      //send the data to the server using Volley library
and set the  SYNC_COLUMN status TRUE

  } else {
      // show the error message
      showMessage()
  }
           }
          moveToNextRecord();
          // move to next record
      }
  }
  //-----------------------------------------------------
```

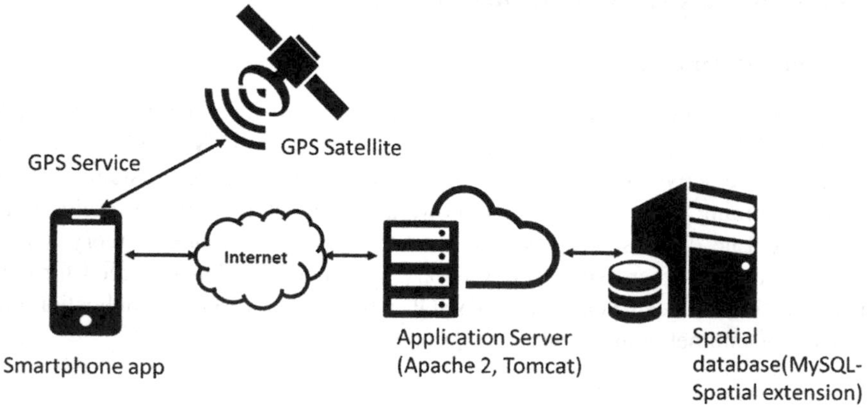

Fig. 1. Application architecture

2.2 Development of the Web-Based Tool

There are three major components of the tool, which include client, application server, and a spatial database server. The web server handles the request and response to the client via the internet. The role of the map server is to provide a map tiles to the map viewer control of the client browser. We have used Google and OSM (Open Street Map) to map tiles service providers for Leaflet map viewer. Figure 2 shows the architecture of Web tool.

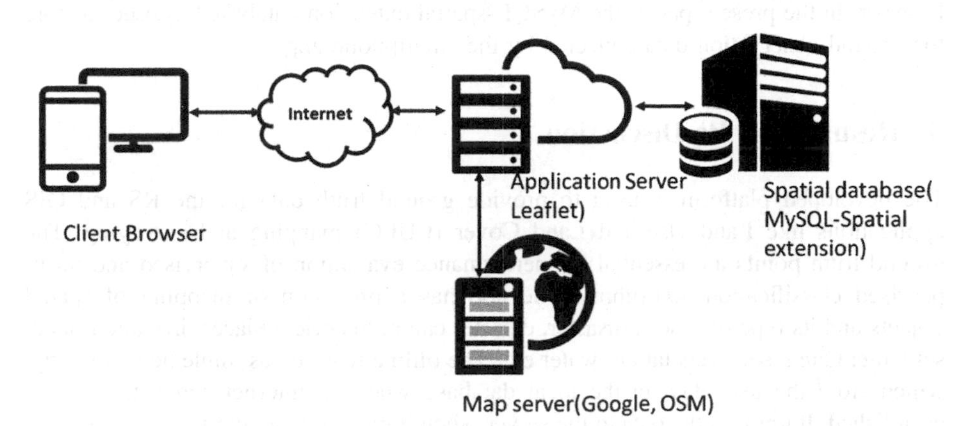

Fig. 2. The architecture of web based tool

2.2.1 Leaflet

The leaflet is an open source Javascript library for web mapping application which is developed by Vladimir Agafonkin, in 2011. It is widely popular because of extremely lightweight (size 38 KB), robust, and supports most mobile and desktop platform [6]. Therefore is used by Microsoft, Facebook, Pinterest, FourSquare, Flickr, The Washington post, GitHub and many more (http://Leafletjs.com). The LeafletJS provides the bunch of plugins and API with good documentation to fulfill the requirement of mapping application based on Open Geospatial Consortium (OGC) standards [15–17].

2.2.2 Apache2 Webserver Component

The Apache HTTP Server ("httpd") is an extremely used Web server on the Internet since April 1996. It is used to manage the request from the client and **serve** the response in term of a web document. It utilizes various protocols like Hyper Text Transfer Protocol (HTTP), HyperText Transfer Protocol over Secure Sockets Layer (HTTPS), and File Transfer Protocol (FTP), a protocol for web page transmission, and files uploading and downloading. Generally, Apache web server is used in combination with MySQL database, HyperText Preprocessor (PHP) and other scripting languages such as Pearl and Python (https://httpd.apache.org/). Moreover, it is a powerful platform for open source development of web applications over the Unix, and Windows platform [18–21].

2.2.3 MySQL-Spatial Extension

The MySQL is a worldwide popular open source relational database management system (RDBMS). It is the main component of the LAMP (Linux, Apache, MySQL, PHP/Pearl/Python) open source web development stack. By following the OGC standards, MySQL has implemented the spatial extension for a generation, storage, and analysis of geometrical data [7].

The phpMyAdmin is an open-source tool developed for MySQL database administration, which provides the bunch of features like create, delete, update database, tables, fields, SQL query, and user permission management through the web browser. In the present paper, the MySQL-spatial extension database has used to store the ground observation data collected by the smartphone app.

3 Results and Its Discussion

The developed platform is used to provide ground truth data for the RS and GIS applications like Land Use and Land Cover (LULC) mapping and surveying. The ground truth points are essential for performance evaluation of supervised and unsupervised classification algorithms. The app has a provision of mapping of spatial objects and its type of assets, disaster, disaster camp, historical places, irrigation, land, settlement area, soil, vegetation, water etc. The offline feature has implemented in-app, which store the user data in the local database when an internet connection is not established. It uploads the data to the server when it received the internet connection so that this application is useful in data collection in remote areas. Figures 3 and 4 shows the screenshots of an android app.

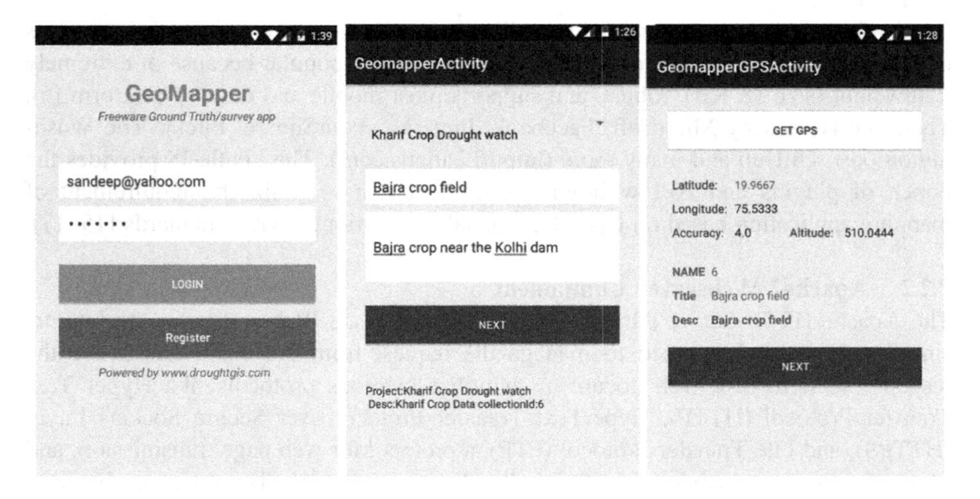

Fig. 3. Interface of GeoMapper android application

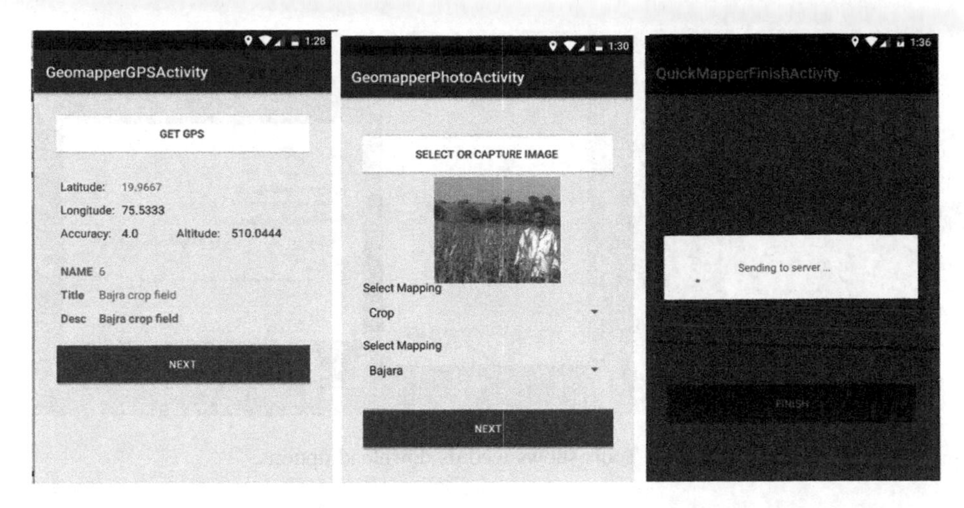

Fig. 4. Interface of GeoMapper android application (continue…)

Besides, we have developed a web-based tool for handling the records at server sides which are uploaded by the android application. It is based on an open-source component like Bootstrap, Leaflet, PHP, MySQL-Spatial extension which is hosted on Linux server (http://www.droughtgis.com). The bootstrap provides a responsive and user-friendly user interface on multiple screen layout. The tool provides excellent features like user, project, records management and visualization of the map along with map markers. Furthermore, the user can download the records in various format like KML and CSV which are compatible with Google Earth and ArcGIS software. The interface testing has carried out in various browsers such as Safari, Chrome, Firefox, and Internet Explorer etc. Figures 5, 6 and 7 illustrates the features of the developed web based tool.

Fig. 5. Leaflet based map viewer shows the GT data in Popup window.

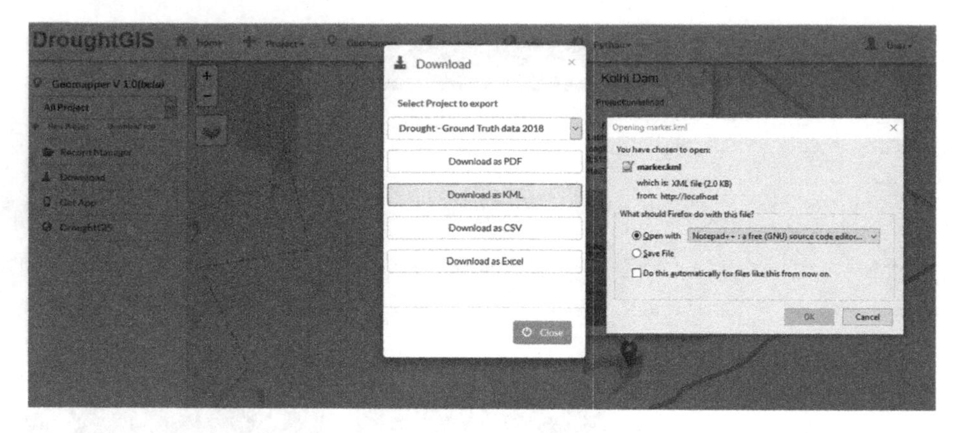

Fig. 6. Tools shows records download options.

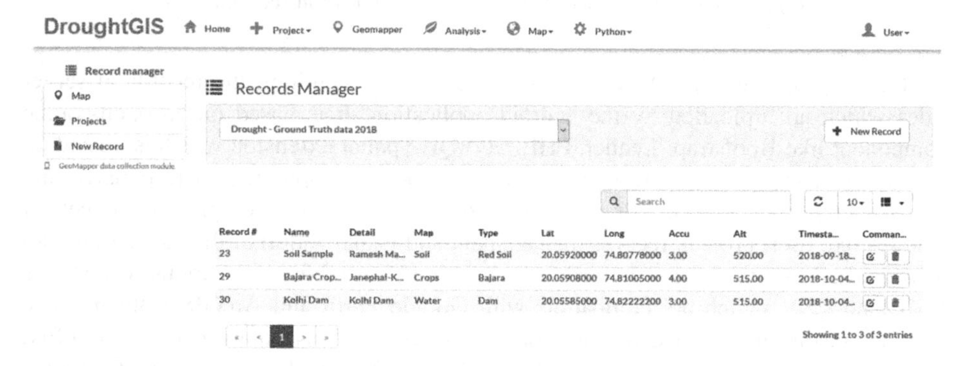

Fig. 7. Web based record manager

4 Conclusions

The open source components have been used for development of the ground truth data acquisition and management. The application has successfully met the requirement that was specified by the researchers. The smartphone (android) app has an excellent feature like online-offline data storage mode, spatial object mapping, and user management. The algorithm has developed for automatic synchronization of records between device and server. In addition, the GPS accuracy enhancement module has integrated into the application which provides good accuracy when GPS signals are weak. The web-based tool was developed for management of project and records generated by user. Moreover, the map viewer is also developed for visualization of ground truth data along with photographs.

Acknowledgments. The authors would like to acknowledge and thanks to UGC, India for granting UGC SAP (II) DRS Phase-I & Phase-II F. No. 3-42/2009 &4-15/2015/DRS-II for Laboratory facility to Dept. of CS & IT, Dr. BAM University, Aurangabad, Maharashtra, India and financial assistance under UGC BSR Fellowship for this work.

References

1. Nusser, S., Miller, L., Clarke, K., Goodchild, M.: Geospatial IT for mobile field data collection. Commun. ACM **46**(1), 1–2 (2003). https://doi.org/10.1145/602421.602446
2. Seebregts, C.J., et al.: Handheld computers for survey and trial data collection in resource-poor settings: development and evaluation of PDACT, a Palm Pilot interviewing system. Int. J. Med. Inform. **78**(11), 721–731 (2009). https://doi.org/10.1016/j.ijmedinf.2008.10.006
3. de Abreu Freirea, C.E., Painhoa, M.: Development of a mobile mapping solution for spatial data. Procedia Technol. **16**, 481–490 (2014)
4. Berger, M., Platzer, M.: Field evaluation of the smartphone-based travel behaviour data collection app "SmartMo". Transp. Res. Procedia **11**, 263–279 (2015)
5. Athanasis, N., Karagiannis, F., Palaiologou, P., Vasilakos, C., Kalabokidis, K.: AEGIS app wildfire information management for windows phone devices. Procedia Comput. Sci. **56**, 544–549 (2015)
6. Neene, V., Kabemba, M.: Development of a mobile GIS property mapping application using mobile cloud computing. (IJACSA) Int. J. Adv. Comput. Sci. Appl. **8**(10) (2017)
7. Chen, D., Shams, S., Carmona-Moreno, C., Leone, A.: Assessment of open source GIS software for water resources management in developing countries. J. Hydro-Environ. Res. **4**, 253–264 (2010)
8. Zhang, H., Yi, S., Wu, Y.: Decision support system and monitoring of eco-agriculture. Energy Procedia **14**, 382–386 (2012)
9. Jordan, R., Eudoxie, G., Maharaj, K., Belfon, R., Bernard, M.: AgriMaps: improving site-specific land management through mobile maps. Comput. Electron. Agric. **123**, 292–296 (2016)
10. Ye, S., Zhu, D., Yao, X., Zhang, N., Fang, S., Li, L.: Development of a highly flexible mobile GIS-based system for collecting arable land quality data. IEEE J. Sel. Top. Appl. Earth Observ. Remote Sens. **7**(11), 4432–4441 (2014)
11. Beaudette, D.E., O'Geen, A.T.: Soil-web: an online soil survey for California, Arizona, and Nevada. Comput. Geosci. **35**, 2119–2128 (2009)
12. Learn It Team, Simple GPS Survey app. https://play.google.com/store/apps/details?id=com.justbtech.app.simplegpssurvey. Accessed 10 Aug 2018
13. Digrasoft UG, My GPS Location app. https://play.google.com/store/apps/details?id=com.digrasoft.mygpslocation. Accessed 10 Aug 2018
14. Olyazadeh, R., Sudmeier-Rieux, K., Jaboyedoff, M., Derron, M.-H., Devkota, S.: An offline–online Web-GIS Android application for fast data acquisition of landslide hazard and risk. Nat. Haz. Earth Syst. Sci. **17**, 549–561 (2017)
15. Wang, L., Pi, R., Zhou, X., Zhou, H.: The construction of off-line map based on OpenStreetMap and leaflet. In: 4th International Conference on Computer, Mechatronics, Control and Electronic Engineering, (ICCMCEE) (2015)
16. Lu, W., Ai, T., Zhang, X., He, Y.: An interactive web mapping visualization of urban air quality monitoring data of China. Atmosphere **8**, 148 (2017). https://doi.org/10.3390/atmos8080148
17. Open Geospatial Consortium. www.opengeospatial.org/standards. Accessed 7 July 2018s

18. Hea, Y., Zhanga, D., Fang, Y.: Development of a mobile post-disaster management system using free and open source technologies. Int. J. Disaster Risk Reduction **25**, 101–110 (2017)
19. Steiniger, S., Hay, G.J.: Free and open source geographic information tools for landscape ecology. Ecol. Inform. **4**, 183–195 (2009)
20. HTTPD - Apache2 Web Server. https://help.ubuntu.com/lts/serverguide/httpd.html. Accessed 1 July 2018
21. Apache HTTP Server Project. https://httpd.apache.org/. Accessed 1 July 2018

Automatic Classification of Normal and Affected Vegetables Based on Back Propagation Neural Network and Machine Vision

Manohar Madgi[1(✉)] and Ajit Danti[2]

[1] K. L. E. Institute of Technology, Hubballi 580030, India
manohar.madgi@gmail.com
[2] Christ (Deemed to be University), Bengaluru 560074, India
ajit.danti@christuniversity.in

Abstract. This article presents a neural network and machine vision-based approach to classify the vegetables as normal or affected. The farmers will have great difficulty if there is a change from one disease control to another. The examination through an open eye to classify the diseases by name is more expensive. The texture and color features are used to identify and classify different vegetables into normal or affected using a neural network and machine vision. The mixture of both the features is proved to be more effective. The results of experiments show that the proposed methodology extensively supports the accuracy in automatic detection of affected and normal vegetables. The applications in packing and grading of vegetables are the outcome of this research article.

Keywords: Vegetable disease · Color features · Texture features · Classifier

1 Introduction

The analysis of vegetables can be useful in many different situations. For instance in the preparation of curry at home or restaurant, in the process of packing and selling in the malls, sorting in the field by the farmer or in the APMC market. The person who is an artist or a scientist or a skilled one can identify the vegetable whether it is affected or not. The spontaneous decision and scientific methods are used for the process of bifurcating vegetables as an affected or a good one. The images of vegetables with signs and symptoms of diseases are used to enhance the depiction of diseases. The pathologists of vegetable can include these digital images in the diagnosis of vegetable diseases. The farmers are worried about the expenditure involved in such activities. The automatic recognition and categorization of diseases based on the indication are useful for the farmers and agricultural scientists. The initial detection of the symptoms of the diseases is the major challenge in the agriculture field. A proper method

K. C. Santosh and R. S. Hegadi (Eds.): RTIP2R 2018, CCIS 1037, pp. 529–537, 2019.
https://doi.org/10.1007/978-981-13-9187-3_47

needs to be developed in this area. The symptoms of the diseases are the main indicators for the diagnosis. The automatic disease diagnosis and control is the major concern of the agricultural scientists. The samples of affected and normal vegetable images are shown in Fig. 1. A literature survey is carried out to know the applications in agriculture/horticulture in general and identification of vegetable diseases in particular. The core of the survey is as follows.

Fig. 1. (a)–(f) Normal vegetables, (g)–(l) Affected vegetables.

Karargyris et al. [1] have described a method to classify digital chest x-rays (CXRs) into tuberculosis (TB) and non-tuberculosis cases using combined shape and texture features. They have applied Air-cavity segmentation and Expected lung anatomy segmentation algorithm on CXRs to segment affected lung portion for TB findings.

Singh and Misra [2] have presented an image segmentation technique to identify and classify the plant leaf diseases. For extracting the features, they considered the color and texture of an image. The color co-occurrence method is used for feature extraction.

Raut and Fulsunge [3] have introduced a technique to find out diseases related to both leaf and fruit. They have used color, texture, morphology and structure feature vectors to detect plant disease. The K-means clustering algorithm with multi SVM is used for identification and classification of diseases.

Mokhtar et al. [4] have described a method to identify and detect unhealthy tomato leaves. The proposed method includes pre-processing, feature extraction and classification. They have used a texture feature to identify tomato leaf state. The SVM classifier with different kernel functions is used for classification. They have achieved 99.83% accuracy over 800 healthy and affected tomato leaves.

Savita and Arora [5] have studied various classification techniques used to classify plant leaf diseases using image processing. They concluded that for the given test sample, the k-nearest-neighbor technique looks to be more suitable and the simplest of all algorithms for class prediction.

Mrunalini and Prashant [6] have presented a technique to identify and classify the different diseases of the affected plants. A machine learning based recognition method is very much useful it saves time, effort and money. The color co-occurrence technique used to obtain the feature set. A neural network used for the automatic detection of the diseases from the leaves.

Arivazhagan et al. [7] have reported the disease identification process which includes structure, a precise threshold value, the green pixel masking and segmentation processes. The texture statistics are computed to replace useful segments. Finally, a classifier is applied on extracted features to classify the diseases.

Anand and Ashwin [8] have presented a system to detect early and correct plant diseases by using an Artificial Neural Network (ANN) and different techniques of image processing. The suggested method extracts feature using a Gabor filter and an ANN classifier to classify. It provides results with an identification percentage to 91. The classifier ANN classifies various diseases plant on combined of texture and color features.

Smita and Niket [9] have implemented a method for plant disease identification by using histogram matching. In plants, the disease emerges on the leaf; hence, the match of a histogram is made based on edge and color features. Layers division technique and edge detection technique are used to divide the bands of RGB image into red, green and blue and also they are used to detect edges of the layered images.

Danti et al. [10] have designed a system for the identification and classification of leafy vegetables using a reduced color feature set. A reverse engineering process is adopted for the reduction of features and has considered the mean and range color features for the classification.

Chaudhary et al. [11] have developed an algorithm for the spot segmentation of the diseases in plant leaves by using image processing techniques. To detect a spot of disease is performed by examining the impact of HSI, YCbCr and CIELAB color space. For smoothing the image used a median filter. Finally, applying the Otsu method to the color component for finding out a diseased spot is done. The background noise is removed by the CIELAB color model.

Rumpf et al. [12] have presented an automatic method for early detection of the plant diseases which are vital for the precision of crop protection. The early detection helps the farmers to avoid a huge loss. The technology support helps the farmers by cutting the cost of the pesticides.

Al-Bashish et al. [13] have developed software for automatic classification and identification of leaf diseases of the plant which is an essential technique to prove benefits in watching crops in large fields. It automatically identifies the signs of the diseases as soon as they arise on the plant leaf.

Danti et al. [14] have proposed color features values of Mean and Range to identify the leafy vegetables. Proposed methodology uses a set of reduced color features. They found that the maximum recognition of 100% and the minimum is 92%.

Bernardeset et al. [15] have presented a system to the analysis of cotton diseases using the extracted features of foliar signs from the images. The system uses wavelet transform energy for extract feature. SVM used for the classification.

The literature review shows that computers are employed for the detection of plant leaf diseases and crop diseases. Not much work is carried out on vegetable diseases. The present work has considered images of leafy and non-leafy vegetables. The images of various categories are subjected to pre-processing, feature extraction and are classified using the ANN. The extracted color and texture features are finally combined and subjected to the classifier to detect normal and an affected vegetable.

This paper has five sections. The second section contains the proposed approach, image capture and resize. Feature extraction is explained in section three. Results and discussion are in section four. Section five concludes the task.

2 Proposed Approach

The suggested method consists of four stages, image capture & resize, feature extraction, feature reduction and classification. Figure 2 shows all these stages.

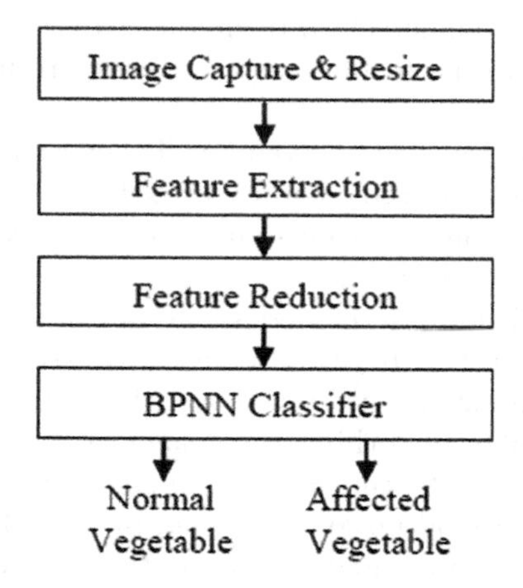

Fig. 2. Feature extraction and classification

2.1 Image Capture and Resize

The six different types of vegetables such as Capsicum, Brinjal, Tomato, Mint leaves, Spinach, White radish leaves of normal and affected are selected from the market and taken for this research work. The images are taken by placing the bulk or bench of vegetable on the plain white background. Images are captured in daylight using a digital camera with a 10-megapixel resolution put at a distance of 40 cms. The obtained images are 3264×2448 pixels. For the computational reason, images are resized to 150×150 pixels. In total, 600 different samples images are cropped from the same vegetables 100 times.

3 Feature Extraction

Humans recognize the vegetables by color, texture, shape and size. Color is the main significant feature for identification of vegetables. The majority of the vegetables are in green color, but green color shade differs from vegetable to vegetable. Another important feature to identify vegetables is texture. The reduced color and texture features set are involved in recognition of vegetables.

3.1 Color Feature Extraction

The RGB color pattern has taken for color features extraction. The components of Red, Green and Blue are broken up from the image. To have the values in the range [0, 1] the Red, Green and Blue color component values of vegetable images are normalized. By the Eqs. (1 to 3) the Hue (H), Saturation (S) and Intensity (I) color features are acquired from the RGB color segments.

$$H = \begin{cases} \theta \text{ \& if } B \leq G \\ 360\text{-}\theta \text{ \& if } B > G \end{cases} \tag{1}$$

$$\theta = cos^{-1}\left\{\frac{R - G/2 - B/2}{[(R-G)^2 + (R-B)(G-B)]^{1/2}}\right\}$$

$$S = 1 - \frac{3}{R+G+B}[min(R, G, B)] \tag{2}$$

$$I = \frac{1}{3}(R+G+B) \tag{3}$$

The images of vegetable have a distribution and disparity in color across the image. The distinguishing features are depicted by the values of Mean, Range and Variance of the input image. 18 color features are selected to describe an image. The Eqs. (4 to 6) are used to obtain the above 18 features.

$$Mean = \sum_{x,y} xP(x, y) \tag{4}$$

$$Variance = \sum_{x,y} (x - \mu)^2 P(x, y) \tag{5}$$

$$Range = Max(p(x, y)) - Min(p(x, y)) \tag{6}$$

The mean and range feature used because they satisfy the reduction of feature policy.

3.2 Texture Feature Extraction

In the samples of the vegetable image, color features overlap, but texture change from one vegetable type to another. In such cases, the texture will be ideal for identification. In this work, the gray level co-occurrence matrix (GLCM) is employed to bring out the analysis. The different kinds of texture features are obtained by using the GLCM. The Eq. (7) compute the co-occurrence matrix. Equations (8 to 12) compute texture features [16] such as Energy, Contract, Cluster Shade, Homogeneity and Cluster Prominence of Hue image component.

$$C = \frac{1}{4}(P_{0^0} + P_{45^0} + P_{90^0} + P_{135^0}) \tag{7}$$

$$Contrast = \sum_{i,j=0}^{N-1} (i,j)^2.C(i,j) \tag{8}$$

$$Energy = \sum_{i,j=0}^{N-1} (i,j)^2 \tag{9}$$

$$LocalHomogeneity = \sum_{i,j=0}^{N-1} C(i,j)/(1+(i+j)^2) \tag{10}$$

$$ClusterShade = \sum_{i,j=0}^{N-1} (i - M_x + j - M_y)^3 C(i,j)) \tag{11}$$

$$ClusterProminence = \sum_{i,j=0}^{N-1} (i - M_x + j - M_y)^4 C(i,j)) \tag{12}$$

3.3 Feature Reduction

A threshold-based approach adopted on color features for the feature reduction purpose. From the color feature set, it is noted that for all type of vegetables, the variance feature values are very low (<0.5) and such features do not contribute for the classification. Hence, the variance feature is discarded. The threshold value is obtained by the experiment. A threshold is set to 0.5 and used to compare each feature. If the feature value is higher than the threshold value, it considered as a healthy feature. Thus 18 color features are reduced to 12 color features.

The texture features Energy, Contract, Cluster Shade, Homogeneity and Cluster Prominence are obtained from GLCM. These five texture features along with 12 color features have contributed for identifying a normal or affected vegetable. Totally 17 features are considered for classification.

3.4 Classifier and Parameters

A three-layer back-propagation neural network implemented in Matlab. The mse set to 0.001. There is 17 node in the input layer and 36 nodes in the hidden layer. The output layer of the two nodes corresponds to the two different vegetable types. 0.01 set as the learning rate.

4 Results and Discussion

About 600 images of six different Indian vegetables such as Capsicum, Brinjal, Tomato, Mint leaves, Spinach, White radish leaves are considered in the approach. On acquired image sample extract useful color and texture features required for separating normal and affected vegetable images using an ANN classifier. The vegetable images are split into testing and training sets. 50% of the vegetable images are from every group used to train the network the remaining 50% serve testing set. The support vector machine is trained and tested by considering all color and texture features. During the experiment, noted that features with very low values do not contribute to the classification. Hence, reduced the feature set to 17. The classification accuracy of vegetables is presented in Fig. 3. The graph reveals, the highest classification accuracy of 100% and the lowest 75%. The classification accuracy is low for spinach and white radish leaves as they are similar in color and texture.

$$Accuracy = \frac{Correctly\ recognized\ test\ image\ samples}{Total\ number\ of\ test\ images} * 100$$

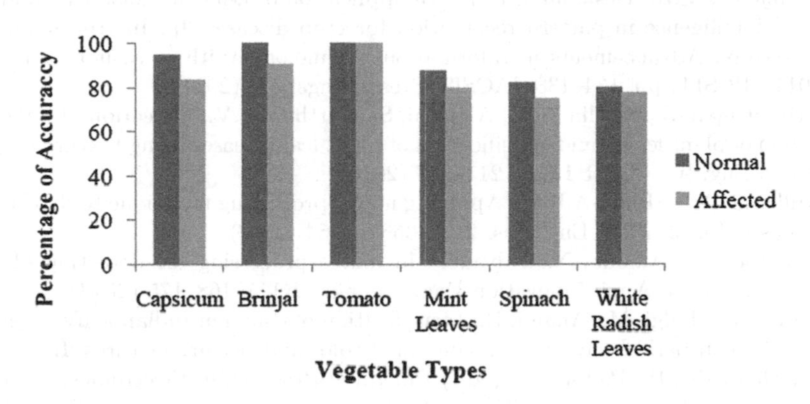

Fig. 3. Classification accuracy of different vegetables.

5 Conclusion

In this paper, BPNN based classifier is proposed to use a combined texture and co-lour feature for identity and classify different vegetables. These features have provided different accuracies for a variety of vegetables. The maximum classification accuracy of 100% is observed with Tomato and Brinjal and a minimum of 75% with spinach. The results encourage developing a good machine-vision system for classification of vegetables. The significance of the work to the present world is an automatic classification of normal and affected vegetables, their grading, in food preparation and sales in the malls.

Acknowledgement. We wish to express gratitude to our beloved Principal, Dr Basavaraj Anami, K.L.E. Institute of Technology, Hubballi for his advice.

References

1. Karargyris, A., et al.: Combination of texture and shape features to detect pulmonary abnormalities in digital chest X-rays. Int. J. Pattern Recognit. Artif. Intell. **11**(1), 99–106 (2016)
2. Singh, V., Misra, A.K.: Detection of plant leaf diseases using image segmentation and soft computing techniques. Inf. Process. Agric. **4**(1), 41–49 (2016)
3. Raut, S., Fulsunge, A.: Plant disease detection in image processing using matlab. Int. J. Innovative Res. Sci. Eng. Technol. **6**(6), 10373–10381 (2017)
4. Mokhtar, U., et al.: SVM-based detection of tomato leaves diseases. In: Filev, D., Filev, D., et al. (eds.) Intelligent Systems 2014. AISC, vol. 323, pp. 641–652. Springer, Cham (2015). https://doi.org/10.1007/978-3-319-11310-4_55
5. Ghaiwat, S.N., Arora, P.: Detection and classification of plant leaf diseases using image processing techniques: a review. Int. J. Recent Adv. Eng. Technol. **2**(3), 1–7 (2014)
6. Badnakhe, M.R., Deshmukh, P.R.: An application of k-means clustering and artificial intelligence in pattern recognition for crop diseases. In: International Conference on Advancements in Information Technology With workshop of ICBMG 2011. IPCSIT, pp. 134–138. IACSIT Press, Singapore (2011)
7. Arivazhagan, S., Newlin, S.R., Ananthi, S., Varthini, S.V.: Detection of unhealthy region of plant leaves and classification of plant leaf diseases using texture features. Agric. Eng. Int. CIGR **15**(1), 211–217 (2013)
8. Kulkarni, A.H., Patil, A.R.K.: Applying image processing technique to detect plant diseases. Int. J. Mod. Eng. Res. **2**(5), 3661–3664 (2012)
9. Naikwadi, S., Amoda, N.: Advances in image processing for detection of plant diseases. Int. J. Appl. Innovation Eng. Manage. **2**(11), 168–175 (2013)
10. Danti, A., Madgi, M., Anami, B.: Identification of common Indian leafy vegetables based on statistical measures on combined color and texture features. In: Sridhar, V., Sheshadri, H., Padma, M. (eds.) Emerging Research in Electronics, Computer Science and Technology. LNEE, vol. 248, pp. 381–389. Springer, New Delhi (2014). https://doi.org/10.1007/978-81-322-1157-0_38
11. Chaudhary, P., Chaudhari, A.K., Cheeran, N.A., Godara, S.: Color transform based approach for disease spot detection on plant leaf. Int. J. Comput. Sci. Telecommun. **3**(6), 65–70 (2012)
12. Rumpf, T., Mahlein, A.-K., Steiner, U., Oerke, E.-C., Dehne, H.-W., Plumer, L.: Early detection and classification of plant diseases with support vector machines based on hyperspectral reflectance. Comput. Electron. Agric. **74**(1), 91–99 (2010)
13. Al-Bashish, D., Braik, M., Bani-Ahmad, S.: Detection and classification of leaf diseases using K-meansbased segmentation and neural-networks-based classification. Inf. Technol. J. **10**(2), 267–275 (2011)
14. Danti, A., Madgi, M., Anami, B.: Mean and range color features based identification of common Indian leafy vegetables. Int. J. Sig. Process. Image Process. Pattern Recognit. **5**(3), 151–160 (2012)

15. Bernardes, A.A., et al.: Identification of foliar diseases in cotton crop. In: Tavares, J., Natal Jorge, R. (eds.) Topics in Medical Image Processing and Computational Vision. LNCVB, vol. 8, pp. 67–84. Springer, Dordrecht (2013). https://doi.org/10.1007/978-94-007-0726-9_4
16. Haralick, R.M., Shanmugam, K., Dinstein, I.: Texture features for image classification. IEEE Trans. Syst. Man Cybern. **3**(6), 610–621 (1973)

Data Mining, Information Retrieval and Applications

Data Mining Learning of Behavioral Pattern of Internet User Students

Aniket Muley(✉) and Atish Tangawade

School of Mathematical Sciences, SRTM University, Nanded, MS, India
aniket.muley@gmail.com, tangawadeatish@gmail.com

Abstract. This study focuses on the students internet use in their personal life. Various aspects has been assumed with the help of data mining technique and tried to obtain some hidden outcomes of student's internet behavior. The special focus is to test the significance based the on gender wise, location wise and different financial income group perspective to discriminate the behavioral pattern. Here, online survey is carrying out and 217 students information is gathered. The random sampling is performed for collection of data. The unsupervised and supervised learning analysis was carried out with SPSS 22.0v software package. The obtained result helps in future planning the direction of appropriate use of internet by students.

Keywords: Educational analytics · Chi-square test · Data mining · Internet

1 Introduction

Today use of internet is enormous part of our life. In today's era most of the people are depending on the internet to do different task viz., searching things, playing games data set that they crave, entertainment, correspondence or connecting to other, online shopping etc. The use of social media and networking services such as, face-book, twitter, Instagram and snap-chat are plays crucial role in student's life. In today's generation most of the learners can't live without internet. It has positive as well as negative impacts on daily life, indirectly effects on the physical fitness, psychological and mental health. Nervousness due to use of internet is important factor that have been found to relate to people's use and attitudes towards the Internet [2]. As the online cognition scale score increased, students' performance of internet activities viz., general information and academic research decreased. Whereas, the performance of interactive and distraction behavior increased viz., chat, transactions, games, downloading, and listening to MP3s. While, positive association was establish between problematic use with loneliness and depression, a negative correlation was originate between problematic use and alleged social support [4,6,7,9,12,25,26]. Some researcher investigated in the survey that, how college students identify academic institutions through internet and he suggested that a disproportionate number of

K. C. Santosh and R. S. Hegadi (Eds.): RTIP2R 2018, CCIS 1037, pp. 541–549, 2019.
https://doi.org/10.1007/978-981-13-9187-3_48

Internet dependent students might be finding among the hard science subjects [5,6,12,13,16]. In a gender-wise study, results reveals that internet use patterns of males mainly for pastime and relaxation purposes, whereas women's primarily use for interpersonal contacts and academics [15,33]. It also found that, the age of students has significant effect on use of internet. Nalwa [18] performed research of school children's internet habit of the age group of 16–18 years. The Davis Online Cognition Scale is used to evaluate pathological internet use. The loneliness measure observed to be significant in dependents scoring is more than independents. Tsitsika et al. [32] evaluates the features and forecaster of excessive use of internet and investigates the prevalence of pathological Internet use among Greek adolescents by Cross-Sectional study. Multivariate regression analysis was used for prediction of parameters. Chou et al. [8] explores study of research on the social effects of Internet addiction. They were chosen Internet use and time, identifiable problems, gender differences, psychological variables and computer attitudes as key factors. Gross [14] focuses study on adolescent Internet use and their expectations in California suburban public schools. It is observed that, teenager boy's and girl's online actions have become more alike. Kim et al. [20] accomplish study which explores the reasons for using Social Networking Sites (SNS) among college students in US and Korea. Their results reveals that, number of friends, social support, information, entertainment, and convenience were identified as primary reasons for using SNSs among college students in both countries. Jenaro et al. [18] identified pathological Internet and cell-phone usage correlates with health as well as psychological, behavioural pattern. A cross sectional plan is implemented and Logistic Regression is used for prediction of parameters. Their study reveals deep internet use is associated with high nervousness; high cell phone use is allied to being female, and having high anxiety and insomnia [18]. Linan et al. [22] studied the data mining learning's for advancement based parameters in academics. Ogata et al. [27] highlighted educational analytics for E-book foundation on higher education with respect to the concept of big data. Ahad et al. [3] focuses on learning analytics for internet on educational things by using deep learning techniques viz., their structural design, confronts and applications. Jang et al. [17] performed a study on the improvement and relevance for encouragement about student centric education. Kimmons and Veletsianos [21] discussed community internet data mining techniques in training plan, enlightening technology as well for online learning. Cooper and Klein [10] investigated college student's online pornography use in view of general dissimilarity and precise variables with their social learning. Raman et al. [30] evaluated the novelty distribution of student's inspiration by adopting programming competition. Data mining can be applied to such databases in order to gain challenging outputs [1,9,11,23,24,31,34]. Data mining helps users extract useful information from large databases. Statistics has the same general uses and results as data mining. Regression is used in statistics quite often it creates models that are predictive of behavior, and these models are built from large stores of historical data [23]. Data mining effectively automates the statistical process, thereby the reliving the users burden. This results in tool that is easier to use. In the educational data mining where we primarily investigate analytics for good insights [1,11,13,14,23,28,34]. Data mining broadly classified into two parts: unsupervised and supervised learning. In unsupervised learning, extraction and exploration of data is performed, while in supervised learning prediction and forecasting of future as well intermediate terminologies with various statistical tools. These insights are in terms of associations, correlations, clusters or even outline. The revealed myths suggested that, their paradigm can be exercise for better resource management. The present study aims to determine behavioral patterns of users of internet students, impact of internet use and awareness about internet. Another objective is to investigate the gender wise significance among university students regarding to their internet use. To study, the students re-log in internet pattern. To study is there any specific time schedule used for internet purpose by students. Also, does it give significance according to their location and use of internet?

2 Research Methodology

In this study, structured questionnaire is prepared with forty seven issues that were administered to S. R. T. M. University campus in Nanded during months of January– February 2018. The survey was accomplished through online Google forms with 217 individuals. The closed institutional environment allowed survey to be administered and taken for data collection purpose [11]. The proposed questionnaire is organized in four sections viz., general information, purpose of Internet use, emotional and physical problems due to its use and its impact on students. In this study, some of the responses were represented in the form of Likert scale. The extracted responses are coded from a point scale of 0 to 4, where: 0 = very often, 1 = often, 2 = sometimes, 3 = rarely, 4 = never. Guttmann's coefficient lambda computes association and a coefficient of predictability [29]. It is a statistical technique that measures the degree of one variable can be accurately predicted with knowledge of the other variable. This quality of predictability is one way of looking at the association between variables [1,11,23,34]. To compute Lambda symmetrical the following Eq. (1) is employed:

$$\lambda = \frac{\sum f_r + \sum f_c - (F_r + F_c)}{[2N - (F_r + F_c)]} \quad (1)$$

Where,
$f_r = Maximum\ frequency\ (row)$
$f_c = Maximum\ frequency\ (column)$
$F_r = Maximum\ marginal\ frequency\ among\ the\ rows$
$F_c = Maximum\ marginal\ frequency\ among\ the\ columns$
$N = Total\ number\ of\ frequency$

The Yule's Q measure is use to evaluate association for variables having only two values [2]. This could be done by using the following Eq. (2):

$$Q = \frac{(ad - bc)}{(ad + bc)} \tag{2}$$

Where, a, b, c, d represents the respective frequency in the 2 × 2 table.

In this study, to accumulate the data and to perform graphical representation is done through MS Excel and to execute inferential statistical investigation SPSS 22.0v software is used [15, 29]. The data mining techniques viz., unsupervised learning for exploration of data (Figs. 1, 2, 3 and 4) and supervised learning methodology were implemented by inferential techniques viz. Chi square test of significance and one-way ANOVA.

3 Result and Discussions

In this study, Fig. 1 represents geographical distribution of students. Clearly, it is observed that, most of the students i.e. 73% of the total students are coming rural part of the areas. Figure 2 reveals that, 33% of the students feels like restless, moody and depressed or get irrigated if they are not having internet access.

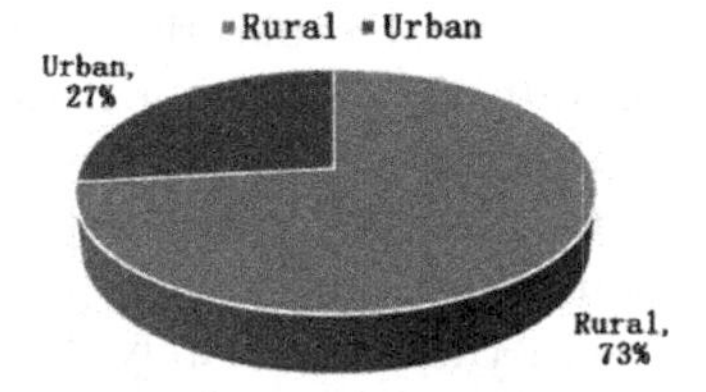

Fig. 1. Students residence location

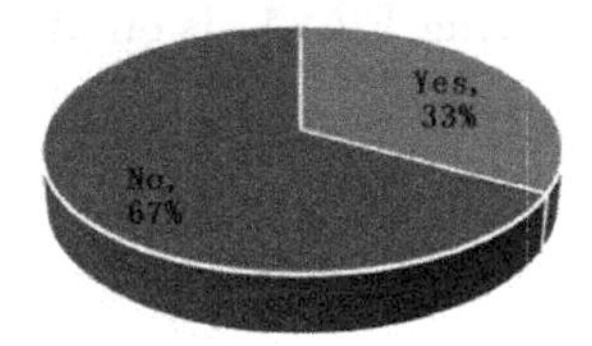

Fig. 2. Students negative feelings

Figure 3 explores the information about to check the gender wise independency about the use of internet. Figure 4 reveals that, 38% of the students performed unsuccessful efforts for controlling, cut back or stop internet use and 62% of them are not retrying to connect it. Here, observed that, the use of internet is independent of gender.

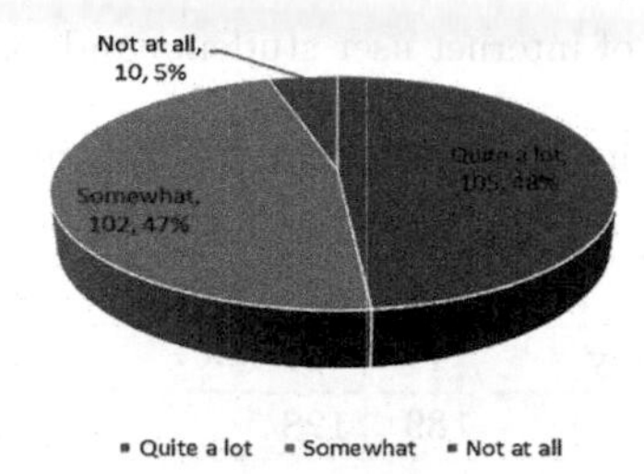

Fig. 3. Students accessing relevant information for study

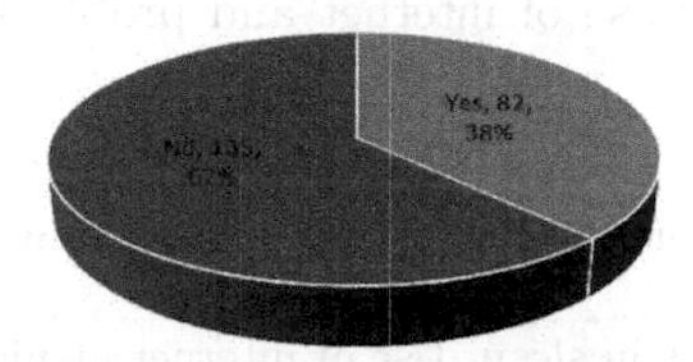

Fig. 4. Students unsuccessful efforts of internet use

Table 1. Gender wise internet users

Gender	Do you use internet?		Total
	Yes	No	
Male	46	1	47
Female	162	8	170
Total	208	9	217

Table 2. ANOVA gender Vs. Use of internet

	Value	df	Asymp. Sig. (2-sided)	Exact Sig. (Two-tailed)	(One-tailed)
Pearson chi square	0.616	1	0.433		
Continuity correction	0.138	1	0.710		
Likelihood ratio	0.712	1	0.399		
Fisher's exact test				0.688	0.383
Linear-by-linear association	0.613	1	0.434		
Valid cases	217				

The results obtained in Table 2 reveals that, Pearson's Chi Square $(x^2) = 0.616$ and p-value is 0.433(>0.05). Hence, the use of internet is independent of gender.

Table 3. Summary of internet user students and avoiding doing work

Use of internet	Avoiding doing work		Total
	Yes	No	
Daily	78	117	195
Not daily	11	11	22
Total	89	128	217

Based on Table 3, $\lambda = 0.9009$ (Eq. 1), we can wrap up that there is a strong association between daily use of internet and problem of avoiding doing work among the students.

Table 4. Frequency distribution of use of internet at night

Disturbance in sleep	Use of internet at night		Total
	Yes	No	
Yes	a = 77	b = 107	184
No	c = 11	d = 22	33
Total	88	129	217

Table 4 illustrates the Yule's Q formulae (Eq. 2) and the obtained result reveals that, the level of association between the problem of disturbance in sleep and students' use of internet at night is 0.18, indicating a moderately low positive relationship.

Table 5. Location Vs. Annual family income

	Value	d.f.	Asymp. Sig. (2-sided)
Pearson chi square	27.988	7	.000
Likelihood ratio	25.493	7	.001
Linear-by-linear association	25.763	1	.000
Valid cases	217		

Table 5 explores that, there is significance among the student's annual family income and their place of living.

Table 6 reveals the significance of peculiar timing for use of internet. The data explore the information that, they use night timing for using surfing social media sites.

In this study, result reveals the insignificance of use of internet among male and female students (Table 1). In gender wise study, student's time spending

Table 6. Location Vs. Annual family income

	Value	d.f	Asymp. Sig. (2-sided)
Pearson chi square	18.460	7	.000
Likelihood ratio	17.574	7	.014
Valid cases	217		

Table 7. Gender Vs. students anticipating re-online again

		Re-online users					Total
		Always	Often	Sometimes	Rarely	Never	
Gender	Male	8	1	28	4	6	47
	Female	16	8	73	32	41	170
	Total	24	9	101	36	47	217

on internet is significantly observed at night (Table 6). It is found that, the students belongs to different income groups were uses the internet insignificantly (Table 5). It is observed that, 46.54% of the students sometimes re-logging to internet, 21.65% of them never login again, 16.58% of the students rarely re-logging, 11.05% of them always making re-loin and 4.14% of them often login to the internet (Table 7). Also, it is observed that, there is insignificant difference among rural and urban resident students internet use.

4 Conclusions

In the present study, numerous societal issues undertaken if prognostication methods are implemented to study the behavioral aspects with their gender, location, financial income can boost student's internet use. Here, we observed that most of these students uses internet for the purpose of communicating or connecting to other and entertainment. Out of 217 students, 29 students thought that, internet has negative effect on their study. The negative effect on their study is observed due to: (i) purpose of the internet use, (ii) get irritated while some is disturbing, (iii) losing sleep due to online much of the time, (iv) suffering problems while using internet, (v) sudden, scream or becomes angry if someone hassle, (vi) performance endure because of the internet. Also, data reveals that 72% student feels like restive, grumpy, miserable or short-tempered as go to slash down or discontinue internet. Around 50% students use internet for getting relevant information for study. The use of internet is observed to be independent of their gender. The problem of avoiding doing work is strongly related with daily use of internet. There is relationship between the problems of disturbance in sleep and students use of internet. In future, if we are focusing on avoiding the use of social media site on working hours, reducing the time on it at night and re-login again then it will be helpful in increasing the performance of students.

References

1. Achana, R.A., Hegadi, R.S., Manjunath, T.N.: A novel data security framework using E-MOD for big data. In: IEEE International WIE Conference on Electrical and Computer Engineering, pp. 546–551 (2015)
2. Adeyemi, O.: Measures of association for research in educational planning and administration. Res. J. Math. Stat. **3**(3), 82–90 (2011)
3. Ahad, M.A., Tripathi, G., Agarwal, P.: Learning analytics for IoE based educational model using deep learning techniques: architecture, challenges and applications. Smart Learn. Environ. **5**(1), 7 (2018)
4. Ali, S., Haider, Z., Munir, F., Khan, H., Ahmed, A.: Factors contributing to the students academic performance: a case study of Islamia University Sub-Campus. Am. J. Educ. Res. **1**(8), 283–289 (2013)
5. Anderson, K.J.: Internet use among college students: an exploratory study. J. Am. Coll. Health **50**(1), 21–26 (2001)
6. Bratti, M., Staffolani, S.: Student time allocation and educational production functions, University of Ancona Department of Economics Working Paper No. 170 (2002)
7. Ceyhan, A.A.: Predictors of problematic internet use on Turkish university students. Cyberpsychol. Behav. **11**(3), 363–366 (2008)
8. Chou, C., Condron, L., Belland, J.C.: A review of the research on internet addiction. Educ. Psychol. Rev. **17**(4), 363–388 (2005)
9. Considine, G., Zappala, G.: Influence of social and economic disadvantage in the academic performance of school students in Australia. J. Sociol. **38**, 129–148 (2002)
10. Cooper, D.T., Klein, J.L.: College students' online pornography use: contrasting general and specific structural variables with social learning variables. Am. J. Crim. Justice. **43**(3), 551–569 (2018)
11. Divya, M., Manjunath, T.N., Hegadi, R.S.: A study on developing analytical model for groundnut pest management using data mining techniques. In: IEEE's International Conference on Computational Intelligence and Communication Networks, pp. 691–696 (2014)
12. Field, A.: Discovering Statistics using R for Windows. Sage publications, Thousand Oaks (2000)
13. Graetz, B.: Socio-economic status in education research and policy in John A., et al., socio-economic status and school education DEET/ACER Canberra. J. Pediatr. Psychol. **20**(2), 205–216 (1995)
14. Gross, E.F.: Adolescent internet use: what we expect, what teens report. J. Appl. Dev. Psychol. **25**(6), 633–649 (2004)
15. Gupta, S.L., Hitesh, G.: SPSS 17.0 for Researchers. International book house, Pvt. Ltd. (2011)
16. Han, J., Pei, J., Kamber, M.: Data Mining: Concepts and Techniques. Elsevier, New York (2011)
17. Jang, Y., Kim, J., Lee, W.: Development and application of internet of things educational tool based on peer to peer network. Peer-to-Peer Network. Appl. **11**(6), 1217–1229 (2018)
18. Jenaro, C., Flores, N., Gómez-Vela, M., González-Gil, F., Caballo, C.: Problematic internet and cell-phone use: psychological, behavioural, and health correlates. Addict. Res. Theor. **15**(3), 309–320 (2007)
19. Joiner, R., Gavin, J., Duffield, J., Brosnan, M., Crook, C., Durndell, A., Lovatt, P.: Gender, internet identification, and internet anxiety: correlates of internet use. Cyber Psychol. Behav. **8**(4), 371–378 (2005)

20. Kim, Y., Sohn, D., Choi, S.M.: Cultural difference in motivations for using social network sites: a comparative study of American and Korean college students. Comput. Hum. Behav. **27**(1), 365–372 (2011)
21. Kimmons, R., Veletsianos, G.: Public internet data mining methods in instructional design, educational technology, and online learning research. TechTrends. **62**(5), 492–500 (2018)
22. Liñán, L.C., Pérez, Á.A.J.: Educational data mining and learning analytics: differences, similarities, and time evolution. Int. J. Educ. Technol. High. Educ. **12**(3), 98–112 (2015)
23. Manjunath, T.N., Hegadi, R.S.: Statistical data quality model for data migration business enterprise. Int. J. Soft Comput. **8**(5), 340–351 (2013)
24. Dunham, M.: Data Mining: Introductory and Advanced Topics. Pearson publications, USA (2002)
25. Metzger, M.J., Flanagin, A.J., Zwarun, L.: College student web use, perceptions of information credibility, and verification behaviour. Comput. Educ. **41**(3), 271–290 (2003)
26. Nalwa, K., Anand, A.P.: Internet addiction in students: a cause of concern. Cyberpsychol. Behav. **6**(6), 653–656 (2003)
27. Ogata, H., Oi, M., Mohri, K., Okubo, F., Shimada, A., Yamada, M., Wang, J., Hirokawa, S.: Learning analytics for e-book-based educational big data in higher education. In: Yasuura, H., Kyung, C.-M., Liu, Y., Lin, Y.-L. (eds.) Smart Sensors at the IoT Frontier, pp. 327–350. Springer, Cham (2017). https://doi.org/10.1007/978-3-319-55345-0_13
28. Özcan, N.K., Buzlu, S.: Internet use and its relation with the psychosocial situation for a sample of university students. Cyber Psychol. Behav. **10**(6), 767–772 (2007)
29. Pritchard, M.E., Wilson, G.S.: Using emotional and social factors to predict student success. J. Coll. Student Dev. **44**(1), 18–28 (2003)
30. Raman, R., Vachharajani, H., Achuthan, K.: Students motivation for adopting programming contests: innovation-diffusion perspective. Educ. Inf. Technol. **23**(5), 1919–1932 (2018)
31. Rokach, L., Maimon, O.: Data Mining with Decision Trees: Theory and Applications. World scientific, Singapore (2014)
32. Tsitsika, A., et al.: Internet use and misuse: a multivariate regression analysis of the predictive factors of internet use among Greek adolescents. Eur. J. Pediatr. **168**(6), 655 (2009)
33. Weiser, E.B.: Gender differences in internet use patterns and Internet application preferences: a two-sample comparison. Cyber Psychol. Behav. **3**(2), 167–178 (2000)
34. Yogish, D., Manjunath, T.N., Hegadi, R.S.: Survey on trends and methods of an intelligent answering system. In: IEEE's International Conference on Electrical, Electronics, Communication, Computer, and Optimization Techniques, pp. 346–353 (2017)

Research Challenges in Big Data Security with Hadoop Platform

M. R. Shrihari[1], T. N. Manjunath[1(✉)], R. A. Archana[2(✉)], and Ravindra S. Hegadi[3(✉)]

[1] Department of ISE, BMS Institute of Technology and Management, Bengaluru, India
shrihari.mr@gmail.com, manju.tn@gmail.com
[2] Bharathiar University, Coimbatore, Tamil Nadu, India
archana.tnm@gmail.com
[3] School of Computational Sciences, Solapur University, Solapur, Maharastra, India
rshehadi@gmail.com

Abstract. Every minute in the internet, enormous volume of data is generating tuning into big data. A new paradigm of data storage and processing is essential. There is extremely no query that Hadoop is an essentially unruly technology. Latest innovation in scalability, appearance, and data dispensation competence has been beating us apiece of few months over the last few years. In this source of ecosystem is the extremely classification of novelty. Big data has distorted data analytics given that extent, presentation, and flexibility that was just not potential a few years ago, at an outlay that was evenly inconceivable. But as Hadoop turn into the new standard of information technology, developers, and security policy are playing grab awake to identify with Hadoop security. Moreover, current security hypothesis and mechanism have been established. In this paper, we discuss lays out a series of recommended security controls for Hadoop along with an access control framework, which enforces access control policies dynamically based on the sensitivity of the data and systemic security, operational security and architecture for data security. A relative study of latest advances in big data for security. A number of prospect information for big data security and privacy methods are discussed.

Keywords: Big data · Hadoop · Security · Privacy

1 Introduction

In the last few years Hadoop has developed as of a effortless disseminated data organization system for successively Map Reduce queries, into a complete application framework for dispensation enormous quantity of data using presently concerning any technique. We are extensive history wondering whether Hadoop is a feasible technology; however Hadoop is now creature embraced both by enterprises on data processing in shared and treating Hadoop as currently a different open source distribution. To survey to show Hadoop is affecting the path of the popular novel consequence development in movement. Hadoop's significance is extensively established and these

K. C. Santosh and R. S. Hegadi (Eds.): RTIP2R 2018, CCIS 1037, pp. 550–560, 2019.
https://doi.org/10.1007/978-981-13-9187-3_49

schemes are here to stay, but currently they necessity remain to shared security and data authority frameworks. Getting Hadoop secure is an essential obstacle the majority information technology and security. A handle on Hadoop security but extra prominently concern accessible information authority and observance controls to the group. Similar too many security projects and inventiveness to secure [1]. The most frequent difficulty from consumers are easy ones like How can secure Hadoop and How map accessible data authority policies to No SQL platforms. Hadoop's efficient development enhancements over the previous few reality have been corresponding by much better security and authority aptitude. Because huge activity in progress to squeeze this technology policy, they pushed the seller of enterprise Hadoop sharing and made security and compliance mechanism of their least supplies. Currently security controls are not only available, but for most requirements more than one option. Hadoop has reached security parity with the relational policy. As of this quick development, a new assessment of Hadoop security is in organizes. In this paper will discuss a concise impression of Hadoop's architecture and emphasize security challenges are built. We will discuss set of strategic and tactical responses to address and challenges. Our objective is to help individuals tasked with Hadoop security concentrate on possibility to the cluster, as well as construct an authority outline to sustain prepared requirements [8].

2 Architecture and Composition

Objective of the Hadoop cluster is together the data and process which is used by the end user for specific purposes. This assist perceptive of the security challenges, the length of with which kind of security can protect them. Data scientists and developers continue to enlarge scheme presentation and scalability, with modified arrangement of open basis and profitable products. To identify with threats and suitable reply, one should primary identify with all the aspects [2]. It is significant to identify with apiece constituent interface. Each constituent offer aggressor an explicit set of prospective expands, even as defenders have a consequent set of option for attack exposure and anticipation. Perceptive architecture and gather work is the primary stair to place jointly your security strategy. The subsequent is a simplify inspection of Apache Hadoop's essential MapReduce system:

Hadoop can be passionately attractive because it scales extremely healthy, preserve be configured to switch a extensive assortment and is extremely expensive compared to older. Hadoop is cheap, fast, and flexible. To perceive why and how it scales, get a look at Hadoop cluster architecture, demonstrate in the above diagram. This architecture encourages extent and presentation. It supports equivalent dispensation; with extra nodes provide 'horizontal' scalability [15]. This architecture is also essentially multi-tenant, sustaining several customer application diagonally solitary or new file collection. There are numerous belongings to a security perspective. Presently various moving component each node converse with its peers to guarantee to facilitate

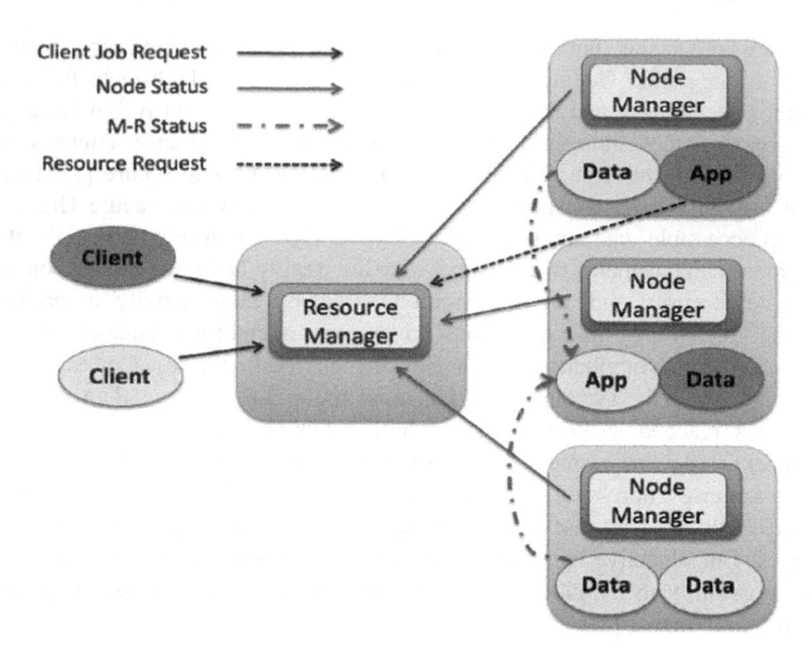

Fig. 1. Hadoop architecture

information is correctly replicated, nodes are online and efficient, storage is optimized, and function needs are individual procedure. Apiece line in the plan is a trust association utilize a announcement protocol (see Fig. 1).

3 The Hadoop Framework

Towards understand Hadoop's agility to must to identify with to facilitate meet preserve be completely adapted. In support of individuals of novel to Hadoop it might assist to believe of the Hadoop framework as a heap, greatly approximating a light heap. Among a jetpack and night-vision googles. The quantity in the direction of which can mix add mechanism is inconceivable. As HBase is frequently implement on top of HDFS, it may decide a dissimilar technique of investigate, such as Solr. It can implement Sqoop to present relational data access or influence Pig intended for elevated intensity MapReduce principle. it preserve choose dissimilar SQL uncertainty engines through Spark, Drill, Impala, and Hive all helpful SQL queries [11]. The modularity propose huge suppleness to accumulate and modify separate to carry out precisely as preferred. It's not currently to facilitate Hadoop preserve finger information dispensation in dissimilar behavior, or that you preserve change its presentation individuality, modify development, or append contribution data parsing. A Hadoop cluster preserve be customized to healthy the precise requirements of your production. We can plan a group to please your usability, scalability, and performance goals [2]. We can adapt it to specific types of data and append component to ease study of certain data sets. Except plasticity convey complexity, an enthusiastic framework of features, occupation and ticket position make security additional complicated. Apiece alternative convey its individual security preference and concern. Each component sprint an explicit description of system, have it's possess understanding and might need self-governing authentication to effort in the group. a lot of piece should employment in tandem here to development data, so each necessitate its own security assessment [7]. It can also be that security, auditing, governance, and configuration management can also be added (see Fig. 2).

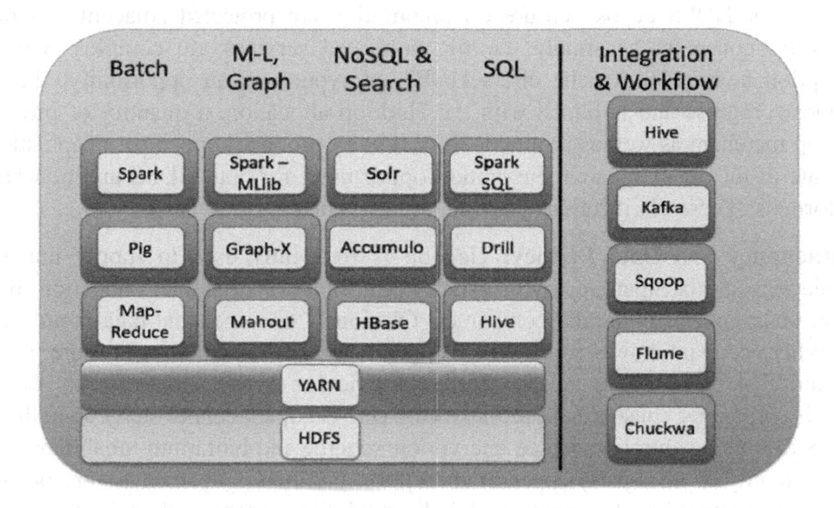

Fig. 2. Hadoop framework

4 Systemic Security

At the present to we have outline exposed the essentials a Hadoop group and one appear like, let's converse threats. We require to believe equally the communications itself and the information inferior organization. Specified the complication of a Hadoop cluster, the mission is closer to securing an entire set of applications [12]. All the features so as to offer elasticity, scalability, presentation, and honesty produce explicit security challenges. Consequently at this time are quantities of detailed aspect of arrangement scheme attacker determination objective [4].

Data Permission and Possession: Responsibility of access roles is vital to mainly to concern RDBMS and data warehouse security system and Hadoop system. Nowadays Hadoop offer complete incorporation by individuality provisions, the length of by role based services to separate awake data access between groups of users. Relational and quasi-relational policy incorporates roles, collection, schemas, make security, and a variety of extra services for preventive consumer admission to subsets of accessible data. To make easy understood, verification and permission need collaboration between the appliance fashionable and the information technology team managing the cluster. Leveraging existing Active Directory services assist extremely by essential user individuality and predefined position might be obtainable for restrictive admission to receptive information [10].

Data at Respite Security: The normal intended for defending data at respite is encryption which defends touching challenge to right of entry of data is external to established relevance interface. With Hadoop system concern regarding community pinching records or honestly analysis records starting disk and encryption by the organizer or HDFS deposit ensure documentation are protected adjacent to straight access by consumer as simply the file equipped services are complete with the encryption keys [11]. Apache offers HDFS encryption as an opportunity; this is a foremost progress and is bunch with the Hadoop allocation. a quantity of profitable Hadoop merchant as well as profitable third revelry products, encompass sophisticated the state of the art in apparent encryption opportunity for both HDFS and non-HDFS file formats. These clarifications provide key organization as well [6].

Multitenancy and Data Privacy: Hadoop is frequently used to supply numerous applications and occupant, every of which might be as of dissimilar collection by one dense, or in general dissimilar companies. Obviously one occupant data is not public with other occupant, however it should execute a security control to make sure privacy. a quantity of dense utilize Access Control Entries or Access Control Lists for both basically organizer consent built to make sure one occupant cannot study an additional data. Still others control are called encryption scheme and build into inhabitant HDFS and a quantity of third-party apparent encryption products. Effectively every occupant has define sector for file groups and individual files where each definite sector is encrypted with a different key to ensure data privacy [13].

Internode Communication: Hadoop having huge popular of distributions (Cassandra, MongoDB, Couchbase, etc.) to converse securely by default and they use unencrypted RPC over TCP/IP. TLS and SSL potential are package in big data distributions, however not constantly used to involving customer application and the cluster reserve administrator, and rarely for internodes communication. This plants information in transfer along with application queries easily reached for examination and interferes [7].

Client Communication: Clients network with the source supervisor and nodes. Opportunity military can be produced to consignment information and clients communicate directly with both source managers and personality data nodes. Concession clients may send spiteful data or relation to services. This eases efficient communication apart from create it tricky to defend nodes from customers, clients from nodes, and even name servers as of nodes. Of inferior quality the allocation of self-organizing nodes is a pitiable fit for security apparatus such as gateways, firewalls, and monitors [7].

Scattered Nodes: One of the key compensation of big data is the previous platitude affecting estimation is cheaper than affecting data. Data is development where assets are obtainable and enable especially equivalent estimation. Regrettably this produce difficult environment by lots of assault surface. With so a lot of stirring parts it is hard to validate stability or security across a extremely dispersed cluster of different platforms. Patching, association organization, node personality and data at rest protection and reliable exploitation for all matter [7].

5 Operational Security

Information technology teams are expecting several frequent tools and services which are universal transversely scheme relevance. So as to contain vulnerability estimation, policy supervision and preserve piece intensity diagonally a compound meeting of sustaining module. Hadoop security has come a extensive system in immediately a few years and most of the frequent subject can now be concentrate on with various time and attempt on the part of Information technology and security teams. The subsequent is an impression of the majority frequent threats to Hadoop (data organization systems in general) beside with prepared control contribution preventive security to secure off frequent attack [10].

Authentication and Permission: Individuality verification are vital to any security attempt to elite determine who have to access to information. Luckily the most gain in Hadoop security has in individuality and access information. It also provides of enterprise Hadoop distributions, we encompass develop commencing defaulting configurations contribution no verification preference to completely incorporated LDAP, Active Directory, Kerberos, and X.509 based options. Through leveraging these potential we protect to use recognized roles for authorization information, and occasionally make longer it to fine-grained consent services similar to Apache Sentry, or convention endorsement record restricted from surrounded by the profession relevance [5].

Organizational Data Access: The majority of the organizations contain platform administrator and Hadoop administrator, both with access to the cluster's records. Towards to supply partition of duties to guarantee officer cannot inspection contented a capability of desirable to separate organizational roles and confine unnecessary access to a lowest quantity [14]. Straight admission to records or data is frequently concentrate on a arrangement of responsibility based-authorization, access organize list, file permissions and segregation of organizational roles such as with separate organizational financial records outlook in different responsibility and recommendation. This grants basic protection but cannot defend legitimate admission to archived. Stronger security necessitate a arrangement of data encryption and key management services, with excellent keys for each function or cluster as present with apparent file or HDFS encryption [2].

Authentication of Applications and Nodes: If a defender preserves to add a new node they organize to the cluster, they preserve information. To authenticate nodes (rather than users) before they can adhere a cluster and most dense we converse with either use X.509 certificates or Kerberos. Both schemes can authenticate users as well but we draw this feature to emphasize the threat of applications or nodes organism extra to the cluster. Consumption of these services brings risks as well. Certificate support personality opportunities implicitly obscure setup and consumption but appropriately organize them can grant strong verification and improve security [5].

Inspection and Classification: If you believe a name has violate your cluster, can you identify it, or outline reverse to the root basis. A diversity of append classification ability are accessible for open foundation and profitable. Leverage of the cluster to build up its personality logs, except several security professionals concern an enemy can face their path by remove or transform log admission. As well note a quantity of classification option does not offer enough information for an examiner to decide precisely. It needs to validate that your logs are configured to confine both the accurate incident types and adequate in order to verify on consumer events [7].

API Security: In a Big data cluster APIs require to be confined as of code and authority insertion, defense excess attacks and the other entire standard web examine attack. This dependability classically domain on the relevance using the cluster. General security control incorporate addition with listing services, plan to API services, strain requests, effort validation and organization policies nodes. a quantity of people control API gateways and fair list acceptable relevance requirements. Another time, a handful of the solutions can help concentrate on API security [11].

6 Security Architectures

Security architectures are extremely cooperative for conceptualizing approach for cluster security and they are very useful to acquire an exchange on information.

6.1 Walled Garden

The most frequent approach today is a “walled garden” security model, related to the ‘moat’ model from mainframe security: consign the complete cluster on its possess network, and securely control consistent access during firewalls or API gateways, by access controls for customer or relevance authentication. The representation provides practically no security within the Hadoop cluster information and communications security depend on the external “protective shell” of the system and relevance that to be enclose it. The benefit is simplicity and any compact can realize this representation by existing tools and skills, exclusive of presentation or efficient degradation to the Hadoop cluster. On the difficulty security is delicate [9]. This representation also does not avoid credentialed users as of exploitation the scheme or performance/change data accumulate in the cluster. For institute not particularly worried about security, this is a simple and cost-effective approach (see Fig. 3).

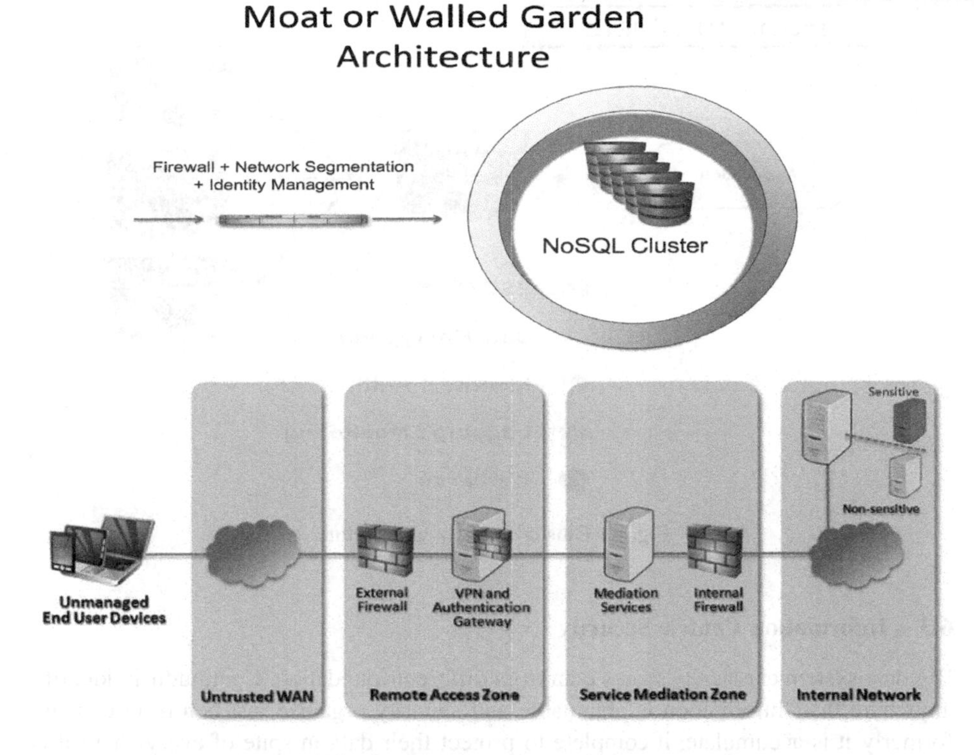

Fig. 3. Walled garden architecture

6.2 Cluster Security

Distinct relational database which utility like black boxes, Hadoop interpretation its intestines to the system. Inter-node statement, imitation and extra cluster function acquire place among several technology, using special types of services. Proposed for the great quantity of effective protection and creation of security into cluster process is critical [7]. The advance leverages security tools built into intermediary harvest incorporated into the NoSQL cluster. This security is complete and construct to be component of the originate cluster design. It includes SSL/TLS for secure communication, Kerberos for node validation, apparent encryption for data at rest security, and identity and endorsement organization. And third-party security utensils can be exclusive [15]. But they preserve successfully secure clusters as of attackers, administrators. It is the most efficient and comprehensive approach to Hadoop security (see Fig. 4).

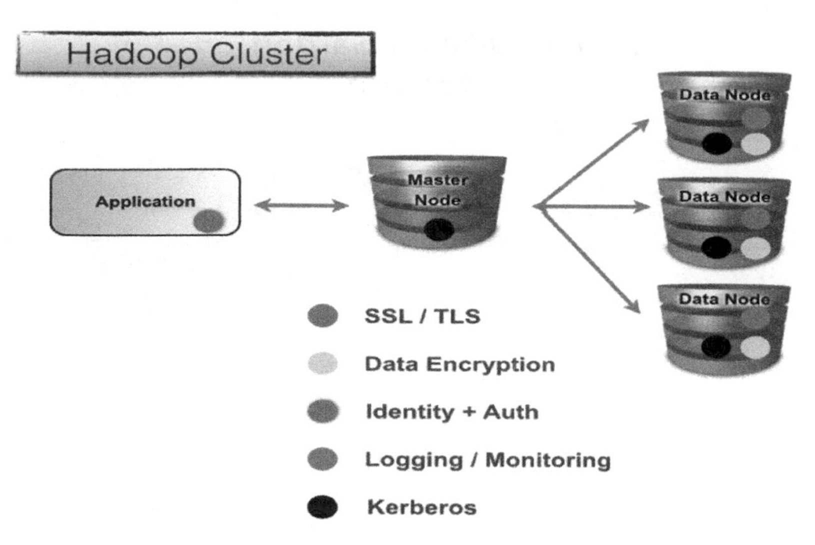

Fig. 4. Cluster security architecture

6.3 Information Centric Security

Big data system classically shares data from different foundation. Continuously identify anywhere their information is obtainable and security organize just before in consign formerly it is accumulate, it complete to protect their data in spite of everywhere it is used. This representation is called information centric security since the controls are part of the data [9]. The indispensable mechanism that support information centric security are tokenization, masking and data element encryption the shortly frequently realize as system Preserving Encryption. An information indication has no natural value. It only orientation the innovative assessment in a indication store. Masking is another fashionable device use to defend information essentials although maintains the collective value of an information set [3]. The information centric security model provides an immense agreement of security once the system so as to development information cannot be exclusively conviction and in cases us doesn't want to share data with users. Fixed masking can be used to entirely remove responsive data as of clusters yet still allocate for significant data analysis. Excluding a information centric security representation require suspicious preparation and implement variety it is additional concerning information lifecycle organization [4]. Small of obliterate sensitive data, this is the most excellent representation and populate a big data group for investigation work assurance its security (see Fig. 5).

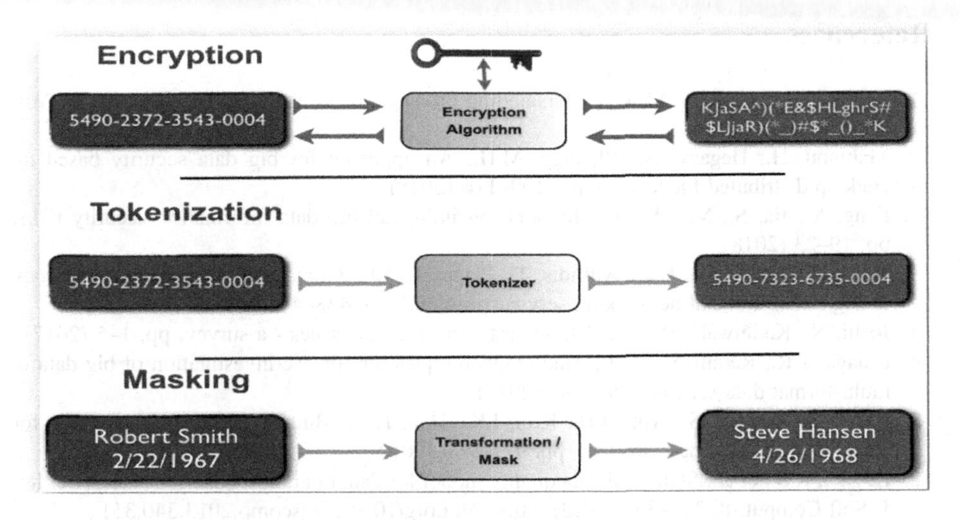

Fig. 5. Information centric security architecture

7 Conclusion

In this paper we presented the big data information and challenges used in the world wide. The security issue is pointed more in order to enhance the security in big data. The issues are provide idea about the big data issues. We preserve improve security in big data by using any one of the approach or by combining these three advance in Hadoop Distributed File System which is the base layer in Hadoop, where it surround huge quantity of blocks. These move toward are found to conquer certain issues occurs in the name node and also in Data node. In potential these advance be as well execute in additional layer of Hadoop Technology. Security mechanism will tell you both assault on business Information technology systems and data contravene are recognized, so with gobs of data underneath organization, Hadoop provides an attractive objective for hackers. Hadoop responsible in a foremost data contravene. Consequently this kind of broadcast tiny credence by Information technology Operations. As insightful information, consumer data, medical histories, intellectual property, and presently concerning all type of information used in venture compute is currently frequently used in Hadoop clusters, One of the foremost change we encompass see over the last couple years has been Hadoop becoming business critical communications. A new, which flows honestly with the origination and Information technology creature tasked with convey existing clusters in procession with enterprise fulfillment necessities. This is challenging of a new set up of Hadoop System security as traditional Information technology systems, so it takes effort to establish security. And additional to produce strategy and information for the observance team. Intended for clusters to need to choose technologies and an exploitation operations.

References

1. Shamsi, J.A., Khojaye, M.A.: Understanding privacy violations in big data systems. IT Prof. **20**(3), 73–81 (2018)
2. Mahmou, H., Hegazy, A., Khafagy, M.H.: An approach for big data security based on Hadoop distributed file system, pp. 109–114 (2018)
3. Feng, X., Jia, S., Mai, S.: The research on industrial big data information security risks, pp. 19–23 (2018)
4. Stergiou, C., Psannis, K.E., Xifilidis, T., Plageras, A.P., Gupta, B.B.: Security and privacy of big data for social networking services in cloud, pp. 438–443 (2018)
5. Joshi, N., Kadhiwala, B.: Big data security and privacy issues - a survey, pp. 1–5 (2017)
6. Balaga, T.R., Reram, S., Pi, L.: Hadoop techniques for concise investigation of big data in multi-format data sets, pp. 490–495 (2017)
7. Lee, J.-H., Kim, Y.S., Kim, J.H., Kim, I.K., Han, K.-J.: Building a big data platform for large-scale security data analysis, pp. 976–980 (2017)
8. Hegadi, R.S., et al.: Statistical data quality model for data migration business enterprise. Int. J. Soft Comput. **8**, 340–351 (2013). https://doi.org/10.3923/ijscomp.2013.340.351
9. Manjunath, T.N., et al.: Data quality assessment model for data migration business enterprise. Int. J. Eng. Technol. (IJET) **5**(1), February–March 2013. ISSN: 0975-4024
10. Behnia, R., Yavuz, A.A., Ozmen, M.O.: High-speed high-security public key encryption with keyword search. In: Livraga, G., Zhu, S. (eds.) DBSec 2017. LNCS, vol. 10359, pp. 365–385. Springer, Cham (2017). https://doi.org/10.1007/978-3-319-61176-1_21
11. Shrihari, M.R., Archana, R.A., Manjunath, T.N., Hegadi, R.S.: A review on different methods to protect big data sets, issue 12, p. 4 (2018)
12. Mehmood, A., Natgunanathan, I., Xiang, Y., Hua, G., Guo, S.: Protection of big data privacy, pp. 2169–3536 (2016). IEEE
13. Hongbing, C., Chunming, R., Kai, H., Weihong, W., Yanyan, L.: Secure big data storage and sharing scheme for cloud tenants. China Commun. **12**(6), 106–115 (2015)
14. Wang, H., Jiang, X., Kambourakis, G.: Special issue on security, privacy and trust in network-based big data. Inf. Sci. Int. J. **318**, 48–50 (2015)
15. Thuraisingham, B.: Big data security and privacy. In: Proceedings of the 5th ACM Conference on Data and Application Security and Privacy, San Antonio, TX, USA, pp. 279–280, 2–4 March 2015

Mining Frequent Patterns in Firewall Logs Using Apriori Algorithm with WEKA

Hajar Esmaeil As-Suhbani(✉) and S. D. Khamitkar

Department of Computational Sciences and Technology, S.R.T.M University, Nanded, India
hajar.esmaeil@gmail.com

Abstract. With the enormous growth of security incidents in computer networks, the network security defense has gained significant attention from the information industry and network community. Firewalls are the first lines of defense for protecting computer networks and important information. They function as routers to connect different network segments together. Furthermore, they considered as the most important elements in the networks used by organizations to enforce their security policy. The security policies of enterprises and companies are implemented as firewall rules. These firewall rules are sensitive and any misconfiguration of them will cause anomalies. The subject of mining of frequent patterns in itemsets of the dataset is considered as one of the most important aspects in data mining technology. Apriori algorithm is the simplest and most powerful association rule mining (ARM) algorithms which can be efficiently used for mining frequent itemsets in the dataset. In this study, we proposed Apriori algorithm on WEKA to extract frequent itemset in the firewall logs to determine the best association rules that ensure the general orientations in the dataset.

Keywords: Firewall logs · Firewall rules · Data mining · Association rule · WEKA · Apriori

1 Introduction

In the era of information technology, computer networks constantly change human society, with the accompanying information and network security issues. The networks' administrators spend a lot of money to purchase network security tools such as anti-virus software and firewalls. Also, they spend a lot of time to ensure the availability and integrity of the networks and retain the confidentiality of network information to prevent attacks from outside or inside the network [1]. Most network devices record large amounts of network security data in the form of logs and alarms, so these data become an important basis for the response, detection, and defense in network security work.

K. C. Santosh and R. S. Hegadi (Eds.): RTIP2R 2018, CCIS 1037, pp. 561–571, 2019.
https://doi.org/10.1007/978-981-13-9187-3_50

Firewalls are the core devices of network security. They considered as the primary devices in computer networks for enforcing and ensuring the security policy of the network and information security. Firewall act as a router that joins multiple network segments together. Managing and customizing firewall rules is a very important aspect in firewall policy, it is extremely complex, costly and error-prone due to the continuous growth of security threats and managing the security rules [2]. The security policies of organizations, enterprises and companies are implemented as ordered firewall rules. Since, most firewalls use addressing information in the packet header [3], they first check the data packets passing the networks and compare them with a list of predetermined logical rules. These rules are implemented by the network administrators or created dynamically based on the status of the active connections. These logical rules are referred to as firewall rules. Firewall policy is a set of ordered rules with accepting or denying actions which compare the packets passing the network with these ordered rules to decide whether the packets should be allowed to pass or should be denied.

In computer networks, systems and devices can generate much information about their states as log files. A log file is a file that is used to track the activities performed by any user in the network. They are generated by several sources such as a specific application, firewalls, service, or an operating system, etc. [2]. In addition, different functioning logs and alert services generate them. Log files are excellent sources that reflect the health status of a system or a network. Systems or devices in order to provide information about actions were performed; errors or problems, and a wide variety of information create log files. By analyzing these information, most of the anomalies and potential security breaches in firewalls and networks can be revealed. In the real network environment, these security data are often massive and sporadic and cannot exist directly as valid security information [4]. These data must be analyzed deeply to find valuable information. The existing database technology can effectively realize the functions of data entry, inquiry and preliminary statistics, but it is very difficult to predict the potential threats and attacks from the massive safety data and predict the future development trend based on the existing data.

In recent years, data mining technology has gained much significant attention in the information technology and also in information industry mainly due to the existence of a massive amount of data that can be widely used, and the ability to draw on and utilize the useful knowledge behind these data [2]. Mining log data is an effective method, but log analysis will face many challenges such as:

1. log data is bulky and beyond the capabilities of the analytical system and analysts, thus destroying the possibility of reaching conclusions. Indeed, the logs could reach several GBs and scale to a few terabytes, necessitating the choice of specialized tools to deal with this "flood".
2. Insufficient data due to different reasons, the key part of the data is easy to be deleted, which makes the log analysis beyond the necessary difficulty.
3. Record a variety of: Need to analyze too many different and dissimilar log sources, in order to come to the truth. The problem is the lack of uniform

auditing standards and the fact that most application logs are in the format developed by the creators, resulting in a number of analytical challenges.
4. Data duplication where different logs refer to the same event, without any indication. This situation is often complicated by the lack of time synchronization between different log sources [5].

This experiment has been implemented using firewall logs which was collected from a Lab in our department using Snort and TWIDS tools [15]. We have proposed Apriori algorithm to find out the frequent patterns hidden in the dataset. We have used only the 6 major attributes/features in firewall logs out of 10 attributes by making some guesses to select the most significant attributes and decide which attribute can be removed. We have used WEKA 3.8 Data Mining tool as an analysis software to analyze the firewall logs, deduce firewall policy rules to obtain the best 10 association rules.

1.1 Related Work

There are many studies were carried out by researchers in which firewall logs were analyzed using data mining techniques.

Ucar and Ozhan [4], proposed some of classification algorithms of machine learning to detect the anomalies in firewall rules. In this study large-scale log files are automatically analyzed. The study was implemented using 5,000,000 logs taken from a firewall which analyzed through the supervised learning in WEKA.

Golnabi et al. [2] processed Linux operating system firewall initially with 10 firewall policy rules to generate firewall log dataset which contained 33,172 instances and they selected only the 7 major attributes in the firewall logs. They tried to analyze and detect the firewall rules anomalies using Association Rule Mining (ARM) and Filtering-Rule Generalization (FRG) algorithms, in addition, they also, detected dominant and decaying rules in the firewall policy.

Saboori et al. [6] proposed a model to improve detecting novel anomaly attacks based on data mining techniques, they illustrated how to use Apriori Algorithm to detect anomalies in the network, and prophesied a novel attack to generate a set of real-time rules for the firewall.

In another study, Al-Shaer et al. [7] designed a tool named "Firewall Policy Advisor" in Java for analyzing firewall policies and they tried to use this tool to detect the anomalies in firewall rules which could exist in centralized and distributed firewalls.

Breier and Brani [8], proposed a method to detect anomalies in firewall logs using data mining techniques to generates the rules dynamically from certain patterns in the sample file.

Caruso et al. [9], proposed K-means and EM algorithms to reveal the network daily checklist template. They used a firewall records which contained 76.702 instances only with 8 attributes. Tanna and Ghodasara [10], used Apriori algorithm in WEKA with minSupport = 50% and min confidence = 50% for mining repeated patterns in enormous dataset.

In another paper, Shrivastava and Panda [11], proposed Apriori algorithm with WEKA tool to generate the best association rule in the dataset.

The main aim of this experiment was to present how can Apriori algorithm be employed for mining frequent patterns in firewall logs on WEKA tool. In this experiment, we proposed the use of Apriori algorithm with 50% of threshold to analyze the firewall logs, deduce firewall policy rules for mining frequent itemsets in firewall logs dataset on WEKA tool [12]. It is a very powerful tool that few machine learning platforms offer for the researchers. For our study, we have used only the major six attributes/features for analyzing firewall logs and later could be used for detecting anomalies in firewall rules.

2 Mining Frequent Patterns in Firewall Logs

The main purpose of our study is to take help of data mining techniques such as feature selection, eliminating redundancies and normalization in order to analyze the firewall logs dataset. For that, we used association rule mining (ARM) to analyze the firewall log, deduce firewall policy rules for mining frequent item sets using Apriori Algorithm with WEKA tool.

2.1 Collecting the Dataset

Logs can be generated by several sources such as firewalls, service, specific software or an operating system, etc. Firewall Rules were processed to generate the firewall log files using Snort IDS [13] and TWIDS [14] tool which have been used to collect the activities of 5 people in a period of one week on "weekday morning" from Wednesday to Thursday: 1st-9th March 2017 from a Lab in our Department, with initially 13 firewall rules as in our previous work [15]. We had seven files in ".csv" format. A record in a log file is a packet, which contained the following attributes: Date, Time, Action (Allow, Deny), Source IP, Source port, Destination IP, Destination port, Protocol and another attributes. A line sample of our firewall logs is shown in Table 1.

Table 1. A sample of firewall Logs

02/03/2017,13:16:35,ALLOW,192.168.137.5,49203,182.50.136.237,80,88800003,5868,UDP

2.2 Dataset Preprocessing

A preprocessing step must be performed before any analysis because the log data cannot be used for analysis and mining purposes in the form as it is stored in the log files. The seven log files have been combined into one log file. Redundant data was removed. Preprocessing involves loading a dataset and manipulating

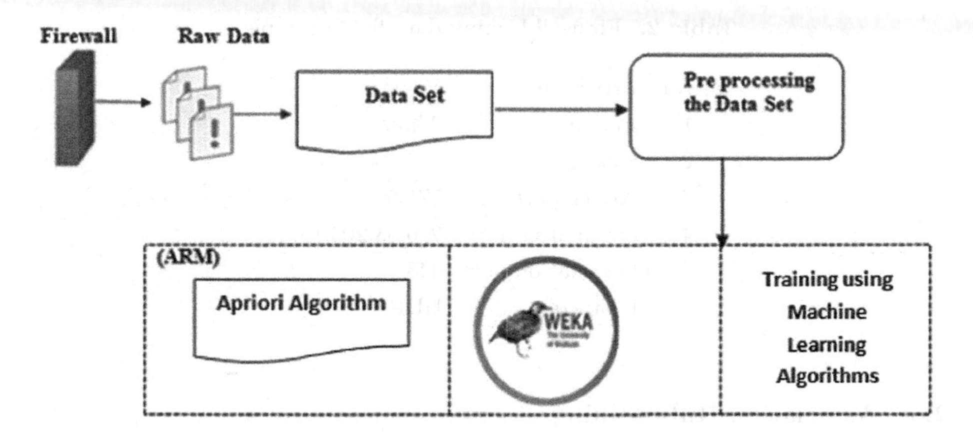

Fig. 1. The framework of the system

the data into a form that you want to work with. This process follows different steps as illustrated in Fig. 1.

In this study, the firewall log files stored on the Gateway PC as a (.CSV) files. Therefore, in the data-preprocessing step, first stores the log records of the firewall through the automatically exported function and then organize the original data into one file which contains the dataset required for the analysis. The testing and learning operations were implemented on the dataset loaded to WEKA and it was observed that the dataset was running correctly.

2.3 Dataset Description

The total number of instances of firewall log file included in training and testing dataset were 94,384 log instances/records with 10 attributes, which was obtained from a Snort IDS and TWIDS tools, initially with 13 firewall policy rules. The log records of the firewall exported through the automatically exported function and then organize the original data into one file which contains the dataset required for the analysis. The data used in the analysis were loaded to WEKA software. The data used in the analysis were loaded to WEKA software. After the dataset has been loaded, testing and learning were implemented properly and the dataset was running effectively.

2.4 Information Gain for Features Selection

For this study, the 10 attributes of the firewall logs dataset were reduced to only the 6 major attributes using the attribute selection filter in WEKA, by making some guesses as to which attribute can be removed. A firewall logs data field sample is shown in Table 2.

Table 2. Firewall logging data field example

No.	Attribute	Value
1	Action	Allow
2	Source IP	192.168.137.5
3	Source port	57127
4	Destination IP	216.58.203.162
5	Destination port	443
6	Protocol	UDP

2.5 Association Rule Mining

Generally, association rule mining (ARM) algorithm is used for discovering the interesting relationships between items and terms in a dataset and find out the frequent associations in large itemsets [16]. The definition of association rules is as follows (Agrawal et al. [17]): If there is an association rule between the two item sets X and Y, they are expressed in the form: X→ Y, where X and Y is a combination of one or more items, and X ∩ Y= ⊘. There are two conditions used to determine the association Whether or not the rule X→Y is established: First, the support is expressed as including (X ∪ Y) the transaction amount. The proportion of material to the total transaction data; The other is confidence, which is expressed as containing The number of transaction data (X ∪ Y) contains the ratio of the number of X transaction data. Support (X ∪Y) is computed using Eq. (1). The formula of confidence (X ∪Y) could be easily obtained using Eq. (2). Where, the total number of transactions in D is represented by $|D|$.

$$Support(X \cup Y) = \frac{|(X \cup Y)|}{|D|} \tag{1}$$

$$confidence(X \rightarrow Y) = \frac{Support(X \cup Y)}{support(X)} \tag{2}$$

2.6 Apriori Algorithm

Apriori is considered as the most basic and simplest algorithms for association rules mining. It is one of the association rule (ARM) algorithms which was proposed by Srikant and Agrawal [17] for mining association rules in datasets. The goal of Apriori algorithm is to find out all the association rules in the dataset having a confidence and support greater than the given thresholds, i.e., the specified minimum confidence and minimum support. Obviously, The two inputs of the Apriori algorithm are the minimum support and the dataset respectively, and the output is the largest item-sets and the best rules drawn from dataset. The process of Apriori algorithm can be described in the following steps:

1. Generates a list of itemsets for all individual items.
2. Scan the transaction records to see which itemsets meet the minimum threshold requirements, and those that do not meet the minimum threshold will be removed.
3. The remaining collections are then combined to produce a set of items containing two elements.
4. Next, rescan the transaction records and remove item sets that do not meet the minimum threshold.
5. This process is repeated until all item sets in the m-th layer containing m elements are removed.

It is important to specify the attributes in the firewall logs dataset as nominal data type to avoid gives their values any significance. In contrast, by using small values for minimum support and minimum confidence, i.e., minimum threshold, Apriori algorithm will find out and generates more rules and vice versa. Therefore, it is important to specify the values of minimum support and minimum confidence according to the number of rules we need to find out from the dataset. Figure 2 illustrated the Pseudo code of Apriori Algorithm.

Input:
- D= Training dataset, the minimum support count threshold.

Output:
- frequent itemsets in the database

steps:
- C_k: Candidate itemset of size k
- L_k: frequent itemset of size k

- L_1 = {frequent items};
- for (k = 1; L k != Ø; k++) do begin
  - C_{k+1} = candidates generated from L k;
  - for each transaction t in database do
    - increment the count of all candidates in C_{k+1} that are contained in t
  - L_{k+1} = candidates in C_{k+1} with min_support
  - end
- return $\cup_k L_k$;

Fig. 2. Apriori algorithm

3 Experimental Analysis

The first step in our experiment was deploying the firewall using Snort IDS and TWIDS in windows 7 platform which have been collected the activities of 5 people in a period of one week on "weekday morning" from Wednesday to Thursday: 1st–9th March 2017 from a Lab in our Department, with initially 13 firewall rules [15]. Then, processed the firewall rules to generate the firewall logs. The total number of instances/records of the dataset included in training and testing stages were 94,384 log instances with 10 attributes/features. The dataset have been trained and tested after that Apriori algorithm have been applied to the trained and tested dataset.

The experiment was evaluated using Apriori algorithm with WEKA on Intel (R) Core(TM) i3CPU@ 2.10 GHz and 4 GB RAM running windows 7. In addition, MATLAB (R2017b) software and Excel program have been used for organization and eliminate the duplications in the dataset. Figure 3 shows a sample of raw logs taken from the firewall logs dataset. Each single record in the firewall log indicates the following corresponding information which includes in the packet header of the packet: Date, Time, Action (Allow, Deny), Source IP, Source port, Destination IP, Destination port, Session ID, Process ID and Protocol (TCP/UDP) as shown in Table 3.

```
08/03/2017,11:37:06,Allow,192.168.137.2,49823,161.69.226.71,443,88800009,5868,UDP
08/03/2017,11:37:06,Allow,192.168.137.2,49824,161.69.226.71,443,88800009,4,UDP
08/03/2017,11:37:06,Allow,192.168.137.2,49825,161.69.226.71,443,88800009,1036,UDP
08/03/2017,11:37:06,Allow,192.168.137.2,49826,161.69.226.71,443,88800009,4,UDP
08/03/2017,11:37:06,Allow,192.168.137.2,60041,216.58.203.162,443,1000009,33,TCP
08/03/2017,11:37:07,Allow,192.168.137.2,49830,104.244.42.193,443,88800009,5292,UDP
08/03/2017,11:37:07,Allow,192.168.137.2,49827,104.244.42.193,443,88800003,5868,UDP
08/03/2017,11:37:07,Allow,192.168.137.2,49829,104.244.42.193,443,88800009,5292,UDP
08/03/2017,11:37:07,Allow,192.168.137.2,49831,104.244.42.193,443,88800009,5292,UDP
08/03/2017,11:37:07,Drop,192.168.137.2,49837,104.113.242.64,443,88800009,848,UDP
08/03/2017,11:37:07,Drop,192.168.137.2,49835,104.113.242.64,443,88800009,4,UDP
08/03/2017,11:37:07,Drop,192.168.137.2,49833,104.113.242.64,443,88800009,1036,UDP
08/03/2017,11:37:07,Drop,192.168.137.2,49838,104.113.242.64,443,88800009,4,UDP
08/03/2017,11:37:07,Drop,192.168.137.2,49834,104.113.242.64,443,88800009,1036,UDP
08/03/2017,11:37:07,Drop,192.168.137.2,49836,104.113.242.64,443,88800009,4,UDP
08/03/2017,11:37:07,Drop,192.168.137.2,49839,104.244.46.135,443,88800009,1036,UDP
08/03/2017,11:37:07,Drop,192.168.137.2,49840,104.244.46.135,443,88800009,4,UDP
08/03/2017,11:37:07,Drop,192.168.137.2,49841,104.244.42.193,443,88800009,1036,UDP
08/03/2017,11:37:07,Allow,192.168.137.2,61072,172.217.27.195,443,88800009,4,TCP
08/03/2017,11:37:07,Allow,192.168.137.2,49828,104.244.42.193,443,1000004,34,TCP
08/03/2017,11:37:07,Allow,192.168.137.2,49832,104.244.42.193,443,88800009,5868,UDP
```

Fig. 3. A sample of firewall log file

The second step of our experiment was processing the dataset of 94,384 records, with 6 attributes to extract the major attributes/features to implement data mining using Apriori algorithm. In order to reduce the number of states for our experiment, some attributes have been removed based on some guesses. Therefore, we considered only the 6 major attributes/features as in the IP packet header. The selected major attributes are: Action (Allow, Deny), Source IP,

Table 3. A list of feature selection by Apriori algorithm

Attributes: 10
Date
Time
Action
Source IP
Source port
Destination IP
Destination port
SID
PID
Protocol (TCP,UDP)

Source port, Destination IP, Destination port, and Protocol (TCP/UDP). The Action field was specified as the class attribute. The result of this experiment was implemented on WEKA with 50% of threshold, i.e., the values of minimum support and minimum confidence were (0.5) for each and the number of association rules was set to 10 rules. The output of association displayed the results of Apriori algorithm in the form of 10 best rules.

Each single record of the firewall logs dataset indicates the following information for each packet: Action (Allow, Drop), source IP address, destination IP, source port, destination port and protocol. Figure 4 presented a sample results of Apriori algorithm association output for the source IP address 192.168.137.2.

Action	'Source IP'	'Source Port'	'Dest IP'	'Dest Port'	Protocol
Allow	192.168.137.2	65485	192.168.137.1	53	UDP
Allow	192.168.137.2	49543	192.168.137.1	443	UDP
Allow	192.168.137.2	49542	192.168.137.1	443	UDP
Allow	192.168.137.2	61308	192.168.137.1	53	TCP
Allow	192.168.137.2	61309	172.217.31.3	443	TCP
Allow	192.168.137.2	63804	192.168.137.1	53	UDP
Allow	192.168.137.2	63805	172.217.31.4	443	UDP
Allow	192.168.137.2	49546	172.217.31.4	443	UDP
Allow	192.168.137.2	49547	172.217.31.4	443	UDP
Allow	192.168.137.2	65159	192.168.137.1	53	UDP
Allow	192.168.137.2	61288	192.168.137.1	53	UDP
Allow	192.168.137.2	61348	192.168.137.1	53	UDP
Allow	192.168.137.2	57118	192.168.137.1	53	UDP
Allow	192.168.137.2	65160	172.217.31.3	443	UDP
Allow	192.168.137.2	57119	172.217.27.194	443	UDP
Allow	192.168.137.2	57120	172.217.26.238	443	UDP
Allow	192.168.137.2	57121	172.217.166.46	443	UDP
Allow	192.168.137.2	49548	172.217.26.227	443	UDP
Allow	192.168.137.2	49549	172.217.166.46	443	UDP
Allow	192.168.137.2	49550	172.217.26.227	443	UDP
Allow	192.168.137.2	49551	172.217.166.46	443	UDP
Allow	192.168.137.2	49545	172.217.31.3	443	TCP
Allow	192.168.137.2	49552	161.69.226.70	443	UDP
Allow	192.168.137.2	57127	216.58.203.162	443	TCP

Fig. 4. Sample of association rules for IP address 192.168.137 after using Apriori

4 Conclusions and Future Works

Firewall is mainly, considered as the most important element in the networks. It is used by organizations to enforce their security policy. However, Managing and customizing firewall rules is a very important aspect in firewall policy, it is extremely complex, costly and error-prone due to the continuous growth of security threats and managing the security rules. WEKA platform allows you to quickly design and run experiments. We can say that Apriori algorithm gives excellent results in mining frequent items in firewall logs. The main aim of this experiment was to present how can Apriori algorithm be employed for mining frequent patterns in firewall logs on WEKA tool. Firewall logs have been analyzed. In addition, we have deduced firewall rules.

The study was done using firewall logs dataset which was collected from a Lab in our department using Snort and TWIDS tools. Apriori algorithm has been used to find out the frequent patterns hidden in the firewall logs dataset. We have used only the major 6 attributes/features out of the overall 10 attributes by making some guesses as to which attribute can be removed. Moreover, the dataset attributes were specified as nominal data type to avoid gives their values any significance.

The result of this experiment was implemented with 50% of threshold, i.e., the values of minimum support and minimum confidence were (0.5) for each and the number of association rules was set to 10 rules. The output of association displayed the results of Apriori algorithm in the form of 10 best rules.

The goal of mining frequent itemset in firewall logs to reduce its number will be implemented in future work due to our research work is in progress for the same.

In future work we will use data mining techniques for detecting anomalies in firewall rules, correct firewall policy rules to make them optimal, generalized and anomaly-free.

References

1. Rizzardi, A.: Security in Internet of Things: networked smart objects. Doctoral thesis, Universitá degli Studi dell'Insubria (2016)
2. Golnabi, K., Min, R.K., Khan, L., Al-Shaer, E.: Analysis of firewall policy rules using data mining techniques. In: 10th IEEE/IFIP Network Operations and Management Symposium, NOMS 2006, vol. 5, pp. 305–315. IEEE (2006). https://doi.org/10.1109/NOMS.2006.1687561. Nagel, W.E., Walter, W.V., Lehner, W. (eds.) Euro-Par 2006. LNCS, vol. 4128, pp. 1148–1158. Springer, Heidelberg (2006). https://doi.org/10.1007/11823285_121
3. Lawal, O.B., Ibitola, A., Longe, O.B.: Analysis and evaluation of network-based intrusion detection and prevention system in an enterprise network using snort freeware. Afr. J. Comput. ICTs. **6**(1), 169–184 (2013)
4. Ucar, E., Ozhan, E.: The analysis of firewall policy through machine learning and data mining. Wirel. Pers. Commun. **96**, 2891 (2017). https://doi.org/10.1007/s11277-017-4330-0
5. Bello-Orgaz, G., Jung, J.J., Camacho, D.: Social big data: recent achievements and new challenges (2015)
6. Saboori, E., Parsazad, S., Sanatkhani, Y.: Automatic firewall rules generator for anomaly detection systems with Apriori algorithm. In: 3rd International Conference on Advanced Computer Theory and Engineering ICACTE, pp. 57–60 (2010)
7. Al-Shaer, E., Hamed, H., Boutaba, R., Hasan, M.: Conflict classification and analysis of distributed firewall policies. IEEE J. Sel. Areas Commun. **23**(10), 2069–2084 (2005). https://doi.org/10.1109/JSAC.2005.854119
8. Breier, J., Branišová, J.: A dynamic rule creation based anomaly detection method for identifying security breaches in log records. Wirel. Pers. Commun. (2015). https://doi.org/10.1007/s11277-015-3128-1

9. Caruso, C., Malerba, D., Papagni, D.: Learning the daily model of network traffic. In: Hacid, MS., Murray, N.V., Raś, Z.W., Tsumoto, S. (eds.) ISMIS 2005. LNCS, vol. 3488, pp. 131–141. Springer, Heidelberg (2005). https://doi.org/10.1007/11425274_14
10. Tanna, P., Ghodasara, Y.: Using Apriori with WEKA for frequent pattern mining. arXiv preprint arXiv:1406.7371 (2014)
11. Shrivastava, A.K., Panda, R.N.: Implementation of Apriori algorithm using WEKA. KIET Int. J. Intell. Comput. Inform. **1**(1), 4 (2014)
12. URL download WEKA: http://www.cs.waikato.ac.nz/ml/weka/
13. Snort. An open source network intrusion detection system. http://www.Snort.org/
14. TWIDS Tool: TWIDS. http://twids.cute.edu.tw/en
15. As-Suhbani, H., Khamitkar, S.D.: Enhancing snort IDS performance using TWIDS for collecting network logs dataset. Int. J. Res. Adv. Eng. Technol. 42–45 (2017). https://doi.org/10.22271/engineering
16. Kotsiantis, S., Kanellopoulos, D.: Association rules mining: a recent overview. GESTS Int. Trans. Comput. Sci. Eng. **32**(1), 71–82 (2006)
17. Agrawal, R., Imielinski, T., Swami, A.: Mining association rules between sets of items in large databases. In: Proceedings of the: Webb. G.I, Association Rules (1993). In Handbook

Application of Decision Tree for Predictive Analysis of Student's Self Satisfaction with Multivariate Parameters

Aniket Muley[1(✉)], Parag Bhalchandra[2], and Govind Kulkarni[2]

[1] School of Mathematical Sciences, SRTM University, Nanded, MS, India
aniket.muley@gmail.com
[2] School of Computational Science, S.R.T.M. University, Nanded, India
srtmun.parag@gmail.com, govindcoolkarni@gmail.com

Abstract. Decision trees are better known for automatic feature selection in predictive analysis. There is no need of variable transformation while processing data in decision tree as such trees work on selection of important variables as parent nodes and they act as split node. An attempt is made herein to enumerate decision tree over customized educational dataset. The work is based on the hypothesis that distinctiveness, routine and sensitivity related variables of students have close association together with self satisfaction and motivation. This paper demonstrates better feature selection using tree analytics. Experimental analytics were carried out with R software package.

Keywords: Educational analytics · Decision trees · Classification · Multivariate analysis

1 Introduction

The educational data mining is one of the up-and-coming areas in the research arena related to academics where traditional and hybrid algorithms are implemented to discover myths from educational data. Numerous applications of educational data mining are in place which also has add-on component of machine learning [4,5]. The core research is student centric as students are always centring attention in educational systems. There has been a significant contemporary research literature available across the internet including the work of Rawat and Malhan [15] and Kamal and Ahuja [11] have tried to associate the impact of non academic variables on academic performance of students using machine learning. Their questionnaire has used factors such as demographic, social and behavioral. Their results have revealed family's financial health, educational profile and communication with teacher were the prime factors or the influential aspects. The other promising work includes the work of Hussain et al. [10] which tried for implementation association rules using this data. Various clustering and classification methods were used for comparison over

K. C. Santosh and R. S. Hegadi (Eds.): RTIP2R 2018, CCIS 1037, pp. 572–579, 2019.
https://doi.org/10.1007/978-981-13-9187-3_51

the dataset through the ORANGE, WEKA and R Studio software. The Singh et al. [17] tried for classification algorithms for diagnosis in medical data. The Ahuja et al. [1] made empirical analysis of various clustering and classification algorithms over an educational dataset and analyzed the same using machine learning. The Angiani et al. [3] made analytical escalations for students' behaviors and their results. These all cited works have encouraged us to analyze educational databases at our end. This work is an illustration of a multiparty interdisciplinary work and the key objective of this study is to implement decision trees for automatic feature selection for some of the variables associated with students. The broader objective is to use data mining functions to study and evaluate student academic data. We hypotheses that there are other non academic factors associated with the satisfaction of the students. The satisfaction is directly proportionate to performance likewise. This could possibly be the reason why higher educational performance is not significant even though Government spend of GDP on higher education [12]. Among other known methods of data mining [6,9], we insisted for a decision tree model. Our primary aim was for classification of important and less important parameters. A decision tree is a data structure based binary tree which is commonly seen as a flow chart. It has branches and nodes. The nodes can be internal or external. The internal nodes are denoted by rectangles, and terminal nodes are denoted by ovals. All internal/non leaf nodes have two or more children nodes. The other related structural philosophy including split, provision for edges, class levelling, learning and classification is borrowed from standard literatures on decision trees [6,16]. The given entropy as appraise of the impurity is then defined. This calculation is called information gain. The section two below summarizes all this considerations.

2 Research Methodology

A database of students from computational sciences was collected using closed questionnaire method. It had 360 records, each having 46 fields [2,8,9]. Experimental algorithms were implemented using Rattle GUI of R software [7,13,14]. The traditional algorithm is implemented in the rpart package. Standard ETL-Extract, Transform and Load [7,13–15] cycle of pre-processing of data was followed. A decision tree model is one of the most common data mining models. It is popular because the resulting model is easy to understand. The algorithms use a recursive partitioning approach. The traditional algorithm is implemented in the rpart package. It is comparable to CART and ID3/C4. The conditional tree algorithm is implemented in the party package. It builds trees in a conditional inference framework.

3 Experimentations and Discussions

Experimentations were carried out to devise out decision tree. The main concern behind experimentations was to find relationship between considered variables. To draw the decision tree, rattle uses rpart command from rpart package.

We obtained the rules associated to the tree as summarized in Table 1. It represents the root node, level of trees, split of the observations, loss of frequency at each level split and corresponding probability values. The * denotes terminal node. Total number of population selected was n = 251.

Table 1. Summary of the decision tree model for classification

Variables	Nodes	Children generated	Probability
Root	251	(64,1)	(0.25498008 0.74501992)
FT FRIEND $\geq$ 1.5	56	(23,1)	(0.41071429 0.58928571)
M.M. < 0.5	12	(1,0)	(0.91666667 0.08333333)
M.M. $\geq$ 0.5	195	(41,1)	(0.27272727 0.72727273)
OWN NOTES $\geq$ 0.5	9	(4,0)	(0.21025641 0.78974359)
M.JOB $\geq$ 4.5	186	(36,1)	(0.55555556 0.44444444)
M.JOB $\leq$ 4.5	7	(1,1)	(0.19354839 0.80645161)

The terminal node variables actually used in the tree construction are: M.JOB, M.M and OWN NOTES. Initially, the root node error is 0.25498 (n = 251). The Table 2 shows the error matrix for the Decision Tree. Tables 3 and 4 can be consulted for validating, training, testing and full dataset models proportions and the observed overall errors.

Table 2. Error matrix

Actual		Predicted		Error
		0	1	
Validation	0	7.5	13.2	63.6
	1	1.9	77.4	2.4
Testing	0	3.6	32.7	90
	1	7.3	56.4	11.4
Training	0	6.4	19.1	75
	1	12.0	72.5	2.7
Full	0	6.1	20.3	76.8
	1	2.8	70.8	3.8

Table 3. The error occurring in the models (%)

Train	Test	Validate	Full
21.1%	40%	15.1%	23.1%

The area under the ROC curve for the rpart model is computed for train set data (70%), test (15%), validation (15%) and full (100%) data. The ROC accuracy of the result was summarized in below Table 4. It results validates simply states the random selection is responsible for the accuracy level of the data.

Table 4. ROC values in the models

Train	Test	Validate	Full
0.6403	0.6613	0.5043	0.62

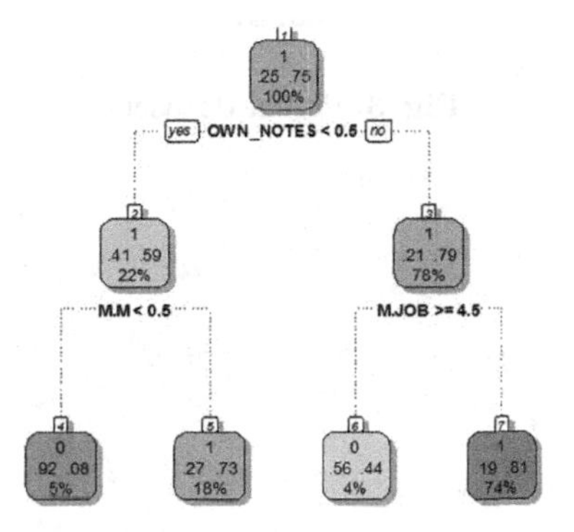

Fig. 1. Decision tree

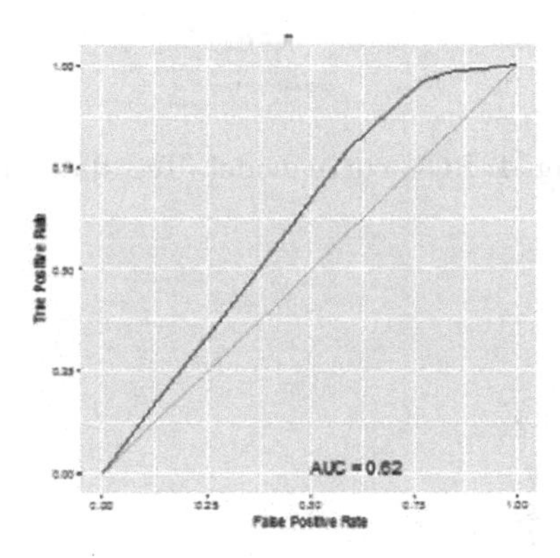

Fig. 2. ROC curve model (Full dataset)

Figures 2, 3, 4 and 5 indicates the area under ROC for given dataset. It simply elaborates the accuracy of the model in the form of probability. As its value closer to 1, it gives more suitability about the proposed model. Figures 6, 7, 8 and 9 represents the precision chart, which explores the precision level of the given decision tree (Fig. 1). Figure 9 it self shows the efficiency of the stated decision tree.

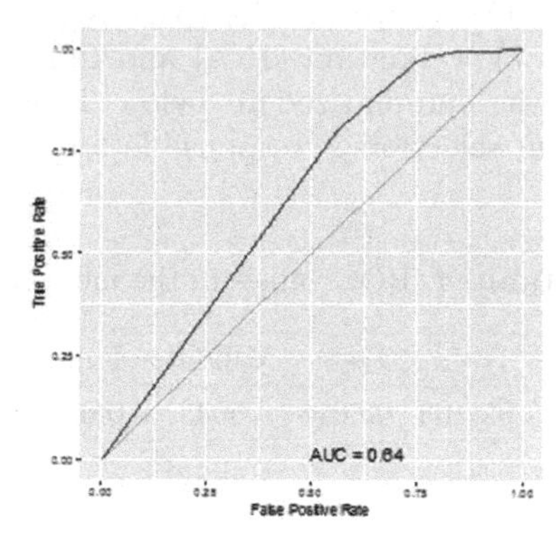

Fig. 3. (Train dataset)

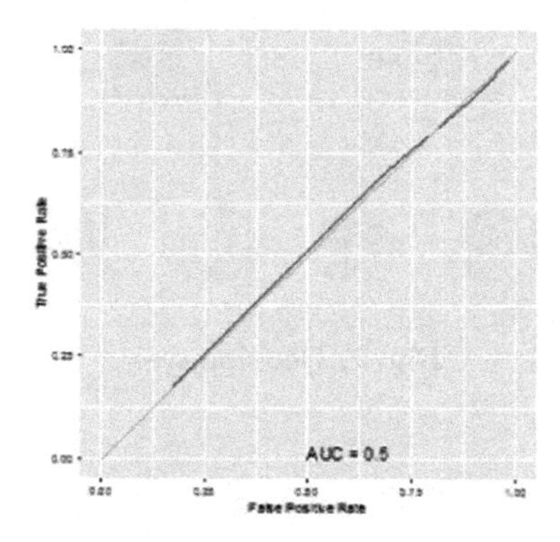

Fig. 4. ROC curve model (Test dataset)

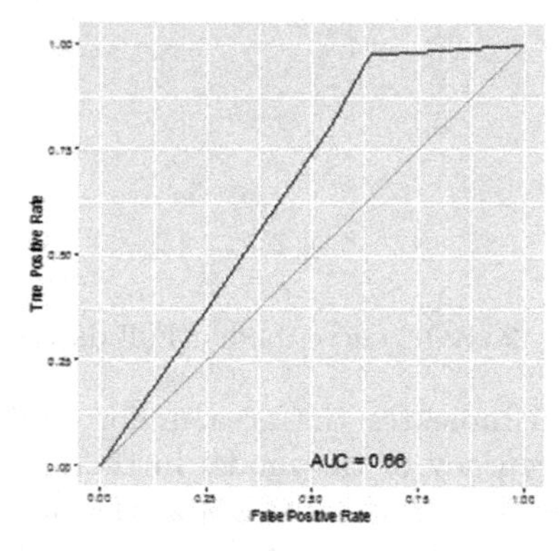

Fig. 5. ROC curve model (Validate dataset)

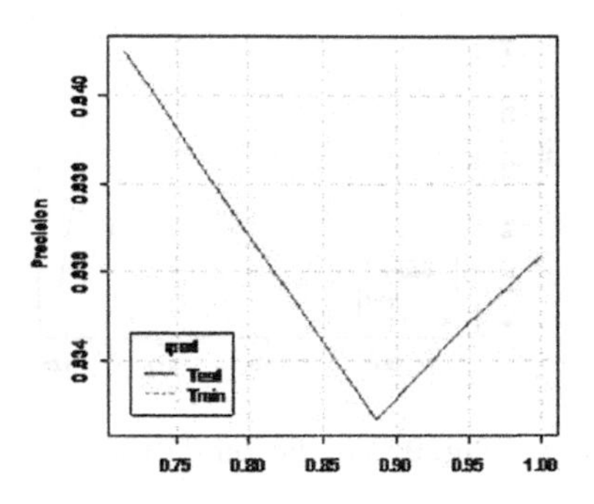

Fig. 6. Full data

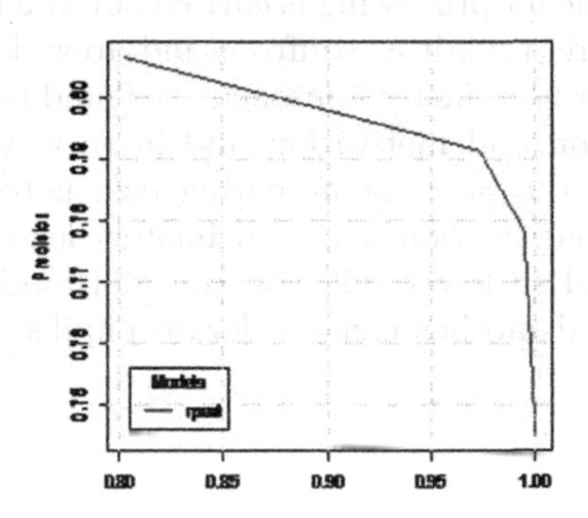

Fig. 7. Train-set

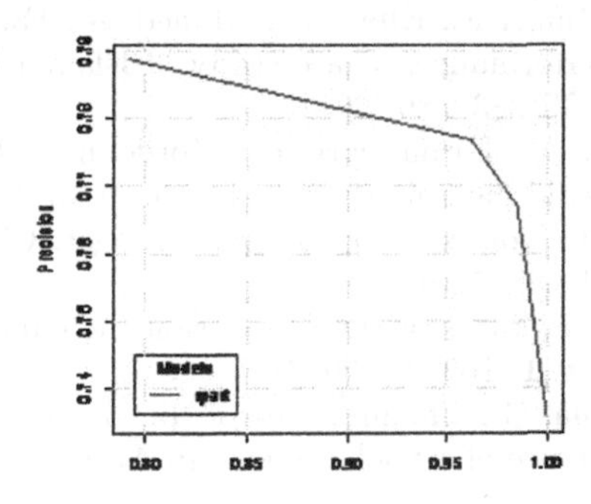

Fig. 8. Test-set

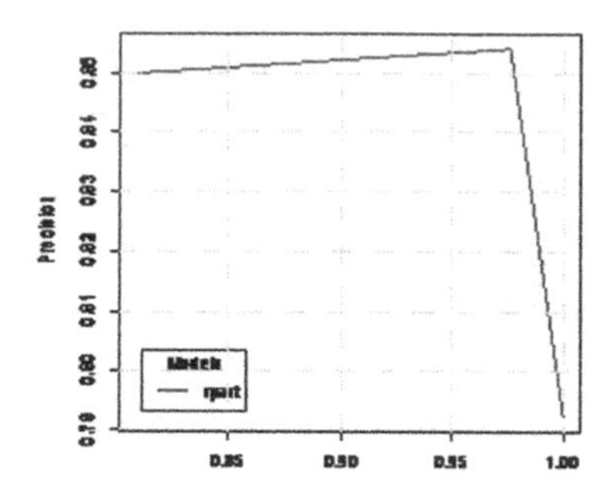

Fig. 9. Validation dataset

4 Conclusions

The application of data mining in educational domain is increasing with rapid fire. Although no significant processing is carried out at our university's end. The study, as a joint venture, took it as confront and drew decision tree for selected parameters. Using interdisciplinary approach, we found multivariate relationship between self satisfaction and motivation and in turn with performance. This study highlights that, the habit of preparing own notes significantly matters and is related to self satisfaction which ultimately leads to good performance of the students. Second such variable boosting self motivation is the mother's education. An educated mother can accelerate pupil's interest in studies and thus in performance.

References

1. Ahuja, R., Jha, A., Maurya, R., Srivastava, R.: Analysis of educational data mining. In: Yadav, N., Yadav, A., Bansal, J.C., Deep, K., Kim, J.H. (eds.) Harmony Search and Nature Inspired Optimization Algorithms. AISC, vol. 741, pp. 897–907. Springer, Singapore (2019). https://doi.org/10.1007/978-981-13-0761-4_85
2. Ali, S., Haider, Z., Munir, F., Khan, H., Ahmed, A.: Factors contributing to the students academic performance: a case study of Islamia University Sub-Campus. Am. J. Educ. Res. **1**(8), 283–289 (2013)
3. Angiani, G., Ferrari, A., Fornacciari, P., Mordonini, M., Tomaiuolo, M.: Real marks analysis for predicting students' performance. In: Di Mascio, T., et al. (eds.) MIS4TEL 2018. AISC, vol. 804, pp. 37–44. Springer, Cham (2019). https://doi.org/10.1007/978-3-319-98872-6_5
4. Bratti, M., Staffolani, S.: Student time allocation and educational production functions. Ann. Econ. Stat. **1**, 103–40 (2013)
5. Considine, G., Zappalà, G.: The influence of social and economic disadvantage in the academic performance of school students in Australia. J. Sociol. **38**(2), 129–48 (2002)
6. Dunham, M.H.: Data Mining: Introductory and Advanced Topics. Pearson Education India (2002)
7. Field, A.: Discovering Statistics Using IBM SPSS Statistics. Sage (2013)

8. Graetz, B.: Socio-economic status in education research and policy in John A., et al. Socio-economic status and School Education DEET/ACER Canberra. J. Pediatr. Psychol. **20**(2), 205–16 (1995)
9. Han, J., Pei, J., Kamber, M.: Data Mining: Concepts and Techniques. Elsevier, Amsterdam (2011)
10. Hussain, S., Atallah, R., Kamsin, A., Hazarika, J.: Classification, clustering and association rule mining in educational datasets using data mining tools: a case study. In: Silhavy, R. (ed.) CSOC2018 2018. AISC, vol. 765, pp. 196–211. Springer, Cham (2019). https://doi.org/10.1007/978-3-319-91192-2_21
11. Kamal, P., Ahuja, S.: Academic performance prediction using data mining techniques: identification of influential factors effecting the academic performance in undergrad professional course. In: Yadav, N., Yadav, A., Bansal, J.C., Deep, K., Kim, J.H. (eds.) Harmony Search and Nature Inspired Optimization Algorithms. AISC, vol. 741, pp. 835–843. Springer, Singapore (2019). https://doi.org/10.1007/978-981-13-0761-4_79
12. Online resources information. https://economictimes.indiatimes.com
13. Pritchard, M.E., Wilson, G.S.: Using emotional and social factors to predict student success. J. Coll. Stud. Dev. **44**(1), 18–28 (2003)
14. R software information. https://www.r-project.org
15. Rawat, K.S., Malhan, I.V.: A hybrid classification method based on machine learning classifiers to predict performance in educational data mining. In: Krishna, C.R., Dutta, M., Kumar, R. (eds.) Proceedings of 2nd International Conference on Communication, Computing and Networking. LNNS, vol. 46, pp. 677–684. Springer, Singapore (2019). https://doi.org/10.1007/978-981-13-1217-5_67
16. Rokach, L., Maimon, O.Z.: Data Mining with Decision Trees: Theory and Applications. World Scientific (2008)
17. Singh, G.K., Jain, R.K., Dubey, P.: Study of classification techniques on medical datasets. In: Iyer, B., Nalbalwar, S.L., Pathak, N.P. (eds.) Computing, Communication and Signal Processing. AISC, vol. 810, pp. 557–565. Springer, Singapore (2019). https://doi.org/10.1007/978-981-13-1513-8_57

Simplified Deterministic Finite Automata Construction Algorithm from Language Specification

Darshan D. Ruikar[1(✉)], Amruta D. Ruikar[2], Suhas G. Kulkarni[3], and Ravindra S. Hegadi[1]

[1] Department of Computer Science, Solapur University, Solapur 413255, Maharastra, India
{ddruikar,rshegadi}@sus.ac.in

[2] Department of Computer Technology, Solapur Education Society's Polytechnic, Solapur 413002, Maharastra, India
amrutashegadar8887@gmail.com

[3] Department of BCA, Prin. K.P. Mangalvedhekar Institute of Management, Career Development & Research, Solapur 413001, Maharastra, India
suhas_gkulkarni@rediffmail.com

Abstract. Deterministic Finite Automata (DFA) has several real-life applications. It is a useful tool in diversified areas like pattern matching, pattern recognition, control theory, text editors, lexical analyzer, and models of software interfaces. Several methods are developed to construct a DFA using Regular Expression (RE), set of strings, derivations in RE and regular grammar. The motivation for this paper is the observation that students find DFA construction as a tedious task in the early stage of learning. This paper provides a simple solution to the standard computation-theory problem that asks a student to construct a DFA for given language description (i.e., validation rules). The presented paper keeps the concerns of beginner level students at the core.

Keywords: Deterministic Finite Automata · Regular expression · Validation rules

1 Introduction

The importance of DFA in computer science education hardly needs justification. Besides being part of almost all computer science curriculum, the DFA has several real-life applications like compiler construction, robot (BOT) behavior, state transition charts, designing digital circuits and network protocols. Further, DFAs are largely used in linguistics, computer graphics, knowledge engineering and game theory. In word processing-based applications like string recognition, string matching, and editing, spell checker and advisor, till the date DFA is the ultimate solution. DFA finds its extensive application in pattern recognition and pattern mating that is to check whether the given pattern 'p' is available in

K. C. Santosh and R. S. Hegadi (Eds.): RTIP2R 2018, CCIS 1037, pp. 580–588, 2019.
https://doi.org/10.1007/978-981-13-9187-3_52

provided text [5]. Apart from this, vast vocabulary set in natural and formal languages can be effectively described by using DFA [7].

According to the formal language theory, a regular expression (RE) and DFA are the two different representations of regular language [4]. RE acts as a generative device, i.e., expansion of RE results in a set of string that set is known as regular language set. Whereas DFA can be used as accepter, validator or recognition system. It informs whether the supplied string belongs to language or not. That is DFA is the tool to solve the membership problem of regular language [11].

Constructing precise and optimum DFA for given language description is a relatively harder problem. In the initial stage, students find difficulty in constructing DFA for a given set of language specification rules [14]. Deciding what steps to follow and its correct order of execution is a time-consuming task, and it is a matter of practice and experience. This paper presents a simple DFA construction algorithm to construct DFA for a given set of validation rules.

The paper is organized as follows: Sect. 2 briefly describes related preliminaries. Other related work to construct DFA is discussed in Sect. 3. Section 4 presents the proposed algorithm and its detailed explanation. Section 5 evaluates the proposed algorithm with the help of counterexamples. Conclusion is presented in Sect. 6.

2 Preliminaries

This section provides a preliminary background on DFA and related terms. All the notations and definitions in this section are followed from [8,9].

The symbol is an abstract entity. Like point in mathematics symbols can not be defined formally. Alphabet is certain set of symbols for some language. Formally it is denoted by $\sum$. The string is a finite sequence of symbols from some alphabet set. Language is set of string over alphabet (represented by $\sum^*$). Regular language, context-free language, context-sensitive language, recursive language and recursively enumerable language are the five different types formal languages defined by Chomsky [6]. In all aforementioned languages, the regular language is a most intersected and restricted language. It can be effectively stated by certain acceptance specifications. A regular expression (RE) is an algebraic representations to represent these specifications. RE is a finite sequence of operators and operands. It has only three operators: '+' (Union), '.' (Concatenation) and '*' (Kleen closure) and operands are input symbols from $\sum$. Each regular expression can be converted to equivalent DFA.

More generally, finite automata are represented by state transition diagram or transition graph. A model of state transition diagram is shown in Fig. 1. It has finite number states and transitions. Transitions from state to state are occurred due to input symbol is chosen from $\sum$. A state transition diagram has two distinct types of states: the final state (represented by double concentric circles) and non-final state (represented by a single circle). Each transition starts its execution from an initial state (represented by a circle with an arrow pointed

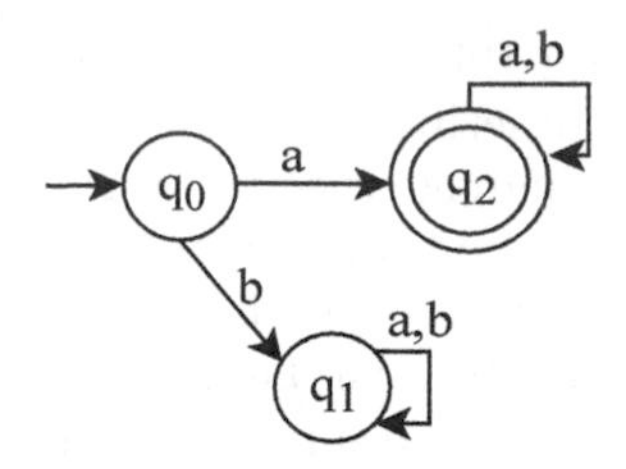

Fig. 1. State transition diagram

towards it. More generally that arrow is labeled by 'start'). During string validation process with DFA, every time a character is consumed, and there is only one specific corresponding state transition. The time complexity of state transition procedure shows linear growth and is proportional to the length of the target character string. At the end of the string, if finite automata halt at final state then that string is valid (i.e., it satisfies all the provided validation rules) or else string is get rejected.

Definition 1. *Deterministic Finite Automata*

DFA is 5 tuple M = (Q, $\sum, \delta, q_0$, F) where

1. Q = finite non-empty set of states
2. $\sum$ = finite non-empty set of input symbols (alphabet set)
3. δ = state transition function ($\delta : Q \times \sum \rightarrow Q$)
4. $q_0 \in Q$ = an initial state
5. $F \subseteq Q$ = finite non-empty set of final states[1]

3 Related Work

Daciuk et al. [7] presented a new algorithm to construct an incremental, minimal, acyclic DFA for language dictionary trie from a set of strings. The proposed algorithm is capable of constructing DFA for both sorted and unsorted list. Berry et al. [3] Presented an elegant algorithm to construct DFA from RE based on derivations and marking of the regular expression. Derivations of RE is responsible for converting RE to DFA and marking scheme is used to identify distinct input symbol from RE. Antimirov et al. [1] presented a partial derivative based RE to DFA conversion algorithm. The presented algorithm is an improved version of the derivative invented by Brzozowski. Bhargava et al. [4] described time efficient method with time complexity O(nlog$_e$n) to convert given RE to DFA without using intermediate NFA. Tie et al. [14] proposed concept and formula for checking the similarity between regular expressions. In addition to this, a grouping algorithm-based DFA construction technique is also presented. To comment on the correctness of proposed techniques, a large number of tested data is

[1] In some references, F is also denoted as A set of accepting states.

collected, and clustering analysis is performed. By confirming results, they concluded that grouping algorithm supervised by similarity checking could generate optimal DFA. Zafar et al. [15] presented a DFA construction procedure from a regular expression. At first accepted regular expression is converted to abstract syntax tree then that tree is converted to transition diagram to construct DFA. The formal specification is described using Z notation and correctness is analyzed by Z/Eves toolset. Along with this, Regular grammar to minimal DFA conversion algorithm is proposed in [8]. The proposed algorithm uses two function of LR parser i.e. CLOSURE( ) and GOTO( ) to construct minimal DFA. In addition to this, two applications: first the ambiguity test for given grammar and second membership problem of regular set are also discussed.

DFA construction and minimization based on learning from example is proposed in [10]. On the other hand, fill climbing with the heuristically guided approach based method is adapted to construct and minimize the DFA in [13]. Adaptive search method is adapted to avoid suboptimal results. The proposed method is demonstrated with the help of fourteen counterexamples are (i.e. seven simple examples and seven its reversal). Constructed DFA correctly accept strings in right list and rejects the stings in the wrong list. Multi-start, random hill climber evolutionary algorithm that uses smart state labeling is implemented in [2] to learn DFA construction through examples.

In the literature, various successful research attempts are made to construct DFA. Most researchers follow one of two alternatives: first to convert given regular expression to DFA with or without intermediate NFA and the second alternative is to train a neural network to simulate the behavior of DFA. However, minimal attempts are made find out some provision to train students how to construct a DFA from given language specification. One such alternative is discussed in the paper.

4 Proposed DFA Construction Algorithm

The proposed DFA construction method is the simple and effective solution for the learners to construct DFA for elementary validation rules. That is validation rule must not contain any intricate words such as, ‘or’, ‘and’, ‘and not having’. For such validation rules, authors have already devised a divide and conquer-based DFA construction algorithm [12].

Algorithm 1 explains the DFA construction process in detail. The algorithm begins it execution by accepting validation rule. As a first step, it prepares two sets L and $\bar{L}$ having valid and invalid strings respectively. Consider the first string in set L as a minimal string that satisfies all validation rules and draws the basic finite using the sixth step in the Algorithm 1. To add remaining transitions for remaining input symbols from each state, the loop in step fourteen is iterated several numbers of times until DFA become complete. The wholly constructed DFA has exactly one transition for each input symbol from every state. That is, if in resultant DFA there are ‘m’ states and ‘n’ input symbols then it must have ‘m × n’ distinct transitions.

Algorithm 1. ConstructDFA (P)

Require: P // *Specification rule*
Ensure: DFA M (Q,$\sum, \delta, q_0$,F)
1: Determine sets L and $\overline{L}$ having some valid and invalid strings according to supplied specification rules respectively;
2: x:= first string L; // *x is minimum string*
3: $|x|$:=number of symbols in x; // $|x|$ *is string length*
4: Minimum number states in DFA := $|x| + 1$;
// *Draw basic Transition diagram having same number states calculated in step 2.*
// *In that first state is initial and last one is final.*
5: i:=1;
6: while (i $\leq |x|$) do
7: Add transition from state i to state i+1;
8: Transition label := x[i];
9: i := i + 1;
10: end while
11: if($\epsilon \in$ L)then
12: initial state is final state;
13: end if
// *To add remaining transitions to each state*
14: do
15: check weather is self loop possible to current state for current input symbol;
16: if no then check weather one step backtracking possible;
17: if no then continue backtracking till possible;
18: if the backtracking leads to reject state never do backtracking;
19: if backtracking is not possible add new state;
20: while (DFA become complete);
21: Return (M (Q,$\sum, \delta, q_0$,F))

The next Section provides the detailed explanation of the proposed algorithm with the help of counterexamples.

5 Evaluation

The correctness of proposed algorithm is illustrated by solving three counterexamples.

Example 1. Construct a DFA over {0, 1}, where all string starts with 0. (i.e. L = {w $\in \sum^*$/w starts with 0 over {0, 1}}).
At first, analyze the given validation rule carefully and construct two sets L and $\bar{L}$, one having valid and other having invalid strings.
L = {0, 00, 000, 01, 011, 0111, 010, 0101, 0011, ...}
$\bar{L}$ = {1, 11, 111, 10, 1000, 101, 1010, 1100, ...}
First string in L is minimum string (i.e. 0) and its length is one. So the minimum number of states in DFA is two. Now draw the basic transition diagram having two states q_0 and q_1 and draw a transition labeled 0 between them. State q_0

is initial whereas state q_1 in final state. To make DFA complete, still three transitions need to be added. These are (1) 1 form state q_0; (2) 0 from state q_1; and (3) 1 from state q_1. As state q_0 is the first state, symbols occurred at state q_0 are considered as the first occurrence of that symbol. According to the validation rule, each string must start with 0. The first occurrence of symbol 1 to state q_0 will violate this rule. Hence the transition labeled one from state q_0 will get rejected. To add this transition, there is need to add a new state q_2 which will treat as reject state[2].

Now for adding second and third transitions, according to loop in step fourteen in Algorithm 1, a direct self loop of input symbols 0 and 1 is possible. Because, as per given validation rule, there is restriction on occurrence of first input symbol in the string, contrary to this, no validation rule exists for last symbol in the sting hence a self-loop is possible for each input symbol at state q_1. The complete transition diagram (i.e. DFA) is shown in Fig. 2.
Let M_1 = (Q, $\sum, \delta, q_0$, F) be the required DFA for given language is as shown in Fig. 2(a).
where Q = $\{q_0, q_1, q_2\}$, $\sum$ = {0, 1}, $q_0 = q_0$, F = $\{q_1\}$ and transition table δ is shown in Fig. 2(b).

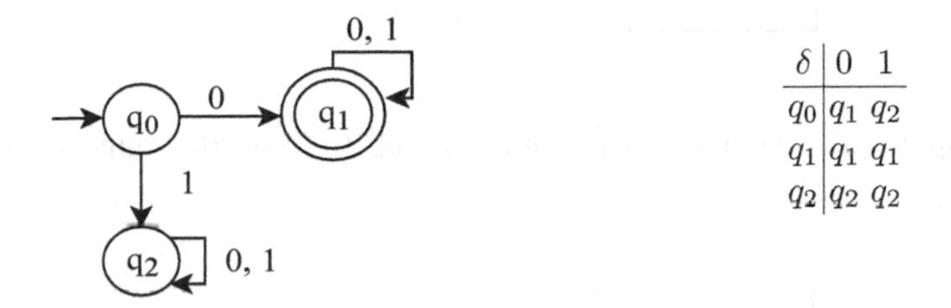

δ	0	1
q_0	q_1	q_2
q_1	q_1	q_1
q_2	q_2	q_2

Fig. 2. (a) Transition diagram for M_1 and (b) Transition table for M_1

Example 2. Construct a DFA over {0, 1}, where all string ends with 1. (i.e. L = {w $\in \sum^*$/w ends with 1 over {0, 1}}).
Same like Example 1, two sets are constructed L and $\bar{L}$, one having valid and other having invalid strings by analyzing the given validation rules.
L = {1, 01, 001, 11, 111, 101, 0101, 0011, ...}
$\bar{L}$ = {0, 00, 000, 10, 1000, 101, 1010, 1100, ...}
Here also, the first string in L is minimum string (i.e. 1), having string length one and the minimum number of states in DFA is two. The basic transition diagram is drawn having two states (q_0, q_1) and a transition is drawn having label 1 between them. To make DFA complete, still three transitions need to be added. These are (1) 0 form state q_0; (2) 0 from state q_1; and (3) 1 from state

[2] Reject state always has self-loop of each input symbols, as it does not show progress (i.e., no transition leads to final state).

q_1. In this example, validation rules are existing at last, opposite to this, no validation rules are exists at beginning, hence a self of symbol 0 can be possible to state q_0. Same like this, self loop of 1 to state q_1 is also valid, because string can be ends with 1. To add remaining transition i.e. 0 from state q_1, self loop is not possible because after adding a self loop DFA will accept the strings those are ends with 0 also, this will violates a given validation rule. Hence to add this transition, according to loop in step fourteen in Algorithm 1 the transition for input symbol 0 from state q_1 will be backtracked to state q_0. The complete transition diagram (i.e. DFA) is shown in Fig. 3.
Let M_2 = (Q, $\sum, \delta, q_0$, F) be the required DFA for given language is as shown in Fig. 3(a).
where Q = $\{q_0, q_1, q_2\}$, $\sum$ = {0, 1}, $q_0 = q_0$, F = $\{q_1\}$ and transition table δ is shown in Fig. 3(b).

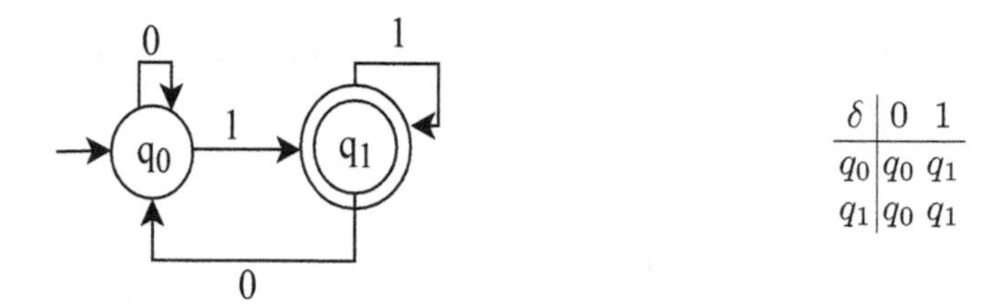

δ	0	1
q_0	q_0	q_1
q_1	q_0	q_1

Fig. 3. (a) Transition diagram for M_2 and (b) Transition table for M_2

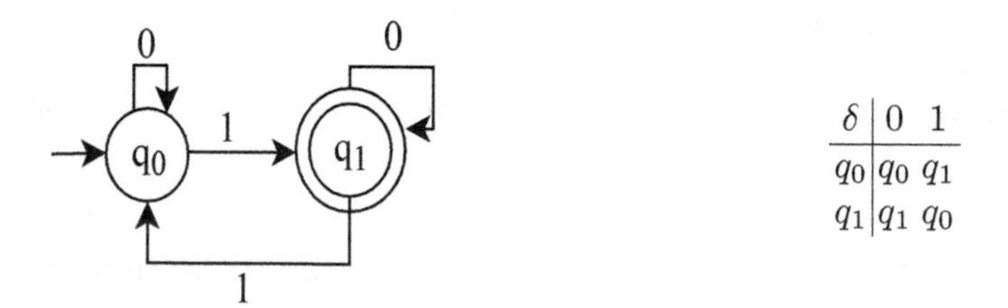

δ	0	1
q_0	q_0	q_1
q_1	q_1	q_0

Fig. 4. (a) Transition diagram for M_3 and (b) Transition table for M_3

Example 3. Construct a DFA over {0, 1}, where all string must have even number of 1. (i.e. L = $\{w \in \sum^* / w$ has even occurrences of 1 over $\{0, 1\}\}$).
In this example zero occurrence of symbol 1 that is empty string (ϵ) is acceptable as zero is treated as even number. Hence to construct DFA for this example, according to step eleven in Algorithm 1 first (i.e. initial) state q_0 itself is final state to accept empty (null) string ϵ. Now the input symbol 1 occurred from

state q_0 will transited to state q_1 as it is first occurrence of symbol 1. Transition for symbol 1 from state q_1 is backtracked to state q_0 as it second (i.e. even) occurrence of symbol 1. In the given validation rule there is no restriction for occurrences of symbol 0 from any state, hence directly a self loop of symbol 0 is added for both states q_0 and q_1. The complete transition diagram (i.e. DFA) is shown in Fig. 4.
Let M_3 = (Q, $\sum, \delta, q_0$, F) be the required DFA for given language is as shown in Fig. 4(a).
where Q = $\{q_0, q_1, q_2\}$, $\sum$ = {0, 1}, $q_0 = q_0$, F = $\{q_1\}$ and transition table δ is shown in Fig. 4(b).

6 Conclusions

We investigated and tried to find a simple solution to the DFA construction problem. Proposed tool is evaluated by solving three counterexamples. We come across few conclusions while solving these counterexamples. These conclusions are (1) if validation rules exist in the beginning then a self-loop can be added for remaining input symbols to the last state; (2) same like above, if validation rules exist at the end then a self-loop can be added for remaining input symbols to the first state; and (3) violation of rules in the beginning results rejection of the transition whereas violation of rules at later states may lead to backtracking.

Along with this, we ask students to use the proposed tool to construct DFA. According to students, feedback proposed tool is simple to construct DFA, and they are satisfied with the tool. In future we are planing to implement the proposed algorithm. In addition this, we are planning to combine proposed algorithm and the algorithm proposed in [12] to develop effective DFA construction technique for any sort of language specification rules.

Acknowledgements. Authors thank the Ministry of Electronics and Information Technology (MeitY), New Delhi for granting Visvesvaraya Ph.D. fellowship through file no. PhD-MLA\4(34)\201-1 Dated: 05/11/2015.

References

1. Antimirov, V.: Partial derivatives of regular expressions and finite automaton constructions. Theoret. Comput. Sci. **155**(2), 291–319 (1996)
2. Ben-David, S., Fisman, D., Ruah, S.: Embedding finite automata within regular expressions. Theoret. Comput. Sci. **404**(3), 202–218 (2008)
3. Berry, G., Sethi, R.: From regular expressions to deterministic automata. Theoret. Comput. Sci. **48**, 117–126 (1986)
4. Bhargava, S., Purohit, G.: Construction of a minimal deterministic finite automaton from a regular expression. Department of Computer Science, Banasthali University (2011)
5. Brzozowski, J.A., Davies, S., Madan, A.: State complexity of pattern matching in finite automata. arXiv preprint arXiv:1806.04645 (2018)

6. Chomsky, N.: On certain formal properties of grammars. Inf. Control **2**(2), 137–167 (1959)
7. Daciuk, J., Mihov, S., Watson, B.W., Watson, R.E.: Incremental construction of minimal acyclic finite-state automata. Comput. Linguist. **26**(1), 3–16 (2000)
8. Kumar, K.S., Malathi, D.: A novel method to construct deterministic finite automata from a given regular grammar. Int. J. Sci. Eng. Res. **6**(3), 106–111 (2015)
9. Martin, J.C.: Introduction to Languages and the Theory of Computation, vol. 4. McGraw-Hill, New York (1991)
10. Parekh, R., Honavar, V.: Learning DFA from simple examples. In: Li, M., Maruoka, A. (eds.) ALT 1997. LNCS, vol. 1316, pp. 116–131. Springer, Heidelberg (1997). https://doi.org/10.1007/3-540-63577-7_39
11. Rabin, M.O., Scott, D.: Finite automata and their decision problems. IBM J. Res. Dev. **3**(2), 114–125 (1959)
12. Ruikar, D.D., Hegadi, R.S.: Simple DFA construction algorithm using divide-and-conquer approach. In: Nagabhushan, P., Guru, D.S., Shekar, B.H., Kumar, Y.H.S. (eds.) Data Analytics and Learning. LNNS, vol. 43, pp. 245–255. Springer, Singapore (2019). https://doi.org/10.1007/978-981-13-2514-4_21
13. Stearns, R.E., Hartmanis, J.: Regularity preserving modifications of regular expressions. Inf. Control **6**(1), 55–69 (1963)
14. Tie, Y., Qiang, X., Jin, H.: A grouping algorithm based on regular expression similarity for DFA construction. In: 2011 IEEE 13th International Conference on Communication Technology (ICCT), pp. 671–674. IEEE (2011)
15. Zafar, N.A., Alsaade, F.: Syntax-tree regular expression based DFA formalconstruction. Intell. Inf. Manage. **4**(04), 138 (2012)

Review on Natural Language Processing Trends and Techniques Using NLTK

Deepa Yogish[1], T. N. Manjunath[2(✉)], and Ravindra S. Hegadi[3(✉)]

[1] VTU-RC-ISE, BMSIT&M, Bangalore, India
deepayogish@gmail.com
[2] ISE, BMSIT&M, Bangalore, India
manju.tn@bmsit.in
[3] School of Computational Sciences, Solapur University, Solapur, India
rshegadi@sus.ac.in

Abstract. In modern age of information explosion, every day millions of gigabytes of data are generated in the form of documents, web pages, e-mail, social media text, blogs etc., so importance of effective and efficient Natural Language Processing techniques become crucial for an information retrieval system, text summarization, sentiment analysis, information extraction, named entity recognition, relationship extraction, social media monitoring, text mining, language translation program, and question answering system. Natural Language Processing is a computational technique applies different levels of linguistic analysis for representing natural language into a useful representation for further processing. NLP is recognized as a challenging task in computer science and artificial intelligence because understanding human natural language is not only depends on the words but how those words are linked together to form precise meaning is also considered. Regardless of language being one of the easiest concepts for human to learn, but for training computers to understand natural language is a difficult task due to the ambiguity of language syntax and semantics. Natural Language processing techniques involves processing documents or text which reduces storage space and also reduces the size of index and understanding the given information which satisfies user's need. NLP techniques improve the performance of the information retrieval efficiency and effective documentation processes. Common dialect handling procedures incorporates tokenization, stop word expulsion, stemming, lemmatization, parts of discourse labeling, lumping and named substance recognizer which enhances execution of NLP applications. The Natural Language Toolkit is the best possible solution for learning the ropes of NLP domain. NLTK, a collection of application packages which encourage researchers and learners in natural language processing, computational linguistics and artificial intelligence.

Keywords: Natural Language Processing (NLP) · Artificial Intelligence (AI) · Information Retrieval (IR) · Natural Language Tool Kit (NLTK)

K. C. Santosh and R. S. Hegadi (Eds.): RTIP2R 2018, CCIS 1037, pp. 589–606, 2019.
https://doi.org/10.1007/978-981-13-9187-3_53

1 Introduction

Natural Language Processing is a territory of software engineering, man-made consciousness, and computational semantics. NLP is related to human-computer interaction leveraged many applications such as document summarization, sentiment analysis, topic modeling, named entity recognition, relationship extraction, text mining, language translation, and question answering system. Natural language processing is influenced by many organizations to improve the effective documentation processes and identify the most relevant information from large data corpus. Many organizations use NLP techniques to optimize customer support by collecting the useful information. NLP improvises the efficiency of text analytics and enhance social media monitoring. For example, banks might implement NLP algorithms to optimize customer support and a large consumer products brand might combine natural language processing and semantic analysis to improve their knowledge management strategies and social media monitoring. Natural Language Processing is a computational technique applies different levels of linguistic analysis for representing natural language into a useful representation for further processing. NLP is recognized as a challenging problem in computer science and artificial intelligence because understanding human natural language is not only depends on the words but how those words are linked together to form precise meaning is also considered. Regardless of language being one of the easiest concepts for humans to learn, but for training computers to understand natural language is a difficult task due to the ambiguity of language.

Developers utilize Natural Language Processing concepts and techniques to organize and structure the data or information to perform different NLP applications which includes:

- **Machine Translation:** Machine translation program which automatically translate text from one form to another form where human and machine can understand. Google has a translation program used as statistical engine which translates each word in a sentence to other readable form by preserving the meaning of each word in a sentence.
- **Email Spam Filtering:** Spam filters have become important as the protection against unwanted email from unknown persons. The issues of spam filters are at the challenging task of NLP technology to extract correct meaning from strings of text in email to avoid spam emails.
- **Information Extraction:** The task of NLP is to provide pertinent information by extracting important entities which are affected by news, opinions and announcements in the form of natural language for financial market decisions.
- **Summarization:** NLP task in summarization is to produce a short summary which contains pertinent information from large amounts of data such as to produce a short summary from large academic articles. Summarization is used to understand customer opinion based on collective data from social media from which a company can decide next product offers.
- **Question Answering system:** Challenging task of NLP in question answering system is to understand Natural language queries and to provide short and

precise answer from large data corpus. Google's efforts in NLP in answering user question has focused in providing relevant documents but still Question Answering system remains a major challenge for search engines to provide exact answer to the user queries.

- **Social Media Monitoring:** Web based life checking gives an extraordinary chance to organizations to recognize what their customers are discussing via web-based networking media stages, web journals, and so forth and to find pertinent data for their future business choices. By communicating with customers, preparing their discussions and basically understanding clients in their very own words, organizations can more likely comprehend their customers' needs and enhance the associations with them.
- **Text Analytics:** Many associations use characteristic dialect preparing to approach content issues and enhance exercises, for example, information administration and huge information examination [12]. Morphological, linguistic [13], syntactic and semantic examinations of dialect empower distinguishing proof and extraction of elements like themes, areas, individuals, association, dates, and so forth and produce the metadata that can be utilized to tag and classify content in the most exact way.

Components of NLP

Natural Language Understanding (NLU) is to understand chunks of text data in the given natural language and to preprocess into a user understandable representation using different level of analysis.

Natural Language Generation (NLG) is to generate text in to human readable language with meaningful phrases and sentences in the natural language using different level of synthesis.

General Steps in Natural Language Processing

Figure 1 depicts the stages of analysis [1] in processing natural language with different level of analysis such as morphological analysis, syntactic analysis, semantic analysis, discourse analysis and pragmatic analysis.

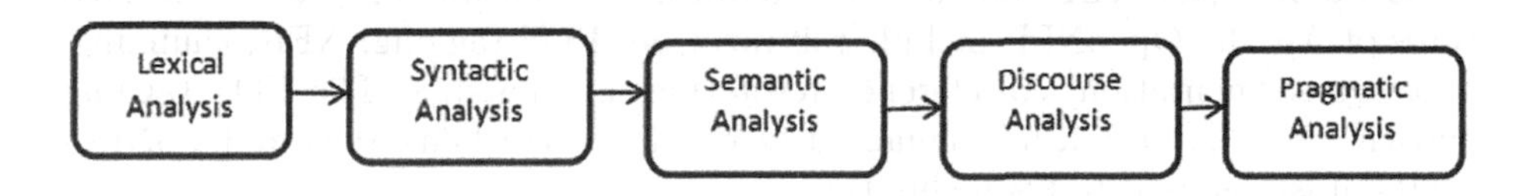

Fig. 1. Stages of analysis in processing natural language

- **Lexical Analysis:** Lexical Analysis involves analyzing the computational meaning of each word in context of document and divides the whole text into paragraphs, sentences and words.
- **Syntactic Analysis:** Syntactic analysis involves analysis of the rules or grammar of the sentence which will give a correct meaning by providing an

order and structure of each sentence in the text. Syntactic analysis is responsible for creation of a syntactic structure through parse tree of the sentence. Syntactic analyzer rejects the sentence if they violate the rules of language. Ex- "Blue dress girl the go the to market".
- **Semantic Analysis:** Semantic analysis means finding out the dictionary meaning of the text for meaningfulness. Sentence is rejected if no semantic meaning. Ex- "colorless - blue dress girl go to the market" meaningless sentence.
- **Discourse Analysis:** To frame remedy importance in a sentence which relies upon the sentence that goes before it and furthermore impacted by the significance of sentences that tail it. Ex-"Blue dress young lady required it" "it" relies on young lady.
- **Pragmatic Analysis:** Pragmatic investigation gets learning from outside presence of mind data and gives intentional utilization of dialect in circumstance. Ex-"Do you recognize what time she goes to showcase?" Interpreted as a demand.

2 Standard NLP Toolkit

Few toolkits are supporting for NLP tasks for the development of many applications. The current challenge is to select the best tool depends on application specific, source of text and performance of NLP tasks. Natural Language Toolkit (NLTK) is a gathering of program modules, libraries and diverse content change procedures for English dialect written in Python programming dialect. NLTK was produced by Steven Bird and Edward Loper at the University of Pennsylvania. The NLTK toolbox is in charge of various NLP assignments, for example, tokenization, stemming, parse tree portrayals, labeling, parsing, chunking and NER. NLTK stuffed with test codes for corpus readers, tokenizers, stemmers, taggers, chunkers, parsers, word net and NER. NLTK utilizes the Penn treebank tokenizer and POS tagger utilizes the Penn treebank tagset. The ACE corpus with a Maximum Entropy prototype is utilized to prepare chunkers and NER modules.

Apache OpenNLP toolkit is a Java library for natural language tasks. Features of Apache OpenNLP include tokenization, POS tagging, NER, chunking, sentence segmentation, co-reference resolution and parsing. The POS tagging and chunking for English language uses the Penn treebank tags and CoNLL-2000 dataset is used to train chunker.

Stanford CoreNLP toolbox is a Java library for normal dialect handling assignments. Stanford CoreNLP is in charge of numerous NLP assignments, for example, labeling, named entity recognizer [NER], co reference goals, assumption examination, bootstrapped design learning and data extraction.

The CoreNLP utilizes a Penn Treebank style tokenization; POS module utilizes Penn Treebank tagset with usage of the Maximum Entropy demonstrate. Conditional Random Field (CRF) display is utilized for NER utilizing CoNLL-2003 dataset.

Pattern is a Python library underpins NLP, web mining and Machine Learning undertakings. Provides modules to Google, Twitter, Wikipedia API, a web crawler, POS labeling, estimation investigation, Word Net, Vector space model, grouping and SVM.

GATE (General Architecture for Text Engineering) created by division of Computer Science, University of Sheffield with License GNU LGPL. Door bolsters NLP undertakings with program modules, for example, tokenizer, sentence splitter, POS tagger, NER and co-reference goals tagger.

TextBlob is a Python library like NLTK gives modules to NLP errands, for example, POS labeling, tokenization, Chunking, assumption investigation, order, interpretation, and some more.

SpaCy is a modern quality programming library written in Python and Cython for cutting edge Natural Language Processing applications. Backings tokenization, NER, POS labeling, parsing and word vectors.

Table 1 shows comparing different NLP toolkits [2] for NLP techniques like tokenization, POS tagging, chunking, NER tagging and stemming. NLTK, OpenNLP and CoreNLP can perform almost all lower-level NLP tasks. We have chosen the NLTK as best tool for research and applications to best of our knowledge. NLTK is the best solution for learning the ropes of NLP domain. Its modular structure helps comprehend the dependencies between components and get the best experience with composing appropriate models for solving certain tasks.

Table 1. Comparison of various NLP toolkits

NLP toolkit	Programming language	Tokenization	POS tagging	Chunking	NER	Stemmer
NLTK	Python	Yes	Yes	Yes	Yes	Yes
Apache OpenNLP	Java	Yes	Yes	Yes	Yes	Yes
Stanford CoreNLP	Java	Yes	Yes	No	Yes	Yes
Pattern	Python	Yes	Yes	Yes	No	No
GATE	Java	Yes	Yes	Yes	Yes	No
Spacy	Python/Cython	Yes	Yes	Yes	Yes	No

3 Natural Language Processing Techniques Using NLTK

NLTK [3] is written in Python which supports many features like transparent syntax, good string-handling features and easy to understand. Python programming as an object oriented language supports interactive interpreter, encapsulated data and code which has extensive library. NLTK uses the treebank word tokenizer and POS tagger utilizes the Penn Treebank tagset which is prepared on the PENN treebank corpus with a Maximum Entropy prototype. Chunking and the NER modules are utilized on ACE corpus with a Maximum Entropy prototype.

3.1 Text Transformation Techniques Using NLTK

NLTK is a gathering of test Python modules for some NLP tasks, for example, tokenization, stop word expulsion, stemming, POS tagging,chunking and NER. Brown Corpus, CoNLL-2000 Chunking Corpus, CMU Pronunciation Dictionary, NIST IEER Corpus, PP Attachment Corpus, Penn Treebank, and the SIL Shoebox corpus design are the corpus tests incorporated into NLTK.

Tokenization. Tokenization is a procedure of producing significant tokens from textual information. Tokenization is a process of identification of tokens within documents which reduces storage space and satisfies user's information need more precisely with reduced search space. Janani [4] investigated the execution of seven open source tokenization tools. For tokenize an archive, numerous tools are accessible. The tools are Nlpdotnet Tokenizer, Mila Tokenizer, NLTK Word Tokenizer, TextBlob Word Tokenizer, MBSP Word Tokenizer, Pattern Word Tokenizer, and Word Tokenization with Python NLTK. Among all above tokenization tools Word Tokenization with Python NLTK is the best tool which gives the best result. This tool can take up to 50000 characters at a time. The text is first tokenized into sentences using the PunktSentence Tokenizer. Then each sentence is tokenized into words using four different word tokenizers such as TreebankWordTokenizer, WordPunctTokenizer, PunctWord-Tokenizer and WhitespaceTokenizer. NLTK supports treebankWordTokenizer which uses regular expressions to tokenize text with Penn Treebank corpus. The text will be segmented into sentences by using method sent_tokenize ( ) and word_tokenize ( ) method is invoked for tokenization.

Stop Word Removal. The most commonly used preprocessing method in NLP application is stop word removal. The main purpose of stop word removal is removing the words that occur commonly across all the documents in the corpus. A generally utilized stop words are "the", "an", "an", "in". Stop-word removal [5] is utilized to enhance the execution of the Information Retrieval System, Text Analytics, Text Summarization and Question-Answering framework. NLTK import stop words and set stop word for English dialect by executing command on python shell as demonstrated as follows. import nltk from nltk.corpus import stopwords set(stopwords.words("English")).

Stemming. Stemming is a most commonly used preprocessing technique in text mining and Information retrieval systems. Stemming process used by information retrieval systems for indexing words by reducing the size of index and search space by increasing retrieval accuracy. Stemming [6] process reduces different grammatical word forms to its root form. The "stem" is acquired by applying an arrangement of standards without thinking about the grammatical feature (POS) with regards to the word event in sentence. Over stemming and under stemming [6] are two blunders in stemming. Over stemming happens when two words with various stems are stemmed to indistinguishable root and known from a false positive. Under-stemming happens when two words that ought to be stemmed to a similar root are not and known as a false negative. The issue of over stemming and under stemming can be decreased by thinking about the language structure, semantics and parts of discourse of a sentence. Figure 2 depicts classification of stemming algorithms for stemming process.

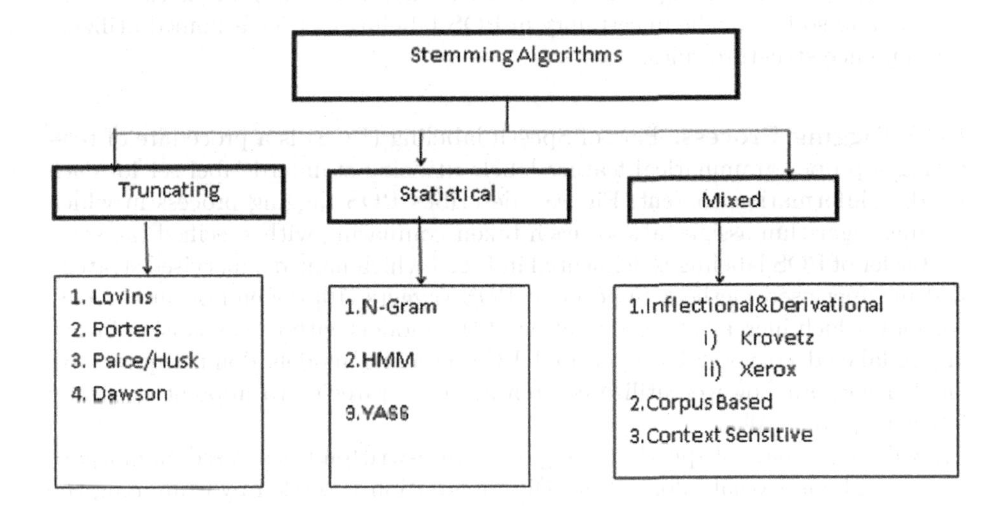

Fig. 2. Classification of stemming algorithms

Truncating Methods (Affix Removal): Truncating methods produces stem by removing affixes of a word based on the some value 'n' that is to consider word till 'n' letters and to discard rest of all alphabets.

Statistical Methods: Statistical stemmers are based on statistical analysis and techniques produces stem by applying few statistical procedure.

Mixed Methods: stemmers in hybrid methods depend on both the inflectional and derivational morphology analysis with a given corpus to develop these types of stemmers.

The Porter's Stemmer is the most widely used by researchers; the basic algorithm will be changed based on their requirements [6]. The Porter stemmer was produced by Martin Porter in 1980 at the University of Cambridge. The Porter stemmer [7] is a setting touchy postfix expulsion calculation comprises of number of direct advances, five or six utilized to create the last stem. The control looks like <condition> <suffix> leads to <new suffix>.

POS Tagging. Part of Speech labeling (POS) is a method of allotting proper grammatical feature tag for each word in a given content. Label set is the arrangement of labels from which the tagger chooses proper descriptor for each word. The coarse-grained label set are N (Noun), V (Verb), ADJ (descriptive word), ADV (Adverb), PREP (Preposition), CONJ (Conjunction) and fine-grained label set are NNOM (Noun-Nominative), NSOC (NounSociative), VFIN (Verb limited), VNFIN (Verb Nonfinite). Generally most taggers utilize just fine grained label set. POS labeling [9] is a normally utilized pre-preparing module for different NLP undertakings like Machine interpretation, Natural dialect content handling, synopsis, Multilingual and cross dialect data recovery, Speech acknowledgment, Artificial knowledge, Parsing, Expert framework. Difficulties of POS labeling, for example, remote words, ambiguities, ungrammatical information and so forth. The uncertainty in POS labeling can be fathomed utilizing the sentence structure rules.

POS Tagging Process. Part of Speech labeling (POS) is a procedure of relegating a proper grammatical feature labels utilizing standard label set for each word in information content. Figure 3 describes POS tagging process in which tagging algorithm assign tags for each token comparing with specified tag set.

Order of POS labeling is delineated in Fig. 4 which named supervised strategy and unsupervised method. Supervised POS taggers [10] depend on pre-labeled corpora, which increment the execution of the models with the expansion in size of pre-labeled corpora. Unsupervised POS labeling models don't require pre-labeled corpora however utilize computational strategies to appoint naturally labels to words [10].

Rule-based part-of-speech tagging uses hand-written rules based on morphological and contextual information. Disambiguation is settled by examining the lexical, semantic, syntax of each word in a sentence, its previous word, its following word and different perspectives. For instance, if the former word is article then the word in question must be a noun. This information is coded in the form of hand written rules in rule based tagger. The rule tagger [8] process is divided into two stages. In first stage, tagger returns a set of possible POS tags by searching words in dictionary and in second stage uses a set of hand-coded rules to assign a single part of speech for each word by removing disambiguate words using linguistic features of word. Examples of rule based tagger are TAGGIT and ENGTWOL. A stochastic approach assigns a tag to word using probability and statistics. Stochastic approaches assign a tag based on the probability of certain tag occurrences in a word sequence and predictions based on past observations. Example of stochastic tagger is CLAWS (constituent likelihood automatic word -tagging system). The hybrid approach [8] is a combination of Rule based approach and statistical approach. Hybrid tagger assign most probable tag to the word using statistical method, if any disambiguate found after assigning tag, then apply grammar rules to change. Hybrid tagger uses rule-based systems to specify tags using hand

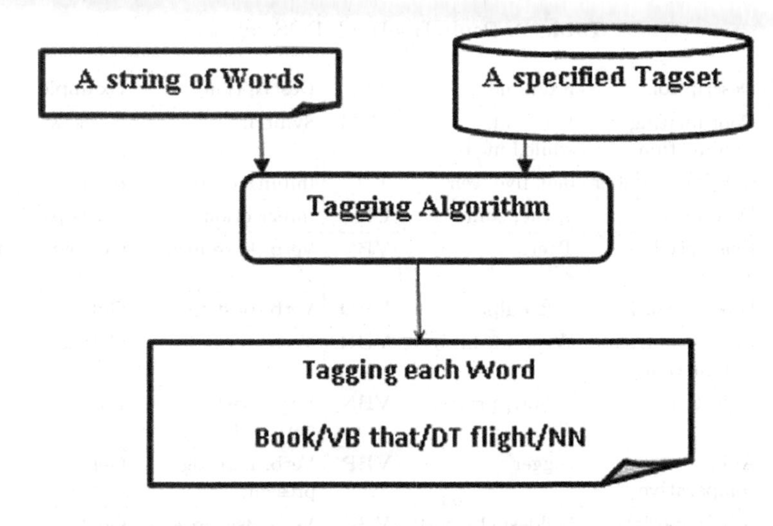

Fig. 3. POS tagging process

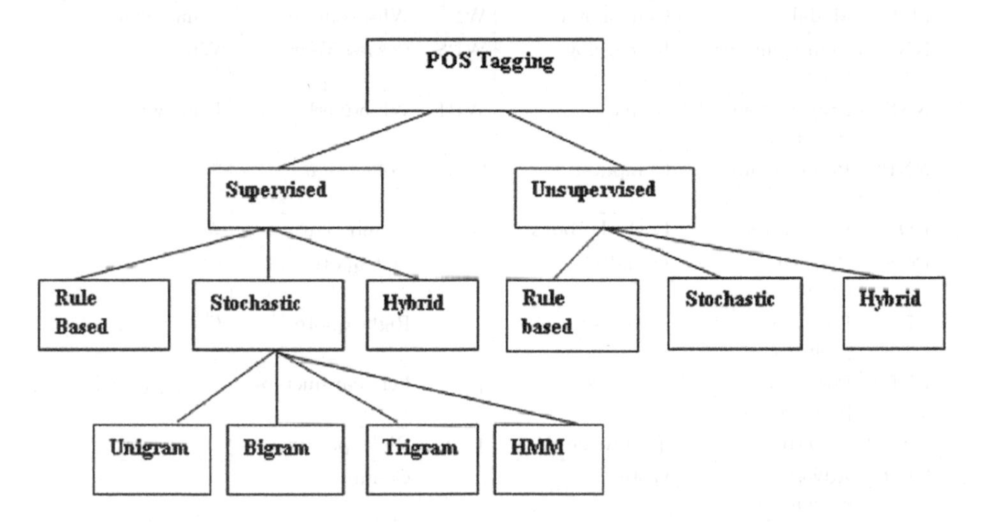

Fig. 4. Classification of POS tagging

written rules and use stochastic systems by applying machine-learning to induce rules from a tagged training corpus automatically. Example of hybrid approach is Brill tagger or transformation based tagger. NLTK POS tagger utilizes the Penn Treebank label set as default and is prepared on the PENN Treebank corpus with a Maximum Entropy prototype. Table 2. Demonstrates the Penn Treebank POS label set.

Table 2. Penn TreeBank POS tag set

Tag	Description	Example	Tag	Description	Example
CC	Coordinating Conjunction	For, and, while,but, or	SYM	Symbol	+, %, &
CD	Cardinal number	one, five, ten	TO	Infinitive ‘to’	to go
DT	Determiner	a, the, some	UH	Interjection	ah, oops
EX	Existential There	There	VB	Verb, base form	Get, write, jump
FW	Foreign word	mea culpa	VBD	Verb, past tense	Got
IN	Preposition/sub-conjunction	With, of, in, by	VBG	Verb, gerund	Getting
JJ	Adjective	Yellow, pretty, old	VBN	Verb, past participle	Eaten
JJR	Adj., comparative	Bigger	VBP	Verb, non 3sg present	Get
JJS	Adj., superlative	Wildest, biggest	VBZ	Verb, 3sg present	Gets
LS	List item marker	1, 2, one	WDT	Wh-determiner	Which, that
MD	Modal	Can, should	WP	Wh-pronoun	What, who
NN	Noun, sing, or mass	Rama, dog	WPS	Possessive wh	Whose
NNS	Proper noun, singular	Ramas	WRB	Wh-adverb	How, where
NNPS	Proper noun, plural	Computers	$	Dollar sign	$
PDT	Predeterminer	Both the boys	#	Pound sign	#
POS	Possessive ending	Friend’s	“	Left quote	(“)
PP	Personal pronoun	I, you, she	”	Right quote	(”)
PRP$	Possessive pronoun	your, his	{	Left parameters	(—, (,{,<)
RB	Adverb	quickly, never	}	Right parameters	(—,),},>)
RBR	Adverb, comparative	Faster	,	Comma	,
RBS	Adverb, superlative	Fastest	.	Sentence-final punc	(. ! ?)
RP	Particle	up, off	:	Mid-sentence punc	(: : ... - -)

Figure 5 shows code snippet for tokenization, stop word removal, stemming and POS tagging process using word_ tokenize ( ) method, importing stop words and setting for English language, using porter stemmer algorithm which instantiate the Porter Stemmer class and call the stem ( ) method using Penn Treebank tagset.

```
from nltk.tokenize import sent_tokenize, word_tokenize from nltk.corpus import stopwords
from nltk.stem import PorterStemmer, WordNetLemmatizer
#read a file and word tokenize
data = "Bengaluru is the capital of Karnataka and also known as Garden city .
        VTU is one of the largest Technological University."

#stop word removal
stopWords =
set(stopwords.words('english')) tokenwords
= word_tokenize(data) stopwordsFiltered=
[]
for w in tokenwords:
   if w not in stopWords:
      stopwordsFiltered.append
      ()
print(stopwordsFiltered)

#stemming using porter
stemmer ps=
PorterStemmer()
for w in
   stopwordsFiltered:
   print(ps.stem(w))

#POS TAGGING using PENN TREEBANK POS TAG SET
tokens_pos= pos_tag(wordsFiltered)
```

Fig. 5. Code snippet for tokenization, stop word removal, stemming and POS tagging process

Figure 6 shows snapshot of tokenization, stop word removal, stemming and POS tagging process.

Lemmatization. Lemmatization is the way toward decreasing a gathering of words into their lemma. Lemmatization delivers the "lemma" of a word after fathom the Parts of Speech and the setting of a word in the given sentence. Lemmatization used in information retrieval application like indexing and searching as a normalization technique. Riddhi et al. [11], discussed five different Lemmatization approaches such as Edit Distance on dictionary algorithm, Morphological analyzer, radix trie, Affix lemmatizer and fixed length truncation approach. Figure 7 shows Lemmatization process instantiate WordNetLemmatizer class using lemmatize ( ) method.

Figure 8 shows snapshot for Lemmatization process using lemmatize ( ) method for noun and verb parts of speech for different inputs.

Word Frequency – Frequency Distribution. After processing text using tokenizion, stop word removal and stemming we can figure the word recurrence that is checking the occasions every token happens in the information corpus. Python object Counter is used to perform word frequency.

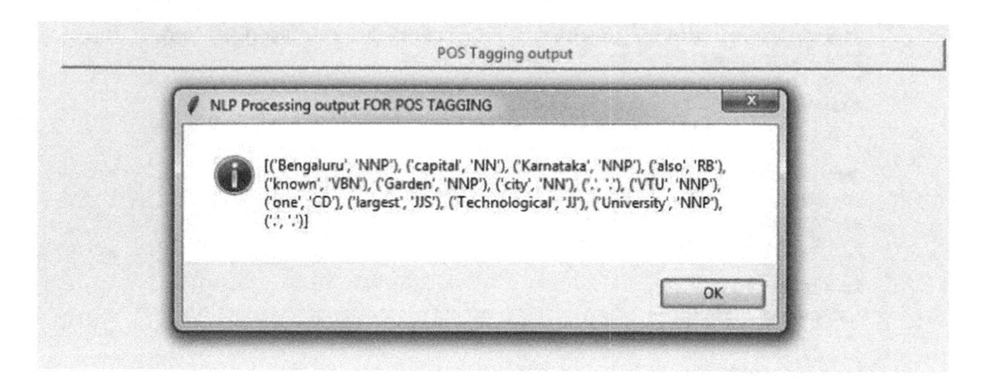

Fig. 6. Snapshot of tokenization, stop word removal, stemming and POS tagging process

```
from nltk.stem import WordNetLemmatizer

input_words = ['writing', 'calves', 'be', 'branded', 'horse', 'randomize',
    'possibly', 'provision', 'hospital', 'kept', 'scratchy', 'code']

# Create lemmatizer object
lemmatizer =
WordNetLemmatizer()

# Create a list of lemmatizer names for display
lemmatizer_names = ['LEMMA-NOUN' , 'LEMMA-VERB'] formatted_text
= '{:>20}' * (len(lemmatizer_names) + 1) print(formatted_text.format('INPUT
WORD', *lemmatizer_names), '\n', '='*40)

# Lemmatize each word and display the output
for word in input_words:
    output = [word, lemmatizer.lemmatize(word, pos='n'),
        lemmatizer.lemmatize(word, pos='v')]
    print(formatted_text.format(*output))
```

Fig. 7. Code snippet for Lemmatization process

Figure 9 shows code snippet for counting frequency of each token in document using counter ( ) method after preprocessing given text using tokenization, stop word removal and stemming.

Figure 10 shows snapshot for counting frequency of each token in document using counter ( ) method.

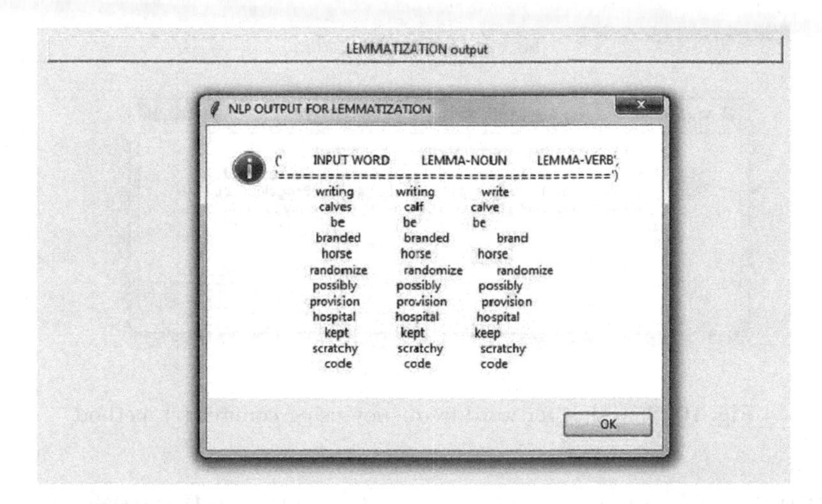

Fig. 8. Snapshot of Lemmatization process for noun and verb POS

Named Entity Recognizer (NER). Named-Entity Recognition (NER) is an essentially connected in errand of data extraction for extricating basic substances, for example, the people, areas, associations, articulations of times, amounts, fiscal qualities, rates, and so on. Named Entity Recognition is a type of content mining that isolates out unstructured content information and finds thing phrases called named elements. Question Answering (QA) framework utilizes NER method to enhance the exactness of Information Retrieval by recovering important archive which contains a response to the client's inquiry.

```
import nltk
from nltk.tokenize import
word_tokenize from collections import
Counter
s= " Bengaluru is the capital ofKarnataka and also known as Garden city .
  VTU is one of the largest Technological University."

#TOKENIZATION
PROCESS tokens
=word_tokenize(s)
#print(tokens)

#counting word frequency
count = Counter(tokens)
print(count.most_common(50
))
```

Fig. 9. Code snippet for word frequency using counter ( ) method

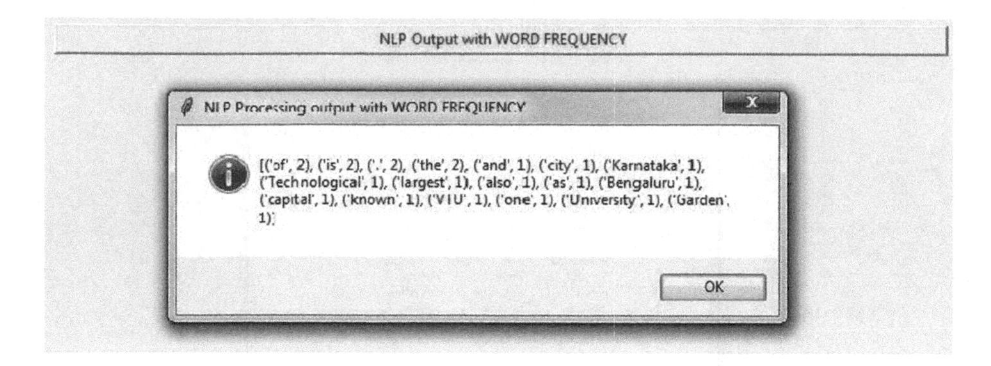

Fig. 10. Snapshot for word frequency using counter ( ) method

NER is performed after pre-processing of records by tokenization, stop word expulsion and Part-Of-Speech(POS) Tagging. Named Entity Recognition is the assignment of recognizing substances in a sentence with labels like a man as PERSON, association as ORG, date as DATE, area as GPE, time as TIME and so on. Figure 11 shows code snippet for named entity recognizer using nltk.ne_chunk( ) function.

Figure 12 shows snapshot of named entity recognizer representing tags as person, location, organization for a given text.

Chunking. Chunking is a procedure of extricating noun or verb phrases from unstructured content. Chunking is applied after POS tagging and provides chunks as output. Chunking utilizes standard arrangement of Chunk labels like Noun Phrase (NP), Verb Phrase (VP), and Adjective Phrase (AP) and so forth. Chunking is essentially utilized when we need to remove data from content, for example, Locations, Person Names and so forth. Chunking is a three stage process, for example, labeling a sentence by POS tagger, Chunk the POS labeled sentence and break down the parse tree to extricate data. In order to create an NP or VP chunker, define a chunk grammar consisting of rules with a single regular expression. In order to chunk, combine the part of speech tags with regular expressions. To shape standard articulation, use meta characters, for example, + = at least one or more events of the former component.

? = zero or one events of the former component.

* = zero or More events of the former component.

. = Any character with the exception of another line.

The part of speech tags are denoted with the "<" and ">" and place regular expressions within the tags to chunk the tagged sentence. Few examples are shown below.

<JJ.?>* = "0 or more of any tense of adjective", <VB.?>* = "0 or more of any tense of verb", <NNP>+ = "One or more proper nouns".

```
import nltk
from nltk.tokenize import sent_tokenize,
word_tokenize from nltk.corpus import stopwords
from nltk.stem import PorterStemmer,
WordNetLemmatizer from nltk import pos_tag

data = "In Bangalore,I like to ride the Metro to visit MG road and some restaurants rated well by Karan."
#print (word_tokenize(data))

#stop word removal
stopWords =
set(stopwords.words('english'))
tokenwords = word_tokenize(data)
stopwordsFiltered = []

for w in tokenwords:
   if w not in stopWords:
      stopwordsFiltered.appen
      d(w)
print(stopwordsFiltered)

#stemming using porter
stemmer ps =
PorterStemmer()
for w in
   stopwordsFiltered:
   print(ps.stem(w))

#POS TAGGING using PENN TREEBANK POS TAG SET
tokens_pos = pos_tag(wordsFiltered)

#Namedentity recognizer
print(nltk.ne_chunk(tokens_p
os))
```

Fig. 11. Code snippet for NER process

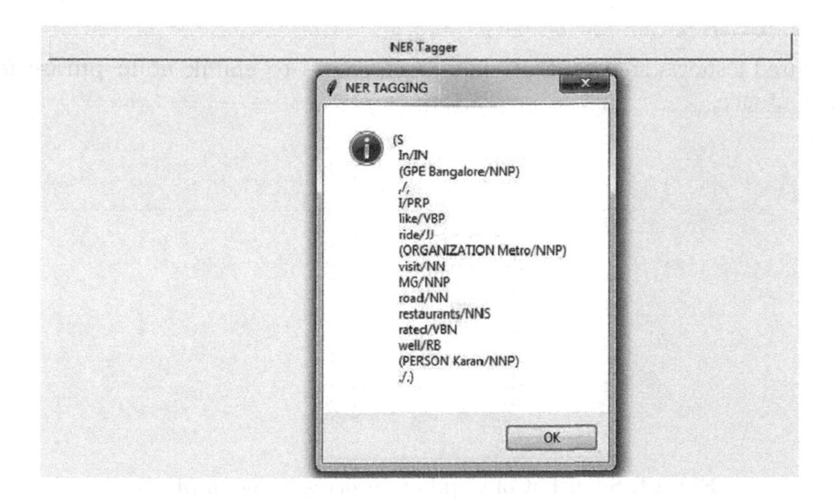

Fig. 12. Snapshot of Named Entity recognizer

Figure 13 illustrates chunking process using regular expression. For a given sentence, preprocess the text using tokenization and POS tagging. Chunk the tagged text using regular expression to noun phrase, verb phrase or adverb phrase.

```
import nltk

def prepareForNLP(text):
  sentences = nltk.sent_tokenize(text)
  sentences = [nltk.word_tokenize(sent) for sent in sentences]
  sentences = [nltk.pos_tag(sent) for sent in sentences] return
  sentences

def chunk(sentence):
  chunkToExtract =
  """

  NP: {<NNP>*}
    {<DT>?<JJ>?<NNS
    >}
    {<NNP><NNP>}""
    "

  parser = nltk.RegexpParser(chunkToExtract)
  result = parser.parse(sentence)

  for subtree in result.subtrees
    (): if subtree.label() ==
    'NP':
      t = subtree
      t = ' '.join(word for word, pos in t.leaves())
      print(t)

sentences = prepareForNLP("Bangalore is garden city. VTU is Technical University.") for
sentence in sentences:
  chunk(sentence)
```

Fig. 13. Code snippet for Chunking process to chunk noun phrase

Figure 14 shows snapshot of chunking process to chunk noun phrase for a given sentence.

Fig. 14. Snapshot of chunking process for noun phrase

4 Conclusion

Huge volume of data is generated every minute through social media platforms which are used by much organization for their business operations. In connection with decision making, making the machine to learn to arrive suitable decisions at the right time. In this regard Natural Language processing plays a vital role in machine learning. Language being one of the easiest things for humans to learn, but for training computers to understand natural language is a difficult problem due to the ambiguity of language syntax and semantics. Information retrieval process provides the world's wealth of information at our fingertips, but still required improvement and challenging task to provide precise answer for answering specific questions posed by humans. Natural language processing is the process of transforming a natural language text into a structured format that a computer can process and produce useful information to users. So pre-processing of natural language is necessary for many applications. We have used NLTK tool kit for natural language preprocessing techniques such as tokenization, stop word removal, stemming, Lemmatization, POS tagging, NER tagging and chunking which reduces search space, reduce index size and improve the accuracy of retrieval process in many applications.

References

1. Swapnil, V., Jayshree, A.: Natural language processing preprocessing techniques. Int. J. Comput. Eng. Appl. **XI**(Special Issue) (2017). http://www.ijcea.com/. ISSN 2321-3469
2. Alexandre, P., Hugo, G.O., Ana, O.A.: Comparing the performance of Different NLP toolkits in formal and social media text. In: 5th Symposium on Languages, Applications and Technologies, Germany (2016).https://doi.org/10.4230/OASIcs, SLATE.2016
3. Steven, B.: NLTK: the natural language toolkit. In: Proceedings of the COLING/ACL Interactive Presentation Sessions, Association for Computational Linguistics, Sydney, pp. 69–72 (2006)
4. Vijayarani, S., Janani, R.: Text mining: open source tokenization tools - an analysis. Adv. Comput. Intell. Int. J. (ACII) **3**(1), 37–47 (2016)
5. Raulji, J.K., Saini, J.R.: Stop-word removal algorithm and its implementation for Sanskrit language. Int. J. Comput. Appl. (0975–8887) **150**(2), 15–17 (2016)
6. Jivani, A.G.: A comparative study of stemming algorithms. Int. J. Comp. Tech. Appl. **2**(6), 1930–1938 (2011). ISSN 2229-6093
7. Anjali, M.K., BabuAnto, P.: Parts of speech taggers for dravidian languages. Int. J. Eng. Trends Technol. (IJETT) **21**(7), 342–347 (2015). https://doi.org/10.14445/22315381/IJETT-V21P263. ISSN 2231-5381
8. Shubhangi, R., Sharvari, G.: Survey of various POS tagging techniques for Indian regional languages. Int. J. Comput. Sci. Inf. Technol. (IJCSIT) **6**(3), 2525–2529 (2015)
9. Mahar, J.A., Qadir, G.: MEMON: rule based part of speech tagging of Sindhi language. In: Proceeding of International Conference on Signal Acquisition and Processing (2010)

10. Mahar, J.A., Memon, G.Q.: Parts of speech taggers for Dravidian languages. Int. J. Eng. Trends Technol. (IJETT) **21**(7), 1933–1938 (2015). ISSN 2231-5381
11. Riddhi, D., Prem, B.: Survey paper of different lemmatization approaches. Int. J. Res. Advent Technol. (2015). (E-ISSN 2321-9637) Special Issue 1st International Conference on Advent Trends in Engineering, Science an d Technology
12. Manjunath, T., Ravindra, N., Hegadi, S.: Statistical data quality model for data migration business enterprise. Int. J. Soft Comput. Medwell J. (2013). ISSN 1816-9503
13. Ruikar, D.D., Hegadi, R.S.: Simple DFA construction algorithm using divide-and-conquer approach. In: Nagabhushan, P., Guru, D.S., Shekar, B.H., Kumar, Y.H.S. (eds.) Data Analytics and Learning. LNNS, vol. 43, pp. 245–255. Springer, Singapore (2019). https://doi.org/10.1007/978-981-13-2514-4_21

A Deep Recursive Neural Network Model for Fine-Grained Opinion Classification

Ramesh S. Wadawadagi[1(✉)] and Veerappa B. Pagi[2]

[1] Department of Computer Applications, Basaveshwar Engineering College, Bagalkot 587 102, India
rswlib@yahoo.co.in

[2] Department of Computer Science and Engineering, Basaveshwar Engineering College, Bagalkot 587 102, India
veereshpagi@gmail.com

Abstract. In recent times, deep neural networks (DNN) have acquired greater significance in providing solutions to many deep learning tasks. Particularly, recursive neural networks (RNN) have been efficiently utilized in exploring semantic compositions for natural language content represented with structured formats (e.g. parse-trees). Despite the fact that RNN are deep in structure, yet they fail to exhibit hierarchical representations observed in traditional deep feed-forward networks (DFNN) and also in revolutionary deep recurrent neural networks (DRcNN). However, the notion of depth can be incorporated through stacking multiple recursive layers, which results in deep recursive neural networks (DRNN). On the other hand, enhanced word spaces offer added benefits in capturing fine-grained semantic regularities. In this paper, we address the problem of fine-grained opinion classification using DRNN and word embeddings. Furthermore, the efficiency of DRNN model is estimated through the conduction of a series of experiments over several opinion datasets. The results report that the proposed DRNN architecture achieves better prediction rate for fine-grained classification when compared with conventional shallow counterparts that employ similar parameters.

Keywords: Deep learning · Recursive neural networks · Fine-grained opinion classification · Compositional semantics

1 Introduction

Investigating semantic compositions in short opinion sentences need more sophisticated supervised training, language models and also evaluation resources. Despite several models proposed in the literature, the efficiency of a binary (positive/negative) opinion classifier has not surpassed 85%. Especially, for the more complicated fine-grained classification model consisting of a neutral class, accuracy reported is not more than 55% [1]. This fact indeed triggers a demand for the construction of new classification systems to cope with many challenges. There have been a surge of advancements in developing and training DNN,

K. C. Santosh and R. S. Hegadi (Eds.): RTIP2R 2018, CCIS 1037, pp. 607–621, 2019.
https://doi.org/10.1007/978-981-13-9187-3_54

and enabled them in natural language processing (NLP) research, for various tasks including simpler ones like POS-tagging and named entity recognition (NER), to the most complex ones such as semantic role labeling (SRL) [2]. In particular, DRNN have shown great potential in exploring semantic compositionality utilized for user-generated content (UGC) with structured representations. However, developments in DRNN are still constrained by insufficiency of huge, task specific and labeled compositionality resources that precisely captures the underlying concept hidden in the opinion data [3,4].

Deep neural architectures are stacked with several hidden layers of nonlinear information processing [5]. They focus on learning feature hierarchies formed by the combination of lower level features into higher level hierarchies, in such a way that, any subsequent layer possibly gains a more common meaning [6]. More specifically, RNN are the generic class of DNN constructed through recursive application of the fixed set of weights over structured inputs (directed positional acyclic graphs) [3,4,7]. Furthermore, RNN are considered as the abstract form of recurrent neural networks (RcNN) [8] that have skewed tree representation. Since, each hidden state in RcNN is defined as a function of all previous hidden states, they are inherently deep in time [9,10]. In order to accomplish depth in space, primitive RcNN could be engineered with stacked representation of multiple recurrent hidden layers, similar to the way how feed-forward networks are stacked in traditional deep networks. Unlike RcNN which are deep in time, RNN possess deep in structure due to the iterative computation of recursive connections. Thus, for the given structural representation of a sentence, RNN recursively yields parent structures in a bottom-up manner through the combination of words to obtain structure for phrases that eventually produces the entire sentence. Subsequently, the sentence level representation is provided to obtain a classification prediction for a given input sentence. In spite of being deep in structure, RNN fails to model hierarchical representation observed in traditional DFNN and also in DRcNN. The notion of depth can be incorporated into traditional RNN through the organization of multiple recursive layers in a stacked presentation. Nevertheless, deep learning hypothesis is built around a fact that hierarchical models are considerably good at representing certain functions than a shallow one [5,11].

The model discussed in this paper is basically a hybrid architecture that combines the features of a deep feed-forward neural network and a recursive network. In particular, a DFNN is extended with an auxiliary structural processing inside each layer. The information is concurrently propagated through each node of a structure (e.g. parse tree) synchronized with each hidden layer. During forward propagation of the recursive network, the weights of structural processing components are shared. On the contrary, each node in the structure feeds its own hidden states successively to its counterpart in the adjacent layer. Figure 1 demonstrates the general idea of the proposed deep recursive neural networks. In such a network architecture, each layer learns certain portion of composition to exploit, and transform partial representation to the subsequent layer to accomplish processing of remaining constituents. To evaluate the performance of a new architecture, a task specific dataset similar to Stanford sentiment treebank (SST) [4] is constructed with the aid of Stanford CoreNLP[1]. The proposed DRNN architecture is then trained on newly constructed treebank dataset. The model outperforms several contemporary methods when evaluated on different metrics with different datasets. The remainder of the article is organized according to the following sections. Section 2 provides a literature review on the contemporary DNN architectures that exploit linguistic features for NLP applications. A detailed discussion on the proposed DRNN model for fine-grained opinion classification is presented in Sect. 3. In Sect. 4, the performance of the proposed DRNN model is evaluated against two opinion datasets. Finally, Sect. 5 renders a brief summary of the research contributions, concluding remarks and a discussion on future work.

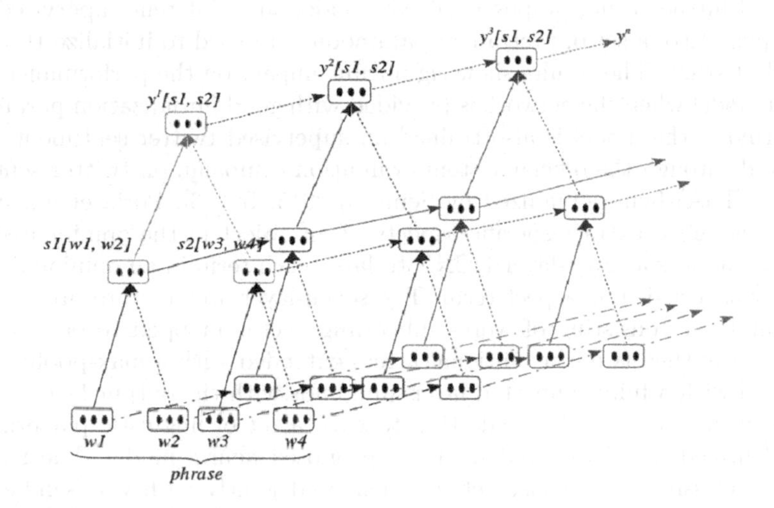

Fig. 1. The three layer architecture of a deep recursive neural network.

2 Related Work

Research on sentiment analysis has gradually evolved from traditional bag-of-words (BOW) model to compositional semantics by the means of applying techniques that preserve meaning and context using DNN. They are categorized into convolutional networks [12–16], recurrent neural networks [17,18,21,22], recursive neural network [4,23], memory augmented networks [24] and other hybrid neural architectures [25–28]. In the following subsections, we provide a review on several DNN models that have been successfully employed for various sentiment analysis tasks.

[1] https://nlp.stanford.edu/sentiment/code.html.

2.1 Deep Convolution Neural Networks

The significant amount of research work in DNN is witnessed for the usage of deep convolutional neural networks (DCNN) to exploit compositional semantics in natural language. DCNN are basically deep multi-layered architectures that use improved regularization technique and generate more training samples, exploiting the existing results [11]. As a matter of fact, a model based on DCNN for twitter sentiment analysis to predict polarities at both sentence and phrase level is discussed in [12]. The work proposes a novel technique for parameter weight initialization of the CNN to assist in training an accurate model without additional parameters. Finally, the approach employs unsupervised neural language model (UNLM) for training initial word-embeddings which are tuned further using proposed DCNN model on a different supervised opinion corpus. Also, a set of pretrained parameters are used to initialize the model in the last stage. The results show significant impact on the performance of the trained model when the network is provided with good initialization parameters. Furthermore, the model is also trained on supervised twitter sentiment corpus presented through the official system evaluation campaign on twitter sentiment analysis (TSA) being organized by Semeval-2015. In [13], Poria et al., present a DNN model to extract specific aspects of an object in the opinionated text. The work adopts a seven-layer DCNN to label each term in an opinion sentence as either aspect or non-aspect term. The seven-layer architecture proceeds into an input layer consisting of word-embedding vector mapped to each word in the sentence; then two convolutional layers extended with a max-pooling layer; continued with a fully connected layer; and ends with the output layer containing one neuron per word. In addition to this, a set of linguistic patterns have been identified and integrated with the proposed neural model. The resulting ensemble classifier thus constructed, when used jointly with word-embeddings yield better accuracy compared to the state-of-the-art DNN models. However, in [14] a multi-lingual aspect extraction and aspect-based sentiment analysis using DCNN is proposed. In this work, the task of aspect extraction is posed as a multi-label classification task that computes probabilities over aspect parameterized by a threshold. Accordingly, the sentiment towards an aspect is estimated by applying convolution over the aspect vector concatenated with each word-embedding. The model achieves decent results across several languages and domains. Furthermore, sarcasm detection (SD) is one of the fundamental task in many NLP applications. More specifically, in sentiment analysis sarcasm can reverse the polarity of a sentence and, hence, degrades the performance of the system. In view of this, a model based on pretrained DCNN for automatic learning of sarcasm features is addressed [15]. The model investigates that, when features extracted from pretrained sentiment models will increase the efficiency of sarcasm detection. Lastly, a novel approach for feature extraction of handwritten text images based on CNNs is presented [16]. The architecture is tailored with three fundamental types of layers, namely, convolutional layer, pooling layer and fully connected layer. Convolutional layer uses a set of kernels to produce parameters and generates an activation map to an output. Then all CNNs take max-pooling operation at their corresponding pooling layer. The model outperforms the previously reported techniques, when considering the complexity and the size of the dataset.

2.2 Deep Recurrent Neural Networks

A recurrent neural network (RcNN) establishes recurrent linkage among the hidden layers. As a consequence, each hidden unit binds itself with remaining nodes within the hidden layer. Since, RcNN perform recursive tasks over each instance of the sequence, the output always depends on previous computations. RcNN are embedded with short memory to store previous computations, and results are used in current processing. When multiple recurrent hidden layers are combined in the network hierarchy, then it becomes a Deep RcNN. Numerous applications based on DRcNN have been proposed in the literature. For example, in [17], different ways to extend a basic RcNN to an efficient DRcNN is presented. The research identifies shallow parts of an RcNN, and each shallow part is replaced with an alternative deep architecture. Then, a novel framework based on neural operators is introduced to interpret these DRcNN. As a result, this deeper variant is effectively utilized for the task of predicting the next symbol in phrase construction. Furthermore, a category of RcNN, named multiplicative RcNN (mRcNN) [18] is proposed as a general model for exploring compositional semantics in languages. Essentially, mRcNNs are sequential models of language, which are similar to matrix-space models, and as a variant of RcNN they implicitly model compositionality. Unlike recursive neural networks, mRcNN does not require parse trees. Computations are being performed sequentially rather than recursively, and performance is independent of accuracy of the parser. The efficiency of the model is evaluated through its application on the task of fine-grained sentiment analysis. Despite its success, RcNN often suffers from vanishing and exploding gradient problems [19]. However, long short-term memory (LSTM) [20] that are aided with additional "forget" gates will overcome the above problems. LSTM allows the error to back-propagate through large number of time steps. Consequently, in [21], a twitter sentiment prediction system based on the notion of LSTM-RcNN is discussed. In this method, interactions among the words are controlled through a compositional function that uses gates and error carousels in the memory structure. Flexibility is achieved through multiplicative operations between word-embeddings, and leads to better compositional results compare to the additive ones. Empirically, the model outperforms many classifiers and feature engineering techniques. On the other hand, RcNN are proved to be powerful for sequential data processing. Nevertheless, end-to-end training mechanisms like connectionist temporal classification can be used to train RcNN for sequence labelling problems. The combined approach with LSTM-RcNN yields fruitful results in many NLP problems. A DRcNN model that combines multiple levels of representation have proved effective with the flexible use of long range context [22]. The work admits deep LSTM-RcNN achieve decent results when trained end-to-end with suitable regularization.

2.3 Deep Recursive Neural Networks

The general idea of deep recursive neural networks (DRNN) is discussed in Sect. 1. Based on the principles of DRNN and parse tree representation, numerous applications have been proposed. In [4], the task of fine-grained sentiment classification that uses binary parse trees as input representation is addressed. The model is constructed by staking multiple recursive layers each on top of other. The performance of the model is empirically determined against several shallow recursive networks. The research also performs qualitative analysis of the model through input perturbation, and examines nearest neighboring phrases of given samples. The results reveal that adding depth to recursive network captures different aspect of the compositionality. Techniques based on fixed-length representations are used in construction of tree-structured recursive neural networks (TreeRNN) to explore compositional semantics in natural language [23]. In this model, two phrases to be compared are handled independently using a pair of recursive networks that use common parameters. The resultant vectors are input to comparison layers to generate features vector. Finally, the result of the output layer is fed into softmax classifier to compute probabilities over different relations. Two variants, plain TreeRNN and tree-structured neural tensor networks (TreeRNTN) have been evaluated for learning to identify logical relationships such as entailment and contradiction using these representations. When evaluated on SICK challenge data, both models perform better compared to contemporary tree based models.

2.4 Memory-Augmented Networks

Recently, memory-augmented networks have gained much interest in DNN that combines neural networks with explicit memory. As a consequence, [24] address the problem of aspect level sentiment classification using deep memory network. Unlike sequential LSTM, these models explicitly determine the significance of each context word while approximating sentiment polarity of an aspect. Further, feature representations are obtained using multi-layers with shared parameters, that in turn are neural attention models over external memories. Each layer in such a model follows content and location based attention, where the network first learns the weight of each context word and then computes continuous text representation. The text representation is the end layer and considered as the feature vector for sentiment classification. Since, each component is differentiable, the network is efficiently trained end-to-end with gradient descent with cross-entropy error as a loss function. The model proved to be computationally efficient and is independent of any parser or sentiment lexicon.

2.5 Hybrid Deep Neural Networks

There is an increased interest in combining methods and techniques from DNN which carry out information fusion, thereby overcoming the constraints of employing a single network model. Several techniques have been proposed in the literature for sentiment analysis, employing several computational views. In [25], a novel recursive recurrent neural network (RRcNN) that models end-to-end decoding for statistical machine translation is proposed. The model combines RNN and RcNN to integrate their respective capabilities. A new framework that combines CNN and RcNN into a single model on top of pretrained word vectors is discussed [26]. The work utilized LSTM as an alternative for pooling layers to reduce loss of detailed, local information and capture long-range dependencies across the input sequence. Furthermore, a composite representation for questions and answers using combination of CNN and bidirectional long short-term memory (biLSTM) is addressed [27]. The distributional sentence model used in this approach maps queries and answers to their distributional vectors to learn the semantic identicalness between the sentences. In [28], an effective hybrid architecture that combines CNN and recursive networks for sentence-level sentiment analysis is proposed. Additionally, the work used transfer learning from a huge document-level labeled sentiment corpus to enhance the word-embeddings.

3 Background

3.1 Recursive Neural Networks

Recursive neural networks (RNN) are non-linear adaptive models that deal with structured information. Given a directed positional acyclic graph (DPAG), transformations are recursively applied to obtain higher level representations learned from previously computed representations of descendants, i.e., fixed set of weights are recursively applied within a structural organization of the input. This allows the network model to include semantic representations, so that each subsequent layer potentially has a more abstract representation. In general, RNN could be applied to any sort of DPAG. However, for the sake of simplicity and efficiency, the usage of RNN is restricted to positional binary trees (e.g. parse-trees) [3,29]. Hence, for a given binary parse-tree representation of a n-gram phrase, with every leaf node comprised of initial word-embeddings, RNN computes the potential parent representations using following state representations.

$$x(v) = f(W_f \times x(ch[v]) + b) \tag{1}$$

where, $ch[v] = \{ch_1[v], ch_2[v]\}$ represents left and right children of v, $W_f = \{W_l, W_r\}$ are the synaptic weights that connect the left and right children to the parent v, and b is a bias vector. Given that W_l and W_r are square matrices, and $ch_1[v]$ and $ch_2[v]$ are not distinguishable from terminal and non-terminal nodes, then representations at terminal nodes and non-terminal nodes lie in same space. Recursive neural model will then computes parent vectors in a bottom up fashion through the combination of two successive subphrases with same semantic space. Accordingly, a task-dependent output layer on the top of representation layer is defined as follows:

$$y(v) = g(W_g \times x(v) + c) \tag{2}$$

where W_g is the output weight matrix and c is the bias vector to the output layer and $y(v)$ is a class prediction for the node v. In the case of fine-grained opinion classification, $y(v)$ will predict a sentiment polarity of the phrase given by the subtree rooted at v. During the course of supervised training, external errors incurred on y are back-propagated from root towards its descendants [29]. Though terminal and non-terminal nodes are assumed to be similar, it is necessary to make the distinction between themselves. To achieve this, a simple parametrization of the weights W is used to decide whether the incoming edge originates from a terminal or non-terminal node [4]. More technically, the idea is illustrated in the following Eq. 3.

$$z(v) = f(W_f^{ch[v]} \times z(ch[v]) + b) \tag{3}$$

If $z(v) = x(v)$ then v is a leaf node, and $z(v) \epsilon \chi$, otherwise $z(v) \epsilon \psi$, and also if v is terminal node then $W^v = W^{xz}$ otherwise $W^v = W^{zz}$. Then, the sets χ and ψ represents vector spaces of words and phrases respectively. Accordingly, the weight matrix W^{xz} corresponds to transformation matrix from word to phrase space and W^{zz} corresponds to transformation matrix from phrase space to itself. Furthermore, the dimension of W^{xz} is $|z| \times |x|$ and W^{zz} is $|z| \times |z|$, meaning that even a large pretrained word vectors and a small number of hidden units could be used without a quadratic dependence on the word vector dimensionality $|x|$.

3.2 Deep Recursive Neural Networks

The above mathematical description proves RNN to be deep in structure. However, they fail to model hierarchical interpretation of data observed in conventional stacked deep learners (e.g. deep feed-forward networks). The fact of depth is understood through the hierarchy among hidden representations, i.e. each hidden layer yields a different representation space and exhibit a more abstract representation of the input space. The notion of depth can be incorporated into traditional RNN by stacking multiple layers of recursive networks. The following Eq. 4 gives the mathematical representation of DRNN.

$$z(v)^{(i)} = f(W_f^{(i)} \times z(ch^{(i)}[v]) + J^{(i)} z(v)^{(i-1)} + b^{(i)}) \tag{4}$$

where i represents the indices of the multiple stacked layers, $J^{(i)}$ is the weight matrix that binds i^{th} hidden layer to the $(i+1)^{th}$ hidden layer, $W_f^{(i)}$ and $b^{(i)}$ are defined as in Eq. 1 within each layer i. Furthermore, it is implicit that in proposed deep architecture, distinction between terminal and non-terminal nodes is required only for the first layer. Hence, mapping is performed using two separate $J^{(i=2)}$ for terminal and non-terminal nodes as $J^{xz(i=2)}$ and $J^{zz(i=2)}$ respectively. In consequence, except the first layer all nodes above this layer are represented in the same space. Finally, class prediction can be obtained by connecting the output layer to the final hidden layer using the following Eq. 5.

$$y(v) = g(W_g \times z(v)^{(l)} + c) \tag{5}$$

where l is the total number of layers. This enables the network to express more abstract information at the final layer which helps supervised decision.

4 Applications

In consideration of the above background knowledge, numerous application areas of DRNN have been identified in the literature. The following list provides some suggested applications in the field of NLP.

Syntactic Parsing: Syntactic parsing is known to be a fundamental aspect of any NLP task. It plays a key role in providing an interface between a linguistic expression and its meaning. Much of the work is tackled with word-category disambiguation (NP, VP etc.), but this kind of representation often fails to capture the rich semantic linguistic information. A DRNN can be used for both parsing natural language text as well as learning vector space models (VSM) for hidden phrases and gives syntactic information to predict phrase structure trees accurately.

Document Summarization: Document summarization aims at creating short text summary from a larger text corpus through the selection of prominent phrases in the documents. Numerous approaches have been proposed for ranking the sentences using domain specific features. The most crucial step in document summarization is to keep track of information coverage and redundancy among the sentences. Learning techniques based on DRNN are proven to be successful at abstractive summarization that generates a summary using language models, but not from transcribing literatim from the document.

Paraphrase Detection: DRNN can also be used effectively in paraphrase detection, where understanding of deep semantics is essential to obtain high accuracy. Paraphrase is defined as the process of rephrasing text or paragraphs in different way. Paraphrase detection is important for applications such as summarization, plagiarism detection, text segmentation, information extraction and question answering, etc.

Word-Embeddings: Distributed representations of words in a vector space, i.e. word-embeddings often useful in learning techniques to improve the performance by grouping identical terms. The word-embeddings obtained using DRNN are very efficient because the learned vectors explicitly encode many linguistic regularities and patterns.

Language Modelling and Prediction: A language model studies the likelihood of occurrence of a word based on previously used sequence of words in a sentence. Deep architectures are preferred methods for developing statistical language models because they use word-embeddings and large context of recently observed words when making predictions.

Machine Translation: Deep networks are also preferred in machine translation, where a source language gets translated into target language. However, the fundamental difference between machine translation and language modeling is that the output starts only after the complete input has been fed into the network.

5 Experimental Setup

To demonstrate the effectiveness of the proposed model, a series of experiments have been conducted, from preparation of datasets to selection of activation functions and network training. In the following subsections, we demonstrate various steps involved in the process of training and evaluating the proposed DRNN architecture.

5.1 Preparation of Datasets

The performance evaluation of the proposed model is examined with two opinion treebank datasets. One such is recently published Stanford sentiment treebank (SST) [4] dataset, which includes real-valued sentiment labels for 215,154 phrases in the form of parse trees of 11,855 opinion statements, with an average length of 19.1 tokens per sentence. The sentiment labels are then mapped to an ordered integer label between 0 to 4, using thresholding to make the supervised task as a five-class classification problem. However, SST is constructed for demonstrating generic fine-grained sentiment classification and may not be suitable for task-specific problems. Hence, a task-specific treebank dataset similar to Stanford sentiment treebank (SST) [4] is prepared with the help of Stanford CoreNLP library. The new treebank dataset includes parse-trees of 7550 opinion statements with respective sentiment labels of several product reviews collected form e-commerce websites such as amazon, imbd and yelp. The DRNN architecture is then trained using single training-validation-test set partitioning on both the SST and newly constructed treebank dataset in two different trials.

5.2 Activation Functions

Based on the research work reported in [30], employing rectifier units for training DNN without pre-training step yields better results. Hence, for hidden layers a rectified linear unit (ReLU), $f(x) = max\{0, x\}$ is employed that returns *0* if it receives negative input, and returns unaltered for positive values of x. And a standard softmax activation, $g(x) = \frac{e^{x}}{\sum_{k=1}^{K} e^{x_k}}$ is used for the output layer. The output of the softmax function is equivalent to a categorical probability distribution, i.e. the probability that any of the classes are true. Experimental results reveals, utilization of rectifier activation units will improve the performance of the network and bring faster convergence.

5.3 Word-Embeddings

Semantic vector space representations have been successful in capturing fine-grained semantic regularities from natural language text content [31,32]. A semantic vector (word-embedding) is a high-dimensional vector consisting of real-valued numbers. Consequently, these vectors can be used as features in solving many NLP problems. However, several approaches in semantic space use single vector to represent a word. This ignores polysemy, i.e. coexistence of many possible meanings of a word. To overcome this problem, a context vector space model (*context2vec*) [33,34] that adopts multi-prototype to learn multiple embeddings for each word is used. The *context2vec* uses a neural tensor skip-gram model (NTSG) to learn the distributed representations of words and topics. In this work, a pretrained model[2] is used to generate 50 dimensional word vectors of sentential contexts of the target words.

5.4 Regularization of Network

It is evident from the previous studies, DNN face the problem of overfitting when trained with a large number of parameters. To address this problem, a recently proposed dropout technique [35] is adopted. In dropout mechanism, certain units from network are dropped during training. However, the selection of units to drop is taken randomly. Initially, each hidden unit in a network is retained with a probability of 1.0, further, it is tuned between 1.0 to 0.5 over the validation set. Since, dropped units are shared, same units of the hidden layer are dropped at each node.

5.5 Training the Networks

The proposed DRNN is trained using stochastic gradient descent (SGD) with a fixed learning rate of 0.01. To update training parameters, a diagonal variant of AdaGrad [36] is applied. AdaGrad performs smaller updates, adapts the learning rate to the parameters, and gives a smooth and faster convergence. Weights are updated regularly after processing a mini-batch of twenty sentences. Number of epochs for training is set to 200. Lastly, the recursive weight matrix Whh within a layer is initialized as $0.5I + \epsilon$, where I is the identity matrix and ϵ is a small uniformly random noise. This indicates, the representation of each node is approximately initialized to the average of its two descendants. The values of depth are tested with l = *3, 4 and 5* and width is varied from ($|h|$ = 100) to ($|h|$ = 200) as suggested in [4]. The width of a network represents number of hidden units in each hidden layer. Furthermore, pre-training step is ignored, because, DNN are always trained using supervised error signal.

[2] https://github.com/orenmel/context2vec.

6 Results and Discussion

The proposed DRNN model is evaluated on both fine-grained opinion classification (five-class problem), as well as binary classification (positive and negative). However, there is no separate training conducted for binary classification. The network trained for fine-grained classification alone is used, which further decodes five-dimensional posterior probability vector into a binary prediction. The experimental results show that DRNN when trained with $l = 4$ and $|h| = 175$ outperforms the other shallow counterparts for the task of fine-grained classification. However, with various depth and width values DRNN even yields better results for binary classification. In addition to this, the model is compared with several previous works. Baselines include naive bayes (NB) that employs bi-grams and the recursive neural tensor network (RNTN) where the composition is defined as a bilinear tensor product, then a matrix-vector RNN (MV-RNN) in which every word is assigned a matrix-vector pair, and variant of deep convolution networks (CNN, DCNN). Table 1 depicts, a comparative analysis of the

Table 1. Performance comparison of DRNN with other methods

Methodology	Fine-grained	Binary
NB [3]	43.2	82.4
MV-RNN [38]	44.4	82.9
RNTN [3]	45.7	85.4
DCNN [37]	48.5	86.8
Paragraph-Vec [39]	48.7	87.8
CNN-LSTM (word2vec) [39]	51.5	89.5
Proposed DRNN (context2vec, SST)	51.2	87.7
Proposed DRNN (context2vec, OwnTB)	**52.4**	88.3

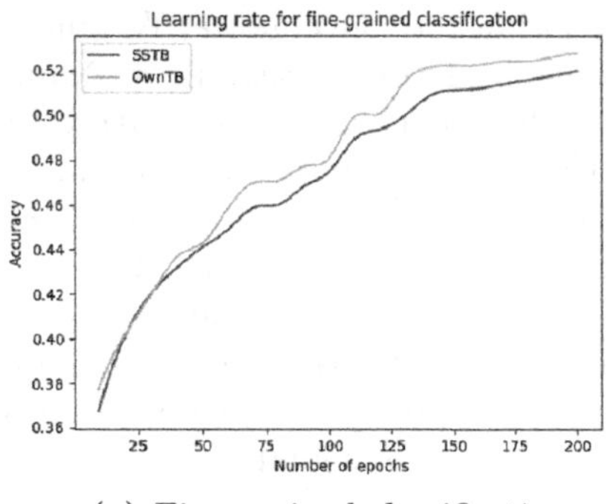

(a) Fine-grained classification

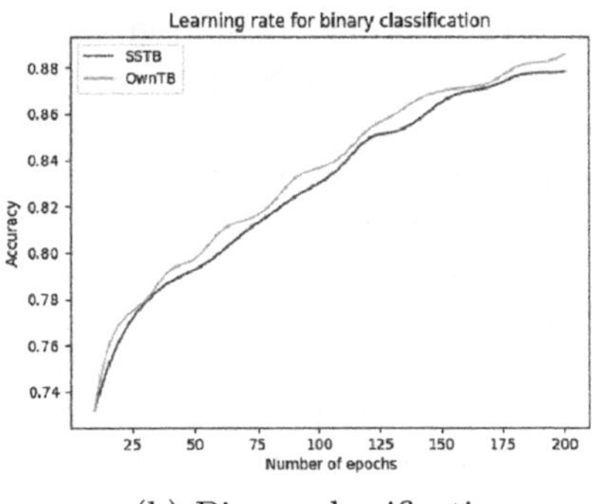

(b) Binary classification

Fig. 2. Learning rate of DRNN model against number of epochs.

proposed DRNN model and other related works. Figure 2(a) and (b) demonstrate the behavior of DRNN trained with two different datasets for fine-grained and binary classifications respectively. The experiment is being conducted for running 200 epochs for training the model. The curves on the plot illustrates the rate at which the model receives training as the number of epochs increases. It is also revealed from the plots that the model exhibits better results when it is trained on our own tree-bank dataset in either cases.

7 Conclusion and Future Directions

In this paper, a deep recursive neural network (DRNN) model that uses stacked framework of multiple recursive neural networks is presented. The proposed architecture is employed for the task of fine-grained opinion classification of online customer reviews. The performance of the model is evaluated against two datasets; a standard Stanford sentiment tree-bank and our own tree-bank dataset. Further, a comparative analysis of several contemporary approaches with the proposed model is performed. Experimental results reveal that DRNN outperform the baselines, accomplishing best performance towards fine-grained opinion classification. As a future enhancement, there is much scope for exploring insights on the functionalities of hidden layers and their benefits. How to tune network models for more generalized classification tasks is always an open issue. Learning our own word vectors with task specific learning corpus can also improve the performance of the model. Validating task-specific deep neural networks require much attention from research community.

References

1. Wang, H., Can, D., Kazemzadeh, A., Bar, F., Narayanan, S.: A system for real-time Twitter sentiment analysis of 2012 US presidential election cycle. In: Proceedings of the ACL 2012 System Demonstrations, pp. 115–120 (2012)
2. Liu, W., Wang, Z., Liu, X., Zeng, N., Liu, Y., Alsaadi, F.E.: A survey of deep neural network architectures and their applications. Neurocomputing **234**, 11–26 (2017)
3. Socher, R., et al.: Recursive deep models for semantic compositionality over a sentiment treebank. In: Proceedings of the Conference on Empirical Methods in Natural Language Processing, EMNLP 2013 (2013)
4. Irsoy, O., Cardie, C.: Deep recursive neural networks for compositionality in language. In: Advances in Neural Information Processing Systems, pp. 1–4 (2014)
5. Bengio, Y.: Learning deep architectures for AI. Found. Trends Mach. Learn. **2**(1), 1–127 (2009)
6. Glorot, X., Bengio, Y.: Understanding the difficulty of training deep feedforward neural networks. In: Proceedings of the Thirteenth International Conference on Artificial Intelligence and Statistics, pp. 249–256 (2010)
7. Gori, M., Maggini, M., Sarti, L.: A recursive neural network model for processing directed acyclic graphs with labeled edges. In: Proceedings of the International Joint Conference on Neural Networks (2003)

8. Elman, J.L.: Finding structure in time. Cogn. Sci. **14**(2), 179–221 (1990)
9. Pascanu, R., Gulcehre, C., Cho, K., Bengio, Y.: How to construct deep recurrent neural networks. arXiv preprint arXiv:1312.6026 (2013)
10. Pascanu, R., Mikolov, T., Bengio, Y.: On the difficulty of training recurrent neural networks. In: ICML-2013 (2013)
11. LeCun, Y., Bengio, Y., Hinton, G.: Deep learning. Nature **521**, 436–444 (2015)
12. Severyn, A., Moschitti, A.: Twitter sentiment analysis with deep convolutional neural networks. In: Proceedings of the 38th International ACM SIGIR Conference on Research and Development in Information Retrieval (SIGIR-15), pp. 959–962 (2015)
13. Poria, S., Cambria, E., Gelbukh, A.: Aspect extraction for opinion mining with a deep convolutional neural network. Knowl.-Based Syst. **108**, 42–49 (2016)
14. Ruder, S., Ghaffari, P., Breslin, J.G.: INSIGHT-1 at SemEval-2016 Task 5: deep learning for multilingual aspect-based sentiment analysis. In: Proceedings of SemEval-2016, pp. 330–336 (2016)
15. Poria, S., Cambria, E., Hazarika, D., Vij, P.: A deeper look into sarcastic tweets using deep convolutional neural networks. In: Proceedings of COLING 2016 (2016)
16. Ukil, S., Ghosh, S., Obaidullah, Sk.Md., Santosh, K.C., Roy, K., Das, N.: Deep learning for word-level handwritten Indic script identification. Cornell University Library (arXiv:1801.01627) (2018)
17. Elman, J.L.: Finding structure in time. Cogn. Sci. **14**(2), 179–211 (1990)
18. Pascanu, R., Gulcehre, C., Cho, K., Bengio, Y.: How to construct deep recurrent neural networks. In: Proceedings of ICLR 2014, pp. 1–10 (2014)
19. İrsoy, O., Cardie, C.: Modeling compositionality with multiplicative recurrent neural networks. In: Proceedings of ICLR 2015 (2015)
20. Young, T., Hazarika, D., Poria, S., Cambria, E.: Recent trends in deep learning based natural language processing. Comput. Intell. Mag. **13**, 55–75 (2018)
21. Hochreiter, S., Schmidhuber, J.: Long short-term memory. Neural Comput. **9**(8), 1735–1780 (1997)
22. Wang, X., Liu, Y., Sun, C., Wang, B., Wang, X.: Predicting polarities of tweets by composing word-embeddings with long short-term memory. In: Proceedings of the 53rd Annual Meeting of the Association for Computational Linguistics and the 7th International Joint Conference on Natural Language Processing, pp. 1343–1353 (2015)
23. Graves, A., Mohamed, A., Hinton, G.: Speech recognition with deep recurrent neural networks. In: 2013 IEEE International Conference on Acoustics, Speech and Signal Processing (2013)
24. Bowman, S.R., Potts, C., Manning, C.D.: Recursive neural networks can learn logical semantics. In: Proceedings of the third Workshop on Continuous Vector Space Models and their Compositionality (2015)
25. Tang, D., Qin, B., Liu, T.: Aspect level sentiment classification with deep memory network. In: Proceeding of EMNLP 2016 (2016)
26. Liu, S., Yang, N., Li, M., Zhou, M.: A recursive recurrent neural network for statistical machine translation. In: Proceedings of the 52nd Annual Meeting of the Association for Computational Linguistics, pp. 1491–1500 (2014)
27. Hassan, A., Mahmood, A.: Convolutional recurrent deep learning model for sentence classification. IEEE Access **6**, 13949–13957 (2018)
28. Li, Z., Huang, J., Zhou, Z., Zhang, H., Chang, S., Huang, Z.: LSTM-based deep learning models for answer ranking. In: 2016 IEEE First International Conference on Data Science in Cyberspace (DSC) (2016)

29. Van, V.D., Thai, T., Nghiem, M.-Q.: Combining convolution and recursive neural networks for sentiment analysis. In: Proceedings of the Eighth International Symposium on Information and Communication Technology 2017. ACM (2018)
30. Socher, R., Lin, C.C.-Y., Ng, A.Y., Manning, C.D.: Parsing natural scenes and natural language with recursive neural networks. In: Proceedings of the 28th International Conference on International Conference on Machine Learning (ICML 2011), pp. 129–136 (2011)
31. Glorot, X., Bordes, A., Bengio, Y.: Deep sparse rectifier networks. In: Proceedings of the 14th International Conference on Artificial Intelligence and Statistics, vol. 15, pp. 315–323 (2011)
32. Mikolov, T., Sutskever, I., Chen, K., Corrado, G., Dean, J.: Distributed representations of Words and Phrases and their compositionality. arXiv:1310.4546, pp. 1–9 (2013)
33. Pennington, J., Socher, R., Manning, C.D.: GloVe: global vectors for word representation. In: Proceedings of the 2014 Conference on Empirical Methods in Natural Language Processing (EMNLP), pp. 1532–1543 (2014)
34. Liu, P., Qiu, X., Huang, X.: Learning context-sensitive word-embeddings with neural tensor skip-gram model. In: Proceedings of the 24th International Conference on Artificial Intelligence (IJCAI 2015), pp. 1284–1290 (2015)
35. Melamud, O., Goldberger, J., Dagan, I.: Context2vec: learning generic context embedding with bidirectional LSTM. In: Proceedings of the 20th SIGNLL Conference on Computational Natural Language Learning (CoNLL), pp 51–61 (2016)
36. Srivastava, N.: Dropout: a simple way to prevent neural networks from overfitting. J. Mach. Learn. Res. **15**(1), 1929–1958 (2014)
37. Duchi, J., Hazan, E., Singer, Y.: Adaptive subgradient methods for online learning and stochastic optimization. J. Mach. Learn. Res. **12**, 2121–2159 (2011)
38. Socher, R., Huval, B., Manning, C.D., Ng, A.Y.: Semantic compositionality through recursive matrix-vector spaces. In: Proceedings of the 2012 Joint Conference on Empirical Methods in Natural Language Processing and Computational Natural Language Learning (EMNLP-CoNLL 2012), pp. 1201–1211 (2012)
39. Kalchbrenner, N., Grefenstette, E., Blunsom, P.: A convolutional neural network for modelling sentences. arXiv preprint arXiv:1404.2188 (2014)

Effective Emoticon Based Framework for Sentimental Analysis of Web Data

Shoieb Ahamed[1(✉)] and Ajit Danti[2]

[1] Department of Computer Science, Government First Grade College, Soraba, Shimoga, Karnataka, India
shoiabahmed@gmail.com
[2] Department of Computer Science and Engineering, Christ(Deemed to be University), Bangalore, Karnataka, India
ajit.danti@christuniversity.in

Abstract. The Explosive development in the social media domain has created a platform for mass generation of textual and emoticon based web data from micro blogging sites. Sentimental Analysis refers to analysis of sentiments or emotions from such heterogeneous reviews are the present urge of the market. Thus, an effective emoticon based framework is proposed which generates scores of both textual and emoticons into seven layered categories using SentiWordNet and weighs performance of various machine learning techniques like SVM/SMO, K-Nearest Neighbor (IBK), Multilayer Perception (MLP) and Naive Bayes (NB). Using Jsoup crawler input reviews are obtained and processed with initial preprocessing model for emoticons and text data followed by stemming and POS tagger. Projected framework is investigated on college and hospital dataset obtaining upper attainment level by Kappa statistic metrics having 98.4% correctness and lesses bug value. Proposed Framework showcases greater competence score with lesser FP Rate based on weighted average of correctness measures. The investigational outcomes are tested on training data with Ten-Fold cross validation. The outcome reveals that suggested emoticon based framework for the task of Sentimental analysis can be efficaciously applied in online decision job.

Keywords: Sentiment analysis · Opinion mining · Emoticon · SentiWordNet

1 Introduction

Development in web technology and their rapid advances have produced path for generation of vast size of data in the web for internet users. Sentimental analysis refers to method of obtaining attitude of any person relating to a specific theme from a vast group of opinions or reviews which are openly presented in web. Thus mining or analysis of sentiment is most needed. Assume a web user needs to perform buying of goods, get details from people in sense of their opinions. Its a cumbersome task to derive a decision from these huge available reviews in

K. C. Santosh and R. S. Hegadi (Eds.): RTIP2R 2018, CCIS 1037, pp. 622–633, 2019.
https://doi.org/10.1007/978-981-13-9187-3_55

web. This demand for an automated system to mine the goodness and badness of a product which will be effective and efficient for the web users for making the task of decision making in online buying and selling.

Internet is a proven platform for sharing opinions, online learning and transferring thoughts. Social Web interface sites like Twitter, Google+ Facebook and so on have swiftly made popularity as they allow public to prompt and share onea^{TM}s views about issues, make conversation with various communities, or post messages across the world. Now-a-days most online reviews, posts, comment and tweets are posted in the emoticons ways such as (e.g. Happy emoticons: ":-)", ":)", "=)", ":D" and Sad emoticons: ":-(", ":(", "=(", ";("). Use of emoticons allows an online user to showcase his emotions about a topic, product, issue in a more graphical and easy way. Thus in the task of sentimental analysis of online reviews, emoticons are as important as a textual review is in an online data. The proposed work highlights the importance of emoticons and considers them as a vital feature for sentimental analysis of web data. In the proposed work a seven level classification of sentiments is done using SentiWordNet as a lexical resource.

Dataset refers to all the needed information that is vital for the task of experimentation of any system. In this work dataset is created by collecting the reviews from the twitter social site. For experimentation the web domains used in this work include various colleges and hospitals in Karnataka state.

2 Literature Survey

various work has been conducted in the area of opinion or sentimental analysis. Pang et al., [8] has suggested sentiment analysis approach with different machine learning algorithms for the task of categorizing movie opinions by implementing POS Tagger and Scored using SentiWordNet into three categories like negative, positive and neutral. Dang et al., [2] have proposed a method called lexicon-enhanced that represents a group of sentiment words and datasets like Books, Digital cameras, Electronics, DVDs. For experimentation they have used three Sentiment features and Ten-fold cross validation evaluation is performed.

Li et al., [5] has instrumented with a multiple domain based framework. DVD, BOOKS Electronics and Kitchen from Amazon are used as Multi-Domain Dataset. Here in task of pre-processing stop words were removed and case conversion is done. Term frequency for weighting features was used. Linearly-Separable manner framework is used. For the task of classification LIBLINEAR SVM model is used.

Santosh et al., [1] have suggested categorization learning method that used noise with label depending on many way visions. This method gets cases to label with upper influence. In another model, When a case A is said to have upper influence it selects a case with good agree factor than a disagree factor. These approaches are compared on different datasets, which has got good accuracy compared to other methods.

Vajda et al., [10] have proposed a quick approach using kNN to classify patterns. A generic strategy to speed up the process of distance calculation, training data is clustered. Then an iterative method is used to select data closest to the cluster centers and far from the cluster centers. This method has got efficiency by reducing up to 71.

Li et al., [6] have deviced a new clustering method to be worked on Ka"mean model. Here positive and negative clusters are created for the documents. Here weighting was applied. Last outcome is obtained with recursive working of said process. Movies data is collected to perform experimentation which has 1000 negative and positive opinions with an accuracy of 77.17.

Liang et al., [4] proposes a micro blogs sentiment obtaining approach implementing maintained learning which does the task of extracting, understanding with categorization of opinion tweets i.e., mining opinions. for selecting the data feature chi-square is used with M-inclusion. over the period of 2012's November to 2013's January cell phone, Films dataset is used in the implementation of this work. 91

Vaidya et al., [9] has suggested am sentiment analysis method incorporating a improved version of analysis of data at three levels of classification using preprocessing techniques.

3 Proposed Methodology

An effective emoticon based framework is proposed in this work that deploys SentiWordNet to generate emoticon and textual tweet counts and labelling them into seven levels as strong-negative, weak-negative, negative, positive, strong-positive, weak-positive and neutral tweets.

College and hospitals Online web tweets both textual and emoticons are examined. For the proposed seven class of classification levels the generated score values are applied with various machine learning technique of IBK - K-Nearest Neighbor, SMO/SVM - Support Vector Machine, Multilayer Perception (MLP) and Naïve Bayes (NB) which have resulted in superior results. Since emoticons are considered for sentimental analysis of reviews, the proposed framework is best suited for online users to take online decisions.

Figure 1 depicts twitter's online college reviews SentiWordNet's output obtained by the proposed emoticon framework. Every statement in output showcase a tweet reviews numerical score with its related score type.

Projected framework is experimented with twitters tweets on college and hospital domains those are extracted from the Jsoup Crawler. The projected emoticon enabled framework for the task of sentimental analysis is depicted in Fig. 2.

A large amount of reviews from twitter^{TM}s portal are obtained by implementing the web crawler named - JSoup. This JSoup will spawns pure sentences and excludes every tag data by parsing each page. The finally obtained input reviews are then deposited in a way of textual forms or files.

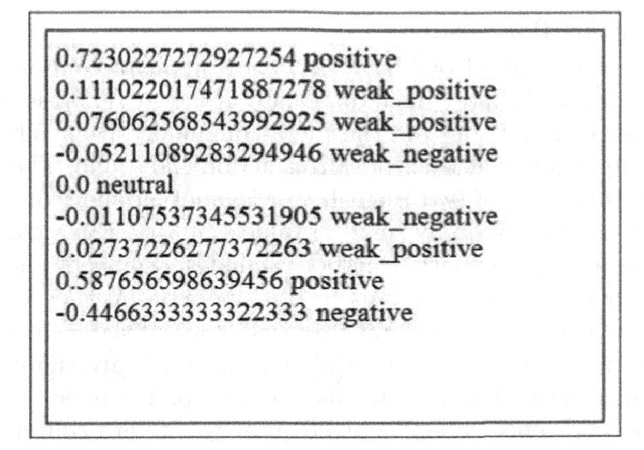

0.7230227272927254 positive
0.111022017471887278 weak_positive
0.076062568543992925 weak_positive
-0.05211089283294946 weak_negative
0.0 neutral
-0.01107537345531905 weak_negative
0.02737226277372263 weak_positive
0.587656598639456 positive
-0.4466333333322333 negative

Fig. 1. Sample output of proposed framework.

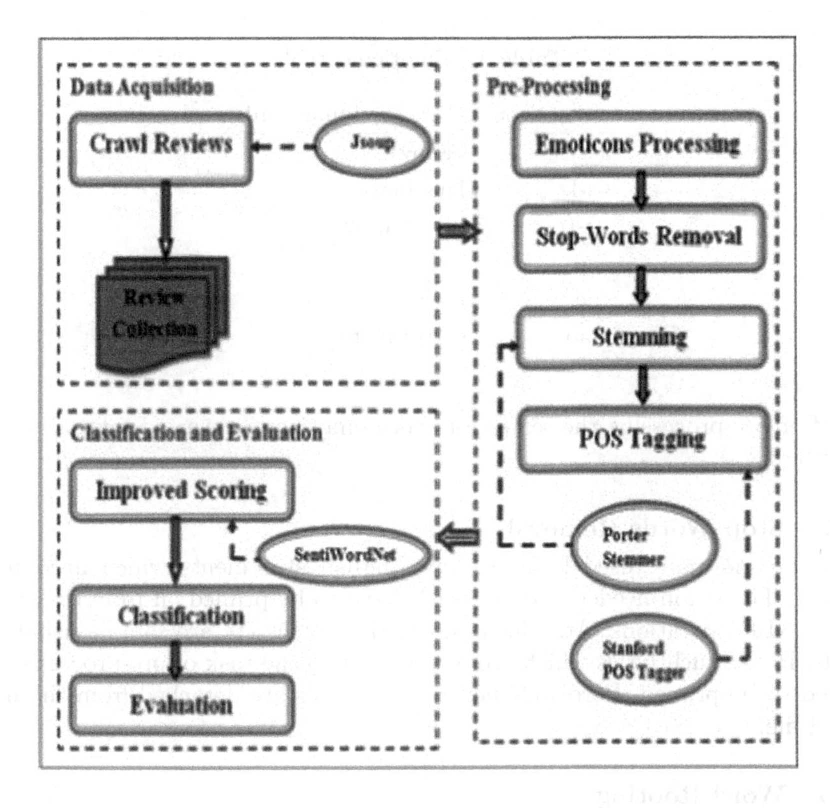

Fig. 2. Proposed emoticon based framework of sentimental analysis.

3.1 Pre-processing

3.1.1 Emoticons Processing

In present technology emoticons have become a mode of communicating emotions. Todaya^{TM}s technology includes many ways to express the emotions through emoticons. Emoticons are the meta-communicative symbolic representation of the expression in absence of words, action and sound. The name Emoticon is the amalgamation of two English words emotion and icon, which means exchanging a textual emotion by small graphical images. Emoticons are created using combination of punctuation marks (includes alphabets and digits also). Ita^{Tm}s a fact that our brain responds to the emoticons as they were real faces. Hence each emoticon is a sign for some textual word. If these emoticons are not given a proper significance in the task of sentimental analysis, our system may end up with inefficient analysis towards the user online reviews. Thus, in the proposed framework once college and hospital tweets are collected from web, every input emoticon is substituted by its matching textual data value. Thus we arrive at enhanced input data reviews from the raw data collected through the web crawler. Below are some examples showing emoticons and its corresponding words (Table 1).

Table 1. Emoticons table

Emoticons	Corresponding word
:-)	happy
:-D	laughing
:-))	very happy
:(	sad
:c	frown
:-o	surprise etc.

After pre-processing the scores for every emoticon converted textual word is generated.

3.1.2 Stop Words Removal

These are normally used lyrics in any English statements which need to be ignored. These non-needed textual words have to be pruned off from the system in any NLP operations. At, which, a, on, the, an, is etc are sample usual stop words. Hence such lyrics which are not needed to the task of pre-processing are needed to be pruned. Here 4027 non-needed lyrics are detached from the input stored file.

3.1.3 Word Rooting

Twitter opinions are usually casual language those having huge variety of web based flaky jargon embedded with in any of the needed comments. These may

represent any of the extra ing formats, or bunch of slang styled terms and so on. Such form of terms have to be rescanned by the automated system and it has to be pruned at its stem level for retaining only the root term of the slang word that can be a more filtered way of keeping the input text ready for the fast processing of the future operations in the projected methodology. This framework deploys Porter Stemming procedure. Following are the stages deployed in the process.
Stage 1: prune the all –ed. -ing and plurals.
Stage 2: take it's changed from i to y station.
Stage 3: Plan for dual suffixes to solos.
Stage 4: handle all the suffixes with full and ness and so on.
Stage 5: get all -ant, -ence, etc.
Stage 6: Rejects the last –e
Following are some examples showing process of stemmer.

- Possesses -possess
- Operatives - operative
- Interesting - interest
- Confess - confess
- Infuriating – Infuriate
- Predicate – predict etc.

3.1.4 Labelling with POS

POS Labeler will be next used in the process to perform the task of connecting all the words with its relevant linkable tag values. These tag values represent the part of speech identifier like rbr, vbz, jj, nn and so on. The stated POS tagging is linked with every scanned words in the comment by Eqs. (1) and (2).

$$T(W_{i,n}) = arg, max_{t1,n} max \; xe^{-x^2} \prod_{i=1}^{n} P(W_i \mid W_{i,i-1}) \tag{1}$$

$$T(W_{i,n}) = arg, max_{t1,n} max \; xe^{-x^2} \prod_{i=1}^{n} P(t_i \mid W_i) \tag{2}$$

Where 'W' refers ot word, 't_i' refers to tag. This tagging scenario is refered to as arrangement of tags in the form of 't_i, n'.

In this labelling ME - Maximum Entropy model is deployed that is same as stochastic tagging that used a Penn Treebank tag set. The Table 2 depicts sample POS tag with labels.

3.2 Classification

The proposed emotion based framework for effective sentimental analysis deploys the community collection SentiWordNet which will issue emoticon containing words into different seven labels as strong- negative class, Negative class, weak-negative class, Neutral Class, weak-positive class, Positive Class and strong-positive class and mark its label values. These obtained label values are very

Table 2. List of POS tags

POS tags	Description
NNS	Noun, plural
NNP	Proper noun, singular
CC	Coordinating conjunction
VB	Verb, base form
VBD	Verb, past tense
RBR	Adverb, comparative
IN	Preposition or subordinating conjunction
JJ	Adjective
JJR	Adjective, Comparative
JJS	Adjective, Superlative
PRP	Personal pronoun

crucial in the working of the projected model of sentiment calculation. The SentiWordNet, a lexical resource, is implemented in pure object based java programming coding skill set to the task of generate the score count of every emotion and textual words in the input data.

The proposed paper uses Stanford POS Tagger. The Table 3 shows SentiWordNet tags and its translation.

Table 3. SentiWordNet tags and its translation to POS tags

SentiWordNet tag	POS tag
a (adjective)	JJ, JJR, JJSl
n (noun)	NN, NNS, NNP, NNPS
v (verb)	VB,VBD, VBG, VBN, VBP VPZ
r (adverb)	RB, RBR, RBS

4 Experimental Results

The proposed framework is investigated with different twitter^{TM}s input data ranging with diverse domains of internet like College tweets data and hospital tweet data. In the course of experimentation count values are determined by obtaining the scores of both textual words and emotions. For the task of classification and sentimental analysis the obtained counts are used.

Input College Review Dataset is equipped by gathering twitter users reviews pertaining to college domain namely KLES College of Engineering and Technology (KLESCET), Maharaja College, PES College, and RVCE Bangalore and hospital domain namely Apollo Hospital, Brookfield hospital and Colombia Asia Hospital Bangalore. Tweets of around 2150 online users are gathered for testing.

Both the datasets emoticon processing is done which generates the textual values for each emotion encountered in the tweets. Further the enhanced input reviews are processed with stop words removal and stemming followed by tagging using StanfordPOS. Using SentiWordNet scores are generated for tagged documents. The results obtained from the experimentation reveal an higher efficacy of the model proposed. Table 4 depicts the Score Count of College twitter reviews datasets.

Table 4. Score count of college datasets.

Class	KLESCET	Maharaja	PES	RV
Strong positive	299	25	120	322
Positive	398	288	599	386
weak positive	179	507	278	200
Neutral	402	310	255	307
Weak negative	101	155	59	47
Negative	21	20	15	12
Strong negative	1	0	1	0

Figure 3 represents the obtained scores values for the College twitter reviews datasets (Table 5).

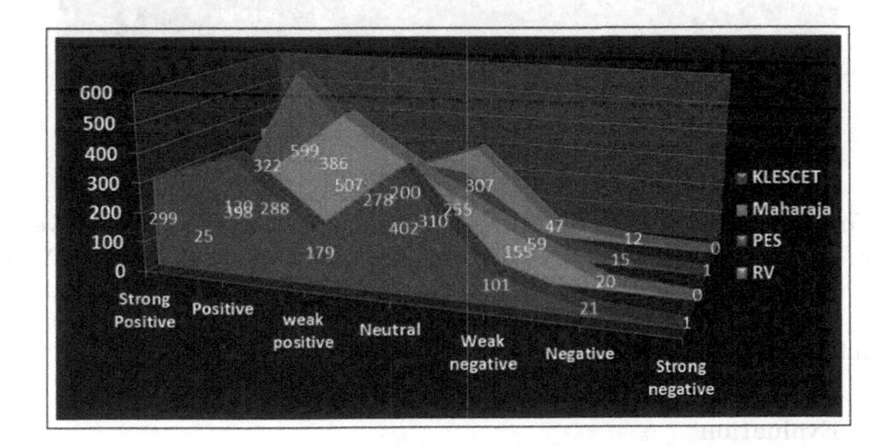

Fig. 3. Representing score of college datasets.

Table 5. Score count of hospital datasets.

Class	Apollo	Brookefield	ColumbiaAsia
Strong positive	190	247	308
Positive	296	201	193
Weak positive	498	468	201
Neutral	300	365	155
Weak negative	200	198	113
Negative	42	21	20
Strong negative	12	20	8

Figure 4 represents the obtained scores values for the Hospital twitter reviews datasets.

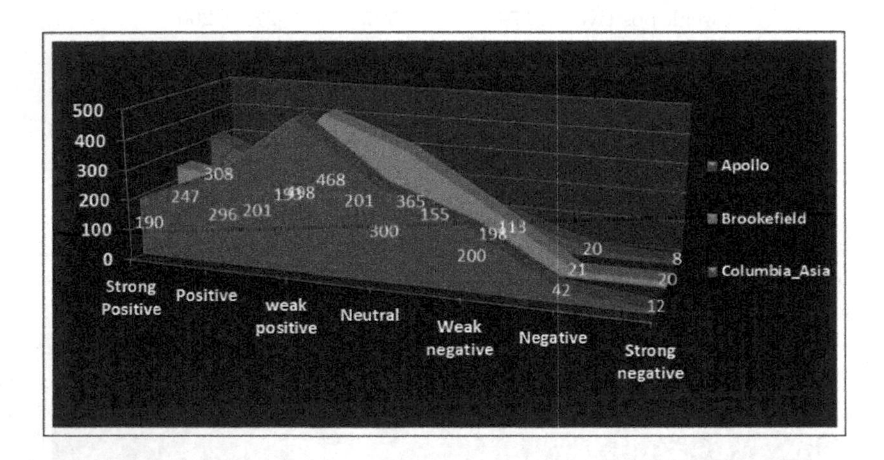

Fig. 4. Representing score of college datasets.

Following supervised learning classification algorithms like Support Vector Machine (SVM/SMO), K-Nearest Neighbor (IBK), Multilayer Perception (MLP) and Naïve Bayes (NB) are applied. Table 6 depict the twitter's college and hospital datasets obtained Precision score gained by various machine learning techniques.

4.1 Evaluation

Table 7 shows the numerically calculated representation of obtained success values when worked with stated statistical metrics.

Table 6. Precision value of different datasets.

Machine learning algorithms	Precision value	
	College	Hospital
Support Vector Machine (SMO)	42.3	55.2
K-Nearest Neighbor (IBK)	56.2	61.3
Multilayer Perception (MLP)	86.2	93.2
Naïve Bayes	84.7	83.3

Table 7. Various accuracy metrics for college tweets.

Sl. no.	Metrics	SVM	IBK	MLP	Naïve Bayes
1	Kappa statistic	43.13%	39.50%	98.40%	81.3.7%
2	Mean absolute error	20.60%	38.40%	09.33%	13.80%
3	Root mean squared error	37.21%	41.21%	18.21%	30.30%
4	Relative absolute error	98.16%	90.16%	31.72%	49.30%
5	Root relative squared error	97.03%	93.12%	43.20%	51.32%

The exactness of the projected evaluation done is judged using the Kappa metrics and is obtained by evaluating the Eq. (3).

$$(kappa)\text{K} = \frac{a_0 - a_e}{1 - a_e} \tag{3}$$

where a_0 is perceived accurateness and a_e is predictable accuracy. Fleiss considers kappa (K) > 0.75 as excellent, 0.40–0.75 as fair to good, and < 0.40 as poor. In this work, the kappa is 0.984, which indicates that this framework has higher efficacy and is an excellent tool task of sentimental analysis.

Means Absolute Error (MAE) which is the Middling absolute variance between classifier forecasted outcome with definite outcome that is obtained from Eq. (4).

$$\text{MAE} = \frac{1}{N}\sum_{i=1}^{N}\left(Desired_i - Actual_i\right) \leq \varepsilon \tag{4}$$

Root Mean Square Error (RMSE) is often handled tool for measuring modifications between value forecasted with definite values that is obtained with the Eq. (5).

$$\text{RMSE} = \sqrt{\frac{1}{N}\sum_{i=1}^{N}\left(Desired_i - Actual_i\right)^2} \leq \varepsilon \tag{5}$$

Table 8 showcases the weighted correctness rate with various correctness metrics such as Receiver Operating Characteristic Curve, Recall, FP Rate, Precision, TP Rate, F-Measure obtained using various machine learning techniques such as Support Vector Machines (SVM/SMO), K-Nearest Neighbour (IBK), Multilayer Perceptron (MLP) and Naïve Bayes.

Table 8. Weighted average rate on various correctness metrics.

Weighted average	TP rate	FP rate	Precision	Recall	F-measure	ROC area
SVM(SMO)	0.453	0.404	0.438	0.419	0.246	0.595
IBK	0.321	0.401	0.311	0.394	0.206	0.502
MLP	0.889	0.098	0.841	0.892	0.879	0.962
Naïve Bayes	0.792	0.186	0.901	0.802	0.732	0.788

Table 8 infers that, projected framework shows upper Precision, F-Measure, TP Rate and Recall Rate and lesser FP Rate by MLP and Naïve Bayes machine learning algorithms, that concludes the projected framework shows greater efficiency which will be a more effective opinion mining tool which can be used by any web client to take any proper decision making task.

5 Conclusion

In the stated work an attempt is made to explore the new methodology for dealing with both the textual as well current running emoticon based comments in the web. Theses emoticons are becoming the most widely used style of expressing the views by any one in internet for showing his feel or views about anything happening in the internet. Projected model is scrutinized on twitter's college and hospital tweets and experimental results showcase the efficiency of the proposed framework with upper accuracy.

References

1. Bouguelia, M.R., Nowaczyk, S., Santosh, K.C., et al.: Agreeing to disagree: active learning with noisy labels without crowdsourcing. Int. J. Mach. Learn. Cybern. **9**, 1307 (2018). https://doi.org/10.1007/s13042-017-0645-0
2. Dang, Y., Zhang, Y., Chen, H.: A lexicon-enhanced method for sentiment classification: an experiment on online product reviews. Intell. Syst. IEEE **25**(4), 46–53 (2010)
3. Virmani, D., Malhotra, V., Tyagi, R.: Sentimental analysis using collaborated opinion mining. Int. J. Soft Comput. Eng. 4(ICCIN-2014) (2014). ISSN 2331–2037
4. Liang, P.-W., Dai, B.-R.: Opinion mining on social media data. In: 14th International Conference on Mobile Data Management (MDM), vol. 2. IEEE (2013)
5. Li, L., et al.: Multi-domain active learning for text classification. In: Proceedings of the 18th ACM SIGKDD International Conference on Knowledge Discovery and Data Mining. ACM (2012)
6. Li, G., Liu, F.: A clustering-based approach on sentiment analysis. In: International Conference on IEEE Intelligent Systems and Knowledge Engineering (ISKE) (2010)
7. Ohana, B., Tierney, B.: Sentiment classification of reviews using SentiWordNet. In: 9th IT&T Conference. Dublin Institute of Technology, Dublin, Ireland, p. 13 (2009)

8. Pang, B., Lee, L., Vaithyanathan, S.: Thumbs up?: sentiment classification using machine learning techniques. In: Proceedings of the ACL-02 Conference on Empirical Methods in Natural Language Processing-Volume 10, pp. 79–86. Association for Computational Linguistics (2002)
9. Vaidya, S., Rafi, M.: An improved SentiWordNet for opinion mining and sentiment analysis. J. Adv. Database Manag. Syst. **1**(2), 1–7 (2014)
10. Vajda, S., Santosh, K.C.: A fast k-nearest neighbor classifier using unsupervised clustering. In: Santosh, K.C., Hangarge, M., Bevilacqua, V., Negi, A. (eds.) RTIP2R 2016. CCIS, vol. 709, pp. 185–193. Springer, Singapore (2017). https://doi.org/10.1007/978-981-10-4859-3_17

Website Analysis: Search Engine Optimization Approach

Vijaykumar Sambhajirao Kumbhar(✉) and Kavita S. Oza(✉)

Department of Computer Science, Shivaji University, Kolhapur, Maharashtra, India
{vsk_csd,kso_csd}@unishivaji.ac.in

Abstract. Search engines play an important role in the popularity of the business. For business to reach masses its website link should appear on the first page of search engine result set. This requires SEO (Search Engine Optimization) focused design of website. Unfortunately website developers do not consider this factor seriously due to this there is lack of search engine optimization tools in designing the websites. One of the crucial website everyone deals with is their bank website and to make bank officials aware about SEO status of their website a dataset of 115 bank website was created for SEO analysis. This analysis resulted in downtrend pattern. Results have shown that banks in India need to focus on search engine optimization while designing their websites. Developers need to consider how Google sort their search results (RankBrain and drain time), CTR (click through rate) etc.

Keywords: Banking sector · SEO · Rstudio · Clustering · Websites

1 Introduction

Nowadays a general resource hub for all the resources like business, education, research, service etc is search engine. Search engines have taken over the conventional searching methods. Almost all business houses, Institutes, financial sectors and Academic Institutes depend on search engines for attracting customers/students/clients. These sectors don't need to invest in marketing by putting advertisements in newspaper or magazines or yellow page directories. Most of the customers are online customers and they search for resources using search engines. This search gives a list of links where the required resource is likely to be found in the form of search result. To get the good business, Company's or Institutes website link should appear in the search result set. Another important aspect associated with search result is the position of the website link in the search result. It is preferred to appear on the first page of search result. For websites links to appear in the search result they need to be search engine optimized (SEO). A set of attributes are associated with websites like contents, links, architecture, social media, trust etc. To make websites search engine friendly these attributes play an important role. Depending on the value of these attributes websites are ranked by the search engines. Role of few attributes is

K. C. Santosh and R. S. Hegadi (Eds.): RTIP2R 2018, CCIS 1037, pp. 634–639, 2019.
https://doi.org/10.1007/978-981-13-9187-3_56

discussed below: Content attribute summarizes the webpage content in one or two sentences. Detail website content need to focus on keywords commonly used by netizens while searching. Keywords should be target customer oriented like a Bank website should have the keywords and their synonyms related to banking sectors commonly used by customers. Search engines uses frequency count of keywords as one of the attribute for ranking. Another interesting attribute of website is Trust. Every business house wants more and more visitors to their website i.e. more and more traffic to their websites. If more people visit a website it creates a reputation for the website depending on the contents of the website. If contents are great then website link can appears at a good position in the search engine result. Traffic to the website can be direct or referral. Website architecture attributes deals with design and planning of website. Websites should be designed in way to increase the page speed as page speed is also one of the attribute used by the search engine. Websites should be planned for fast access i.e. it should not take more time to load. One way to increase loading speed is caching a webpage. Coding in a better way can increase the server speed of web server. URL's should not be messy they should be clean without underscore in them. Clean URL's helps search engines to show relevant URL's in the search result. To make the website SEO friendly www resolve should also be implemented while designing. If the website allows social media sharing option then it drives a good amount of traffic to a website. On a social media if one person likes a website and shares it then most of his/her friends will also visit that website. Social media is also a very good platform for advertising. Links play a key role in popularity of the website as depending on the words in the link user decides whether to click or no. If all the links coming to a webpage are with valid text related to contents of web page then chances of increase in search engine ranking are more.

2 Literature Review

Rehman and et al. reviewed different optimization techniques for individual page pages or the whole website to make them search engine friendly. Also paper analyzes and summaries the core techniques and do the comparative study of the work related to SEO along with guidelines for optimizing the websites [1]. Kumar in his paper introduces the concept of speculative search engine optimization and gave guidelines for how to optimize scholarly literature [2]. Spais examined the possibility of an extension of Bedny's perspective of 'activity' theory as a framework for search engine optimization [3]. Rababah and et al. analyze the influence of local geographical area, in terms of cultural values, and the effect of local society keywords in increasing Website visibility [4]. Main aim of search engines is to provide users with the most relevant output to their queries for this their algorithms are constantly revised and updated. Bedi and et al. shows the importance of few techniques of search engine optimization along with strategies used by Google, Microsoft in support of SEO [5]. Khorsheed and et al. improve the search time of search engines to the maximum extent using the K-Means

Algorithm [6]. Singh and et al. discussed different techniques for improving page rank and image search accuracy along with the comparison of page ranking techniques [7]. Meenakshi Bansal and et al. highlighted the on-page optimization include actual code merged with various languages, keyword placement and keyword density [8]. Hussien identified the most popular techniques used to rank a web page highly in Google [9]. Vignesh and et al. introduces the key concepts of a theoretical framework for search engine optimization [10]. Manral and et al. presents an approach to identify efficient techniques used in Web Search Engine Optimization (SEO) [11]. Poongkode and et al. discussed many optimization tactics used for the development of a website viz. ON Page SEO, OFF Page SEO, Black Hat, White Hat and Grey hat techniques [12]. Singh and et al. studied and compared various algorithms and techniques used for cluster analysis using weka tools with aim to present the comparison of 9 clustering algorithms in terms of their execution time, number of iterations, sum of squared error and log likelihood [13]. Khaimar and et al. discussed about the different techniques and improvements of K-means clustering algorithm along with comparison and limitations [14]. Irani and et al. took survey of various clustering techniques along with limitations and advantages [15].

3 Website Analysis

Present work focuses on banking sector domain for SEO compatibility analysis of websites. Data related to search engine optimization for each bank website is collected using site SEO tool available at http://www.site-seo-analysis.com. Dataset of 115 different banks website is created. Following is sample dataset.

Table 1. Sample of dataset

Sr. no.	Total score	Architecture	Content	Links	Social med	Trust
8	55.8	60	78.7	18.5	100	90
9	23.8	65	21.2	15	0	43.7
10	39	14.4	57.7	18.8	100	33.7
11	49.4	75	70.4	18.6	60	29.4
12	39.7	75	47.5	18.5	100	20.1
13	37.3	55	53	18.8	0	20

Using above mentioned SEO tool we could collect data related to their SEO optimization. As observed in Fig. 1. First attribute is total score which gives over all grading for SEO optimization. Second attribute architecture gives total score of SEO, Architecture attribute gives the score related to how friendly website design is to search engines. Contents of the website should be well tagged so that search engines can index them easily. Number of links coming to website also plays a role in search engine optimization as it indicates how many other websites are directing visitors to your website. Social media is another attribute which can help in popularity of the website if website is shared by more number of visitors on their social network. Trust is related to popularity of the website indicating users visiting website again and again.

4 Experimental Work

Rattle package of Rstudio [16] was used for experimental work. For clustering attribute selection was carried out using weka tool. Attributes which were most significant for clustering were content and link. After selection of attributes dataset was again preprocessed to remove records with maximum null entries so after preprocessing dataset used for analysis had 80 records. Clustering technique was applied on this dataset to cluster data into three categories. Here dataset was clustered into three clusters. Following is the rattle interactive screen image showing three clusters with their size. As observed in Fig. 2 there are three clusters with size 56, 11 and 13 respectively. It can also be observed that sum of square error within clusters is also negligible. Following Table 1 shows cluster assignments with sum of square error and Fig. 3 shows pictorial cluster assignment.

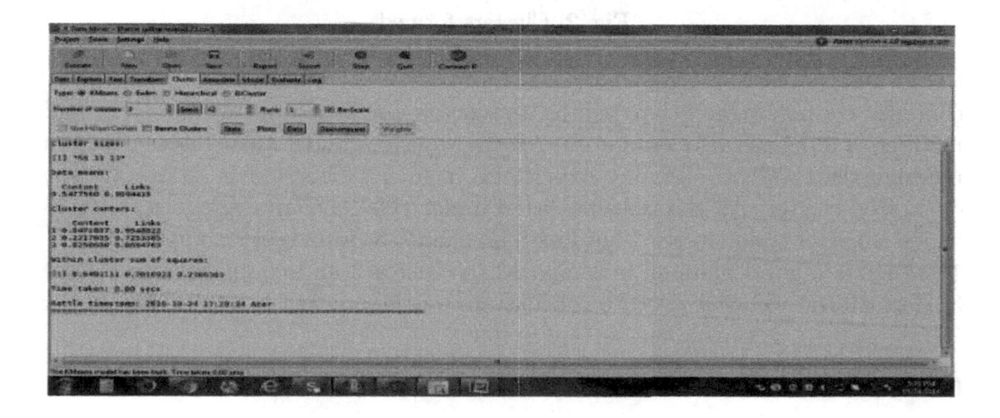

Fig. 1. Clustering using rattle package

5 Cluster Analysis

As indicated in Fig. 3 cluster are distinct and well formed. After analysis of cluster assignment following investigations was carried out. Cluster 3 with 13 records has all the websites which are search engine optimized. Looking at their content attribute its score on average it is 75 with good number of links approximating to 18.5. All the banks whose websites are categorized under Cluster-3 are search engine friendly. Cluster-2 has only 11 websites under it. Here on average

Table 2. Clusters with instances

Cluster no.	No. of instances	Within cluster sum of square error
1	56	0.6493111
2	11	0.7016932
3	13	0.2306303

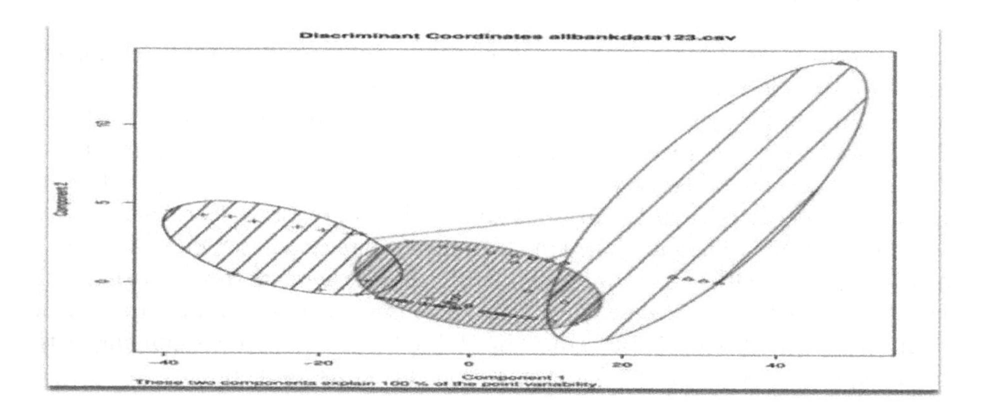

Fig. 2. Clusters formed

contents are not tagged that well as its score is too low, approximately 19 and number of links are also less i.e. 15. All the websites under this cluster needs to redesign their websites keeping seo criteria in focus. Cluster-1 is an average seo optimized cluster. All the websites listed under this have an average score of 48 which is better than cluster 2 but less than cluster 3. Link score is approximately 18 which is a good number as compared to cluster 2 and compatible to cluster 3. Here all the websites need to concentrate on the contents more (Table 2).

6 Conclusion

This is an era of online business. To survive in this online competitors business houses (banks) need to work on their websites as they represent their business online. These websites need to be user friendly and search engine friendly to get good business. There is need for SEO awareness in banking sectors so that they can appear on the first page of search engine results and get noticed by more and more customers. Search engine algorithms are updated very frequently so to keep pace with these one need to update the website regularly by incorporating Google's ranking factor which is currently based on RankBrain algorithm which analyses user interaction with search results, time spent by user on the webpage, click through rate etc.

References

1. Rehman, K., Khan, M.N.A.: The foremost guidelines for achieving higher ranking in search results through search engine optimization. Int. J. Adv. Sci. Technol. **52**, 101–109 (2013)
2. Kumar, A.: Search engine optimization (SEO): technical analysis concepts. Int. J. Emerg. Technol. Adv. Eng. **3**(3), 123–128 (2013). ISO 9001:2008 Certified, ISSN 2250–2459
3. Spais, G.S.: Search Engine Optimization (SEO) as a dynamic online promotion technique: the implications of activity theory for promotion managers. Innov. Mark. **6**(1), 7–24 (2010)
4. Rababah, O., Al-Shboul, M., Al-Zaghoul, F., Ghnemat, R.: Website search engine optimization: geographical and cultural point of view. J. Softw. Eng. Appl. **7**, 1087–1095 (2014)
5. Bedi, G.S., Singh, M.A.: Analysis of Search Engine Optimization (SEO) techniques. Int. J. Adv. Res. Comput. Sci. Softw. Eng. **4**(3), 563–566 (2014). ISSN 2277 128
6. Khorsheed, K.O., Madbouly, M.M., Guirguis, S.K.: Search engine optimization using data mining approach. Int. J. Comput. Eng. Appl. Part I, **IX**(VI), 184–200 (2015)
7. Singh, E.T., Maini, R.: Search engine optimization for improving page rank and image search accuracy. Int. J. Eng. Res. Appl. (IJERA) **3**(4), 1145–1152 (2013). ISSN 2248–9622
8. Bansal, M., Sharma, D.: Improving webpage visibility in search engines by enhancing keyword density using improved on-page optimization technique. Int. J. Comput. Sci. Inf. Technol. **6**(6), 5347–5352 (2015). ISSN 0975–9646
9. Hussien, A.S.: Factors affect search engine optimization. Int. J. Comput. Sci. Netw. Secur. **14**(9), 28–33 (2014)
10. Vignesh, J., Deepa, V.: Search engine optimization to increase website visibility. Int. J. Sci. Res. (IJSR) **3**(2), 425–430 (2014). ISSN (Online) 2319–7064
11. Manral, J., Hossain, M.A.: An innovative approach for online meta search engine optimization. In: The 6th Conference on Software, Knowledge, Information Management and Applications, Chengdu, China, vol. 57, pp. 1–7, 9–11 September 2012 (2012)
12. Poongkode, J.P.S., Nirosha, V.: Study on various search engine optimization techniques. Int. J. Innov. Res. Comput. Commun. Eng. **2**(11), 6738–6742 (2014). (An ISO 3297: 2007 Certified Organization), ISSN (Online) 2320–9801 ISSN (Print) 2320–9798
13. Singh, P., Surya, A.: Performance analysis of clustering algorithms in data mining in weka. Int. J. Adv. Eng. Technol. **7**(6), 1866–1873 (2015). ISSN 2231–1963
14. Khaimar, V., Patil, S.: Efficient clustering of data using improved K-means algorithm: a review. Imp. J. Interdiscip. Res. (IJIR) **2**(1), 226–230 (2016). ISSN 2454–1362
15. Irani, J., Pise, N., Phatak, M.: Clustering techniques and the similarity measures used in clustering: a survey. Int. J. Comput. Appl. **134**(7), 9–14 (2016). ISSN 0975–8887
16. RStudio Team: RStudio: Integrated Development for R. RStudio Inc., Boston, MA (2015). http://www.rstudio.com/. Accessed 23 Mar 2016

Efficient Feature Selection Based on Modified Cuckoo Search Optimization Problem for Classifying Web Text Documents

Ankita Dhar[1(✉)], Niladri Sekhar Dash[2], and Kaushik Roy[1]

[1] Department of Computer Science, West Bengal State University, Kolkata, India
ankita.ankie@gmail.com, kaushik.mrg@gmail.com
[2] Linguistic Research Unit, Indian Statistical Institute, Kolkata, India
ns_dash@yahoo.com

Abstract. The continuous increase of information in the web with varying dimensions is becoming difficult for users to filter and analyse them efficiently as it incorporates redundant and irrelevant terms. Managing, filtering and organizing such huge datasets need the classification of text documents to be performed. Text classification is the process of assigning the text documents to their predefined text categories based on the content. The aim of this paper is to explore Cuckoo search optimization (CSO) problem established from the behaviour of cuckoo birds for selection of relevant features by modifying the algorithm. The revised algorithm is named as modified Cuckoo search (MCS) optimization algorithm that can be proved to be useful for developing an efficient text classification system. The proposed method is generated by combining the ability of MCS with the sharpness of Naive Bayes Multinomial (NBM) algorithm for generating proper feature which increases the rate of success. The approach adopted here is tested on 9000 text documents that cover eight different domains fetched from several web sources and obtains encouraging outcome. The results compared with the results from other well-known approaches for text classification task show the effectiveness of the proposed approach as an automatic Bangla text classification system.

Keywords: Text classification · Meta-heuristic · Modified Cuckoo search · Feature selection · Naive Bayes Multinomial

1 Introduction

The great advancement in information technology field has made it possible to store huge amount of large dimensional data with noisy, redundant and irrelevant terms or features. These datasets are mainly comprised of unstructured data which does not suited properly to any systems. When the crawling is performed over several web sources, it becomes necessary to classify or categorize

K. C. Santosh and R. S. Hegadi (Eds.): RTIP2R 2018, CCIS 1037, pp. 640–651, 2019.
https://doi.org/10.1007/978-981-13-9187-3_57

these unstructured text documents into their respective categories based on some criteria. Maintaining, indexing, organizing and filtering such large amount of documents collectively is not an easy job. To manage such datasets we need to perform the text classification task of related text documents. Information retrieval is becoming an important form of accessing information and text document classification plays a major role by assigning the text documents into their respective text categories or domains from such a large volume of datasets based on their contents. Though various meaningful work has already been done, more efforts need to be paid for certain languages especially for Indian languages to enhance the expertise of the text document classification procedure.

Here, text classification is a multi-step supervised learning process where the text documents are grouped into predefined number of text categories based on their contextual similarity to each other and using template documents as training set to establish the contextual support for each category. This becomes useful while dealing with an unspecified cluster of unstructured text documents. The process begins by carrying out pre-processing techniques such as tokenization and removal of stopwords; then by extracting and selecting the terms followed by weighing them and ultimately a vector representation of the documents is created. This representation is often found to be the closeness between the pairs of terms; and in the classification process the random set of text documents are classified to their respective categories fully or partially.

While dealing with unstructured text documents, it is difficult to select features with the increasing dimension of features. Thus various feature selection techniques are applied for selecting relevant features from the documents. Yang and Deb [23] proposed the meta-heuristic Cuckoo search algorithm which is integrated with various functions and named as Parallel Cuckoo Search Optimization (PCSO). In this paper an evolutionary algorithm that incorporates functions, such as cost and purity, along with the basic concept of Cuckoo search optimization using Naive Bayes Multinomial algorithm is proposed. In this work, Naive Bayes Multinomial is chosen as a classification algorithm because of its advantages over other Bayesian approaches. It is mostly design keeping the nature of text documents in mind and uses a multinomial distribution for all the features being selected.

The rest of the paper is sorted out as follows: The related work along with the application of Cuckoo search optimization problem in text classification task is discussed in Sect. 2. The methodology being adopted is given in Sect. 3 followed by the results analysis in Sect. 4. Lastly, Sect. 5 concludes the paper.

2 Related Works

2.1 Text Classification with Traditional Approaches

The literature survey shows that several works have already been performed for resourceful languages except those languages not having standard datasets, resources and tools. For English text classification, DeySarkar et al. [6] implemented technique based on clustering on 13 databases using NB classifier. Guru and Suhil [9] worked with the feature based on the occurrence of a word in a category along with SVM and KNN for classifying 20 Newsgroup documents. Jin et al. [12] classified two databases using bag of embeddings concept and SGD. Wang et al. [21] performed experiment on two standard databases using tf and t-test methods and obtained quite satisfactory outcome. For categorizing Arabic medical texts, Al-Radaideh and Al-Khateeb [1] applied classifier depending on rules. For categorizing Punjabi documents, Gupta and Gupta [8] worked on hybrid approach by combining NB and Ontology classification. ArunaDevi and Saveeth [3] worked using cfeatures acquired from Mozhi and CIIL texts. Bolaj and Govilkar [4] classified Marathi documents using dictionary along with Modified KNN, MNB, ontological and SVM. For Bangla, N-Gram technique was proposed by Mansur et al. [16] to classify news texts obtained from one-year data of a single newspaper. Mandal and Sen [15] provided the comparison among four algorithms for classifying 1000 news texts with 22,218 tokens and obtained maximum accuracy of 89.14% for SVM. Kabir et al. [13] used SGD for classifying documents from 9 categories and obtained an accuracy of 93.85%. Various classifiers were used for comparison purpose. Islam et al. [11] applied Tf-Idf and Chi square and gained accuracy of 92.57% using SVM.

2.2 Text Classification Using Cuckoo Search Optimization Problem

Redmond et al. [18] proposed two class tweet sentiment classification architecture using hybrid method based on the information from Emoticons and Emojies by keeping the textual interpretation intact. They proposed hybrid approach by compiling several methods such as SVD, Extended Binary Cuckoo Search, SVM and other methods. The experiment shows the approach performs better involving Extended Binary Cuckoo Search and SVM. Aly and Kelleny [2] implemented the Cuckoo Search algorithm for efficiently clustering the text documents. They have used dynamic nests initialized with different surviving selection of initial centroids, updated using Lévy flight and evaluated to choose the best solution. Their method proves to be better than K-means algorithm. Sujana et al. [19] implemented parallel cuckoo search optimization (PCSO) algorithm along with Naive Bayes classifier for selecting feature vector with minimum features from seven different datasets consisting of balanced and imbalanced classes. Their results show higher accuracy compare to other algorithms. Rautray and Balabantaray [17] proposed cuckoo search algorithm for optimal selection of sentences for text summarizer. They mainly focused on content coverage and length in the summaries. The proposed model maintained the inter-sentence and sentence-to-document relationship by using cosine similarity measure. They evaluated their experiment on DUC dataset using ROUGE tool and the result is significant in multi-document summarization.

3 Proposed Methodology

The proposed adopted Bangla text classification method includes several stages: pre-processing that involves tokenization and removal of stopwords followed by applying Tf-Idf-Icf scheme for extracting features referred in Sect. 3.3. Then modified cuckoo search optimization problem is used for selecting features to train the Naive Bayes Multinomial classifier. The outline (Fig. 1) illustrates the detailed description of the overall methodology adopted in this experiment for developing an automatic Bangla text categorization system.

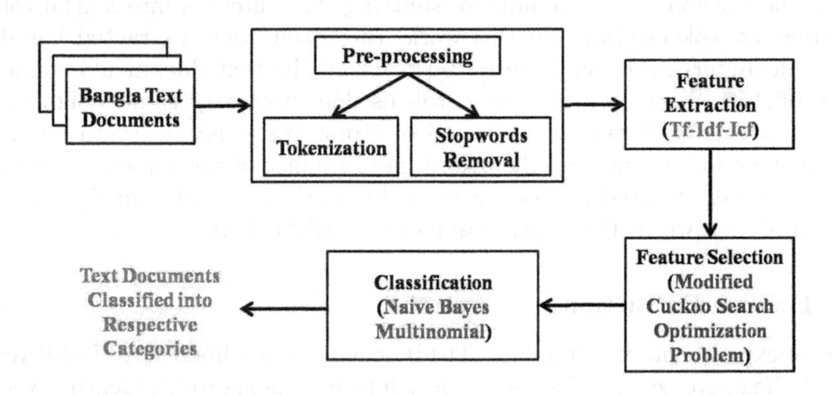

Fig. 1. Outline of the methodology

3.1 Data Collection

Data collection is the most important task for carrying out any experiment. The quality as well as quantity plays a major part in the outcome of any experiment. While developing a database attention needs to be paid so that errors cannot occurs in the further processing. A total of 9000 Bangla texts was acquired from eight different text categories. The detailed distribution of the text documents is provided in Table 1. The links from which the texts were obtained is provided in [7].

Table 1. Text documents distribution for each domain

Domain	Distribution
Business	1100
Entertainment	1100
Food	1000
Medical	1000
State News	1400
Sports	1400
Science & Technology	1000
Travel	1000
Total	**9000**

3.2 Pre-processing of Texts

Prior to the extraction and selection of features from the text documents using various approaches, the task of tokenization and stopwords removal is necessary. Every documents involves sentences that consists series of characters, words and phrases. Thus the sentences are required to be segregated into words to obtain the features. The technique of splitting the sentences into lexical tokens is termed as 'tokenization'. In this work, the tokens were extracted based on 'space' delimiter. Total terms retrieved from all the text documents counts to be **28,06,159**. Since each and every tokens does not carry same weightage to be retained in the feature set, removal of stopwords is necessary to lower the dimension as well as increase the overall performance of the system. The list of words that were treated as stopwords in this work is provided in [7] and after removal of such words the tokens results to be **13,10,119**.

3.3 Feature Extraction

For this experiment, the standard Tf-Idf scheme is modified into Tf-Idf-Icf [7] by including inverse class frequency (Icf) into it to generate the feature vector. The Tf determines the presence of a term i in a document x. The Idf determines the significance of i in x using Eq. 1 where D is the dataset and Df (Document frequency) is the count of texts involving i in it.

$$Idf(i) = log\frac{D}{DF(i)} \tag{1}$$

The Icf determines the presence of i in total C text categories. The Tf-Idf-Icf approach for a term i in a document x is the product of Tf, Idf and Icf.

$$Icf(i) = log\frac{D}{CF(i)} \tag{2}$$

$$Tf - Idf - Icf = Tf(i) * Idf(i) * Icf(i) \tag{3}$$

3.4 Feature Selection

The standard Cuckoo search optimization problem is adapted and modified in this section. The revised Cuckoo search optimization problem is named as Modified Cuckoo Search (MCS) optimization problem to select relevant features from the feature space. The parasitic behaviour of Cuckoo is extremely fascinating. The Cuckoo lay eggs in a host nests and often mimic the exterior characteristics like colour and patterns of host eggs. While their eggs get identified by the host, either it is thrown away or the host simply evacuate its nest. Using this concept, Yang and Deb [23] established a novel Cuckoo Search (CS) optimization algorithm and summarized it using the following order:

1. A nest is selected randomly by a cuckoo to lay eggs.

2. The number of nests is fixed, and the best nests with high quality of eggs will be provided to the next generations.
3. If the cuckoo egg is identified by the host; either the egg is thrown away or host evacuate the nest.

Based on the above mentioned rules, the steps of the Modified Cuckoo Search (MCS) have been encapsulated in the following algorithm.

Input: the feature set, f_z;
Generate n host nests;
While (iteration < max_generation)
Get a cuckoo randomly and replace its solution by performing Levy flights using Eq. 4;

$$z_{new} = z + lLevy(\lambda) \tag{4}$$

where l is the step size and $Levy(\lambda)$ is estimated from a standard normal distribution
Evaluate the purity or fitness with the help of Eq. 5;

$$Purity(F_{z_{new}}) = \frac{\sum_{i=1}^{d}(F_i * w_i) * 100}{N} \tag{5}$$

if $(F_{z_{new}} > F_z)$
Replace current solution by the new nests;
end if
When the worst nests are discarded, new nests are built and cost is calculated by 6;

$$Cost = \frac{\sum_{i=1}^{N}\{\frac{\sum_{q=1}^{C_i} S(c_i,d_{iq})}{C_i}\}}{N} \tag{6}$$

where d_{iq}: is the qth text documents in class i;
c_i: is the centroid of class i;
$S(c_i, d_{iq})$: is the distance between text document d_{iq} and centroid of class c_i;
C_i: Number of text documents in a class;
N: Total number of classes;

The best nests is kept;
Rank the solutions to get the current best;
Provide the current best nest to the next generation;
end while

3.5 Classifier

Naive Bayes Multinomial (NBM) [14,22] algorithm is applied as a model classifier by using Weka [10], an open source classification tool based on the feature set being developed for classifying the Bangla text documents. K fold cross validation is performed where K is chosen to be 5. NBM is mostly designed considering the nature of text documents in mind. It generally computes the class probabilities $P(x|y)$ of a text document using Eq. 7, where $P(t_l|x)$ is the conditional probability of term t_l in text document of category x and t_n is the number of terms in document y. NBM counts the terms and worked with the assessments in it. It considers the multiple presence of a term in a classification task. This variation measures the conditional probability $P(t|x)$ as the relative frequency for a term t of category x using Eq. 8 where N is the training dataset, Z is the set of terms and N_{x_t} is the presence of t in text document of domain x.

$$P(x|y) \propto P(x) \prod_{1 \leq l \leq t_n} P(t_l|x) \tag{7}$$

$$P(t|x) = \frac{N_{x_t}}{\sum_{t' \in Z} N_{x'_t}} \tag{8}$$

4 Result Analysis

The proposed approach was evaluated on the obtained dataset consisting of 9000 Bangla text documents from eight different categories (namely Business, Entertainment, Food, Medical, State news, Sports, Science & Technology and Travel) based on Tf-Idf-Icf and modified Cuckoo search optimization problems using Naive Bayes Multinomial classification algorithm. The tuning parameters used in the proposed MCS optimization problem is provided in following Table 2. Figure 2 shows the change of values of cost (C) and purity (P) along with the iterations. It is evident from the Fig. 2 that with the reducing value of the cost the purity gets increases.

Table 2. Tuning parameters

Parameter	Value
Number of Nests	8
Rate of alien solutions	0.25
Levy flight	2
Number of iterations	100
Minimum fitness	0.9
Lower bound	0
Upper bound	1

The domain-wise accuracy achieved after performing classification is provided in Fig. 3. The fall of accuracy in the adopted methodology is due to the fact that the occurrence of terms in the text documents of some domains have much more similarity with the presence of the terms in the contents of the text documents of

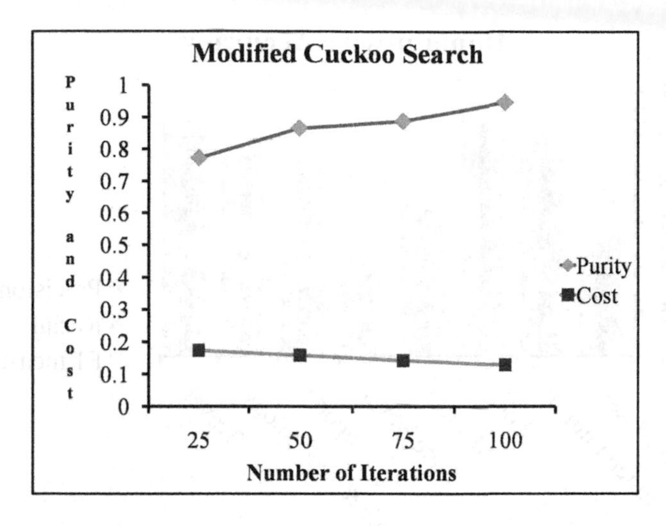

Fig. 2. Purity and Cost values for various iterations

other domains which leads to the misclassification. The rate of correctly classified text documents as well as the rate of misclassification for each domain is given in Table 3 calculated using Eqs. 9 and 10.

$$Classification = \frac{Correctly\_classified\_documents}{Total\_documents} * 100\% \quad (9)$$

$$Misclassification = \frac{Incorrectly\_classified\_documents}{Total\_documents} * 100\% \quad (10)$$

Table 3. Rate of classification and misclassification for each domain

Domain	Classification	Misclassification
Business	96.63%	3.36%
Entertainment	99.00%	1.00%
Food	99.60%	0.40%
Medical	97.30%	2.70%
State News	96.14%	3.86%
Sports	98.36%	1.64%
Science & Technology	98.40%	1.60%
Travel	99.20%	0.80%
Average	**98.00%**	**2.00%**

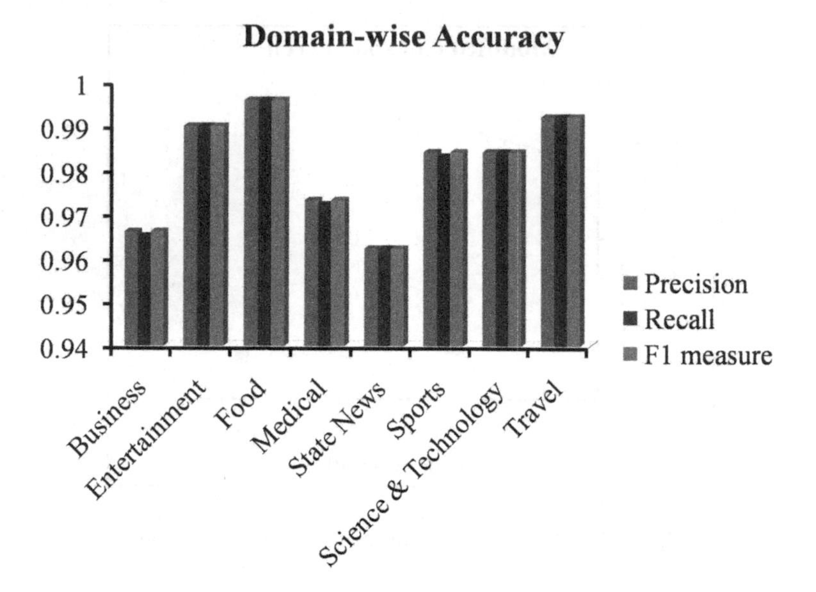

Fig. 3. Domain-wise accuracy

4.1 Statistical Significance Test

The result of the proposed methodology has been compared with some other commonly used classification algorithms namely Decision Tress (J48), LibSVM, Naive Bayes (NB) and RIPPER. The NBM classifier outperforms all other classification algorithms; validated by the Friedman test [6]. For comparing the results of the classifiers, Friedman's non-parametric test is applied as it does not require any assumption about the model being used in the experiment. For this test, the classifiers (k) and datasets (N) were set to 5 and 5 respectively. On each parts the accuracies are measured and a rank R_j^i is provided. The mean rank has been calculated using Eq. 11. From the comparison of the mean ranks, it has been observed that NBM has a better mean rank (rank is assigned in ascending order i.e. top rank: 1.00) and the mean ranks for all the classifiers are presented in the following Table 4. The Friedman Statistic χ_F^2 [6] is estimated using Eq. 12. The details of the statistics is provided in Table 5 where N, df and Asymp. Sig. represents the dataset, degree of freedom and the significance level.

$$R_j = \frac{1}{N}\sum_i R_j^i \tag{11}$$

$$\chi_F^2 = \frac{12N}{k(k+1)}\Big[\sum_j R_j^2 - \frac{k(k+1)^2}{4}\Big] \tag{12}$$

Table 4. Mean rank for all classifiers

Classifiers	Mean rank
NBM	1.0
J48	3.6
LibSVM	2.0
NB	5.0
RIPPER	3.4

Table 5. Friedman test statistics

N	5
Chi-Square Test	19.040000
Degree of Freedom (df)	4
Asymptotic Significance level	0.05
p-value	0.000772

4.2 Comparison with Existing Methods

The proposed algorithm was compared with existing methods, based on the feature extraction and selection approaches used in various researches. The approaches is evaluated on the obtained database consisting of 9000 Bangla text documents and classified using Naive Bayes Multinomial classifier. The obtained accuracy for all the methods is illustrated in Table 6 given below.

Table 6. Comparison with earlier works in Bangla

Reference	Used feature	Size of data	Accuracy (in %)
Mandal and Sen [15]	Tf-Idf	9000	94.27
Kabir et al. [13]	Tf-Idf	9000	95.97
Islam et al. [11]	Tf-Idf and Chi square	9000	95.78
Present work	**Tf-Idf-Icf + MCS**	**9000**	**98.00**

5 Conclusion

This paper presents an approach for feature selection which modifies the basic concept of Cuckoo search optimization problem by incorporating few functions to it. It can be observed that the optimization problem works well for categorizing Bangla documents. In future this experiment can be carried out in expanded text documents as well as domains or text categories. We also plan to use active learning [5] and clustering-based approaches like fast KNN as proposed in [20] on a larger database and observe the performance of our system. Also new hybrid approaches can be introduced in this domain. From this experiment it can also be observed that among all the classification algorithm Naive Bayes Multinomial outperforms for classifying documents with multiple domain.

Acknowledgement. One of the authors thank DST for the INSPIRE fellowship and also thank various links provided in [7] from which the data has been collected.

References

1. Al-Radaideh, Q.A., Al-Khateeb, S.S.: An associative rule-based classifier for Arabic medical text. Int. J. Knowl. Eng. Data Min. **03**, 255–273 (2015)
2. Aly, W., Kelleny, H.A.: Adaptation of Cuckoo search for documents clustering. Int. J. Comput. Appl. Technol. **86**, 4–10 (2014)
3. ArunaDevi, K., Saveeth, R.: A novel approach on tamil text classification using C-Feature. Int. J. Sci. Res. Dev. **2**, 343–345 (2014)
4. Bolaj, P., Govilkar, S.: Text classification for Marathi documents using supervised learning methods. Int. J. Comput. Appl. **155**, 6–10 (2016)
5. Bouguelia, M.R., Nowaczyk, S., Santosh, K.C., Verikas, A.: Agreeing to disagree: active learning with noisy labels without crowdsourcing. Int. J. Mach. Learn. Cybern. **9**, 1307–1319 (2018)
6. DeySarkar, S., Goswami, S., Agarwal, A., Akhtar, J.: A novel feature selection technique for text classification using Naive Bayes. Int. Sch. Res. Not. **2014**, 10 (2014)
7. Dhar, A., Dash, N.S., Roy, K.: Categorization of bangla web text documents based on TF-IDF-ICF text analysis scheme. In: Mandal, J.K., Sinha, D. (eds.) CSI 2018. CCIS, vol. 836, pp. 477–484. Springer, Singapore (2018). https://doi.org/10.1007/978-981-13-1343-1_39
8. Gupta, N., Gupta, V.: Punjabi text classification using Naive Bayes, centroid and hybrid approach. In: Proceedings of the 3rd Workshop on South and South East Asian Natural Language Processing, pp. 109–122 (2012)
9. Guru, D.S., Suhil, M.: A novel term_ class relevance measure for text categorization. In: Proceedings of International Conference on Advanced Computing Technologies and Applications, pp. 13–22 (2015)
10. Hall, M., Frank, E., Holmes, G., Pfahringer, B., Reutemann, P., Witten, I.H.: The WEKA data mining software: an update. SIGKDD Explor. **11**, 10–18 (2009)
11. Islam, Md.S., Jubayer, F.E.Md., Ahmed, S.I.: A support vector machine mixed with TF-IDF algorithm to categorize Bengali document. In: Proceedings of International Conference on Electrical, Computer and Communication Engineering, pp. 191–196 (2017)
12. Jin, P., Zhang, Y., Chen, X., Xia, Y.: Bag-of-embeddings for text classification. In: Proceedings of the 25th International Joint Conference on Artificial Intelligence, pp. 2824–2830 (2016)
13. Kabir, F., Siddique, S., Kotwal, M.R.A., Huda, M.N.: Bangla text document categorization using stochastic gradient descent (SGD) classifier. In: Proceedings of International Conference on Cognitive Computing and Information Processing, pp. 1–4 (2015)
14. Kim, S., Han, K., Rim, H., Myaeng, S.: Some effective techniques for Naive Bayes text classification. IEEE Trans. Knowl. Data Eng. **18**, 1457–1466 (2006)

15. Mandal, A.K., Sen, R.: Supervised learning methods for Bangla web document categorization. Int. J. Artif. Intell. Appl. **05**, 93–105 (2014)
16. Mansur, M., UzZaman, N., Khan, M.: Analysis of N-gram based text categorization for Bangla in a Newspaper Corpus. In: Proceedings of International Conference on Computer and Information Technology, p. 08 (2006)
17. Rautray, R., Balabantaray, R.C.: CSTS: cuckoo search based model for text summarization. In: Dash, S.S., Vijayakumar, K., Panigrahi, B.K., Das, S. (eds.) Artificial Intelligence and Evolutionary Computations in Engineering Systems. AISC, vol. 517, pp. 141–150. Springer, Singapore (2017). https://doi.org/10.1007/978-981-10-3174-8_13
18. Redmond, M., Salesi, S., Cosma, G.: A novel approach based on an extended cuckoo search algorithm for the classification of tweets which contain Emoticon and Emoji. In: Proceedings of IEEE International Conference on Knowledge Engineering and Applications, pp. 13–19 (2017)
19. Sujana, T.S., Rao, N.M.S., Reddy, R.S.: An efficient feature selection using parallel cuckoo search and Naive Bayes classifier. In: Proceedings of IEEE International Conference on Networks & Advances in Computational Technologies, pp. 167–172 (2017)
20. Vajda, S., Santosh, K.C.: A fast k-nearest neighbor classifier using unsupervised clustering. In: Santosh, K.C., Hangarge, M., Bevilacqua, V., Negi, A. (eds.) RTIP2R 2016. CCIS, vol. 709, pp. 185–193. Springer, Singapore (2017). https://doi.org/10.1007/978-981-10-4859-3_17
21. Wang, D., Zhang, H., Liu, R., Lv, W.: Feature selection based on term frequency and T-Test for text categorization. In: Proceedings of the ACM International Conference on Information and Knowledge Management, pp. 1482–1486 (2012)
22. Wilbur, W.J., Kim, W.: The ineffectiveness of within-document term frequency in text classification. Inf. Retrieval **12**, 509–525 (2009)
23. Yang, X.S., Deb, S.: Cuckoo search via Levy flights. World Congress on Nature & Biologically Inspired Computing, pp. 210–214 (2009)

Detection of Sarcasm from Consumer Sentiments on Social Media About Luxury Brands

V. Haripriya[1(✉)] and Poornima G. Patil[2(✉)]

[1] Department of MCA, Jain Deemed-to-be University, Bangalore, India
haripriyav07@gmail.com
[2] Department of MCA, Visvesvaraya Technological University, Belagavi, India
poornima_g_patil@gmail.com

Abstract. Social media sites act as a platform for customers to express their opinions/sentiments on brands and products. The opinion of the customers in social media in case of luxury brands plays a great role in improving the sales by building a better brand strategy. Most of the existing analysis used by the luxury brand industry ignores the importance of sarcasm analysis. A common type of sarcasm that is given in the form of opinion is positive sentiments, which contain a negative meaning. This paper studies the scope of Lexicon based approach, K-means and Naïve Bayes for analyzing the sarcastic opinion and analyzing the impact of these algorithms in recognition of sarcasm, which has a negative context for analyzing the luxury, brand data.

Keywords: Machine learning · Sarcasm · Branding strategy · K-means · Naïve Bayes · Lexicon based

1 Introduction

Since the industrial revolution, businesses continually look for ways and means to remain competitive in the market. In today's era of new technologies of information and communication, internet makes it possible for businesses to reach out to immense number of prospects around the world in a wink of an eye. However, to covert these prospects to customers, businesses need to exercise a certain amount of creativity and ingenuity. Branding is the process, which consists for a business to create a unique name and image of its products or services into consumers' minds. Lake et al. [1] mentioned that branding constitutes the procedures that businesses implement in order to retain people attention. Baker et al. [2] classified branding with two essential approaches: emotional branding and rational branding. Sarcasm detection is considered as the challenging factor in the sentiment analysis since it gives a gap between the intended meaning and the literal meaning of the sentences. This ambiguity poses challenge to the industries in making proper decisions regarding branding. Lexicon based approach is

K. C. Santosh and R. S. Hegadi (Eds.): RTIP2R 2018, CCIS 1037, pp. 652–667, 2019.
https://doi.org/10.1007/978-981-13-9187-3_58

used for sentiment analysis. Naïve Bayes and K-Means clustering have used to detect the sarcasm in the proposed work. Machine learning, Lexicon Based and Rule based approaches are the commonly used for analyzing the sentiments in data. Machine learning approaches like Bayesian Networks Naive Bayes Classification Maximum Entropy etc. are used in creation and training of models in domain-specific contents. But the drawback of these approaches is that training becomes very difficult in case of application of new data. This problem occurs as labeling of each data is costly and time consuming. The commonly used Lexicon based approaches are Dictionary based approach; Corpus based approach Ensemble Approaches, etc. This approach can cover unlimited words but it lacks availability of linguistics resources. Rule based approaches are based on rule chaining for decision making. When the number of rules increases and the rule chain expands the decision making becomes a complex process. A simplest classification algorithm is k-Nearest Neighbor algorithm used for classification and regression. KNN is generally known as Lazy algorithm since distance calculation is difficult for large data set. To make the process of distance calculation faster Vajda and Santosh [3] have proposed a generic method by using KNN classification. Authors have been used an iterative method to find percentage of the data which are nearer to the cluster centers and could obtained an efficiency of the method by reducing the classification speed by 71% by maintaining the same performance.

A semi-supervised machine-learning algorithm, which helps to efficiently interact with user's queries to get the expected output, is Active Learning. A new active method has been proposed by Mohamed-Rafik Bouguelia et al. [4] for managing labeled noise without the need to depend crowdsourcing and two strategies also have been proposed to select the instances based on the maximum influence on the learned model and the second strategy which helps to choose instances which are highly affected by the modification in the learned model. A comparison also has been made on these two strategies.

2 Related Works

Consumer perceptions are considered as the major factor of brand equity and therefore marketing strategy. Culotta et al. [5] suggested that to get a clear idea of the brand's strengths and weakness these perceptions can be segmented into various attributes such as eco-friendliness, nutrition, and luxury. The large volume of user-generated content (UGC) on social media is loaded with potential value for brand managers. Since it was difficult to fetch the answer for brand management questions a framework has introduced by Liu et al. [6] that derives latent brand topics and classifies brand sentiments by using text mining and sentiment analysis. A quantitative research has targeted females to find out the impact of self-concept and life style of customers on purchases of luxury goods and obtained that there is a positive significant impact by Naz et al. [7]. Sentiment of the users can be identified using various techniques but the major challenge is Sarcasm detection in writing. Various studies have been done on sarcasm detection. Mohd et al. [8] proposed feature extraction model which used bilingual texts to detect the sarcasm on Facebook comments. The authors could obtain F-measurement score of 0.852 by combining various features. Parsing-based lexicon generation algorithm and the occurrence of the interjection words have proposed by Bharti et al. [9] detect the sarcasm on Twitter data and a comparison has been done with the existing approach. Francesco et al. [10] to detect sarcasm in Twitter data and the results shows that the proposed model is easy to implement when compared to the existing system since it does not include word pattern as features has been developed a novel computational model. A model has been developed by Riloff et al. [11] to detect sarcasm using bootstrapping algorithms, which helps to provide a list of positive sentiment phrases and negative phrases in the sarcastic tweets automatically. Luxury brands essentially reflect the identity, image, position, personality, equity, experience, differentiation, communication, gap, and the extension of their companies suggested by Sharma et al. [13]. An examination has done by Rodrigues et al. [14] on the behavior of luxury brand purchasing customers, considering that the luxury brands including solid consumer emotions. Phau et al. [15] mentioned that the perception of a brand personality depends more on people traits and emotions than on the skill and ability with which marketers and advertisers present the brand to people. A comprehensive study has been done by Yang et al. [16] on the relationship between an individual's personal traits and his/her brand preferences. Chakraborty et al. [17] have done an analysis done on four factors such as interpersonal influence, brand image, brand consciousness and different demographic components and found Interpersonal influence has been maximum priority by the respondents than other factors. A theory of planned behavior framework has been developed by Jain et al. [18] to analyze the purchasing behavior for luxury fashion goods. An expert-predefined lexicon has been used by Mostafa et al. [19] to evaluate consumer sentiments towards the famous brands and the results indicate a generally positive consumer sentiment.

3 Proposed Research Methodology

The proposed methodology classifies the sentiments expressed by customers in textual contents as Positive, Negative or Neutral and Sarcastic. When the user communicates with an industry or brand, like a reply or comments, the sentiment in the related content likely demonstrates this user's sentiment towards that industry or brand said by Hu et al. [12]. These reviews and comments are subjected to the lexicon-based analysis to determine the sentiments expressed by people toward luxury brands. K-means and Naïve Bayes algorithms are used to detect the sarcastic comments from the positive sentiments. The proposed model is represented in Fig. 1, which gives step-by-step procedure involves in the sarcasm detection process. The classified output is shown in the form of charts.

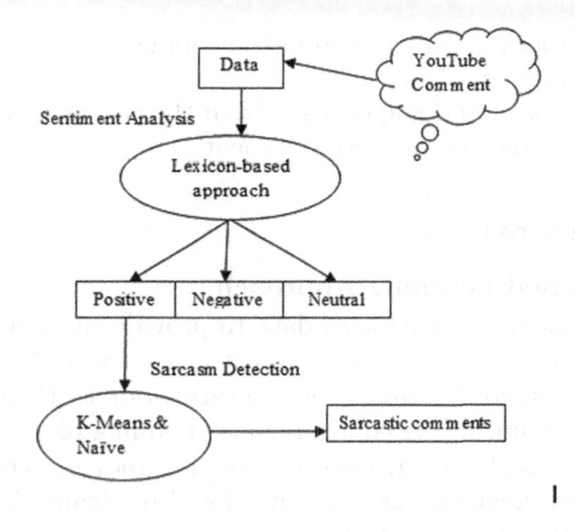

Fig. 1. Proposed sarcasm detection model

3.1 Data and Preprocessing

Preprocessing social media data is a difficult task because the collected data are noisy, and sometimes it violates the rules of English grammar. Main steps involved in the proposed work are Collections of data, Data preprocessing, classification and clustering.

3.1.1 Data Collection

The customers' opinion on luxury brand data have been collected from YouTube on three luxury brands: Armani, Zara and Gucci of various seasons like Spring Summer campaign for Armani, Falls winter and Spring summer data of Gucci and Autumn winter reviews of Zara for the years 2017 to 2018. The reviews have been collected using YouTube Comment Scraper and the data generated are in the .csv format. The total dataset used for the proposed study is 2314 comments where 347 comments on Zara, 88 comments on Armani and 1879 comments on Gucci.

3.1.2 Data Cleaning

This step involves the process of denoising the data. Social media data is a rich container of user-generated data with lots of irrelevant information and inconsistent data with special characters. This affects the quality of data analysis outcomes. Various tools also can be used for data cleaning such as Data Cleaner and XLTools.net Data Cleaning, as an add-in product for Excel. For the proposed work Data Cleaner has used as an add-in to the excel sheet to remove the noisy data.

1. Collect comments using comment scrapper.
2. Save the comments in a file to denoise.

3. Remove all the special characters from the comments.
4. Remove all the punctuations.
5. Remove all the spaces and empty fields from the comments.
6. Add a full stop at the end of every comment.

3.2 Learning Approaches

3.2.1 Unsupervised Learning Approach

This approach works on the unlabeled data to provide the structure and it is a function used to compare the features of a text against word lexicons in discriminatory representation to determine the polarity prior to their use. Unlabeled data are provided for unsupervised algorithms. Training of data is not required in case of unsupervised learning, however; the accuracy of the results depends on the resource used. Lexicon based algorithm and K-Means classifier algorithm have been used in the proposed study.

3.2.1.1 The Lexicon Based Algorithm: The lexicon based algorithm mainly used in sentiment analysis. It is built on a collection of known and precompiled sentiment terms. This algorithm classifies the sentiment expressed by textual contents as negative, neutral or positive.

Steps of the lexicon based algorithm: To determine the sentiment expressed by users in the textual content, the lexicon-based algorithm performs a series of steps:

Split textual content into micro-phrase: This operation consists of breaking the social media comments based to the splitting cues. A splitting cue might be an adverb, a conjunction or punctuation.

Split micro-phrase into terms: This operation consists of breaking micro-phrases into individual terms. Each word of the micro-phrases is considered as a standalone component.

Determine the score of each and every term: Based on the lexicon resource in use, a score is attributed to each term. The score could be any real number.

Compute the polarity of each micro-phrase: Based on the score of each term, the polarity of each micro-phrase is computed. The polarity of each micro-phrase is function of the variant method used.

Compute the polarity of the entire textual content: The polarity of the entire textual content is computed based on the polarity of every micro-phrase.

Determine the sentiment of the entire textual content: The sentiment of the textual content is determined based on the sign of the polarity number: negative for (−), neutral for (0) and positive for (+).

Example: I liked this dress, it looks elegant. Here, the sentence is divided into two micro-phrases. The first micro-phrase is *"I liked this dress"* and the second one is *"it looks elegant"*. The comma is used as the splitting cue to divide the comments into two parts. Next step, each micro-phrase has to be divided into individual terms like: *"I", "liked", "this", "dress" and "it", "looks", "elegant"*. Stop-words such as *"I", "this", "looks"* etc. have to be removed from the micro-phrase. Third step, polarity has to be assigned to each words from the lexicon resource used. Example: the term *"liked"* has the polarity 75 and the term *"elegant"* has the polarity 90. Once the polarity has been assigned for each word, the polarities of the entire micro-phrase have to be calculated. As the forth step, the polarity of the entire comments have to determine by using variant methods in lexicon based and based on the results obtained, the sentiments of the comments can be calculated. The obtained score has been used to assign the sentiments to the comments as Positive if the results is '+'ive number and negative if it is '−'ive value and 0 is considered as Neutral.

3.2.1.2 K-means Clustering Algorithm: K-means is one of the oldest and mostly used clustering techniques. It is mainly used for spatial analysis. K-means clustering algorithm operation consists of grouping data into specific clusters. These clusters are formed based on features available into the data. Each datum is assigned to a cluster based on feature similarity. However, many techniques are available to validate K, the number of cluster. These techniques include cross validation, information criteria, information theoretic jump method, silhouette method and G-means algorithm.

3.2.2 Supervised Learning

Supervised learning approach is a data mining activity that consists of revealing the algorithm (the functions) that a feature vector went through for a particular supervisory signal to be produced. The algorithm gets trained with the help of sample data from the same domain as the initial pair in order to be consolidated and optimized. The consolidated and optimized algorithm gets generalized so as to be able to generate exact labeled data from an input vector. Naïve Bayes Classifier has been used as supervised learning algorithm in this study.

3.2.2.1 Naïve Bayes Classifier Algorithm: Naïve Bayes classification is a fast and easy algorithm with good classification accuracy. Naïve Bayes mainly involves two main data sets: the feature matrix and the response vector. The former is considered as the input and the later as output of the algorithm. Naïve Bayes is said to be naïve because of its principle that advocates that:

1. Each feature input is independent
2. Each input feature is given the same weight and consideration, and equally contributes to the outcome.

Naïve Bayes algorithm complies with the Bayes' theorem. This theorem enables to find the probability of an event to occur given the probability of another event that has previously occurred. Bayes' theorem is given in the Eq. (1):

$$P(Y|X) = \frac{p(X|Y)p(Y)}{P(X)} \tag{1}$$

where, y is a class variable and X is a dependent feature vector of size n. $X = (x1, x2, x3, \ldots xn)$

3.3 Implementation of the Algorithms

The implementation of algorithms is done by using Python.

3.3.1 Sentiment Analysis Using Lexicon-Based Approach

To test the scenario and simulate the experiment the users' comments have been analyzed using the lexicon-based algorithm. The opinions are extracted from YouTube. Within each set, the comments were split into Positive, Negative and Neutral and TextBlob has been used for finding out the polarity of the text from the opinion corpus. TextBlob is a python in-built library for text processing. All the three sentiments features are determined from the comments. Total numbers of positive, negative and neutral comments have been calculated as shown in Figs. 3, 4 and 5. The below given screen shot shows how the polarity is been calculated using Lexicon-based approach. In the Fig. 2 the polarity generated is −0.8 that is considered as Negative sentiment. The same calculations are done for Positive and Neutral.

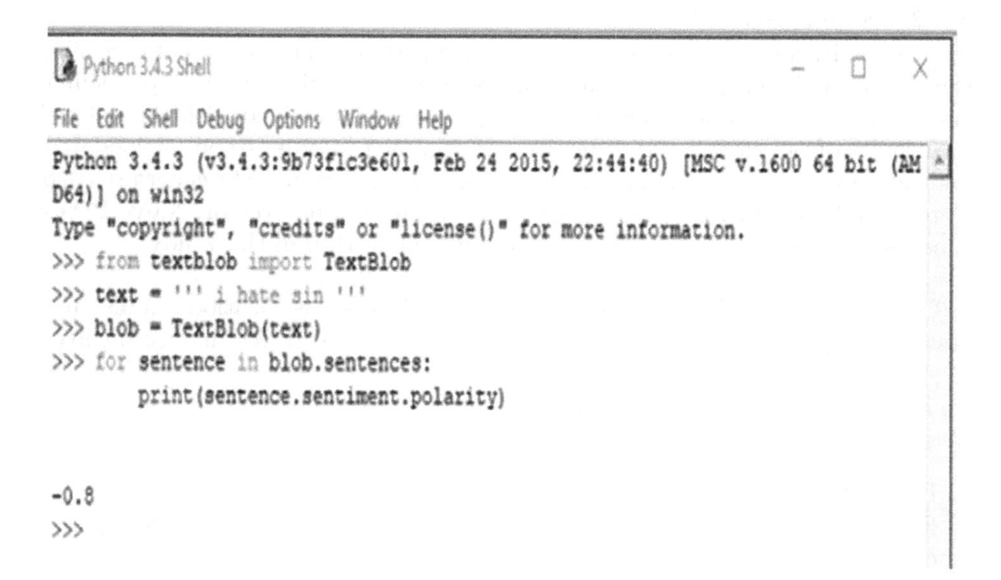

Fig. 2. Sample screenshot of polarity calculation

The results obtained from the Lexicon based approach have Positive, Negative and Neutral comments. The positive comments generated from the lexicon based approach are given as the input to Naïve Bayes classifier and K-Means clustering.

3.3.2 Sarcasm Detection Using Naïve Bayes Classifier

Naïve Bayes algorithm has been used in this study for the purpose of classification. The main objective of using this algorithm is to detect the sarcastic comments from the positive comments derived from the Lexicon based approach. The preprocessed dataset is split into training set and testing set in the ratio 60:40 where 60% of the data are assigned for training set and 40% is assigned for testing set. The algorithm is trained using the training data and tested by using test dataset.

According to the Bayes theorem mentioned in Eq. 1 the probability of positive sarcastic and positive comments have been computed by Eqs. (2) and (3):

$$P(PositiveComments|Comments) = \frac{P(comments|PositiveComments)P(PositiveComments)}{P(Comments)} \quad (2)$$

$$P(Sarcastic|Comments) = \frac{P(comments|Sarcastic)P(Sarcastic)}{P(Comments)} \quad (3)$$

Consider an example *"I like shopping"*. Given the dataset, the comment *"I like shopping"* has to be classified as positive comments or sarcastic comments by using Bayes theorem.

1. P(Comments | Positive Comments) can be calculated by considering the total number of times the word *"I"*, *"like"*, and *"shopping"* occurs in the positive comments divided by the total number of unique words present in the positive comments
2. P(Positve Comments) is the total number of positive comments divided by the total number of comments in the dataset.

The same procedure is repeated for finding the probability of Sarcastic comments. Comparison takes place between two generated results from Eqs. (2) and (3) and the highest probability value will be considered as the result. Based on the calculation, the example *"I like shopping"* is classified as positive comment.

3.3.3 Sarcasm Detection Using K-Means Clustering

The steps involved for the K-Means clustering in the proposed study are as follows:

Step 1: As k-means algorithm accepts only numerical data as parameter the first step is to convert both the columns into numerical columns with the help of LabelEncoder object in sklearn.preprocessing module of python.
writing the numerical data into binarized_sentiment.csv

data.to_csv("binarized_sentiment.csv")
In this step the binarization of the sentiment column takes place and which contains only two possible binary values '0' for 'sarcastic' and '1' indicates 'positive' comments. Dummification of comments attribute is very challenging as it contains distinct text comments of different length. So again Label Encoder has been used to encode the comments into numerical values.

Step 2: Added one more column called 'counter' into newly generated dataset stored in binarized_sentiment.csv file. This column does not have any significant role for further processing. It has been added just to increase the dimension by one as K-means algorithm does not accept 1D array as input. This counter column can be dropped after obtaining the result of K-means algorithm which is a cluster column.

Step 3: Standardize the data to make all the data in the same scale so that the data with large scale will not dominate the data with small scale. Standardizing data is one of the first steps to be performed in K-means algorithm as it does not have any built in feature scaling capability after pre-processing. This uses dissimilarity measures like Euclidean distance or cosine dissimilarity to measure the distance between the data point and the centroid and then the data point is assigned to cluster or the centroid with the minimum distance that is the closest centroid. This has been done by using standard scalar object in sklearn.preprocessing.StandardScaler.

Step 4: Principal Component Analysis (PCA) is applied on the data to speed up the model performance and reduce the dimension. By reducing the dimensions of the data it makes it less complex to perform clustering and visualization.

Step 5: After scaling the data, the dataset consist of comments, sentiments and counter attributes in numerical format and the data have been stored in numerical.csv file and the dataset is read from this numerical.csv file, stored in pandas data frame object, and passed as a parameter to K-means algorithm: kmeans_model.fit(scaled_data)

Internally kmeans++ algorithm calculates the distance between the data points and assigns the data point to the nearest centroid.
There are 2 centroid points, so the initial centroid point is calculated randomly and then mean is computed to reposition the centroid each time until the model is converged.

Step 6: Once the K-means model is in converged state the labels_ attribute of K-means object has been used to obtain the cluster column and add as a cluster column to the data set in the original file named sentiment.csv. Now the sentiment.csv file contains 3 columns such as comments, sentiments in text form and cluster column which specifies which cluster the observation belongs to. Cluster 0 for positive comments and cluster 1 for sarcastic comments.
Clustered data are stored in clustred_sentiment.csv file which can be used for further analysis.

Step 7: Scatter plot and histogram have been used for representing how K-means clusters the data graphically so that users can have solid understanding of how data have been grouped together based on their similarities.

4 Results and Discussions

Lexicon-based approach have used in the proposed study to analyze the sentiments of the customers on three luxury brands: Armani, Zara and Gucci. The data collected for two corresponding years from 2017 to 2018. The implementation has divided into two steps: the first step is sentiment analysis to determine the Positive, Negative or Neutral sentiment of the customers and the second step is Sarcasm detection from the positive comments. Table 1 represents the brief analysis of the sentiments of opinion. The opinions are classified into Positive, Negative and Neutral comments. The values are represented in "%". The opinions are classified for three brands: Gucci, Zara and Armani. Table 2 represents the number of Sarcastic comments and Positive Comments obtained as a result of Naïve Bayes and K-Means algorithms and the same results are generated in the form of charts.

Table 1. Sentiment analysis of luxury brands for the years 2017–2018

Brand name	Total	Positive	Negative	Neutral
Gucci	1879	526	559	794
Zara	347	84	107	156
Armani	88	50	11	27

Table 2. Sarcasm detection from Positive comments

Brand name	Sarcastic comments	Positive comments
Gucci	488	38
Zara	80	46
Armani	36	14

Figures 3, 4 and 5 represents the sentiment analysis results of above mentioned brands for the years 2017 to 2018 for both men and women and the results are represented in the form of charts.

Figures 6, 7 and 8 represents the results of sarcasm detection from the positive comments of sentiment analysis for two years from 2017 to 2018 for three brands: Gucci, Zara and Armani using Naïve Bayes Classification. The results show the Sarcastic comments are more when compared to Positive comments.

Figures 9, 10 and 11 represents the results of K-Means clustering for sarcasm detection from the positive comments of lexicon based approach for three brands: Gucci, Zara and Armani. The results have two clusters one is for Sarcastic and the other cluster is for Positive Comments. The results are given in the form of scatter plot and histogram. The results of this study reveal that the K-means algorithm has to be improved better in such a way that the error rates can be reduced even with less number of iterations. In K-Means clustering when the

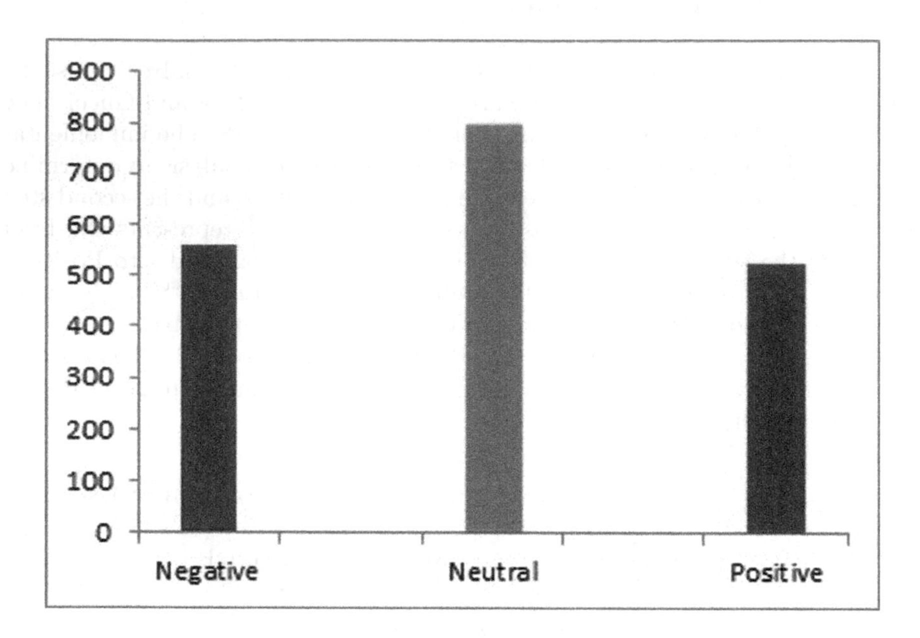

Fig. 3. Result of sentiment analysis - Gucci

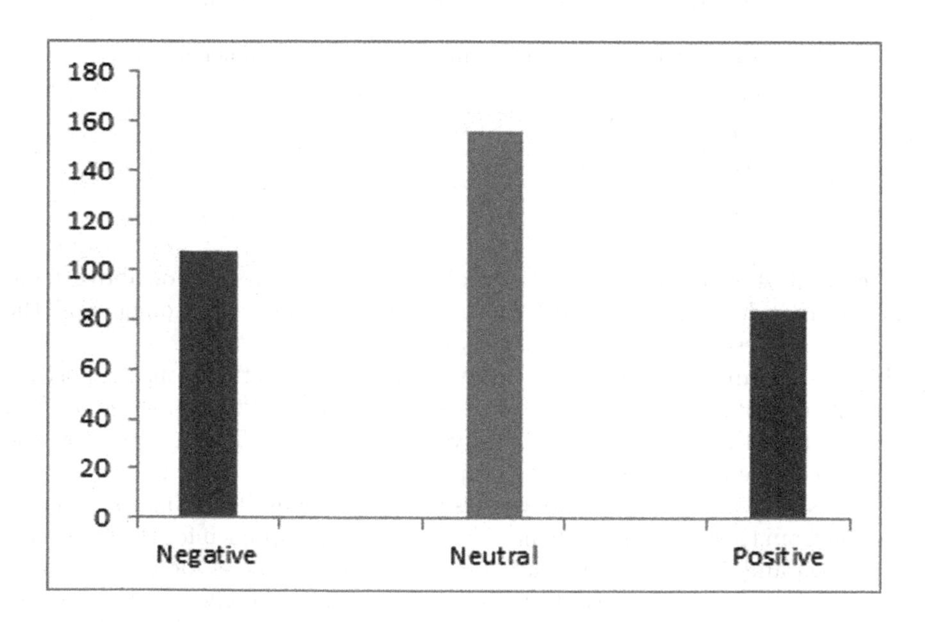

Fig. 4. Result of sentiment analysis - Zara

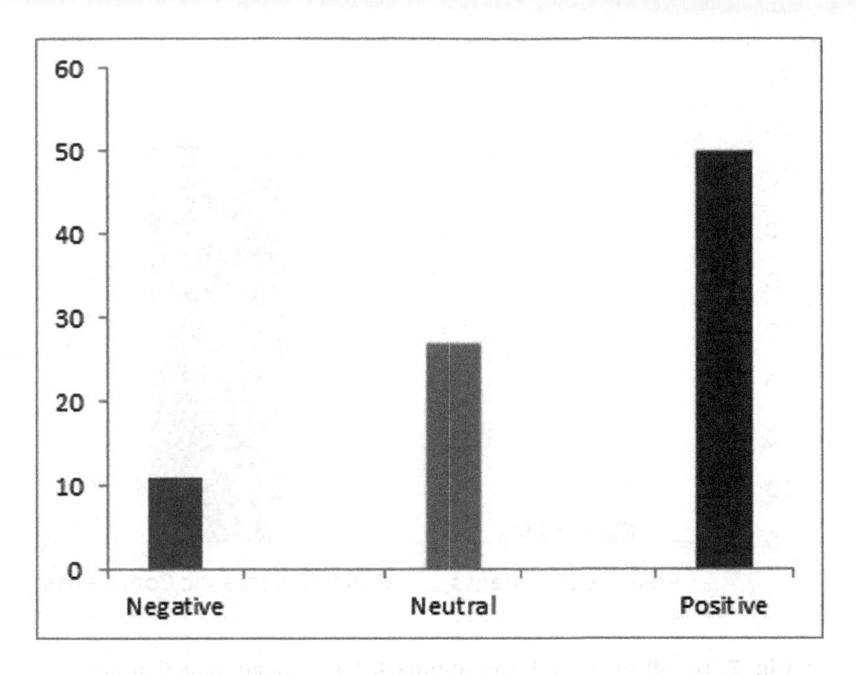

Fig. 5. Result of sentiment analysis - Armani

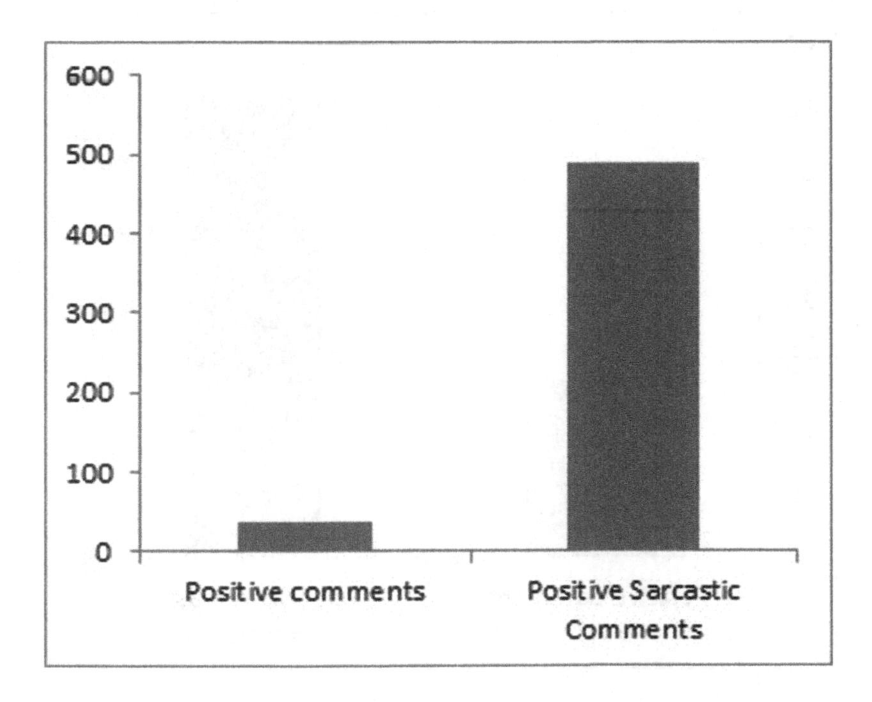

Fig. 6. Result of Naïve Bayes approach for sarcasm detection-Gucci

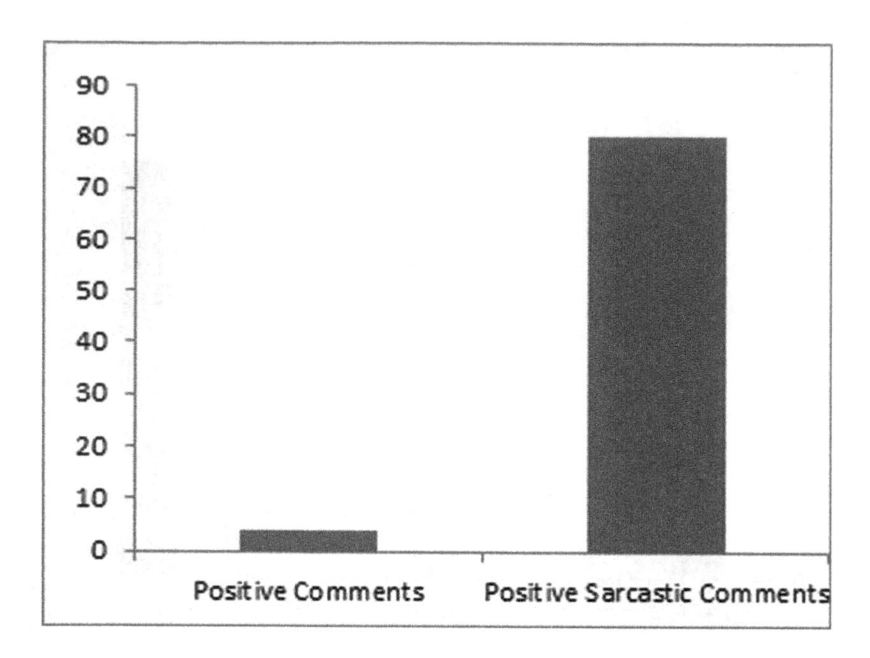

Fig. 7. Result of Naïve Bayes approach for sarcasm detection-Zara

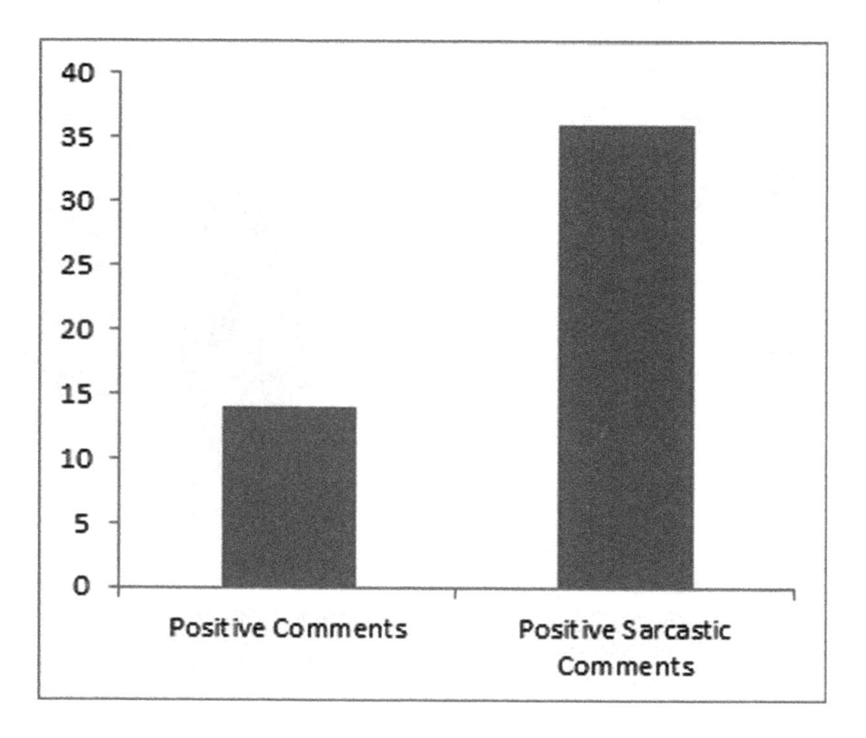

Fig. 8. Result of Naïve Bayes approach for sarcasm detection-Armani

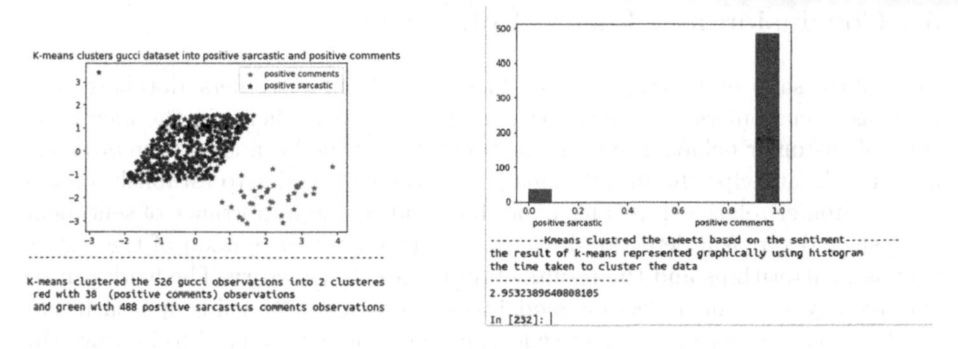

Fig. 9. Results of K-means clustering in scatter plot and histogram - Gucci

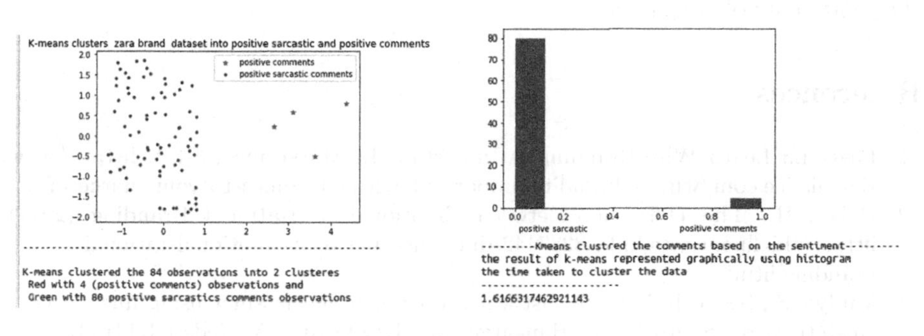

Fig. 10. Results of K-means clustering in scatter plot and histogram - Zara

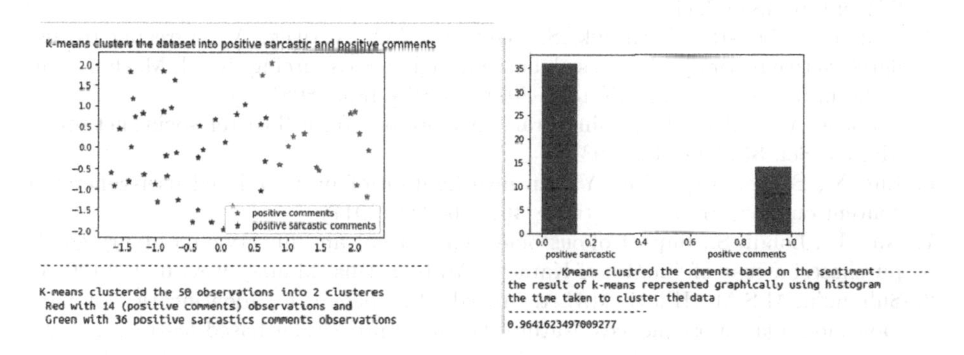

Fig. 11. Results of K-means clustering in scatter plot and histogram - Armani

size of the data is larger it takes more time to cluster the data and in some of the cases the distance between the cluster and the centroid is high thereby error rates can't be reduced. Naïve Bayes classification requires labeling of data whereas K-Means works on unlabeled data. The proposed study analyses that Naïve Bayes works on textual data and in case of K-Means it require numerical data for clustering.

5 Conclusion and Future Enhancement

Most of the sales of luxury brand products come from customers' database. The analysis of customers' sentiments that may be sarcastic helps in the identifications of customer behavior and satisfaction towards the brands and the products and thereby it helps the brand managers or the companies to establish a long-term customer relationship. This paper has studied the importance of sentiment analysis and sarcasm detection by applying the customer opinion of the luxury brands in algorithms and the results obtained are satisfactory. The work can be extended by designing a decision model based on machine learning approach that can be domain independent. Active learning method can be used to improve the effectiveness of the annotation and also for the creation of corpora for analyzing the sentiments of consumers.

References

1. Lake, L.: Learn Why Branding Is Important In Marketing (2018). http://www.thebalance.com/why-is-branding-important-when-it-comes-to-your-marketing
2. Baker, H.: The Difference Between Emotional & Rational Branding (2017). https://bizfluent.com/info-8287442-difference-between-emotional-rational-branding.html
3. Vaidya, S., Santosh, K.C.: A fast k-nearest neighbor classifier using unsupervised clustering. In: Santosh, K., Hangarge, M., Bevilacqua, V. (eds.) RTIP2R 2016. CCIS, vol. 709, pp. 185–193. Springer, Heidelberg (2016). https://doi.org/10.1007/978-981-10-4859-3_17
4. Bouguelia, M. -R., Nowaczyk, S., Santosh, K.C., Verikas, A.: Agreeing to disagree: active learning with noisy labels without crowdsourcing. Int. J. Mach. Learn. Cybern. **9**, 1307–1319. ISSN 1868–8071, E-ISSN 1868–808X,
5. Culotta, A., Cutler, J.: Mining brand perceptions from Twitter social networks. Market. Sci. **35**, 343–362 (1981)
6. Liu, X., Burns, A.C., Hou, Y.: An investigation of brand-related user-generated content on Twitter. J. Advertising **46**, 236–247 (2017)
7. Naz, U., Lohdi, S.: Impact of customer self concept and life style on luxury goods purchases: a case of females of Karachi. Arab. J. Bus. Manag. Rev. **6**, 192 (2016)
8. Suhaimin, M.S.M., Hijazi, M.H.A., Alfred, R., Coenen, F.: Identification of common molecular subsequences. Natural language processing based features for sarcasm detection: an investigation using bilingual social media texts. In: 8th International Conference on Information Technology (ICIT). IEEE (2017)
9. Bharti, S.K., Vachha, B., Pradhan, R.K., Babu, K.S., Jena, S.K.: Sarcastic sentiment detection in tweets streamed in real time: a bigdata approach. Digit. Commun. Netw. http://dx.doi.org/10.1016/j.dcan.2016.06.002
10. Barbieri, F., Saggion, H., Ronzano, F.: Modelling sarcasm in Twitter, a novel approach. In: Sentiment and Social Media Analysis, pp. 50–58. Association for Computational Linguistics, Baltimore (2014)
11. Riloff, E., Qadir, A., Surve, P., De Silva, L., Gilbert, N., Huang, R.: Sarcasm as contrast between a positive sentiment and negative situation. In: Conference on Empirical Methods in Natural Language Processing, pp. 704–714. Association for Computational Linguistics, Seattle (2013)

12. Hu, G., Bhargava, P., Fuhrmann, S., Ellinger, S., Spasojevic, N.: Analyzing users' sentiment towards popular consumer industries and brands on Twitter. arXiv preprint arXiv:1709.07434 (2017)
13. Sharma, E.: 10 Branding Elements And What They Mean (2015). http://www.brandanew.co/10-branding-elements-and-what-they-mean/
14. Rodrigues, P., Costa, P.: Why consumers buy Luxury Brand? Conference proceeding (2016). https://www.researchgate.net/publication/306600838
15. Phau, I., Lau, K.C.: Brand personality and consumer self-expression: single or dual carriageway. J. Brand Manag. **8**, 428–444 (2001)
16. Yang, C., Pan, S., Mahmud, J., Yang, H., Srinivasan, P.: Using personal traits for brand preference prediction. In: Proceedings of the Conference on Empirical Methods in Natural Language Processing, pp. 86–96 (2015). https://doi.org/10.1016/0022-2836(81)90087-5
17. Chakraborty, S., Sheppard, L.: An explanatory study on indian young consumers' luxury consumption: the underlying relationship of interpersonal influence, brand image, brand consciousness and demographic components with luxury brand purchase decision. Int. J. Curr. Eng. Technol. **6**, 622–634 (2016)
18. Jain, S., Khan, M.N., Mishra, S.: Understanding consumer behavior regarding luxury fashion goods in India based on the theory of planned behavior. J. Asia Bus. Stud. **11**, 4–21 (2017)
19. Mostafa, M.M.: More than words: social networks' text mining for consumer brand sentiments. Expert Syst. Appl. **40**, 4241–4251 (2013)

Question Answer Based Chart Summarization

Aditi Deshpande and Namrata Mahender(✉)

Department of Computer Science, Dr. BAMU University, Aurangabad 431005, Maharashtra, India
aditi.deshpande96@gmail.com, nam.mah@gmail.com

Abstract. To summarize documents worths to summationof the main points. A summarization is this kind of summing up. Elementary school book reports are big on summarization. To provide a comprehensible declaration of the significant points is nothing but summarization. In current years, natural language processing (NLP) has stimulated to statisticall base. Many tribulations in NLP, e.g., parsing, word sense disambiguation, and involuntary paraphrasing. In recent times, robust graph-based methods for NLP is also a lot of scope, e.g., in clustering of words and attachments of prepositional phrase. In proposed paper, we will take in account of graph-based summarization techniques, approaches used for that etc. We will talk about how arbitrary traversing on images of graphs can help in making of question answer based summarization. In current exploration work, question answer based graph summarization system for Bar Graph is shown. The extraction procedure is completely computerized using image processing and text recognition methods. The extracted information can be used to improve the indexing component for bar charts and get better exploration results. After generating questions, questions are rank the according to frequency or priority and answer of the ranked question is summary of given input.

Keywords: Extraction of chart data · Question answer generation system

1 Introduction

Summarization is the procedure of display the original ideas of the text in as less words as possible. It can be done theoretically, verbally, through play, through song, in groups and independently. Extensive range of research in the area of summarization shows that it is well liked and broadly used techniques in knowledge field. There is widespread research that shows that summarization is among the top most effective teaching strategies in the history of education. Summarization is a ephemeral and precise illustration of input text such that the output covers the utmost essential conceptions of the source in a abbreviated fashion.

Text summarization is procedure of extraction of important information from source text and presents that in form of short summary. To select information from

K. C. Santosh and R. S. Hegadi (Eds.): RTIP2R 2018, CCIS 1037, pp. 668–677, 2019.
https://doi.org/10.1007/978-981-13-9187-3_59

large amount of information from various sources is difficult for human beings. Automatic summarization is the process of utilizing ready to use software on given text for production of summary of given text with an idea that summary holds unique and important aspects of the original text. Query - Based Summarization technique consists only answered the thing which is asked by the user. These queries are generally natural language queries or keywords which are relevant to a particular area under discussion. Automatic graph summarization is nothing but produce a Question Answer based graph summary. To provide a comprehensible declaration of the significant points is nothing but summarization.

In current years, natural language processing has stimulated to statistical base. Many tribulations in NLP, e.g., parsing, word sense disambiguation, and involuntary paraphrasing. In recent times, robust graph-based methods for NLP is also a lot of scope, e.g., in clustering of words and attachments of prepositional phrase. In this research work, we review graph summarization mostly from a methodological perspective, answering how we can algorithmically obtain summaries of graph data. Most of the work has been done on binary tree graphs and search trees and charts etc. We are creating a summary of bar graph using automated question answering system.

We propose to use an Data Inference Algorithm as a preprocessing step for text recognition with a common ocr system. From last 300 years, charts, graphs, and other visual data has been the important part in communicating the quantitative information. In proposed paper, we evaluate a process of data extraction from bar charts. We use automated extraction process in image processing. For the recognition of textual or arithmetical objects of every text section in the charts a Tesseract OCR is used.

2 Literature Survey

Prasad et al. [21] has categorize bitmap images of charts into five levels. All this types form a chain of operations in preprocessing, calculation of global curve saliency and features used for local image segmentation. The hinder of insight of technical graphs has been addressed in an variety of studies. Scrupulous data types i.e. nominal, ordinal, or quantitative positioned by optical variables in charts are explained in detail by Mackinlay's APT [19]. He also extended his work in support of collection of mechanized picture designs [20]. Polaris is a structure designed for breeding lesser multiples [13] display to built on consumers query on multi-dimensional data table given by Stolte et al.'s [16]. Zhou and Tan [15] developed process of combining boundary tracing with the Hough transform for extraction of bars from charts. Charts having bar, lines are used for generating edge maps by Huang et al., they designed rules and vectorized these edge maps for extraction of marks [5,17,18]. Yang et al. [14] built a system considering Huangs's procedure for mechanically creating vector representations of charts.

2.1 Review of Related Work

Author and Year	Paper title	Method	Rule	Limitations
Chaddha et al. [9]; Gupta (1994); Schaar-Mitrea and de With (1998) and LeBourgeois (1997)	Text Segmentation in Mixed-mode type Images, Systems and Computers Textual loom for Indexing and repossession of Image And Video	Text detection method using Texture Features	Frequent color transition on image	Computationally complex when there are many edges
Sato et al. (1998) [15]	Digital News collection needs Video OCR, Proc. of IEEE practicum on Content based admittance of Image and Video Databases	Segmentation Methods using Character Contours	Measure exact character boundaries	Sow & may generate incomplete or distorted contour
Yu et al. (1995) and Lee and Kankanhalli (1995) [13]	Text recognition In difficult Color Images, Pattern Recognition, Zoning methods for handwritten character recognition	Character Recognition Methods Using Pixel Connectivity	Use cc-component	Fails to extract all the characters when there are joined or broken
Forssen Erik and Lowe (2007) and Matas et al. (2002) [20]	Figure Descriptors For MSER, vigorous large baseline stero from MSER	MSER method	Remove non-text regions, threshold value needed	
Hase and Shinokawa [19]; Yoneda and Suen (2001) and Sakai et al. (1986) [5]	Extraction of strings of characters from Color Documents, Pattern Recognition,	Edge (text boundaries, strong edges), Compression	Use kernel matrix, threshold value needed	Garbage values taken

3 Working of Proposed System

The proposed system contains three sections: graphical-component extraction section, text extraction section, and the data inference section. The system is only build for 2-D bar chart as source image and reproduces the data values as output. We have described the working of all modules below:

3.1 Data Extraction and Mark Extraction

The existing accomplishment stresses on uses of marks for data extraction form bar and similar types of chart. We observed that chart images from the source of internet are commonly condensed and noisy. To decrease such drawbacks, the bilateral filter is firstly [12] to each bar in the image. The bilateral filter helps in smoothing the left small changes in color but retains sharp edges in the image. It is observed that marks are identified and tried to fit to fitting models of various shapes for extraction in mark based extraction. In our work a mapping is established between data space and image space i.e. link labels with marks and finally the data relevant with each mark is extracted.

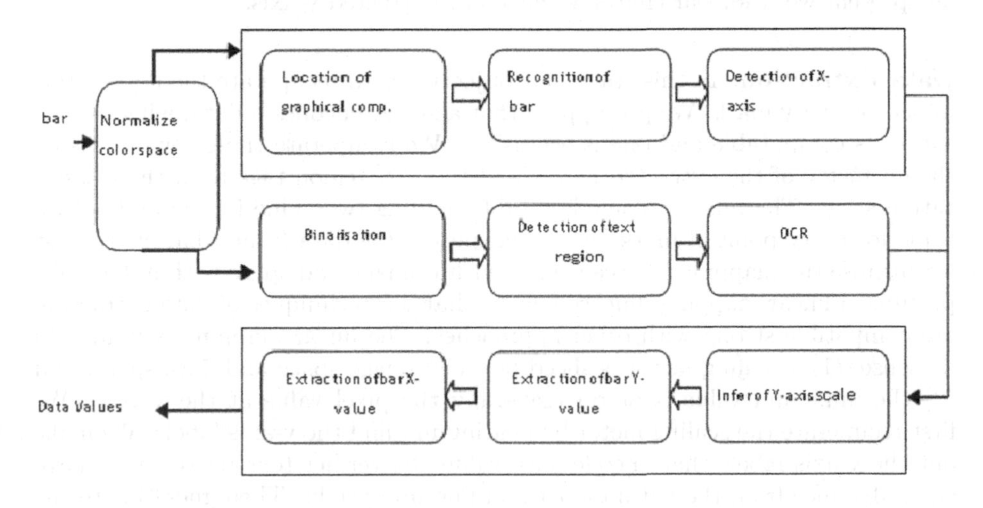

Fig. 1. Data extraction system overview

Extraction of Marks. Here first the marks are identified, to extract bars and axis. The bars are recognized by mapping it with rectangular fitting models. Then the location of those bars are used to extract X and Y axis [11]. Discover rectangular shapes. First from the filtered image we extract connected components by assemblage adjacent pixels of similar color, i.e., which are with an L2 norm less than 0.04 in normalized RGB space. By computing how much every component fills its bounding box we recognize rectangular connected components we classify the component as a rectangle [9]. We follow Raba [11] for component extraction, that is if component pixels fill more than 90% of the bounding box, Then: extract, otherwise: remove the component. Even small and thin rectangular components (with width or height less than 2 pixels), are rejected as such small components does not resemble bars but are image artifacts. However, some of these components do stand for very small bars, and in the extraction procedure, those are called back later [9].

Remove Background Side Rectangles. We observed that the bars are mostly colored in darker shades compared to background bars or rectangles. The true bars mostly hold red, blue or orange color while the side rectangles are lighter in shades mostly white. The background colour is used to compare each rectangle those similar in shades are discarded. This procedure is followed by traversing next from the rectangle to pint where the colour of pixel differs from border pixel or if not then a minimum of 5 pixels next to the rectangle [11]. With colour demarcation bar orientation is also important as many of times bars are found tilted horizontally or vertically. To correct this we assumed base line as y-axis, then histogram of the chart is drawn, using histogram equalization process bars are normalized. These bars are located and used for extraction. In our present work for bar charts we have not extracted y-axis.

Data Extraction. In this stage, our objective is to recuperate the data determined by every mark. We presuppose that a smear encodes a data, where single aspect is computable and one is ostensible. We recuperate these rows by using the geometry of the extracted marks and the text region tags from the classification stage. The final outcome is a table holding two value i.e. ID and a data tuple for every point of mark [9]. To recuperate the data from a bar chart, first we surmise the mapping between space within image and space within data. We presume a linear mapping, but we believe that our techniques of data extraction are compatible strong with other approaches. The linear mapping is unique in its sense: (1) a scaling factor is sliced between image space and data space, and (2) the minimum value is origin (generally the pixel value at the x-axis). We first recuperate the scaling factor by bearing in mind the y-axis labels. We make out the y-axis labels that encode data values by verdict text regions that have equal distance from the leftmost bar and line up upright. Then guesstimate the scaling factor using each pair of value labels. We guess the labels were improved by our text extraction procedure in the classification stage. The pixel distance between labels "6" and "10" is di = 60 pixels, and the estimated scaling factor is $60/(10-6) = 15$ pixels/data unit. We compute the scaling factor for each pair of labels and take the median as our final estimate. For this chart, the median scaling factor across all label pairs is 12.5 pixels/data unit. Then we find the minimum value. We begin with the y-axis label vertically closest to the x-axis. If this label is "0", the minimum value is set to 0, else re-calculate the minimum value using alike approach to compute the scaling factors. For each label, we estimate a minimum value considering the y-distance (in pixels) from the label's center to the x-axis, and the chart scaling factor [9]. The chart's minimum value is set to the median of these minimum values estimated. The median minimum for the chart is kept −0.2. Finally, a supposed appropriate value is associated to each bar with its adjoining label underneath the baseline x-axis [9].

3.2 Preprocessing and Normalization

Preprocessing is needed to remove the redundant data and will focus on contrast of background. Extraction of y-title is not possible in pass, it need second pass by rotating image in −900. As we have done extraction process, we are getting some garbage values in the numerical data extraction like 00 or 1g instead of 10. Whatever the unneeded garbage data is out we directly delete it. And where the recognition problem is occurred because of the quality of image, sometimes it considered g instead of 0. At this stage we have to assume it is 0 not g. The below standardizations are prepared concerning the charts [11]: • Charts should be 2D charts. • The fill color for the bars is a rock-solid color, to a certain extent of any pattern. • The y-axis should follow a linear scale, rather than logarithmic scale. • Alignment of Y-axis is on the left, alignment of x-axis is on the foot scale of the chart. Some preprocessing is done with the chart image, such as color normalization and noise filtering. Noise eliminating filters are used to the image for text identification in graphics [4]. Noise deduction filters such as bilateral filters are applied. To transitory of chart to the extraction section only pre-processing step is changes the color-space as a priority. The chart should be converted into grayscale version is needed for the text extraction module [11].

4 Experiments

We have done the experiments on a total 10 bar charts, those are taken from the net images sets [9]. OCR tool requires two pass for extraction of horizontal data and vertical data i.e. In the first pass OCR extracts all x-values, title of graph, legends if they are presented in horizontal form. And in the second pass it extracts the y-scale value i.e. all remaining vertical scale values. The method shows the result of bars almost 80% in the first pass of image charts; if the bars were identified and located hundred percent, then also the almost 20% of y-scaling of these charts wasn't evaluated properly because of the drawback of the OCR tool. Because of the image resolution and cleanness of the bars the maximum accuracies for the bar charts gained from the higher resolution images. Similarly the y-scale evaluated correctly for most charts apart from when results of OCR are inaccurate. For some images, more enrichment of the text region is obligatory to get concluding values for the chart data. Figure 2 Shows a sample input chart shows regions we are to be extracted. Additionally, some text improvement methods can be used to raise the precision of the results of the OCR tool. This should develop the exactness of the y-scale extraction and the axes titles extraction.

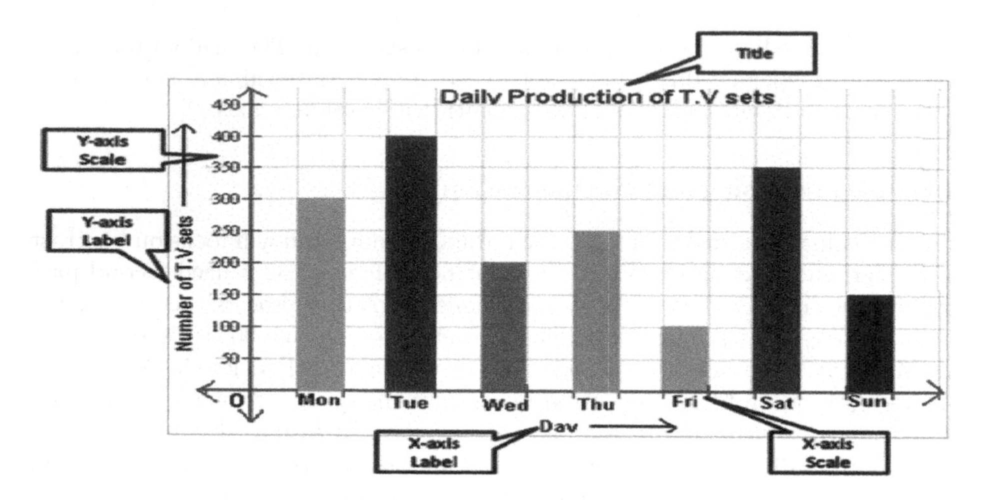

Fig. 2. Extraction of bar component

5 Question Answer Generation System

Automatic question generation is sub field of natural language processing. Questions are used to reveal the informational needs. Graph based methods are used for both single document and multi-document summarization. The answer is request or response to particular query. Answer is a thing that is done to deal with or as a response or reaction, statement or situation. An answer to a particular question may express various form like, definition manner, list format, factoid based, descriptive manner etc. We have performed the five point summary on the graph to extract particular values. We have extracted the points of Minimum and Maximum values of X and Y-axis. We have also extract the minimum and maximum values of intersect points. Then extract the median point. Minimum quartile value that is Q1, and maximum quartile value Q3. In the quartile values we calculate the 25% for Q1, 75% for Q3 is extracted after median point. We also generate the questions on the basis of color of bar. As the x ticks are extracted, depends on that we can generate questions like specific x-tick shows which color? And vice- versa. Finally we put all above calculated values in one matrix. Then ask query and answer will be the extracted values in the matrix.

6 Sample Question and Answer

QUESTIONS ANSWERS
Q.1 what data shows on x-axis? - Days in week
Q.2 how much gap between each data Intervals? - 5° gap between each data intervals.
Q.3 what data shows on Y-axis? - Temperature in Farenheight

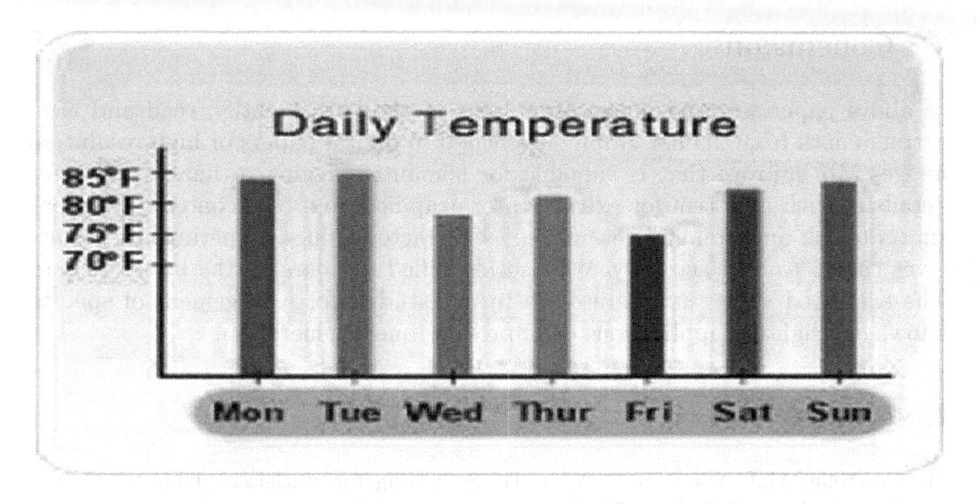

Fig. 3. Sample input image

Q.4 what is the title of graph? - Daily Temperature
Q.5 what is five point summary of this graph? - Min(Y)-73,
- Max(Y)-84, median(x)-Thru, median(Y)-77.5,
- Q1(x)-72.5, Q3(x)-82

Table 1. Experimental Results

Sample images	Q1. Min(x)	Q2. Min(Y)	Q3. Max(x)	Q4. Max(Y)	Q5. Median(x)	Q5. Median(Y)	Q6. Q1 (25%)	Q7. Q3 (75%)	1st x tick color	Green color x tick
Img1	0	0	-	70	-	35	15	55	Blue	Kiwi
Img2	0	0	2008	30000	2004	15000	6000	24000	Green	2001
Img3	0	0	Others	200	Haryana	100	40	160	Red	India
Img4	0	0	Sun	450	250	Thu	100	350	Yellow	Mon
Img5	0	0	Crackers	35	Ice-cream	17.5	7	28	Blue	Lollypop
Img6	0	0	Dec	100k	July	50k	20k	80k	Orange	Jan
Img7	0	0	-	.1	-	.05	.03	.07	Pink	-
Img8	0	0	Others	50	Calcium	25	6.5	43.7	Purple	Vitamin
Img9	0	0	Match5	70	Match3	35	15	55	Yellow	Match2
Img10	0	0	2010	350	2005	175	43.7	306.5	Fast yellow	2001
Img11	0	0	7	20	3.2	10	2.5	17.5	Chocolaty	1
Img12	0	0	Salad	175	Grilled chees	90	40	140	Fast red	Burger
Img13	0	0	Jazz	14	-	6.5	3	11	green	Piano

We have done experiment on above 13 number of images. And will get answers as given in the chart. The method gives the results, on average 75% of bars in the first pass of image charts. The y-scaling of about 20% of these charts was not evaluated properly because of the inaccuracies of the OCR tool. Due to the image superiority and cleanness of the bars, The highest accuracies is for the bar charts obtained from the high resolution images; Though the bars were identified and situated correctly.

7 Conclusion

In above paper we used image processing methods to identify, read and then extracts data from 2D bar graph entrenched in digital papers or high resolution images. To improve that is valuable for semantic tagging of images. We give details of such a system for extraction of graphical and text contents from bar charts to ask question and get answer as extracted values. Experimental results gives the 75 to 80% accuracy. We also identified the bars on the basis of color. The extracted values can be used in future studies for enlargement of specific knowledge sighting applications or improving query generation.

References

1. Cafarella, M.J., Chen, S.Z., Adar, E.: Searching for statistical diagrams. Front. Eng. Natl. Acad. Eng. 69-78 (2011)
2. Adar, E., Chen, Z., Cafarella, M.: Diagramflyer: a search engine for data-driven diagrams. In: Proceedings of the 24th International Conference on World Wide Web Companion, International World Wide Web Conferences Steering Committee, pp. 183–186 (2015)
3. Chester, D., Elzer, S.: Getting computers to see information graphics so users do not have to. In: Hacid, M.-S., Murray, N.V., Raś, Z.W., Tsumoto, S. (eds.) ISMIS 2005. LNCS (LNAI), vol. 3488, pp. 660–668. Springer, Heidelberg (2005). https://doi.org/10.1007/11425274_68
4. Kasturi, R., Fletcher, L.A.: A robust algorithm for text string separation from mixed text/graphics images. IEEE Trans. Pattern Anal. Mach. Intell. **10**, 910–918 (1988)
5. Huang, W., Tan, C.L., Leow, W.K.: Model-based chart image recognition. In: Lladós, J., Kwon, Y.-B. (eds.) GREC 2003. LNCS, vol. 3088, pp. 87–99. Springer, Heidelberg (2004). https://doi.org/10.1007/978-3-540-25977-0_8
6. Leow, W.K., Huang, W., Tan, C.L.: Associating text and graphics for scientific chart understanding. In: 2005 Eighth International Conference on Document Analysis and Recognition, Proceedings. IEEE, pp. 580–584 (2005)
7. Giles, C.L., Liu, Y., Mitra, P., Bai, K.: Automatic extraction of table metadata from digital documents. In: Proceedings of the 6th ACM/IEEE-CS Joint Conference on Digital Libraries, pp. 339-340. ACM (2006)
8. Atlas. http://atlas.qz.com/ (2015)
9. Fei-Fei, L., Agrawala, M., Savva, M., Kong, N., Chhajta, A., Heer, J.: Revision: automated classification, analysis and redesign of chart images. In: Proceedings of the 24th Annual ACM Symposium on User Interface Software and Technology, pp. 393–402. ACM (2011)
10. Fomina, Y., Vassilieva, N.: Text detection in chart images. Pattern Recognit. Image Anal. **23**, 139–144 (2013)
11. Al-Zaidy, R.A., Giles, C.L.: Automatic extraction of data from bar charts. In: Proceedings of the International Conference on Knowledge Capture, K-CAP, pp. 30:1–30:4 (2015)
12. Tomasi, C., Manduchi, R.: Bilateral filtering for gray and color images. In: ICCV, pp. 839–846, January 1998
13. Tufte, E.R.: The Visual Display of Quantitative Information. Graphics Press (1983)

14. Yang, L., Huang, W., Tan, C.L.: Semi-automatic ground truth generation for chart image recognition. In: Bunke, H., Spitz, A.L. (eds.) DAS 2006. LNCS, vol. 3872, pp. 324–335. Springer, Heidelberg (2006). https://doi.org/10.1007/11669487_29
15. Zhou, Y.P., Tan, C.L.: Hough technique for bar charts detection and recognition in document images. In: International Conference on Image Processing, pp. 605–608, September 2000
16. Stolte, C., Tang, D., Hanrahan, P.: Polaris: a system for query, analysis, and visualization of multidimensional relational databases. IEEE Trans. Vis. Comput. Graph. **8**(1), 52–65 (2002)
17. Huang, W., Tan, C.L.: A system for understanding imaged infographics and its applications. In: Proceedings of the ACM Symposium on Document Engineering, DocEng 2007, pp. 9–18. ACM, New York (2007)
18. Liu, R., Huang, W., Tan, C.L.: Extraction of vectorized graphical information from scientific chart images. In: Document Analysis & Recognition (ICDAR), pp. 521–525 (2007)
19. Mackinlay, J.D.: Automating the design of graphical presentations of relational information. ACM Trans. Graph. **5**(2), 110–141 (1986)
20. Mackinlay, J.D., Hanrahan, P., Stolte, C.: Show me: automatic presentation for visual analysis. IEEE Trans. Vis. Comput. Graph. **13**(6), 1137–1144 (2007)
21. Santosh, K.C., Wendling, L.: Character recognition based on non-linear multi-projection profiles measure. Front. Comput. Sci. **9**, 678–690 (2014)
22. Santosh, K.C., Lamiroy, B.: Character recognition based on DTW-Radon. IAPR, International Conference on Document Analysis and Recognition (ICDAR), pp. 264–268. IEEE (2013)
23. Santosh, K.C., Aafaque, A.: Line segment-based stitched multipanel figure separation for effective biomedical CBIR. Int. J. Pattern Recognit. Artif. Intell. (IJPRAI) **31**(6), 1–18 (2017)
24. Ghosh, S., Lahiri, D., Bhowmik, S., Kavallieratou, E., Sarkar, R.: g-DICE: graph mining based document Information Content Exploitation. Int. J. Doc. Anal. Recognit. (IJDAR) **18**(04), 337–355 (2018)

Online Behavior Patterns of Terrorists: Past and Present

Dhanashree Deshpande[1(✉)], Shrinivas Deshpande[2], and Vilas Thakare[3]

[1] SGBAU, Amravati, India
dhanashree.rani@gmail.com
[2] Department of Computer Science and Technology, DCPE, Amravati, India
shrinivasdeshpande68@gmail.com
[3] Department of Computer Science and Engineering, SGBAU, Amravati, India
vilthakare@yahoo.co.in

Abstract. Terrorists have shifted their activities of planning and coordination of attacks to internet by using web technologies. They are educated enough to create their websites, communicate through apps. They use online technologies to fulfill their objectives such as distribution of materials, overcome obstacles, communication, recruitment, fundraising, coordination, publicity, etc. Terrorists recruit required people through social media, websites, video games, etc. It has become easier for them to prepare for attacks with the help of modern form of terrorism. The paper explained various behavior patterns of terrorists from past to present. They have increased their number of activities since 1970 till the date. The activities are represented from year 1970 to 2016. The records are collected from Kaggle site. It is studied and analyzed. The terrorism affected badly in every country and its activities increase day by day. In this paper, the efforts took place to identify terrorist's way of online behavior.

Keywords: Behavior patterns · Terrorism · Online behavior · Cyber terrorism · Old and new terrorism

1 Introduction

Terrorism is originated from Latin word 'terrere' that means to frighten. The modern form of terrorism was obtained in 1793 and 1794 during the Reign of Terror in France [4]. Global terrorism consequences give rise to suffering, an untold amount of death, and horror. Whole world is facing the threat of terrorism. There are most dangerous groups such as ISIS, Al-Qaeda, Taliban, Boko Haram, Lashkar-E-Toba, Tehrik-I-Taliban, Hezbollah, Al-Shabaab, Hamas, Farc, Kurdistan Worker's Party, Al-Nusra Front, Naxalites, Irish Republican Army, the Lord's Resistance Army. Terrorists use the internet mainly to communicate and to spread propaganda. Cyber terrorists want to hide their intentions of logical attacks and act anonymously on internet [3]. On World Wide Web, many web sites provide information about how to generate explosive weapons. There are four thousand searching keywords related to terrorist and handbook found on the Google search engine. Terrorists use the Internet as the main medium to get aware the procedure

K. C. Santosh and R. S. Hegadi (Eds.): RTIP2R 2018, CCIS 1037, pp. 678–688, 2019.
https://doi.org/10.1007/978-981-13-9187-3_60

of bombs. They also plan and coordinate specific attacks through the various internet services. Terrorism is psychological warfare and run as a campaign on the Internet. Al-Qaeda was the first terrorist group to use the internet to expand their activities and ISIS brought modernization into terrorism with their use of social media. Al Qaeda combines advanced communication technologies and multimedia propaganda to create a sophisticated psychological warfare [5]. Terrorists use and follow various online tools and technologies for planning and preparation of attacks. It helps them to coordinate the interested people and reach up to them. Although training is in the physical form but communication and propaganda are done through the online services. In this paper, while analyzing terrorist's online behavior patterns, the events of terrorism took place are studied since year 1970 to 2016 and represented graphically. It could be possible to analyze the database for global terrorism but here in this paper, analysis is done by focusing on one country that is India. Terrorist's past and present behavior pattern is studied and differentiated.

The paper is organized as below. In Sect. 2, behavior patterns of terrorists have been explained. In Sect. 3, difference between old and new terrorism is explained. In Sect. 4, the ways of terrorist identification are explained. The terrorist's detail database is analyzed in Sect. 5. The paper is concluded in Sect. 6.

2 Terrorist's Behavior Patterns

Terrorists use Internet services to plan and coordinate various terror attacks [6]. They use message boards, chat rooms to communicate with each other within the groups, to share the information, recruitment, to coordinate the various attacks, to spread the propaganda, and to raise the funds in order to run these vulnerable activities. Their own websites offer tutorials and learning materials in order to teach the process of building bombs, firing surface to air missiles, shooting at soldiers, sneaking into Iraq from abroad countries etc. They host propaganda videos that helps to expand the recruitment and fundraising networks. Their own encryption tools and techniques are available for security. It includes the technique known as steganography, which hides the messages in graph files. They also use dead dropping that transmit the information through saved drafts in email account and get accessible to anyone with the known password of that particular email id. The Internet provides information about potential recruits and donors.

2.1 Communication

Terrorist's recruitment takes place on social media, messaging apps, websites. Recruiters are experts in determining the personalities that are easily manipulated for the tasks. They try to reach out up to each member in group by various services available on social media such as friend requests, following each other, and sending messages. Recruiters also use video games to target the members to become their followers. They also use chatting services such as WhatsApp for this purpose. It was reported that the e-mail communication service was used to coordinate the attacks in the investigation of 9/11. The interception of Voice-over-IP calls is more challenging as compared to the interception of regular phone calls. There are different ways terrorists use the internet for recruitment.

2.1.1 Social Media

Terrorists use websites, Twitter, Facebook, YouTube, RapidShare, Instagram, Tumblr online message apps, chat rooms, al-Qaeda's Inspire magazines, forum to recruit, radicalize and raise funds. ISIS is fluent in this approach. Twitter has suspended more than 1,000 accounts which are suspected for terrorist links. ISIS had found succeeded in creating 700,000 accounts discussing the terrorist group. They directly upload videos of them firing weapons and attacking towns. Al-Qaeda is using internet since two decades. It is spreading its significance among the Muslim community in Europe through social media. Al-Qaeda uses the internet to distribute materials anonymously or 'meet in dark spaces'. The Taliban group is active on Twitter since May 2011 and has many thousands of followers. Taliban tweets frequently.

2.1.2 Building Websites

Only 12 terrorists' vulnerable websites were active in 1998. Currently, this number raised to 7,000 active web sites. All terrorists' organizations own and regularly use at least one website. It is found that many terrorist organizations use more than one website in different languages. During the process of recruitment, these websites contain the messages such as 'the west is extremely aggressive, threatening towards Islam, the only way to treat them is through violence, so the option to counter the west is only jihad'. Many websites give information about killing or targeting specific person such as foreign journalists [5]. Their websites are not comparatively appropriate and hosted on servers provided by freely available public ISPs [7]. The websites appear and disappear continuously. They modify website's pattern and design frequently. These websites reappear within few hours or days [8]. The candidates who are interested to become their disciples get the video of propaganda and Pal Talk communication software. They use peer level close networks and different torrent online services for communication [3].

2.1.3 Online Video Games

Terrorists use online video games as part of their preparation for real attacks. These games simulate the real-world situation by changing character in virtual world. Islamic terrorist groups have released these online video games for teens and young adults. They gave children the goal for killing President George Bush. Hezbollah announced two games such as Special Force and its version-2 which indicate fighting with the Israeli military.

2.1.4 Indoctrinating Children with Cartoons

Both Hamas, Al Fateh, and al-Qaeda put the learning videos on websites in the form of Disney style animation. Their purpose was to recruit the children. In this animation, Al Fateh used cartoons to implement violence against Israel. In Yemen, Al-Qaeda released a cartoons short film in which young boys in dressed participated in terrorist plots and raids.

2.1.5 Targeting Prospective Recruits

Young people search and navigate web sites belongs to their community, traditions, heritage, ideologies associated with their group. Terrorists use to target these susceptible members of society. They take benefits of these feelings of weakness, loneliness, need for belonging. Terrorists diverts their minds and recruits these people for their tasks.

2.1.6 Instructing Procedure to Carry Out Attacks

Terrorists use online technologies to train the potential recruits in order to build explosives, execute specific attacks, obtain firearms and to join their organization.

2.2 Collecting Information About Targets

Terrorists access sensitive information about important targets from web and perform illegal activities such as blackmailing governments and businesses. Many such footprints were found on websites and removed after the attack on World Trade Center.

2.3 Financial Support

Terrorists gather economic support and trade stocks through the internet which is equivalent to any other organization. The Chechen Republic provided procedure on their website for donation to the Chechen breakaway from Russia. Terrorists also do cyber frauds to generate finance. The IRA website had provided a page link on which visitors could able to donate through credit card [5]. Approximately, dollar 1.6 million illegal funds are generated through 1,400 credit cards [1].

2.4 Extortion

Financial institutions are main target of extortion. Terrorists extort institutions in the form of threatening. They threaten to use the internet to commit cyber-attacks.

2.5 Publicity

Terrorists use the internet to do publicity. For example, while bombing was going on in Afghanistan, Osama bin Laden broadcasted a call via television and the internet to kill the Americans. Publicity is the first priority of terrorists and internet does that.

2.6 Global Freedom

Terrorists do coordination and communication of attacks from other countries. They able to operate from another friendly nation against a remote target. They use technology to terrorize people psychologically.

2.7 Secrecy

Terrorists send and post encrypted messages over the internet. When government were busy in checking these encrypted messages then the coordinators of the September 11 attacks were communicating in emails.

2.8 Propaganda

Terrorist groups spread their propaganda by using social media platforms, and publishing a magazine. The goal of propaganda is to appeal and convince their thoughts and ideology to attract more followers. There is no any specific website of terrorists such as "ISIS.com" or "alQaeda.org" but still this propaganda is available on internet. ISIS's media campaign is different and more brutal from other terrorist groups such as Al Qaeda in the social media. ISIS frequently posts bombing videos, beheadings, and targets children for these executions. ISIS mainly uses social media such as twitter, Facebook, YouTube, and share propaganda campaigns to recruit youths and radicalize them. Al Shabaab also uses these social media to for these campaigns. The Klu Klux Klan used twitter account to spread their messages. Al Qaeda and ISIS publish online magazines many times in a year which contains interviews, comments on target country's events, instructions on bomb making activity, other tips on successful execution of attacks etc. ISIS mainly focuses on the points of recruitment and spreading their extremist religion through magazine.

3 Difference Between Old and New Terrorism

In old terrorism, terrorist's activities planned and coordinated in small organizations whereas in modern it is planned and coordinated in network organizations. Nationalism remained a strong motivator of terrorism. Now a days, nationalism is mixed with religious themes. In old days, nature of terrorism was easy to detect and not safe whereas now a days, online terrorism in safe, beneficial and hard to detect compared to actions of traditional terrorism. Traditional terrorists recruited and converted young stray persons of Muslim background to become suicide bombers whereas modern terrorism recruit the right people. Terrorists were not well educated and they were not potential suicide bombers but now, new form of terrorism appears to be addressed solely to well-educated persons and not to suicide bombers in a traditional sense. Terrorists did not have knowledge of IT. They did not well versed with western culture. The recruited people for terrorist activities are highly educated, very skilled, born in Europe and USA with citizenship. They understand the western culture. They are not potential suicide bombers. They are competent with knowledge of IT. Terrorism had a well-defined geographical 'centre of gravity' to which all the group's activity could be related. Some of the new terrorists have no single permanent geographical point of reference. Terrorists often killed civilians and occasionally produced large numbers of casualties. In new terrorism, mass-casualty attacks against civilians have become routine [2, 3].

4 Identification of Terrorist

Traditional investigation methods were used to identify the person physically that was in control of the subscriber account. It was possible to identify the defendant by tracing the static IP address which is used to access an email account under surveillance.

Relevant ISP links the IP address to a subscriber account which is used by multiples of a household including the terrorist. Investigators intercepted the data traffic for the subscriber's account and were able to established links between IP address and activity on the jihadist website. Terrorists created multiple connections to the website's discussion forum. Investigators were able to correlate the times for these connections by increasing internet data volume linked to the defendant's personal e-mail account. Ping and Traceout commands may be used. Ping is used to send a signal to computer to check whether it is connected to internet at a given time. Traceout shows the path between two computers in network which is helpful to determine the physical location. Trojan horse or Remote Administration Trojans (RATs) are used covertly to collect information and to enable remote control over the compromised machine. Keystroke monitoring tool could be installed to monitor and record keyboard activities. Keystroke loggers are helpful in obtaining information regarding passwords, communications, website and local activity. Sniffers used to gather information directly from a network. It provides information of the source and content of communications. The computer's storage discs and RAM may contain important evidences of used programs by the suspects [1].

5 Analysis of Terrorism in India

Global terrorism's database has been analyzed. The database contains terrorism occurred from year 1970 to 2016. Total number of attributes in database are 135 with 1,70,350 records. It contains attributes such as year, country, region, state, city, latitude, longitude, attack type, target of attack, group names of terrorists, weapon type, number of killed etc. The database is analyzed and few of the factors represented in graphical format as below.

5.1 Global Terror Attacks in Countries

The graph shows total number of terrorist's attacks in different countries within the year 1970–2016. The countries which have faced more than 300 number of attacks are considered. The maximum number of terrorist's attacks happened in Iraq then Pakistan, Afghanistan, and India. In Iraq, 22,130 attacks took place and in India, it is 10,978 attacks.

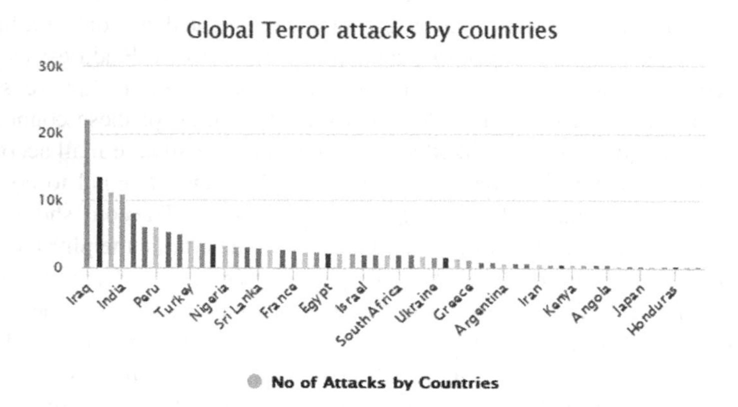

5.2 India Terror Attacks by Year

The graph shows total number of terrorist's attacks in India within the year 1970–2016. The maximum number of terrorist's attacks happened in 2016 i.e. 1,019 attacks. In 1977, there was only 1 attack occurred. The number of attacks has increased so much since then.

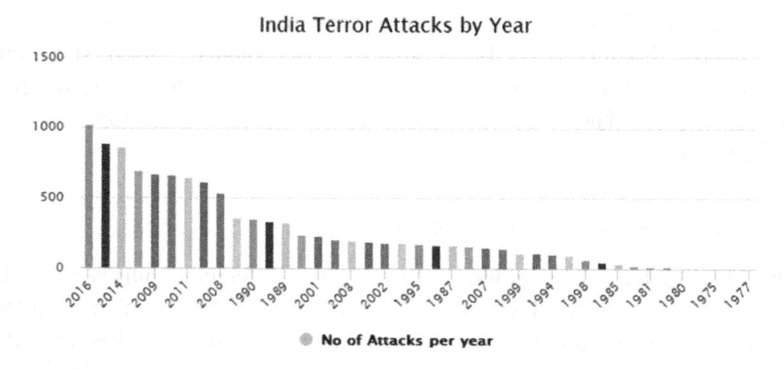

5.3 India Terror Attacks by States

The graph shows total number of terrorist's attacks in different states of India within the year 1970–2016. The maximum number of terrorist's attacks happened in Jammu and Kashmir i.e. 2,197 attacks then in Assam, Manipur, Punjab, Chhattisgarh etc. In Puducherry, Sikkim and Goa, lesser number of attacks occurred.

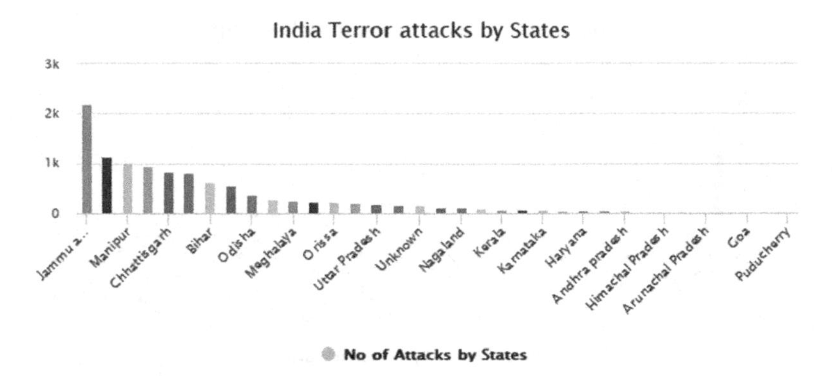

5.4 India Terror Attacks by City

The graph shows total number of terrorist's attacks in different cities of India within the year 1970–2016. The number of attacks more than 25 are considered. The maximum number of terrorist's attacks happened in Srinagar i.e. 637 attacks then in Imphal, New Delhi, Amritsar, Sopore etc. In Malkangiri, Lalgarh and Ahmedabad, Mumbai lesser number of attacks occurred as compared to other cities.

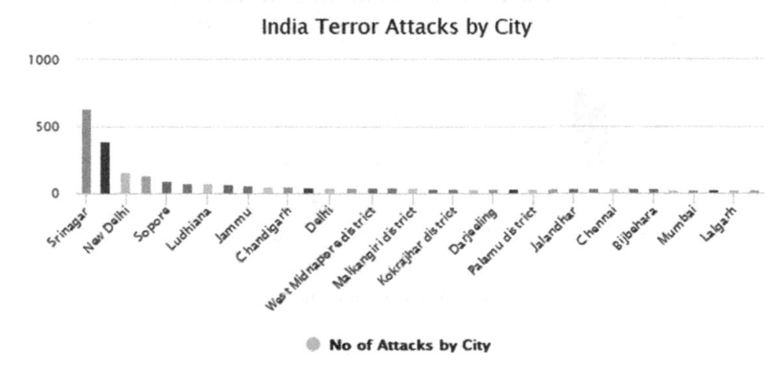

5.5 India Terror Attacks by Groups of Terrorists

The graph shows total number of terrorist's attacks in India by group of terrorists within the year 1970–2016. The number of attacks more than 35 are considered. The maximum number of terrorist's attacks coordinated and implemented by Communist party (CPI-Maoist) i.e. 1,765 attacks then Sikh extremists, United liberation front of Assam, NDFB, etc.

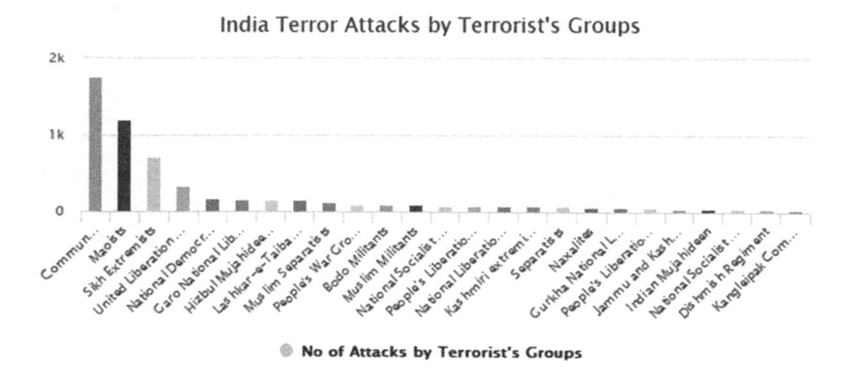

5.6 India Terror Attacks by Attack Types

The graph shows total number of terrorist's attacks in India by attack types within the year 1970–2016. The number of attacks more than 35 are considered. The maximum number of terrorist's attacks done through Bombing/Explosion i.e. 4,516 then through Armed assault, assassination, kidnapping etc.

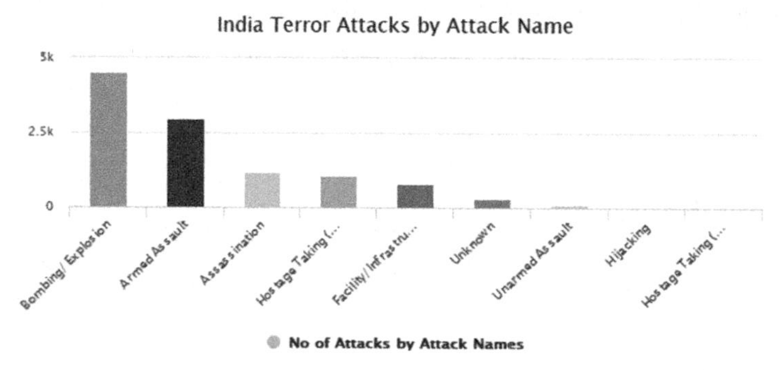

5.7 India Terror Attacks by Target

The graph shows total number of terrorist's attacks in India by keeping various targets within the year 1970–2016. The maximum number of terrorist's attacks done on the private citizens & property i.e. 2,904 attacks then on police, government (general), business, military etc.

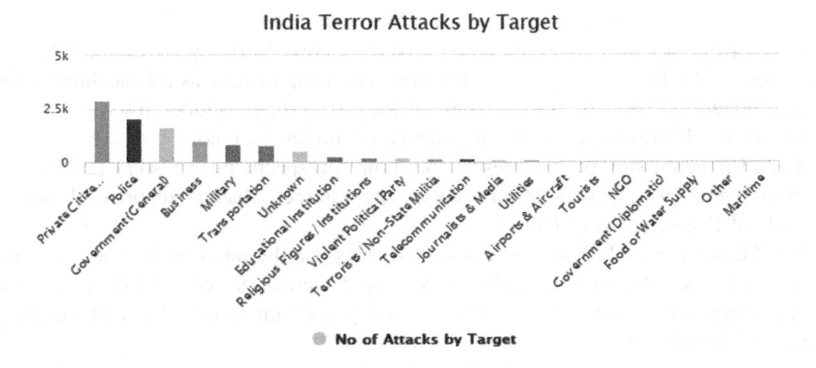

6 Conclusion

The research is focused on study and analysis of online terrorist's behavior patterns. Due to growth of technology many activities are implemented through internet. In this paper, terrorist's past and present behavior is analyzed. In this behavior patterns, terrorists are using online technology for planning and coordination purpose. They use social media, various apps, web sites, online video games, forums for communication. They recruit interested people, motivate them, do publicity online, provides procedure of how to use weapons, procedure of bombs through their online handbooks, also they provide videos for this task. The terrorist's database from 1970 to 2016 is studied and analyzed. It contains all activities of global terrorists. Though the database contains all the countries, the work is focused on India. Maximum number of terrorist's attacks occurred in Iraq. India is on fourth rank. Maximum number of attacks occurred in India in year 2016 as compared to other years whereas in 1972, only 1 attack took place. These numbers indicates that there is tremendous growth in attacks due to increase in technology. Maximum number of attacks occurred in Jammu and Kashmir whereas Puducherry, Goa and Sikkim are lesser in these attacks. Srinagar city is at highest rank, then Imphal, New Delhi, Amritsar etc. Attacks occurred due to Bombing/Explosion are more in number. The group of Maoists terrorists done maximum attacks. Private citizens remain main target for the attacks. By this analysis, it is found that the terrorist's activities are increasing day by day. This study mainly focused on the online behavior patterns of terrorists so that in future online suspicious user identification focusing on terrorism will become easier. Future work can be focused on further analysis of database focusing all countries, number of people killed, weapons used etc.

References

1. The use of internet for terrorist purposes, UNODC, New York, pp. 1–158 (2012)
2. Neumann, P.R.: Terrorism in the 21st Century, The Rule of Law as a Guideline for German policy, Friedrich-Ebert-Stiftung, Compass 2020, pp. 1–16, December 2008
3. Heickero, R.: Terrorism online and the change of modus operandi
4. Mannik, E.: Terrorism: its past, present and future prospects, pp. 151–171 (2007)
5. Weimann, G.: How modern terrorism uses the internet. Special report of United States Institute of Peace, pp. 1–12 (2004)
6. Cohen-Almagor, R.: Jihad online: how do terrorists use the internet? In: Campos Freire, F., Rúas Araújo, X., Martínez Fernández, V.A., López García, X. (eds.) Media and Metamedia Management. AISC, vol. 503, pp. 55–66. Springer, Cham (2017). https://doi.org/10.1007/978-3-319-46068-0_8
7. Qin, J., Zhou, Y., Reid, E., Lai, G., Chen, H.: Analyzing terror campaigns on the internet: technical sophistication, content richness, and web interactivity. Int. J. Hum. Comput. Stud. **65**, 71–84 (2006)
8. Goth, G.: Terror on the internet: a complex issue, and getting harder. IEEE Distrib. Syst. Online **9**(3), no. 0803, 1–4 (2008)

Author Impression on Dependent Authors Using Wordtovector Method

Maheshkumar B. Landge(✉), Ramesh R. Naik(✉), and C. Namrata Mahender(✉)

Department of CS and IT, Dr. B.A.M. University, Aurangabad (MS), India
Maheshkumar.landge@gmail.com, nam.mah@gmail.com, ramesh.naik31@yahoo.com

Abstract. This paper provides an introduction to types of authorship analysis which is important in many applications of NLU, QA, Plagiarism detection etc. Author profiling helps to identify the traits of an author in the given text/texts, which finally leads to predict whether those traits are present in other text that reflects the important characteristics of the original author. The original author normally observed to have some sort of his impact or impression on dependent writers or authors The main focus of this paper is identifying the weight of impact of original author on dependent author. Word to vector technique is been used in this work to identify impact of original author on dependent authors.

Keywords: Author impression · Authorship analysis · Wordtovector

1 Introduction

Plagiarism is finding the similarities between two texts. Plagiarism means an act or instance of using or closely copying the language and thoughts of another author without authorization and the presentation of that author's work as one's own without acknowledging the original author [1]. There are two types of plagiarism detection techniques (1) Extrinsic plagiarism detection (2) Intrinsic plagiarism detection.

In extrinsic plagiarism detection reference corpus is used to identify plagiarism and in intrinsic plagiarism detection author writing style is used to detect plagiarism [9]. The paper contains five parts: first part covers introduction to plagiarism, second part contains authorship analysis and its types, third part covers new feature of author profiling i.e. author impression, fourth part covers text corpus description, fifth part covers methodology description, and result shows percentage of author impact for three poems.

2 Authorship Analysis

In authorship analysis the important characteristics are counted on part of writing to identify the authorship of the given text [2]. Authorship analysis having three types: (1) Authorship Identification, (2) Authorship characterization, (3) Similarity detection.

K. C. Santosh and R. S. Hegadi (Eds.): RTIP2R 2018, CCIS 1037, pp. 689–695, 2019.
https://doi.org/10.1007/978-981-13-9187-3_61

2.1 Authorship Identification

Authorship identification tries to find the similarity of a part of writing produced by a specific author by examining other works by the same author.

2.2 Authorship Characterization

Authorship characterization produces an author profile based on the traits reviewed on the various works of the author.

2.3 Similarity Detection

Similarity detection compares multiple pieces of writing and decides whether they were written by a single author without actually identifying the author [2].

In intrinsic plagiarism detection stylometry technique is used to identify stylistically different segments in a text. Some features used for stylometry are sentence length, frequencies of words, word length [3, 4]. We have used stylometry feature extraction technique for our work. Stylometry feature extraction technique is effective for detecting plagiarism. Also we will consider Stylometric features and author writing style for the purpose of uniqueness. Because every author has unique style of writing. So the robustness of our tool will increase and it will be useful for linguistic researchers.

3 Author Impression

In our present work we have invented a new feature of author profiling i.e. author impression. To illustrate exactly what we want to extract with this trait is the author's impact on the authors who uses his work. In writing literature we come across the concept of dominant impression, which is the author's wish of creation of an environment to

उघडोनी नेत्र पाहे जंव पुढें । तंव दृष्टी पडे पांडुरंग ॥ १ ॥
देखिली पंढरी ध्याना तेची येत जयराम दिसत दृष्टीपुढें ॥ २ ॥
ब्राह्मण स्वप्नांत देखिला तो जाण । त्याची आठवण मनीं वाहे ॥ ३ ॥
न दिसे आणिक नेत्रांपुढें जाण । नामाचें स्मरण मनीं राहे ॥ ४ ॥
पूर्वील हरिकथा आयकिल्या होत्या । त्या मनीं मागुत्या आठवती ॥ ५ ॥
तुकोबाची पदें अद्वैत प्रसिद्ध । त्यांचा अनुवाद चित्त झुरवी ॥ ६ ॥
ऐसीं ज्याचीं पदें तो मज भेटतां । जीवास या होतां तोष बहु ॥ ७ ॥
तुकोबाचा छंद लागला मनासी । ऐकतां पदांसी कथेमाजीं ॥ ८ ॥
तुकोबाची भेटी होईल ते क्षण । वैकुंठासमान होये मज ॥ ९ ॥
तुकोबाची ऐकेन कानीं हरिकथा । होय तैसें चित्ता समाधान ॥ १० ॥
तुकोबाचें ध्यान करोनी अंतरीं । राहे त्याभीतरीं देहामाजीं ॥ ११ ॥
बहिणी म्हणे तुका सद्गुरु सहोदर । भेटतां अपार सुख होय ॥ १२ ॥

Fig. 1. Sample of abhangas written by SantBahinabai [5]

express what is wanted by the author that the reader should feel or get affected while reading the author's work. Such impression can be seen as a form emotion, inspiration. Marathi literature generally shows the author or poet impressions in their writings, like abhangas written SantBahinabai. The abhangas of SantBahinabai shows her impression in her writing in ways, Firstly Her name comes in abhangas. From that we come to know that was written by SantBahinabai and secondly due to the similar household context she uses to tell how much she is devoted or what a person should do to let fair life (Fig. 1).

The same way the authors who work on a particular known author. The original author shows some impact or impression on dependent writers or authors. So this work tries to focus on the aspect of finding those impacts of well-known writers on general dependent authors.

4 Text Corpus

As this work starts with development of text corpus because the corpus required for our purpose is not available. We created Marathi text corpus using three poems from 2nd standard text book of Marathi and collected explanatory summaries from different users.

Poem 1	Poem 2	Poem 3
आनंदाने नाचूया ! चला गड्यानो खेळूया ! आनंदाने नाचूया ! बाग चिमुकली फुले उमलली सुंदर फुलली फुलासारखे होऊया फुलाभोवती नाचूया ‖ फुलाभोवती भुंगे फिरती गाणी गाती तसेच आपण गाऊया ताल धरोनी नाचूया ‖ आभाळाचे तहेतहेचे रंग मजेचे चला चला रे पाहूया आनंदाने नाचूया ‖ – के. नारखेडे	पावसा, पावसा ये ! ये! झरझर झरझर आला वारा, टपटप टपटप पडल्या गारा. धड्म धड्म ढगांचा बाजा, कोकीळ म्हणतो पाऊसराजा. पावशा म्हणतो पाणी पाणी, धरती म्हणते द्या न्हाऊनी. नदी-नाल्यांना आला पूर, केरकचरा गेला दूर. माती झाली चिंब चिंब, आता उगवतील हिरवे कोंब. गडबड करते वीजराणी, आम्ही गातो पाऊसगाणी. पावसा, पावसा ये ! ये ! पीकपाणी खूप दे! --उषाकिरण आत्राम	गवताचं पात वाऱ्यावर डोलत, डोलताना म्हणत खेळायला चला. झऱ्यातलं पाणी खळाखळा हसत,हसताना म्हणत पळायला चला. निळनिळ पाखरू आंब्यावर गात,गाताना म्हणत नाचायला चला. झिम्मड पावसात गारांची बरसात,बरसात म्हणते वेचायला चला. छोटासा मोती लपाछपी खेळतो,धावताना म्हणतो शिवायला चला. मनीच पिल्लू पायाशी लोळत,लोळताना म्हणत जेवायला चला. --कुसुमाग्रज

Fig. 2. Sample of three Marathi poems

We have taken summary of poems written by users. The text corpus contains 900 summary files from 300 authors for three Marathi poems.

We have considered three Marathi poems namely "Anandane nachuya", "Pavsa pavsa ye ye", and "Gavtache pate" (Figs. 2 and 3).

कवी के. नारखेडे यांनी निसर्गातील प्रसंगाचे वर्णन करून त्यांच्यासोबत हर्ष ने नाचायला सांगितले आहे. छोटयाशा बागेत फुललेल्या सुंदर फुलांसारखे होऊन, फुलांभोवती गाणी गात फिरणाऱ्या भून्यासोबत ताल धरून तसेच आकाशातील विविध गमतीदार रंग पाहून नृत्य करणे सांगितले आहे.

Fig. 3. Sample of summary written by user

5 Methodology

See Fig. 4.

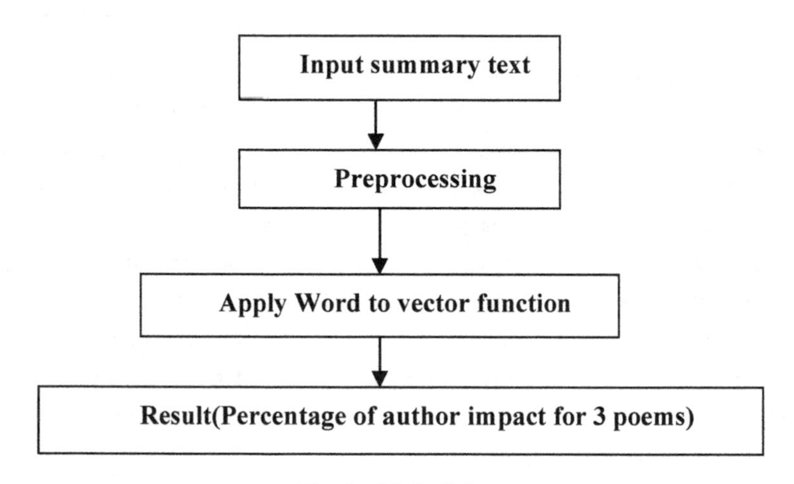

Fig. 4. Methodology

5.1 Input

In first step the summary file written by author is taken as input to system.

5.2 Preprocessing

In preprocessing first step is taking input file of summary written by user. Then in second step tokenization process is applied on text file. The tokenized data passed to word to vector function.

5.3 Word to Vector

word2vec is a tool developed by Mikolov, Sutskever, Chen, Corrado, and Dean et al. at Google [6]. Gensim provides a Python reimplementation of word2vec [7]. Word vectors encode semantic meaning and capture many different degrees of similarity [8]. word2vec calculates similarity between words. Word2vec is widely used in natural Language processing. Word2vec Mikolov, To-mas, [10] is a word embedding method that takes a corpus of words as input and produces vectors as output. word2vec having two models which are continues bag of words and Skip-gram. The difference between them is the word order, which is followed in Skip-gram and ignored in continues bag of words Rong [11]. In this paper, we apply the gensim Rehurek and Sojka [7] which is a python library to help us implement the Word2Vec. We first build a vocabulary from the entire training data. After training, each word attaches a vector and it generates word vectors. Finally, overall percentage of author impact for three poems is calculated.

5.4 Result

In above Tables 1, 2 and 3 there are ten sample summary files in each table, in which

गवताचपात, झर्यातिलपाणी, पाखरू, मोती, पिल्लु, वारा, कोकीळ, धरती, कचरा, कोंब, फुले, भृंगे

Words are present. These are some specific words, which are not replaceable with other words because it specifies perfect meaning. The presence of these words in dependent author file is denoted by 1 and 0 represents absence of that word. These words shows author's impact and which are used by dependent authors. One of the way which we can also term authors impression as authors impact on others writing, this can help for multiple application like in literature we can find dominate authors, expression expressed by authors even the plagiarism can be detected. The result from Tables 1, 2 and 3 shows the summary written by users is having impact of main poet of all three poems.

Table 1. A sample of ten Author users for first Marathi poem 'gavtach pate'

File_name	गवताच पात	झर्यातिल पाणी	पाखरू	मोती	पिल्लु
F1.txt	1	1	1	1	1
F2.txt	1	1	1	1	1
F3.txt	1	1	1	1	1
F4.txt	0	0	1	0	1
F5.txt	0	0	0	0	0
F6.txt	1	1	1	1	1
F7.txt	1	1	1	1	1
F8.txt	1	1	0	1	1
F9.txt	0	0	0	0	0
F10.txt	0	0	0	0	0

Table 2. A sample of ten Author users for second Marathi poem 'pavasa pavsa yeye'

File_name	वारा	कोकीळ	धरती	कचरा	कोंब
F1.txt	0	0	0	0	0
F2.txt	0	0	0	0	0
F3.txt	0	0	0	0	0
F4.txt	1	1	1	0	1
F5.txt	0	1	1	1	1
F6.txt	0	1	1	1	1
F7.txt	1	1	1	1	1
F8.txt	1	1	1	1	1
F9.txt	0	1	1	1	1
F10.txt	1	1	1	1	0

Table 3. A sample of ten Author users for third Marathi poem 'Anandane nachuya'

File_name	फुले	भुंगे
F1.txt	1	1
F2.txt	1	1
F3.txt	1	1
F4.txt	1	1
F5.txt	1	1
F6.txt	1	0
F7.txt	1	1
F8.txt	1	1
F9.txt	1	1
F10.txt	1	1

Table 4. Overall percentage of author impact for three poems

Sr. no.	Poem name	Percentage of dependent author impact	Average percentage of author impact for all poems
1	"Gavtacha pate"	62%	72.33%
2	"Pavsa Pavsa Ye Ye"	60%	
3	"Anandane nachuya"	95%	

The above Table 4 shows percentage of dependent author impact is higher for "Anandane nachuya" i.e. 95%. The average percentage of author impact for all poems is 72.33%.

6 Conclusion

Every person is having unique style of writing. They uses that style in his writing documents, emails, blogs. The impact of writing is present in his writing. The impact of main author comes in writing of dependent authors when dependent author follows main author. In this paper we created Marathi text corpus which contains 900 summary files of 3 Marathi poems. The summary is written by 300 different authors for three poems. Wordtovec technique used to find similarity of words used as the original author by the dependent author. The percentage for dependent author impact for poem “Gavtache pate” is 62%, for second poem “Pavsa Pavsa Ye Ye” the impact percentage is 60% and for third poem “Anandane nachuya” the impact percentage is 95%. The average percentage of author impact for all poems is 72.33%.

Acknowledgement. Authors would like to acknowledge and thanks to CSRI DST Major Project sanctioned No. SR/CSRI/71/2015(G), Computational and Psycholinguistic Research Lab Facility supporting to this work and Department of Computer Science and Information Technology, Dr. Babasaheb Ambedkar Marathwada University, Aurangabad, Maharashtra, India.

References

1. What is plagiarism? (2018). http://www.plagiarism.org/ (plagiarism definition). Accessed 8 Aug 2018
2. Zheng, R., et al.: A framework for authorship identification of online messages: writing-style features and classification techniques. J. Am. Soc. Inform. Sci. Technol. **57**(3), 378–393 (2006)
3. Brennan, M., et al.: Adversarial stylometry. ACM Trans. Inf. Syst. Secur. **15**(3), 12:1–12:22 (2012)
4. Krause, M., et al.: A behavioral biometrics based authentication method for MOOC’s that is robust against imitation attempts. In: L@S 2014 Proceedings of the First ACM Conference on Learning@ Scale Conference (Work in Progress), Atlanta, GA, USA, pp. 201–202. ACM Press (2014)
5. SantBahinabaicheabhang (2018). http://www.transliteral.org/pages/z70612224312/view. Accessed 01 Aug 2018
6. Mikolovet, T., et al.: Efficient estimation of word representations in vector space. CoRR, vol. abs/1301.3781 (2013)
7. Rehurek, R., et al.: Software framework for topic modeling with large corpora. In: Proceedings of the LREC 2010 Workshop on New Challenges for NLP Frameworks, Valletta, Malta, pp. 45–50 (2010)
8. Levy, O., et al.: Dependency-based word embeddings. In: Proceedings of the 52nd Annual Meeting of the Association for Computational Linguistics (Volume 2: Short Papers), Baltimore, Maryland, pp. 302–308 (2014)
9. Clough, P.: Plagiarism in natural and programming languages: an overview of current tools and technologies. Technical report, University of Sheffeld, Sheffeld, UK, June 2000
10. Mikolov, T., et al.: Discovering word senses from text using random indexing. arXiv preprint arXiv:1301.3781 (2013)
11. Rong, X.: word2vec parameter learning explained. arXiv preprint arXiv:1411.2738 (2014)

Correlation Based Real-Time Data Analysis of Graduate Students Behaviour

Shankar M. Patil[1(✉)] and A. K. Malik[2]

[1] Department of Computer Engineering, Singhania university, Pacheri, India
smpatil2k@gmail.com
[2] Department of Mathematics, BK Birla Institute of Engineering and Technology, Pilani, India
ajendermalik@gmail.com

Abstract. Engineering institutes nurture the mind sets of a nation. Knowledge is an intangible asset of the organization. Quest for increased performance and reliability has made it imperative to develop techniques for utilization of resources to make informed decisions. The contradiction is represented with engineering intake and outcomes. So there is need for planning better adaptive strategies of student's behaviour to increase productivity. In this paper, research work has been carried out to find behaviour patterns and exemplify the quality of educational system. Our approach illustrates the behaviour of engineering students in various dimensions. Specifically, it determines the association among academics and behaviour for general evaluation of students' in engineering institutes. This evaluation helps to address common problems which keep away students from functioning productively in class. The correlation evaluation with ten parameters and 31 supporting questions gives real life, real time analysis to make strategic planning, action and development for betterment of students and engineering institutes.

Keywords: Students behaviour · Informed decision · Behaviour pattern · Real time analysis · Engineering institute

1 Introduction

Engineering institutes have the capability to describe their own set of behavioural potential. There is no universal set of behaviours that can be commonly regarded as demanding. Behavioural expectations approaches to promoting optimistic behaviour and knowledge of behavioural potential should be made accessible in a college. There are several possible influences on student behaviour and lots of factors that can direct to behaviour that is difficult for institutes to compact through. Behaviour patterns would facilitate in pleasing essential steps to prevent/promote this behaviour according to expectations and make informed decisions. On the other hand, if there are some contrary performance patterns in the similar field below dissimilar situations with students, this would identify for awareness.

K. C. Santosh and R. S. Hegadi (Eds.): RTIP2R 2018, CCIS 1037, pp. 696–706, 2019.
https://doi.org/10.1007/978-981-13-9187-3_62

Our approach is to provide the key solution to the institute with different issues. They would be able to achieve the behavioural pattern of each individual student as well as of the entire class in a specific institute. This paper deals with finding behavioural attributes of the entire class or institute. It would include module for collecting the individual opinion from the students to the behavioural pattern extraction, the entire processing of the behavioural attributes is explained in this research paper. In this work, we make an attempt to discuss a system that analysis and provides us with the behaviour of the students. We have conducted a survey among students of the engineering institute. The survey collects the feedback from students which will then be processed to get the appropriate and relevant data of all the behaviour parameters. This result would be beneficial for the institutes to know each aspect of the individual attributes which would be needed to be improved for their overall performance. The analysis of student's behaviour can provide an effective reference for campus construction, as well as the cultivation and management of student's life.

A total number of 334 student's data is collected from engineering college, including ten different attributes for the research work. In general, the findings of this work are likely to be representative for the higher education institutions, which produce graduates each year in the India.

The contents of the paper are organized as follows, Sect. 2 includes Literature review, Sect. 3 method, Sect. 4 explains Data Analysis Model for Student Behaviour, Sect. 5 Result and discussion, Sect. 6 mentioned conclusion.

2 Literature Review

Dass [2], described a work done to comprehend alliance of a variety of magnitude of behaviour of school kids to resolve Anand Niketan school management's problem. School management's problems exasperated with get higher in number of students and faculty along with bigger struggle from other schools in the neighbourhood. The school management attention of getting better a number of their heart procedure to allow improved understanding of their students' performance and providing individual concentration to students. Association rule mining is used for student's behavioural analysis. The system helps to understand students' strong and weak areas on the basis of behavioural analysis. Behavioural attributes are useful along with academic parameters to find out performance pattern of a particular student. Main et al. [6] identified the role of marks in foremost decision and most important switching behaviour, their investigation identifies factors allied with switching within engineering, at the same time, examines how students opportunities regarding upcoming results may manipulate foremost alternative. Kometani et al. [7] proposed a coaching behaviour blueprint form that can approximate instruction behaviour from student opinion of every class. In their research, they identified correspondences among class improvement and instruction behaviours from instantaneous student evaluations performed throughout the class. Fan et al. [8], proposed a network based technique to discover the association between the attainment of a student and his learning partners. For this study they have gathered the information from the digital campus card and study the relationship between book borrowing behaviour and educational achievements, at the same time overwhelming ability of sharing and campus movement trajectory from two different departments in a Chinese university.

Spies [9], proposed integrated large information insights with computerized and assisted processes associated to key client touch points to get better the client experience. The author has suggested how to progress their business performance, using exclusive tactic designed to choose the accurate key excellence indicators, construct precise key business purpose "method," forecast client behaviour and eventually recognize which factors are influencing the majority. To develop customer understanding, service providers must be capable to successfully determine and sculpt customer experience. This means that facility providers must also know what matters generally to their customers.

As Godfrey explains, varying traditions requires "not just a change in behaviour and practices, but sufficient experience and strengthening to those behaviours and practices to change values and educational norms and finally the instinctively held faith and assumptions that are the spirit of civilization [15]".

3 Method

3.1 Context of the Population Studied in This Work

Engineering student's data is used for the study. For this data, survey was conducted through online Google form to collect the responses of the students. Response data of first year, second, third and final year students is in a excel sheet. Question wise analysis is carried out for each class. The response analysis of the survey results are, first year students 35%, second, third and final year 25%, 56% and 40% responses respectively.

To investigate the student behaviour pattern data of 334 students is acquired through Google survey form. Analysis focuses on the change in correlation among the different ten attributes, using the low, Medium and high correlation value for each attributes measurement. The correlation value greater than +0.70 is called very strong relationship and +0.40 to +0.69 strong positive, +0.30 to +0.39 moderate, +0.20 to +0.29 weak positive, +0.01 to +0.19 no or negligible relationship and 0 no relationship, also similar in negative relationship. Python open source software is used for statistical analysis. We developed and implemented algorithm to compute correlation of ten different attributes of the class behaviour. Pearson correlation coefficient is used to compute correlation between different attributes. Our algorithm compute average of the collected data from the survey form and finally, the correlation is computed of all attributes.

3.2 Population Studied in This Work

The student behaviour is very important aspect in today's world for every educational institute. Analysis of behavioural attributes give idea about lagging strong or weak attributes of students, so institute can carry out various measure to improve quality of students. This work focuses on the 334 students of the engineering college from 2016–17 academic year. The collected data is sufficient to calculate the student's behaviour from ten different attributes. A survey is conducted that consists of different questions related to engineering. We find the behaviour pattern of engineering students by making them answer to survey questions present in the survey form. This work is helpful to find out the strong and the weak areas of each attributes of the Class/Batch in the institute, for knowing the student behaviour of the engineering college.

3.3 Research Question and Attributes Used in This Work

In this work, the research survey questions are designed based on the defined attributes of the research work. The ten attributes are used to know the engineering college students behaviour. The survey Questions are given in the Table 1. The one to four questions are defined for each attributes of this study.

4 Modelling Behaviour of Engineering Graduates

To find out the behaviour pattern of engineering student's from various aspects of collected data, correlation coefficient is used. The result of Pearson correlation symmetric matrix is shown in Table 2.

4.1 Employability

The perspective for aptitude attribute was testing the ability of engineering graduates against employability as this is the first mandatory criteria for campus selection. While analysing the real time collected data,16% students in the cluster of poor performance. So, considering institute requirement and departmental efforts they must motivate these students to achieve the goal of recruitment, as, nowadays the first enquiry while counselling engineering graduates is job offerings.

We found in this study is, the final year students aptitude knowledge is good compared to the lower classes. The relationship between the attributes is a positive and medium level, and P value is less than 0.001, in other words there is evidence for associations between these two variables.

4.2 Work Capabilities

Class discussion will enhance critical thinking skill of the learners. Class participation is one of the ten parameters to evaluate and analyze student behaviour against attentiveness, focused and enthusiastic to learn co-curricular activities during teaching learning process. It has been analyzed considering students focus for subject contents, his answering ability and courage to ask difficulties. It has been found that 74% students are attentive and comparable during question

Table 1. Survey Questions of Students class behaviour

Attributes	Questions
Aptitude (AP)	1. Which one of the following is not a prime number?
	2. Find Odd one: 16, 25, 36, 72, 144, 196, and 225
	3. Find the correctly spell word
	4. Choose the alternative which is closely resembles the mirror image of the given combination
Attendance(AT)	1. What do you think about the minimum attendance criterion?
Class Participation (CP)	1. I feel comfortable answering questions in class
	2. If I don't think my answer is 100% correct still I will say it during class discussions
	3. Students in my class room support their peers when they give a wrong answer
	4. Do you encourage weekly seminars and quizzes and related co-curricular activities in class?
Communication Skills (CS)	1. When communicating with others I pay attention to non
	verbal signals - body language, facial expression, and gestures?
	2. I am often thinking of something witty to say in response while listening
Emotional Intelligence (EI)	1. I do not become defensive when criticized
	2. I utilize criticism and other feedback for growth
	3. I try to see things from another's perspective
	4. I recognize how my behaviour affects others
Innovation (IV)	1. Would you like to do some new, out-of-box projects based on some new language or new trend?
Leadership (LD)	1. When assigning tasks I consider people skills and interests
	2. I am highly motivated because I know How to become successful
	3. Time spent worrying about team moral, is time that wasted
	4. My actions show people what I want from them
Punctuality (PN)	1. Do You believe that students who are late for classes are likely to be late to other events in their daily life?
	2. Do You think unpunctual students will grow up to be unpunctual at their workplace?
	3. Do you believe unpunctuality is a problem that needs addressing and one that could make the country more efficient if we all learnt to be more punctual?
Responsibility (RS)	1. Do I regularly set goals for things I want to accomplish?
	2. Do I manage my time and set priorities to make sure I do the important things first?
	3. Do I know that I am responsible for what I do and that I must accept the consequences of my behaviour?
	4. Do I keep track of my college assignments and keep a daily study plan?
Subject Knowledge (AP)	1. SQL stands for
	2. How long is an IPv6 address?
	3. Which of the following is the 1's complement of 1010?
	4. From what location are the 1st computer instructions available on boot up?

Table 2. Pearson correlation matrix of ten different attributes

Attribute	Ap	AT	CP	CS	EI	IV	LD	PN	RS	SK
Aptitude	1	−0.019	0.07	0.073	0.12	0.17	0.11	0.036	0.16	0.29
Attendance	−0.019	1	0.089	−0.22	0.11	−0.19	0.06	0.12	0.066	−0.12
Class participation	0.07	0.089	1	0.19	0.37	0.26	0.39	0.21	0.34	0.0043
Communication skills	0.073	−0.22	0.19	1	0.18	0.26	0.24	0.041	0.2	0.053
Emotional intelligence	0.12	0.11	0.37	0.18	1	0.36	0.35	0.33	0.38	0.1
Innovation	0.17	−0.019	0.26	0.26	0.36	1	0.28	0.28	0.39	0.25
Leadership	0.11	0.06	0.39	0.24	0.35	0.28	1	0.29	0.46	−0.012
Punctuality	0.036	0.12	0.21	0.041	0.33	0.28	0.29	1	0.3	−0.012
Responsibility	0.16	0.066	0.34	0.2	0.38	0.39	0.46	0.3	1	0.14
Subject knowledge	0.29	−0.12	0.0043	0.053	0.1	0.25	−0.012	−0.012	0.14	1

answer session but 12% students are more focused and participate in debate against learning process. These 12% students may lead to project handling capabilities in future. The relationship between class participation, responsibility and emotional intelligence towards leadership is positive, medium and high level correlation; there is an evidence of association between the variables of leadership and class participation. The analysis investigated collective behaviours of the all classes are between medium level correlation towards responsibility, Leadership and emotional intelligence.

The final year students class participation is low and Second year class participations is high and notable to other classes. First year and third year class students shown interest in the leadership skill in the class participation, the investigated correlation values are 0.38 and 0.52. Second year class students are always interested to do the activities in the college, so it shows in the result, correlation coefficient is high level.

4.3 State-of-the-art Development

Intelligence is an ability to stand tall when been criticized. It does not allow being defensive as it has emotional origin and makes decision based on scenario. The emotional intelligence parameter for engineering students have been tasted against four criteria's and got results which are indications of growth when being criticized. This is because most of students took it in positive sense for their benefits and not only for criticism. One more reason was, students started thinking and respecting others perspective than own opinion. They got many angles to tackle the same problem and may be easiest way to get rid of most painful experience. Few students approach got very interesting perspective than others, in which they play role and become more responsible towards their goal setting.

Table 3 Shows that, Second year students have noticeable innovative correlation value and highly responsible in the class participation. The correlation coefficient value for the second year class is very strong towards responsibility.

Final year student's shows negligible relationship in emotional intelligence and class participation but their innovation result shows that, they are good in innovation and also responsible students in class. Students who actively participate in class are emotionally intelligent. The relationship between Emotional Intelligence towards Innovation and Responsibility is positive and medium level correlation; there is a strong evidence of association between the Emotional Intelligence and Responsibility.

Table 3. Pearson correlation of emotional intelligence Vs innovation and responsibility

Attribute	Class	Emotional Intelligence	Class participation	Innovation	Responsibility
Emotional Intelligence	Collective	1	0.37	0.36	0.38
Emotional Intelligence	F.E	1	0.34	0.29	0.27
Emotional Intelligence	S.E	1	0.54	0.44	0.55
Emotional Intelligence	T.E	1	0.30	0.21	0.54
Emotional Intelligence	B.E	1	0.17	0.39	0.48

4.4 Learning Craft

Analyzing 334 tuples against attendance the general remark can be drawn as, 85% students are interested to attend more than 75% lectures. While digging more into questions, it seems that, students save time, cost, efforts and more rely on self learning after listening it once from teacher.

Table 2 shows that, there is negative correlation in the attendance towards communication skill, aptitude and subject knowledge. It means, communication skill and Aptitude are not dependent on attendance. So, this clearly indicates that attendance have negative correlation towards learning. Only class participation and attendance shows the positive relationship, means without attendance class participation is not possible.

The investigation shows that, students who are more attentive in classes generally have low communication skills and students who are more attentive in classes generally are more punctual than others in the activities. Students who are more attentive are generally less innovative and students who actively participate in extracurricular activities have better communication skills.

4.5 Creative Innovation

The data shows the students active class participation and involvement leads to innovative thinking. The soft skills are used to exchange and conceptualization of thought processes. It is a way of knowledge sharing and gives rise to informed decision making. It helps students to keep pace with rapidly changing technology the innovative ability. Communication is used to share knowledge in student community. This conclusion is based on groups study during examination. His base of exchanging ideas while studying, the data shows active class participation with good communication ability leads with innovative ideas.

Results identifies that, first and second year class attendance is good but third year onwards it decreases. All classes are punctual about their studies, second year students show their attitude towards leadership so their class participation and responsibility is high, it is noticeable.

Students who participate more in extracurricular activities have good leadership skills. Students who are innovative are generally taking more responsibilities.

5 Results and Discussion

We focused on the evaluation behaviour attributes of class participation, aptitude, communication skill, innovation, leadership, emotional intelligence and responsibility towards leadership, learning, responsibility and subject knowledge. For comparison with the absolute values of evaluation, we also investigated the correlation between different attributes. Figure 1 shows, the differences in evaluation attributes of the classes and their behaviour pattern.

By considering various aspects of student's behaviour a complete survey has been carried out in an engineering college. For this survey Students population is undergraduate programs of various departments. Our result shows the correlation of the ten different attributes of students from FE, SE, TE and BE.

From above results it is concluded that, the final year class attributes such as aptitude, innovation and subject knowledge correlation is high. The communication skill, class participation and subject knowledge attributes are good of all the classes, it also shows that, student's correlation is strongly positive with attributes like aptitude, Innovation and subject Knowledge. In comparison, for the first year and third year, the score is high in the innovation attribute. Also this graph shows the final year student are less punctual than the other classes, their relationship is less comparatively to the other classes. The reason is that, the final year students are more focused on the placement in the college campus, so they focused on aptitude rather than other activities in the class room. Therefore, the final year students aptitude attribute behaviour is higher than the lower classes. Another observation is that the first year students losing subject knowledge, due to late admission in the college and less time is getting them to study in the first year, also the new environment and location they are taking more time to adjust.

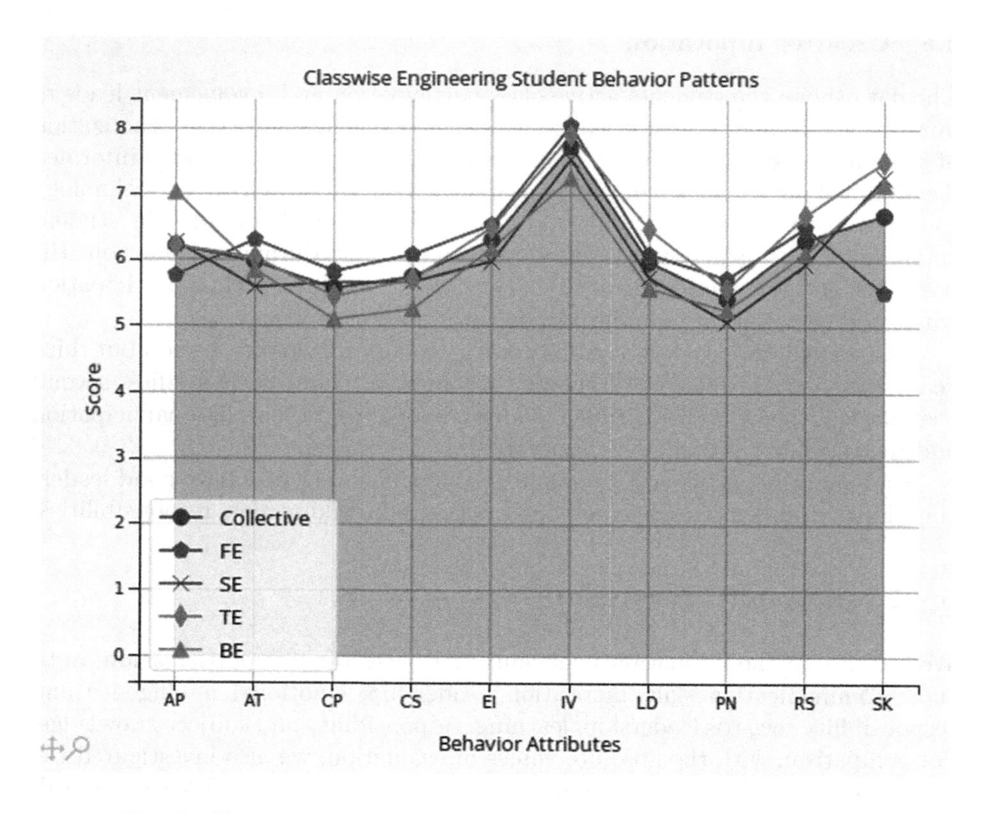

Fig. 1. Class-wise student's behaviour pattern of different attributes.

5.1 Disscussion

We found that, the students of engineering college are viewing strength in behavioural attributes such as class Participation, Innovation, Leadership, Responsibility, Emotional Intelligence, and Subject Knowledge. While, many more of the students are weak in attributes such as Aptitude, Attendance and Communication skills. The correlation coefficient between these attributes in Pearson correlation are found and shown in the Table 2. For example, we found that students, who are Innovative in class, are also good in class participation and Leadership. Similarly, for some behavioural attributes are showing weaknesses.

It was observed that the attributes such as Subject Knowledge, Class Participation, Responsibility, Leadership and Innovation go hand in hand with each other and the attributes such as Attendance, Emotional Intelligence, Innovation does not adhere to a positive correlation as they oppose trajectories.

The management needs to understand and apply this knowledge of extracted behaviour patterns so as to add in their decision making process and will know where special attention needs to be given. This paper can help institute to improve the management decision while developing necessary policies.

6 Conclusion

1847, the establishment of first engineering college in India and since then the graph of establishment of new colleges and courses is ever increasing. Every year India crops 1.5 million engineering graduates and as many as 97% wants employment. However, only 3% have suitable skills to be employed in software, and only 7% can handle core engineering tasks. This statistics gives rise to analysing engineering graduate's behaviour to cultivate their mind-sets for productive outcome. The fertile mind of engineering graduates must be fostered with path breaking thinking and action oriented learning.

This research investigated changes in correlation of student behaviour of different attributes. While improving engineering graduates job offerings, the institute and departmental efforts can motivate students to achieve the goal of recruitment. The result showed that class participation is one of the parameters used to evaluate and analyze student behaviour against attentiveness, focused and enthusiastic to learn co-curricular activities during teaching learning process. It has been analyzed considering students focus for subject contents, his answering ability and courage to ask difficulties.

The emotional intelligence parameter for engineering students have been tasted against four criteria's and got results which are indications of growth when being criticized. This is because most of students took it in positive sense for their benefits and not only for criticism, in which they play role and become more responsible towards their goal setting. The analyzed attentive result seems that, students save time, cost, efforts and more rely on self learning after listening it once from teacher.

The result showed that correlation between student behaviour and classes, changes their attribute correlation values as per individual class. Providing this information change of behaviour pattern to the institute for further decision making and development.

References

1. Lord, S.M., Layton, R.A., Ohland, M.W.: Multi-institution study of student demographics and outcomes in electrical and computer engineering in the USA. IEEE Trans. Educ. **58**(3) (2015). https://doi.org/10.1016/0022-2836(81)90087-5
2. Dass, R.: Using association rule mining for behavioural analysis of school students: a case from India. In: Proceedings of the 42nd Hawaii International Conference on System Sciences. IEEE (2009). https://doi.org/10.1007/11823285_121

3. Oskouei, R.J.: Identifying students' behaviours related to internet usage patterns. In: T4E 2010. IEEE (2010)
4. Peter Salhofer, G.: Analysing student behaviour in CS courses a case-study on detecting and preventing cheating. In: 2017 IEEE Global Engineering Education Conference (EDUCON), 25–28 April 2017
5. Srour, S.E., Luo, J., Goda, K., Mine, T.: Correlation of grade prediction performance with characteristics of lesson subject. In: 15th International Conference on Advanced Learning Technologies. IEEE (2015)
6. Main, J.B., Mumford, K.J., Ohland, M.W.: Examining the influence f engineering students course grades on major choice and major switching behaviour. Int. J. Eng. Educ. **31**(6(A)), 1468–1475 (2015)
7. Kometani, Y., Tomoto, T., Akakura, T., Nagaoka, K.: Correlation between teaching behaviour and real time student evaluations. In: IEEE Computer Society, IEEE sixth International conference on Technology for Education (2014)
8. Fan, S., Li, P., Liu, T., Chen, Y.: Population behaviour analysis of Chinese university students via digital campus cards. In: 2015 IEEE 15th International Conference on Data Mining Workshops, IEEE Computer Society (2015)
9. Spies, J., T'Joens, Y., Dragnea, R., Spencer, P., Pilippart, L.: Using big data to improve customer experience and business performance. Bell Labs Tech. J. **18**, 3–17 (2014)
10. Cápay, M., Mesárošová, M., Balogh, Z.: Analysis of Students' Behaviour in E-Learning System (2011)
11. Benko, L'., Reichel, J., Munk, M.: Analysis of Student Behaviour in Virtual Learning Environment Depending on Student Assessments (2015)
12. Dodge, P.R.: Managing school behaviour: a qualitative case study. Iowa State University, Graduate theses and Dissertations (2011)
13. Pan, J., Jun, Y.: Design and realization of college service center system based on MVC. In: College of Communication and Information Engineering, Xi'an University of Science and Technology, Xi'an, China. 2014, IEEE Workshop on Advanced Research and Technology in Industry Applications (WARTIA) (2014)
14. Norasiah, M.A., Norhayati, A.: Intelligent student information system. In: 4th National Conference on Telecommunication Technology Proceedings, Shah Alam, Malaysia (2003)
15. Godfrey, E.: Cultures within cultures: welcoming or unwelcoming for women? In: Proceedings of ASEE Annual Conference, Honolulu, HI, USA (2007)

Compression Based Modeling for Classification of Text Documents

S. N. Bharath Bhushan[1(✉)] and Ajit Danti[2]

[1] Department of Computer Science and Engineering, Sahyadri College of Engineering & Management, Mangaluru, India
sn.bharath@gmail.com

[2] Faculty of Engineering-CSE, Christ (Deemed to be University), Kengeri Campus, Bangalore 560074, India
ajitdanti@yahoo.com

Abstract. Classification of text data one of the well known, interesting research topic in computer science and knowledge engineering. This research article, address the classification of text files issue using lzw text compression algorithms. LZW is a lossless compression technique which requires two pass on the input data. These two passes are treated separately as training stage and text stage for classification of text data. The proposed compression based classification technique is tested on publically available datasets. Results of the experiments shows the effectiveness of the proposed algorithm.

Keywords: Text classification · LZW text compression · Compressed representation

1 Introduction

Nowadays text files are the common platform for information exchange, sentiment sharing. It is because of the huge production of electronic documents because of internet and web. These text files productions is increasing on a daily basis but data structure and algorithms to handling them and understand them is still constant. Due to this, there is huge scope for research in text mining and allied areas like sentiment analysis, patent classification and clustering of text data.

Text mining applications are designed based on the features extracted, representation method and classification method adapted. In the beginning, conventional text processing techniques are adapted to read the text files. Then, feature extraction techniques are applied for the construction of knowledge base. Here, free running text files need to be represented in a machine understandable mode and then relevant features are extracted from the text files. Later supervised learning algorithm is adapted for assigning class information to the query document.

K. C. Santosh and R. S. Hegadi (Eds.): RTIP2R 2018, CCIS 1037, pp. 707–715, 2019.
https://doi.org/10.1007/978-981-13-9187-3_63

Many researchers have found solutions to classification of text document problem. The traditional technique is construction of knowledge base from the extracted features, using this a class label will be assigned to the query sample Q_D. Here the algorithm will be trained with the support from the extracted features during training stage. Generally, in most of the text mining application frequency of occurrence of terms are considered as the features using which knowledge base will be constructed. Classification of text data is process of assigning a boolean value say zero or one to all the test samples to the number of classes defined in the dataset. Basically it is a approximation function between training and testing say FunctionTrain :Domain and Document $\rightarrow$ 0 or 1 with another FunctionTest :Domain and Document $\rightarrow$ 0 or 1 in such way that, these FunctionTrain and FunctionTest will be similar to the maximum level. Here the function FunctionTest will be classifier [1,2].

One of the important entities of text classification algorithms is parameter for computing the similarity between train and text functions since the efficiency of text mining application is mainly depends on this. There are different and well known metrics are there for computing the degree of similarity between two text files. In literature, well known measures like cosine similarity, kullback-leibler divergence, euclidean distance and many other approaches are explored by many researchers to compute the degree of closeness between two text documents [3,4]. In this research article, new method will be proposed for proximity using LZW compression technique. Lempel-Ziv-Welch (LZW) is one of the popular dictionary based lossless text compression technique. The working principle of lzw is explained in algorithm – 1 of Sect. 3. The main motivation here to choose LZW technique is, it require two pass on the input data. In the first pass, algorithm construct the dictionary and actual compression is achieved in the second pass. So, here we used first pass as the training stage and second stage as the classification stage of text files. The other remaining parts of the article is arranged as follows,

In Sect. 2 literature survey of different representations and state of the art classification approaches are presented. In Sect. 3 discuss the proposed proximity measure for text document classification. Section 4 presents the details of the experimentation conducted. Article will be concluded in Sect. 5.

2 Literature Survey

Text mining application are having stages like preprocessing and document representation which are most important steps in recent machine learning systems. The performance of these are having direct impact on the recognition of topics and assigning class label to the test documents. Since this research article is discuss the problem related to text classification, state of the art document representation methods accessible are here.

We can find many intuitive text document representation schemes in the literature. These can be broadly categorized into vector – space model, distributed text representation and co-occurrence representation model. Representation like vector – space model handles terms as independent entities. Due to this, problems like generation huge dimensional feature space and it cannot capture the correlation with the adjacent terms in the document which leads to loss of semantic information of the document [2]. To tackle this issue, many document level representation models have been proposed such as D-VSM [5], N-gram representation, LSI and LPI [6]. In term co-occurrence model, features related to particular topic and terms index (position in the document) are be considered for the construction of topic specific dictionary. This universal representation can also be extended to graph based representation, where topics are represented as nodes and relationships are represented as the edges connecting the nodes. Another popular representation is known as bag of topics representation for text samples. Here terms related to topics are indexed based on the bag of topics representation. This representation can also be extended to features space also. Topic modelling was proposed by [7], it is based on the fact that a text sample contain several topics and the terms will be present in various topics. This representation will reduce the dimension of the features space and also enable high level features for topic understanding and document representation. In Bengio et al. 2003 multilayer neural network architecture is used for construction of bigram language model for obtaining word vectors. Some other techniques are based on compression techniques, SVM, regression-based approach and artificial neural network [2]. In literature we can find many approaches for classification of text documents. These approaches include naïve bayes [8] nearest neighbor [8], decision trees [9], support vector machines [1,10] and neural network [11–13]) approaches.

3 Proposed Model

Current section of the article introduces you the novel similarity metric using text compression algorithms for text mining applications. Text processing is one of the most discussed domain of computer science which has many applications like classification text documents, sentiment analysis, document clustering and patent analysis. All these applications have few stages as common and all these are pictorially shown in Fig. 1. In these applications, two things which contribute a lot is representation model for text data and selection of prominent feature. Generally bag-of-words (BoW) and vector space representation models are treated as the most popular representation models for text data. After representing the documents, next issue will be selection of similarity or dissimilarity measure for text data. Understanding the importance proximity measure for developing an efficient application, here we are proposing a compression based similarity metric for effective classification of text documents. The proposed similarity metric has the capability capturing important document features like terms and frequency of terms for obtaining the similarity between text data.

Compression as a Similarity Metric: Many researchers have already proposed classification of text files using compression algorithms. But the commonality among these algorithms are, text data is represented using compressed representation not on regular representation of the text data. Classification of text documents using compression model is achievable only statistical reappearance of terms (features) in the input data. Here this article, address the classification of text files issue using lzw text compression algorithms. LZW is a lossless compression technique which requires two pass on the input data. These two passes are treated separately as training stage and text stage for classification of text data. During first pass, text compression algorithms, will identify the terms from the input data and construct the dictionary. Later based on the dictionary test file will be compressed. Based on the compression ration class information will be assigned to the test file.

$Class_a = [doc_1^a, doc_2^a, doc_3^a]$ and
$Class_b = [doc_1^b, doc_2^b, doc_3^b]$

Where $doc_i^j = [f_{i1}^j, f_{i2}^j, f_{i3}^j, ..., f_{im}^j.]$

The function $Proximity_Compression$ will define models say $model_a$ and $model_b$ as mentioned like above classes during the training stage. These models are again considered for compressing test file. This situation is illustrated in the following example.

3.1 LZW Based Model Construction

LZW - Lempel–Ziv–Welch compression technique is a universal lossless compression technique. LZW compression algorithm will an index dictionary during the first pass, of the input data [14,15]. In our algorithm it is treated as training stage, where all the dominant features of the input data is taken into account. In second pass on the input data, dictionaries from different domains are considered for actual compression. This process in our algorithm is treated as classification stage. Since compression of the actual input depend on the selected dictionary, there will be huge variation in the compression ratio. It is clear from the observation and experiments, for some dictionary compression ratio will be very since it contains some terms (features) from the domain. This process is clearly explained in the following illustration.

Illustration 1: Let us consider two class data as follows and these are denoted as $Class_{education}$ and $Class_{computer}$
$Class_{education-1}$ = [Education begins at home.]
$Class_{education-2}$ = [Education is the key to success.]
$Class_{computer-1}$ = [Computer is a machine.]
$Class_{computer-2}$ = [Computer programs are written by computer programmers.]

Conventional text mining techniques are adapted and stopwords ($is, was, where, ..$)are removed from the input collection. The transformed data will be as follows
$Class_{education-1}$ = [education,begins,home.]
$Class_{education-2}$ = [education,success.]
$Class_{computer-1}$ = [Computer,machine.]
$Class_{computer-2}$ = [Computer,programs,written,computer,programmers.]

Two samples from each domain will be concatenated to form a single file. Later a word level LZW compression will be applied. As a result, two models say model education and computer will be created.

$Model_{education-1}$ = [education,begin,home,success,1]
$Model_{computer-2}$ = [computer,programs,written,programmers,1]

Since the term education is present in all the sample, the term will be replaced by its index as show above. This will be applicable all the other classes of the database.

Consider a test sample Q_D = [education helps to mold character].

For $Model_{education}$ = [1 helps to mold character].
For $Model_{computer}$ = [education helps to mold character].
(Stop words are eliminated.)

It is clear from the above example, the LZW compression technique will provide best result based on the model selected. Means, test sample Q_D will have more compression ratio for model education and there will be no change for the remaining model. Hence we can come to a conclusion that the query document can be assigned with the class name ***Education***.

Algorithm - 1

Algorithm : LZW text compression
Input : Collection of uncompressed text documents
Output : Collection of compressed text documents, String table
Method :
Step 1 : Initialize table to contain single character strings.
Step 2 : Prefix string $\omega \leftarrow$ Read first input character
Step 3 : K $\leftarrow$ Read next input character.
If no such K (input exhausted) : code
(ω) - output; EXIT.
Step 4 : If ωK exists in string table : ωK- ω ; repeat 3;
Step 5 : else ω K not in string table : code (ω) - output;
Step 6 : ω K - string table;
Step 7 : K - ω ; repeat Step.
Algorithm End

4 Experimentation

The proposed algorithms are evaluated using following datasets.
First Data Corpus - Vehicle Wikipedia data set: Dataset Size = 440 documents from 04 classes.
Second Data Corpus- Google-Newsgroup dataset: Dataset Size = 1000 documents from 10 classes.

Third Data Corpus - Twenty mini newsgroups: Dataset Size = 2000 documents from 20 classes.
Fourth Data Corpus - Twenty newsgroups: Dataset Size = 20000 documents from 20 classes.
Experiments = Two Sets.

Set 1 = 40%Train and 60%Test
Set 2 = 60%Train and 40%Test

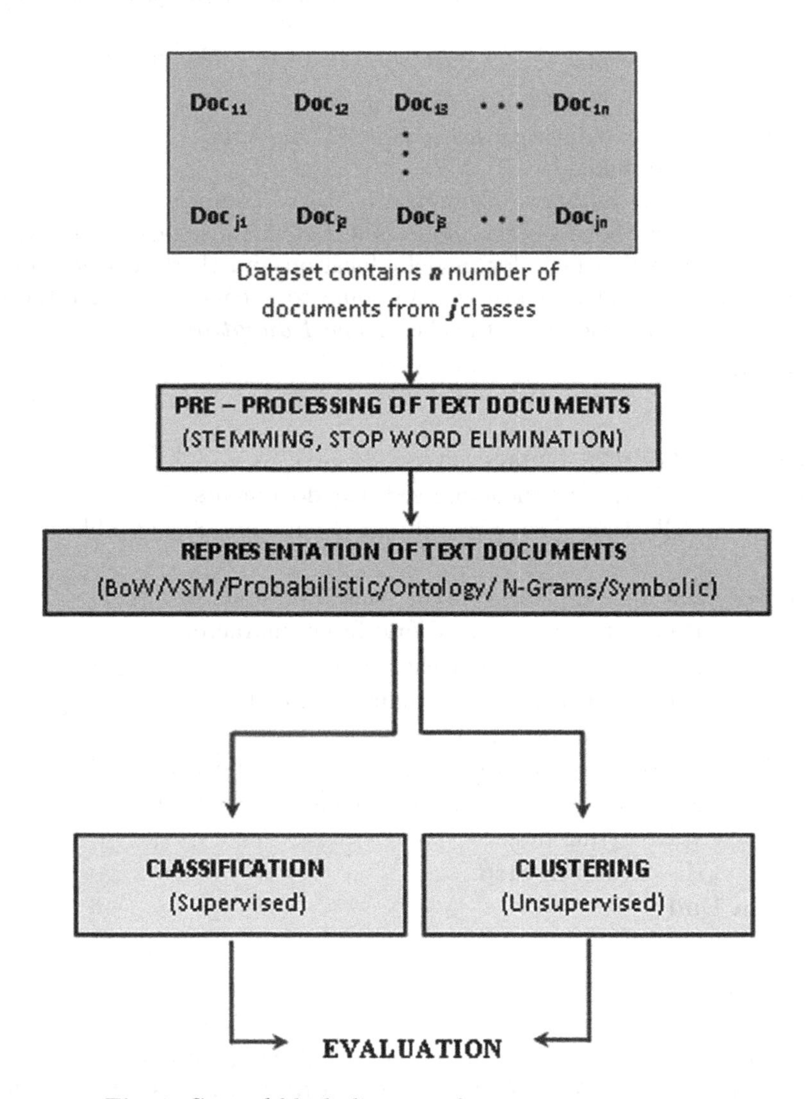

Fig. 1. General block diagram of text processing system

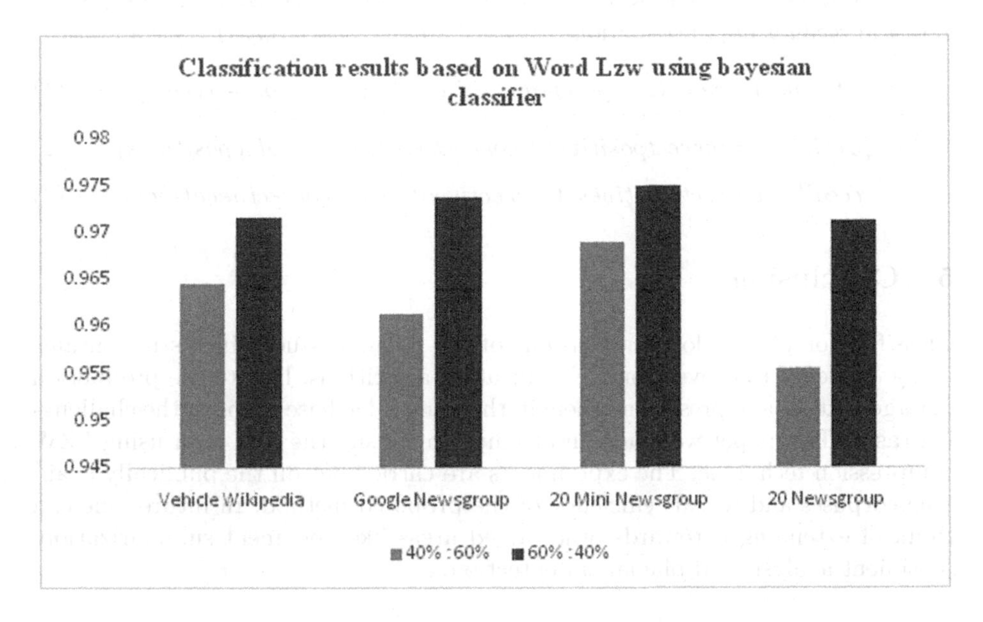

Fig. 2. Classification results based on word Lzw using bayesian classifier

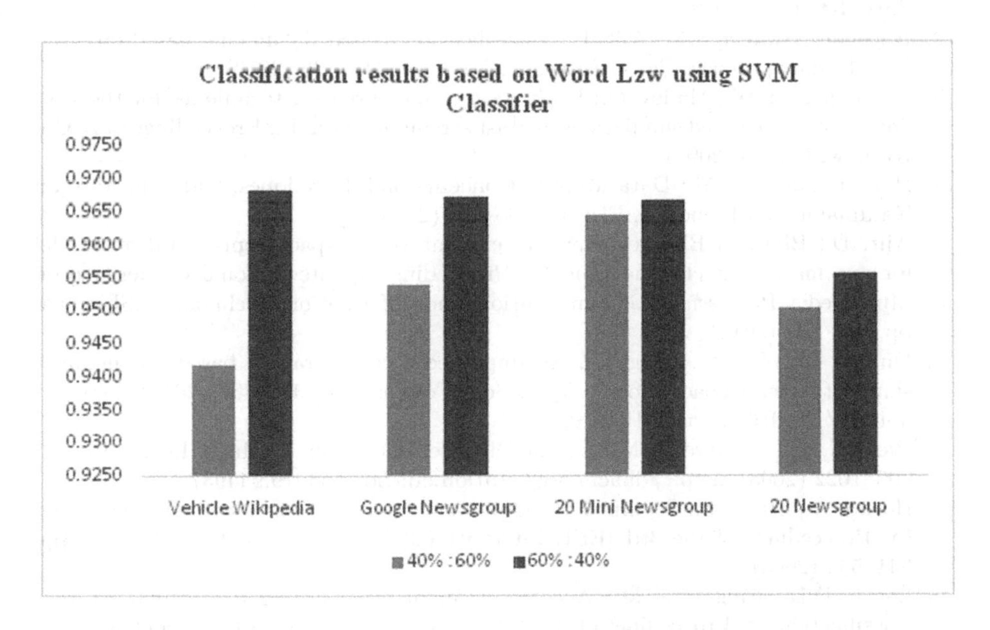

Fig. 3. Classification results based on word Lzw using SVM Classifier

The proposed model is evaluated by well-known metrics like f-Measure, precision and recall using the equations Eqs. 1, 2 and 3 respectively (Figs. 2 and 3).

$$f - measure = (2 \times precision \times Recall)/(precision + recall) \tag{1}$$

$$precision = correctpositives/(correctpositives + falsepositives) \tag{2}$$

$$recall = correctpositives/(correctpositives + correctnegatives) \tag{3}$$

5 Conclusion

Classification of text documents is one of the popular issue, which still contains many challenge to conventional classification algorithms. Due to the production of large text data, representing them in the knowledge base is one of the challenging tasks. This paper we made an attempt to classify the text data using LZW compression technique. The experiments are carried out on the publically available corpuses and reveals efficiency of the proposed method. In future, one can think of extending it towards other allied areas like document summarization, sentiment analysis and plagiarism detection.

References

1. Sebastiani, F.: Machine learning in automated text categorization. ACM Comput. Surv. **34**, 1–47 (2002)
2. Bhushan Bharath, S.N., Ajit, D.: Classification of text documents based on score level fusion approach. Pattern Recogn. Lett. **94**, 118–126 (2017)
3. Schoenharl, T.W., Madey, G.: Evaluation of measurement techniques for the validation of agent-based simulations against streaming data. In: Proceedings of ICCS, Kraków, Poland (2008)
4. Han, J., Kamber, M.: Data Mining: Concepts and Techniques, 2nd edn. Morgan Kaufmann, San Francisco, Elsevier, Boston (2006)
5. Ajit, D., Bhushan Bharath, S.N.: Document vector space representation model for automatic text classification. In: Proceedings of International Conference on Multimedia Processing, Communication and Information Technology, Shimoga, pp. 338–344 (2013)
6. Du, Y., LiuW, L.X., Peng, G.: An improved focused crawler based on semantic similarity vector space model. Appl. Soft Comput. **36**, 392–407 (2015). https://doi.org/10.1016/j.asoc.2015.07.026
7. Blei, D., Ng, A., Jordan, M.: Latent Dirichlet allocation. J. Mach. Learn. Res. **3**, 993–1022 (2003). http://dl.acm.org/citation.cfm?id=944919.944937
8. Hotho, A., Staab, S., Stumme, G.: Ontologies improve text document clustering. In: Proceedings of the 3rd IEEE International Conference on Data Mining, pp. 541–544 (2003)
9. Lewis, D.D., Ringuette, M.: A comparison of two learning algorithms for text classification. In: Proceedings of the 3rd Annual Symposium on Document Analysis and Information Retrieval, pp. 81–93 (1998)

10. Joachims, T.: Text categorization with support vector machines: learning with many relevant features. In: Nédellec, C., Rouveirol, C. (eds.) ECML 1998. LNCS, vol. 1398, pp. 137–142. Springer, Heidelberg (1998). https://doi.org/10.1007/BFb0026683
11. Donald, S.: Probabilistic neural networks. J. Neural Networks **3**(1), 109–118 (1990)
12. Patra, A., Singh, D.: Neural network approach for text classification using relevance factor as term weighing method. Int. J. Comput. Appl. **68**(17), 37–41 (2013)
13. Ajit, D., Bharath, B.: Classification of text documents using integer representation and regression: an integrated approach. Spec. Issue IIOAB Scopus Indexed J. **7**(2), 45–50 (2016)
14. Bharath Bhushan, S.N., Danti, A.: Classification of compressed and uncompressed text documents. Future Gener. Comput. Syst. **88**, 614–623 (2018)
15. Bharath Bhushan, S.N., Danti, A.: Comparative study of clustering algorithms on compressed text data. Int. J. Comput. Eng. Appl. **XII**(I), 182–190 (2018)

Detecting Depression in Social Media Posts Using Machine Learning

Abhilash Biradar(✉) and S. G. Totad

Department of Computer Science, B.V.B College of Engineering, Hubli 580031, India
I.biradar.abhilash@gmail.com

Abstract. The utilization of Social Networking Sites (SNS) like Twitter is expanding quickly and particularly by the more youthful age. The profit capacity of SNS enables us to express their interests, emotions and offer their day by day schedule. SNS sites such as Twitter allow for constant investigation of user behaviour. Such examples are important for the psychological research network to comprehend the periods and area of most prominent interest. Worlds fourth biggest disease depression has turned out to be a standout amongst the most huge research subject. We propose a system which uses tweets as source of data and SentiStrength sentiment analysis to create a training data for our system and a Back Propagation Neural Network (BPNN) model to classify the given tweets into depressed or not depressed categories.

Keywords: Back Propagation Neural Network (BPNN) · Depression · Machine learning · Social Network Sites (SNS) · SentiStrength · Twitter

1 Introduction

Social Network Sites (SNS) like Facebook and Twitter have changed the ways in which people represent their opinions, communicate with others and share their knowledge. This yields a ceaseless stream of a mind-boggling measure of information containing hints of profitable data, like, individuals' sentiments and opinions. For example, About 22% of adults online use Twitter and more than 500 million tweets are sent each day [1].

Depression has been a hot topic for psychology research. With the rise of the SNS, Many psychologists turn their views on web media from traditional case studies. They detect depressed users with psychological information and observe their online features [2,3]. Most of the perception concerns every person's practices and little attention is paid to interaction between users.

A statistics made by e-marketer K [4] has shown that Number of active monthly users on Twitter from the third quarter of 2017 worldwide, is averaged at 330 million, and the number is increasing rapidly. Most people use social media to express their feelings, emotions and their daily activities. Rich working groups have demonstrated that social media is an open space for many to express

K. C. Santosh and R. S. Hegadi (Eds.): RTIP2R 2018, CCIS 1037, pp. 716–725, 2019.
https://doi.org/10.1007/978-981-13-9187-3_64

their negative emotions by displaying information that reflects emotions [5]. Many researchers are constantly demonstrating that social media can be used successfully to maintain the mental health of the people.

From the social media user profile, we can collect all information about a person's mood, activities, sleeping hours, thinking styles, interactions and helplessness. We will obtain a complete picture of the natural behavior of the user by analyzing the social media posts of the user. Such behavioral attributes in social media show depression symptoms that can be used to predict whether the user is depressed or not. Depressed user friends and parents can track user depression using the proposed tool, which saves time for the depressed user to deal with major problems.

2 Related Work

One can call Twitter a microblog or a social site. This is called because the main activity is to post short status updates via the web or mobile phone (tweets). Twitter is also a social network site because members have a personal information profile page and can connect to other members by following them and accessing their content. It seems to be used every day to share information and describe minor activities [8]. Around 80% of users in Twitter post update on what they do to followers, while the rest of them focus on information [9].

In the sentiment analysis method used by Thelwall et al. [10] for short informational text, A random sample of comments by Myspace was taken by examining the profiles of all 15th member who joined on 18 June 2007, up to 40000. The standard SentiStrength algorithm here has done better than standard machine learning methods. When their performance was improved by the use of subsumption and information gain feature reduction, the difference in the result was not statistically different. A slightly modified SentiStrength version was significantly better than the improved methods of machine learning.

Depression detection has been a popular research on online social media platforms or SNS. A depression prediction model proposed by Aldarwish [6] is developed using RapidMiner. Both classifiers, SVM classifier, and the Naive Bayes are tested using number of processes. The proposed model contained seven main operators of two datasets. The first set of data is the training data set containing the 2073 depressed posts manually trained and the second set of data consisting of users of SNS posts. The datasets here are manually trained which limits the number of training samples that can be used which makes it difficult to modify or update the dataset which makes the tool less adaptive.

The other method which has been developed by Husain [7], It begins by collecting User Generated Content (UGC) from social media on Facebook. They then labeled the words from the posts that refer to either user in the training phase are depressed or not depressed. After training, text classification algorithms are used to assign a test to one of the classes by using the support vector machine classifier (SVM). Tokenization, conversion of lower case and word stemming and all pre-processing operations are carried out on the text before it is converted into vector space. The developed tool is restricted only to Facebook users and users are labeled manually during the training phase, and in SNS such as Facebook, there is no much data available about one's mood or mental health as he/she is more likely to share pictures and events.

3 Methodology

In this work, to monitor a user's mood from their daily tweets and users at risk of depression, a harmonic analysis technique is developed which uses the textual cues from daily tweets. The proposed system is mainly composed of four modules Data collection, Sentiment analysis, Machine learning model, and the classification of the tweets Fig. 1 Methodology.

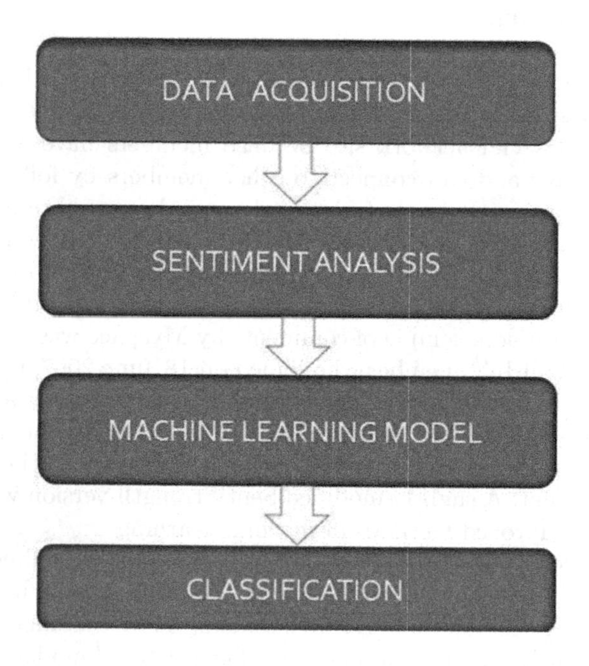

Fig. 1. Methodology

The tweets that are collected using the Twitter API, undergo a sentiment analysis and a sentiment is assigned to each tweet which is used as training data (Table 3). Classification is made to grouped tweets which is test data for the proposed classification model.

3.1 Data Acquisition

This method is for analyzing the emotional feeling expressed on Twitter, an online social media service that allows the user to post and read short text messages (tweets) of 140 characters. The vast share of Twitter accounts are openly accessible, permitting any enlisted or unregistered user to see posted tweets this stands out from other community stages which have diverse protection settings. Twitter offers an open API (Application Program Interface) which empowers automatic utilization of distributed tweets as they happen. So as to get to Twitter Streaming API, we have to get 4 snippets of data from Twitter: API key, API secret, Access token and Access token secret, which can be picked up by making a tweeter account. We will use a Python library called Tweepy for the Twitter Streaming API interface and download the information. Tweets can be collected using specific criteria such as keywords, which is used while occupying the data for classification. There are numerous sorts of burdensome issue, and each sort has its own one of a kind manifestations. The most widely recognized kind of Depression is called Major Depressive Disorder (MDD), which interfaces most with the capacity to eat, study, work, rest, and having a ton of fun. Frances et al. has shown to analyze Major Depressive Disorder, the patient will have at least five of the accompanying nine side effects amid the time of about fourteen days and almost consistently. The main side effect is having a discouraged state of mind a large portion of the day. The second is losing enthusiasm for all exercises. The third manifestation is weight reduction or weight gain and dozes excessively. The fourth side effect is body fomentation or impediment. The fifth is feeling tired or loss of vitality. The sixth is the sentiment of blame or uselessness. The seventh side effect is discovering fixation, considering or settling on a choice turns into a troublesome undertaking. The eight manifestation is inconvenience having rest or rest excessively. The ninth and last manifestation is thinking about death. As based on the symptoms of Major Depressive Disorder (MDD). The suitable keywords were (a) depression (b) sadness (c) tired (d) guilt (e) suicidal (f) anxiety and (g) mental health for the information regarding mind state and some random tweets. Hence the tweets were acquired using the above keywords (Table 2).

Table 1. Tweets acquisition keywords

Keyword	Number of tweets acquired
Depression	27,900
Sadness	5,000
Tired	5,000
Guilt	5,000
Suicidal	5,000
Anxiety	10,000
Mental health	2,500
Random	1,000
	61,400

Table 2. Tweets example

Keyword	Example
Depression	Someone please be my friend and come hangout with me tonight?? Not to party or be crazy, but just hangout. Have conversation. Bring me out of my depression ... any takers?
Sadness	Worst feeling ever. Sadness turns into depression
Tired guilt	Im just so tired of everything and i'm tired of feeling all this guilt that i buried inside of my ribcage
Suicidal	I want love and support#suicidal
Anxiety	Yo, anxiety, I would appreciate it if you would stop making me bite the fuck outta my lip because I literally have NO FUCKIN LIP LEFT AND THIS SHIT BLEEDIN
Mental health	Why am i still sad over someone who quite literally hurt me in the worst possible way and made my mental health worse

3.2 Sentiment Analysis

Sentiment assessment is of use for examining online communication, because it enables emotional measurements in online texts automatically. There are developed algorithms that automatically detect sentiment in text. We use the sentiment analysis to create the training data for the Back Propagation Neural Network (BPNN) model.

There are many developed Sentiment Analysis tool out there, but SentiStrength Sentiment Analysis is best to use for short informal text [10] Based on its key features that are (a) A word list of sentiments with judgements of human polarity and strength. (b) Algorithm for spelling correction. (c) To strengthen or weaken the emotion of the following words of sentiment, a booster word list is used. (d) a list of languages is used to identify the feelings of a few common phrases. This overrides the strengths of individual sentiment words. (e) The list of negative words is used to invert the following words of emotion. (f) At least two repeated words added to words boost the feeling of words by 1. (h) To identify additional feelings, an emoticon list of polarities is used. (i) Exclamation mark sentences have a minimum positive strength of 2. (j) Repeated punctuation with one or more exclamation marks increases the strength of the word of feeling immediately preceding by 1. (k) Negative feeling is ignored in questions (l) a training algorithm that optimizes the strengths of feeling words and potentially changes polarity as well.

The data collected by the various keywords are Split into one of two parts: (a) sample for training and (b) sample for testing, from the collected 60,400 tweets approximately 50,000 tweets are selected as training samples which are collectively selected tweets of the different keywords used and the remaining tweets are used as the test samples. The 50,000 training samples are given input to the sentiment analysis tool of SentiStrength which gives the positive and negative sentiment to the tweets, based on the sentiment assigned to the samples they are categorized or labelled into depressed or not depressed. And thus the training samples for our model have generated automatically.

Table 3. Sentiment analysis example

Input	Output
Someone please be my friend and come hangout with me tonight?? Not to party or be crazy, but just hangout. Have conversation. Bring me out of my depression ... any takers?	Positive sentiment is 2 Negative Sentiment is -4 Analysis: Someone[0] please[1] be[0] my[0] friend[0] and[0] come[0] hangout[0] with[0] me[0] tonight[0] [[Sentence=-1,2=word max, 1-5]] Not[0] to[0] party[0] or[0] be[0] crazy[0] but[0] just[0] hangout[0] [[Sentence=-1,1=word max, 1-5]] Have[0] conversation[0] [[Sentence=-1,1=word max, 1-5]] Bring[0] me[0] out[0] of[0] my[0] depression[-3] [[Sentence=-4,1=word max, 1-5]] any[0] takers[0] [[Sentence=-1,1=word max, 1-5]][[[2,-4 max of sentences]]]
and i'm tired of feeling all this guilt that i buried inside of my ribcage	Positive sentiment is 1 Negative sentiment is -4 Analysis: and[0] i'm[0] tired[0] of[0] feeling[0] all[0] this[0] guilt[-3] that[0] i[0] buried[0] inside[0] of[0] my[0] ribcage[0] [[Sentence=-4,1=word max, 1-5]][[[1,-4 max of sentences]]]

3.3 Machine Learning Model

Machine learning algorithm in our proposed system is used for text classification. Text classification is a method that defines the category automatically based on text actual content. Mostly with the boom of the Internet, a classification system is urgently needed Which can help individuals assemble information comprehensively. Researchers have shown that many techniques have been introduced to improve the accuracy of the classifications. However, it is still complicated after using these techniques and requires a strong text classification classifier. Several classifiers have been used to categorize texts such as Vector Space Models, SVM, Nave Bayes, and so on [12]. The Neural Network has demonstrated very prominent and interesting results amongst the many methods. As per a comparison, study was done by Wang et al. [13] among the three Neural Networks that are the Backpropagation (BP) Competitive, and the Radial Basis Function (RBF). The Backpropagation (BP) network is among the most commonly used models in artificial neural network patterns and its mature backpropagation mechanism has been proven to be very effective in text classification.

In the proposed system we work with twitter dataset which is a dataset which contains approximately 50,000 training samples and 10,000 testing samples which are categorized and collected using the specific keyword as in Table 1. The tweets are in English word format, not in a numerical form and needs to be converted to a feature vector.

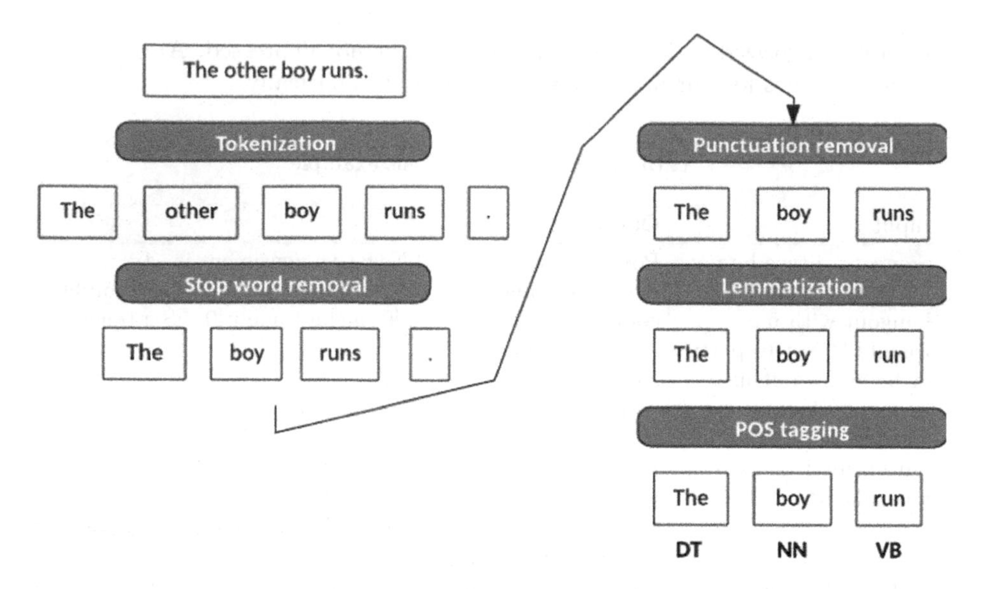

Fig. 2. Pre-processing example

We will use the NLTK (Natural Language Toolkit) to help us with the pre-processing of data. Our main concern is the word tokenizer and the lemmatizer. For us, words tokenizers is used to separate words. A lemmatizer extracts comparative words to transforms them into a similar word. It helps us to keep our samples much smaller without losing too much value (Fig. 2). And after processing all the tweets we will arrive with a bag of words, if any word occurs less than 800 but more than 20 times in the dataset, it is added to a unique word list or else called as a lexicon. Then we create a zero training vector the same size as a unique word list. Now we pass all the words in the sample sentence and, if they are in the unique word list, we make the value of that index equal to 1 in the training vector, the data is converted in a vector of features.

• For example consider a unique word list as [cats, dogs, spoon, television] and training sample is "It's raining cats and dogs", at First generate our training vector 1×4: [0 0 0 0], Since cats (index: 0) and dogs (index: 1) are in the list unique words. So, our new training feature vector is [1 1 0 0].

We take our input feature vector and send it to hidden layers the system has a three hidden layer each of 500 nodes making it a Deep Neural Network. We compare the output with the intended output. To determine the difference in

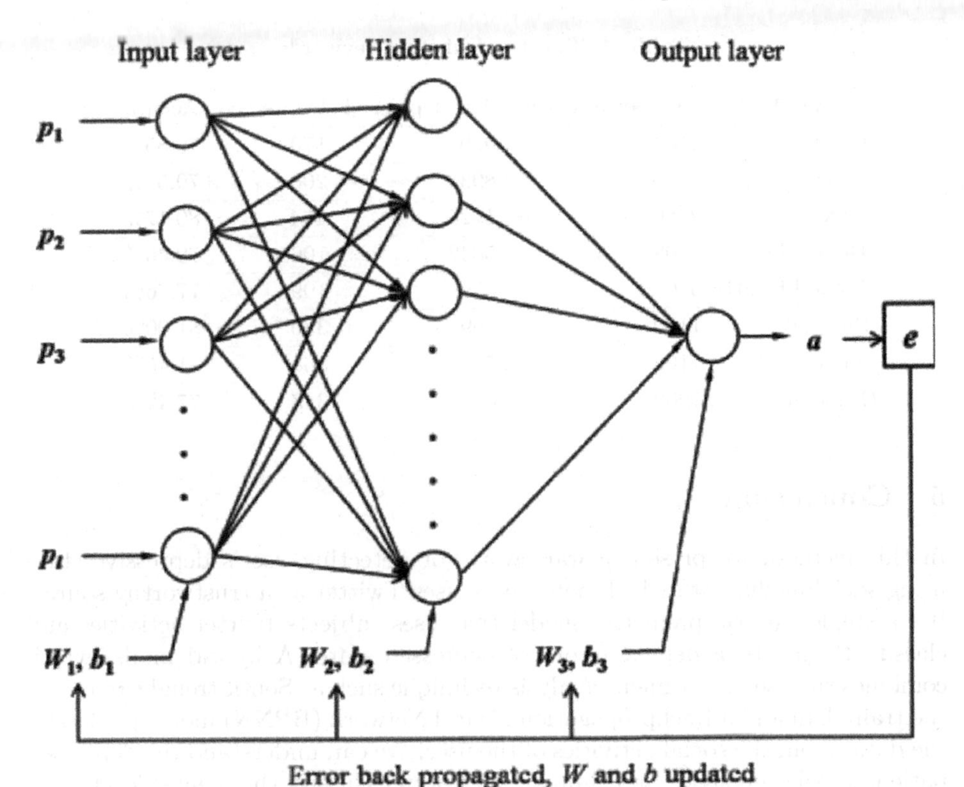

Fig. 3. Back Propagation Neural Network

results, a cost function is used. Finally, An optimizer function is used, in this case Adam Optimizer, to minimize costs. The act that we send data directly through the neural network means that we operate a neural feed-forward network. The weights adjustments and base backward on errors is a back propagation (Fig. 3 Back Propagation Neural Network). This cycle of feed forward and back propagation is called an epoch. And the proposed system goes through 15 epoch cycles to arrive at the result.

4 Classification

The test samples of the data are also categorized using the specific keywords as in Table 1. And this each group of test samples are classified using the proposed model, and thus result of all the test samples are noted in Table 4.

Table 4. Tweets acquisition keywords

Keywords	Number of tweets	Not depressed	Depressed	Accuracy
Guilt	1100	920	170	79.85%
Sadness	1100	893	206	79.56%
Anxiety	2200	1725	474	80.17%
Depression	6908	5842	1065	79.46%
Mental health	550	350	199	77.76%
Suicidal	1100	769	330	81.00%
Tired	1100	731	368	81.46%
Random	1000	656	343	77.42%

5 Conclusion

In this method, we present a framework for detecting user's depressive state using social media data which potentially uses Twitter as a trustworthy source. It's a single, centric predictive model that uses subjects twitter activities and classify them into a depressed or not-depressed state. A hybrid model which combines the use of sentiment analysis technique such as SentiStrenght to create the train data and a Backpropagation Neural Network (BPNN) model to classify the data. From the social activities of the user, We can understand the depressed patient's social behavior and thinking and better classify the mental level.

References

1. Duggan, M., Ellison, N.B., Lampe, C., Madden, M.: Social Media Update 2014. Pew Research Center (2015)
2. Ramirez-Esparza, N., Chung, C.K., Kacewicz, E., Pennebaker, J.W.: The psychology of word use in depression forums in English and in Spanish: testing two text analytic approaches. In: Proceedings of the International Conference on Weblogs and Social Media, pp. 102–108. AAAI Press, Menlo Park (2008)
3. Moreno, M.A., et al.: Feeling bad on Facebook: depression disclosures by college students on social networking site. Depression Anxiety **28**(6), 447–455 (2011)
4. Numbers of monthly active Twitter users in 4th Quarter 2017 — Statistic (2017). https://www.statista.com/statistics/282087/number-of-monthly-active-twitter-users/. Accessed 21 Jan 2018
5. Jalonen, H.: An arena for venting negative emotions. In: International Conference on Communication, Media, Technology and Design, Istanbul, Turkey, pp. 224–230 (2014)
6. Aldarwish, M.M., Ahmed, H.F.: Predicting depression levels using social media posts. In: 2017 IEEE 13th International Symposium on Autonomous Decentralized Systems (2017)
7. Hussain, J., et al.: SNS based predictive model for depression. In: Geissbühler, A., Demongeot, J., Mokhtari, M., Abdulrazak, B., Aloulou, H. (eds.) ICOST 2015. LNCS, vol. 9102, pp. 349–354. Springer, Cham (2015). https://doi.org/10.1007/978-3-319-19312-0_34

8. Java, A., Song, X., Finin, T., Tseng, B.: Why we Twitter: understanding microblogging usage and communities. In: Proceedings of the 9th WebKDD and 1st SNA-KDD 2007 Workshop on Web Mining and Social Network Analysis, pp. 56–65. ACM Press, New York (2007)
9. Naaman, M., Boase, J., Lai, C.-H.: Is it really about me? Message content in social awareness streams. In: Proceedings of the 2010 ACM Conference on Computer Supported Cooperative Work. ACM Press, New York (2010)
10. Thelwall, M., Buckley, K., Paltoglou, G., Cai, D., Kappas, A.: Sentiment strength detection in short informal text. J. Am. Soc. Inform. Sci. Technol. **61**(12), 2544–2558 (2010)
11. Frances, A., Pincus, H., First, M.: Major depressive episode. In: The Diagnostic and Statistical Manual of Mental Disorders: DSM-IV, 4th edn., pp. 326–327. American Psychiatric Association, Washington, D.C. (1994)
12. Pang, J., Bu, D., Bai, S.: Research and implementation of text categorization system based on VSM. Appl. Res. Comput. **9**, 23–26 (2003)
13. Wang, Z., He, Y., Jiang, M.: A comparison among three Neural Network for text classification. In: Proceedings of the 2006 International Conference on Signal Processing (2006)

Big-Five Personality Traits Based on Four Main Methods

P. Hima(✉) and M. Shanmugam(✉)

Vignan's Foundation for Science, Technology and Research,
Guntur 522 213, Andhra Pradesh, India
himasyamala137@gmail.com, shaninfo247@gmail.com

Abstract. A cognitive structure focused to explain diverse behaviors of human regarding fixed and quantifiable features is called personality. Personality shows effects on various customs, traditions and daily routines. Big-five personality traits which is five-factor method contains agreeableness, extraversion, conscientiousness, openness, and neuroticism. This paper is a study of personality traits of humans based on usage of mobile apps, social media, handwriting analysis and facial expressions. By using this four methods people can be categorized into these five personality traits. This personality traits categorization will give best results in the areas of medical and marketing.

Keywords: Personality traits · Mobile apps · Social media · Facial expressions · Handwriting

1 Introduction

The Big-Five personality features factor model contains agreeableness, extraversion, conscientiousness, openness, and neuroticism. Neuroticism is being anxious, depressed and worried. Agreeableness is being courteous, trusting, tolerant and cooperative. Extraversion is being cordial, communicative, and agile. Conscientiousness is being careful, organized and responsible. Openness to experience is imaginative, curious and independent. The effect on the particular person's process of making important decisions is based on behavioral features [10]. The Big-Five characteristic features are one of the methods to define the behavioral features impact on human regarding with this model. Outputs showed an important part in forecasting human behavior based on this model and also individual interest. Extraversion, conscientiousness and neuroticism factors show more impact on human behavior. This model enlarged towards predicting behaviors in each and every field. Providing communication interface between users and electronic devices to classify users by making electronic devices as user's brain according to characteristic features. The output of this model is favorable as they are applied over applications. Despite area or field, this model is giving best results to detect the features or characteristics of a particular person. In this mainly discussed four methods based on this model they are handwriting,

K. C. Santosh and R. S. Hegadi (Eds.): RTIP2R 2018, CCIS 1037, pp. 726–734, 2019.
https://doi.org/10.1007/978-981-13-9187-3_65

facial expressions, social media and mobile apps. These four methods will be clustered their individual categorization based on this model [5].

2 Personality Traits Based on Four Methods

2.1 Mobile Apps

To predict behavioral features of mobile operators an application was built and launched. This application contains several questions to gather information about the particular person. Through this application, apps which are launched and frequently used also known [4]. This method outputs the effect of smart phone applications on the behavioral features of a particular person. Smart phone usage is increasing day by day and everyone is using apps and a maximum number of apps are available in the app stores. So predicting the behavioral features from apps usage may give the best accuracy. Through application program interface data regarding app can be gathered from the android os devices. This API consist of the detailed list of applications resides or launched in the smart phone and also the name of the package, launched time and the app category. The personality test is an app which gathers all the details and other than application details it also gathers the personal basic details of the particular user in the form of multiple questions. This app outputs the behavioral features through the graph of a particular person when all the process is completed. While the app is opened an unspecified, distinctive and individual identification number created and showed on the application window. Simultaneously in the back end of the application, a function started to gather application details from mobile and store for further process. In the front end after completion of answers by operator, those will also store in the database. After submission of answers, they cannot modified. The stored details of applications and the answers will be analyzed and gives the output. For one device only once this test can be taken. To prevent multiple entries of answers from the same person and not to avail the test to other persons other than the owner of the mobile [12]. Extraversion: These type of people are cordial, active, communicative and agile. These people have a number of friends and they will communicate through social media says that these type of persons frequently uses messaging and social applications. From the study of past research gives additional confirmation to determine this kind of people. These type of people will tell their experiences to others, which are the main functions of social networking apps. Neuroticism: People are anxious, depressed, nervous and worried. These type of persons afraid of using novel methods and assistance as a result of that there is a gradual decrement in internet usage. However, empirical studies showed contradictory results. Past research has discovered shopping provides psychologically unsecured persons to set ones to the state of mind. People expressed their views about these type of people that they spare their time with mobile social applications to be a part of a group to acquire information. These type of people to stay away from isolation they spare their time in applications which are related to social. Conscientiousness: These type of persons are well ordered, restraint, alert, dedicated and well grounded.

These type of properties leads to achieve heights in their chosen field, they will be successful in every profession. These type of people always have a vision towards success so they will not waste their valuable time on applications which will waste their time and they feel like these application divert them from their desires. These type of people does not use any kind of applications which are consuming more time. The people who are not conscientious they use these type of applications which are interesting to spend time. This is the reason that this kind of people is not coordinated with innovations. The other people will spend more time on the creation of new patterns modifying the images [4]. Agreeableness: These type of persons are polite, unbiased, unsuspicious and productive. They will easily habituate to spare some time with trending applications because they are Open-minded and humane. An agreeable people found to be determined in investigating unauthorized Web sites that are not user-friendly and difficult to follow. Although, these type of persons will have a number of interactions with social media as the social media is used to share information and make friends online rather than direct interactions. All the others also habituated to interact with these people through online or through phone calls and messages rather than interacting with them directly. Openness to Experience: These type of persons are creative, progressive, not depending and experimental. These type of persons will experience a different type of situations and they will find the way to handle them, so they are more interested towards novel methods and they will easily inherit them. In short messaging and social media apps: these people started using the applications when they launched while others not yet known regarding the apps. From the past research, these type of people adapted to spend more time in online. However, as the innovation spread out quickly, everyone was using social media frequently. Because by nature persons do not belong to this type, frequently spending time in online decreased the percentage in this kind in the whole population. Accordingly, research on openness based on usage of social apps like Twitter, WhatsApp, and Facebook. Thus researchers state that the impact in this type of people regarding usage in online is not that much extent, mainly when so many people use the apps frequently. Because of this certainty, the effect of the usage of different applications is not that much extent in this kind of people. Smart phone applications are universal, without the cost and created to use everyone. The factors which are influencing people usage of mobile apps are different from past research. Accordingly, this survey analyzes past research and to give a conclusion about usage of a variety of mobile apps and that effects on the personality traits. There is no technical classification of mobile apps. Consequently, Google Play Store, which is the vast mobile app store is taken as the reference for the present app classification. There are so many varieties of applications in Google Play Store. Few apps highly favored mobile operators for example music, community and shopping applications, other applications contain fewer operators, for example, medical and comic applications. This survey concentrates few classified applications which frequently in use between, which are having scope to coordinate with the stated behaviors, as discussed by past researchers. Those are games, music and video apps, photography apps (filtering, polishing, and editing), shopping apps, finance apps (expense management, mobile banking, and budget planning), social apps (chat, tweet and sharing), and personalized apps (ring tones changing, wallpapers changing and font changing).

2.2 Handwriting

Graphology is a scientific method of handwriting analysis which identifies, evaluates and understands the personality from the strokes and structures disclosed in handwriting. Handwriting tells the accurate personality together with emotional output, distress, integrity, protection and more behaviors [15]. This cannot report scrutiny and contains analyzing the written script from that particular person character is decided. Handwriting Analysis is generally made in reference to brain writing analysis. Every single behavior entitled to a neurological brain pattern [13]. Every pattern makes a distinct effect on nerves and which shows the same movement who is having the same type of behavior. At the time of writing, minute activities happen without realizing. Every single behavior is based on discussed specifications either stroke or activity [17]. The method graphology recognizes every activity occur in the script to give a detailed explanation of the certain personality trait. The personality traits disclosed by the pen pressure, the baseline, the slope of the writing and the letter 't' originate in every single script explained clearly. Pen pressure, baseline slope of the writing and T-bar height with respect to the stem are input data for Rule-Base and gives output as a particular person behavior according to specifications [12]. Assessment regarding baseline through a method called polygonalization and assessment regarding pen pressure through a method call grey level threshold. T-bar height regarding stem evaluated through a method called template matching. Through Template matching only the slope of writing is evaluated. These specifications from a handwriting tell correct information about the particular person [13]. The handwriting analyzed through baseline in one's handwriting tells accurate details about the particular person. The line through which writing flows of a particular person in the handwriting is called the Baseline. The three main common baselines elevate in the handwriting of anyone are ascending (Optimistic), descending (Pessimistic) and level [13]. The writing pressure is the important specification in a handwriting. The depth of feeling represents the quantity of pressure applied to the paper while writing which is named as emotional intensity [15]. From the pen pressure, particular persons categorized into three types. They are a light writer, medium writer, and heavy writer. Light writer: A particular person can suffer from disturbing events which are not affected seriously. These type of events do not make that much effect on that particular person. Medium writer: A particular person perceives emotional events for an average time span. He effects with a moderate state of emotional potency. Heavy writer: A particular writer has extending and long-lasting experiences. The writer can excuse but never fail to remember it. Every single event effect on him intensely. Personality trait based on position of the T-bar 1. High Self Esteem: The T-bar does not touch the stem but near the stem location. This tells trust, intention and the capacity to plan further, desires, believes in his decision and altogether well dignified. These are the main specifications to victory and satisfaction. 2. Average Self Esteem: The T-bar touched and moved over the stem and does not reach near the closed region. The person with average self-esteem is experimental, which is common among successful people. 3. Low Self Esteem: The T-bar moved outward the stem with minimum height. The writer avoids non-success due to opposing the problems. That person mainly ends up with uncertain events and relationships for long-lasting and realizes defects with himself. This particular person is hardly fortunate enough in his own ways without being affected by his desires. 4. Dreamer: The T-bar moved outward the stem. The individual's aims to desire top position are not related to real life. These people repeatedly tell about what they actually interested to do instead of making it possible. The slope of the writing slant in the handwriting can correctly predict the emotional reactions [9]. Writing slant Personality traits Extreme Left: Self-contained; introvert. Left: Adore being alone; emotions cannot be expressed. Vertical: Discernment acts; follows the head, not the heart. Right: Emotional highs and low; Extravert. Extreme Right: Intense mood changes; follows the heart, not the head. Openness to experience: Light writer, T-bar crossed above the stem people come under this category as they are curious. Conscientiousness: T-bar crossed very high, Vertical people will come under this category as they are self-disciplined and organized. Extraversion: Light writer, T-bar crossed the middle zone, Right will come under this category as they are outgoing and talkative. Agreeableness: Medium writer, the Extreme left will come under this category as they are compassionate. Neuroticism: T-bar The T-bar moved outward the stem with minimum height, Heavy writer, Left will come under this category as they are nervous and lack of emotional stability.

2.3 Facial

To determine the Big-Five personality types includes so many questions, which will take inappropriate time span and which is not used frequently. In this paper, this part focuses to survey a new unused method which predicts BFPT through facial expressions which obtained from FACS. At the end in this which shows the relation between the action units which contain utmost extremities of expressions of the face through behaviors of the particular person. Additionally, this method provides the good result to estimate openness behavior, extraversion and neuroticism show actual outputs in few minutes in comparison with the time taken by questionnaire method [5]. To estimate the facial expressions, FACS is one of the scientific methods which usually estimates the untold intentions through micro-expressions [6]. Based on the past research that each emotion occurs through a particular muscle of the face, thus converts into particular AU which gives detailed data regarding the state of emotion about a particular object. Every single face movement contains a particular set of AU which will either additive or non-additive. For every single Action unit determining a multistate face model, the face is divided into five parts: Eyes, Cheeks, Brows, Lips, and Wrinkles. In the phase of training, the model is obtained with few frames which includes movements of face in distinct data responses. Few videos build to predict the six emotions (joy, disgust, happiness, anger, fear, and surprise) [11]. Every single video contain less amount of length; thus the model was instructed in the particular order of certain amount of frames. Every frame carried towards initial part to identify the face through the algorithm which is VJFD [5]. Detected face is further divided into five parts through that algorithm. Each divided part in this training frame obtains similar categories from that category it will state whether AU is present or absent and also the effectiveness of the level. Mentioned classifiers are already trained and they gave the utmost accuracy. From the past description, the presence of AU classified from I to V to build a consolidated function which gives the behavior map. Every single part in behavior map carried towards modification based on results neural network, to lower the rate of the error backpropagation process is used [5]. Outputs from neural network measured with fetched one from model which is a questionnaire. The model prepared to move into phase of testing when this process is completed. In this testing phase real-time information obtains from this model subject. Each obtained frame passed to VJFD algorithm [14], later the divided part carried to blocks, which provides whether AU is present or not after that checking the effect. The behavior maps structured from neural network through limited rows to limit the processing time, every row is obtained from this process of steps that predict respective behavior through these phases. Openness to experience: Adventurous, emotional, unusual ideology, appreciation for art will come under curious. Conscientiousness: Depend on others, self-disciplined, aim for achievement, fondness over planned behavior who avoids spontaneous behavior will come under organized. Extraversion: confident, prompt by others company, positive emotions, talkativeness will come under this category. Agreeableness: compassionate, not suspicious, helping nature, and tempered will come under this category. Neuroticism: nervous, unpleasant emotions will experience easily, lack of instinct control will come under this category [3].

2.4 Social Media

Every single person has wide variety of personality, hobbies, interests and so many other things which will determine them as unique person [8]. All these differences can classify or categorize into a different groups, from that industries or business partners can be useful to improve the advertisements and promotions, measuring job performance and other functions effectively [16]. Almost all people are using social media. Additionally, frequently they post something on social media which associates to their interest [7]. From this we can be determined to know their personality type they belong. For every person who is active in social media, if the psychological behavior is correct and certain then we can estimate the person's personality trait even though they have not posted in the internet, this gives advantage for marketing and for other purposes [2]. It's beneficial if we know others personality. Though, at present the perfect process to predict others personality is mainly by using psychological test, but this test is taking a lot of time and power. The past research has proved that data in social media can return actual personality, now a days it is not effectively used. Effective digital marketing can be developed by predicting the others personality. However, social media followed so many methods to develop the efficiency of marketing. Although, all the methods are not that much effective to develop the efficiency of the marketing by using the predicted personality through their posted information in the social media [2]. The present approach which impacts more on the short-term marketing and mainly based on the history or cookies of particular person on the internet. However past research concentrated on English as the language to do personality prediction. There is still no proper approach to do the personality prediction based on several methods. Combining all the data from social media like Facebook, Twitter, LinkedIn, Instagram we can categorize people's personality without any questionnaire which is time and energy consuming [1]. The algorithm which simple, powerful and accurate is Naive Bayes classifier algorithm which can be used for machine learning approach [8]. This approach is used because of its capacity to predict situation and eliminate few unwanted words in a phrase. From this type of automated personality classifier, constructive and productive marketing will be obtained. Each person has a certain personality, as different person has different taste or interest in some type of advertisements or certain promotions. This psychological test obtained from this prediction can be used for faster prediction of candidate's psychology for employment. HR department can gain candidates performance measure for the employment through this psychological test which is obtained from the social media data. This helps the HR department to save more time and to concentrate on next process. Facebook. Twitter, LinkedIn, and Instagram are the online social networking sites through which a lot of researches were performed to predict personality based on public data, online social networking applications and behavior in social media towards friends list and followers list. Organized extracting of knowledge from the social media content can provide researchers the capacity to predict behavior accurately. This survey provides researchers with a brief description of different methods used for studies and research focusing on predicting user personality and behavior using online social networking content. The below table shows the personality traits according to the behavior of social media.

Personality traits	Behavior on social media
Extraversion	A frequent user of social media
	More use of social media components
	Number of Facebook Friends or Twitter followers
	More Social media groups
Neuroticism	Spend more time on social media apps, More use of Facebook wall
	Less use of private message
	Share more information
Agreeableness	Number of friends
Openness	More use of Facebook or Twitter for communication
	Number of components
	More knowledge of features
Conscientiousness	Limited social media activity

3 Conclusion

Personality traits are usually categorized by an extensive study, which included so many aspects those are price and capacity. Accordingly, based on these four methods we can predict personality traits of a person. Personality traits are crucially affected by social media, mobile apps. Particularly, neuroticism influence more on social media and mobile apps. There are so many directions to affect the people who are categorized under these behaviors. On the other hand conscientiousness individuals having less impact on social media and mobile apps based on their behavior. Personality traits are important to determine for further applications based on medical and marketing. The discussed behaviors given a model to know and mold our behavior.

References

1. Yogish, D., Manjunath, T.N., Hegadi, R.S.: Survey on trends and methods of an intelligent answering system. In: 2017 International Conference on Electrical Electronics, Communication, Computer, and Optimization Techniques (ICEECCOT), pp. 346–353. IEEE (2017)
2. Laleh, A., Shahram, R.: Analyzing Facebook activities for personality recognition. In: 2017 16th IEEE International Conference on Machine Learning and Applications (ICMLA), Cancun, Mexico, pp. 960–964. IEEE (2017)
3. Pramodh, K.C., Vijayalata, Y.: Automatic personality recognition of authors using big five factor model. In: IEEE International Conference on Advances in Computer Applications (ICACA), pp. 32–37. IEEE (2016)
4. Xu, R., Frey, R.M., Fleisch, E., Ilic, A.: Understanding the impact of personality traits on mobile app adoption-Insights from a large-scale field study. Comput. Hum. Behav. **62**, 244–256 (2016)

5. Gavrilescu, M.: Study on determining the Big-Five personality traits of an individual based on facial expressions. In: E-Health and Bioengineering Conference (EHB), pp. 1–6. IEEE (2015)
6. Abadi, M.K., Correa, J.A.M., Wache, J., Yang, H., Patras, I., Sebe, N.: Inference of personality traits and affect schedule by analysis of spontaneous reactions to affective videos. In: 2015 11th IEEE International Conference and Workshops on Automatic Face and Gesture Recognition (FG), vol. 1, pp. 1–8. IEEE (2015)
7. Achana, R.A., Hegadi, R.S., Manjunath, T.N.: A novel data security framework using E-MOD for big data. In: 2015 IEEE International WIE Conference on Electrical and Computer Engineering (WIECON-ECE), pp. 546–551. IEEE (2015)
8. Markovikj, D., Gievska, S., Kosinski, M., Stillwell, D.: Mining Facebook data for predictive personality modeling. In: Proceedings of the 7th International AAAI Conference on Weblogs and Social Media (ICWSM 2013), Boston, MA, USA, pp. 23–26 (2013)
9. Djamal, E.C., Darmawati, R., Ramdlan, S.N.: Application image processing to predict personality based on structure of handwriting and signature. In: 2013 International Conference on Computer, Control, Informatics and Its Applications (IC3INA), pp. 163–168. IEEE (2013)
10. Chittaranjan, G., Jan, B., Gatica-Perez, D.: Who's who with big-five: analyzing and classifying personality traits with smartphones. In: 2011 15th Annual International Symposium on Wearable Computers (ISWC), pp. 29–36. IEEE (2011)
11. Zhan, Y.Z., Cheng, K.Y., Chen, Y.B., Wen, C.J.: A new classifier for facial expression recognition: fuzzy buried Markov model. J. Comput. Sci. Technol. **25**(3), 641–650 (2010)
12. Prasad, S., Singh, V.K., Sapre, A.: Handwriting analysis based on segmentation method for prediction of human personality using support vector machine. Int. J. Comput. Appl. **8**(12), 25–29 (2010)
13. Champa, H.N., AnandaKumar, K.R.: Automated human behavior prediction through handwriting analysis. In: 2010 First International Conference on Integrated Intelligent Computing (ICIIC), pp. 160–165. IEEE (2010)
14. Valstar, M., Pantic, M.: Fully automatic facial action unit detection and temporal analysis. In: Conference on Computer Vision and Pattern Recognition Workshop, CVPRW 2006, p. 149. IEEE (2006)
15. Mogharreban, N., Rahimi, S., Sabharwal, M.: A combined crisp and fuzzy approach for handwriting analysis. In: IEEE Annual Meeting of the Fuzzy Information, Processing NAFIPS 2004, vol. 1, pp. 351–356. IEEE (2004)
16. Manjunath, T.N., Hegadi, R.S.: Data quality assessment model for data migration business enterprise. Int. J. Eng. Technol. (IJET) **5**(1), 101–109 (2013)
17. Madhvanath, S., Govindaraju, V.: The role of holistic paradigms in handwritten word recognition. IEEE Trans. Pattern Anal. Mach. Intell. **23**(2), 149–164 (2001)

A Review on Utilizing Bio-Mimetics in Solving Localization Problem in Wireless Sensor Networks

R. I. Malar and M. Shanmugam(✉)

Vignan's Foundation for Science, Technology and Research,
Guntur 522 213, Andhra Pradesh, India
igneshia40@gmail.com, shaninfo247@gmail.com

Abstract. Wireless Sensor Networks (WSNs) have the feasibility to connect the physical world with the virtual world by framing a network of sensors. The supreme function of a sensor network is to collect and forward data to the destination. Applications based on WSNs needs location knowledge about randomly deployed nodes. Localization of these nodes is the basic problem in WSNs. Several types of research have been done so far, using various strategies to improve the network performance as well as energy efficiency and communications effectiveness of WSNs. Among the strategies used, algorithms inspired by natural behaviours of a group of organisms like butterflies, fireflies, grey wolf, etc., showed higher efficiency in locating the nodes. In this survey, some of the inherent nature inspired localization algorithms are briefly discussed. Also, some other collective behaviours which can be used to develop localization algorithms are also explained.

Keywords: Sensor networks · Bio inspired · Localization

1 Introduction

A Wireless Sensor Network (WSN) consists of a lot of sensor nodes. These sensor nodes have the potential for sensing, computation and wireless communication [1]. Owing to its vigorous functionalities and low cost it is widely used in the fields of environmental observation, military monitoring, disaster relief and building monitoring [2]. The performance of WSNs has a pivotal impact on how precisely the sensor nodes are localized. Sensor Localization Information is used in self-configuration as well as the configuration of networks in deciding the location where the event has taken place, assisting traffic routing, tracking moving target and providing the network geographic coverage [3]. In most applications of sensor networks, the information collected by the sensors will be absurd unless the location from where the information is procured is known. Hypothetically, a localization measurement device, such as GPS can be used to locate a sensor node, but it is not practical to use GPS on each sensor node because it is so expensive [4]. Moreover, GPS will not be pertinent to indoor and underground

K. C. Santosh and R. S. Hegadi (Eds.): RTIP2R 2018, CCIS 1037, pp. 735–745, 2019.
https://doi.org/10.1007/978-981-13-9187-3_66

deployments as well as dense forests [5]. So many localization methods have been developed to solve this problem. All localization techniques assume that GPS is installed with only a few nodes, which are called anchor nodes [6]. Other sensors can connect with nearby sensors and measure differences between them using algorithms like Time of Arrival (TOA), Received Signal Strength (RSS), etc. [7] Another way to solve localization problem is by algorithms based on natural behavior of animals like Particle Swarm Optimization Algorithm (PSO), Bacteria Foraging Algorithm (BFA), Cuckoo Search Algorithm (CSA), Bat Algorithm (BA), and Modified Bat Algorithm (MBA), which will be discussed briefly in this survey. All these exploratory algorithms are potential techniques for solving the node localization issues [8]. In this survey, we also attempt to propose some animal behavior for development of efficient network localization algorithms.

2 Localization Algorithms Developed Based on Animal Behaviours

In this section, the localization algorithms that are based on different behaviours of animals, birds, and microorganisms are explained briefly. Each algorithm has its advantages and disadvantages.

2.1 Butterfly Optimization Algorithm

A nature-inspired meta-heuristic optimization algorithm was developed based on the food foraging behaviour of butterflies. Butterflies have chemoreceptors all over their body which can sense fragrance and used to identify the location of food source. As per Butterfly Optimization Algorithm (BOA), butterfly act as search agents, performing an optimized search of another butterfly with more intense fragrance than its own. When the butterfly changes its position, its intensity diverges, leading to a random local search by another butterfly [9]. This BOA algorithm has been enhanced by introducing chaotic maps to improve the performance efficiency of convergence speed and local optima avoidance [10]. Further, it was improved for numerical optimization [11], global optimization [10] and mechanical design optimization [13]. BOA for localization has been compared with other algorithms like FA [14,15], PSO [16] regarding performance efficiency, errors, computing time and localized nodes [14]. Simulation results clearly showed that the proposed scheme performs more consistent and accurate location nodes than existing PSO and FA based Localization schemes.

2.2 Firefly Algorithm

Fireflies have a social behaviour of generating various flashing patterns to search, communicate and choose their mating partner [10,14,16,17,49]. This behaviour of flashing was utilized for developing a meta-heuristic algorithm for localization based on three rules [17,18]. Irrespective of their gender, a firefly can get attracted towards other firefly located nearby, as all fireflies are assumed as unisexual. Fireflies are attracted to each other based on their brightness. Fly with less brightness will get attracted towards the fly that exerts more brightness. The brightness of a firefly is determined by the location of the objective function. FA compared with other algorithms like CS, BA and found to have the better speed of convergence, efficient and quick optimization and powerful local search [19].

2.3 Bat Algorithm

Bats have a fascinating ability to find their prey in complete darkness using echolocation [20]. The Bat Algorithm (BA) was developed based on the echolocation characteristics with the following rules [21] (i) The distance is sensed by bats using echolocation (ii) Bats can vary the wavelength and intensity based on the distance of the target. (iii) The loudness intensity is varied from a high value to a minimum one. Several improvements have been made in BA to increase its performance efficiency [22], global search [23], job scheduling problems [24], various engineering applications [25], etc. This algorithm has also been enhanced for solving node localization problem in WSN in a distributed environment [26].

Modified Bat Algorithm. The original bat algorithm [21] is modified to improve its efficiency, using bacteria foraging strategies [41]. In this algorithm, the bat movement selection is decided by the fitness function. The bat movement will be swimming when it move towards optimized fitness function value, otherwise, it follows bacterial chemotactic movement [28].

2.4 Cuckoo Search Algorithm

Cuckoos have a reproduction strategy of laying eggs in the nests of other species, as well as it makes use of Levy flights search pattern, based on which the Cuckoo Search Algorithm (CSA) has been developed [29]. After laying eggs, the cuckoo will remove the eggs of other species to increase the probability of hatching of own eggs. To avoid the detection of cuckoo eggs by the host species in their nest, female cuckoos use the strategy of mimicking the colour and patterns of host eggs [29,30]. The rules followed to develop CS algorithm are as follows: (i) Each cuckoo lays only one egg at a time and dumps its egg in the randomly chosen nest; (ii) The offspring is developed in the best nests with good quality of eggs; (iii) The available host nests are fixed, and the single egg laid by the cuckoo is discovered by the host species. Sean Walton et al. [31] modified the CS algorithm and developed MCS. It has better convergence speed than CS [32]. An adaptive CS algorithm was developed for global optimization [33].Further, this algorithm was modified to solve flow shop scheduling problems [34–36], Knapsack problems [37], neural network training [38], travelling salesman problem [39], node localization problem in WSN [40].

2.5 Bacterial Foraging Algorithm

Bacterial foraging is the strategy used by several bacteria (e.g., E.coli) to move towards nutrient-rich medium which will favour them in cell division. The movement is initiated as swimming, and if required, it will be followed by a tumbling movement based on the nutrient location [41]. Bacterial Foraging Algorithm (BFA) is developed based on these swimming and tumbling movements hybrid algorithm [42] was developed with Genetic Algorithm and BFA for global optimization. It also has its diverse applications like image segmentation [43], Automatic circle detection on digital images [44]. It was also used to solve the issue of node localization after deployment [45]. Simulation studies showed that BFA locates the nodes with more accuracy. However, the high computing time and memory requirement reduce the potential application of BFA [17].

2.6 Chicken Swarm Optimization

Chickens are social and domestic birds that live together as flocks. They have the ability to recognize their flock member among 100 or more chickens, though they are separated for months. They communicate through a range of different sounds like clucks, chrisps, giggles, etc. They have the capability to make decisions based on experiences, trial and error method. The flock is constituted based on hierarchy, in which the most dominant one is the rooster that controls the flock. Other dominant chickens are considered as hens and are tightly attached with the rooster, while the weaker ones are chicks. The rooster calls their flock members when the food source is available and warns the chicks during predator or trespasser invasion into their territory. The chicks are always around their mother and move as a swarm towards the food source. Based on this social behaviour of chicken, Chicken Swarm Optimization algorithm was developed following three concepts [46] (i) In the swarm of chickens, the hierarchy is purely based on dominating character. (ii) The order of hierarchy remains unchanged for a defined time, however, gets updated in stipulated timings. (iii) The flock members are guided by rooster in food search, and it prevents the stealing of their food by other flock members.

An improved Chicken Swarm Optimization [47] was proposed because of premature convergence for high-dimensional complex problems. Chen et al. [48] combined Penalty function with improved chicken swarm optimization. To solve the node localization problem in WSN, a new algorithm based on Chicken Swarm Optimization(CSO) and mobile at he anchor [49] was established, and the overall results show that CSO algorithm has better positioning accuracy.

2.7 Grey Wolf Algorithm

Grey wolf, also known as timber wolf is native to North America and Eurasia. They are considered as apex predators, next to humans and tigers. They always live in as a group of 5 to 12 wolves and have a hierarchical order among them based on dominance. By hierarchy and hunting strategy, they are classified as alpha, beta, delta and omega wolves [50]. The dominant ones are alpha wolves, which makes all decisions and solutions for the welfare of the group. Beta wolves assist the alpha wolves in making a decision and hold the in-charge position in the absence of alpha wolves. Next, the delta wolves serve as protectors or guards for the groups. Also they hunt and behave as caretakers for omega wolves, the submissive ones. Omega wolves are the scapegoats and follow the instructions of higher order wolves and also helps to solve internal problems [50, 51]. While hunting the prey, the group follows three strategies: they track down the prey, chase and approach it in the group; the group then encircles the prey and start harassing it to stop its movement; finally, the group attacks the prey [51].

Based on these group behaviours of grey wolves, in particular, hunting strategy, Grey Wolf Optimization (GWO) algorithm was developed. In GWO, the first hunting strategy of grey wolves, searching for prey serves as the exploration phase in a global search space. The other two strategies are used in exploitation phase to provide an optimal solution in local search space. Jitkongchuen et al. improved the performance of GWO by integrating invasion based migration operation in which there are more packs compared to GFO and migration between them [51]. *GWO was found to provide minimum errors, high location accuracy, fast convergence rate and less computation time in comparison with other algorithms like PSO and MBA when implemented in node localization issue* [52].

2.8 Particle Swarm Optimization Algorithm

Particle Swarming Optimization (PSO) algorithm was developed by considering each bird in a flock or each fish in the group (school) as an individual particle in the search space and area around the target food source represents the search space. Each particle has its distance from the target, which is defined as the cost function of the issue. Each particle in the space has its self pbest (particle best), and the particle which is nearer to the target is called gbest (global best). The entire group of particles will move towards the direction in which the particle with gbest is moving. The search for new gbest will keep on repetition, considering the individual particle's updated position and velocity, until the condition is terminated [53].

The above-discussed algorithms are compared for their parameters, efficiency, and accuracy in the following table (Table 1).

3 Animal Behaviours for Future Prespectives

Apart from the animal behaviours studied so far, several group behaviours of other animals can be used to develop localization algorithms which may provide an improved solution to node localization problem. Some of such animal behaviours are discussed in this section.

Table 1. Comparative Analysis of some bio-inspired Algorithms

Type of Algorithm	Based on	Localization error (%)	Convergence rate	Success rate	Accuracy	Other Domains using this Algorithm
BOA [9]	Food foraging behavior of the butterfly	0.183942	1.43 s	75	Better Localization accuracy than FA and PSO	Numerical Optimization, Global Optimization
FA [9]	Flashing behavior of firefly	0.291862	2.73 s	75	High convergence rate and gives good solutions in minimum number of iterations	Stochastic test functions and design Optimization, Levy flights and Global Optimization, Solving non-convex economic dispatch problems
BA [20,27]	Echolocation behavior of bat	0.0035	5.22 s	37	Good accuracy but convergence rate and success rate is not so good	Feature selection, Traveling salesman problem
MBA [52]	Added the bacteria foraging behavior with bat behavior	0.5205[52]	1.75 s	25	Better convergence rate and success rate than BA	Data warehousing, neural networks
CSA [19,40]	Brood parasitism of cuckoo	0.266	More run time than FA and BA	35	Better Localization accuracy than PSO and implementation is more simpler	To solve the structural optimization problem, neural networks, Traveling salesman problem
GWO [52]	hierarchy and Social hunting behavior	0.7771	1.98	185	identifies a maximum number of unknown positions with minimum Localization error compared to PSO and MBA	Feature Selection, Multi-criterion Optimization
PSO [52]	Birds flocking behavior	0.5357	3.03	142	Estimates positions of node in less time but takes more computation time	Multi-robot path planning, spacecraft reentry trajectory

3.1 Maze Solving by Slime Mould

Slime mould is a common name used to represent different single cellular eukaryotic organisms that aggregate together to form reproductive structures [54]. One such organism is Physarum polycephalum, which is capable of identifying the shortest and optimal path in labyrinth maze When this slime was put in a maze that has food sources at entry and exit, it covered and colonized the whole maze surface to utilize the source. Through communication from all the ends, the slime mould started withdrawing from dead ends and lengthy paths to form a single short and optimal path to reach the food source. The shortest path identified by the slime mould is one of the four likely solutions [55]. After this study, several researches showed that slime mould is able to solve extremely complex mazes [56]. In all these studies, the slime mould starts with a global search of all paths in the maze, following swarm intelligence and ends with an optimized path through communication between its units [57].

Voting by Painted Dogs. African wild dog, also known as African painted dogs, similar to humans has a democratic kind of behaviours like voting to make group decisions. These dogs sneeze to make a decision of taking rest or start hunting, and it is based on the dominance of the dog which starts the rally. For dominant dogs, only two to three sneezes are enough to make the pack start, but the subordinate ones have to sneeze a lot. Based on the voting of the pack members, the decision will be taken to hunt, lead, path determination, locomotion, etc. [58]

Co-Operative Hunting by Whales. The Pack Ice killer whales express a group hunting behaviour to attack a crabeater seal sitting on floating ice sheets. The group is usually comprised of seven whales, and they swam 100 m away from the ice floe, suddenly turn and swam towards the floe creating a wave that broke the floe, leaving the seal into the water. This behaviour is known as wave-wash hunting. The type of movement towards floe, and the intensity of waves created are based on the size of the ice floe. Initially, the ice floe will be analysed by the spy whales for the presence of seals [59], size of the floe, etc. Apart from wave-washing, the whales also use several strategies to push off the seal into the water [60].

Uprising of a New Queen in Bee Hive. Usually, the bees in a hive decide to have a new queen for them under three inevitable situations: 1. The existing queen gets old or exhibits poor governance; 2. The queen is missing or dead or left the hive; 3. The bees want to develop a new hive. During such a circumstance, the bees chose some female eggs randomly and grew them in bigger queen cells. They will be fed with nutrient-rich food (Royal Jelly), while the other larvae will be fed with just honey. Usually, both types of larvae are fed with royal jelly during their first three days, but only the larvae selected to be a queen will be continued with royal jelly. They mature faster than other larvae and transform into queens. The first emerging virgin queen roams in the hive and communicates with its sister queens through "tooting" sound, to which the sisters respond by "quacking." Based on the response, the first queen then visits their cells and kills them. After this, according to the purpose of their rising, either they will stay in the hive or leave with a group of bees to make their hive [61].

4 Conclusion

In this paper, the natural behaviours of animals and other organisms that were used to develop algorithms for solving node localization issue in WSNs were reviewed. In addition, some new behaviour has been added that can be studied for their potentiality to be applied for network localization. This review adds collective knowledge that gives insights into bio-inspired algorithms and their importance in developing potential network localization algorithms.

References

1. Kumar, A., Shwe, H.Y., Wong, K.J., Chong, P.H.J.: Location-based routing protocols for wireless sensor networks: A survey. Wireless Sens. Netw. **9**, 25–72 (2017)
2. Rashid, B., Rehmani, M.H.: Applications of wireless sensor networks for urban areas: A survey. J. Netw. Comput, Appl. **60**, 192–219 (2016)
3. Patwari, N., Ash, J.N., Kyperountas, S., Hero, A.O., Moses, R.L., Corral, N.S.: Locating the nodes: Cooperative localization in wireless sensor networks. IEEE Sig. Process. Mag. **22**, 54–69 (2005)
4. Chong, C.-Y., Kumar, S.P.: Sensor networks: Evolution, opportunities, and challenges. Proc. IEEE **91**(8), 1247–1256 (2013)
5. Rawat, P., Singh, K.D., Chaouchi, H., Bonnin, J.M.: Wireless sensor networks: A survey on recent developments and potential synergies. J. Super Comput. **68**(1), 353–393 (2014)
6. Kuriakose, J., Joshi, S., Vikram Raju, R., Kilaru, A.: A review on localization in wireless sensor networks. In: Thampi, S.M., Gelbukh, A., Mukhopadhyay, J. (eds.) Advances in Signal Processing and Intelligent Recognition Systems. AISC, vol. 264, pp. 599–610. Springer, Cham (2014). https://doi.org/10.1007/978-3-319-04960-1_52
7. Chaczko, Z., Klempous, R., Nikodem, J., Nikodem, M.: Methods of sensors localization in wireless sensor networks. In: IEEE International Conference and Workshops on the Engineering of Computer-Based Systems, ECBS 2007, Tucson, AZ, pp. 26–29 (2007)
8. Kulkarni, R.V., Venayagamoorthy, G.K., Cheng, M.X.: Bio-inspired node localization in wireless sensor networks. In: IEEE International Conference on Systems, Man and Cybernetics, San Antonio, pp. 205–210 (2009)
9. Arora, S., Singh, S.: Node localization in wireless sensor networks using butterfly optimization algorithm. Arab. J. Sci. Eng. **42**(8), 3325–3335 (2017)
10. Arora, S., Singh, S.: An improved butterfly optimization algorithm with chaos. J. Intell. Fuzzy Syst. **32**(1), 1079–1088 (2017)
11. Arora, S., Singh, S.: An effective hybrid butterfly optimization algorithm with artificial bee colony for numerical optimization. Int. J. Interact. Multimed. Artif. Intell **4**(4), 14 (2017)

12. Arora, S., Singh, S.: An improved butterfly optimization algorithm for global optimization. Adv. Sci. Eng. Med. **8**(9), 711–717 (2016)
13. Arora, S., Singh, S., Yetilmezsoy, K.: A modified butterfly optimization algorithm for mechanical design optimization problems. J. Braz. Soc. Mech. Sci. Eng. **40**(1), 21 (2018)
14. Yang, X.-S.: Engineering Optimization An Introduction with Metaheuristic Applications. 2nd edn. John Wiley and Sons INC, Hoboken, New Jersey (2010)
15. Yang, X.-S., He, X.: Firefly algorithm: Recent advances and applications. Int. J. Swarm Intell. **1**(1), 36–50 (2013)
16. Gopakumar, A., Jacob, L.: Localization in wireless sensor networks using Particle Swarm Optimization. In: IET International Conference on Wireless Mobile and Multimedia Networks, Beijing, China, pp. 227–230 (2008)
17. Harikrishnan, R., Jawahar Senthil Kumar, V., Sridevi Ponmalar, P.: Firefly algorithm approach for localization in wireless sensor networks. In: Nagar, A., Mohapatra, D.P., Chaki, N. (eds.) Proceedings of 3rd International Conference on Advanced Computing, Networking and Informatics. SIST, vol. 44, pp. 209–214. Springer, New Delhi (2016). https://doi.org/10.1007/978-81-322-2529-4_21
18. Bingnan, P., Zhang, H., Pei, T., Wang, H.: Firefly algorithm optimization based WSN localization algorithm. In: International Conference on Information and Communication Technologies, Xi'an, China, pp. 26–5 (2015)
19. Arora, S., Singh, S.: A conceptual comparison of firefly algorithm, bat algorithm, and cuckoo search. In: International Conference on Control Computing Communication and Materials (ICCCCM), Allahabad, India, pp. 1–4. IEEE (2013)
20. Goyal, S., Patterh, M.S.: Wireless sensor network localization based on BAT algorithm. Int. J. Emerg. Technol. Comput. Appl. Sci. **3**(192), 507–512 (2013)
21. Yang, X.-S., He, X.: Bat algorithm: Literature review and applications. Int. J. Bio-Inspired Comput. **5**(3), 141–149 (2013)
22. Gandomi, A.H., Yang, X.S.: Chaotic bat algorithm. J. Comput. Sci. **5**(2), 224–232 (2014)
23. Rezaee Jordehi, A.: Chaotic bat swarm optimization (CBSO). Appln. Soft. Comput. **26**, 523–530 (2015)
24. Dao, T.-K., Pan, T.-S., Nguyen, T., Pan, J.-S.: Parallel bat algorithm for optimizing makespan in job scheduling problems. J. Intell. Manuf. **29**(2), 451–462 (2015)
25. Jayabarathi, T., Raghunathan, T., Gandomi, A.H.: The bat algorithm, variants and some practical engineering applications: A review. In: Yang, X.-S. (ed.) Nature-Inspired Algorithms and Applied Optimization. SCI, vol. 744, pp. 313–330. Springer, Cham (2018). https://doi.org/10.1007/978-3-319-67669-2_14
26. Mihoubi, M., Rahmoun, A., Lorenz, P., Lasla, N.: An effective bat algorithm for node localization in a distributed wireless sensor network. Secur. Priv. **1**(1), e7 (2018)
27. Goyal, S., Patterh, M.S.: Modified bat algorithm for localization of wireless sensor network. Wireless Pers. Commun. **86**(2), 657–670 (2016)
28. Yılmaz, S., Ugur Kucuksille, E., Cengiz, Y.: Modified bat algorithm. Elektronika ir Elektrotechnika **20**(2), 71–78 (2014)
29. Yang, X.S., Deb, S.: Cuckoo search via Lévy flights. In: Proceedings of World Congress on Nature and Biologically Inspired Computing (NaBIC 2009), pp. 210–214. IEEE (2009)
30. Rajabioun, R.: Cuckoo optimization algorithm. Appl. Soft Comput. **11**(8), 5508–5518 (2011)

31. Walton, S., Hassan, O., Morgan, K., Rowan Brown, M.: A review of the development and applications of the cuckoo search algorithm. In: Swarm Intelligence and Bio-Inspired Computation Theory and Applications, pp. 257–271 (2013)
32. Walton, S., Hassan, O., Morgan, K., Rowan Brown, M.: Modified cuckoo search: A new gradient-free optimization algorithm chaos. Solitons and Fractals **44**(9), 710–718 (2011)
33. Mareli, M., Tawla, B.: An adaptive Cuckoo search algorithm for optimization. Appl. Comput. Inf. **14**(2), 107–115 (2018)
34. Marichelvam, M.K., Prabaharan, T., Yang, X.-S.: Improved cuckoo search algorithm for hybrid flow shop scheduling problems to minimize makespan. Appl. Soft Comput. **19**, 93–101 (2014)
35. Marichelvam, M.K.: An improved hybrid Cuckoo Search (IHCS) metaheuristics algorithm for permutation flow shop scheduling problems. Int. J. Bio-Inspired Comput. **4**(4), 200–205 (2012)
36. Li, X., Yin, M.: A hybrid cuckoo search via Lévy flights for the permutation flow shop scheduling problem. Int. J. Prod. Res. **51**(16), 4732–4754 (2013)
37. Gherboudj, A., Layeb, A., Chikhi, S.: Solving 0–1 knapsack problems by a discrete binary version of the cuckoo search algorithm. Int. J. Bio-Inspired Comput. **4**(4), 229–236 (2012)
38. Valian, E., Mohanna, S., Tavakoli, S.: Improved cuckoo search algorithm for feed-forward neural network training. Int. J. Artif. Intell. Appl. **2**(3), 36–43 (2011)
39. Ouaarab, A., Ahiod., B., Yang, X.-S.: Discrete cuckoo search algorithm for the traveling salesman problem. Neural Computing and Applications **24**(7-8), 1659–1669 (2014)
40. Goyal, S., Patterh, M.S.: Wireless sensor network localization based on cuckoo search algorithm. Wireless Pers. Commun. **79**(1), 223–234 (2014)
41. Passino, K.: Biomimicry of bacterial foraging for distributed optimization and control. IEEE Control Syst. Mag **22**(3), 52–67 (2002)
42. Kim, D.H., Abraham, A., Cho, J.H.: A hybrid genetic algorithm and bacterial foraging approach for global optimization. Inf. Sci. **177**(18), 3918–3937 (2007)
43. Sathya, P.D., Kayalvizhi, R.: Modified bacterial foraging algorithm based multilevel thresholding for image segmentation. Eng. Appl. Artif. Intell. **24**(4), 595–615 (2011)
44. Dasgupta, S., Das, S., Biswas, A., Abraham, A.: Automatic circle detection on digital images with an adaptive bacterial foraging algorithm. Soft Comput. **14**(11), 1151–1164 (2011)
45. Kulkarni, R.V., Ganesh Kumar, V.: Bio-inspired algorithms for autonomous deployment and localization of sensor nodes. IEEE Trans. Syst. Man Cybern. Part C (Appl. Rev.) **40**(6), 663–675 (2010)
46. Meng, X., Liu, Y., Gao, X., Zhang, H.: A new bio-inspired algorithm: Chicken swarm optimization. In: Tan, Y., Shi, Y., Coello, C.A.C. (eds.) ICSI 2014. LNCS, vol. 8794, pp. 86–94. Springer, Cham (2014). https://doi.org/10.1007/978-3-319-11857-4_10
47. Wu, D., Kong, F., Gao, W., Shen, Y., Ji, Z.: Improved chicken swarm optimization. In: 2015 IEEE International Conference on Cyber Technology in Automation, Control, and Intelligent Systems (CYBER), Shenyang, China, pp. 681–686. IEEE (2015)
48. Chen, Y.L., He, P.L., Zhang, Y.H.: Combining penalty function with modified chicken swarm optimization for constrained optimization. Adv. Intell. Syst. Res. **126**, 1899–1907 (2015)

49. Al Shayokh, M., Shin, S.Y.: Bio-inspired distributed WSN localization based on chicken swarm optimization. Wireless Pers. Commun. **97**(4), 5691–5706 (2017)
50. Mirjalili, S., Mirjalili, S.M., Lewis, A.: Grey wolf optimizer. Adv. Eng. Softw. **69**, 46–61 (2014)
51. Jitkongchuen, D., Phaidang, P., Pongtawevirat, P.: Grey wolf optimization algorithm with invasion-based migration operation. In: 2016 IEEE/ACIS 15th International Conference on Computer and Information Science (ICIS), Okayama, Japan, pp. 1–5. IEEE (2016)
52. Rajakumar, R., Amudhavel, J., Dhavachelvan, P., Vengattaraman, T.: GWO-LPWSN: Grey wolf optimization algorithm for node localization problem in wireless sensor networks. J. Comput. Netw. Commun. (2017)
53. Chen, G.-C., Yu, J.-S.: Particle swarm optimization algorithm. Inf. Cont.-Shenyang **34**, 318 (2005)
54. Schmickl, T., Crailsheim, K.: A navigation algorithm for swarm robotics inspired by slime mold aggregation. In: Şahin, E., Spears, W.M., Winfield, A.F.T. (eds.) SR 2006. LNCS, vol. 4433, pp. 1–13. Springer, Heidelberg (2007). https://doi.org/10.1007/978-3-540-71541-2_1
55. Nakagaki, T., Yamada, H., Toth, A.: Path finding by tube Morphogenesis in an amoeboid organism. Biophys. Chem. **92**(1), 47–52 (2001)
56. Nakagaki, T.: Smart behavior of true slime mold in a labyrinth. Res. Microbiol. **152**(9), 767–770 (2001)
57. Li, K., Torres, C.E., Thomas, K., Rossi, L.F., Shen, C.C.: Slime mold inspired routing protocols for wireless sensor networks. Swarm Intell. **5**(3–4), 183–223 (2011)
58. Conradt, L., Roper, T.J.: Consensus decision making in animals. Trends Ecol. Evol. **20**(8), 449–456 (2005)
59. Pitman, R.L., Durban, J.W.: Cooperative hunting behavior, prey selectivity and prey handling by pack ice killer whales (Orcinus orca), type B, Antarctic Peninsula waters. Marine Mammal Sci. **28**(1), 16–36 (2012)
60. Visser, I.N., Smith, T.G., Bullock, I.D., Green, G.D., Carlsson, O.G.L., Imberti, S.: Antarctic peninsula killer whales (Orcinus orca) hunt seals and a penguin on floating ice. Marine Mammal Sci. **24**(1), 225–234 (2008)
61. Tarpy, D.R., Gilley, D.C., Seeley, T.D.: Levels of selection in a social insect: A review of conflict and cooperation during honey bee (Apis mellifera) queen replacement. Behav. Ecol. Sociobiol. **55**(6), 513–523 (2004)

Correction to: Recent Trends in Image Processing and Pattern Recognition

K. C. Santosh and Ravindra S. Hegadi

Correction to: K. C. Santosh and R. S. Hegadi (Eds.): *Recent Trends in Image Processing and Pattern Recognition*, CCIS 1037, https://doi.org/10.1007/978-981-13-9187-3

In the originally published version, the names of the two Authors on pages 108, 149, and 159 were incorrect. The names have been corrected as "AKM Shahariar Azad Rabby" and "Syed Akhter Hossain".

The updated version of these chapters can be found at
https://doi.org/10.1007/978-981-13-9187-3_10
https://doi.org/10.1007/978-981-13-9187-3_14
https://doi.org/10.1007/978-981-13-9187-3_15

K. C. Santosh and R. S. Hegadi (Eds.): RTIP2R 2018, CCIS 1037, p. C1, 2019.
https://doi.org/10.1007/978-981-13-9187-3_67

Author Index

Printed in the United States
By Bookmasters